ADVANCED ENGINEERING MATHEMATICS

THE PRINDLE, WEBER & SCHMIDT SERIES IN MATHEMATICS

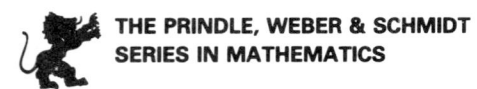

Althoen and Bumcrot, *Introduction to Discrete Mathematics*
Boye, Kavanaugh, and Williams, *Elementary Algebra*
Boye, Kavanaugh, and Williams, *Intermediate Algebra*
Burden and Faires, *Numerical Analysis, Fourth Edition*
Cass and O'Connor, *Fundamentals with Elements of Algebra*
Cullen, *Linear Algebra and Differential Equations, Second Edition*
Dick and Patton, *Calculus, Volume I*
Dick and Patton, *Calculus, Volume II*
Dick and Patton, *Technology in Calculus: A Sourcebook of Activities*
Eves, *In Mathematical Circles*
Eves, *Mathematical Circles Adieu*
Eves, *Mathematical Circles Squared*
Eves, *Return to Mathematical Circles*
Fletcher, Hoyle, and Patty, *Foundations of Discrete Mathematics*
Fletcher and Patty, *Foundations of Higher Mathematics, Second Edition*
Gantner and Gantner, *Trigonometry*
Geltner and Peterson, *Geometry for College Students, Second Edition*
Gilbert and Gilbert, *Elements of Modern Algebra, Third Edition*
Gobran, *Beginning Algebra, Fifth Edition*
Gobran, *Intermediate Algebra, Fourth Edition*
Gordon, *Calculus and the Computer*
Hall, *Algebra for College Students*
Hall, *Beginning Algebra*
Hall, *College Algebra with Applications, Third Edition*
Hall, *Intermediate Algebra*
Hartfiel and Hobbs, *Elementary Linear Algebra*
Humi and Miller, *Boundary-Value Problems and Partial Differential Equations*
Kaufmann, *Algebra for College Students, Fourth Edition*
Kaufmann, *Algebra with Trigonometry for College Students, Third Edition*
Kaufmann, *College Algebra, Second Edition*
Kaufmann, *College Algebra and Trigonometry, Second Edition*
Kaufmann, *Elementary Algebra for College Students, Fourth Edition*
Kaufmann, *Intermediate Algebra for College Students, Fourth Edition*
Kaufmann, *Precalculus, Second Edition*
Kaufmann, *Trigonometry*
Kennedy and Green, *Prealgebra for College Students*
Laufer, *Discrete Mathematics and Applied Modern Algebra*
Nicholson, *Elementary Linear Algebra with Applications, Second Edition*
Pence, *Calculus Activities for Graphic Calculators*
Pence, *Calculus Activities for the TI-81 Graphic Calculator*
Plybon, *An Introduction to Applied Numerical Analysis*
Powers, *Elementary Differential Equations*
Powers, *Elementary Differential Equations with Boundary-Value Problems*
Proga, *Arithmetic and Algebra, Third Edition*
Proga, *Basic Mathematics, Third Edition*
Rice and Strange, *Plane Trigonometry, Sixth Edition*
Schelin and Bange, *Mathematical Analysis for Business and Economics, Second Edition*
Strnad, *Introductory Algebra*
Swokowski, *Algebra and Trigonometry with Analytic Geometry, Seventh Edition*
Swokowski, *Calculus, Fifth Edition*
Swokowski, *Calculus, Fifth Edition (Late Trigonometry Version)*
Swokowski, *Calculus of a Single Variable*
Swokowski, *Fundamentals of College Algebra, Seventh Edition*
Swokowski, *Fundamentals of College Algebra and Trigonometry, Seventh Edition*
Swokowski, *Fundamentals of Trigonometry, Seventh Edition*
Swokowski, *Precalculus: Functions and Graphs, Sixth Edition*
Tan, *Applied Calculus, Second Edition*
Tan, *Applied Finite Mathematics, Third Edition*
Tan, *Calculus for the Managerial, Life, and Social Sciences, Second Edition*
Tan, *College Mathematics, Second Edition*
Trim, *Applied Partial Differential Equations*
Venit and Bishop, *Elementary Linear Algebra, Alternate Second Edition*
Venit and Bishop, *Elementary Linear Algebra, Third Edition*
Wiggins, *Problem Solver for Finite Mathematics and Calculus*
Willard, *Calculus and Its Applications, Second Edition*
Wood and Capell, *Arithmetic*
Wood and Capell, *Intermediate Algebra*
Wood, Capell, and Hall, *Developmental Mathematics, Fourth Edition*
Zill, *A First Course in Differential Equations with Applications, Fourth Edition*
Zill, *Calculus, Third Edition*
Zill, *Differential Equations with Boundary-Value Problems, Second Edition*
Zill and Cullen, *Advanced Engineering Mathematics*

THE PRINDLE, WEBER & SCHMIDT SERIES IN ADVANCED MATHEMATICS

Brabenec, *Introduction to Real Analysis*
Ehrlich, *Fundamental Concepts of Abstract Algebra*
Eves, *Foundations and Fundamental Concepts of Mathematics, Third Edition*
Keisler, *Elementary Calculus: An Infinitesimal Approach, Second Edition*
Kirkwood, *An Introduction to Real Analysis*
Ruckle, *Modern Analysis: Measure Theory and Functional Analysis with Applications*
Sieradski, *An Introduction to Topology and Homotopy*

ADVANCED ENGINEERING MATHEMATICS

DENNIS G. ZILL
Loyola Marymount University

MICHAEL R. CULLEN
Loyola Marymount University

PWS PUBLISHING COMPANY
Boston

PWS PUBLISHING COMPANY
20 Park Plaza, Boston, MA 02116-4324

Copyright © 1992 by PWS Publishing Company

All rights reserved. No part of this book may be reproduced, stored in a retrieval system, or transcribed, in any form or by any means—electronic, mechanical, photocopying, recording, or otherwise—without the written permission of PWS Publishing Company.

PWS Publishing Company is a division of Wadsworth, Inc.

I(T)P ™

International Thomson Publishing
The trademark ITP is used under license

Library of Congress Cataloging-in-Publication Data
Zill, Dennis G.
   Advanced engineering mathematics/Dennis G. Zill, Michael R. Cullen.
     p. cm.
   Includes index.
   ISBN 0-534-92800-5
   1. Engineering mathematics. I. Cullen., Michael R. II. Title.
TA330.Z55 1992   91-24071
620′.001′51—dc20   CIP

Part Opening Photo Credits:

Part I: San Francisco, California, Bay Bridge, (c) Peter Menzel/Stock Boston (p. 1); *Part II:* YF-22 Advanced Tactical Fighter (ATF), courtesy of Lockheed/Boeing (p. 321); *Part III:* Collapse of the Nimitz Freeway, photo by Jayne Kamin-Oncea, Copyright 1989, Los Angeles Times (p.617): *Part IV:* NASA space shuttle "Atlantis," photo courtesy of Space Transportation & Systems Group, Rockwell International Corporation (p. 723); *Part V:* Cray Y-MP/832 (supercomputer), photo by Paul Shambroom, courtesy of Cray Research, Inc. (p. 843); *Part VI:* Fast Flux Test Facility, photo courtesy of Department of Energy (p. 929).

This book is printed on recycled, acid-free paper.

Sponsoring Editor: Steve Quigley
Developmental Editor: Barbara Lovenvirth
Production Coordinator/Cover Designer: Robine Andrau
Interior Designer: Julia Gecha
Cover Photo: © Rob Atkins/The Image Bank
Interior Illustrator: Network Graphics
Typesetter: Polyglot Pte Ltd
Cover Printer: Henry N. Sawyer Company
Printer and Binder: Arcata Graphics/Hawkins

Printed in the United States of America
    95 96 — 10 9 8 7

CONTENTS

Preface xiii

Part I
ORDINARY DIFFERENTIAL EQUATIONS 1

1 INTRODUCTION TO DIFFERENTIAL EQUATIONS 3

- 1.1 Basic Definitions and Terminology 4
- 1.2 Differential Equation of a Family of Curves 13
- 1.3 Mathematical Models 17
  - Summary 32
  - Chapter 1 Review Exercises 32

2 FIRST-ORDER DIFFERENTIAL EQUATIONS 34

- 2.1 Preliminary Theory 35
- 2.2 Separable Variables 39
- 2.3 Homogeneous Equations 46
- 2.4 Exact Equations 52
- 2.5 Linear Equations 58
- [O] 2.6 Equations of Bernoulli, Ricatti, and Clairaut 66
- [O] 2.7 Substitutions 70
- [O] 2.8 Picard's Method 73
- 2.9 Orthogonal Trajectories 76
- 2.10 Applications of Linear Equations 81
- 2.11 Applications of Nonlinear Equations 92
  - Summary 101
  - Chapter 2 Review Exercises 102

v

3 LINEAR DIFFERENTIAL EQUATIONS OF HIGHER ORDER 105

- 3.1 Preliminary Theory 106
  - 3.1.1 Initial-Value and Boundary-Value Problems 106
  - 3.1.2 Linear Dependence and Linear Independence 110
  - 3.1.3 Solutions of Linear Equations 114
- 3.2 Constructing a Second Solution from a Known Solution 126
- 3.3 Homogeneous Linear Equations with Constant Coefficients 131
- 3.4 Undetermined Coefficients 139
- 3.5 Differential Operators and Undetermined Coefficients Revisited 148
  - 3.5.1 Differential Operators 148
  - [O] 3.5.2 An Alternative Approach to Undetermined Coefficients 152
- 3.6 Variation of Parameters 157
- 3.7 Systems of Linear Differential Equations with Constant Coefficients 164
- 3.8 Simple Harmonic Motion 172
- 3.9 Damped Motion 180
- 3.10 Forced Motion 189
- 3.11 Electric Circuits and Other Analogous Systems 198
  - Summary 204
  - Chapter 3 Review Exercises 206

4 LAPLACE TRANSFORM 208

- 4.1 Laplace Transform 209
- 4.2 Inverse Transform 217
- 4.3 Operational Properties 224
  - 4.3.1 Translation Theorems and Derivatives of a Transform 224
  - 4.3.2 Transforms of Derivatives and Integrals 233
  - 4.3.3 Transform of a Periodic Function 237
- 4.4 Applications 242
- 4.5 Dirac Delta Function 255
- 4.6 Systems of Differential Equations 260
  - Summary 267
  - Chapter 4 Review Exercises 268

5 DIFFERENTIAL EQUATIONS WITH VARIABLE COEFFICIENTS 270

- 5.1 Cauchy-Euler Equation 271
- 5.2 Power Series Solutions 278
- 5.3 Solutions About Singular Points 290
- 5.4 Two Special Equations 307

5.4.1 Solution of Bessel's Equation 307
5.4.2 Solution of Legendre's Equation 313
Summary 318
Chapter 5 Review Exercises 319

Part II
VECTORS, MATRICES, AND VECTOR CALCULUS 321

6 VECTORS 323

6.1 Vectors in the Plane 324
6.2 Vectors in Space 332
6.3 The Dot Product 339
6.4 The Cross Product 348
6.5 Lines and Planes in 3-Space 356
6.6 Vector Spaces 365
Summary 373
Chapter 6 Review Exercises 374

7 MATRICES 376

7.1 Matrix Algebra 377
7.2 Systems of Linear Algebraic Equations 387
7.3 Determinants 400
7.4 Properties of Determinants 406
7.5 Inverse of a Matrix 414
7.5.1 Finding the Inverse 414
7.5.2 Using the Inverse to Solve Systems 421
7.6 Cramer's Rule 426
7.7 The Eigenvalue Problem 430
7.8 Orthogonal Matrices 437
7.9 Diagonalization 444
7.10 Cryptography 454
7.11 An Error-Correcting Code 458
7.12 Method of Least Squares 465
Summary 469
Chapter 7 Review Exercises 469

8 VECTOR CALCULUS 473

8.1 Vector Functions 474
8.2 Motion on a Curve; Velocity and Acceleration 482
8.3 Curvature; Components of Acceleration 488
8.4 Functions of Several Variables; Chain Rule 494
8.5 The Directional Derivative 502

	8.6	Tangent Planes and Normal Lines 510
	8.7	Divergence and Curl 515
	8.8	Line Integrals 521
	8.9	Line Integrals Independent of Path 533
[O]	8.10	Review of Double Integrals 542
[O]	8.11	Double Integrals in Polar Coordinates 553
	8.12	Green's Theorem 559
	8.13	Surface Integrals 566
	8.14	Stokes' Theorem 575
[O]	8.15	Review of Triple Integrals 582
	8.16	Divergence Theorem 597
	8.17	Change of Variables in Multiple Integrals 605
		Summary 613
		Chapter 8 Review Exercises 614

Part III
SYSTEMS OF DIFFERENTIAL EQUATIONS 617

9 SYSTEMS OF LINEAR FIRST-ORDER DIFFERENTIAL EQUATIONS 619

	9.1	Systems in Normal Form 620
	9.2	Matrix Functions 625
	9.3	Preliminary Theory 629
	9.4	Homogeneous Linear Systems 641
		9.4.1 Distinct Real Eigenvalues 641
		9.4.2 Complex Eigenvalues 645
		9.4.3 Repeated Eigenvalues 649
	9.5	Solution by Diagonalization 656
	9.6	Nonhomogeneous Linear Systems 658
		9.6.1 Undetermined Coefficients 658
		9.6.2 Variation of Parameters 661
		9.6.3 Diagonalization 664
[O]	9.7	Matrix Exponential 667
		Summary 670
		Chapter 9 Review Exercises 671

10 PLANE AUTONOMOUS SYSTEMS AND STABILITY 673

	10.1	Autonomous Systems, Critical Points, and Periodic Solutions 674
	10.2	Stability and Linear Systems 680
	10.3	Linearization and Local Stability 688
	10.4	Applications of Autonomous Systems 698
[O]	10.5	Periodic Solutions, Limit Cycles, and Global Stability 707
		Summary 718
		Chapter 10 Review Exercises 720

Part IV
FOURIER SERIES AND BOUNDARY-VALUE PROBLEMS 723

11 ORTHOGONAL FUNCTIONS AND FOURIER SERIES 725

- 11.1 Orthogonal Functions 726
- 11.2 Fourier Series 732
- 11.3 Fourier Cosine and Sine Series 737
- 11.4 Sturm-Liouville Problem 746
- 11.5 Bessel and Legendre Series 755
  - 11.5.1 Fourier-Bessel Series 755
  - 11.5.2 Fourier-Legendre Series 759
  - Summary 762
  - Chapter 11 Review Exercises 763

12 BOUNDARY-VALUE PROBLEMS IN RECTANGULAR COORDINATES 765

- 12.1 Separable Partial Differential Equations 766
- 12.2 Classical Equations and Boundary-Value Problems 771
- 12.3 Heat Equation 777
- 12.4 Wave Equation 780
- 12.5 Laplace's Equation 784
- 12.6 Nonhomogeneous Equations and Boundary Conditions 788
- 12.7 Use of Generalized Fourier Series 791
- [O] 12.8 Boundary-Value Problems Involving Fourier Series in Two Variables 795
  - Summary 798
  - Chapter 12 Review Exercises 799

13 BOUNDARY-VALUE PROBLEMS IN OTHER COORDINATE SYSTEMS 801

- 13.1 Problems Involving Laplace's Equation in Polar Coordinates 802
- 13.2 Problems in Polar and Cylindrical Coordinates: Bessel Functions 807
- 13.3 Problems in Spherical Coordinates: Legendre Polynomials 813
  - Summary 815
  - Chapter 13 Review Exercises 816

14 INTEGRAL TRANSFORM METHOD 818

- 14.1 Error Function 819
- 14.2 Applications of the Laplace Transform 821

14.3	Fourier Integral	828
14.4	Fourier Transforms	834
	Summary	841
	Chapter 14 Review Exercises	842

Part V
NUMERICAL ANALYSIS 843

15 NUMERICAL METHODS 845

15.1	Newton's Method	846
15.2	Approximate Integration	854
15.3	Direction Fields	866
15.4	The Euler Methods	871
15.5	The Three-Term Taylor Method	877
15.6	The Runge-Kutta Method	880
15.7	Multistep Methods, Errors	885
15.8	Higher-Order Equations and Systems	888
15.9	Second-Order Boundary-Value Problems	892
15.10	Numerical Methods for Partial Differential Equations: Elliptic Equations	896
15.11	Numerical Methods for Partial Differential Equations: Parabolic Equations	903
15.12	Numerical Methods for Partial Differential Equations: Hyperbolic Equations	911
15.13	Approximation of Eigenvalues	917
	Summary	925
	Chapter 15 Review Exercises	926

Part VI
COMPLEX ANALYSIS 929

16 FUNCTIONS OF A COMPLEX VARIABLE 931

16.1	Complex Numbers	932
16.2	Polar Form of Complex Numbers; Powers and Roots	936
16.3	Set of Points in the Complex Plane	942
16.4	Functions of a Complex Variable; Analyticity	945
16.5	Cauchy-Riemann Equations	952
16.6	Exponential and Logarithmic Functions	958
	16.6.1 Exponential Function 958	
	16.6.2 Logarithmic Function 962	
16.7	Trigonometric and Hyperbolic Functions	967
16.8	Inverse Trigonometric and Hyperbolic Functions	971
	Summary	974
	Chapter 16 Review Exercises	975

17 INTEGRATION IN THE COMPLEX PLANE 977

- 17.1 Contour Integrals 978
- 17.2 Cauchy-Goursat Theorem 984
- 17.3 Independence of Path 990
- 17.4 Cauchy's Integral Formula 996
  - Summary 1003
  - Chapter 17 Review Exercises 1004

18 SERIES AND RESIDUES 1006

- 18.1 Sequences and Series 1007
- 18.2 Taylor Series 1013
- 18.3 Laurent Series 1020
- 18.4 Zeros and Poles 1028
- 18.5 Residues and Residue Theorem 1032
- 18.6 Evaluation of Real Integrals 1039
  - Summary 1047
  - Chapter 18 Review Exercises 1048

19 CONFORMAL MAPPINGS AND APPLICATIONS 1050

- 19.1 Complex Functions as Mappings 1051
- 19.2 Conformal Mapping and the Dirichlet Problem 1056
- 19.3 Linear Fractional Transformations 1066
- 19.4 Schwarz-Christoffel Transformations 1073
- [O] 19.5 Poisson Integral Formulas 1080
- 19.6 Applications 1086
  - Summary 1095
  - Chapter 19 Review Exercises 1096

APPENDICES A-1

- **Appendix I** Gamma Function A-3
- **Appendix II** Table of Laplace Transforms A-6
- **Appendix III** Conformal Mappings A-9
- **Appendix IV** BASIC Programs for Numerical Methods in Chapter 15 A-20

Answers to Odd-Numbered Problems A-29

Index A-93

14.3 Fourier Integral 828
14.4 Fourier Transforms 834
Summary 841
Chapter 14 Review Exercises 842

Part V
NUMERICAL ANALYSIS 843

15 NUMERICAL METHODS 845

15.1 Newton's Method 846
15.2 Approximate Integration 854
15.3 Direction Fields 866
15.4 The Euler Methods 871
15.5 The Three-Term Taylor Method 877
15.6 The Runge-Kutta Method 880
15.7 Multistep Methods, Errors 885
15.8 Higher-Order Equations and Systems 888
15.9 Second-Order Boundary-Value Problems 892
15.10 Numerical Methods for Partial Differential Equations: Elliptic Equations 896
15.11 Numerical Methods for Partial Differential Equations: Parabolic Equations 903
15.12 Numerical Methods for Partial Differential Equations: Hyperbolic Equations 911
15.13 Approximation of Eigenvalues 917
Summary 925
Chapter 15 Review Exercises 926

Part VI
COMPLEX ANALYSIS 929

16 FUNCTIONS OF A COMPLEX VARIABLE 931

16.1 Complex Numbers 932
16.2 Polar Form of Complex Numbers; Powers and Roots 936
16.3 Set of Points in the Complex Plane 942
16.4 Functions of a Complex Variable; Analyticity 945
16.5 Cauchy-Riemann Equations 952
16.6 Exponential and Logarithmic Functions 958
  16.6.1 Exponential Function 958
  16.6.2 Logarithmic Function 962
16.7 Trigonometric and Hyperbolic Functions 967
16.8 Inverse Trigonometric and Hyperbolic Functions 971
Summary 974
Chapter 16 Review Exercises 975

PART III

Systems of Differential Equations

9 SYSTEMS OF LINEAR FIRST-ORDER DIFFERENTIAL EQUATIONS

10 PLANE AUTONOMOUS SYSTEMS AND STABILITY

9

SYSTEMS OF LINEAR FIRST-ORDER DIFFERENTIAL EQUATIONS

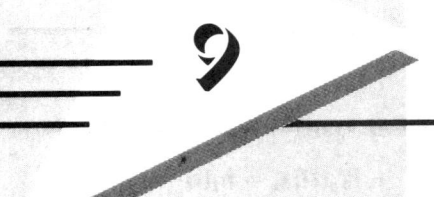

TOPICS TO REVIEW

solution of linear first-order differential equations (2.5)
matrix algebra (7.1)
inverse of a matrix (7.5)
eigenvalues and eigenvectors of a matrix (7.7)

IMPORTANT CONCEPTS

linear normal form
homogeneous system
nonhomogeneous system
solution vector
superposition principle
linear dependence
linear independence
Wronskian
fundamental set of solutions
general solution
fundamental matrix
solution by diagonalization
undetermined coefficients
variation of parameters

9.1 Systems in Normal Form
9.2 Matrix Functions
9.3 Preliminary Theory
9.4 Homogeneous Linear Systems
9.5 Solution by Diagonalization
9.6 Nonhomogeneous Linear Systems
[O] 9.7 Matrix Exponential
Summary
Chapter 9 Review Exercises

INTRODUCTION

In Sections 3.7 and 4.6 we examined two methods for solving systems of linear ordinary differential equations. The differential equations in those systems were first-order, second-order, and so on. Our focus in this chapter will be on the general theory and systematic solution of simultaneous linear first-order differential equations, or simply **linear systems**, in which all coefficients are constants. The discussion that follows will rely heavily on the material in Chapter 7.

9.1 SYSTEMS IN NORMAL FORM

In Sections 3.7 and 4.6 we dealt with linear systems that were of the form

$$P_{11}(D)x_1 + P_{12}(D)x_2 + \cdots + P_{1n}(D)x_n = b_1(t)$$
$$P_{21}(D)x_1 + P_{22}(D)x_2 + \cdots + P_{2n}(D)x_n = b_2(t)$$
$$\vdots \qquad \vdots \qquad \qquad \vdots \qquad \qquad \vdots \qquad (1)$$
$$P_{n1}(D)x_1 + P_{n2}(D)x_2 + \cdots + P_{nn}(D)x_n = b_n(t)$$

where the P_{ij} were polynomials in the differential operator D. However, the study of systems of *first-order* differential equations

$$\frac{dx_1}{dt} = g_1(t, x_1, x_2, \ldots, x_n)$$
$$\frac{dx_2}{dt} = g_2(t, x_1, x_2, \ldots, x_n)$$
$$\vdots \qquad \vdots \qquad (2)$$
$$\frac{dx_n}{dt} = g_n(t, x_1, x_2, \ldots, x_n)$$

is particularly important in advanced mathematics since every nth-order differential equation

$$y^{(n)} = F(t, y, y', \ldots, y^{(n-1)})$$

as well as most systems of differential equations, can be reduced to form (2).

Linear Normal Form Of course, a system such as (2) need not be linear and need not have constant coefficients. Consequently, the system may not be readily solvable, if at all. In the remaining sections of this chapter we shall be interested in only a particular, but important, case of (2)—namely, those systems having the linear **normal**, or **canonical**, form

$$\frac{dx_1}{dt} = a_{11}(t)x_1 + a_{12}(t)x_2 + \cdots + a_{1n}(t)x_n + f_1(t)$$
$$\frac{dx_2}{dt} = a_{21}(t)x_1 + a_{22}(t)x_2 + \cdots + a_{2n}(t)x_n + f_2(t) \qquad (3)$$
$$\vdots$$
$$\frac{dx_n}{dt} = a_{n1}(t)x_1 + a_{n2}(t)x_2 + \cdots + a_{nn}(t)x_n + f_n(t)$$

where the coefficients a_{ij} and the f_i are functions continuous on a common interval I. When $f_i(t) = 0$, $i = 1, 2, \ldots, n$, the system (3) is said to be **homogeneous**; otherwise it is called **nonhomogeneous**.

9.1 Systems in Normal Form

We shall now show that every linear nth-order differential equation can be reduced to a linear system having the normal form (3).

Equation to a System Suppose a linear nth-order differential equation is first written as

$$\frac{d^n y}{dt^n} = -\frac{a_0}{a_n} y - \frac{a_1}{a_n} y' - \cdots - \frac{a_{n-1}}{a_n} y^{(n-1)} + f(t) \qquad (4)$$

If we then introduce the variables

$$y = x_1, \quad y' = x_2, \quad y'' = x_3, \quad \ldots, \quad y^{(n-1)}(x) = x_n \qquad (5)$$

it follows that $y' = x_1' = x_2$, $y'' = x_2' = x_3, \ldots, y^{(n-1)} = x_{n-1}' = x_n$, and $y^{(n)} = x_n'$. Hence, from (4) and (5) we obtain

$$\begin{aligned} x_1' &= x_2 \\ x_2' &= x_3 \\ x_3' &= x_4 \\ &\vdots \\ x_{n-1}' &= x_n \\ x_n' &= -\frac{a_0}{a_n} x_1 - \frac{a_1}{a_n} x_2 - \cdots - \frac{a_{n-1}}{a_n} x_n + f(t) \end{aligned} \qquad (6)$$

Inspection of (6) reveals that it has the same form as (3).

EXAMPLE 1 Reduce the third-order equation

$$2y''' - 6y'' + 4y' + y = \sin t$$

to the normal form (3).

Solution Write the differential equation as

$$y''' = -\frac{1}{2} y - 2y' + 3y'' + \frac{1}{2} \sin t$$

and then let $y = x_1$, $y' = x_2$, $y'' = x_3$. Since

$$\begin{aligned} x_1' &= y' = x_2 \\ x_2' &= y'' = x_3 \\ x_3' &= y''' \end{aligned}$$

we find

$$\begin{aligned} x_1' &= x_2 \\ x_2' &= x_3 \\ x_3' &= -\frac{1}{2} x_1 - 2x_2 + 3x_3 + \frac{1}{2} \sin t \end{aligned}$$ □

Systems Reduced to Normal Form Using a procedure similar to that just outlined, we can reduce *most* systems of the linear form (1) to the linear normal form (3). To accomplish this it is first necessary to solve the system for the highest-order derivative of each dependent variable. As we shall see, this may not always be possible.

EXAMPLE 2 Reduce
$$(D^2 - D + 5)x + 2D^2y = e^t$$
$$-2x + (D^2 + 2)y = 3t^2$$

to the normal form (3).

Solution Write the system as
$$D^2x + 2D^2y = e^t - 5x + Dx$$
$$D^2y = 3t^2 + 2x - 2y$$

and then eliminate D^2y by multiplying the second equation by 2 and subtracting. We have
$$D^2x = e^t - 6t^2 - 9x + 4y + Dx$$

Since the second equation of the system already expresses the highest-order derivative of y in terms of the remaining functions, we are in a position to introduce new variables. If we let
$$Dx = u \quad \text{and} \quad Dy = v$$

the expressions for D^2x and D^2y become, respectively,
$$Du = e^t - 6t^2 - 9x + 4y + u$$
$$Dv = 3t^2 + 2x - 2y$$

Thus, the original system can be written in the normal form
$$Dx = u$$
$$Dy = v$$
$$Du = -9x + 4y + u + e^t - 6t^2$$
$$Dv = 2x - 2y + 3t^2$$
□

Degenerate Systems Those systems of differential equations of form (1) that cannot be reduced to a linear system in normal form are said to be **degenerate**. For example, it is a straightforward matter to show that it is impossible to solve the system
$$(D + 1)x + (D + 1)y = 0$$
$$2Dx + (2D + 1)y = 0 \tag{7}$$

for the highest derivative of each variable; hence, the system is degenerate.*

---

*This does *not* mean that the system does not have a solution (see Problem 21).

9.1 Systems in Normal Form

The reader may be wondering why anyone would want to convert a single differential equation to a system of equations, or for that matter a system of differential equations to an even larger system. While we are not in a position to completely justify their importance, suffice it to say that these procedures are more than a theoretical exercise. There are times when it is actually desirable to work with a system rather than with one equation. In the numerical analysis of differential equations, almost all computational algorithms are established for first-order equations. Since these algorithms can be generalized directly to systems, to compute numerically, say, a second-order equation, we could reduce it to a system of two first-order equations (see Chapter 15).

A linear system such as (3) also arises naturally in some physical applications. The following example will illustrate a homogeneous system in two dependent variables.

EXAMPLE 3 Tank A contains 50 gallons of water in which 25 pounds of salt are dissolved. A second tank, B, contains 50 gallons of pure water. Liquid is pumped in and out of the tanks at the rates shown in Figure 9.1. Derive the differential equations that describe the number of pounds $x_1(t)$ and $x_2(t)$ of salt at any time in tanks A and B, respectively.

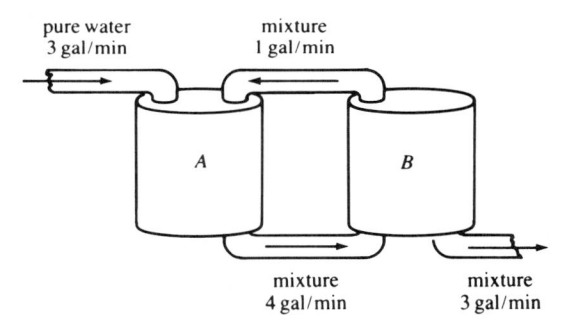

FIGURE 9.1

Solution By an analysis similar to that used in Section 3.2, we see that the net rate of change in $x_1(t)$ in lb/min is

$$\frac{dx_1}{dt} = \overbrace{(3 \text{ gal/min}) \cdot (0 \text{ lb/gal}) + (1 \text{ gal/min}) \cdot \left(\frac{x_2}{50} \text{ lb/gal}\right)}^{\text{input}} - \overbrace{(4 \text{ gal/min}) \cdot \left(\frac{x_1}{50} \text{ lb/gal}\right)}^{\text{output}}$$

$$= -\frac{2}{25} x_1 + \frac{1}{50} x_2$$

In addition, we find that the net rate of change in $x_2(t)$ is

$$\frac{dx_2}{dt} = 4 \cdot \frac{x_1}{50} - 3 \cdot \frac{x_2}{50} - 1 \cdot \frac{x_2}{50} = \frac{2}{25} x_1 - \frac{2}{25} x_2$$

Thus, we obtain the first-order system

$$\frac{dx_1}{dt} = -\frac{2}{25}x_1 + \frac{1}{50}x_2$$
$$\frac{dx_2}{dt} = \frac{2}{25}x_1 - \frac{2}{25}x_2$$
(8)

Observe that the foregoing system is accompanied by the initial conditions $x_1(0) = 25$, $x_2(0) = 0$. □

It is left as an exercise to solve (8) by the Laplace transform.

EXERCISES 9.1 Answers to odd-numbered problems begin on page A-55.

In Problems 1–8 rewrite the given differential equation as a system in normal form (3).

1. $y'' - 3y' + 4y = \sin 3t$

2. $2\dfrac{d^2y}{dt^2} + 4\dfrac{dy}{dt} - 5y = 0$

3. $y''' - 3y'' + 6y' - 10y = t^2 + 1$

4. $4y''' + y = e^t$

5. $\dfrac{d^4y}{dt^4} - 2\dfrac{d^2y}{dt^2} + 4\dfrac{dy}{dt} + y = t$

6. $2\dfrac{d^4y}{dt^4} + \dfrac{d^3y}{dt^3} - 8y = 10$

7. $(t + 1)y'' = ty$

8. $t^2 y'' + ty' + (t^2 - 4)y = 0$

In Problems 9–16 rewrite, if possible, the given system in the normal form (3).

9. $x' + 4x - y' = 7t$
   $x' \quad\;\; + y' - 2y = 3t$

10. $x'' + y' = 1$
    $x'' + y' = -1$

11. $(D - 1)x - Dy = t^2$
    $x + Dy = 5t - 2$

12. $x'' - 2y'' = \sin t$
    $x'' + y'' = \cos t$

13. $(2D + 1)x - 2Dy = 4$
    $Dx - Dy = e^t$

14. $m_1 x_1'' = -k_1 x_1 + k_2(x_2 - x_1)$
    $m_2 x_2'' = -k_2(x_2 - x_1)$

15. $\dfrac{d^3x}{dt^3} = 4x - 3\dfrac{d^2x}{dt^2} + 4\dfrac{dy}{dt}$
    $\dfrac{d^2y}{dt^2} = 10t^2 - 4\dfrac{dx}{dt} + 3\dfrac{dy}{dt}$

16. $D^2x + \quad\quad Dy = 4t$
    $-D^2x + (D + 1)y = 6t^2 + 10$

17. Use the Laplace transform to solve system (8) subject to $x_1(0) = 25$ and $x_2(0) = 0$.

18. Consider two tanks A and B with liquid being pumped in and out at the same rates as given in Example 3. What is the system of differential equations if, instead of pure water, a brine solution containing 2 lb of salt per gallon is pumped into tank A?

19. Using the information given in Figure 9.2, derive the system of differential equations describing the number of pounds of salt x_1, x_2, and x_3 at any time in tanks A, B, and C, respectively.

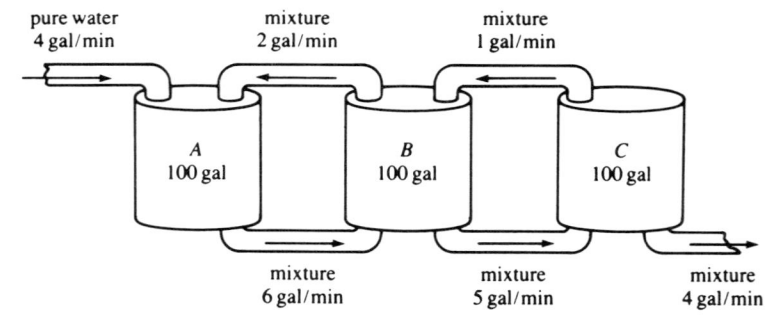

FIGURE 9.2

20. Consider the first-order system

$$(a_1 D - b_1)x + (a_2 D - b_2)y = 0$$
$$(a_3 D - b_3)x + (a_4 D - b_4)y = 0$$

where the a_i are nonzero constants. Determine a condition on the a_i such that the system is degenerate.

21. Verify that the degenerate system (7) possesses the solution $x(t) = c_1 e^{-t}$, $y(t) = -2c_1 e^{-t}$.

9.2 MATRIX FUNCTIONS

In this chapter matrices whose entries are functions will play an important role. If the entries $a_{ij}(t)$ are functions defined on some common interval I, then

$$\mathbf{A}(t) = [a_{ij}(t)]_{m \times n}$$

is called a **matrix function** or a **matrix-valued function**. In other words, a matrix function is a rule of correspondence that assigns to each value of t in I a unique $m \times n$ matrix $\mathbf{A}(t)$. If each $a_{ij}(t)$ is continuous, then $\mathbf{A}(t)$ is said to be continuous. The derivative and integral of a matrix function $\mathbf{A}(t)$ are defined as follows.

DEFINITION 9.1 Derivative of a Matrix Function

If $\mathbf{A}(t) = [a_{ij}(t)]_{m \times n}$ is a matrix whose entries are functions differentiable on a common interval I, then

$$\frac{d\mathbf{A}}{dt} = \left[\frac{d}{dt} a_{ij} \right]_{m \times n}$$

DEFINITION 9.2 Integral of a Matrix Function

If $\mathbf{A}(t) = [a_{ij}(t)]_{m \times n}$ is a matrix whose entries are functions continuous on a common interval I containing t and t_0, then

$$\int_{t_0}^{t} \mathbf{A}(s)\, ds = \left[\int_{t_0}^{t} a_{ij}(s)\, ds \right]_{m \times n}$$

In other words, to differentiate (integrate) a matrix function we simply differentiate (integrate) each entry of the matrix. The derivative of a matrix function is also denoted by $\mathbf{A}'(t)$.

EXAMPLE 1 For $\mathbf{A}(t) = \begin{bmatrix} \sin 2t \\ e^{3t} \\ 8t - 1 \end{bmatrix}$ compute (a) $\mathbf{A}'(t)$ and (b) $\int_0^t \mathbf{A}(s)\, ds$.

Solution (a) $\mathbf{A}'(t) = \begin{bmatrix} \dfrac{d}{dt}\sin 2t \\ \dfrac{d}{dt}e^{3t} \\ \dfrac{d}{dt}(8t-1) \end{bmatrix} = \begin{bmatrix} 2\cos 2t \\ 3e^{3t} \\ 8 \end{bmatrix}$

(b) $\int_0^t \mathbf{A}(s)\,ds = \begin{bmatrix} \int_0^t \sin 2s\,ds \\ \int_0^t e^{3s}\,ds \\ \int_0^t (8s-1)\,ds \end{bmatrix} = \begin{bmatrix} -\tfrac{1}{2}\cos 2t + \tfrac{1}{2} \\ \tfrac{1}{3}e^{3t} - \tfrac{1}{3} \\ 4t^2 - t \end{bmatrix}$ □

The rules for differentiating and integrating matrix functions are similar to those for ordinary functions:

$$\frac{d}{dt}\mathbf{CA} = \mathbf{C}\frac{d\mathbf{A}}{dt}, \quad \mathbf{C} \text{ a matrix of constants} \tag{1}$$

$$\frac{d}{dt}(\mathbf{A} + \mathbf{B}) = \frac{d\mathbf{A}}{dt} + \frac{d\mathbf{B}}{dt} \tag{2}$$

$$\frac{d}{dt}\mathbf{AB} = \mathbf{A}\frac{d\mathbf{B}}{dt} + \frac{d\mathbf{A}}{dt}\mathbf{B} \tag{3}$$

$$\int_a^b (\mathbf{A}(t) + \mathbf{B}(t))\,dt = \int_a^b \mathbf{A}(t)\,dt + \int_a^b \mathbf{B}(t)\,dt \tag{4}$$

The result in (3) is, of course, the matrix analogue of the product rule. Note, however, that the order of multiplication in (1) and (3) must be strictly observed, since matrix multiplication is not commutative. In (4) we also assume that the entries in both $\mathbf{A}(t)$ and $\mathbf{B}(t)$ are integrable on the interval $[a, b]$.

Matrix Form of a System If $\mathbf{X}$, $\mathbf{A}(t)$, and $\mathbf{F}(t)$ denote the matrix functions

$$\mathbf{X} = \begin{bmatrix} x_1(t) \\ x_2(t) \\ \vdots \\ x_n(t) \end{bmatrix} \quad \mathbf{A}(t) = \begin{bmatrix} a_{11}(t) & a_{12}(t) & \cdots & a_{1n}(t) \\ a_{21}(t) & a_{22}(t) & \cdots & a_{2n}(t) \\ \vdots & \vdots & & \vdots \\ a_{n1}(t) & a_{n2}(t) & \cdots & a_{nn}(t) \end{bmatrix} \quad \mathbf{F}(t) = \begin{bmatrix} f_1(t) \\ f_2(t) \\ \vdots \\ f_n(t) \end{bmatrix}$$

then the system of linear first-order differential equations

$$\frac{dx_1}{dt} = a_{11}(t)x_1 + a_{12}(t)x_2 + \cdots + a_{1n}(t)x_n + f_1(t)$$

$$\frac{dx_2}{dt} = a_{21}(t)x_1 + a_{22}(t)x_2 + \cdots + a_{2n}(t)x_n + f_2(t)$$

$$\vdots$$

$$\frac{dx_n}{dt} = a_{n1}(t)x_1 + a_{n2}(t)x_2 + \cdots + a_{nn}(t)x_n + f_n(t)$$

9.2 Matrix Functions

can be written as

$$\frac{d}{dt}\begin{bmatrix} x_1 \\ x_2 \\ \vdots \\ x_n \end{bmatrix} = \begin{bmatrix} a_{11}(t) & a_{12}(t) & \cdots & a_{1n}(t) \\ a_{21}(t) & a_{22}(t) & \cdots & a_{2n}(t) \\ \vdots & & & \vdots \\ a_{n1}(t) & a_{n2}(t) & \cdots & a_{nn}(t) \end{bmatrix} \begin{bmatrix} x_1 \\ x_2 \\ \vdots \\ x_n \end{bmatrix} + \begin{bmatrix} f_1(t) \\ f_2(t) \\ \vdots \\ f_n(t) \end{bmatrix}$$

or simply
$$\frac{d\mathbf{X}}{dt} = \mathbf{A}(t)\mathbf{X} + \mathbf{F}(t) \qquad (5)$$

If the system is homogeneous, (5) becomes

$$\frac{d\mathbf{X}}{dt} = \mathbf{A}(t)\mathbf{X} \qquad (6)$$

Equations (5) and (6) are also written as $\mathbf{X}' = \mathbf{A}\mathbf{X} + \mathbf{F}$ and $\mathbf{X}' = \mathbf{A}\mathbf{X}$, respectively.

EXAMPLE 2 In matrix terms the nonhomogeneous system

$$\frac{dx}{dt} = -2x + 5y + e^t - 2t$$
$$\frac{dy}{dt} = 4x - 3y + 10t$$

can be written as

$$\frac{d\mathbf{X}}{dt} = \begin{bmatrix} -2 & 5 \\ 4 & -3 \end{bmatrix}\mathbf{X} + \begin{bmatrix} e^t - 2t \\ 10t \end{bmatrix} \qquad \text{where } \mathbf{X} = \begin{bmatrix} x \\ y \end{bmatrix} \qquad \square$$

EXAMPLE 3 The matrix form of the homogeneous system

$$\frac{dx}{dt} = 2x - 3y$$
$$\frac{dy}{dt} = 6x + 5y$$

is
$$\frac{d\mathbf{X}}{dt} = \begin{bmatrix} 2 & -3 \\ 6 & 5 \end{bmatrix}\mathbf{X} \qquad \text{where } \mathbf{X} = \begin{bmatrix} x \\ y \end{bmatrix} \qquad \square$$

EXERCISES 9.2 Answers to odd-numbered problems begin on page A-55.

In Problems 1–6 find $\dfrac{d\mathbf{X}}{dt}$.

1. $\mathbf{X} = \begin{bmatrix} 5e^{-t} \\ 2e^{-t} \\ -7e^{-t} \end{bmatrix}$

2. $\mathbf{X} = \begin{bmatrix} \frac{1}{2}\sin 2t - 4\cos 2t \\ -3\sin 2t + 5\cos 2t \end{bmatrix}$

3. $\mathbf{X} = 2\begin{bmatrix} 1 \\ -1 \end{bmatrix} e^{2t} + 4\begin{bmatrix} 2 \\ 1 \end{bmatrix} e^{-3t}$

4. $\mathbf{X} = \begin{bmatrix} 5te^{2t} \\ t\sin 3t \end{bmatrix}$

5. $\mathbf{X} = \begin{bmatrix} 2\cos t \\ -\sin t \end{bmatrix} e^t \ln|\cos t|$

6. $\mathbf{X} = \begin{bmatrix} 2\cos t \\ -\sin t \end{bmatrix} e^t + \begin{bmatrix} 2\sin t \\ \cos t \end{bmatrix} e^t$

7. Let $\mathbf{A}(t) = \begin{bmatrix} e^{4t} & \cos \pi t \\ 2t & 3t^2 - 1 \end{bmatrix}$.

   Find (a) $\dfrac{d\mathbf{A}}{dt}$, (b) $\displaystyle\int_0^2 \mathbf{A}(t)\, dt$, (c) $\displaystyle\int_0^t \mathbf{A}(s)\, ds$.

8. Let $\mathbf{A}(t) = \begin{bmatrix} 1 & 3t \\ t^2+1 & \\ t^2 & t \end{bmatrix}$ and $\mathbf{B}(t) = \begin{bmatrix} 6t & 2 \\ \frac{1}{t} & 4t \end{bmatrix}$.

   Find (a) $\dfrac{d\mathbf{A}}{dt}$, (b) $\dfrac{d\mathbf{B}}{dt}$, (c) $\displaystyle\int_0^1 \mathbf{A}(t)\, dt$, (d) $\displaystyle\int_1^2 \mathbf{B}(t)\, dt$,

   (e) $\mathbf{A}(t)\mathbf{B}(t)$, (f) $\dfrac{d}{dt}\mathbf{A}(t)\mathbf{B}(t)$, (g) $\displaystyle\int_1^t \mathbf{A}(s)\mathbf{B}(s)\, ds$.

In Problems 9–14 write the given system in matrix form.

9. $\dfrac{dx}{dt} = 3x - 5y$

   $\dfrac{dy}{dt} = 4x + 8y$

10. $\dfrac{dx}{dt} = 4x - 7y$

    $\dfrac{dy}{dt} = 5x$

11. $\dfrac{dx}{dt} = -3x + 4y - 9z$

    $\dfrac{dy}{dt} = 6x - y$

    $\dfrac{dz}{dt} = 10x + 4y + 3z$

12. $\dfrac{dx}{dt} = x - y$

    $\dfrac{dy}{dt} = x + 2z$

    $\dfrac{dz}{dt} = -x + z$

13. $\dfrac{dx}{dt} = x - y + z + t - 1$

    $\dfrac{dy}{dt} = 2x + y - z - 3t^2$

    $\dfrac{dz}{dt} = x + y + z + t^2 - t + 2$

14. $\dfrac{dx}{dt} = -3x + 4y + e^{-t}\sin 2t$

    $\dfrac{dy}{dt} = 5x + 9y + 4e^{-t}\cos 2t$

In Problems 15–18 write the given system without the use of matrices.

15. $\mathbf{X}' = \begin{bmatrix} 4 & 2 \\ -1 & 3 \end{bmatrix} \mathbf{X} + \begin{bmatrix} 1 \\ -1 \end{bmatrix} e^t$

16. $\mathbf{X}' = \begin{bmatrix} 7 & 5 & -9 \\ 4 & 1 & 1 \\ 0 & -2 & 3 \end{bmatrix} \mathbf{X} + \begin{bmatrix} 0 \\ 2 \\ 1 \end{bmatrix} e^{5t} - \begin{bmatrix} 8 \\ 0 \\ 3 \end{bmatrix} e^{-2t}$

17. $\dfrac{d}{dt}\begin{bmatrix} x \\ y \\ z \end{bmatrix} = \begin{bmatrix} 1 & -1 & 2 \\ 3 & -4 & 1 \\ -2 & 5 & 6 \end{bmatrix} \begin{bmatrix} x \\ y \\ z \end{bmatrix} + \begin{bmatrix} 1 \\ 2 \\ 2 \end{bmatrix} e^{-t} - \begin{bmatrix} 3 \\ -1 \\ 1 \end{bmatrix} t$

18. $\dfrac{d}{dt}\begin{bmatrix} x \\ y \end{bmatrix} = \begin{bmatrix} 3 & -7 \\ 1 & 1 \end{bmatrix} \begin{bmatrix} x \\ y \end{bmatrix} + \begin{bmatrix} 4 \\ 8 \end{bmatrix} \sin t + \begin{bmatrix} t - 4 \\ 2t + 1 \end{bmatrix} e^{4t}$

In Problems 19 and 20 show that the given matrix is nonsingular for every real value of t. Find $\mathbf{A}^{-1}(t)$.

19. $\mathbf{A}(t) = \begin{bmatrix} 2e^{-t} & e^{4t} \\ 4e^{-t} & 3e^{4t} \end{bmatrix}$

20. $\mathbf{A}(t) = \begin{bmatrix} 2e^t \sin t & -2e^t \cos t \\ e^t \cos t & e^t \sin t \end{bmatrix}$

9.3 PRELIMINARY THEORY

Our goal in this chapter is to examine a systematic procedure for solving linear first-order systems in the normal form $\mathbf{X}' = \mathbf{AX} + \mathbf{F}(t)$. Before examining this method, which is based on finding the eigenvalues and eigenvectors of the coefficient matrix $\mathbf{A}$, we need some preliminary theory.

DEFINITION 9.3 Solution Vector

A **solution vector** on an interval I is any column matrix

$$\mathbf{X} = \begin{bmatrix} x_1(t) \\ x_2(t) \\ \vdots \\ x_n(t) \end{bmatrix}$$

whose entries are differentiable functions satisfying the system $\mathbf{X}' = \mathbf{AX} + \mathbf{F}(t)$ on the interval.

EXAMPLE 1 Verify that

$$\mathbf{X}_1 = \begin{bmatrix} 1 \\ -1 \end{bmatrix} e^{-2t} = \begin{bmatrix} e^{-2t} \\ -e^{-2t} \end{bmatrix} \quad \text{and} \quad \mathbf{X}_2 = \begin{bmatrix} 3 \\ 5 \end{bmatrix} e^{6t} = \begin{bmatrix} 3e^{6t} \\ 5e^{6t} \end{bmatrix}$$

are solutions of
$$\mathbf{X}' = \begin{bmatrix} 1 & 3 \\ 5 & 3 \end{bmatrix} \mathbf{X} \tag{1}$$

on the interval $(-\infty, \infty)$.

Solution We have
$$\mathbf{X}'_1 = \begin{bmatrix} -2e^{-2t} \\ 2e^{-2t} \end{bmatrix}$$

and $\quad \mathbf{AX}_1 = \begin{bmatrix} 1 & 3 \\ 5 & 3 \end{bmatrix} \begin{bmatrix} e^{-2t} \\ -e^{-2t} \end{bmatrix} = \begin{bmatrix} e^{-2t} - 3e^{-2t} \\ 5e^{-2t} - 3e^{-2t} \end{bmatrix} = \begin{bmatrix} -2e^{-2t} \\ 2e^{-2t} \end{bmatrix} = \mathbf{X}'_1$

Now $\quad \mathbf{X}'_2 = \begin{bmatrix} 18e^{6t} \\ 30e^{5t} \end{bmatrix}$

and $\quad \mathbf{AX}_2 = \begin{bmatrix} 1 & 3 \\ 5 & 3 \end{bmatrix} \begin{bmatrix} 3e^{6t} \\ 5e^{6t} \end{bmatrix} = \begin{bmatrix} 3e^{6t} + 15e^{6t} \\ 15e^{6t} + 15e^{6t} \end{bmatrix} = \begin{bmatrix} 18e^{6t} \\ 30e^{6t} \end{bmatrix} = \mathbf{X}'_2$ □

Much of the theory of systems of n linear first-order differential equations is similar to that of linear nth-order differential equations.

Initial-Value Problem   Let t_0 denote a point on an interval I and

$$\mathbf{X}(t_0) = \begin{bmatrix} x_1(t_0) \\ x_2(t_0) \\ \vdots \\ x_n(t_0) \end{bmatrix} \quad \text{and} \quad \mathbf{X}_0 = \begin{bmatrix} \gamma_1 \\ \gamma_2 \\ \vdots \\ \gamma_n \end{bmatrix}$$

where the γ_i, $i = 1, 2, \ldots, n$, are given constants. Then the problem

$$\boxed{\begin{aligned} \text{Solve:} \quad & \frac{d\mathbf{X}}{dt} = \mathbf{A}(t)\mathbf{X} + \mathbf{F}(t) \\ \text{Subject to:} \quad & \mathbf{X}(t_0) = \mathbf{X}_0 \end{aligned}} \quad (2)$$

is an **initial-value problem** on the interval.

THEOREM 9.1   **Existence of a Unique Solution**

Let the entries of the matrices $\mathbf{A}(t)$ and $\mathbf{F}(t)$ be functions continuous on a common interval I that contains the point t_0. Then there exists a unique solution of the initial-value problem (2) on the interval.

Homogeneous Systems   In the next several definitions and theorems we will be concerned with only homogeneous systems. Without stating it, we shall always assume that the a_{ij} and the f_i are continuous functions of t on some common interval I.

Superposition Principle   The following result is a **superposition principle** for solutions of linear systems.

THEOREM 9.2   **Superposition Principle**

Let $\mathbf{X}_1, \mathbf{X}_2, \ldots, \mathbf{X}_k$ be a set of solution vectors of the homogeneous system $\mathbf{X}' = \mathbf{A}\mathbf{X}$ on an interval I. Then the linear combination

$$\mathbf{X} = c_1\mathbf{X}_1 + c_2\mathbf{X}_2 + \cdots + c_k\mathbf{X}_k$$

where the c_i, $i = 1, 2, \ldots, k$, are arbitrary constants, is also a solution on the interval.

It follows from Theorem 9.2 that a constant multiple of any solution vector of a homogeneous system of linear first-order differential equations is also a solution.

9.3 Preliminary Theory

EXAMPLE 2 One solution of the system

$$\mathbf{X}' = \begin{bmatrix} 1 & 0 & 1 \\ 1 & 1 & 0 \\ -2 & 0 & -1 \end{bmatrix} \mathbf{X} \tag{3}$$

is

$$\mathbf{X}_1 = \begin{bmatrix} \cos t \\ -\tfrac{1}{2}\cos t + \tfrac{1}{2}\sin t \\ -\cos t - \sin t \end{bmatrix}$$

For any constant c_1 the vector $\mathbf{X} = c_1 \mathbf{X}_1$ is also a solution, since

$$\frac{d\mathbf{X}}{dt} = \begin{bmatrix} -c_1 \sin t \\ \tfrac{1}{2}c_1 \sin t + \tfrac{1}{2}c_1 \cos t \\ c_1 \sin t - c_1 \cos t \end{bmatrix}$$

and

$$\mathbf{AX} = \begin{bmatrix} 1 & 0 & 1 \\ 1 & 1 & 0 \\ -2 & 0 & -1 \end{bmatrix} \begin{bmatrix} c_1 \cos t \\ -\tfrac{1}{2}c_1 \cos t + \tfrac{1}{2}c_1 \sin t \\ -c_1 \cos t - c_1 \sin t \end{bmatrix} = \begin{bmatrix} -c_1 \sin t \\ \tfrac{1}{2}c_1 \cos t + \tfrac{1}{2}c_1 \sin t \\ -c_1 \cos t + c_1 \sin t \end{bmatrix}$$

Inspection of the resulting matrices shows that $\mathbf{X}' = \mathbf{AX}$. □

EXAMPLE 3 Consider the system (3) of Example 2. If

$$\mathbf{X}_2 = \begin{bmatrix} 0 \\ e^t \\ 0 \end{bmatrix} \quad \text{then} \quad \mathbf{X}'_2 = \begin{bmatrix} 0 \\ e^t \\ 0 \end{bmatrix}$$

and

$$\mathbf{AX}_2 = \begin{bmatrix} 1 & 0 & 1 \\ 1 & 1 & 0 \\ -2 & 0 & -1 \end{bmatrix} \begin{bmatrix} 0 \\ e^t \\ 0 \end{bmatrix} = \begin{bmatrix} 0 \\ e^t \\ 0 \end{bmatrix} = \mathbf{X}'_2$$

Thus, we see that $\mathbf{X}_2$ is also a solution vector of (3). By the superposition principle, the linear combination

$$\mathbf{X} = c_1 \mathbf{X}_1 + c_2 \mathbf{X}_2 = c_1 \begin{bmatrix} \cos t \\ -\tfrac{1}{2}\cos t + \tfrac{1}{2}\sin t \\ -\cos t - \sin t \end{bmatrix} + c_2 \begin{bmatrix} 0 \\ e^t \\ 0 \end{bmatrix}$$

is yet another solution of the system. □

Linear Independence We are primarily interested in linearly independent solutions of homogeneous systems.

> **DEFINITION 9.4 Linear Dependence/Independence**
>
> Let $X_1, X_2, \ldots, X_k$ be a set of solution vectors of the homogeneous system $X' = AX$ on an interval I. We say that the set is **linearly dependent** on the interval if there exist constants $c_1, c_2, \ldots, c_k$, not all zero, such that
>
> $$c_1 X_1 + c_2 X_2 + \cdots + c_k X_k = 0$$
>
> for every t in the interval. If the set of vectors is not linearly dependent on the interval, it is said to be **linearly independent**.

The case when $k = 2$ should be clear; two solution vectors X_1 and X_2 are linearly dependent if one is a constant multiple of the other, and conversely. For $k > 2$, a set of solution vectors is linearly dependent if we can express at least one solution vector as a nontrivial linear combination of the remaining vectors.

EXAMPLE 4 It can be verified that

$$X_1 = \begin{bmatrix} 3 \\ 1 \end{bmatrix} e^t \quad \text{and} \quad X_2 = \begin{bmatrix} 1 \\ 1 \end{bmatrix} e^{-t}$$

are solution vectors of the system

$$X' = \begin{bmatrix} 2 & -3 \\ 1 & -2 \end{bmatrix} X \tag{4}$$

Now X_1 and X_2 are linearly independent on the interval $(-\infty, \infty)$, since

$$c_1 X_1 + c_2 X_2 = 0 \quad \text{or} \quad c_1 \begin{bmatrix} 3 \\ 1 \end{bmatrix} e^t + c_2 \begin{bmatrix} 1 \\ 1 \end{bmatrix} e^{-t} = \begin{bmatrix} 0 \\ 0 \end{bmatrix}$$

is equivalent to

$$3 c_1 e^t + c_2 e^{-t} = 0$$
$$c_1 e^t + c_2 e^{-t} = 0$$

Solving this system for c_1 and c_2 immediately yields $c_1 = 0$ and $c_2 = 0$. □

EXAMPLE 5 The vector $X_3 = \begin{bmatrix} e^t + \cosh t \\ \cosh t \end{bmatrix}$ is also a solution of the system (4). However, X_1, X_2, and X_3 are linearly dependent, since

$$X_3 = \frac{1}{2} X_1 + \frac{1}{2} X_2$$

□

Wronskian As in our earlier consideration of the theory of a single ordinary differential equation, we can introduce the concept of the **Wronskian** determinant as a test for linear independence. We state the following theorem without proof:

THEOREM 9.3   Criterion for Linear Independence

Let $\quad \mathbf{X}_1 = \begin{bmatrix} x_{11} \\ x_{21} \\ \vdots \\ x_{n1} \end{bmatrix}, \quad \mathbf{X}_2 = \begin{bmatrix} x_{12} \\ x_{22} \\ \vdots \\ x_{n2} \end{bmatrix}, \quad \ldots, \quad \mathbf{X}_n = \begin{bmatrix} x_{1n} \\ x_{2n} \\ \vdots \\ x_{nn} \end{bmatrix}$

be n solution vectors of the homogeneous system $\mathbf{X}' = \mathbf{AX}$ on an interval I. A necessary and sufficient condition that the set of solutions be linearly independent is that the Wronskian

$$W(\mathbf{X}_1, \mathbf{X}_2, \ldots, \mathbf{X}_n) = \begin{vmatrix} x_{11} & x_{12} & \cdots & x_{1n} \\ x_{21} & x_{22} & \cdots & x_{2n} \\ \vdots & & & \vdots \\ x_{n1} & x_{n2} & \cdots & x_{nn} \end{vmatrix} \neq 0 \quad (5)$$

for every t in I.

In fact it can be shown that if $\mathbf{X}_1, \mathbf{X}_2, \ldots, \mathbf{X}_n$ are solution vectors of $\mathbf{X}' = \mathbf{AX}$, then for every t in I either

$$W(\mathbf{X}_1, \mathbf{X}_2, \ldots, \mathbf{X}_n) \neq 0 \quad \text{or} \quad W(\mathbf{X}_1, \mathbf{X}_2, \ldots, \mathbf{X}_n) = 0$$

Thus, if we can show that $W \neq 0$ for some t_0 in I, then $W \neq 0$ for every t and hence the solutions are linearly independent on the interval.

Notice that, unlike our previous definition of the Wronskian, the determinant (5) does not involve differentiation.

EXAMPLE 6   In Example 1 we saw that

$$\mathbf{X}_1 = \begin{bmatrix} 1 \\ -1 \end{bmatrix} e^{-2t} \quad \text{and} \quad \mathbf{X}_2 = \begin{bmatrix} 3 \\ 5 \end{bmatrix} e^{6t}$$

are solutions of the system (1). Clearly, $\mathbf{X}_1$ and $\mathbf{X}_2$ are linearly independent on $(-\infty, \infty)$ since neither vector is a constant multiple of the other. In addition, we have

$$W(\mathbf{X}_1, \mathbf{X}_2) = \begin{vmatrix} e^{-2t} & 3e^{6t} \\ -e^{-2t} & 5e^{6t} \end{vmatrix} = 8e^{4t} \neq 0$$

for all real values of t.   □

Fundamental Set of Solutions

> **DEFINITION 9.5 Fundamental Set**
>
> Any set $X_1, X_2, \ldots, X_n$ of n linearly independent solution vectors of the homogeneous system $X' = AX$ on an interval I is said to be a **fundamental set of solutions** on the interval.

> **THEOREM 9.4 Existence of a Fundamental Set**
>
> There exists a fundamental set of solutions for the homogeneous system $X' = AX$ on an interval I.

> **DEFINITION 9.6 General Solution—Homogeneous System**
>
> Let $X_1, X_2, \ldots, X_n$ be a fundamental set of solutions of the homogeneous system $X' = AX$ on an interval I. The **general solution** of the system on the interval is defined to be
>
> $$X = c_1 X_1 + c_2 X_2 + \cdots + c_n X_n$$
>
> where the c_i, $i = 1, 2, \ldots, n$, are arbitrary constants.

Although we shall not give the proof, it can be shown that, for appropriate choices of the constants $c_1, c_2, \ldots, c_n$, *any* solution of $X' = AX$ on the interval I can be obtained from the general solution.

EXAMPLE 7 From Example 6 we know that

$$X_1 = \begin{bmatrix} 1 \\ -1 \end{bmatrix} e^{-2t} \quad \text{and} \quad X_2 = \begin{bmatrix} 3 \\ 5 \end{bmatrix} e^{6t}$$

are linearly independent solutions of (1) on $(-\infty, \infty)$. Hence, X_1 and X_2 form a fundamental set of solutions on the interval. The general solution of the system on the interval is then

$$X = c_1 X_1 + c_2 X_2 = c_1 \begin{bmatrix} 1 \\ -1 \end{bmatrix} e^{-2t} + c_2 \begin{bmatrix} 3 \\ 5 \end{bmatrix} e^{6t} \qquad (6)$$

EXAMPLE 8 The vectors

$$\mathbf{X}_1 = \begin{bmatrix} \cos t \\ -\tfrac{1}{2}\cos t + \tfrac{1}{2}\sin t \\ -\cos t - \sin t \end{bmatrix} \quad \mathbf{X}_2 = \begin{bmatrix} 0 \\ 1 \\ 0 \end{bmatrix} e^t \quad \mathbf{X}_3 = \begin{bmatrix} \sin t \\ -\tfrac{1}{2}\sin t - \tfrac{1}{2}\cos t \\ -\sin t + \cos t \end{bmatrix}$$

are solutions of the system (3)* in Example 2. Now

$$W(\mathbf{X}_1, \mathbf{X}_2, \mathbf{X}_3) = \begin{vmatrix} \cos t & 0 & \sin t \\ -\tfrac{1}{2}\cos t + \tfrac{1}{2}\sin t & e^t & -\tfrac{1}{2}\sin t - \tfrac{1}{2}\cos t \\ -\cos t - \sin t & 0 & -\sin t + \cos t \end{vmatrix}$$

$$= e^t \begin{vmatrix} \cos t & \sin t \\ -\cos t - \sin t & -\sin t + \cos t \end{vmatrix} = e^t \neq 0$$

for all real values of t. We conclude that $\mathbf{X}_1$, $\mathbf{X}_2$, and $\mathbf{X}_3$ form a fundamental set of solutions on $(-\infty, \infty)$. Thus, the general solution of the system on the interval is

$$\mathbf{X} = c_1 \mathbf{X}_1 + c_2 \mathbf{X}_2 + c_3 \mathbf{X}_3 = c_1 \begin{bmatrix} \cos t \\ -\tfrac{1}{2}\cos t + \tfrac{1}{2}\sin t \\ -\cos t - \sin t \end{bmatrix} + c_2 \begin{bmatrix} 0 \\ 1 \\ 0 \end{bmatrix} e^t + c_3 \begin{bmatrix} \sin t \\ -\tfrac{1}{2}\sin t - \tfrac{1}{2}\cos t \\ -\sin t + \cos t \end{bmatrix} \quad \square$$

Nonhomogeneous Systems For nonhomogeneous systems a **particular solution** $\mathbf{X}_p$ on an interval I is any vector, free of arbitrary parameters, whose entries are functions satisfying the system $\mathbf{X}' = \mathbf{AX} + \mathbf{F}(t)$.

EXAMPLE 9 Verify that the vector $\mathbf{X}_p = \begin{bmatrix} 3t - 4 \\ -5t + 6 \end{bmatrix}$ is a particular solution of the nonhomogeneous system

$$\mathbf{X}' = \begin{bmatrix} 1 & 3 \\ 5 & 3 \end{bmatrix} \mathbf{X} + \begin{bmatrix} 12t - 11 \\ -3 \end{bmatrix} \tag{7}$$

on the interval $(-\infty, \infty)$.

Solution We have $\mathbf{X}'_p = \begin{bmatrix} 3 \\ -5 \end{bmatrix}$ and

$$\begin{bmatrix} 1 & 3 \\ 5 & 3 \end{bmatrix} \mathbf{X}_p + \begin{bmatrix} 12t - 11 \\ -3 \end{bmatrix} = \begin{bmatrix} 1 & 3 \\ 5 & 3 \end{bmatrix} \begin{bmatrix} 3t - 4 \\ -5t + 6 \end{bmatrix} + \begin{bmatrix} 12t - 11 \\ -3 \end{bmatrix}$$

$$= \begin{bmatrix} (3t - 4) + 3(-5t + 6) \\ 5(3t - 4) + 3(-5t + 6) \end{bmatrix} + \begin{bmatrix} 12t - 11 \\ -3 \end{bmatrix}$$

$$= \begin{bmatrix} -12t + 14 \\ -2 \end{bmatrix} + \begin{bmatrix} 12t - 11 \\ -3 \end{bmatrix} = \begin{bmatrix} 3 \\ -5 \end{bmatrix} = \mathbf{X}'_p$$

$\square$

---

*On page 631 it was verified that $\mathbf{X}_1$ and $\mathbf{X}_2$ are solutions; it is left as an exercise to demonstrate that $\mathbf{X}_3$ is also a solution.

THEOREM 9.5

Let $X_1, X_2, \ldots, X_k$ be a set of solution vectors of the homogeneous system $X' = AX$ on an interval I, and let X_p be any solution vector of the nonhomogeneous system $X' = AX + F(t)$ on the same interval. Then

$$X = c_1 X_1 + c_2 X_2 + \cdots + c_k X_k + X_p$$

is also a solution of the nonhomogeneous system on the interval for any constants $c_1, c_2, \ldots, c_k$.

DEFINITION 9.7 General Solution—Nonhomogeneous System

Let X_p be a given solution of the nonhomogeneous system $X' = AX + F(t)$ on an interval I, and let

$$X_c = c_1 X_1 + c_2 X_2 + \cdots + c_n X_n$$

denote the general solution on the same interval of the corresponding homogeneous system $X' = AX$. The **general solution** of the nonhomogeneous system on the interval is defined to be

$$X = X_c + X_p$$

The general solution X_c of the homogeneous system is called the **complementary function** of the nonhomogeneous system.

EXAMPLE 10 In Example 9 it was verified that a particular solution of the nonhomogeneous system (7) on $(-\infty, \infty)$ is

$$X_p = \begin{bmatrix} 3t - 4 \\ -5t + 6 \end{bmatrix}$$

In Example 7 we saw that the complementary function of (7) on the same interval, or general solution

$$X' = \begin{bmatrix} 1 & 3 \\ 5 & 3 \end{bmatrix} X$$

was

$$X_c = c_1 \begin{bmatrix} 1 \\ -1 \end{bmatrix} e^{-2t} + c_2 \begin{bmatrix} 3 \\ 5 \end{bmatrix} e^{6t}$$

Hence by Definition 9.7,

$$X = X_c + X_p = c_1 \begin{bmatrix} 1 \\ -1 \end{bmatrix} e^{-2t} + c_2 \begin{bmatrix} 3 \\ 5 \end{bmatrix} e^{6t} + \begin{bmatrix} 3t - 4 \\ -5t + 6 \end{bmatrix}$$

is the general solution of (7) on $(-\infty, \infty)$. □

As one might expect, if **X** is *any* solution of the nonhomogeneous system $\mathbf{X}' = \mathbf{AX} + \mathbf{F}(t)$ on an interval I, then it is always possible to find appropriate constants $c_1, c_2, \ldots, c_n$ so that **X** can be obtained from the general solution.

A Fundamental Matrix If $\mathbf{X}_1, \mathbf{X}_2, \ldots, \mathbf{X}_n$ is a fundamental set of solutions of the homogeneous system $\mathbf{X}' = \mathbf{AX}$ on an interval I, then its general solution on the interval is

$$\mathbf{X} = c_1 \mathbf{X}_1 + c_2 \mathbf{X}_2 + \cdots + c_n \mathbf{X}_n$$

$$= c_1 \begin{bmatrix} x_{11} \\ x_{21} \\ \vdots \\ x_{n1} \end{bmatrix} + c_2 \begin{bmatrix} x_{12} \\ x_{22} \\ \vdots \\ x_{n2} \end{bmatrix} + \cdots + c_n \begin{bmatrix} x_{1n} \\ x_{2n} \\ \vdots \\ x_{nn} \end{bmatrix} = \begin{bmatrix} c_1 x_{11} + c_2 x_{12} + \cdots + c_n x_{1n} \\ c_1 x_{21} + c_2 x_{22} + \cdots + c_n x_{2n} \\ \vdots \\ c_1 x_{n1} + c_2 x_{n2} + \cdots + c_n x_{nn} \end{bmatrix} \quad (8)$$

Observe that (8) can be written as the matrix product

$$\mathbf{X} = \begin{bmatrix} x_{11} & x_{12} & \cdots & x_{1n} \\ x_{21} & x_{22} & \cdots & x_{2n} \\ \vdots & & & \vdots \\ x_{n1} & x_{n2} & \cdots & x_{nn} \end{bmatrix} \begin{bmatrix} c_1 \\ c_2 \\ \vdots \\ c_n \end{bmatrix} \quad (9)$$

We are led to the following definition:

DEFINITION 9.8 Fundamental Matrix

Let $\quad \mathbf{X}_1 = \begin{bmatrix} x_{11} \\ x_{21} \\ \vdots \\ x_{n1} \end{bmatrix}, \quad \mathbf{X}_2 = \begin{bmatrix} x_{12} \\ x_{22} \\ \vdots \\ x_{n2} \end{bmatrix}, \quad \ldots, \quad \mathbf{X}_n = \begin{bmatrix} x_{1n} \\ x_{2n} \\ \vdots \\ x_{nn} \end{bmatrix}$

be a fundamental set of n solution vectors of the homogeneous system $\mathbf{X}' = \mathbf{AX}$ on an interval I. The matrix

$$\mathbf{\Phi}(t) = \begin{bmatrix} x_{11} & x_{12} & \cdots & x_{1n} \\ x_{21} & x_{22} & \cdots & x_{2n} \\ \vdots & & & \vdots \\ x_{n1} & x_{n2} & \cdots & x_{nn} \end{bmatrix}$$

is said to be a **fundamental matrix** of the system on the interval.

EXAMPLE 11 The vectors

$$\mathbf{X}_1 = \begin{bmatrix} 1 \\ -1 \end{bmatrix} e^{-2t} = \begin{bmatrix} e^{-2t} \\ -e^{-2t} \end{bmatrix} \quad \text{and} \quad \mathbf{X}_2 = \begin{bmatrix} 3 \\ 5 \end{bmatrix} e^{6t} = \begin{bmatrix} 3e^{6t} \\ 5e^{6t} \end{bmatrix}$$

have been shown to form a fundamental set of solutions of the system (1) on $(-\infty, \infty)$. A fundamental matrix of the system on the interval is then

$$\Phi(t) = \begin{bmatrix} e^{-2t} & 3e^{6t} \\ -e^{-2t} & 5e^{6t} \end{bmatrix} \qquad (10)$$

□

The result given in (9) states that the general solution of any homogeneous system $\mathbf{X}' = \mathbf{A}(t)\mathbf{X}$ can always be written in terms of a fundamental matrix of the system: $\mathbf{X} = \Phi(t)\mathbf{C}$, where $\mathbf{C}$ is an $n \times 1$ column vector of arbitrary constants.

EXAMPLE 12 The general solution given in (6) can be written as

$$\mathbf{X} = \begin{bmatrix} e^{-2t} & 3e^{6t} \\ -e^{-2t} & 5e^{6t} \end{bmatrix} \begin{bmatrix} c_1 \\ c_2 \end{bmatrix} \qquad □$$

Furthermore, to say that $\mathbf{X} = \Phi(t)\mathbf{C}$ is a solution of $\mathbf{X}' = \mathbf{A}(t)\mathbf{X}$, we mean

$$\Phi'(t)\mathbf{C} = \mathbf{A}(t)\Phi(t)\mathbf{C} \quad \text{or} \quad (\Phi'(t) - \mathbf{A}(t)\Phi(t))\mathbf{C} = \mathbf{0}$$

Since the last equation is to hold for every t in the interval I and for every possible column matrix of constants $\mathbf{C}$, we must have

$$\Phi'(t) - \mathbf{A}(t)\Phi(t) = \mathbf{0}$$

or
$$\Phi'(t) = \mathbf{A}(t)\Phi(t) \qquad (11)$$

This result will be useful in Section 9.6.

Fundamental Matrix Is Nonsingular Comparison of (5) in Theorem 9.3 and Definition 9.8 shows that $\det \Phi(t)$ is the same as the Wronskian $W(\mathbf{X}_1, \mathbf{X}_2, \ldots, \mathbf{X}_n)$.* Hence the linear independence of the columns of $\Phi(t)$ on an interval I guarantees that $\det \Phi(t) \neq 0$ for every t in the interval; that is, $\Phi(t)$ is nonsingular on the interval.

THEOREM 9.6 A Fundamental Matrix Has an Inverse

Let $\Phi(t)$ be a fundamental matrix of the homogeneous system $\mathbf{X}' = \mathbf{A}\mathbf{X}$ on an interval I. Then $\Phi^{-1}(t)$ exists for every value of t in the interval.

---

*For this reason some texts call $\Phi(t)$ a *Wronski matrix*.

9.3 Preliminary Theory

EXAMPLE 13 For the fundamental matrix given in (10) we see that $\det \mathbf{\Phi}(t) = 8e^{4t}$. It then follows from (4) of Section 7.5 that

$$\mathbf{\Phi}^{-1}(t) = \frac{1}{8e^{4t}} \begin{bmatrix} 5e^{6t} & -3e^{6t} \\ e^{-2t} & e^{-2t} \end{bmatrix} = \begin{bmatrix} \frac{5}{8}e^{2t} & -\frac{3}{8}e^{2t} \\ \frac{1}{8}e^{-6t} & \frac{1}{8}e^{-6t} \end{bmatrix} \qquad \square$$

Special Matrix In some instances it is convenient to form another special $n \times n$ matrix, a matrix in which the column vectors $\mathbf{V}_i$ are solutions of $\mathbf{X}' = \mathbf{A}(t)\mathbf{X}$ that satisfy the conditions

$$\mathbf{V}_1(t_0) = \begin{bmatrix} 1 \\ 0 \\ \vdots \\ 0 \end{bmatrix}, \quad \mathbf{V}_2(t_0) = \begin{bmatrix} 0 \\ 1 \\ \vdots \\ 0 \end{bmatrix}, \quad \ldots, \quad \mathbf{V}_n(t_0) = \begin{bmatrix} 0 \\ 0 \\ \vdots \\ 1 \end{bmatrix} \qquad (12)$$

Here t_0 is an arbitrary chosen point in the interval on which the general solution of the system is defined. We shall denote this special matrix by the symbol $\mathbf{\Psi}(t)$. Observe that $\mathbf{\Psi}(t)$ has the property

$$\mathbf{\Psi}(t_0) = \begin{bmatrix} 1 & 0 & 0 & \cdots & 0 \\ 0 & 1 & 0 & \cdots & 0 \\ \vdots & & & & \vdots \\ 0 & 0 & 0 & \cdots & 1 \end{bmatrix} = \mathbf{I} \qquad (13)$$

where $\mathbf{I}$ is the $n \times n$ multiplicative identity.

EXAMPLE 14 Find the matrix $\mathbf{\Psi}(t)$ satisfying $\mathbf{\Psi}(0) = \mathbf{I}$ for the system given in (1).

Solution From (6) we know that the general solution of (1) is given by

$$\mathbf{X} = c_1 \begin{bmatrix} 1 \\ -1 \end{bmatrix} e^{-2t} + c_2 \begin{bmatrix} 3 \\ 5 \end{bmatrix} e^{6t}$$

When $t = 0$ we first solve for constants c_1 and c_2 such that

$$c_1 \begin{bmatrix} 1 \\ -1 \end{bmatrix} + c_2 \begin{bmatrix} 3 \\ 5 \end{bmatrix} = \begin{bmatrix} 1 \\ 0 \end{bmatrix} \quad \text{or} \quad \begin{array}{l} c_1 + 3c_2 = 1 \\ -c_1 + 5c_2 = 0 \end{array}$$

We find that $c_1 = \frac{5}{8}$ and $c_2 = \frac{1}{8}$. Hence we define the vector $\mathbf{V}_1$ to be the linear combination.

$$\mathbf{V}_1 = \frac{5}{8} \begin{bmatrix} 1 \\ -1 \end{bmatrix} e^{-2t} + \frac{1}{8} \begin{bmatrix} 3 \\ 5 \end{bmatrix} e^{6t}$$

Again when $t = 0$ we wish to find another pair of constants c_1 and c_2 for which

$$c_1 \begin{bmatrix} 1 \\ -1 \end{bmatrix} + c_2 \begin{bmatrix} 3 \\ 5 \end{bmatrix} = \begin{bmatrix} 0 \\ 1 \end{bmatrix} \quad \text{or} \quad \begin{array}{l} c_1 + 3c_2 = 0 \\ -c_1 + 5c_2 = 1 \end{array}$$

In this case we find $c_1 = -\frac{3}{8}$ and $c_2 = \frac{1}{8}$. We then define

$$V_2 = -\frac{3}{8}\begin{bmatrix} 1 \\ -1 \end{bmatrix}e^{-2t} + \frac{1}{8}\begin{bmatrix} 3 \\ 5 \end{bmatrix}e^{6t}$$

Hence, $\Psi(t) = \begin{bmatrix} \frac{5}{8}e^{-2t} + \frac{3}{8}e^{6t} & -\frac{3}{8}e^{-2t} + \frac{3}{8}e^{6t} \\ -\frac{5}{8}e^{-2t} + \frac{5}{8}e^{6t} & \frac{3}{8}e^{-2t} + \frac{5}{8}e^{6t} \end{bmatrix}$ (14)

Observe that $\Psi(0) = \begin{bmatrix} 1 & 0 \\ 0 & 1 \end{bmatrix} = I$. □

Note in the preceding example that since the columns of $\Psi(t)$ are linear combinations of the solutions of $X' = A(t)X$, we know from the superposition principle that each column is a solution of the system.

$\Psi(t)$ Is a Fundamental Matrix

From (13) it is seen that det $\Psi(t_0) \neq 0$, and hence we conclude from Theorem 9.3 that the columns of $\Psi(t)$ are linearly independent on the interval under consideration. Therefore $\Psi(t)$ is a fundamental matrix. Also, it follows from Theorem 9.1 that $\Psi(t)$ is the unique matrix satisfying the condition $\Psi(t_0) = I$. Last, the fundamental matrices $\Phi(t)$ and $\Psi(t)$ are related by

$$\Psi(t) = \Phi(t)\Phi^{-1}(t_0) \qquad (15)$$

Equation (15) provides an alternative method for determining $\Psi(t)$.

The answer to why anyone would want to form an obviously complicated looking fundamental matrix such as (14) will be answered in Sections 9.6 and 9.7.

EXERCISES 9.3
Answers to odd-numbered problems begin on page A-55.

In Problems 1–6 verify that the vector X is a solution of the given system.

1. $\dfrac{dx}{dt} = 3x - 4y$
   $\dfrac{dy}{dt} = 4x - 7y$; $X = \begin{bmatrix} 1 \\ 2 \end{bmatrix}e^{-5t}$

2. $\dfrac{dx}{dt} = -2x + 5y$
   $\dfrac{dy}{dt} = -2x + 4y$; $X = \begin{bmatrix} 5\cos t \\ 3\cos t - \sin t \end{bmatrix}e^t$

3. $X' = \begin{bmatrix} -1 & \frac{1}{4} \\ 1 & -1 \end{bmatrix}X$; $X = \begin{bmatrix} -1 \\ 2 \end{bmatrix}e^{-3t/2}$

4. $X' = \begin{bmatrix} 2 & 1 \\ -1 & 0 \end{bmatrix}X$; $X = \begin{bmatrix} 1 \\ 3 \end{bmatrix}e^t + \begin{bmatrix} 4 \\ -4 \end{bmatrix}te^t$

5. $\dfrac{dX}{dt} = \begin{bmatrix} 1 & 2 & 1 \\ 6 & -1 & 0 \\ -1 & -2 & -1 \end{bmatrix}X$; $X = \begin{bmatrix} 1 \\ 6 \\ -13 \end{bmatrix}$

6. $X' = \begin{bmatrix} 1 & 0 & 1 \\ 1 & 1 & 0 \\ -2 & 0 & -1 \end{bmatrix}X$; $X = \begin{bmatrix} \sin t \\ -\frac{1}{2}\sin t - \frac{1}{2}\cos t \\ -\sin t + \cos t \end{bmatrix}$

In Problems 7–10 the given vectors are solutions of a system $X' = AX$. Determine whether the vectors form a fundamental set on $(-\infty, \infty)$.

7. $X_1 = \begin{bmatrix} 1 \\ 1 \end{bmatrix}e^{-2t}$, $X_2 = \begin{bmatrix} 1 \\ -1 \end{bmatrix}e^{-6t}$

8. $X_1 = \begin{bmatrix} 1 \\ -1 \end{bmatrix}e^t$, $X_2 = \begin{bmatrix} 2 \\ 6 \end{bmatrix}e^t + \begin{bmatrix} 8 \\ -8 \end{bmatrix}te^t$

9. $\mathbf{X}_1 = \begin{bmatrix} 1 \\ -2 \\ 4 \end{bmatrix} + t\begin{bmatrix} 1 \\ 2 \\ 2 \end{bmatrix}$, $\mathbf{X}_2 = \begin{bmatrix} 1 \\ -2 \\ 4 \end{bmatrix}$, $\mathbf{X}_3 = \begin{bmatrix} 3 \\ -6 \\ 12 \end{bmatrix} + t\begin{bmatrix} 2 \\ 4 \\ 4 \end{bmatrix}$

10. $\mathbf{X}_1 = \begin{bmatrix} 1 \\ 6 \\ -13 \end{bmatrix}$, $\mathbf{X}_2 = \begin{bmatrix} 1 \\ -2 \\ -1 \end{bmatrix}e^{-4t}$, $\mathbf{X}_3 = \begin{bmatrix} 2 \\ 3 \\ -2 \end{bmatrix}e^{3t}$

In Problems 11–14 verify that the vector $\mathbf{X}_p$ is a particular solution of the given system.

11. $\dfrac{dx}{dt} = x + 4y + 2t - 7$

$\dfrac{dy}{dt} = 3x + 2y - 4t - 18$; $\mathbf{X}_p = \begin{bmatrix} 2 \\ -1 \end{bmatrix}t + \begin{bmatrix} 5 \\ 1 \end{bmatrix}$

12. $\mathbf{X}' = \begin{bmatrix} 2 & 1 \\ 1 & -1 \end{bmatrix}\mathbf{X} + \begin{bmatrix} -5 \\ 2 \end{bmatrix}$; $\mathbf{X}_p = \begin{bmatrix} 1 \\ 3 \end{bmatrix}$

13. $\mathbf{X}' = \begin{bmatrix} 2 & 1 \\ 3 & 4 \end{bmatrix}\mathbf{X} - \begin{bmatrix} 1 \\ 7 \end{bmatrix}e^t$; $\mathbf{X}_p = \begin{bmatrix} 1 \\ 1 \end{bmatrix}e^t + \begin{bmatrix} 1 \\ -1 \end{bmatrix}te^t$

14. $\mathbf{X}' = \begin{bmatrix} 1 & 2 & 3 \\ -4 & 2 & 0 \\ -6 & 1 & 0 \end{bmatrix}\mathbf{X} + \begin{bmatrix} -1 \\ 4 \\ 3 \end{bmatrix}\sin 3t$; $\mathbf{X}_p = \begin{bmatrix} \sin 3t \\ 0 \\ \cos 3t \end{bmatrix}$

15. Prove that the general solution of

$$\mathbf{X}' = \begin{bmatrix} 0 & 6 & 0 \\ 1 & 0 & 1 \\ 1 & 1 & 0 \end{bmatrix}\mathbf{X}$$

on the interval $(-\infty, \infty)$ is

$$\mathbf{X} = c_1\begin{bmatrix} 6 \\ -1 \\ -5 \end{bmatrix}e^{-t} + c_2\begin{bmatrix} -3 \\ 1 \\ 1 \end{bmatrix}e^{-2t} + c_3\begin{bmatrix} 2 \\ 1 \\ 1 \end{bmatrix}e^{3t}$$

16. Prove that the general solution of

$$\mathbf{X}' = \begin{bmatrix} -1 & -1 \\ -1 & 1 \end{bmatrix}\mathbf{X} + \begin{bmatrix} 1 \\ 1 \end{bmatrix}t^2 + \begin{bmatrix} 4 \\ -6 \end{bmatrix}t + \begin{bmatrix} -1 \\ 5 \end{bmatrix}$$

on the interval $(-\infty, \infty)$ is

$$\mathbf{X} = c_1\begin{bmatrix} 1 \\ -1-\sqrt{2} \end{bmatrix}e^{\sqrt{2}t} + c_2\begin{bmatrix} 1 \\ -1+\sqrt{2} \end{bmatrix}e^{-\sqrt{2}t}$$

$$+ \begin{bmatrix} 1 \\ 0 \end{bmatrix}t^2 + \begin{bmatrix} -2 \\ 4 \end{bmatrix}t + \begin{bmatrix} 1 \\ 0 \end{bmatrix}$$

In Problems 17–20 the indicated column vectors form a fundamental set of solutions for the given system on $(-\infty, \infty)$. Form a fundamental matrix $\Phi(t)$ and compute $\Phi^{-1}(t)$.

17. $\mathbf{X}' = \begin{bmatrix} 4 & 1 \\ 6 & 5 \end{bmatrix}\mathbf{X}$; $\mathbf{X}_1 = \begin{bmatrix} 1 \\ -2 \end{bmatrix}e^{2t}$, $\mathbf{X}_2 = \begin{bmatrix} 1 \\ 3 \end{bmatrix}e^{7t}$

18. $\mathbf{X}' = \begin{bmatrix} 2 & 3 \\ 3 & 2 \end{bmatrix}\mathbf{X}$; $\mathbf{X}_1 = \begin{bmatrix} -1 \\ 1 \end{bmatrix}e^{-t}$, $\mathbf{X}_2 = \begin{bmatrix} 1 \\ 1 \end{bmatrix}e^{5t}$

19. $\mathbf{X}' = \begin{bmatrix} 4 & 1 \\ -9 & -2 \end{bmatrix}\mathbf{X}$; $\mathbf{X}_1 = \begin{bmatrix} -1 \\ 3 \end{bmatrix}e^t$,

$\mathbf{X}_2 = \begin{bmatrix} -1 \\ 3 \end{bmatrix}te^t + \begin{bmatrix} 0 \\ -1 \end{bmatrix}e^t$

20. $\mathbf{X}' = \begin{bmatrix} 3 & -2 \\ 5 & -3 \end{bmatrix}\mathbf{X}$; $\mathbf{X}_1 = \begin{bmatrix} 2\cos t \\ 3\cos t + \sin t \end{bmatrix}$,

$\mathbf{X}_2 = \begin{bmatrix} -2\sin t \\ \cos t - 3\sin t \end{bmatrix}$

21. Find the fundamental matrix $\Psi(t)$ satisfying $\Psi(0) = \mathbf{I}$ for the system given in Problem 17.

22. Find the fundamental matrix $\Psi(t)$ satisfying $\Psi(0) = \mathbf{I}$ for the system given in Problem 18.

23. Find the fundamental matrix $\Psi(t)$ satisfying $\Psi(0) = \mathbf{I}$ for the system given in Problem 19.

24. Find the fundamental matrix $\Psi(t)$ satisfying $\Psi(\pi/2) = \mathbf{I}$ for the system given in Problem 20.

25. If $\mathbf{X} = \Phi(t)\mathbf{C}$ is the general solution of $\mathbf{X}' = \mathbf{AX}$, show that the solution of the initial-value problem $\mathbf{X}' = \mathbf{AX}$, $\mathbf{X}(t_0) = \mathbf{X}_0$ is $\mathbf{X} = \Phi(t)\Phi^{-1}(t_0)\mathbf{X}_0$.

26. Show that the solution of the initial-value problem given in Problem 25 is also given by $\mathbf{X} = \Psi(t)\mathbf{X}_0$.

27. Show that $\Psi(t) = \Phi(t)\Phi^{-1}(t_0)$. [*Hint:* Compare Problems 25 and 26.]

9.4 HOMOGENEOUS LINEAR SYSTEMS

9.4.1 Distinct Real Eigenvalues

For the remainder of this chapter we shall be concerned with only linear systems with real constant coefficients.

We saw in Example 7 of the preceding section that the general solution of the homogeneous system

$$\frac{dx}{dt} = x + 3y$$

$$\frac{dy}{dt} = 5x + 3y$$

is

$$\mathbf{X} = c_1 \begin{bmatrix} 1 \\ -1 \end{bmatrix} e^{-2t} + c_2 \begin{bmatrix} 3 \\ 5 \end{bmatrix} e^{6t}$$

Since both solution vectors have the basic form

$$\mathbf{X}_i = \begin{bmatrix} k_1 \\ k_2 \end{bmatrix} e^{\lambda_i t}, \quad i = 1, 2$$

k_1 and k_2 constants, we are prompted to ask whether we can always find a solution of the form

$$\mathbf{X} = \begin{bmatrix} k_1 \\ k_2 \\ \vdots \\ k_n \end{bmatrix} e^{\lambda t} = \mathbf{K} e^{\lambda t} \quad (1)$$

for the general homogeneous linear first-order system

$$\mathbf{X}' = \mathbf{A}\mathbf{X} \quad (2)$$

where $\mathbf{A}$ is an $n \times n$ matrix of constants.

Eigenvalues and Eigenvectors

If (1) is to be a solution vector of (2), then $\mathbf{X}' = \mathbf{K}\lambda e^{\lambda t}$ so that the system becomes

$$\mathbf{K}\lambda e^{\lambda t} = \mathbf{A}\mathbf{K} e^{\lambda t}$$

After dividing out $e^{\lambda t}$ and rearranging, we obtain

$$\mathbf{A}\mathbf{K} = \lambda \mathbf{K}$$

or

$$(\mathbf{A} - \lambda \mathbf{I})\mathbf{K} = \mathbf{0} \quad (3)$$

The last equation is equivalent to the simultaneous algebraic equations (3) of Section 7.7. To find a nontrivial solution $\mathbf{X}$ of (2) we must find a nontrivial vector $\mathbf{K}$ satisfying (3). But in order that (3) have nontrivial solutions, we must have

$$\det(\mathbf{A} - \lambda \mathbf{I}) = 0$$

The latter equation is recognized as the characteristic equation of the matrix $\mathbf{A}$. In other words, $\mathbf{X} = \mathbf{K}e^{\lambda t}$ will be a solution of the system of differential equations (2) if and only if λ is an **eigenvalue** of $\mathbf{A}$ and $\mathbf{K}$ is an **eigenvector** corresponding to λ.

When the $n \times n$ matrix $\mathbf{A}$ possesses n distinct real eigenvalues $\lambda_1, \lambda_2, \ldots, \lambda_n$, then a set of n linearly independent eigenvectors $\mathbf{K}_1, \mathbf{K}_2, \ldots, \mathbf{K}_n$ can

9.4 Homogeneous Linear Systems

always be found and

$$X_1 = K_1 e^{\lambda_1 t}, \quad X_2 = K_2 e^{\lambda_2 t}, \quad \ldots, \quad X_n = K_n e^{\lambda_n t}$$

is a fundamental set of solutions of (2) on $(-\infty, \infty)$.

> **THEOREM 9.7 General Solution**
>
> Let $\lambda_1, \lambda_2, \ldots, \lambda_n$ be n distinct real eigenvalues of the coefficient matrix **A** of the homogeneous system (2) and let $K_1, K_2, \ldots, K_n$ be the corresponding eigenvectors. Then the general solution of (2) on the interval $(-\infty, \infty)$ is given by
>
> $$X = c_1 K_1 e^{\lambda_1 t} + c_2 K_2 e^{\lambda_2 t} + \cdots + c_n K_n e^{\lambda_n t}$$

EXAMPLE 1 Solve

$$\frac{dx}{dt} = 2x + 3y$$
$$\frac{dy}{dt} = 2x + y \tag{4}$$

Solution We first find the eigenvalues and eigenvectors of the matrix of coefficients. The characteristic equation is

$$\det(\mathbf{A} - \lambda \mathbf{I}) = \begin{vmatrix} 2 - \lambda & 3 \\ 2 & 1 - \lambda \end{vmatrix} = \lambda^2 - 3\lambda - 4 = 0$$

Since $\lambda^2 - 3\lambda - 4 = (\lambda + 1)(\lambda - 4)$, we see that the eigenvalues are $\lambda_1 = -1$ and $\lambda_2 = 4$.

Now for $\lambda_1 = -1$, (3) is equivalent to

$$3k_1 + 3k_2 = 0$$
$$2k_1 + 2k_2 = 0$$

Thus, $k_1 = -k_2$. When we select $k_2 = -1$, the related eigenvector is

$$\mathbf{K}_1 = \begin{bmatrix} 1 \\ -1 \end{bmatrix}$$

For $\lambda_2 = 4$ we have

$$-2k_1 + 3k_2 = 0$$
$$2k_1 - 3k_2 = 0$$

so that $k_1 = 3k_2/2$, and therefore with $k_2 = 2$, the corresponding eigenvector is

$$\mathbf{K}_2 = \begin{bmatrix} 3 \\ 2 \end{bmatrix}$$

Since the matrix of coefficients $\mathbf{A}$ is a 2×2 matrix and since we have found two linearly independent solutions of (4):

$$\mathbf{X}_1 = \begin{bmatrix} 1 \\ -1 \end{bmatrix} e^{-t} \quad \text{and} \quad \mathbf{X}_2 = \begin{bmatrix} 3 \\ 2 \end{bmatrix} e^{4t}$$

we conclude that the general solution of the system is

$$\mathbf{X} = c_1 \mathbf{X}_1 + c_2 \mathbf{X}_2 = c_1 \begin{bmatrix} 1 \\ -1 \end{bmatrix} e^{-t} + c_2 \begin{bmatrix} 3 \\ 2 \end{bmatrix} e^{4t} \tag{5}$$

□

For the sake of review, the reader should keep firmly in mind that a solution of a system of first-order differential equations, when written in terms of matrices, is simply an alternative to the method we used in Section 3.7—namely, listing the individual functions and the relationships between the constants. By adding the vectors given in (5), we obtain

$$\begin{bmatrix} x(t) \\ y(t) \end{bmatrix} = \begin{bmatrix} c_1 e^{-t} + 3c_2 e^{4t} \\ -c_1 e^{-t} + 2c_2 e^{4t} \end{bmatrix}$$

and this in turn yields the more familiar statement

$$x(t) = c_1 e^{-t} + 3c_2 e^{4t}$$
$$y(t) = -c_1 e^{-t} + 2c_2 e^{4t}$$

EXAMPLE 2 Solve

$$\frac{dx}{dt} = -4x + y + z$$
$$\frac{dy}{dt} = x + 5y - z \tag{6}$$
$$\frac{dz}{dt} = y - 3z$$

Solution Using the cofactors of the third row, we find

$$\det(\mathbf{A} - \lambda \mathbf{I}) = \begin{vmatrix} -4-\lambda & 1 & 1 \\ 1 & 5-\lambda & -1 \\ 0 & 1 & -3-\lambda \end{vmatrix} = -(\lambda+3)(\lambda+4)(\lambda-5) = 0$$

and so the eigenvalues are $\lambda_1 = -3$, $\lambda_2 = -4$, $\lambda_3 = 5$.

Now for $\lambda_1 = -3$, Gauss–Jordan elimination gives

$$(\mathbf{A} + 3\mathbf{I} \,|\, \mathbf{0}) = \begin{bmatrix} -1 & 1 & 1 & | & 0 \\ 1 & 8 & -1 & | & 0 \\ 0 & 1 & 0 & | & 0 \end{bmatrix} \implies \begin{bmatrix} 1 & 0 & -1 & | & 0 \\ 0 & 1 & 0 & | & 0 \\ 0 & 0 & 0 & | & 0 \end{bmatrix}$$

Therefore $k_1 = k_3$, $k_2 = 0$. The choice $k_3 = 1$ gives the eigenvector

$$\mathbf{K}_1 = \begin{bmatrix} 1 \\ 0 \\ 1 \end{bmatrix} \tag{7}$$

9.4 Homogeneous Linear Systems

Similarly, for $\lambda_2 = -4$,

$$(\mathbf{A} + 4\mathbf{I} \mid \mathbf{0}) = \begin{bmatrix} 0 & 1 & 1 & \mid & 0 \\ 1 & 9 & -1 & \mid & 0 \\ 0 & 1 & 1 & \mid & 0 \end{bmatrix} \implies \begin{bmatrix} 1 & 0 & -10 & \mid & 0 \\ 0 & 1 & 1 & \mid & 0 \\ 0 & 0 & 0 & \mid & 0 \end{bmatrix}$$

implies $k_1 = 10k_3$, $k_2 = -k_3$. Choosing $k_3 = 1$ gives the second eigenvector:

$$\mathbf{K}_2 = \begin{bmatrix} 10 \\ -1 \\ 1 \end{bmatrix} \quad (8)$$

Finally, when $\lambda_3 = 5$, the augmented matrices

$$(\mathbf{A} - 5\mathbf{I} \mid \mathbf{0}) = \begin{bmatrix} -9 & 1 & 1 & \mid & 0 \\ 1 & 0 & -1 & \mid & 0 \\ 0 & 1 & -8 & \mid & 0 \end{bmatrix} \implies \begin{bmatrix} 1 & 0 & -1 & \mid & 0 \\ 0 & 1 & -8 & \mid & 0 \\ 0 & 0 & 0 & \mid & 0 \end{bmatrix}$$

yield

$$\mathbf{K}_3 = \begin{bmatrix} 1 \\ 8 \\ 1 \end{bmatrix} \quad (9)$$

Multiplying the vectors (7), (8), and (9) by e^{-3t}, e^{-4t}, and e^{5t}, respectively, gives three solutions of (6):

$$\mathbf{X}_1 = \begin{bmatrix} 1 \\ 0 \\ 1 \end{bmatrix} e^{-3t} \quad \mathbf{X}_2 = \begin{bmatrix} 10 \\ -1 \\ 1 \end{bmatrix} e^{-4t} \quad \mathbf{X}_3 = \begin{bmatrix} 1 \\ 8 \\ 1 \end{bmatrix} e^{5t}$$

The general solution of the system is then

$$\mathbf{X} = c_1 \begin{bmatrix} 1 \\ 0 \\ 1 \end{bmatrix} e^{-3t} + c_2 \begin{bmatrix} 10 \\ -1 \\ 1 \end{bmatrix} e^{-4t} + c_3 \begin{bmatrix} 1 \\ 8 \\ 1 \end{bmatrix} e^{5t} \quad \square$$

9.4.2 Complex Eigenvalues

If $\quad \lambda_1 = \alpha + i\beta \quad$ and $\quad \lambda_2 = \alpha - i\beta, \quad i^2 = -1$

are complex eigenvalues of the coefficient matrix $\mathbf{A}$, we can then certainly expect their corresponding eigenvectors to also have complex entries.*

For example, the characteristic equation of the system

$$\begin{aligned} \frac{dx}{dt} &= 6x - y \\ \frac{dy}{dt} &= 5x + 4y \end{aligned} \quad (10)$$

---

*When the characteristic equation has real coefficients, complex eigenvalues will always appear in conjugate pairs.

is $\quad \det(\mathbf{A} - \lambda \mathbf{I}) = \begin{vmatrix} 6 - \lambda & -1 \\ 5 & 4 - \lambda \end{vmatrix} = \lambda^2 - 10\lambda + 29 = 0$

From the quadratic formula we find $\lambda_1 = 5 + 2i$, $\lambda_2 = 5 - 2i$.
Now for $\lambda_1 = 5 + 2i$ we must solve

$$(1 - 2i)k_1 - k_2 = 0$$
$$5k_1 - (1 + 2i)k_2 = 0$$

Since $k_2 = (1 - 2i)k_1$,* it follows, after choosing $k_1 = 1$, that one eigenvector is

$$\mathbf{K}_1 = \begin{bmatrix} 1 \\ 1 - 2i \end{bmatrix}$$

Similarly, for $\lambda_2 = 5 - 2i$ we find the other eigenvector to be

$$\mathbf{K}_2 = \begin{bmatrix} 1 \\ 1 + 2i \end{bmatrix}$$

Consequently two solutions of (10) are

$$\mathbf{X}_1 = \begin{bmatrix} 1 \\ 1 - 2i \end{bmatrix} e^{(5+2i)t} \quad \text{and} \quad \mathbf{X}_2 = \begin{bmatrix} 1 \\ 1 + 2i \end{bmatrix} e^{(5-2i)t}$$

By the superposition principle another solution is

$$\mathbf{X} = c_1 \begin{bmatrix} 1 \\ 1 - 2i \end{bmatrix} e^{(5+2i)t} + c_2 \begin{bmatrix} 1 \\ 1 + 2i \end{bmatrix} e^{(5-2i)t} \tag{11}$$

Note that the entries in $\mathbf{K}_2$ corresponding to λ_2 are the conjugates of the entries in $\mathbf{K}_1$ corresponding to λ_1. The conjugate of λ_1 is, of course, λ_2. We write this as $\lambda_2 = \bar{\lambda}_1$ and $\mathbf{K}_2 = \bar{\mathbf{K}}_1$. We have illustrated the following general result:

THEOREM 9.8

Let $\mathbf{A}$ be the coefficient matrix having real entries of the homogeneous system (2), and let $\mathbf{K}_1$ be an eigenvector corresponding to the complex eigenvalue $\lambda_1 = \alpha + i\beta$, α and β real. Then

$$\mathbf{X}_1 = \mathbf{K}_1 e^{\lambda_1 t} \quad \text{and} \quad \mathbf{X}_2 = \bar{\mathbf{K}}_1 e^{\bar{\lambda}_1 t}$$

are solutions of (2).

---

*Note that the second equation is simply $(1 + 2i)$ times the first.

9.4 Homogeneous Linear Systems

It is desirable and relatively easy to rewrite a solution such as (11) in terms of real functions. Since

$$x = c_1 e^{(5+2i)t} + c_2 e^{(5-2i)t}$$
$$y = c_1(1-2i)e^{(5+2i)t} + c_2(1+2i)e^{(5-2i)t}$$

it follows from Euler's formula that

$$x = e^{5t}[c_1 e^{2it} + c_2 e^{-2it}]$$
$$= e^{5t}[(c_1 + c_2)\cos 2t + (c_1 i - c_2 i)\sin 2t]$$
$$y = e^{5t}[(c_1(1-2i) + c_2(1+2i))\cos 2t + (c_1 i(1-2i) - c_2 i(1+2i))\sin 2t]$$
$$= e^{5t}[(c_1+c_2) - 2(c_1 i - c_2 i)]\cos 2t + e^{5t}[2(c_1+c_2) + (c_1 i - c_2 i)]\sin 2t$$

If we replace $c_1 + c_2$ by C_1 and $c_1 i - c_2 i$ by C_2, then

$$x = e^{5t}[C_1 \cos 2t + C_2 \sin 2t]$$
$$y = e^{5t}[C_1 - 2C_2]\cos 2t + e^{5t}[2C_1 + C_2]\sin 2t$$

or, in terms of vectors,

$$\mathbf{X} = \begin{bmatrix} x \\ y \end{bmatrix} = C_1 \begin{bmatrix} \cos 2t \\ \cos 2t + 2\sin 2t \end{bmatrix} e^{5t} + C_2 \begin{bmatrix} \sin 2t \\ -2\cos 2t + \sin 2t \end{bmatrix} e^{5t} \quad (12)$$

Here, of course, it can be verified that each vector in (12) is a solution of (10). In addition, the solutions are linearly independent on the interval $(-\infty, \infty)$. We may further assume that C_1 and C_2 are completely arbitrary and real. Thus, (12) is the general solution of (10).

The foregoing process can be generalized. Let $\mathbf{K}_1$ be an eigenvector of the matrix $\mathbf{A}$ corresponding to the complex eigenvalue $\lambda_1 = \alpha + i\beta$. Then $\mathbf{X}_1$ and $\mathbf{X}_2$ in Theorem 9.8 can be written as

$$\mathbf{K}_1 e^{\lambda_1 t} = \mathbf{K}_1 e^{\alpha t} e^{i\beta t} = \mathbf{K}_1 e^{\alpha t}(\cos \beta t + i \sin \beta t)$$
$$\bar{\mathbf{K}}_1 e^{\bar{\lambda}_1 t} = \bar{\mathbf{K}}_1 e^{\alpha t} e^{-i\beta t} = \bar{\mathbf{K}}_1 e^{\alpha t}(\cos \beta t - i \sin \beta t)$$

The foregoing equations then yield

$$\frac{1}{2}(\mathbf{K}_1 e^{\lambda_1 t} + \bar{\mathbf{K}}_1 e^{\bar{\lambda}_1 t}) = \frac{1}{2}(\mathbf{K}_1 + \bar{\mathbf{K}}_1)e^{\alpha t} \cos \beta t - \frac{i}{2}(-\mathbf{K}_1 + \bar{\mathbf{K}}_1)e^{\alpha t} \sin \beta t$$

$$\frac{i}{2}(-\mathbf{K}_1 e^{\lambda_1 t} + \bar{\mathbf{K}}_1 e^{\bar{\lambda}_1 t}) = \frac{i}{2}(-\mathbf{K}_1 + \bar{\mathbf{K}}_1)e^{\alpha t} \cos \beta t + \frac{1}{2}(\mathbf{K}_1 + \bar{\mathbf{K}}_1)e^{\alpha t} \sin \beta t$$

For *any* complex number $z = a + ib$, we note that $\frac{1}{2}(z + \bar{z}) = a$ and $\frac{i}{2}(-z + \bar{z}) = b$ are *real* numbers. Therefore, the entries in the column vectors $\frac{1}{2}(\mathbf{K}_1 + \bar{\mathbf{K}}_1)$ and $\frac{i}{2}(-\mathbf{K}_1 + \bar{\mathbf{K}}_1)$ are real numbers. By defining

$$\mathbf{B}_1 = \frac{1}{2}[\mathbf{K}_1 + \bar{\mathbf{K}}_1] \quad \text{and} \quad \mathbf{B}_2 = \frac{i}{2}[-\mathbf{K}_1 + \bar{\mathbf{K}}_1] \quad (13)$$

we are led to the following theorem:

> **THEOREM 9.9**
>
> Let $\lambda_1 = \alpha + i\beta$ be a complex eigenvalue of the coefficient matrix **A** in the homogeneous system (2) and let $\mathbf{B}_1$ and $\mathbf{B}_2$ denote the column vectors defined in (13). Then
>
> $$\begin{aligned} \mathbf{X}_1 &= (\mathbf{B}_1 \cos \beta t - \mathbf{B}_2 \sin \beta t)e^{\alpha t} \\ \mathbf{X}_2 &= (\mathbf{B}_2 \cos \beta t + \mathbf{B}_1 \sin \beta t)e^{\alpha t} \end{aligned} \qquad (14)$$
>
> are linearly independent solutions of (2) on $(-\infty, \infty)$.

The matrices $\mathbf{B}_1$ and $\mathbf{B}_2$ in (13) are often denoted by

$$\mathbf{B}_1 = \text{Re}(\mathbf{K}_1) \quad \text{and} \quad \mathbf{B}_2 = \text{Im}(\mathbf{K}_1) \qquad (15)$$

since these vectors are, in turn, the *real* and *imaginary* parts of the eigenvector $\mathbf{K}_1$. For example, (12) follows from (14) with

$$\mathbf{K}_1 = \begin{bmatrix} 1 \\ 1 - 2i \end{bmatrix} = \begin{bmatrix} 1 \\ 1 \end{bmatrix} + i \begin{bmatrix} 0 \\ -2 \end{bmatrix}$$

$$\mathbf{B}_1 = \text{Re}(\mathbf{K}_1) = \begin{bmatrix} 1 \\ 1 \end{bmatrix} \quad \text{and} \quad \mathbf{B}_2 = \text{Im}(\mathbf{K}_1) = \begin{bmatrix} 0 \\ -2 \end{bmatrix}$$

EXAMPLE 3 Solve $\mathbf{X}' = \begin{bmatrix} 2 & 8 \\ -1 & -2 \end{bmatrix} \mathbf{X}$.

Solution First we obtain the eigenvalues from

$$\det(\mathbf{A} - \lambda \mathbf{I}) = \begin{vmatrix} 2 - \lambda & 8 \\ -1 & -2 - \lambda \end{vmatrix} = \lambda^2 + 4 = 0$$

Thus, the eigenvalues are $\lambda_1 = 2i$ and $\lambda_2 = \bar{\lambda}_1 = -2i$. For λ_1 we see that the system

$$\begin{aligned} (2 - 2i)k_1 + 8k_2 &= 0 \\ -k_1 + (-2 - 2i)k_2 &= 0 \end{aligned}$$

gives $k_1 = -(2 + 2i)k_2$. By choosing $k_2 = -1$, we get

$$\mathbf{K}_1 = \begin{bmatrix} 2 + 2i \\ -1 \end{bmatrix} = \begin{bmatrix} 2 \\ -1 \end{bmatrix} + i \begin{bmatrix} 2 \\ 0 \end{bmatrix}$$

Now from (15) we form

$$\mathbf{B}_1 = \text{Re}(\mathbf{K}_1) = \begin{bmatrix} 2 \\ -1 \end{bmatrix} \quad \text{and} \quad \mathbf{B}_2 = \text{Im}(\mathbf{K}_1) = \begin{bmatrix} 2 \\ 0 \end{bmatrix}$$

Since $\alpha = 0$, it follows from (14) that the general solution of the system is

$$\mathbf{X} = c_1 \left\{ \begin{bmatrix} 2 \\ -1 \end{bmatrix} \cos 2t - \begin{bmatrix} 2 \\ 0 \end{bmatrix} \sin 2t \right\} + c_2 \left\{ \begin{bmatrix} 2 \\ 0 \end{bmatrix} \cos 2t + \begin{bmatrix} 2 \\ -1 \end{bmatrix} \sin 2t \right\}$$

$$= c_1 \begin{bmatrix} 2\cos 2t - 2\sin 2t \\ -\cos 2t \end{bmatrix} + c_2 \begin{bmatrix} 2\cos 2t + 2\sin 2t \\ -\sin 2t \end{bmatrix} \qquad \square$$

EXAMPLE 4 Solve $\mathbf{X}' = \begin{bmatrix} 1 & 2 \\ -\frac{1}{2} & 1 \end{bmatrix} \mathbf{X}$.

Solution The solutions of the characteristic equation

$$\det(\mathbf{A} - \lambda \mathbf{I}) = \begin{vmatrix} 1-\lambda & 2 \\ -\frac{1}{2} & 1-\lambda \end{vmatrix} = \lambda^2 - 2\lambda + 2 = 0$$

are $\qquad \lambda_1 = 1 + i \quad \text{and} \quad \lambda_2 = \bar{\lambda}_1 = 1 - i$

Now an eigenvector associated with λ_1 is

$$\mathbf{K}_1 = \begin{bmatrix} 2 \\ i \end{bmatrix} = \begin{bmatrix} 2 \\ 0 \end{bmatrix} + i \begin{bmatrix} 0 \\ 1 \end{bmatrix}$$

From (15) we find

$$\mathbf{B}_1 = \begin{bmatrix} 2 \\ 0 \end{bmatrix} \quad \text{and} \quad \mathbf{B}_2 = \begin{bmatrix} 0 \\ 1 \end{bmatrix}$$

Thus (14) gives

$$\mathbf{X} = c_1 \left\{ \begin{bmatrix} 2 \\ 0 \end{bmatrix} \cos t - \begin{bmatrix} 0 \\ 1 \end{bmatrix} \sin t \right\} e^t + c_2 \left\{ \begin{bmatrix} 0 \\ 1 \end{bmatrix} \cos t + \begin{bmatrix} 2 \\ 0 \end{bmatrix} \sin t \right\} e^t$$

$$= c_1 \begin{bmatrix} 2\cos t \\ -\sin t \end{bmatrix} e^t + c_2 \begin{bmatrix} 2\sin t \\ \cos t \end{bmatrix} e^t \qquad \square$$

Alternative Method When $\mathbf{A}$ is a 2×2 matrix having a complex eigenvalue $\lambda = \alpha + i\beta$, the general solution of the system can also be obtained from the assumption

$$\mathbf{X} = \begin{bmatrix} c_1 \\ c_2 \end{bmatrix} e^{\alpha t} \sin \beta t + \begin{bmatrix} c_3 \\ c_4 \end{bmatrix} e^{\alpha t} \cos \beta t$$

and then the substitution of $x(t)$ and $y(t)$ into one of the equations of the original system.

9.4.3 Repeated Eigenvalues

Up to this point we have not considered the case in which some of the n eigenvalues $\lambda_1, \lambda_2, \ldots, \lambda_n$ of an $n \times n$ matrix are repeated. For example, the characteristic equation of the coefficient matrix in

$$\mathbf{X}' = \begin{bmatrix} 3 & -18 \\ 2 & -9 \end{bmatrix} \mathbf{X} \qquad (16)$$

is readily shown to be $(\lambda + 3)^2 = 0$, and therefore $\lambda_1 = \lambda_2 = -3$ is a root of *multiplicity 2*. Now for this value we find the single eigenvector

$$\mathbf{K}_1 = \begin{bmatrix} 3 \\ 1 \end{bmatrix}$$

and so one solution of (16) is

$$\mathbf{X}_1 = \begin{bmatrix} 3 \\ 1 \end{bmatrix} e^{-3t} \qquad (17)$$

But since we are obviously interested in forming the general solution of the system, we need to pursue the question of finding a second solution.

In general, if m is a positive integer and $(\lambda - \lambda_1)^m$ is a factor of the characteristic equation while $(\lambda - \lambda_1)^{m+1}$ is not a factor, then λ_1 is said to be an **eigenvalue of multiplicity** m. We distinguish two possibilities:

(*i*) For some $n \times n$ matrices $\mathbf{A}$ it may be possible to find m linearly independent eigenvectors $\mathbf{K}_1, \mathbf{K}_2, \ldots, \mathbf{K}_m$ corresponding to an eigenvalue λ_1 of multiplicity $m \leq n$. In this case the general solution of the system contains the linear combination

$$c_1 \mathbf{K}_1 e^{\lambda_1 t} + c_2 \mathbf{K}_2 e^{\lambda_1 t} + \cdots + c_m \mathbf{K}_m e^{\lambda_1 t}$$

(*ii*) If there is only one eigenvector corresponding to the eigenvalue λ_1 of multiplicity m, then m linearly independent solutions of the form

$$\mathbf{X}_1 = \mathbf{K}_{11} e^{\lambda_1 t}$$
$$\mathbf{X}_2 = \mathbf{K}_{21} t e^{\lambda_1 t} + \mathbf{K}_{22} e^{\lambda_1 t}$$
$$\vdots$$
$$\mathbf{X}_m = \mathbf{K}_{m1} \frac{t^{m-1}}{(m-1)!} e^{\lambda_1 t} + \mathbf{K}_{m2} \frac{t^{m-2}}{(m-2)!} e^{\lambda_1 t} + \cdots + \mathbf{K}_{mm} e^{\lambda_1 t}$$

where $\mathbf{K}_{ij}$ are column vectors, can always be found.

Eigenvalue of Multiplicity 2 We begin by considering eigenvalues of multiplicity 2. In the first example we shall illustrate a matrix for which we can find two distinct eigenvectors corresponding to a double eigenvalue.

EXAMPLE 5 Solve $\mathbf{X}' = \begin{bmatrix} 1 & -2 & 2 \\ -2 & 1 & -2 \\ 2 & -2 & 1 \end{bmatrix} \mathbf{X}$.

9.4 Homogeneous Linear Systems

Solution Expanding the determinant in the characteristic equation

$$\det(\mathbf{A} - \lambda\mathbf{I}) = \begin{vmatrix} 1-\lambda & -2 & 2 \\ -2 & 1-\lambda & -2 \\ 2 & -2 & 1-\lambda \end{vmatrix} = 0$$

yields $-(\lambda+1)^2(\lambda-5) = 0$. We see that $\lambda_1 = \lambda_2 = -1$ and $\lambda_3 = 5$.

For $\lambda_1 = -1$, Gauss–Jordan elimination gives immediately

$$(\mathbf{A}+\mathbf{I}|\mathbf{0}) = \begin{bmatrix} 2 & -2 & 2 & | & 0 \\ -2 & 2 & -2 & | & 0 \\ 2 & -2 & 2 & | & 0 \end{bmatrix} \Longrightarrow \begin{bmatrix} 1 & -1 & 1 & | & 0 \\ 0 & 0 & 0 & | & 0 \\ 0 & 0 & 0 & | & 0 \end{bmatrix}$$

From $k_1 - k_2 + k_3 = 0$ we can express, say, k_1 in terms of k_2 and k_3. By choosing $k_2 = 1$ and $k_3 = 0$ in $k_1 = k_2 - k_3$, we obtain $k_1 = 1$ and so one eigenvector is

$$\mathbf{K}_1 = \begin{bmatrix} 1 \\ 1 \\ 0 \end{bmatrix}$$

But the choice $k_2 = 1$, $k_3 = 1$ implies $k_1 = 0$. Hence a second eigenvector is

$$\mathbf{K}_2 = \begin{bmatrix} 0 \\ 1 \\ 1 \end{bmatrix}$$

Since neither eigenvector is a constant multiple of the other, we have found, corresponding to the same eigenvalue, two linearly independent solutions

$$\mathbf{X}_1 = \begin{bmatrix} 1 \\ 1 \\ 0 \end{bmatrix} e^{-t} \quad \text{and} \quad \mathbf{X}_2 = \begin{bmatrix} 0 \\ 1 \\ 1 \end{bmatrix} e^{-t}$$

Last, for $\lambda_3 = 5$, the reduction

$$(\mathbf{A}-5\mathbf{I}|\mathbf{0}) = \begin{bmatrix} -4 & -2 & 2 & | & 0 \\ -2 & -4 & -2 & | & 0 \\ 2 & -2 & -4 & | & 0 \end{bmatrix} \Longrightarrow \begin{bmatrix} 1 & 0 & -1 & | & 0 \\ 0 & 1 & 1 & | & 0 \\ 0 & 0 & 0 & | & 0 \end{bmatrix}$$

implies $k_1 = k_3$ and $k_2 = -k_3$. Picking $k_3 = 1$ gives $k_1 = 1$, $k_2 = -1$ and thus a third eigenvector is

$$\mathbf{K}_3 = \begin{bmatrix} 1 \\ -1 \\ 1 \end{bmatrix}$$

We conclude that the general solution of the system is

$$\mathbf{X} = c_1 \begin{bmatrix} 1 \\ 1 \\ 0 \end{bmatrix} e^{-t} + c_2 \begin{bmatrix} 0 \\ 1 \\ 1 \end{bmatrix} e^{-t} + c_3 \begin{bmatrix} 1 \\ -1 \\ 1 \end{bmatrix} e^{5t} \quad \square$$

Second Solution Now suppose that λ_1 is an eigenvalue of multiplicity 2 and that there is only one eigenvector associated with this value. A second solution can be found of the form

$$\mathbf{X}_2 = \mathbf{K}te^{\lambda_1 t} + \mathbf{P}e^{\lambda_1 t} \tag{18}$$

where

$$\mathbf{K} = \begin{bmatrix} k_1 \\ k_2 \\ \vdots \\ k_n \end{bmatrix} \quad \text{and} \quad \mathbf{P} = \begin{bmatrix} p_1 \\ p_2 \\ \vdots \\ p_n \end{bmatrix}$$

To see this, we substitute (18) into the system $\mathbf{X}' = \mathbf{A}\mathbf{X}$ and simplify:

$$(\mathbf{A}\mathbf{K} - \lambda_1 \mathbf{K})te^{\lambda_1 t} + (\mathbf{A}\mathbf{P} - \lambda_1 \mathbf{P} - \mathbf{K})e^{\lambda_1 t} = \mathbf{0}$$

Since this last equation is to hold for all values of t, we must have

$$(\mathbf{A} - \lambda_1 \mathbf{I})\mathbf{K} = \mathbf{0} \tag{19}$$

and

$$(\mathbf{A} - \lambda_1 \mathbf{I})\mathbf{P} = \mathbf{K} \tag{20}$$

The first equation (19) simply states that $\mathbf{K}$ must be an eigenvector of $\mathbf{A}$ associated with λ_1. By solving (19), we find one solution $\mathbf{X}_1 = \mathbf{K}e^{\lambda_1 t}$. To find the second solution $\mathbf{X}_2$ we need only solve the additional system (20) for the vector $\mathbf{P}$.

EXAMPLE 6 Find the general solution of the system given in (16).

Solution From (17) we know that $\lambda_1 = -3$ and that one solution is

$$\mathbf{X}_1 = \begin{bmatrix} 3 \\ 1 \end{bmatrix} e^{-3t}$$

Identifying $\mathbf{K} = \begin{bmatrix} 3 \\ 1 \end{bmatrix}$ and $\mathbf{P} = \begin{bmatrix} p_1 \\ p_2 \end{bmatrix}$, we know from (20) that we must now solve

$$(\mathbf{A} + 3\mathbf{I})\mathbf{P} = \mathbf{K} \quad \text{or} \quad \begin{bmatrix} 6 & -18 \\ 2 & -6 \end{bmatrix} \begin{bmatrix} p_1 \\ p_2 \end{bmatrix} = \begin{bmatrix} 3 \\ 1 \end{bmatrix}$$

Multiplying out this last expression gives

$$6p_1 - 18p_2 = 3$$
$$2p_1 - 6p_2 = 1$$

Since this system is obviously equivalent to one equation, we have an infinite number of choices for p_1 and p_2. For example, by choosing $p_1 = 1$, we find $p_2 = \frac{1}{6}$. However, for simplicity, we shall choose $p_1 = \frac{1}{2}$ so that $p_2 = 0$. Hence, $\mathbf{P} = \begin{bmatrix} \frac{1}{2} \\ 0 \end{bmatrix}$. Thus from (18) we find

$$\mathbf{X}_2 = \begin{bmatrix} 3 \\ 1 \end{bmatrix} te^{-3t} + \begin{bmatrix} \frac{1}{2} \\ 0 \end{bmatrix} e^{-3t}$$

9.4 Homogeneous Linear Systems

The general solution of (16) is then

$$\mathbf{X} = c_1 \begin{bmatrix} 3 \\ 1 \end{bmatrix} e^{-3t} + c_2 \left\{ \begin{bmatrix} 3 \\ 1 \end{bmatrix} te^{-3t} + \begin{bmatrix} \frac{1}{2} \\ 0 \end{bmatrix} e^{-3t} \right\} \qquad \square$$

Eigenvalue of Multiplicity 3 When a matrix **A** has only one eigenvector associated with an eigenvalue λ_1 of multiplicity 3, we can find a second solution of form (18) and a third solution of the form

$$\mathbf{X}_3 = \mathbf{K} \frac{t^2}{2} e^{\lambda_1 t} + \mathbf{P} t e^{\lambda_1 t} + \mathbf{Q} e^{\lambda_1 t} \tag{21}$$

where

$$\mathbf{K} = \begin{bmatrix} k_1 \\ k_2 \\ \vdots \\ k_n \end{bmatrix} \quad \mathbf{P} = \begin{bmatrix} p_1 \\ p_2 \\ \vdots \\ p_n \end{bmatrix} \quad \mathbf{Q} = \begin{bmatrix} q_1 \\ q_2 \\ \vdots \\ q_n \end{bmatrix}$$

By substituting (21) into the system $\mathbf{X}' = \mathbf{AX}$, we find the column vectors **K**, **P**, and **Q** must satisfy

$$(\mathbf{A} - \lambda_1 \mathbf{I})\mathbf{K} = \mathbf{0} \tag{22}$$
$$(\mathbf{A} - \lambda_1 \mathbf{I})\mathbf{P} = \mathbf{K} \tag{23}$$
and
$$(\mathbf{A} - \lambda_1 \mathbf{I})\mathbf{Q} = \mathbf{P} \tag{24}$$

Of course the solutions of (22) and (23) can be utilized in the formulation of the solutions $\mathbf{X}_1$ and $\mathbf{X}_2$.

EXAMPLE 7 Solve $\mathbf{X}' = \begin{bmatrix} 2 & 1 & 6 \\ 0 & 2 & 5 \\ 0 & 0 & 2 \end{bmatrix} \mathbf{X}$.

Solution The characteristic equation $(\lambda - 2)^3 = 0$ shows that $\lambda_1 = 2$ is an eigenvalue of multiplicity 3. In succession we find that a solution of

$$(\mathbf{A} - 2\mathbf{I})\mathbf{K} = \mathbf{0} \quad \text{is} \quad \mathbf{K} = \begin{bmatrix} 1 \\ 0 \\ 0 \end{bmatrix}$$

a solution of

$$(\mathbf{A} - 2\mathbf{I})\mathbf{P} = \mathbf{K} \quad \text{is} \quad \mathbf{P} = \begin{bmatrix} 0 \\ 1 \\ 0 \end{bmatrix}$$

and finally a solution of

$$(\mathbf{A} - 2\mathbf{I})\mathbf{Q} = \mathbf{P} \quad \text{is} \quad \mathbf{Q} = \begin{bmatrix} 0 \\ -\frac{6}{5} \\ \frac{1}{5} \end{bmatrix}$$

We see from (18) and (21) that the general solution of the system is

$$\mathbf{X} = c_1 \begin{bmatrix} 1 \\ 0 \\ 0 \end{bmatrix} e^{2t} + c_2 \left\{ \begin{bmatrix} 1 \\ 0 \\ 0 \end{bmatrix} te^{2t} + \begin{bmatrix} 0 \\ 1 \\ 0 \end{bmatrix} e^{2t} \right\} + c_3 \left\{ \begin{bmatrix} 1 \\ 0 \\ 0 \end{bmatrix} \frac{t^2}{2} e^{2t} + \begin{bmatrix} 0 \\ 1 \\ 0 \end{bmatrix} te^{2t} + \begin{bmatrix} 0 \\ -\frac{6}{5} \\ \frac{1}{5} \end{bmatrix} e^{2t} \right\} \quad \square$$

EXERCISES 9.4 Answers to odd-numbered problems begin on page A-56.

[9.4.1]

In Problems 1–12 find the general solution of the given system.

1. $\dfrac{dx}{dt} = x + 2y$
$\dfrac{dy}{dt} = 4x + 3y$

2. $\dfrac{dx}{dt} = 2y$
$\dfrac{dy}{dt} = 8x$

3. $\dfrac{dx}{dt} = -4x + 2y$
$\dfrac{dy}{dt} = -\frac{5}{2}x + 2y$

4. $\dfrac{dx}{dt} = \frac{1}{2}x + 9y$
$\dfrac{dy}{dt} = \frac{1}{2}x + 2y$

5. $\mathbf{X}' = \begin{bmatrix} 10 & -5 \\ 8 & -12 \end{bmatrix} \mathbf{X}$

6. $\mathbf{X}' = \begin{bmatrix} -6 & 2 \\ -3 & 1 \end{bmatrix} \mathbf{X}$

7. $\dfrac{dx}{dt} = x + y - z$
$\dfrac{dy}{dt} = 2y$
$\dfrac{dz}{dt} = y - z$

8. $\dfrac{dx}{dt} = 2x - 7y$
$\dfrac{dy}{dt} = 5x + 10y + 4z$
$\dfrac{dz}{dt} = 5y + 2z$

9. $\mathbf{X}' = \begin{bmatrix} -1 & 1 & 0 \\ 1 & 2 & 1 \\ 0 & 3 & -1 \end{bmatrix} \mathbf{X}$

10. $\mathbf{X}' = \begin{bmatrix} 1 & 0 & 1 \\ 0 & 1 & 0 \\ 1 & 0 & 1 \end{bmatrix} \mathbf{X}$

11. $\mathbf{X}' = \begin{bmatrix} -1 & -1 & 0 \\ \frac{3}{4} & -\frac{3}{2} & 3 \\ \frac{1}{8} & \frac{1}{4} & -\frac{1}{2} \end{bmatrix} \mathbf{X}$

12. $\mathbf{X}' = \begin{bmatrix} -1 & 4 & 2 \\ 4 & -1 & -2 \\ 0 & 0 & 6 \end{bmatrix} \mathbf{X}$

In Problems 13 and 14 solve the given system subject to the indicated initial condition.

13. $\mathbf{X}' = \begin{bmatrix} \frac{1}{2} & 0 \\ 1 & -\frac{1}{2} \end{bmatrix} \mathbf{X}, \; \mathbf{X}(0) = \begin{bmatrix} 3 \\ 5 \end{bmatrix}$

14. $\mathbf{X}' = \begin{bmatrix} 1 & 1 & 4 \\ 0 & 2 & 0 \\ 1 & 1 & 1 \end{bmatrix} \mathbf{X}, \; \mathbf{X}(0) = \begin{bmatrix} 1 \\ 3 \\ 0 \end{bmatrix}$

[9.4.2]

In Problems 15–26 find the general solution of the given system.

15. $\dfrac{dx}{dt} = 6x - y$
$\dfrac{dy}{dt} = 5x + 2y$

16. $\dfrac{dx}{dt} = x + y$
$\dfrac{dy}{dt} = -2x - y$

17. $\dfrac{dx}{dt} = 5x + y$
$\dfrac{dy}{dt} = -2x + 3y$

18. $\dfrac{dx}{dt} = 4x + 5y$
$\dfrac{dy}{dt} = -2x + 6y$

19. $\mathbf{X}' = \begin{bmatrix} 4 & -5 \\ 5 & -4 \end{bmatrix} \mathbf{X}$

20. $\mathbf{X}' = \begin{bmatrix} 1 & -8 \\ 1 & -3 \end{bmatrix} \mathbf{X}$

21. $\dfrac{dx}{dt} = z$
$\dfrac{dy}{dt} = -z$
$\dfrac{dz}{dt} = y$

22. $\dfrac{dx}{dt} = 2x + y + 2z$
$\dfrac{dy}{dt} = 3x + 6z$
$\dfrac{dz}{dt} = -4x - 3z$

23. $\mathbf{X}' = \begin{bmatrix} 1 & -1 & 2 \\ -1 & 1 & 0 \\ -1 & 0 & 1 \end{bmatrix} \mathbf{X}$

24. $\mathbf{X}' = \begin{bmatrix} 4 & 0 & 1 \\ 0 & 6 & 0 \\ -4 & 0 & 4 \end{bmatrix} \mathbf{X}$

25. $\mathbf{X}' = \begin{bmatrix} 2 & 5 & 1 \\ -5 & -6 & 4 \\ 0 & 0 & 2 \end{bmatrix} \mathbf{X}$

26. $\mathbf{X}' = \begin{bmatrix} 2 & 4 & 4 \\ -1 & -2 & 0 \\ -1 & 0 & -2 \end{bmatrix} \mathbf{X}$

In Problems 27 and 28 solve the given system subject to the indicated initial condition.

27. $\mathbf{X}' = \begin{bmatrix} 1 & -12 & -14 \\ 1 & 2 & -3 \\ 1 & 1 & -2 \end{bmatrix} \mathbf{X}, \ \mathbf{X}(0) = \begin{bmatrix} 4 \\ 6 \\ -7 \end{bmatrix}$

28. $\mathbf{X}' = \begin{bmatrix} 6 & -1 \\ 5 & 4 \end{bmatrix} \mathbf{X}, \ \mathbf{X}(0) = \begin{bmatrix} -2 \\ 8 \end{bmatrix}$

[9.4.3]

In Problems 29–38 find the general solution of the given system.

29. $\dfrac{dx}{dt} = 3x - y$
    $\dfrac{dy}{dt} = 9x - 3y$

30. $\dfrac{dx}{dt} = -6x + 5y$
    $\dfrac{dy}{dt} = -5x + 4y$

31. $\dfrac{dx}{dt} = -x + 3y$
    $\dfrac{dy}{dt} = -3x + 5y$

32. $\dfrac{dx}{dt} = 12x - 9y$
    $\dfrac{dy}{dt} = 4x$

33. $\dfrac{dx}{dt} = 3x - y - z$
    $\dfrac{dy}{dt} = x + y - z$
    $\dfrac{dz}{dt} = x - y + z$

34. $\dfrac{dx}{dt} = 3x + 2y + 4z$
    $\dfrac{dy}{dt} = 2x \quad + 2z$
    $\dfrac{dz}{dt} = 4x + 2y + 3z$

35. $\mathbf{X}' = \begin{bmatrix} 5 & -4 & 0 \\ 1 & 0 & 2 \\ 0 & 2 & 5 \end{bmatrix} \mathbf{X}$

36. $\mathbf{X}' = \begin{bmatrix} 1 & 0 & 0 \\ 0 & 3 & 1 \\ 0 & -1 & 1 \end{bmatrix} \mathbf{X}$

37. $\mathbf{X}' = \begin{bmatrix} 1 & 0 & 0 \\ 2 & 2 & -1 \\ 0 & 1 & 0 \end{bmatrix} \mathbf{X}$

38. $\mathbf{X}' = \begin{bmatrix} 4 & 1 & 0 \\ 0 & 4 & 1 \\ 0 & 0 & 4 \end{bmatrix} \mathbf{X}$

In Problems 39 and 40 solve the given system subject to the indicated initial condition.

39. $\mathbf{X}' = \begin{bmatrix} 2 & 4 \\ -1 & 6 \end{bmatrix} \mathbf{X}, \ \mathbf{X}(0) = \begin{bmatrix} -1 \\ 6 \end{bmatrix}$

40. $\mathbf{X}' = \begin{bmatrix} 0 & 0 & 1 \\ 0 & 1 & 0 \\ 1 & 0 & 0 \end{bmatrix} \mathbf{X}, \ \mathbf{X}(0) = \begin{bmatrix} 1 \\ 2 \\ 5 \end{bmatrix}$

If $\mathbf{\Phi}(t)$ is a fundamental matrix of the system, then the initial-value problem $\mathbf{X}' = \mathbf{AX}$, $\mathbf{X}(t_0) = \mathbf{X}_0$ has the solution $\mathbf{X} = \mathbf{\Phi}(t)\mathbf{\Phi}^{-1}(t_0)\mathbf{X}_0$ (see Problem 25 in Exercises 9.3). In Problems 41 and 42 use this result to solve the given system subject to the indicated initial condition.

41. $\mathbf{X}' = \begin{bmatrix} 4 & 3 \\ 3 & -4 \end{bmatrix} \mathbf{X}, \ \mathbf{X}(0) = \begin{bmatrix} 1 \\ 1 \end{bmatrix}$

42. $\mathbf{X}' = \begin{bmatrix} -\frac{2}{25} & \frac{1}{50} \\ \frac{2}{25} & -\frac{2}{25} \end{bmatrix} \mathbf{X}, \ \mathbf{X}(0) = \begin{bmatrix} 25 \\ 0 \end{bmatrix}$

In Problems 43 and 44 find a solution of the given system of the form $\mathbf{X} = t^\lambda \mathbf{K}$, $t > 0$, where $\mathbf{K}$ is a column vector of constants.

43. $t\mathbf{X}' = \begin{bmatrix} 1 & 3 \\ -1 & 5 \end{bmatrix} \mathbf{X}$

44. $t\mathbf{X}' = \begin{bmatrix} 2 & -2 \\ 2 & 7 \end{bmatrix} \mathbf{X}$

45. Consider the system of differential equations describing the motion of two masses m_1 and m_2 connected to two attached springs as shown in Figure 9.3:

$$m_1 x_1'' = -k_1 x_1 + k_2(x_2 - x_1)$$
$$m_2 x_2'' = -k_2(x_2 - x_1)$$

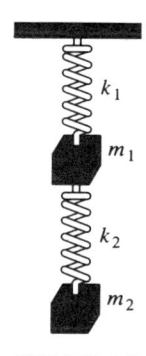

FIGURE 9.3

(a) Show that this system of linear second-order differential equations can be written as the matrix equation $\mathbf{X}'' = \mathbf{AX}$, where

$$\mathbf{X} = \begin{bmatrix} x_1 \\ x_2 \end{bmatrix} \quad \text{and} \quad \mathbf{A} = \begin{bmatrix} -\dfrac{k_1 + k_2}{m_1} & \dfrac{k_2}{m_1} \\ \dfrac{k_2}{m_2} & -\dfrac{k_2}{m_2} \end{bmatrix}$$

(b) If a solution is assumed of the form $\mathbf{X} = \mathbf{K}e^{\omega t}$, show that $\mathbf{X}'' = \mathbf{AX}$ yields

$$(\mathbf{A} - \lambda \mathbf{I})\mathbf{K} = 0 \quad \text{where} \quad \lambda = \omega^2$$

(c) Show that if $m_1 = 1$, $m_2 = 1$, $k_1 = 3$, and $k_2 = 2$, a solution of the system is

$$\mathbf{X} = c_1 \begin{bmatrix} 1 \\ 2 \end{bmatrix} e^{it} + c_2 \begin{bmatrix} 1 \\ 2 \end{bmatrix} e^{-it} + c_3 \begin{bmatrix} -2 \\ 1 \end{bmatrix} e^{\sqrt{6}it} + c_4 \begin{bmatrix} -2 \\ 1 \end{bmatrix} e^{-\sqrt{6}it}$$

(d) Show that the solution in part (c) can be written as

$$\begin{bmatrix} x_1 \\ x_2 \end{bmatrix} = b_1 \begin{bmatrix} 1 \\ 2 \end{bmatrix} \cos t + b_2 \begin{bmatrix} 1 \\ 2 \end{bmatrix} \sin t$$
$$+ b_3 \begin{bmatrix} -2 \\ 1 \end{bmatrix} \cos \sqrt{6}t + b_4 \begin{bmatrix} -2 \\ 1 \end{bmatrix} \sin \sqrt{6}t$$

9.5 SOLUTION BY DIAGONALIZATION

In this section we are going to consider an alternative method for solving a homogeneous system $\mathbf{X}' = \mathbf{AX}$ of linear first-order differential equations. This method is applicable to such a system whenever the coefficient matrix $\mathbf{A}$ is diagonalizable.

A homogeneous system $\mathbf{X}' = \mathbf{AX}$,

$$\begin{bmatrix} x_1' \\ x_2' \\ \vdots \\ x_n' \end{bmatrix} = \begin{bmatrix} a_{11} & a_{12} & \cdots & a_{1n} \\ a_{21} & a_{22} & \cdots & a_{2n} \\ \vdots & & & \vdots \\ a_{n1} & a_{n2} & \cdots & a_{nn} \end{bmatrix} \begin{bmatrix} x_1 \\ x_2 \\ \vdots \\ x_n \end{bmatrix} \qquad (1)$$

in which each x_i' is expressed as a linear combination of $x_1, x_2, \ldots, x_n$, is said to be **coupled**. If the coefficient matrix $\mathbf{A}$ is diagonalizable, then the system can be **uncoupled** in that each x_i' can be expressed solely in terms of x_i.

If the matrix $\mathbf{A}$ has n linearly independent eigenvectors, then we know from Theorem 7.27 that we can find a matrix $\mathbf{P}$ such that $\mathbf{P}^{-1}\mathbf{AP} = \mathbf{D}$, where $\mathbf{D}$ is a diagonal matrix. If we make the substitution $\mathbf{X} = \mathbf{PY}$ in the system $\mathbf{X}' = \mathbf{AX}$, then

$$\mathbf{PY}' = \mathbf{APY} \quad \text{or} \quad \mathbf{Y}' = \mathbf{P}^{-1}\mathbf{APY} \quad \text{or} \quad \mathbf{Y}' = \mathbf{DY} \qquad (2)$$

The last equation in (2) is the same as

$$\begin{bmatrix} y_1' \\ y_2' \\ \vdots \\ y_n' \end{bmatrix} = \begin{bmatrix} \lambda_1 & 0 & 0 & \cdots & 0 \\ 0 & \lambda_2 & 0 & \cdots & 0 \\ \vdots & & & & \vdots \\ 0 & 0 & 0 & \cdots & \lambda_n \end{bmatrix} \begin{bmatrix} y_1 \\ y_2 \\ \vdots \\ y_n \end{bmatrix} \qquad (3)$$

Since $\mathbf{D}$ is a diagonal matrix, an inspection of (3) reveals that this new system is uncoupled; each differential equation in the system is of the form $y_i' = \lambda_i y_i$, where $i = 1, 2, \ldots, n$. The solution of each of these linear equations is $y_i = c_i e^{\lambda_i t}$, where $i = 1, 2, \ldots, n$. Hence, the general solution of (3) can be written as the column vector

$$\mathbf{Y} = \begin{bmatrix} c_1 e^{\lambda_1 t} \\ c_2 e^{\lambda_2 t} \\ \vdots \\ c_n e^{\lambda_n t} \end{bmatrix} \qquad (4)$$

9.5 Solution by Diagonalization

Since we now know Y and since the matrix P can be constructed from the eigenvectors of A, the general solution of the original system $X' = AX$ is obtained from $X = PY$.

EXAMPLE 1 Solve $X' = \begin{bmatrix} -2 & -1 & 8 \\ 0 & -3 & 8 \\ 0 & -4 & 9 \end{bmatrix} X$ by diagonalization.

Solution We begin by finding the eigenvalues and corresponding eigenvectors of the coefficient matrix. From $\det(A - \lambda I) = -(\lambda + 2)(\lambda - 1)(\lambda - 5)$ we get $\lambda_1 = -2$, $\lambda_2 = 1$, and $\lambda_3 = 5$. Since the eigenvalues are distinct, the eigenvectors are linearly independent. Solving $(A - \lambda_i I)K = 0$ for $i = 1, 2,$ and 3 gives, respectively,

$$K_1 = \begin{bmatrix} 1 \\ 0 \\ 0 \end{bmatrix} \quad K_2 = \begin{bmatrix} 2 \\ 2 \\ 1 \end{bmatrix} \quad K_3 = \begin{bmatrix} 1 \\ 1 \\ 1 \end{bmatrix} \tag{5}$$

Thus, a matrix that diagonalizes the coefficient matrix is

$$P = \begin{bmatrix} 1 & 2 & 1 \\ 0 & 2 & 1 \\ 0 & 1 & 1 \end{bmatrix}$$

The entries on the main diagonal of D are the eigenvalues of A corresponding to the order in which the eigenvectors appear in P:

$$D = \begin{bmatrix} -2 & 0 & 0 \\ 0 & 1 & 0 \\ 0 & 0 & 5 \end{bmatrix}$$

As we have shown above, the substitution $X = PY$ in $X' = AX$ gives the uncoupled system $Y' = DY$. The general solution of this last system is immediate:

$$Y = \begin{bmatrix} c_1 e^{-2t} \\ c_2 e^{t} \\ c_3 e^{5t} \end{bmatrix}$$

Hence, the solution of the given system is

$$X = PY = \begin{bmatrix} 1 & 2 & 1 \\ 0 & 2 & 1 \\ 0 & 1 & 1 \end{bmatrix} \begin{bmatrix} c_1 e^{-2t} \\ c_2 e^{t} \\ c_3 e^{5t} \end{bmatrix} = \begin{bmatrix} c_1 e^{-2t} + 2c_2 e^{t} + c_3 e^{5t} \\ 2c_2 e^{t} + c_3 e^{5t} \\ c_2 e^{t} + c_3 e^{5t} \end{bmatrix} \tag{6} \quad \square$$

Note that (6) can be written in the usual manner by expressing the last matrix as a sum:

$$c_1 \begin{bmatrix} 1 \\ 0 \\ 0 \end{bmatrix} e^{-2t} + c_2 \begin{bmatrix} 2 \\ 2 \\ 1 \end{bmatrix} e^{t} + c_3 \begin{bmatrix} 1 \\ 1 \\ 1 \end{bmatrix} e^{5t}$$

Solution by diagonalization will always work provided we can find n linearly independent eigenvectors of the $n \times n$ matrix $\mathbf{A}$; the eigenvalues of $\mathbf{A}$ could be real and distinct, complex, or repeated. The method fails when $\mathbf{A}$ has repeated eigenvalues and n linearly independent eigenvectors cannot be found. Of course, in this last situation $\mathbf{A}$ is not diagonalizable.

Since we have to find eigenvalues and eigenvenctors of $\mathbf{A}$, this method is essentially equivalent to the procedure presented in the last section.

In the next section we shall see that diagonalization can also be used to solve nonhomogeneous systems $\mathbf{X}' = \mathbf{AX} + \mathbf{F}(t)$.

EXERCISES 9.5
Answers to odd-numbered problems begin on page A-56.

In Problems 1–10 use diagonalization to solve the given system.

1. $\mathbf{X}' = \begin{bmatrix} 5 & 6 \\ 3 & -2 \end{bmatrix} \mathbf{X}$

2. $\mathbf{X}' = \begin{bmatrix} \frac{1}{2} & \frac{1}{2} \\ \frac{1}{2} & \frac{1}{2} \end{bmatrix} \mathbf{X}$

3. $\mathbf{X}' = \begin{bmatrix} 1 & \frac{1}{4} \\ 1 & 1 \end{bmatrix}$

4. $\mathbf{X}' = \begin{bmatrix} 1 & 1 \\ 1 & -1 \end{bmatrix} \mathbf{X}$

5. $\mathbf{X}' = \begin{bmatrix} -1 & 3 & 0 \\ 3 & -1 & 0 \\ -2 & -2 & 6 \end{bmatrix} \mathbf{X}$

6. $\mathbf{X}' = \begin{bmatrix} 1 & 1 & 2 \\ 1 & 2 & 1 \\ 2 & 1 & 1 \end{bmatrix} \mathbf{X}$

7. $\mathbf{X}' = \begin{bmatrix} 1 & -1 & -1 \\ -1 & 1 & -1 \\ -1 & -1 & 1 \end{bmatrix} \mathbf{X}$

8. $\mathbf{X}' = \begin{bmatrix} 1 & 1 & 1 & 1 \\ 1 & 1 & 1 & 1 \\ 1 & 1 & 1 & 1 \\ 1 & 1 & 1 & 1 \end{bmatrix} \mathbf{X}$

9. $\mathbf{X}' = \begin{bmatrix} -3 & 2 & 2 \\ -6 & 5 & 2 \\ -7 & 4 & 4 \end{bmatrix} \mathbf{X}$

10. $\mathbf{X}' = \begin{bmatrix} 0 & 2 & 0 \\ 2 & 0 & 2 \\ 0 & 2 & 0 \end{bmatrix} \mathbf{X}$

9.6 NONHOMOGENEOUS LINEAR SYSTEMS

9.6.1 Undetermined Coefficients

The methods of **undetermined coefficients** and **variation of parameters** can both be adapted to the solution of a nonhomogeneous linear system $\mathbf{X}' = \mathbf{AX} + \mathbf{F}(t)$. Of these two methods, variation of parameters is the more powerful technique. However, there are a few instances when the method of undetermined coefficients gives a quick means of finding a particular solution $\mathbf{X}_p$.

9.6 Nonhomogeneous Linear Systems

EXAMPLE 1 Solve the system $\mathbf{X}' = \begin{bmatrix} -1 & 2 \\ -1 & 1 \end{bmatrix} \mathbf{X} + \begin{bmatrix} -8 \\ 3 \end{bmatrix}$ on $(-\infty, \infty)$.

Solution We first solve the homogeneous system

$$\mathbf{X}' = \begin{bmatrix} -1 & 2 \\ -1 & 1 \end{bmatrix} \mathbf{X}$$

The characteristic equation

$$\det(\mathbf{A} - \lambda \mathbf{I}) = \begin{vmatrix} -1-\lambda & 2 \\ -1 & 1-\lambda \end{vmatrix} = \lambda^2 + 1 = 0$$

yields the complex eigenvalues $\lambda_1 = i$ and $\lambda_2 = \bar{\lambda}_1 = -i$. By the procedures of the last section, we find

$$\mathbf{X}_c = c_1 \begin{bmatrix} \cos t + \sin t \\ \cos t \end{bmatrix} + c_2 \begin{bmatrix} \cos t - \sin t \\ -\sin t \end{bmatrix}$$

Now since $\mathbf{F}(t)$ is a constant vector, we shall assume a constant particular solution vector $\mathbf{X}_p = \begin{bmatrix} a_1 \\ b_1 \end{bmatrix}$. Substituting this latter assumption into the original system leads to

$$0 = -a_1 + 2b_1 - 8$$
$$0 = -a_1 + b_1 + 3$$

Solving this system of algebraic equations gives $a_1 = 14$ and $b_1 = 11$, and so $\mathbf{X}_p = \begin{bmatrix} 14 \\ 11 \end{bmatrix}$. The general solution of the system is

$$\mathbf{X} = c_1 \begin{bmatrix} \cos t + \sin t \\ \cos t \end{bmatrix} + c_2 \begin{bmatrix} \cos t - \sin t \\ -\sin t \end{bmatrix} + \begin{bmatrix} 14 \\ 11 \end{bmatrix} \qquad \square$$

EXAMPLE 2 Solve the system $\dfrac{dx}{dt} = 6x + y + 6t$ on $(-\infty, \infty)$.

$$\frac{dy}{dt} = 4x + 3y - 10t + 4$$

Solution We first solve the homogeneous system

$$\frac{dx}{dt} = 6x + y$$

$$\frac{dy}{dt} = 4x + 3y$$

by the method of Section 9.4. The eigenvalues are determined from

$$\det(\mathbf{A} - \lambda\mathbf{I}) = \begin{vmatrix} 6-\lambda & 1 \\ 4 & 3-\lambda \end{vmatrix} = \lambda^2 - 9\lambda + 14 = 0$$

Since $\lambda^2 - 9\lambda + 14 = (\lambda - 2)(\lambda - 7)$, we have $\lambda_1 = 2$ and $\lambda_2 = 7$. It is then easily verified that the respective eigenvectors of the coefficient matrix are

$$\mathbf{K}_1 = \begin{bmatrix} 1 \\ -4 \end{bmatrix} \quad \text{and} \quad \mathbf{K}_2 = \begin{bmatrix} 1 \\ 1 \end{bmatrix}$$

Consequently the complementary function is

$$\mathbf{X}_c = c_1 \begin{bmatrix} 1 \\ -4 \end{bmatrix} e^{2t} + c_2 \begin{bmatrix} 1 \\ 1 \end{bmatrix} e^{7t}$$

Because $\mathbf{F}(t)$ can be written as

$$\mathbf{F}(t) = \begin{bmatrix} 6 \\ -10 \end{bmatrix} t + \begin{bmatrix} 0 \\ 4 \end{bmatrix}$$

we shall try to find a particular solution of the system possessing the *same* form:

$$\mathbf{X}_p = \begin{bmatrix} a_2 \\ b_2 \end{bmatrix} t + \begin{bmatrix} a_1 \\ b_1 \end{bmatrix}$$

In matrix terms we must have

$$\mathbf{X}_p' = \begin{bmatrix} 6 & 1 \\ 4 & 3 \end{bmatrix} \mathbf{X}_p + \begin{bmatrix} 6 \\ -10 \end{bmatrix} t + \begin{bmatrix} 0 \\ 4 \end{bmatrix}$$

or

$$\begin{bmatrix} a_2 \\ b_2 \end{bmatrix} = \begin{bmatrix} 6 & 1 \\ 4 & 3 \end{bmatrix} \left(\begin{bmatrix} a_2 \\ b_2 \end{bmatrix} t + \begin{bmatrix} a_1 \\ b_1 \end{bmatrix} \right) + \begin{bmatrix} 6 \\ -10 \end{bmatrix} t + \begin{bmatrix} 0 \\ 4 \end{bmatrix}$$

$$\begin{bmatrix} 0 \\ 0 \end{bmatrix} = \begin{bmatrix} (6a_2 + b_2 + 6)t + 6a_1 + b_1 - a_2 \\ (4a_2 + 3b_2 - 10)t + 4a_1 + 3b_1 - b_2 + 4 \end{bmatrix}$$

From this last identity we conclude that

$$\begin{aligned} 6a_2 + b_2 + 6 &= 0 \\ 4a_2 + 3b_2 - 10 &= 0 \end{aligned} \quad \text{and} \quad \begin{aligned} 6a_1 + b_1 - a_2 &= 0 \\ 4a_1 + 3b_1 - b_2 + 4 &= 0 \end{aligned}$$

Solving the first two equations simultaneously yields $a_2 = -2$ and $b_2 = 6$. Substituting these values into the last two equations and solving for a_1 and b_1 give $a_1 = -\frac{4}{7}$, $b_1 = \frac{10}{7}$. It follows, therefore, that a particular solution vector is

$$\mathbf{X}_p = \begin{bmatrix} -2 \\ 6 \end{bmatrix} t + \begin{bmatrix} -\frac{4}{7} \\ \frac{10}{7} \end{bmatrix}$$

and so the general solution of the system on $(-\infty, \infty)$ is

$$\mathbf{X} = \mathbf{X}_c + \mathbf{X}_p = c_1 \begin{bmatrix} 1 \\ -4 \end{bmatrix} e^{2t} + c_2 \begin{bmatrix} 1 \\ 1 \end{bmatrix} e^{7t} + \begin{bmatrix} -2 \\ 6 \end{bmatrix} t + \begin{bmatrix} -\frac{4}{7} \\ \frac{10}{7} \end{bmatrix} \quad \square$$

EXAMPLE 3 Determine the form of the particular solution vector $\mathbf{X}_p$ for

$$\frac{dx}{dt} = 5x + 3y - 2e^{-t} + 1$$

$$\frac{dy}{dt} = -x + y + e^{-t} - 5t + 7$$

Solution Proceeding in the usual manner, we find

$$\mathbf{X}_c = c_1 \begin{bmatrix} 1 \\ -1 \end{bmatrix} e^{2t} + c_2 \begin{bmatrix} 3 \\ -1 \end{bmatrix} e^{4t}$$

Now since
$$\mathbf{F}(t) = \begin{bmatrix} -2 \\ 1 \end{bmatrix} e^{-t} + \begin{bmatrix} 0 \\ -5 \end{bmatrix} t + \begin{bmatrix} 1 \\ 7 \end{bmatrix}$$

we assume a particular solution of the form

$$\mathbf{X}_p = \begin{bmatrix} a_3 \\ b_3 \end{bmatrix} e^{-t} + \begin{bmatrix} a_2 \\ b_2 \end{bmatrix} t + \begin{bmatrix} a_1 \\ b_1 \end{bmatrix} \qquad \square$$

The method of undetermined coefficients is not as simple as the last three examples seem to indicate. As in Section 3.4, the method can be applied only when the entries in the matrix $\mathbf{F}(t)$ are constants, polynomials, exponential functions, sines and cosines, or finite sums and products of these functions. There are further difficulties. The assumption for $\mathbf{X}_p$ is actually predicated on a prior knowledge of the complementary function $\mathbf{X}_c$. For example, if $\mathbf{F}(t)$ is a constant vector and $\lambda = 0$ is an eigenvalue of multiplicity 1, then $\mathbf{X}_c$ contains a constant vector. In this case $\mathbf{X}_p$ is *not* a constant vector as in Example 1, but rather

$$\mathbf{X}_p = \begin{bmatrix} a_2 \\ b_2 \end{bmatrix} t + \begin{bmatrix} a_1 \\ b_1 \end{bmatrix}$$

Similarly, in Example 3, if we replace e^{-t} in $\mathbf{F}(t)$ by e^{2t} ($\lambda = 2$ is an eigenvalue), then the correct form of the particular solution is

$$\mathbf{X}_p = \begin{bmatrix} a_4 \\ b_4 \end{bmatrix} te^{2t} + \begin{bmatrix} a_3 \\ b_3 \end{bmatrix} e^{2t} + \begin{bmatrix} a_2 \\ b_2 \end{bmatrix} t + \begin{bmatrix} a_1 \\ b_1 \end{bmatrix}$$

Rather than pursue these difficulties, we shall turn our attention now to the method of variation of parameters.

9.6.2 Variation of Parameters

In Section 9.3 we saw that the general solution of a homogeneous system $\mathbf{X}' = \mathbf{A}\mathbf{X}$ can be written as the product

$$\mathbf{X} = \boldsymbol{\Phi}(t)\mathbf{C}$$

where $\boldsymbol{\Phi}(t)$ is a fundamental matrix of the system and $\mathbf{C}$ is an $n \times 1$ column vector of constants. As in the procedure of Section 3.6, we ask whether it is possible to replace **C** by a column matrix of functions.

$$\mathbf{U}(t) = \begin{bmatrix} u_1(t) \\ u_2(t) \\ \vdots \\ u_n(t) \end{bmatrix}$$

so that
$$\mathbf{X}_p = \mathbf{\Phi}(t)\mathbf{U}(t) \tag{1}$$

is a particular solution of the nonhomogeneous system

$$\mathbf{X}' = \mathbf{A}\mathbf{X} + \mathbf{F}(t) \tag{2}$$

By the product rule, the derivative of (1) is

$$\mathbf{X}'_p = \mathbf{\Phi}(t)\mathbf{U}'(t) + \mathbf{\Phi}'(t)\mathbf{U}(t) \tag{3}$$

Substituting (3) and (1) into (2) gives

$$\mathbf{\Phi}(t)\mathbf{U}'(t) + \mathbf{\Phi}'(t)\mathbf{U}(t) = \mathbf{A}\mathbf{\Phi}(t)\mathbf{U}(t) + \mathbf{F}(t) \tag{4}$$

Now recall from (11) of Section 9.3 that $\mathbf{\Phi}'(t) = \mathbf{A}\mathbf{\Phi}(t)$. Thus (4) becomes

$$\mathbf{\Phi}(t)\mathbf{U}'(t) + \mathbf{A}\mathbf{\Phi}(t)\mathbf{U}(t) = \mathbf{A}\mathbf{\Phi}(t)\mathbf{U}(t) + \mathbf{F}(t)$$

or
$$\mathbf{\Phi}(t)\mathbf{U}'(t) = \mathbf{F}(t) \tag{5}$$

Multiplying both sides of (5) by $\mathbf{\Phi}^{-1}(t)$ gives

$$\mathbf{U}'(t) = \mathbf{\Phi}^{-1}(t)\mathbf{F}(t) \quad \text{or} \quad \mathbf{U}(t) = \int \mathbf{\Phi}^{-1}(t)\mathbf{F}(t)\,dt$$

Hence by assumption (1) we conclude that a particular solution of (2) is given by

$$\boxed{\mathbf{X}_p = \mathbf{\Phi}(t) \int \mathbf{\Phi}^{-1}(t)\mathbf{F}(t)\,dt} \tag{6}$$

To calculate the indefinite integral of the column matrix $\mathbf{\Phi}^{-1}(t)\mathbf{F}(t)$ in (6), we integrate each entry. Thus, the general solution of the system (2) is $\mathbf{X} = \mathbf{X}_c + \mathbf{X}_p$ or

$$\boxed{\mathbf{X} = \mathbf{\Phi}(t)\mathbf{C} + \mathbf{\Phi}(t) \int \mathbf{\Phi}^{-1}(t)\mathbf{F}(t)\,dt} \tag{7}$$

EXAMPLE 4 Find the general solution of the nonhomogeneous system

$$\mathbf{X}' = \begin{bmatrix} -3 & 1 \\ 2 & -4 \end{bmatrix} \mathbf{X} + \begin{bmatrix} 3t \\ e^{-t} \end{bmatrix} \tag{8}$$

on the interval $(-\infty, \infty)$.

9.6 Nonhomogeneous Linear Systems

Solution We first solve the homogeneous system

$$\mathbf{X}' = \begin{bmatrix} -3 & 1 \\ 2 & -4 \end{bmatrix} \mathbf{X} \qquad (9)$$

The characteristic equation of the coefficient matrix is

$$\det(\mathbf{A} - \lambda \mathbf{I}) = \begin{vmatrix} -3 - \lambda & 1 \\ 2 & -4 - \lambda \end{vmatrix} = (\lambda + 2)(\lambda + 5) = 0$$

and so the eigenvalues are $\lambda_1 = -2$ and $\lambda_2 = -5$. By the usual method we find that the eigenvectors corresponding to λ_1 and λ_2, are, respectively,

$$\mathbf{K}_1 = \begin{bmatrix} 1 \\ 1 \end{bmatrix} \quad \text{and} \quad \mathbf{K}_2 = \begin{bmatrix} 1 \\ -2 \end{bmatrix}$$

The solution vectors of the system (9) are then

$$\mathbf{X}_1 = \begin{bmatrix} 1 \\ 1 \end{bmatrix} e^{-2t} \quad \text{and} \quad \mathbf{X}_2 = \begin{bmatrix} 1 \\ -2 \end{bmatrix} e^{-5t}$$

Next we form

$$\mathbf{\Phi}(t) = \begin{bmatrix} e^{-2t} & e^{-5t} \\ e^{-2t} & -2e^{-5t} \end{bmatrix} \quad \text{and} \quad \mathbf{\Phi}^{-1}(t) = \begin{bmatrix} \tfrac{2}{3}e^{2t} & \tfrac{1}{3}e^{2t} \\ \tfrac{1}{3}e^{5t} & -\tfrac{1}{3}e^{5t} \end{bmatrix}$$

From (6) we then obtain

$$\mathbf{X}_p = \mathbf{\Phi}(t) \int \mathbf{\Phi}^{-1}(t) \mathbf{F}(t)\, dt = \begin{bmatrix} e^{-2t} & e^{-5t} \\ e^{-2t} & -2e^{-5t} \end{bmatrix} \int \begin{bmatrix} \tfrac{2}{3}e^{2t} & \tfrac{1}{3}e^{2t} \\ \tfrac{1}{3}e^{5t} & -\tfrac{1}{3}e^{5t} \end{bmatrix} \begin{bmatrix} 3t \\ e^{-t} \end{bmatrix} dt$$

$$= \begin{bmatrix} e^{-2t} & e^{-5t} \\ e^{-2t} & -2e^{-5t} \end{bmatrix} \int \begin{bmatrix} 2te^{2t} + \tfrac{1}{3}e^{t} \\ te^{5t} - \tfrac{1}{3}e^{4t} \end{bmatrix} dt$$

$$= \begin{bmatrix} e^{-2t} & e^{-5t} \\ e^{-2t} & -2e^{-5t} \end{bmatrix} \begin{bmatrix} te^{2t} - \tfrac{1}{2}e^{2t} + \tfrac{1}{3}e^{t} \\ \tfrac{1}{5}te^{5t} - \tfrac{1}{25}e^{5t} - \tfrac{1}{12}e^{4t} \end{bmatrix}$$

$$= \begin{bmatrix} \tfrac{6}{5}t - \tfrac{27}{50} + \tfrac{1}{4}e^{-t} \\ \tfrac{3}{5}t - \tfrac{21}{50} + \tfrac{1}{2}e^{-t} \end{bmatrix}$$

Hence, from (7) the general solution of (8) on the interval is

$$\mathbf{X} = \begin{bmatrix} e^{-2t} & e^{-5t} \\ e^{-2t} & -2e^{-5t} \end{bmatrix} \begin{bmatrix} c_1 \\ c_2 \end{bmatrix} + \begin{bmatrix} \tfrac{6}{5}t - \tfrac{27}{50} + \tfrac{1}{4}e^{-t} \\ \tfrac{3}{5}t - \tfrac{21}{50} + \tfrac{1}{2}e^{-t} \end{bmatrix}$$

$$= c_1 \begin{bmatrix} 1 \\ 1 \end{bmatrix} e^{-2t} + c_2 \begin{bmatrix} 1 \\ -2 \end{bmatrix} e^{-5t} + \begin{bmatrix} \tfrac{6}{5} \\ \tfrac{3}{5} \end{bmatrix} t - \begin{bmatrix} \tfrac{27}{50} \\ \tfrac{21}{50} \end{bmatrix} + \begin{bmatrix} \tfrac{1}{4} \\ \tfrac{1}{2} \end{bmatrix} e^{-t} \qquad \square$$

The general solution of (2) on an interval can be written in the alternative form

$$\mathbf{X} = \mathbf{\Phi}(t)\mathbf{C} + \mathbf{\Phi}(t) \int_{t_0}^{t} \mathbf{\Phi}^{-1}(s)\mathbf{F}(s)\, ds \qquad (10)$$

where t and t_0 are points in the interval. This last form is useful in solving (2) subject to an initial condition $\mathbf{X}(t_0) = \mathbf{X}_0$. Substituting $t = t_0$ in (10) yields

$$\mathbf{X}_0 = \mathbf{\Phi}(t_0)\mathbf{C}$$

from which we see immediately that $\mathbf{C} = \mathbf{\Phi}^{-1}(t_0)\mathbf{X}_0$. We conclude that the solution of the initial-value problem is given by

$$\mathbf{X} = \mathbf{\Phi}(t)\mathbf{\Phi}^{-1}(t_0)\mathbf{X}_0 + \mathbf{\Phi}(t)\int_{t_0}^{t} \mathbf{\Phi}^{-1}(s)\mathbf{F}(s)\,ds \qquad (11)$$

Recall from Section 9.3 that an alternative way of forming a fundamental matrix is to choose its column vectors $\mathbf{V}_i$ in such a manner that

$$\mathbf{V}_1(t_0) = \begin{bmatrix} 1 \\ 0 \\ \vdots \\ 0 \end{bmatrix}, \quad \mathbf{V}_2(t_0) = \begin{bmatrix} 0 \\ 1 \\ \vdots \\ 0 \end{bmatrix}, \quad \ldots, \quad \mathbf{V}_n(t_0) = \begin{bmatrix} 0 \\ 0 \\ \vdots \\ 1 \end{bmatrix} \qquad (12)$$

This fundamental matrix is denoted by $\mathbf{\Psi}(t)$. As a consequence of (12) we know that $\mathbf{\Psi}(t)$ has the property

$$\mathbf{\Psi}(t_0) = \mathbf{I} \qquad (13)$$

But since $\mathbf{\Psi}(t)$ is nonsingular for all values of t in an interval, (13) implies

$$\mathbf{\Psi}^{-1}(t_0) = \mathbf{I} \qquad (14)$$

Thus, when $\mathbf{\Psi}(t)$ is used rather than $\mathbf{\Phi}(t)$, it follows from (14) that (11) can be written as

$$\mathbf{X} = \mathbf{\Psi}(t)\mathbf{X}_0 + \mathbf{\Psi}(t)\int_{t_0}^{t} \mathbf{\Psi}^{-1}(s)\mathbf{F}(s)\,ds \qquad (15)$$

9.6.3 Diagonalization

As in Section 9.5, if the coefficient matrix $\mathbf{A}$ possesses n linearly independent eigenvectors, then we can use diagonalization to uncouple the system $\mathbf{X}' = \mathbf{AX} + \mathbf{F}(t)$.

Suppose $\mathbf{P}$ is the matrix such that $\mathbf{P}^{-1}\mathbf{AP} = \mathbf{D}$, where $\mathbf{D}$ is a diagonal matrix. Substituting $\mathbf{X} = \mathbf{PY}$ into the nonhomogeneous system $\mathbf{X}' = \mathbf{AX} + \mathbf{F}(t)$ gives

$$\mathbf{PY}' = \mathbf{APY} + \mathbf{F} \quad \text{or} \quad \mathbf{Y}' = \mathbf{P}^{-1}\mathbf{APY} + \mathbf{P}^{-1}\mathbf{F} \quad \text{or} \quad \mathbf{Y}' = \mathbf{DY} + \mathbf{G} \qquad (16)$$

In the last equation in (16), $\mathbf{G} = \mathbf{P}^{-1}\mathbf{F}$ is a column vector. So each differential equation in this new system has the form $y_i' = \lambda_i y_i + g_i(t)$, $i = 1, 2, \ldots, n$. But notice that, unlike the procedure for solving a homogeneous system $\mathbf{X}' = \mathbf{AX}$, we now are required to compute the inverse of the matrix $\mathbf{P}$.

EXAMPLE 5 Solve the system $X' = \begin{bmatrix} 4 & 2 \\ 2 & 1 \end{bmatrix} X + \begin{bmatrix} 3e^t \\ e^t \end{bmatrix}$ by diagonalization.

Solution The eigenvalues and corresponding eigenvectors of the coefficient matrix are found to be $\lambda_1 = 0$, $\lambda_2 = 5$, $K_1 = \begin{bmatrix} 1 \\ -2 \end{bmatrix}$, $K_2 = \begin{bmatrix} 2 \\ 1 \end{bmatrix}$. Thus, we find

$$P = \begin{bmatrix} 1 & 2 \\ -2 & 1 \end{bmatrix} \quad \text{and} \quad P^{-1} = \begin{bmatrix} \frac{1}{5} & -\frac{2}{5} \\ \frac{2}{5} & \frac{1}{5} \end{bmatrix}$$

From the substitution $X = PY$ and

$$P^{-1}F = \begin{bmatrix} \frac{1}{5} & -\frac{2}{5} \\ \frac{2}{5} & \frac{1}{5} \end{bmatrix} \begin{bmatrix} 3e^t \\ e^t \end{bmatrix} = \begin{bmatrix} \frac{1}{5}e^t \\ \frac{7}{5}e^t \end{bmatrix}$$

the uncoupled system is

$$Y' = \begin{bmatrix} 0 & 0 \\ 0 & 5 \end{bmatrix} Y + \begin{bmatrix} \frac{1}{5}e^t \\ \frac{7}{5}e^t \end{bmatrix}$$

The solutions of the two differential equations

$$y_1' = \frac{1}{5} e^t \quad \text{and} \quad y_2' = 5y_2 + \frac{7}{5} e^t$$

are, respectively, $\quad y_1 = \frac{1}{5} e^t + c_1 \quad$ and $\quad y_2 = -\frac{7}{20} e^t + c_2 e^{5t}$

Hence, the solution of the original system is

$$X = PY = \begin{bmatrix} 1 & 2 \\ -2 & 1 \end{bmatrix} \begin{bmatrix} \frac{1}{5}e^t + c_1 \\ -\frac{7}{20}e^t + c_2 e^{5t} \end{bmatrix} = \begin{bmatrix} -\frac{1}{2}e^t + c_1 + 2c_2 e^{5t} \\ -\frac{3}{4}e^t - 2c_1 + c_2 e^{5t} \end{bmatrix} \quad (17)$$

Written in the usual manner, (17) is

$$X = c_1 \begin{bmatrix} 1 \\ -2 \end{bmatrix} + c_2 \begin{bmatrix} 2 \\ 1 \end{bmatrix} e^{5t} + \begin{bmatrix} -\frac{1}{2} \\ -\frac{3}{4} \end{bmatrix} e^t \quad \square$$

EXERCISES 9.6 *Answers to odd-numbered problems begin on page A-57.*

[9.6.1]

In Problems 1–8 use the method of undetermined coefficients to solve the given system.

1. $\dfrac{dx}{dt} = 2x + 3y - 7$
   $\dfrac{dy}{dt} = -x - 2y + 5$

2. $\dfrac{dx}{dt} = 5x + 9y + 2$
   $\dfrac{dy}{dt} = -x + 11y + 6$

3. $\dfrac{dx}{dt} = x + 3y - 2t^2$
   $\dfrac{dy}{dt} = 3x + y + t + 5$

4. $\dfrac{dx}{dt} = x - 4y + 4t + 9e^{6t}$
   $\dfrac{dy}{dt} = 4x + y - t + e^{6t}$

5. $X' = \begin{bmatrix} 4 & \frac{1}{3} \\ 9 & 6 \end{bmatrix} X + \begin{bmatrix} -3 \\ 10 \end{bmatrix} e^t$

6. $X' = \begin{bmatrix} -1 & 5 \\ -1 & 1 \end{bmatrix} X + \begin{bmatrix} \sin t \\ -2 \cos t \end{bmatrix}$

7. $X' = \begin{bmatrix} 1 & 1 & 1 \\ 0 & 2 & 3 \\ 0 & 0 & 5 \end{bmatrix} X + \begin{bmatrix} 1 \\ -1 \\ 2 \end{bmatrix} e^{4t}$

8. $X' = \begin{bmatrix} 0 & 0 & 5 \\ 0 & 5 & 0 \\ 5 & 0 & 0 \end{bmatrix} X + \begin{bmatrix} 5 \\ -10 \\ 40 \end{bmatrix}$

9. Solve $X' = \begin{bmatrix} -1 & -2 \\ 3 & 4 \end{bmatrix} X + \begin{bmatrix} 3 \\ 3 \end{bmatrix}$ subject to $X(0) = \begin{bmatrix} -4 \\ 5 \end{bmatrix}$.

10. (a) Show that the system of differential equations for the currents $i_2(t)$ and $i_3(t)$ in the electrical network shown in Figure 9.4 is

$$\frac{d}{dt} \begin{bmatrix} i_2 \\ i_3 \end{bmatrix} = \begin{bmatrix} -R_1/L_1 & -R_1/L_1 \\ -R_1/L_2 & -(R_1+R_2)/L_2 \end{bmatrix} \begin{bmatrix} i_2 \\ i_3 \end{bmatrix} + \begin{bmatrix} E/L_1 \\ E/L_2 \end{bmatrix}$$

(b) Solve the system in part (a) if $R_1 = 2$ ohms, $R_2 = 3$ ohms, $L_1 = 1$ henry, $L_2 = 1$ henry, $E = 60$ volts, $i_2(0) = 0$, and $i_3(0) = 0$.

(c) Determine the current $i_1(t)$.

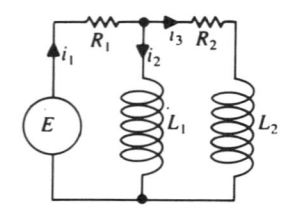

FIGURE 9.4

[9.6.2]

In Problems 11–30 use variation of parameters to solve the given system.

11. $\dfrac{dx}{dt} = 3x - 3y + 4$
$\dfrac{dy}{dt} = 2x - 2y - 1$

12. $\dfrac{dx}{dt} = 2x - y$
$\dfrac{dy}{dt} = 3x - 2y + 4t$

13. $X' = \begin{bmatrix} 3 & -5 \\ \frac{3}{4} & -1 \end{bmatrix} X + \begin{bmatrix} 1 \\ -1 \end{bmatrix} e^{t/2}$

14. $X' = \begin{bmatrix} 2 & -1 \\ 4 & 2 \end{bmatrix} X + \begin{bmatrix} \sin 2t \\ 2 \cos 2t \end{bmatrix} e^{2t}$

15. $X' = \begin{bmatrix} 0 & 2 \\ -1 & 3 \end{bmatrix} X + \begin{bmatrix} 1 \\ -1 \end{bmatrix} e^t$

16. $X' = \begin{bmatrix} 0 & 2 \\ -1 & 3 \end{bmatrix} X + \begin{bmatrix} 2 \\ e^{-3t} \end{bmatrix}$

17. $X' = \begin{bmatrix} 1 & 8 \\ 1 & -1 \end{bmatrix} X + \begin{bmatrix} 12 \\ 12 \end{bmatrix} t$

18. $X' = \begin{bmatrix} 1 & 8 \\ 1 & -1 \end{bmatrix} X + \begin{bmatrix} e^{-t} \\ te^t \end{bmatrix}$

19. $X' = \begin{bmatrix} 3 & 2 \\ -2 & -1 \end{bmatrix} X + \begin{bmatrix} 2e^{-t} \\ e^{-t} \end{bmatrix}$

20. $X' = \begin{bmatrix} 3 & 2 \\ -2 & -1 \end{bmatrix} X + \begin{bmatrix} 1 \\ 1 \end{bmatrix}$

21. $X' = \begin{bmatrix} 0 & -1 \\ 1 & 0 \end{bmatrix} X + \begin{bmatrix} \sec t \\ 0 \end{bmatrix}$

22. $X' = \begin{bmatrix} 1 & -1 \\ 1 & 1 \end{bmatrix} X + \begin{bmatrix} 3 \\ 3 \end{bmatrix} e^t$

23. $X' = \begin{bmatrix} 1 & -1 \\ 1 & 1 \end{bmatrix} X + \begin{bmatrix} \cos t \\ \sin t \end{bmatrix} e^t$

24. $X' = \begin{bmatrix} 2 & -2 \\ 8 & -6 \end{bmatrix} X + \begin{bmatrix} 1 \\ 3 \end{bmatrix} \dfrac{e^{-2t}}{t}$

25. $X' = \begin{bmatrix} 0 & 1 \\ -1 & 0 \end{bmatrix} X + \begin{bmatrix} 0 \\ \sec t \tan t \end{bmatrix}$

26. $X' = \begin{bmatrix} 0 & 1 \\ -1 & 0 \end{bmatrix} X + \begin{bmatrix} 1 \\ \cot t \end{bmatrix}$

27. $X' = \begin{bmatrix} 1 & 2 \\ -\frac{1}{2} & 1 \end{bmatrix} X + \begin{bmatrix} \csc t \\ \sec t \end{bmatrix} e^t$

28. $X' = \begin{bmatrix} 1 & -2 \\ 1 & -1 \end{bmatrix} X + \begin{bmatrix} \tan t \\ 1 \end{bmatrix}$

29. $X' = \begin{bmatrix} 1 & 1 & 0 \\ 1 & 1 & 0 \\ 0 & 0 & 3 \end{bmatrix} X + \begin{bmatrix} e^t \\ e^{2t} \\ te^{3t} \end{bmatrix}$

30. $X' = \begin{bmatrix} 3 & -1 & -1 \\ 1 & 1 & -1 \\ 1 & -1 & 1 \end{bmatrix} X + \begin{bmatrix} 0 \\ t \\ 2e^t \end{bmatrix}$

In Problems 31 and 32 use (11) to solve the given system subject to the indicated initial condition.

31. $X' = \begin{bmatrix} 3 & -1 \\ -1 & 3 \end{bmatrix} X + \begin{bmatrix} 4e^{2t} \\ 4e^{4t} \end{bmatrix}$, $X(0) = \begin{bmatrix} 1 \\ 1 \end{bmatrix}$

32. $X' = \begin{bmatrix} 1 & -1 \\ 1 & -1 \end{bmatrix} X + \begin{bmatrix} 1/t \\ 1/t \end{bmatrix}$, $X(1) = \begin{bmatrix} 2 \\ -1 \end{bmatrix}$

In Problems 33 and 34 use (15) to solve the given system subject to the indicated initial condition. Use the results of Problems 21 and 24 of Section 9.3.

33. $\mathbf{X}' = \begin{bmatrix} 4 & 1 \\ 6 & 5 \end{bmatrix} \mathbf{X} + \begin{bmatrix} 50e^{7t} \\ 0 \end{bmatrix}, \mathbf{X}(0) = \begin{bmatrix} 5 \\ -5 \end{bmatrix}$

34. $\mathbf{X}' = \begin{bmatrix} 3 & -2 \\ 5 & -3 \end{bmatrix} \mathbf{X} + \begin{bmatrix} 2 \\ 3 \end{bmatrix}, \mathbf{X}(\pi/2) = \begin{bmatrix} 0 \\ 0 \end{bmatrix}$

35. (a) Show that the system of differential equations for the currents $i_1(t)$ and $i_2(t)$ in the electrical network shown in Figure 9.5 is

$$\frac{d}{dt}\begin{bmatrix} i_1 \\ i_2 \end{bmatrix} = \begin{bmatrix} -(R_1 + R_2)/L_2 & R_2/L_2 \\ R_2/L_1 & -R_2/L_1 \end{bmatrix}\begin{bmatrix} i_1 \\ i_2 \end{bmatrix} + \begin{bmatrix} E/L_2 \\ 0 \end{bmatrix}$$

(b) Solve the system in part (a) if $R_1 = 8$ ohms, $R_2 = 3$ ohms, $L_1 = 1$ henry, $L_2 = 1$ henry, $E(t) = 100 \sin t$ volts, $i_1(0) = 0$, and $i_2(0) = 0$.

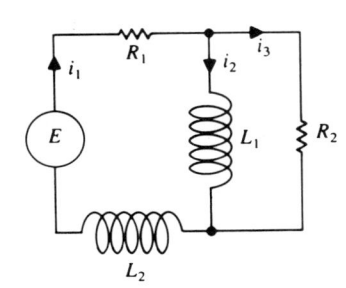

FIGURE 9.5

36. For the electrical network in Figure 9.6 it can be shown that

$$\frac{d}{dt}\begin{bmatrix} i_1 \\ q \end{bmatrix} = \begin{bmatrix} -\dfrac{R_1 R_2}{L(R_1 + R_2)} & -\dfrac{R_1}{CL(R_1 + R_2)} \\ \dfrac{R_1}{R_1 + R_2} & -\dfrac{1}{C(R_1 + R_2)} \end{bmatrix}\begin{bmatrix} i_1 \\ q \end{bmatrix} + \begin{bmatrix} E/L \\ 0 \end{bmatrix}$$

Solve the system if $R_1 = 50$ ohms, $R_2 = 100$ ohms, $L = 5$ henrys, $C = 0.004$ farad, and $E = 10$ volts.

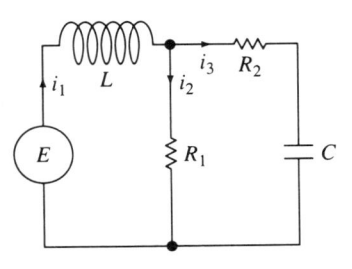

FIGURE 9.6

[9.6.3]

In Problems 37–40 use diagonalization to solve the given system.

37. $\mathbf{X}' = \begin{bmatrix} 5 & -2 \\ 21 & -8 \end{bmatrix}\mathbf{X} + \begin{bmatrix} 6 \\ 4 \end{bmatrix}$

38. $\mathbf{X}' = \begin{bmatrix} 1 & 3 \\ 2 & 2 \end{bmatrix}\mathbf{X} + \begin{bmatrix} e^t \\ e^t \end{bmatrix}$

39. $\mathbf{X}' = \begin{bmatrix} 5 & 5 \\ 5 & 5 \end{bmatrix}\mathbf{X} + \begin{bmatrix} 2t \\ 8 \end{bmatrix}$

40. $\mathbf{X}' = \begin{bmatrix} 0 & 1 \\ 1 & 0 \end{bmatrix}\mathbf{X} + \begin{bmatrix} 4 \\ 8e^{-2t} \end{bmatrix}$

[O] 9.7 MATRIX EXPONENTIAL

Matrices can be utilized in an entirely different manner to solve a homogeneous system of linear first-order differential equations.

Recall that the simple linear first-order differential equation

$$x' = ax$$

where a is a constant, has the general solution

$$x = ce^{at}$$

It seems natural then to ask whether we can define a matrix exponential $e^{t\mathbf{A}}$ so that the homogeneous system

$$\mathbf{X}' = \mathbf{A}\mathbf{X}$$

where **A** is an $n \times n$ matrix of constants, has a solution

$$\mathbf{X} = e^{t\mathbf{A}}\mathbf{C} \qquad (1)$$

Since **C** is to be an $n \times 1$ column vector of arbitrary constants, we want $e^{t\mathbf{A}}$ to be an $n \times n$ matrix. While the complete development of the meaning of the **matrix exponential** would necessitate a more thorough investigation of matrix algebra, one means of computing $e^{t\mathbf{A}}$ is given in the following definition:

DEFINITION 9.9 Matrix Exponential

For any $n \times n$ matrix **A** whose entries are constants,

$$e^{t\mathbf{A}} = \sum_{n=0}^{\infty} \frac{(t\mathbf{A})^n}{n!} = \mathbf{I} + t\mathbf{A} + \frac{t^2}{2!}\mathbf{A}^2 + \frac{t^3}{3!}\mathbf{A}^3 + \cdots \qquad (2)$$

It can be shown that the series given in (2) converges to an $n \times n$ matrix for every value of t. Also, $\mathbf{A}^2 = \mathbf{A}\mathbf{A}$, $\mathbf{A}^3 = \mathbf{A}(\mathbf{A}^2)$, and so on.

Now the general solution of the single differential equation

$$x' = ax + f(t)$$

where a is a constant, can be expressed as

$$x = x_c + x_p = ce^{at} + e^{at}\int_{t_0}^{t} e^{-as}f(s)\,ds$$

For systems of linear first-order differential equations, it can be shown that the general solution of

$$\mathbf{X}' = \mathbf{A}\mathbf{X} + \mathbf{F}(t)$$

where **A** is an $n \times n$ matrix of constants, is

$$\mathbf{X} = \mathbf{X}_c + \mathbf{X}_p = e^{t\mathbf{A}}\mathbf{C} + e^{t\mathbf{A}}\int_{t_0}^{t} e^{-s\mathbf{A}}\mathbf{F}(s)\,ds \qquad (3)$$

The matrix exponential $e^{t\mathbf{A}}$ is always nonsingular and $e^{-s\mathbf{A}} = (e^{s\mathbf{A}})^{-1}$. In practice, $e^{-s\mathbf{A}}$ can be obtained from $e^{t\mathbf{A}}$ by replacing t by $-s$.

Additional Properties

From (2) it is seen that

$$e^{\mathbf{0}} = \mathbf{I} \qquad (4)$$

Also, formal termwise differentiation of (2) shows that

$$\frac{d}{dt}e^{t\mathbf{A}} = \mathbf{A}e^{t\mathbf{A}} \qquad (5)$$

9.7 Matrix Exponential

If we denote the matrix exponential by $\Psi(t)$, then (5) and (4) are equivalent to
$$\Psi'(t) = A\Psi(t) \tag{6}$$
and
$$\Psi(0) = I \tag{7}$$

respectively. The notation here is chosen deliberately. Comparing (6) with (11) of Section 9.3 reveals that e^{tA} is a fundamental matrix of the system $X' = AX$. It is precisely this formulation of the fundamental matrix that was discussed on page 639 of Section 9.3.

By multiplying the series defining e^{tA} and e^{-sA}, we can prove that

$$e^{tA}e^{-sA} = e^{(t-s)A} \quad \text{or equivalently} \quad \Psi(t)\Psi^{-1}(s) = \Psi(t-s)*$$

This last result enables us to relate (3) to (10) of the preceding section:

$$X = e^{tA}C + \int_{t_0}^{t} e^{tA}e^{-sA}F(s)\,ds$$

$$= e^{tA}C + \int_{t_0}^{t} e^{(t-s)A}F(s)\,ds \tag{8}$$

$$X = \Psi(t)C + \int_{t_0}^{t} \Psi(t-s)F(s)\,ds \tag{9}$$

Equation (9) possesses a form simpler than (10) of Section 9.6. In other words, there is no need to compute Ψ^{-1}; we need only replace t by $t - s$ in $\Psi(t)$.

EXERCISES 9.7
Answers to odd-numbered problems begin on page A-57.

In Problems 1 and 2 use (2) to compute e^{tA} and e^{-tA}.

1. $A = \begin{bmatrix} 0 & 1 \\ 1 & 0 \end{bmatrix}$ **2.** $A = \begin{bmatrix} 1 & 0 \\ 0 & 2 \end{bmatrix}$

In Problems 3 and 4 use (1) and Problems 1 and 2 to find the general solution of the given system.

3. $X' = \begin{bmatrix} 0 & 1 \\ 1 & 0 \end{bmatrix} X$ **4.** $X' = \begin{bmatrix} 1 & 0 \\ 0 & 2 \end{bmatrix} X$

In Problems 5–8 use (3) and Problems 3 and 4 to find the general solution of the given system.

5. $X' = \begin{bmatrix} 0 & 1 \\ 1 & 0 \end{bmatrix} X + \begin{bmatrix} 1 \\ 1 \end{bmatrix}$

6. $X' = \begin{bmatrix} 0 & 1 \\ 1 & 0 \end{bmatrix} X + \begin{bmatrix} \cosh t \\ \sinh t \end{bmatrix}$

7. $X' = \begin{bmatrix} 1 & 0 \\ 0 & 2 \end{bmatrix} X + \begin{bmatrix} t \\ e^{4t} \end{bmatrix}$

8. $X' = \begin{bmatrix} 1 & 0 \\ 0 & 2 \end{bmatrix} X + \begin{bmatrix} 3 \\ -1 \end{bmatrix}$

9. If the matrix A can be diagonalized, then $P^{-1}AP = D$ or $A = PDP^{-1}$. Use this last result and (2) to show that $e^{tA} = Pe^{tD}P^{-1}$.

10. Use

$$D = \begin{bmatrix} \lambda_1 & 0 & \cdots & 0 \\ 0 & \lambda_2 & \cdots & 0 \\ \vdots & & & \vdots \\ 0 & 0 & \cdots & \lambda_n \end{bmatrix}$$

---

*Although $e^{tA}e^{-sA} = e^{(t-s)A}$, it is interesting to note that $e^{A}e^{B}$ is, in general, not the same as e^{A+B} for $n \times n$ matrices A and B.

and (2) to show that

$$e^{tD} = \begin{bmatrix} e^{\lambda_1 t} & 0 & \cdots & 0 \\ 0 & e^{\lambda_2 t} & \cdots & 0 \\ \vdots & & & \vdots \\ 0 & 0 & \cdots & e^{\lambda_n t} \end{bmatrix}$$

In Problems 11 and 12 use the results of Problems 9 and 10 to solve the given system.

11. $X' = \begin{bmatrix} 2 & 1 \\ -3 & 6 \end{bmatrix} X$ **12.** $X' = \begin{bmatrix} 2 & 1 \\ 1 & 2 \end{bmatrix} X$

SUMMARY

A first-order system in **normal form** in *two* dependent variables is any system

$$\begin{aligned} \frac{dx}{dt} &= a_{11}(t)x + a_{12}(t)y + f_1(t) \\ \frac{dy}{dt} &= a_{21}(t)x + a_{22}(t)y + f_2(t) \end{aligned} \quad (1)$$

where the coefficients $a_{ij}(t)$, $f_1(t)$, and $f_2(t)$ are continuous on some common interval I. When $f_1(t) = 0$, $f_2(t) = 0$, the system is said to be **homogeneous**; otherwise it is **nonhomogeneous**. Any linear second-order differential equation can be expressed in this form.

Using matrices, we can write the system (1) compactly as

$$\frac{d\mathbf{X}}{dt} = \mathbf{A}(t)\mathbf{X} + \mathbf{F}(t) \quad (2)$$

where

$$\mathbf{X} = \begin{bmatrix} x \\ y \end{bmatrix} \quad \mathbf{A}(t) = \begin{bmatrix} a_{11}(t) & a_{12}(t) \\ a_{21}(t) & a_{22}(t) \end{bmatrix} \quad \mathbf{F}(t) = \begin{bmatrix} f_1(t) \\ f_2(t) \end{bmatrix}$$

The **general solution of the homogeneous system** in two dependent variables

$$\frac{d\mathbf{X}}{dt} = \mathbf{A}(t)\mathbf{X} \quad (3)$$

is defined to be the linear combination

$$\mathbf{X} = c_1\mathbf{X}_1 + c_2\mathbf{X}_2 \quad (4)$$

where $\mathbf{X}_1$ and $\mathbf{X}_2$ form a **fundamental set of solutions** of (3) on I. The **general solution of the nonhomogeneous system** (2) is defined to be

$$\mathbf{X} = \mathbf{X}_c + \mathbf{X}_p$$

where $\mathbf{X}_c$ is defined by (4) and $\mathbf{X}_p$ is *any* particular solution vector of (2).

To solve a homogeneous system (3) we determine the **eigenvalues** of the coefficient matrix $\mathbf{A}$ and then find the corresponding **eigenvectors**.

To solve a nonhomogeneous system we first solve the associated homogeneous system. A particular solution vector $\mathbf{X}_p$ of the nonhomogeneous system is found by **undetermined coefficients, variation of parameters,** or **diagonalization**.

A **fundamental matrix** of a homogeneous system (3) in two dependent variables is defined to be

$$\mathbf{\Phi}(t) = \begin{bmatrix} x_{11} & x_{12} \\ x_{21} & x_{22} \end{bmatrix} \tag{5}$$

The columns in (5) are obtained from two linearly independent solution vectors $\mathbf{X}_1$ and $\mathbf{X}_2$ of (3). In terms of matrices, the method of variation of parameters leads to a particular solution given by

$$\mathbf{X}_p = \mathbf{\Phi}(t) \int \mathbf{\Phi}^{-1}(t) \mathbf{F}(t) \, dt$$

The general solution of (2) on an interval is

$$\mathbf{X} = \mathbf{\Phi}(t)\mathbf{C} + \mathbf{\Phi}(t) \int \mathbf{\Phi}^{-1}(t) \mathbf{F}(t) \, dt$$

where $\mathbf{C}$ is a column matrix containing two arbitrary constants and the matrix $\mathbf{\Phi}^{-1}(t)$ is the **multiplicative inverse** of $\mathbf{\Phi}(t)$.

CHAPTER 9 REVIEW EXERCISES
Answers to odd-numbered problems begin on page A-57.

1. (a) Write as one column matrix $\mathbf{X}$:

$$\begin{bmatrix} 3 & 1 & 1 \\ -1 & 2 & -1 \\ 0 & -2 & 4 \end{bmatrix} \begin{bmatrix} t \\ t^2 \\ t^3 \end{bmatrix} - \begin{bmatrix} 2 \\ -2 \\ -1 \end{bmatrix} + 2t \begin{bmatrix} 1 \\ 0 \\ 4 \end{bmatrix} + 2t^2 \begin{bmatrix} 1 \\ -1 \\ 7 \end{bmatrix}$$

(b) Find $d\mathbf{X}/dt$.

2. Write the differential equation

$$3y^{(4)} - 5y'' + 9y = 6e^t - 2t$$

as a system of first-order equations in linear normal form.

3. Write the system

$$(2D^2 + D)y - D^2 x = \ln t$$

$$D^2 y + (D + 1)x = 5t - 2$$

as a system of first-order equations in linear normal form.

4. Verify that the general solution of the system

$$\frac{dx}{dt} = y$$

$$\frac{dy}{dt} = -x + 2y - 2\cos t$$

on the interval $(-\infty, \infty)$ is

$$\mathbf{X} = c_1 \begin{bmatrix} 1 \\ 1 \end{bmatrix} e^t + c_2 \left\{ \begin{bmatrix} 1 \\ 1 \end{bmatrix} te^t + \begin{bmatrix} 0 \\ 1 \end{bmatrix} e^t \right\} + \begin{bmatrix} \sin t \\ \cos t \end{bmatrix}$$

In Problems 5–14 solve the given system.

5. $\dfrac{dx}{dt} = 2x + y$

$\dfrac{dy}{dt} = -x$

6. $\dfrac{dx}{dt} = -4x + 2y$

$\dfrac{dy}{dt} = 2x - 4y$

7. $\mathbf{X}' = \begin{bmatrix} 1 & 2 \\ -2 & 1 \end{bmatrix} \mathbf{X}$

8. $\mathbf{X}' = \begin{bmatrix} -2 & 5 \\ -2 & 4 \end{bmatrix} \mathbf{X}$

9. $\mathbf{X}' = \begin{bmatrix} 1 & 1 & 1 \\ 1 & 1 & 1 \\ 1 & 1 & 1 \end{bmatrix} \mathbf{X}$

10. $\mathbf{X}' = \begin{bmatrix} 1 & -1 & 1 \\ 0 & 1 & 3 \\ 4 & 3 & 1 \end{bmatrix} \mathbf{X}$

11. $\mathbf{X}' = \begin{bmatrix} 2 & 8 \\ 0 & 4 \end{bmatrix} \mathbf{X} + \begin{bmatrix} 2 \\ 16t \end{bmatrix}$

12. $\dfrac{dx}{dt} = x + 2y$

$\dfrac{dy}{dt} = -\dfrac{1}{2}x + y + e^t \tan t$

13. $\mathbf{X}' = \begin{bmatrix} -1 & 1 \\ -2 & 1 \end{bmatrix} \mathbf{X} + \begin{bmatrix} 1 \\ \cot t \end{bmatrix}$

14. $\mathbf{X}' = \begin{bmatrix} 3 & 1 \\ -1 & 1 \end{bmatrix} \mathbf{X} + \begin{bmatrix} -2 \\ 1 \end{bmatrix} e^{2t}$

10

PLANE AUTONOMOUS SYSTEMS AND STABILITY

TOPICS TO REVIEW

eigenvalues and eigenvectors of a matrix (7.7)
writing a differential equation as a system (9.1)
homogeneous linear systems of differential equations (9.3)
nonhomogeneous linear systems of differential equations (9.3)
eigenvalue–eigenvector method (9.4)

IMPORTANT CONCEPTS

autonomous systems
critical or stationary point
arc
periodic solution or cycle
stable node
unstable node
saddle point
degenerate node
spiral points
stable critical point
unstable critical point
linearization
Jacobian matrix
phase-plane method
global stability
invariant region
limit cycle

10.1 Autonomous Systems, Critical Points, and Periodic Solutions
10.2 Stability of Linear Systems
10.3 Linearization and Local Stability
10.4 Applications of Autonomous Systems
[O] 10.5 Periodic Solutions, Limit Cycles, and Global Stability
Summary
Chapter 10 Review Exercises

INTRODUCTION

In Chapter 9 we concentrated on techniques for solving systems of linear differential equations of the form $\mathbf{X}' = \mathbf{AX} + \mathbf{F}(t)$. When the system of differential equations is not linear, it is usually not possible to find solutions in terms of elementary functions. In this chapter we will demonstrate that valuable information about the geometric nature of solutions can be obtained by first analyzing special constant solutions called **critical points** and then searching for periodic solutions called **limit cycles**. The important concept of **stability** will be introduced and illustrated with examples from physics and biology.

10.1 AUTONOMOUS SYSTEMS, CRITICAL POINTS, AND PERIODIC SOLUTIONS

Terminology and Notation A system of first-order differential equations is called **autonomous** when it can be written in the form

$$\frac{dx_1}{dt} = g_1(x_1, x_2, \ldots, x_n)$$
$$\frac{dx_2}{dt} = g_2(x_1, x_2, \ldots, x_n)$$
$$\vdots \qquad \vdots$$
$$\frac{dx_n}{dt} = g_n(x_1, x_2, \ldots, x_n)$$
(1)

The variable t does *not* appear explicitly on the right-hand side of each differential equation. (Compare with (2) in Section 9.1.)

EXAMPLE 1 The system of differential equations

$$\frac{dx_1}{dt} = x_1 - 3x_2 + t^2$$
$$\frac{dx_2}{dt} = x_1 \sin(x_2 t)$$

is *not* autonomous because of the presence of t^2 and $\sin(x_2 t)$ on the right-hand sides. □

If $\mathbf{X}(t)$ and $\mathbf{g}(\mathbf{X})$ denote the respective column vectors

$$\mathbf{X}(t) = \begin{bmatrix} x_1(t) \\ x_2(t) \\ \vdots \\ x_n(t) \end{bmatrix} \qquad \mathbf{g}(\mathbf{X}) = \begin{bmatrix} g_1(x_1, x_2, \ldots, x_n) \\ g_2(x_1, x_2, \ldots, x_n) \\ \vdots \\ g_n(x_1, x_2, \ldots, x_n) \end{bmatrix}$$

then the autonomous system (1) may be written in the compact **column vector form** $\mathbf{X}' = \mathbf{g}(\mathbf{X})$. The homogeneous linear system $\mathbf{X}' = \mathbf{A}\mathbf{X}$ studied in Sections 9.3 and 9.4 is an important special case.

In this chapter it will be convenient to write (1) using row vectors. If we let

$$\mathbf{X}(t) = (x_1(t), x_2(t), \ldots, x_n(t))$$

and $\quad \mathbf{g}(\mathbf{X}) = (g_1(x_1, x_2, \ldots, x_n), g_2(x_1, x_2, \ldots, x_n), \ldots, g_n(x_1, x_2, \ldots, x_n))$

then the autonomous system (1) may also be written in the compact **row vector form** $\mathbf{X}' = \mathbf{g}(\mathbf{X})$. It should be clear from the context whether we are using column or row vector form, and therefore we will not distinguish between $\mathbf{X}$

10.1 Autonomous Systems, Critical Points, and Periodic Solutions

and $\mathbf{X}^T$, the transpose of $\mathbf{X}$. In particular, when $n = 2$, it is convenient to use row vector form and write an initial condition as $\mathbf{X}(0) = (x_0, y_0)$.

When the variable t is interpreted as time, we can refer to a solution $\mathbf{X}(t)$ as the **state of the system** at time t. Using this terminology, we know that a system of differential equations is autonomous when the rate $\mathbf{X}'(t)$ at which the system changes depends only on the system's present state $\mathbf{X}(t)$. The linear system $\mathbf{X}' = \mathbf{AX} + \mathbf{F}(t)$ studied in Chapter 9 is then autonomous when $\mathbf{F}(t)$ is constant.

Note that when $n = 1$, an autonomous differential equation takes the simple form $dx/dt = g(x)$. Explicit solutions can be constructed, since this differential equation is *separable*, and we will make use of this fact to illustrate the concepts in this chapter.

Vector Field Interpretation When $n = 2$, the system is called a **plane autonomous system**. We write the system as

$$\frac{dx}{dt} = P(x, y)$$

$$\frac{dy}{dt} = Q(x, y)$$

The vector $\mathbf{V}(x, y) = (P(x, y), Q(x, y))$ defines a **vector field** in a region of the plane, and a solution to the system may be interpreted as the resulting path of a particle as it moves through the region. To be more specific, let $\mathbf{V}(x, y) = (P(x, y), Q(x, y))$ denote the velocity of a stream at position (x, y), and suppose that a small particle (such as a cork) is released at a position (x_0, y_0) in the stream. If $\mathbf{X}(t) = (x(t), y(t))$ denotes the position of the particle at time t, then $\mathbf{X}'(t) = (x'(t), y'(t))$ is the velocity vector $\mathbf{v}$. When external forces are not present and frictional forces are neglected, the velocity of the particle at time t is the velocity of the stream at position $\mathbf{X}(t)$:

$$\mathbf{X}'(t) = \mathbf{V}(x(t), y(t))$$

that is,
$$\frac{dx}{dt} = P(x(t), y(t))$$

$$\frac{dy}{dt} = Q(x(t), y(t))$$

Thus, the path of the particle is the solution to the system which satisfies the initial condition $\mathbf{X}(0) = (x_0, y_0)$. We will frequently call on this simple interpretation of a plane autonomous system to illustrate new concepts.

EXAMPLE 2 A vector field for the steady-state flow of a fluid around a cylinder of radius 1 is given by

$$\mathbf{V}(x, y) = V_0 \left(1 - \frac{x^2 - y^2}{(x^2 + y^2)^2}, \frac{-2xy}{(x^2 + y^2)^2} \right)$$

where V_0 is the speed of the fluid far from the cylinder. If a small cork is released at $(-3, 1)$, the path $\mathbf{X}(t) = (x(t), y(t))$ of the cork satisfies the plane autonomous system

$$\frac{dx}{dt} = V_0\left(1 - \frac{x^2 - y^2}{(x^2 + y^2)^2}\right)$$
$$\frac{dy}{dt} = V_0\left(\frac{-2xy}{(x^2 + y^2)^2}\right)$$

subject to the initial condition $\mathbf{X}(0) = (-3, 1)$. See Figure 10.1 for an illustration of flow (left to right) around a cylinder. □

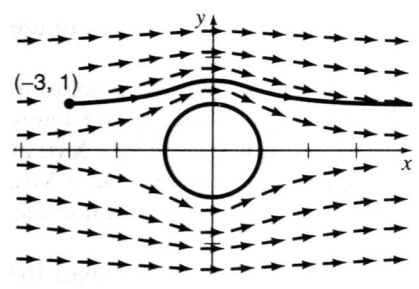

FIGURE 10.1

Any second-order nonlinear differential equation $x'' = g(x, x')$, can be written as a plane autonomous system. When we set $y = x'$, the equation becomes

$$x' = y$$
$$y' = g(x, y)$$

EXAMPLE 3 In Example 3 in Section 1.3 we showed that the displacement angle θ for a pendulum satisfies the second-order nonlinear differential equation

$$\frac{d^2\theta}{dt^2} + \frac{g}{l}\sin\theta = 0$$

If we let $x = \theta$ and $y = \theta'$, this second-order differential equation may be rewritten as the plane autonomous system

$$x' = y$$
$$y' = -\frac{g}{l}\sin x$$ □

Types of Solutions If $P(x, y)$, $Q(x, y)$, and the first-order partial derivatives $\partial P/\partial x$, $\partial P/\partial y$, $\partial Q/\partial x$, and $\partial Q/\partial y$ are continuous in a region R of the plane, then the solutions of the plane autonomous system

$$\frac{dx}{dt} = P(x, y) \qquad \frac{dy}{dt} = Q(x, y)$$

are of three basic types:

(i) A constant solution $x(t) = x_0$, $y(t) = y_0$ (or $\mathbf{X}(t) = \mathbf{X}_0$ for all t): A constant solution is called a **critical** or **stationary point**. When the particle is placed at a critical point $\mathbf{X}_0$ (that is $\mathbf{X}(0) = \mathbf{X}_0$), it remains there indefinitely. Note that since $\mathbf{X}'(t) = \mathbf{0}$, a critical point is a solution of the system of algebraic equations

$$P(x, y) = 0 \qquad Q(x, y) = 0$$

10.1 Autonomous Systems, Critical Points, and Periodic Solutions

(ii) A solution $x = x(t)$, $y = y(t)$ that defines an **arc**, a plane curve that does *not* cross itself: Thus, the curve in Figure 10.2(a) can be a solution of a plane autonomous system, whereas the curve in Figure 10.2(b) cannot be a solution.

(iii) **A periodic solution** $x = x(t)$, $y = y(t)$: A periodic solution is called a **cycle**. If p is the period of the solution, then $\mathbf{X}(t + p) = \mathbf{X}(t)$ and a particle placed on the curve at $\mathbf{X}_0$ will cycle around the curve and return to $\mathbf{X}_0$ in p units of time (see Figure 10.3).

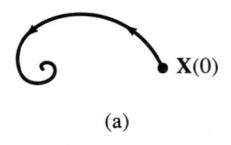

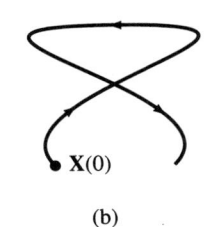

FIGURE 10.2

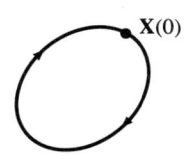

FIGURE 10.3

EXAMPLE 4 Find all critical points of each of the following plane autonomous systems:

(a) $x' = -x + y$ (b) $x' = x^2 + y^2 - 6$ (c) $x' = 0.01x(100 - x - y)$
$\ y' = x - y$ $\ y' = x^2 - y$ $\ y' = 0.05y(60 - y - 0.2x)$

Solution We find the critical points by setting the right-hand sides of the differential equations equal to zero.

(a) The solution to the system

$$-x + y = 0$$
$$x - y = 0$$

consists of all points on the line $y = x$. Thus, there are infinitely many critical points.

(b) To solve the system

$$x^2 + y^2 - 6 = 0$$
$$x^2 - y = 0$$

we substitute the second equation $x^2 = y$ into the first equation to obtain $y^2 + y - 6 = (y + 3)(y - 2) = 0$. If $y = -3$, then $x^2 = -3$, and so there are no real solutions. If $y = 2$, then $x = \pm\sqrt{2}$, and so the critical points are $(\sqrt{2}, 2)$ and $(-\sqrt{2}, 2)$.

(c) Finding the critical points requires a careful consideration of cases. The equation $0.01x(100 - x - y) = 0$ implies $x = 0$ or $x + y = 100$. If $x = 0$, then substituting in $0.05y(60 - y - 0.2x) = 0$, we have $y(60 - y) = 0$. Thus, $y = 0$ or 60, and so $(0, 0)$ and $(0, 60)$ are critical points. If $x + y = 100$, then $0 = y(60 - y - 0.2(100 - y)) = y(40 - 0.8y)$. It follows that $y = 0$ or 50, and so $(100, 0)$ and $(50, 50)$ are critical points. □

When the plane autonomous system is linear, we can use the methods in Chapter 9 to investigate solutions.

EXAMPLE 5 Determine whether the given linear system possesses a periodic solution:

(a) $x' = 2x + 8y$ (b) $x' = x + 2y$
$\ y' = -x - 2y$ $\ y' = -\frac{1}{2}x + y$

In each case sketch the graph of the solution that satisfies $\mathbf{X}(0) = (2, 0)$.

Solution (a) In Example 3 in Section 9.4 we used the eigenvalue–eigenvector method to show that

$$x = c_1(2\cos 2t - 2\sin 2t) + c_2(2\cos 2t + 2\sin 2t)$$
$$y = c_1(-\cos 2t) - c_2 \sin 2t$$

Thus, every solution is periodic with period $p = \pi$. The solution satisfying $\mathbf{X}(0) = (2, 0)$ is

$$x = 2\cos 2t + 2\sin 2t$$
$$y = -\sin 2t$$

This solution generates the ellipse shown in Figure 10.4(a).

(b) In Example 4 in Section 9.4 we used the eigenvalue–eigenvector method to show that

$$x = c_1(2e^t \cos t) + c_2(2e^t \sin t)$$
$$y = c_1(-e^t \sin t) + c_2(e^t \cos t)$$

Because of the presence of e^t in the general solution, there are no periodic solutions (that is, cycles). The solution satisfying $\mathbf{X}(0) = (2, 0)$ is

$$x = 2e^t \cos t$$
$$y = -e^t \sin t$$

This curve (called an **unstable spiral**) is shown in Figure 10.4(b). □

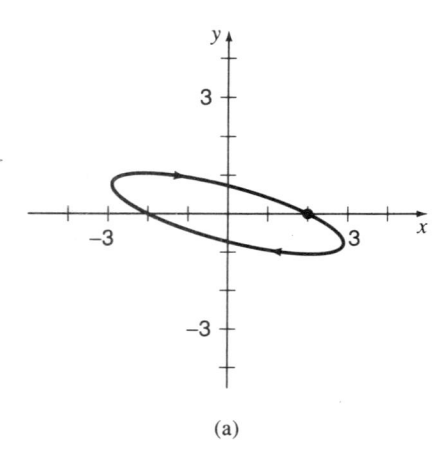

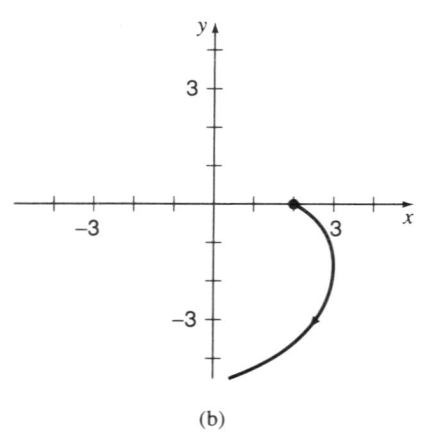

FIGURE 10.4

Changing to Polar Coordinates Except for the case of constant solutions, it is usually not possible to find explicit expressions for the solutions of a *nonlinear* autonomous system. Some nonlinear systems, however, can be solved by changing to polar coordinates. From the formulas $r^2 = x^2 + y^2$ and $\theta = \tan^{-1}(y/x)$, we obtain

$$\boxed{\begin{aligned} \frac{dr}{dt} &= \frac{1}{r}\left(x\frac{dx}{dt} + y\frac{dy}{dt}\right) \\ \frac{d\theta}{dt} &= \frac{1}{r^2}\left(-y\frac{dx}{dt} + x\frac{dy}{dt}\right). \end{aligned}} \qquad (2)$$

EXAMPLE 6 Find the solution of the nonlinear plane autonomous system

$$x' = -y - x\sqrt{x^2 + y^2}$$
$$y' = x - y\sqrt{x^2 + y^2}$$

satisfying the initial condition $\mathbf{X}(0) = (3, 3)$.

10.1 Autonomous Systems, Critical Points, and Periodic Solutions

FIGURE 10.5

Solution From the expressions for dr/dt and $d\theta/dt$ in (2), we obtain

$$\frac{dr}{dt} = \frac{1}{r}[x(-y - xr) + y(x - yr)] = -r^2$$

$$\frac{d\theta}{dt} = \frac{1}{r^2}[-y(-y - xr) + x(x - yr)] = 1$$

with $r(0) = 3\sqrt{2}$ and $\theta(0) = \pi/4$. When we use separation of variables, it follows that the general solution of the system is

$$r = \frac{1}{t + c_1} \qquad \theta = t + c_2$$

for $r \neq 0$. Applying the initial conditions then gives $r = 1/(t + \sqrt{2}/6)$ and $\theta = t + \pi/4$. The spiral $r = 1/(\theta + \sqrt{2}/6 - \pi/4)$ is sketched in Figure 10.5. □

EXAMPLE 7 When expressed in polar coordinates, a plane autonomous system takes the form

$$\frac{dr}{dt} = 0.5(3 - r)$$

$$\frac{d\theta}{dt} = 1$$

Find and sketch the solutions satisfying $X(0) = (3, 0)$ and $X(0) = (0, 1)$.

Solution Applying separation of variables to $dr/dt = 0.5(3 - r)$ and integrating $d\theta/dt = 1$ lead to the general solution

$$r = 3 + c_1 e^{-0.5t} \qquad \theta = t + c_2$$

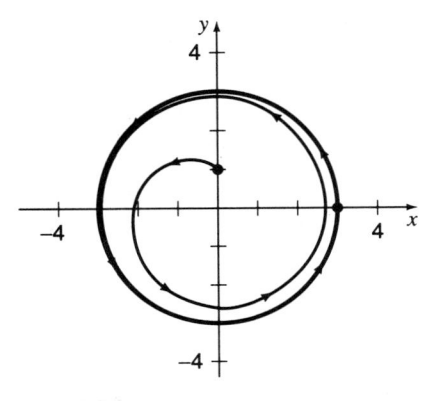

FIGURE 10.6

If $X(0) = (3, 0)$, it follows that $c_1 = c_2 = 0$, and so $r = 3$ and $\theta = t$. Hence, $x = r \cos \theta = 3 \cos t$ and $y = r \sin \theta = 3 \sin t$, and so the solution is periodic. If $X(0) = (0, 1)$, $r = 1$ and $\theta = \pi/2$. Then $c_1 = -2$ and $c_2 = \pi/2$. The solution curve is the spiral $r = 3 - 2e^{-0.5(\theta - \pi/2)}$. Note that as $t \to \infty$, θ increases without bound and r approaches 3. Both solutions are shown in Figure 10.6. □

EXERCISES 10.1   Answers to odd-numbered problems begin on page A-58.

In Problems 1–6 write the given nonlinear second-order differential equation as a plane autonomous system. Find all critical points of the resulting system.

1. $x'' + 9 \sin x = 0$
2. $x'' + (x')^2 + 2x = 0$
3. $x'' + x'(1 - x^3) - x^2 = 0$
4. $x'' + 4\dfrac{x}{1 + x^2} + 2x' = 0$
5. $x'' + x = \varepsilon x^3$ for $\varepsilon > 0$
6. $x'' + x - \varepsilon x|x| = 0$ for $\varepsilon > 0$

In Problems 7–16 find all critical points of the given plane autonomous system.

7. $x' = x + xy$
   $y' = -y - xy$
8. $x' = y^2 - x$
   $y' = x^2 - y$

9. $x' = 3x^2 - 4y$
$y' = x - y$

10. $x' = x^3 - y$
$y' = x - y^3$

11. $x' = x(10 - x - \frac{1}{2}y)$
$y' = y(16 - y - x)$

12. $x' = -2x + y + 10$
$y' = 2x - y - 15\dfrac{y}{y+5}$

13. $x' = x^2 e^y$
$y' = y(e^x - 1)$

14. $x' = \sin y$
$y' = e^{x-y} - 1$

15. $x' = x(1 - x^2 - 3y^2)$
$y' = y(3 - x^2 - 3y^2)$

16. $x' = -x(4 - y^2)$
$y' = 4y(1 - x^2)$

In Problems 17–22 use the given linear plane autonomous system (taken from Exercises 9.4).

(a) Find the general solution and determine whether there are periodic solutions.

(b) Find the solution satisfying the given initial condition.

[O] (c) With the aid of a graphics calculator or graphing software, sketch the solution in part (b) and indicate the direction in which the curve is traversed.

17. $\dfrac{x' = x + 2y}{y' = 4x + 3y}$, $\mathbf{X}(0) = (2, -2)$ (Exercises 9.4, Problem 1)

18. $\dfrac{x' = -6x + 2y}{y' = -3x + y}$, $\mathbf{X}(0) = (3, 4)$ (Exercises 9.4, Problem 6)

19. $\dfrac{x' = 4x - 5y}{y' = 5x - 4y}$, $\mathbf{X}(0) = (4, 5)$ (Exercises 9.4, Problem 19)

20. $\dfrac{x' = x + y}{y' = -2x - y}$, $\mathbf{X}(0) = (-2, 2)$ (Exercises 9.4, Problem 16)

21. $\dfrac{x' = 5x + y}{y' = -2x + 3y}$, $\mathbf{X}(0) = (-1, 2)$ (Exercises 9.4, Problem 17)

22. $\dfrac{x' = x - 8y}{y' = x - 3y}$, $\mathbf{X}(0) = (2, 1)$ (Exercises 9.4, Problem 20)

In Problems 23–26 solve the given nonlinear plane autonomous system by changing to polar coordinates. Describe the geometric behavior of the solution that satisfies the given initial condition(s).

23. $x' = -y - x(x^2 + y^2)^2$
$y' = x - y(x^2 + y^2)^2$, $\mathbf{X}(0) = (4, 0)$

24. $x' = y + x(x^2 + y^2)$
$y' = -x + y(x^2 + y^2)$, $\mathbf{X}(0) = (4, 0)$

25. $x' = -y + x(1 - x^2 - y^2)$
$y' = x + y(1 - x^2 - y^2)$, $\mathbf{X}(0) = (1, 0)$ and $\mathbf{X}(0) = (2, 0)$
[*Hint*: The resulting differential equation for r is a Bernoulli differential equation. See Section 2.6.]

26. $x' = y - \dfrac{x}{\sqrt{x^2 + y^2}}(4 - x^2 - y^2)$

$y' = -x - \dfrac{y}{\sqrt{x^2 + y^2}}(4 - x^2 - y^2)$,

$\mathbf{X}(0) = (1, 0)$ and $\mathbf{X}(0) = (2, 0)$
[*Hint*: See Example 6 in Section 2.2.]

27. If $z = f(x, y)$ is a function with continuous first partial derivatives in a region R, then a flow $\mathbf{V}(x, y) = (P(x, y), Q(x, y))$ in R may be defined by letting $P(x, y) = -(\partial f/\partial y)(x, y)$ and $Q(x, y) = (\partial f/\partial x)(x, y)$. Show that if $\mathbf{X}(t) = (x(t), y(t))$ is a solution of the plane autonomous system

$$x' = P(x, y)$$
$$y' = Q(x, y)$$

then $f(x(t), y(t)) = c$ for some constant c. Thus, a solution curve lies on a level curve of f. [*Hint*: Use the chain rule to compute $(d/dt)f(x(t), y(t))$.]

10.2 STABILITY OF LINEAR SYSTEMS

In Section 10.1 we pointed out that the plane autonomous system

$$\frac{dx}{dt} = P(x, y) \qquad \frac{dy}{dt} = Q(x, y)$$

gives rise to a vector field $\mathbf{V}(x, y) = (P(x, y), Q(x, y))$, and a solution $\mathbf{X} = \mathbf{X}(t)$ may be interpreted as the resulting path of a particle which is initially placed at position $\mathbf{X}(0) = \mathbf{X}_0$. If $\mathbf{X}_0$ is a critical point, the particle remains stationary. If $\mathbf{X}_0$ is placed near a critical point $\mathbf{X}_1$, however, we ask the following questions:

(*i*) Will the particle return to the critical point? More precisely, if $\mathbf{X} = \mathbf{X}(t)$ is the solution that satisfies $\mathbf{X}(0) = \mathbf{X}_0$, then is $\lim_{t \to \infty} \mathbf{X}(t) = \mathbf{X}_1$?

10.2 Stability of Linear Systems

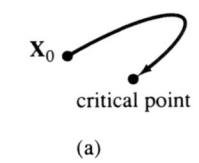

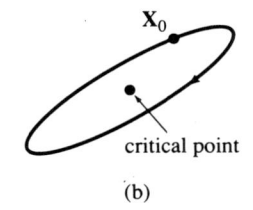

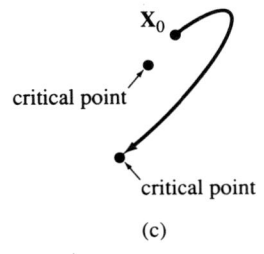

FIGURE 10.7

(*ii*) If the particle does *not* return to the critical point, does it remain close to the critical point or move away from the critical point? It is conceivable, for example, that the particle may simply circle the critical point or even return to a different critical point. See Figure 10.7.

If in some neighborhood of the critical point, case (a) or (b) in Figure 10.7 *always* occurs, then we call the critical point **locally stable**. If, however, an initial value $\mathbf{X}_0$ that results in behavior similar to case (c) can be found in *any* given neighborhood, then we call the critical point **unstable**. These concepts will be made more precise in Section 10.3, where questions (*i*) and (*ii*) will be investigated for nonlinear systems.

Stability Analysis   We will first investigate these two stability questions for linear plane autonomous systems and lay the foundation for Section 10.3. The solution methods of Chapter 9 enable us to give a careful geometric analysis of the solutions to

$$\begin{aligned} x' &= ax + by \\ y' &= cx + dy \end{aligned} \quad (1)$$

in terms of the eigenvalues and eigenvectors of $\mathbf{A} = \begin{bmatrix} a & b \\ c & d \end{bmatrix}$. To ensure that $\mathbf{X}_0 = (0, 0)$ is the only critical point, we will assume that the determinant $\Delta = ad - bc \neq 0$. If $\tau = a + d$ is the **trace*** of the matrix $\mathbf{A}$, then the characteristic equation $\det(\mathbf{A} - \lambda \mathbf{I}) = 0$ can be rewritten as

$$\lambda^2 - \tau \lambda + \Delta = 0$$

Therefore, the eigenvalues of $\mathbf{A}$ are $\lambda = (\tau \pm \sqrt{\tau^2 - 4\Delta})/2$, and the usual three cases for these roots occur according to whether $\tau^2 - 4\Delta$ is positive, negative, or zero.

CASE I   Real Distinct Eigenvalues ($\tau^2 - 4\Delta > 0$)

According to Theorem 9.7 in Section 9.4, the general solution of (1) is given by

$$\mathbf{X}(t) = c_1 \mathbf{K}_1 e^{\lambda_1 t} + c_2 \mathbf{K}_2 e^{\lambda_2 t} \quad (2)$$

where λ_1 and λ_2 are the eigenvalues and $\mathbf{K}_1$ and $\mathbf{K}_2$ are the corresponding eigenvectors. Note that $\mathbf{X}(t)$ can also be written as

$$\mathbf{X}(t) = e^{\lambda_1 t}[c_1 \mathbf{K}_1 + c_2 \mathbf{K}_2 e^{(\lambda_2 - \lambda_1)t}] \quad (3)$$

(a) Both eigenvalues negative ($\tau^2 - 4\Delta > 0$, $\tau < 0$, and $\Delta > 0$)

Stable Node   From (2), it follows that $\lim_{t \to \infty} \mathbf{X}(t) = \mathbf{0}$. If we assume that $\lambda_2 < \lambda_1$, then $\lambda_2 - \lambda_1 < 0$, and so we may conclude from (3) that $\mathbf{X}(t) \approx c_1 \mathbf{K}_1 e^{\lambda_1 t}$ for large values of t. When $c_1 \neq 0$, $\mathbf{X}(t)$ will approach $\mathbf{0}$ from one of

---

*In general, if $\mathbf{A}$ is an $n \times n$ matrix, then the **trace** of $\mathbf{A}$ is the sum of the main diagonal entries.

the two directions determined by the eigenvector $\mathbf{K}_1$ corresponding to λ_1. If $c_1 = 0$, then $\mathbf{X}(t) = c_2\mathbf{K}_2 e^{\lambda_2 t}$ and $\mathbf{X}(t)$ approaches **0** along the line determined by the eigenvector $\mathbf{K}_2$. Figure 10.8 shows a collection of solution curves around the origin. A critical point is called a **stable node** when both eigenvalues are negative.

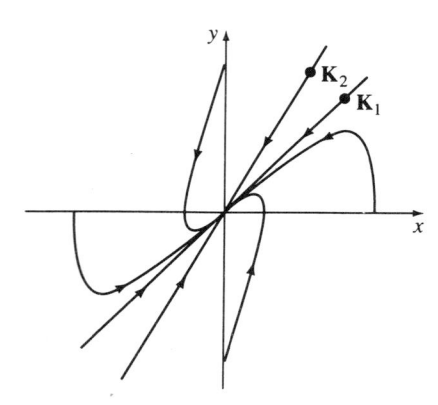

FIGURE 10.8

(b) Both eigenvalues positive ($\tau^2 - 4\Delta > 0$, $\tau > 0$, and $\Delta > 0$)

Unstable Node The analysis for this case is similar to part (a). Again from (1), $\mathbf{X}(t)$ becomes unbounded as t increases. Moreover, from (2), $\mathbf{X}(t)$ becomes unbounded in one of the directions determined by the eigenvector $\mathbf{K}_1$ (when $c_1 \neq 0$) or along the line determined by the eigenvector $\mathbf{K}_2$ (when $c_1 = 0$). Figure 10.9 shows a typical collection of solution curves. This type of critical point corresponding to the case when both eigenvalues are positive is called an **unstable node**.

(c) Eigenvalues with opposite signs ($\tau^2 - 4\Delta > 0$ and $\Delta < 0$)

Saddle Point The analysis of the solutions is identical to part (b) with one exception. When $c_1 = 0$, $\mathbf{X}(t) = c_2\mathbf{K}_2 e^{\lambda_2 t}$ and, since $\lambda_2 < 0$, $\mathbf{X}(t)$ will approach **0** along the line determined by the eigenvector $\mathbf{K}_2$. If $\mathbf{X}(0)$ does not lie on the line determined by $\mathbf{K}_2$, the line determined by $\mathbf{K}_1$ serves as an asymptote for $\mathbf{X}(t)$. This unstable critical point is called a **saddle point**. See Figure 10.10.

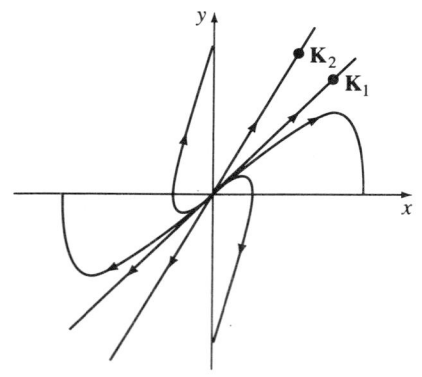

FIGURE 10.9

EXAMPLE 1 Classify the critical point $(0, 0)$ of each linear system $\mathbf{X}' = \mathbf{AX}$ as a stable node, unstable node, or saddle point.

(a) $\mathbf{A} = \begin{bmatrix} 2 & 3 \\ 2 & 1 \end{bmatrix}$ (b) $\mathbf{A} = \begin{bmatrix} -10 & 6 \\ 15 & -19 \end{bmatrix}$

In each case discuss the nature of the solutions in a neighborhood of $(0, 0)$.

Solution (a) Since the trace τ is 3 and the determinant $\Delta = -4$, the eigenvalues are

$$\lambda = \frac{\tau \pm \sqrt{\tau^2 - 4\Delta}}{2} = \frac{3 \pm \sqrt{3^2 - 4(-4)}}{2}$$

$$= \frac{3 \pm 5}{2} = 4 \text{ or } -1$$

The eigenvalues have opposite signs and so $(0, 0)$ is a saddle point. It is not hard to show (see Example 1 in Section 9.4) that the eigenvectors corresponding to $\lambda_1 = 4$ and $\lambda_2 = -1$ are

$$\mathbf{K}_1 = \begin{bmatrix} 3 \\ 2 \end{bmatrix} \quad \text{and} \quad \mathbf{K}_2 = \begin{bmatrix} 1 \\ -1 \end{bmatrix}$$

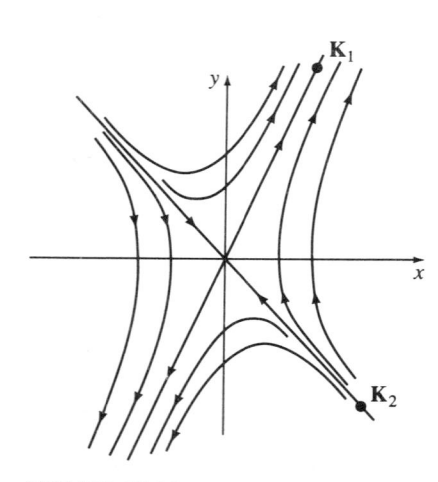

FIGURE 10.10

respectively. If $\mathbf{X}(0) = \mathbf{X}_0$ lies on the line $y = -x$, $\mathbf{X}(t)$ will approach **0**. For any other initial condition, $\mathbf{X}(t)$ will become unbounded in the directions determined by $\mathbf{K}_1$. In other words, the line $y = \frac{2}{3}x$ serves as an asymptote for all these solution curves.

(b) From $\tau = -29$ and $\Delta = 100$ it follows that the eigenvalues of $\mathbf{A}$ are $\lambda_1 = -4$ and $\lambda_2 = -25$. Both eigenvalues are negative, so $(0, 0)$ is in this case a stable node. Since the eigenvectors corresponding to $\lambda_1 = -4$ and $\lambda_2 = -5$ are

$$\mathbf{K}_1 = \begin{bmatrix} 1 \\ 1 \end{bmatrix} \quad \text{and} \quad \mathbf{K}_2 = \begin{bmatrix} 2 \\ -5 \end{bmatrix}$$

respectively, it follows that all solutions approach $\mathbf{0}$ from the direction defined by $\mathbf{K}_1$ except those solutions for which $\mathbf{X}(0) = \mathbf{X}_0$ lies on the line $y = -\frac{5}{2}x$ determined by $\mathbf{K}_2$. These solutions approach $\mathbf{0}$ along $y = -\frac{5}{2}x$. □

CASE II A Repeated Real Eigenvalue ($\tau^2 - 4\Delta = 0$)

Degenerate Nodes   Recall from Section 9.4 that the general solution takes on one of two different forms depending on whether one or two linearly independent eigenvectors can be found for the repeated eigenvalue λ_1.

(a) Two linearly independent eigenvectors
If $\mathbf{K}_1$ and $\mathbf{K}_2$ are two linearly eigenvectors corresponding to λ_1, then the general solution is given by

$$\begin{aligned}\mathbf{X}(t) &= c_1 \mathbf{K}_1 e^{\lambda_1 t} + c_2 \mathbf{K}_2 e^{\lambda_1 t} \\ &= (c_1 \mathbf{K}_1 + c_2 \mathbf{K}_2)e^{\lambda_1 t}\end{aligned}$$

If $\lambda_1 < 0$, $\mathbf{X}(t)$ approaches $\mathbf{0}$ along the line determined by the vector $c_1 \mathbf{K}_1 + c_2 \mathbf{K}_2$, and the critical point is called a **degenerate stable node** (see Figure 10.11(a)). The arrows in Figure 10.11(a) are reversed when $\lambda_1 > 0$, and we have a **degenerate unstable node**.

(a)   (b)

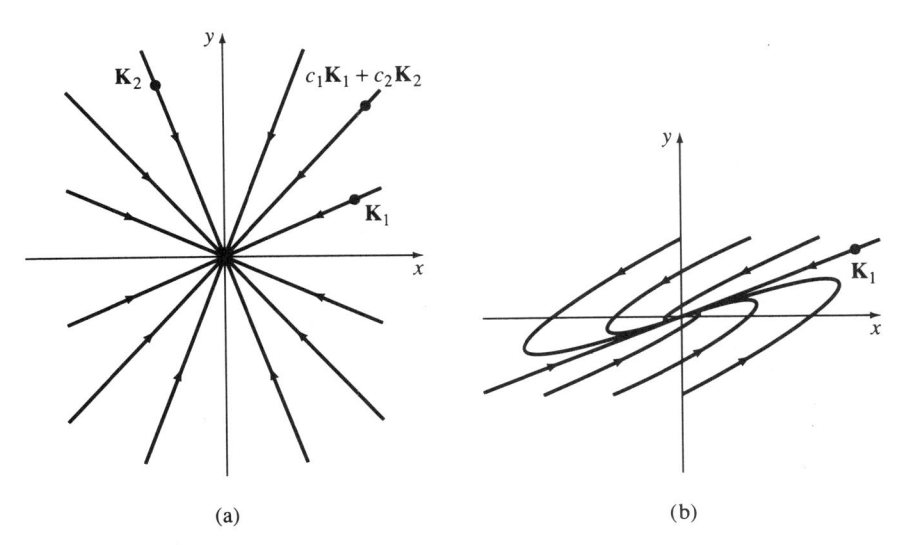

FIGURE 10.11

(b) A single linearly independent eigenvector

When only a single linearly independent eigenvector $\mathbf{K}_1$ exists, the general solution is given by

$$\mathbf{X}(t) = c_1 \mathbf{K}_1 e^{\lambda_1 t} + c_2(\mathbf{K}_1 t e^{\lambda_1 t} + \mathbf{P} e^{\lambda_1 t})$$

where $(\mathbf{A} - \lambda_1 \mathbf{I})\mathbf{P} = \mathbf{K}_1$ (see Section 9.4, (18)–(20)), and the solution may be rewritten as

$$\mathbf{X}(t) = t e^{\lambda_1 t}\left[c_2 \mathbf{K}_1 + \frac{c_1}{t}\mathbf{K}_1 + \frac{c_2}{t}\mathbf{P}\right]$$

If $\lambda_1 < 0$, $\lim_{t \to \infty} t e^{\lambda_1 t} = 0$ and it follows that $\mathbf{X}(t)$ approaches $\mathbf{0}$ in one of the directions determined by the vector $\mathbf{K}_1$ (see Figure 10.11(b)). The critical point is again called a **degenerate stable node**. When $\lambda_1 > 0$, the solutions look like those in Figure 10.11(b) with the arrows reversed, and the line determined by $\mathbf{K}_1$ is an asymptote for *all* solutions. The critical point is again called a **degenerate unstable node**.

CASE III Complex Eigenvalues ($\tau^2 - 4\Delta < 0$)

If $\lambda_1 = \alpha + i\beta$ and $\bar{\lambda}_1 = \alpha - i\beta$ are complex eigenvalues and $\mathbf{K}_1 = \mathbf{B}_1 + i\mathbf{B}_2$ is a complex eigenvector corresponding to λ_1, the general solution can be written as $\mathbf{X}(t) = c_1 \mathbf{X}_1(t) + c_2 \mathbf{X}_2(t)$, where

$$\mathbf{X}_1(t) = (\mathbf{B}_1 \cos \beta t - \mathbf{B}_2 \sin \beta t)e^{\alpha t}$$
$$\mathbf{X}_2(t) = (\mathbf{B}_2 \cos \beta t + \mathbf{B}_1 \sin \beta t)e^{\alpha t}$$

(see (14) in Section 9.4). A solution can therefore be written in the form

$$\begin{aligned} x(t) &= e^{\alpha t}(c_{11} \cos \beta t + c_{12} \sin \beta t) \\ y(t) &= e^{\alpha t}(c_{21} \cos \beta t + c_{22} \sin \beta t) \end{aligned} \quad (4)$$

and when $\alpha = 0$, we have

$$\begin{aligned} x(t) &= c_{11} \cos \beta t + c_{12} \sin \beta t \\ y(t) &= c_{21} \cos \beta t + c_{22} \sin \beta t \end{aligned} \quad (5)$$

(a) Pure imaginary roots ($\tau^2 - 4\Delta < 0$, $\tau = 0$)

Center When $\alpha = 0$, the eigenvalues are pure imaginary and, from (5), all solutions are periodic with period $p = 2\pi/\beta$. Notice that if both c_{12} and c_{21} happened to be 0, then (5) would reduce to

$$\begin{aligned} x(t) &= c_{11} \cos \beta t \\ y(t) &= c_{22} \sin \beta t \end{aligned}$$

which is a standard parametric representation for an ellipse. By solving the system of equations in (4) for $\cos \beta t$ and $\sin \beta t$ and using the identity $\sin^2 \beta t + \cos^2 \beta t = 1$, it is possible to show that *all solutions are ellipses* with center at the origin. The critical point $(0, 0)$ is called a **center**, and Figure 10.12 shows a typical collection of solution curves. The ellipses are either *all* traversed in the clockwise direction or all traversed in the counterclockwise direction.

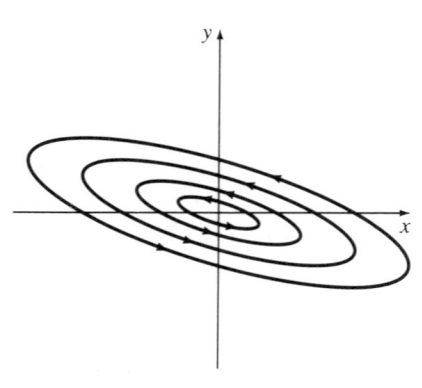

FIGURE 10.12

(b) Nonzero real part ($\tau^2 - 4\Delta < 0$, $\tau \neq 0$)

Spiral Points   When $\alpha \neq 0$, the effect of the term $e^{\alpha t}$ in (4) is similar to the effect of the exponential term in the analysis of damped motion given in Section 3.9. When $\alpha < 0$, $e^{\alpha t} \to 0$ and the elliptic-like solution spirals closer and closer to the origin. The critical point is called a **stable spiral point**. When $\alpha > 0$, the effect is the opposite. An elliptic-like solution is driven farther and farther from the origin, and the critical point is called an **unstable spiral point** (see Figure 10.13).

(a) stable spiral point   (b) unstable spiral point

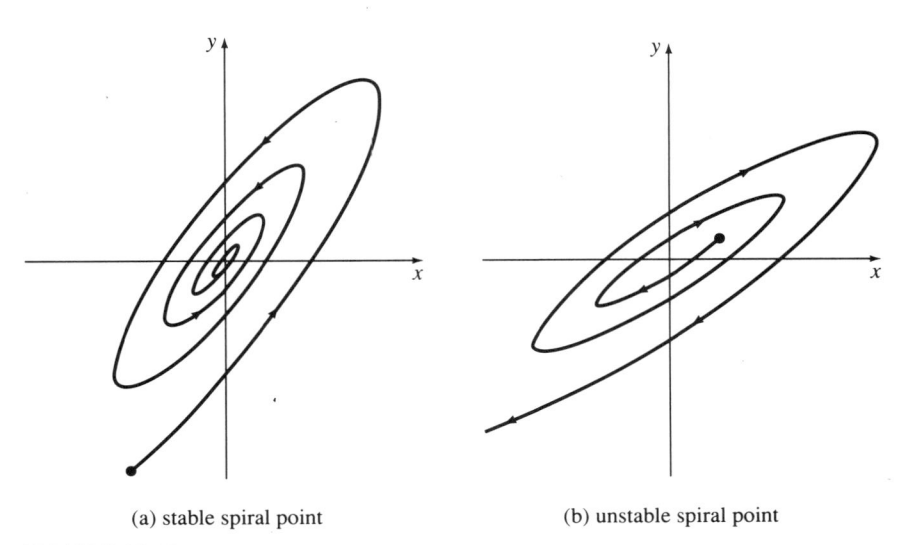

FIGURE 10.13

EXAMPLE 2   Classify the critical point $(0, 0)$ of each linear system $\mathbf{X}' = \mathbf{AX}$:

(a) $\mathbf{A} = \begin{bmatrix} 3 & -18 \\ 2 & -9 \end{bmatrix}$ (b) $\mathbf{A} = \begin{bmatrix} -1 & 2 \\ -1 & 1 \end{bmatrix}$

In each case discuss the nature of the solution that satisfies $\mathbf{X}(0) = (1, 0)$. Determine parametric equations for each solution.

Solution   (a) Since $\tau = -6$ and $\Delta = 9$, the characteristic polynomial is $\lambda^2 + 6\lambda + 9 = (\lambda + 3)^2$, and so $(0, 0)$ is a degenerate stable node. For the repeated eigenvalue $\lambda = -3$, we find a single eigenvector $\mathbf{K}_1 = \begin{bmatrix} 3 \\ 1 \end{bmatrix}$, and so the solution $\mathbf{X}(t)$ which satisfies $\mathbf{X}(0) = (1, 0)$ approaches $(0, 0)$ from the direction specified by the line $y = x/3$.

(b) Since $\tau = 0$ and $\Delta = 1$, the eigenvalues are $\lambda = \pm i$, and so $(0, 0)$ is a center. The solution $\mathbf{X}(t)$ which satisfies $\mathbf{X}(0) = (1, 0)$ is an ellipse which circles the origin every 2π units of time.

From Example 6 in Section 9.4 the general solution of the system in part (a) is

$$\mathbf{X}(t) = c_1 \begin{bmatrix} 3 \\ 1 \end{bmatrix} e^{-3t} + c_2 \left\{ \begin{bmatrix} 3 \\ 1 \end{bmatrix} t e^{-3t} + \begin{bmatrix} \frac{1}{2} \\ 0 \end{bmatrix} e^{-3t} \right\}$$

Using the initial condition gives $c_1 = 0$ and $c_2 = 2$, and so $x = (6t + 1)e^{-3t}$ and $y = 2te^{-3t}$ are parametric equations for the solution.

From Example 1 in Section 9.6 the general solution of the system in part (b) is

$$\mathbf{X}(t) = c_1 \begin{bmatrix} \cos t + \sin t \\ \cos t \end{bmatrix} + c_2 \begin{bmatrix} \cos t - \sin t \\ -\sin t \end{bmatrix}$$

Using the initial condition gives $c_1 = 0$ and $c_2 = 1$, and so $x = \cos t - \sin t$ and $y = -\sin t$ are parametric equations for the ellipse. Note that $y < 0$ for small positive values of t. Therefore the ellipse is traversed in the clockwise direction.

The solutions of parts (a) and (b) are shown in Figure 10.14(a) and (b), respectively. □

Figure 10.15 illustrates the classification of critical points ($\Delta = \det \mathbf{A}$ and $\tau = \text{trace } \mathbf{A}$) and thus conveniently summarizes the results of this section. The general geometric nature of the solutions can be determined by computing the trace and determinant of $\mathbf{A}$. In practice, graphs of the solutions are most easily obtained *not* by constructing explicit eigenvalue–eigenvector solutions but rather by generating the solutions numerically using a method such as the Runge–Kutta method for first-order systems (see Section 15.6).

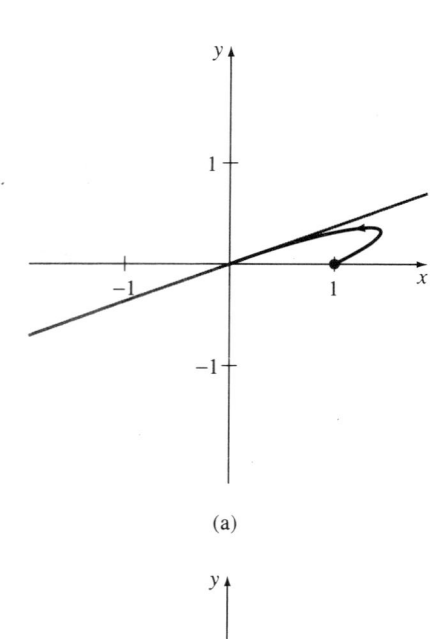

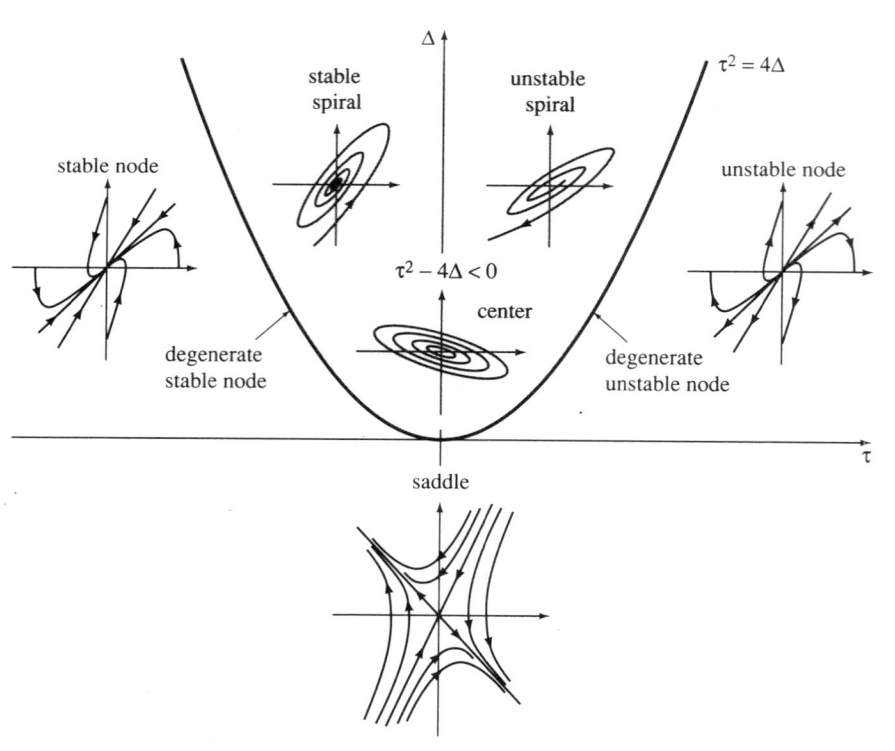

FIGURE 10.14

FIGURE 10.15

EXAMPLE 3 Classify the critical point $(0, 0)$ of each linear system $\mathbf{X}' = \mathbf{AX}$:

(a) $\mathbf{A} = \begin{bmatrix} 1.01 & 3.10 \\ -1.10 & -1.02 \end{bmatrix}$ (b) $\mathbf{A} = \begin{bmatrix} -a\hat{x} & -ab\hat{x} \\ -cd\hat{y} & -d\hat{y} \end{bmatrix}$

for positive constants $a, b, c, d, \hat{x}$, and $\hat{y}$.

Solution (a) For the matrix, $\tau = -0.01$, $\Delta = 2.3798$, and so $\tau^2 - 4\Delta < 0$. Using Figure 10.15, we see that $(0, 0)$ is a stable spiral point.

(b) The matrix arises from the Lotka–Volterra competition model we will study in Section 10.4. Since $\tau = -(a\hat{x} + d\hat{y})$ and all constants in the matrix are positive, $\tau < 0$. The determinant may be written as $\Delta = ad\hat{x}\hat{y}(1 - bc)$. If $bc > 1$, then $\Delta < 0$ and the critical point is a saddle. If $bc < 1$, then $\Delta > 0$ and the critical point is either a stable node, a degenerate stable node, or a stable spiral point. In all three of these cases, $\lim_{t \to \infty} \mathbf{X}(t) = \mathbf{0}$. □

We can now answer each of the questions posed at the beginning of Section 10.2 for the linear plane autonomous system

$$x' = ax + by$$
$$y' = cx + dy$$

with $ad - bc \neq 0$. The answers are summarized in the following theorem:

THEOREM 10.1 Stability Criteria for Linear Systems

For a linear plane autonomous system, let $\mathbf{X} = \mathbf{X}(t)$ denote the solution which satisfies the initial condition $\mathbf{X}(0) = \mathbf{X}_0$, where $\mathbf{X}_0 \neq \mathbf{0}$.

(i) The $\lim_{t \to \infty} \mathbf{X}(t) = \mathbf{0}$ if and only if the eigenvalues of $\mathbf{A}$ have negative real parts. This occurs when $\Delta > 0$ and $\tau < 0$.

(ii) $\mathbf{X}(t)$ is periodic if and only if the eigenvalues of $\mathbf{A}$ are pure imaginary. This occurs when $\Delta > 0$ and $\tau = 0$.

(iii) In all other cases, given any neighborhood of the origin, there is at least one $\mathbf{X}_0$ in the neighborhood for which $\mathbf{X}(t)$ becomes unbounded as t increases.

EXERCISES 10.2 Answers to odd-numbered problems begin on page A-58.

In Problems 1–8 the general solution of the linear system $\mathbf{X}' = \mathbf{AX}$ is given.

(a) In each case discuss the nature of the solutions in a neighborhood of $(0, 0)$.

[O] (b) With the aid of a graphics calculator or graphing software, sketch the solution that satisfies the initial condition $\mathbf{X}(0) = (1, 1)$.

1. $\mathbf{A} = \begin{bmatrix} -2 & -2 \\ -2 & -5 \end{bmatrix}$, $\mathbf{X}(t) = c_1 \begin{bmatrix} 2 \\ -1 \end{bmatrix} e^{-t} + c_2 \begin{bmatrix} 1 \\ 2 \end{bmatrix} e^{-6t}$

2. $\mathbf{A} = \begin{bmatrix} -1 & -2 \\ 3 & 4 \end{bmatrix}$, $\mathbf{X}(t) = c_1 \begin{bmatrix} 1 \\ -1 \end{bmatrix} e^{t} + c_2 \begin{bmatrix} -4 \\ 6 \end{bmatrix} e^{2t}$

3. $\mathbf{A} = \begin{bmatrix} 1 & -1 \\ 1 & 1 \end{bmatrix}$, $\mathbf{X}(t) = e^{t} \left\{ c_1 \begin{bmatrix} -\sin t \\ \cos t \end{bmatrix} + c_2 \begin{bmatrix} \cos t \\ \sin t \end{bmatrix} \right\}$

4. $\mathbf{A} = \begin{bmatrix} -1 & -4 \\ 1 & -1 \end{bmatrix}$,

$\mathbf{X}(t) = e^{-t}\left\{c_1 \begin{bmatrix} 2\cos 2t \\ \sin 2t \end{bmatrix} + c_2 \begin{bmatrix} -2\sin 2t \\ \cos 2t \end{bmatrix}\right\}$

5. $\mathbf{A} = \begin{bmatrix} -6 & 5 \\ -5 & 4 \end{bmatrix}$,

$\mathbf{X}(t) = c_1 \begin{bmatrix} 1 \\ 1 \end{bmatrix}e^{-t} + c_2\left\{\begin{bmatrix} 1 \\ 1 \end{bmatrix}te^{-t} + \begin{bmatrix} 0 \\ \frac{1}{5} \end{bmatrix}e^{-t}\right\}$

6. $\mathbf{A} = \begin{bmatrix} 2 & 4 \\ -1 & 6 \end{bmatrix}$,

$\mathbf{X}(t) = c_1 \begin{bmatrix} 2 \\ 1 \end{bmatrix}e^{4t} + c_2\left\{\begin{bmatrix} 2 \\ 1 \end{bmatrix}te^{4t} + \begin{bmatrix} 1 \\ 1 \end{bmatrix}e^{4t}\right\}$

7. $\mathbf{A} = \begin{bmatrix} 2 & -1 \\ 3 & -2 \end{bmatrix}$, $\mathbf{X}(t) = c_1 \begin{bmatrix} 1 \\ 1 \end{bmatrix}e^t + c_2 \begin{bmatrix} 1 \\ 3 \end{bmatrix}e^{-t}$

8. $\mathbf{A} = \begin{bmatrix} -1 & 5 \\ -1 & 1 \end{bmatrix}$,

$\mathbf{X}(t) = c_1 \begin{bmatrix} 5\cos 2t \\ \cos 2t - 2\sin 2t \end{bmatrix} + c_2 \begin{bmatrix} 5\sin 2t \\ 2\cos 2t + \sin 2t \end{bmatrix}$

In Problems 9–16 classify the critical point $(0, 0)$ of the given linear system by computing the trace τ and determinant Δ and using Figure 10.15.

9. $x' = -5x + 3y$
   $y' = 2x + 7y$

10. $x' = -5x + 3y$
    $y' = 2x - 7y$

11. $x' = -5x + 3y$
    $y' = -2x + 5y$

12. $x' = -5x + 3y$
    $y' = -7x + 4y$

13. $x' = -\frac{3}{2}x + \frac{1}{4}y$
    $y' = -x - \frac{1}{2}y$

14. $x' = \frac{3}{2}x + \frac{1}{4}y$
    $y' = -x + \frac{1}{2}y$

15. $x' = 0.02x - 0.11y$
    $y' = 0.10x - 0.05y$

16. $x' = 0.03x + 0.01y$
    $y' = -0.01x + 0.05y$

17. Determine conditions on the real constant μ so that $(0, 0)$ is a center for the linear system

$$x' = -\mu x + y$$
$$y' = -x + \mu y$$

18. Determine a condition on the real constant μ so that $(0, 0)$ is a stable spiral point of the linear system

$$x' = y$$
$$y' = -x + \mu y$$

19. Show that $(0, 0)$ is always an unstable critical point of the linear system

$$x' = \mu x + y$$
$$y' = -x + y$$

where μ is a real constant and $\mu \neq -1$. When is $(0, 0)$ an unstable saddle point? When is $(0, 0)$ an unstable spiral point?

20. Let $\mathbf{X} = \mathbf{X}(t)$ be the solution of the linear system

$$x' = \alpha x - \beta y$$
$$y' = \beta x + \alpha y$$

that satisfies the initial condition $\mathbf{X}(0) = \mathbf{X}_0$. Determine conditions on the real constants α and β that will ensure

$$\lim_{t \to \infty} \mathbf{X}(t) = (0, 0)$$

Can $(0, 0)$ be a node or saddle point?

21. Show that the nonhomogeneous linear system $\mathbf{X}' = \mathbf{AX} + \mathbf{F}$ has a unique critical point $\mathbf{X}_1$ when $\Delta = \det \mathbf{A} \neq 0$. Conclude that if $\mathbf{X} = \mathbf{X}(t)$ is a solution of the nonhomogeneous system, $\tau < 0$ and $\Delta > 0$,

$$\lim_{t \to \infty} \mathbf{X}(t) = \mathbf{X}_1$$

[Hint: $\mathbf{X}(t) = \mathbf{X}_c(t) + \mathbf{X}_1$.]

22. In Example 3(b), show that $(0, 0)$ is a stable node when $bc < 1$.

10.3 LINEARIZATION AND LOCAL STABILITY

We shall start this section by refining the stability concepts introduced in Section 10.2 in such a way that they will apply to nonlinear autonomous systems as well. While the linear system $\mathbf{X}' = \mathbf{AX}$ had only one critical point when $\det \mathbf{A} \neq 0$, we saw in Section 10.1 that a nonlinear system may have many critical points. We therefore cannot expect that a particle placed initially at $\mathbf{X}_0$ will remain near a given critical point $\mathbf{X}_1$ unless $\mathbf{X}_0$ has been placed sufficiently close to $\mathbf{X}_1$ to begin with. The particle might well be driven to a second critical point $\mathbf{X}_2$.

10.3 Linearization and Local Stability

To emphasize this idea, we consider the physical system shown in Figure 10.16 in which a bead slides along the curve $z = f(x)$ under the influence of gravity alone. We will show in Section 10.4 that the x-coordinate of the bead satisfies a second-order nonlinear differential equation $x'' = g(x, x')$ and, therefore, when $y = x'$, satisfies the nonlinear autonomous system

$$x' = y$$
$$y' = g(x, y)$$

If the bead is positioned at $P = (x, f(x))$ and given zero initial velocity, the bead will remain at P provided $f'(x) = 0$. If the bead is placed near the critical point located at $x = x_1$, it will remain near $x = x_1$ only if its initial velocity does not drive it over the "hump" at $x = x_2$ toward the critical point located at $x = x_3$. Therefore, $\mathbf{X}(0) = (x(0), x'(0))$ must be near $(x_1, 0)$.

In the following definition we will denote the distance between two points $\mathbf{X}$ and $\mathbf{Y}$ by $|\mathbf{X} - \mathbf{Y}|$.

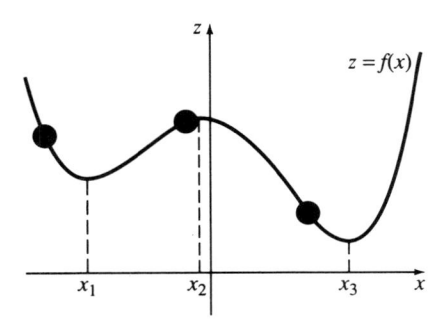

FIGURE 10.16

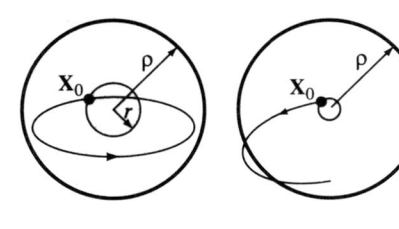

DEFINITION 10.1 Stable Critical Points

Let $\mathbf{X}_1$ be a critical point of an autonomous system, and let $\mathbf{X} = \mathbf{X}(t)$ denote the solution which satisfies the initial condition $\mathbf{X}(0) = \mathbf{X}_0$, where $\mathbf{X}_0 \neq \mathbf{X}_1$. We say that $\mathbf{X}_1$ is a **stable critical point** in this case: Given any radius $\rho > 0$, there is a corresponding radius $r > 0$ such that if initial position $\mathbf{X}_0$ satisfies $|\mathbf{X}_0 - \mathbf{X}_1| < r$, then the corresponding solution $\mathbf{X}(t)$ satisfies $|\mathbf{X}(t) - \mathbf{X}_1| < \rho$ for all $t > 0$. If, in addition, $\lim_{t \to \infty} \mathbf{X}(t) = \mathbf{X}_1$ whenever $|\mathbf{X}_0 - \mathbf{X}_1| < r$, we call $\mathbf{X}_1$ an **asymptotically stable critical point**.

This definition is illustrated in Figure 10.17(a). Given any disk of radius ρ about the critical point $\mathbf{X}_1$, a solution will remain inside the disk provided $\mathbf{X}(0) = \mathbf{X}_0$ is selected sufficiently close to $\mathbf{X}_1$. It is *not* necessary that a solution approach the critical point in order for $\mathbf{X}_1$ to be stable. Stable nodes, stable spiral points, and centers are examples of stable critical points for linear systems. To emphasize that $\mathbf{X}_0$ must be selected close to $\mathbf{X}_1$, the terminology **locally stable critical point** is also used.

By negating Definition 10.1, we obtain the following definition of an unstable critical point:

(a) stable critical point
(b) unstable critical point

FIGURE 10.17

DEFINITION 10.2 Unstable Critical Point

Let $\mathbf{X}_1$ be a critical point of an autonomous system, and let $\mathbf{X} = \mathbf{X}(t)$ denote the solution which satisfies the initial condition $\mathbf{X}(0) = \mathbf{X}_0$, where $\mathbf{X}_0 \neq \mathbf{X}_1$. We say that $\mathbf{X}_1$ is an **unstable critical point** in this case: There is a disk of radius $\rho > 0$ with the property that, for any $r > 0$, there is an initial position $\mathbf{X}_0$ which satisfies $|\mathbf{X}_0 - \mathbf{X}_1| < r$, yet the corresponding solution $\mathbf{X}(t)$ satisfies $|\mathbf{X}(t) - \mathbf{X}_1| \geq \rho$ for at least one $t > 0$.

If a critical point X_1 is unstable, no matter how small the neighborhood about X_1, an initial position X_0 can always be found which results in the solution leaving some disk of radius ρ at some time t (see Figure 10.17(b)). Therefore, unstable nodes, unstable spiral points, and saddles are examples of unstable critical points for linear systems. In Figure 10.16, the critical point $(x_2, 0)$ is unstable. The slightest displacement or initial velocity will result in the bead sliding away from the point $(x_2, f(x_2))$.

EXAMPLE 1 Show that $(0, 0)$ is a stable critical point of the nonlinear plane autonomous system

$$x' = -y - x\sqrt{x^2 + y^2}$$
$$y' = x - y\sqrt{x^2 + y^2}$$

considered in Example 6 of Section 10.1.

Solution In Example 6 in Section 10.1 we showed that in polar coordinates,

$$r = \frac{1}{t + c_1} \qquad \theta = t + c_2$$

is the general solution of the system. If $X(0) = (r_0, \theta_0)$ is the initial condition in polar coordinates, then

$$r = \frac{r_0}{r_0 t + 1} \qquad \theta = t + \theta_0$$

Note that $r \leq r_0$ for $t \geq 0$, and r approaches $(0, 0)$ as t increases. Therefore, given $\rho > 0$, a solution which starts less than ρ units from $(0, 0)$ remains within ρ units of the origin for all $t \geq 0$. Hence, the critical point $(0, 0)$ is stable and is in fact asymptotically stable.   □

EXAMPLE 2 When expressed in polar coordinates, a plane autonomous system takes the form

$$\frac{dr}{dt} = 0.05r(3 - r)$$
$$\frac{d\theta}{dt} = -1$$

Show that $(0, 0)$ is an unstable critical point.

Solution Since $x = r \cos \theta$ and $y = r \sin \theta$, we have

$$\frac{dx}{dt} = -r \sin \theta \frac{d\theta}{dt} + \frac{dr}{dt} \cos \theta$$
$$\frac{dy}{dt} = r \cos \theta \frac{d\theta}{dt} + \frac{dr}{dt} \sin \theta$$

Note that $dr/dt = 0$ when $r = 0$, and we can conclude that $(0, 0)$ is a critical point.

The differential equation $dr/dt = 0.05r(3 - r)$ is a logistic equation which can be solved using either separation of variables or (6) in Section 2.11. If $r(0) = r_0$ and $r_0 \neq 0$, then

$$r = \frac{3}{1 + c_0 e^{-0.15t}}$$

where $c_0 = (3 - r_0)/r_0$. Since

$$\lim_{t \to \infty} \frac{3}{1 + c_0 e^{-0.15t}} = 3$$

it follows that no matter how close to $(0, 0)$ a solution starts, the solution will leave a disk of radius 1 about the origin. Therefore, $(0, 0)$ is an unstable critical point. □

Linearization It is rarely possible to determine the stability of a critical point of a nonlinear system by finding explicit solutions as in Examples 1 and 2. Instead we replace the term $\mathbf{g}(\mathbf{X})$ in the original autonomous system $\mathbf{X}' = \mathbf{g}(\mathbf{X})$ by a linear term $\mathbf{A}(\mathbf{X} - \mathbf{X}_1)$ that most closely approximates $\mathbf{g}(\mathbf{X})$ in a neighborhood of $\mathbf{X}_1$. This replacement process, called **linearization**, will be illustrated first for the first-order differential equation $x' = g(x)$.

An equation of the tangent line to the curve $y = g(x)$ at $x = x_1$ is $y = g(x_1) + g'(x_1)(x - x_1)$, and if x_1 is a critical point of $x' = g(x)$, we have

$$x' = g(x) \approx g'(x_1)(x - x_1)$$

The general solution to the linear differential equation $x' = g'(x_1)(x - x_1)$ is $y = x_1 + ce^{\lambda_1 t}$ where $\lambda_1 = g'(x_1)$, and so, if $g'(x_1) < 0$, $x(t)$ will approach x_1. The following theorem asserts that the same behavior occurs in the original differential equation provided $x(0) = x_0$ is selected close enough to x_1:

THEOREM 10.2 Stability Criteria For $x' = g(x)$

Let x_1 be a critical point of the autonomous differential equation $x' = g(x)$, where g is differentiable at x_1. Then:

(i) If $g'(x_1) < 0$, then x_1 is an asymptotically stable critical point.

(ii) If $g'(x_1) > 0$, then x_1 is an unstable critical point.

EXAMPLE 3 Without solving explicitly, analyze the critical points of the logistic differential equation $x' = rx\left(\dfrac{K - x}{K}\right)$, where r and K are positive constants.

Solution The two critical points are $x = 0$ and $x = K$. Since $g'(x) = \frac{r}{K}(K - 2x)$, $g'(0) = r$ and $g'(K) = -r$. We may therefore conclude that $x = 0$ is an unstable critical point and $x = K$ is an asymptotically stable critical point. □

Jacobian Matrix A similar analysis may be carried out for a plane autonomous system. An equation of the tangent plane to the surface $z = g(x, y)$ at $\mathbf{X}_1 = (x_1, y_1)$ is

$$z = g(x_1, y_1) + \left.\frac{\partial g}{\partial x}\right|_{(x_1, y_1)}(x - x_1) + \left.\frac{\partial g}{\partial y}\right|_{(x_1, y_1)}(y - y_1)$$

and $g(x, y)$ may be approximated by its tangent plane in a neighborhood of $\mathbf{X}_1$. When $\mathbf{X}_1$ is a critical point of a plane autonomous system, $P(x_1, y_1) = Q(x_1, y_1) = 0$ and we have

$$x' = P(x, y) \approx \left.\frac{\partial P}{\partial x}\right|_{(x_1, y_1)}(x - x_1) + \left.\frac{\partial P}{\partial y}\right|_{(x_1, y_1)}(y - y_1)$$

$$y' = Q(x, y) \approx \left.\frac{\partial Q}{\partial x}\right|_{(x_1, y_1)}(x - x_1) + \left.\frac{\partial Q}{\partial y}\right|_{(x_1, y_1)}(y - y_1)$$

The original system $\mathbf{X}' = \mathbf{g}(\mathbf{X})$ may be approximated in a neighborhood of the critical point $\mathbf{X}_1$ by the linear system $\mathbf{X}' = \mathbf{A}(\mathbf{X} - \mathbf{X}_1)$, where

$$\mathbf{A} = \begin{bmatrix} \left.\dfrac{\partial P}{\partial x}\right|_{(x_1, y_1)} & \left.\dfrac{\partial P}{\partial y}\right|_{(x_1, y_1)} \\ \left.\dfrac{\partial Q}{\partial x}\right|_{(x_1, y_1)} & \left.\dfrac{\partial Q}{\partial y}\right|_{(x_1, y_1)} \end{bmatrix}$$

This matrix is called the **Jacobian matrix** at $\mathbf{X}_1$ and will be denoted by $\mathbf{g}'(\mathbf{X}_1)$. If we let $\mathbf{H} = \mathbf{X} - \mathbf{X}_1$, then the linear system $\mathbf{X}' = \mathbf{A}(\mathbf{X} - \mathbf{X}_1)$ becomes $\mathbf{H}' = \mathbf{AH}$, which is the form of the linear system analyzed in Section 10.2. The critical point $\mathbf{X} = \mathbf{X}_1$ for $\mathbf{X}' = \mathbf{A}(\mathbf{X} - \mathbf{X}_1)$ now corresponds to the critical point $\mathbf{H} = \mathbf{0}$ for $\mathbf{H}' = \mathbf{AH}$. If the eigenvalues of $\mathbf{A}$ have negative real parts, then $\mathbf{0}$ is an asymptotically stable critical point for $\mathbf{H}' = \mathbf{AH}$. If there is an eigenvalue with a positive real part, then $\mathbf{H} = \mathbf{0}$ is an unstable critical point. The following theorem asserts that the same conclusions can be made for the critical point $\mathbf{X}_1$ of the original system:

THEOREM 10.3 Stability Criteria For Plane Autonomous Systems

Let $\mathbf{X}_1$ be a critical point of the plane autonomous system $\mathbf{X}' = \mathbf{g}(\mathbf{X})$, where $P(x, y)$ and $Q(x, y)$ have continuous first-order partial derivatives in a neighborhood of $\mathbf{X}_1$. Then:

10.3 Linearization and Local Stability

> (i) If the eigenvalues of $\mathbf{A} = \mathbf{g}'(\mathbf{X}_1)$ have negative real parts, then $\mathbf{X}_1$ is an asymptotically stable critical point.
>
> (ii) If $\mathbf{A} = \mathbf{g}'(\mathbf{X}_1)$ has an eigenvalue with a positive real part, then $\mathbf{X}_1$ is an unstable critical point.

EXAMPLE 4 Classify (if possible) the critical points of each plane autonomous system as stable or unstable.

(a) $x' = x^2 + y^2 - 6$
 $y' = x^2 - y$

(b) $x' = 0.01x(100 - x - y)$
 $y' = 0.05y(60 - y - 0.2x)$

Solution The critical points of each system were determined in Example 4 in Section 10.1.

(a) The critical points are $(\sqrt{2}, 2)$ and $(-\sqrt{2}, 2)$, the Jacobian matrix is

$$\mathbf{g}'(\mathbf{X}) = \begin{bmatrix} 2x & 2y \\ 2x & -1 \end{bmatrix}$$

and so $\mathbf{A}_1 = \mathbf{g}'((\sqrt{2}, 2)) = \begin{bmatrix} 2\sqrt{2} & 4 \\ 2\sqrt{2} & -1 \end{bmatrix}$ and $\mathbf{A}_2 = \mathbf{g}'((-\sqrt{2}, 2)) = \begin{bmatrix} -2\sqrt{2} & 4 \\ -2\sqrt{2} & -1 \end{bmatrix}$. Since the determinant of $\mathbf{A}_1$ is negative, $\mathbf{A}_1$ has a positive real eigenvalue. Therefore, $(\sqrt{2}, 2)$ is an unstable critical point. Matrix $\mathbf{A}_2$ has a positive determinant and a negative trace, and so both eigenvalues have negative real parts. It follows that $(-\sqrt{2}, 2)$ is a stable critical point.

(b) The critical points are $(0, 0)$, $(0, 60)$, $(100, 0)$, and $(50, 50)$, the Jacobian matrix is

$$\mathbf{g}'(\mathbf{X}) = \begin{bmatrix} 0.01(100 - 2x - y) & -0.01x \\ -0.01y & 0.05(60 - 2y - 0.2x) \end{bmatrix}$$

and so

$\mathbf{A}_1 = \mathbf{g}'((0, 0)) = \begin{bmatrix} 1 & 0 \\ 0 & 3 \end{bmatrix}$ $\mathbf{A}_2 = \mathbf{g}'((0, 60)) = \begin{bmatrix} 0.4 & 0 \\ -0.6 & -3 \end{bmatrix}$

$\mathbf{A}_3 = \mathbf{g}'((100, 0)) = \begin{bmatrix} -1 & -1 \\ 0 & 2 \end{bmatrix}$ $\mathbf{A}_4 = \mathbf{g}'((50, 50)) = \begin{bmatrix} -0.5 & -0.5 \\ -0.5 & -2.5 \end{bmatrix}$

Since the matrix $\mathbf{A}_1$ has a positive determinant and a positive trace, both eigenvalues have positive real parts. Therefore, $(0, 0)$ is an unstable critical point. The determinants of matrices $\mathbf{A}_2$ and $\mathbf{A}_3$ are negative, and so in each case one of the eigenvalues is positive. Therefore, both $(0, 60)$ and $(100, 0)$ are unstable critical points. Since the matrix $\mathbf{A}_4$ has a positive determinant and a negative trace, $(50, 50)$ is a stable critical point. □

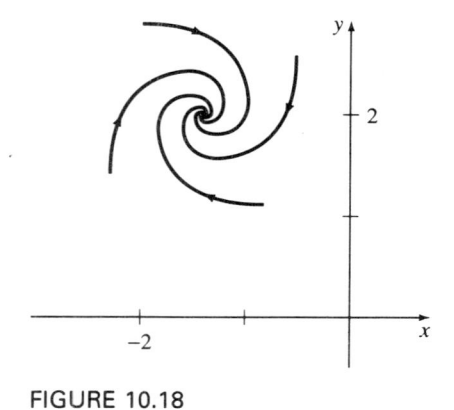

694 CHAPTER 10 Plane Autonomous Systems and Stability

In Example 4 we did not compute $\tau^2 - 4\Delta$ (as in Section 10.2) and attempt to classify the critical points as stable nodes, stable spiral points, saddle points, and so on. For example, for $\mathbf{X}_1 = (-\sqrt{2}, 2)$ in Example 4(a), $\tau^2 - 4\Delta < 0$, and, if the system were linear, we would be able to conclude that $\mathbf{X}_1$ is a stable spiral point. Figure 10.18 shows several solution curves near $\mathbf{X}_1$ that were obtained numerically, and each solution does *appear* to spiral in toward the critical point.

Classifying Critical Points   It is natural to ask whether we can infer more geometric information about the solutions near a critical point $\mathbf{X}_1$ of a nonlinear autonomous system from an analysis of the critical point of the corresponding linear system. The answer is summarized in Figure 10.19, but

FIGURE 10.18

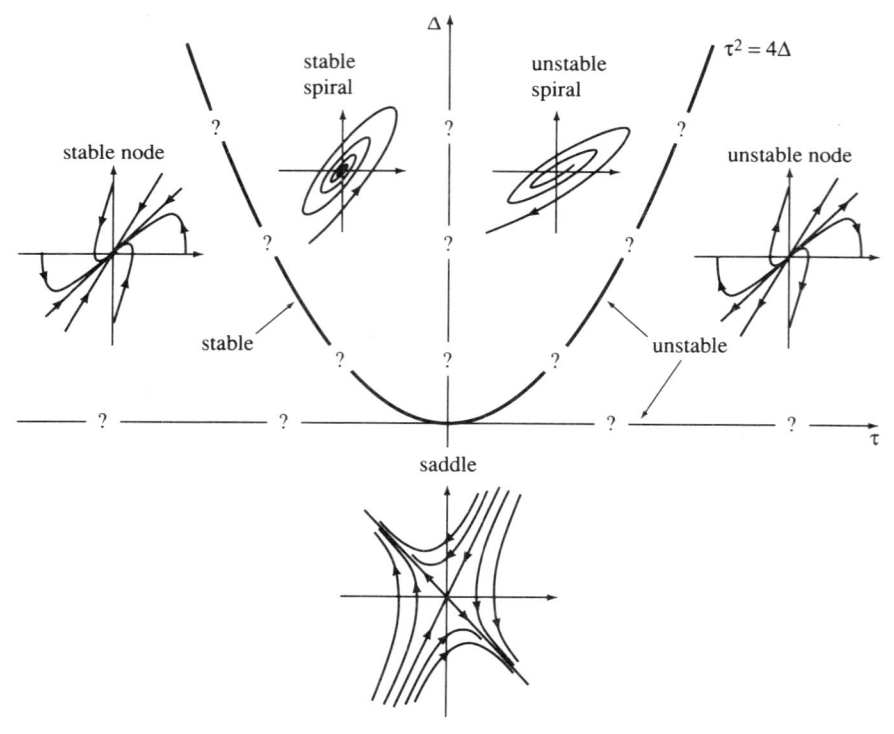

FIGURE 10.19

the student should note the following comments:

(*i*) In five separate cases (stable node, stable spiral point, unstable spiral point, unstable node, and saddle), the critical point may be categorized like the critical point in the corresponding linear system. *The solutions have the same general geometric features as the solutions to the linear system*, and the smaller the neighborhood about $\mathbf{X}_1$, the closer the resemblance.

(*ii*) If $\tau^2 = 4\Delta$ and $\tau > 0$, the critical point X_1 is unstable, but in this borderline case *we are not yet able to decide whether X_1 is an unstable spiral, unstable node, or degenerate unstable node.* Likewise, if $\tau^2 = 4\Delta$ and $\tau < 0$, the critical point X_1 is stable but may be either a stable spiral, a stable node, or a degenerate stable node.

(*iii*) If $\tau = 0$ and $\Delta > 0$, the eigenvalues of $A = g'(X)$ are pure imaginary, and in this borderline case X_1 may be either a stable spiral, an unstable spiral, or a center. *It is therefore not yet possible to determine whether X_1 is stable or unstable.*

EXAMPLE 5 Classify each critical point of the plane autonomous system in Example 4(*b*) as a stable node, a stable spiral point, an unstable spiral point, an unstable node, or a saddle.

Solution For the matrix A_1 corresponding to $(0, 0)$, $\Delta = 3$, $\tau = 4$, and so $\tau^2 - 4\Delta = 4$. Therefore, $(0, 0)$ is an unstable node. The critical points $(0, 60)$ and $(100, 0)$ are saddles, since $\Delta < 0$ in both cases. For matrix A_4, $\Delta > 0$, $\tau < 0$, and $\tau^2 - 4\Delta > 0$. It follows that $(50, 50)$ is a stable node. □

EXAMPLE 6 The plane autonomous system corresponding to the second-order nonlinear differential equation $x'' + x - x^3 = 0$ is

$$x' = y$$
$$y' = x^3 - x$$

Find and classify (if possible) the critical points.

Solution Since $x^3 - x = x(x^2 - 1)$, the critical points are $(0, 0)$, $(1, 0)$, and $(-1, 0)$. The corresponding Jacobian matrices are

$$A_1 = g'((0, 0)) = \begin{bmatrix} 0 & 1 \\ -1 & 0 \end{bmatrix} \qquad A_2 = g'((1, 0)) = g'((-1, 0)) = \begin{bmatrix} 0 & 1 \\ 2 & 0 \end{bmatrix}$$

Since det $A_2 < 0$, the critical points $(1, 0)$ and $(-1, 0)$ are both saddle points. The eigenvalues of matrix A_1 are $\pm i$, and according to comment (*iii*), the status of the critical point at $(0, 0)$ remains in doubt. It may be either a stable spiral, an unstable spiral, or a center. □

Phase-Plane Method The linearization method, when successful, can provide useful information on the local behavior of solutions near critical points. It is of little help if we are interested in solutions whose initial position $X(0) = X_0$ is not close to a critical point or if we wish to obtain a global view of the family of solution curves. The *phase-plane method* is based on the fact that

$$\frac{dy}{dx} = \frac{dy/dt}{dx/dt} = \frac{Q(x, y)}{P(x, y)}$$

It attempts to find y as a function of x using one of many methods available for solving first-order differential equations (see Chapter 2). As we will show in Example 7, the method can sometimes be used to decide whether a critical point like (0, 0) in Example 6 is a stable spiral, an unstable spiral, or a center.

EXAMPLE 7 Use the phase-plane method to determine the nature of the solutions to $x'' + x - x^3 = 0$ in a neighborhood of (0, 0).

Solution If we let $dx/dt = y$, then $dy/dt = x^3 - x$. From this we obtain the first-order differential equation

$$\frac{dy}{dx} = \frac{dy/dt}{dx/dt} = \frac{x^3 - x}{y}$$

which can be solved by separation of variables:

$$\int y\, dy = \int (x^3 - x)\, dx$$

$$\frac{y^2}{2} = \frac{x^4}{4} - \frac{x^2}{2} + c = \frac{x^4 - 2x^2 + 1}{4} + \left(c - \frac{1}{4}\right)$$

The solution may be written as $y^2 = (x^2 - 1)^2/2 + c_0$. If $\mathbf{X}(0) = (x_0, 0)$, where $0 < x_0 < 1$, then $c_0 = -(x_0^2 - 1)^2/2$, and so

$$y^2 = \frac{(x^2 - 1)^2}{2} - \frac{(x_0^2 - 1)^2}{2}$$

$$= \frac{(2 - x^2 - x_0^2)(x_0^2 - x^2)}{2}$$

Note that $y = 0$ when $x = -x_0$. In addition, the right-hand side is positive when $-x_0 < x < x_0$, and so each x has *two* corresponding values of y. The solution $\mathbf{X} = \mathbf{X}(t)$ which satisfies $\mathbf{X}(0) = (x_0, 0)$ is therefore periodic and so (0, 0) is a center.

Figure 10.20 shows a family of solution curves. The original plane autonomous system was used to determine the directions of the solutions. □

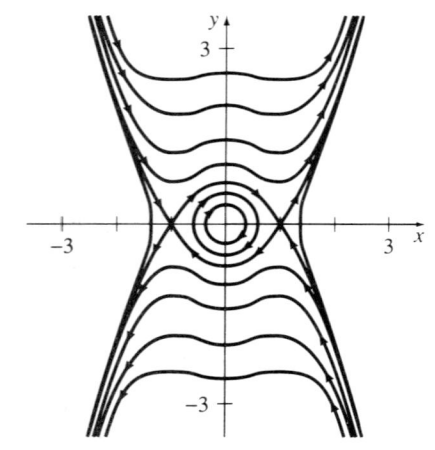

FIGURE 10.20

EXERCISES 10.3   *Answers to odd-numbered problems begin on page A-59.*

1. Show that (0, 0) is an asymptotically stable critical point of the nonlinear autonomous system

$$x' = \alpha x - \beta y + y^2$$
$$y' = \beta x + \alpha y - xy$$

when $\alpha < 0$ and an unstable critical point when $\alpha > 0$. [*Hint:* Switch to polar coordinates.]

2. When expressed in polar coordinates, a plane autonomous system takes the form

$$\frac{dr}{dt} = \alpha r(5 - r)$$

$$\frac{d\theta}{dt} = -1$$

Show that (0, 0) is an asymptotically stable critical point if and only if $\alpha < 0$.

In Problems 3–10, without solving explicitly, classify the critical points of the given first-order autonomous differential equation as either asymptotically stable or unstable. All constants are assumed to be positive.

3. $\dfrac{dx}{dt} = kx(n + 1 - x)$

4. $\dfrac{dx}{dt} = -kx \ln\left(\dfrac{x}{K}\right), \ x > 0$

5. $\dfrac{dT}{dt} = k(T - T_0)$ 6. $m\dfrac{dv}{dt} = mg - kv$

7. $\dfrac{dx}{dt} = k(\alpha - x)(\beta - x)$
   $(\alpha > \beta)$

8. $\dfrac{dx}{dt} = k(\alpha - x)(\beta - x)(\gamma - x)$
   $(\alpha > \beta > \gamma)$

9. $\dfrac{dP}{dt} = P(a - bP)(1 - cP^{-1}), \ P > 0$
   $(a < bc)$

10. $\dfrac{dA}{dt} = k\sqrt{A}(K - \sqrt{A}), \ A > 0$

In Problems 11–20 classify (if possible) each critical point of the given plane autonomous system as a stable node, a stable spiral point, an unstable spiral point, an unstable node, or a saddle point.

11. $x' = 1 - 2xy$
    $y' = 2xy - y$

12. $x' = x^2 - y^2 - 1$
    $y' = 2y$

13. $x' = y - x^2 + 2$
    $y' = x^2 - xy$

14. $x' = 2x - y^2$
    $y' = -y + xy$

15. $x' = -3x + y^2 + 2$
    $y' = x^2 - y^2$

16. $x' = xy - 3y - 4$
    $y' = y^2 - x^2$

17. $x' = -2xy$
    $y' = y - x + xy - y^3$

18. $x' = x(1 - x^2 - 3y^2)$
    $y' = y(3 - x^2 - 3y^2)$

19. $x' = x(10 - x - \tfrac{1}{2}y)$
    $y' = y(16 - y - x)$

20. $x' = -2x + y + 10$
    $y' = 2x - y - 15\dfrac{y}{y + 5}$

In Problems 21–26 classify (if possible) each critical point of the given second-order differential equation as a stable node, a stable spiral point, an unstable spiral point, an unstable node, or a saddle.

21. $\theta'' = (\cos \theta - 0.5) \sin \theta, \ |\theta| < \pi$

22. $x'' + x = [\tfrac{1}{2} - 3(x')^2]x' - x^2$

23. $x'' + x'(1 - x^3) - x^2 = 0$

24. $x'' + 4\dfrac{x}{1 + x^2} + 2x' = 0$

25. $x'' + x = \varepsilon x^3 \quad$ for $\varepsilon > 0$

26. $x'' + x - \varepsilon x|x| = 0 \quad$ for $\varepsilon > 0$

[Hint: $\dfrac{d}{dx} x|x| = 2|x|$]

27. Show that the second-order nonlinear differential equation

$$(1 + \alpha^2 x^2)x'' + [\beta + \alpha^2(x')^2]x = 0$$

has a saddle point at (0, 0) when $\beta < 0$.

28. Show that the plane autonomous system

$$x' = -\alpha x + xy$$
$$y' = 1 - \beta y - x^2$$

has a unique critical point when $\alpha\beta > 1$ and that this critical point is stable when $\beta > 0$.

29. (a) Show that the plane autonomous system

$$x' = -x + y - x^3$$
$$y' = -x - y + y^2$$

has two critical points by sketching the graphs of $-x + y - x^3 = 0$ and $-x - y + y^2 = 0$. Classify the critical point at (0, 0).

(b) Show that the second critical point $\mathbf{X}_1 = (0.88054, 1.56327)$ is a saddle.

30. (a) Show that (0, 0) is the only critical point of the Raleigh differential equation

$$x'' + \varepsilon\left[\tfrac{1}{3}(x')^3 - x'\right] + x = 0$$

(b) Show that (0, 0) is unstable when $\varepsilon > 0$. When is (0, 0) an unstable spiral point?

(c) Show that (0, 0) is stable when $\varepsilon < 0$. When is (0, 0) a stable spiral point?

(d) Show that (0, 0) is a center when $\varepsilon = 0$.

31. Use the phase-plane method to show that (0, 0) is a center of the second-order nonlinear differential equation $x'' + 2x^3 = 0$.

32. Use the phase-plane method to show that the solution of the second-order nonlinear differential equation $x'' + 2x - x^2 = 0$ that satisfies $x(0) = 1$ and $x'(0) = 0$ is periodic.

33. (a) Find the critical points of the plane autonomous system

    $$x' = 2xy$$
    $$y' = 1 - x^2 + y^2$$

    and show that linearization gives no information about the nature of these critical points.

    (b) Use the phase-plane method to show that the critical points in part (a) are both centers. [*Hint:* Let $u = y^2/x$ and show that $(x-c)^2 + y^2 = c^2 - 1$.]

34. The origin is the only critical point of the second-order nonlinear differential equation $x'' + (x')^2 + x = 0$.

    (a) Show that the phase-plane method leads to the Bernoulli differential equation $dy/dx = -y - xy^{-1}$.

    (b) Show that the solution satisfying $x(0) = \frac{1}{2}$ and $x'(0) = 0$ is not periodic.

35. A solution of the second-order nonlinear differential equation $x'' + x - x^3 = 0$ satisfies $x(0) = 0$ and $x'(0) = v_0$. Use the phase-plane method to determine when the resulting solution is periodic. [*Hint:* See Example 7.]

36. The nonlinear differential equation $x'' + x = 1 + \varepsilon x^2$ arises in the analysis of planetary motion using relativity theory. Classify (if possible) all critical points of the corresponding plane autonomous system.

37. When a nonlinear capacitor is present in an *L-C-R* circuit, the voltage drop is no longer given by q/C but is more accurately described by $\alpha q + \beta q^3$, where α and β are constants and $\alpha > 0$. Differential equation (3) of Section 3.11 for the free circuit is then replaced by

$$L \frac{d^2q}{dt^2} + R \frac{dq}{dt} + \alpha q + \beta q^3 = 0$$

Find and classify all critical points of this nonlinear differential equation. [*Hint:* Consider two cases $\beta > 0$ and $\beta < 0$.]

10.4 APPLICATIONS OF AUTONOMOUS SYSTEMS

A plane autonomous system can serve as a mathematical model for diverse physical phenomena. These include problems in steady-state fluid dynamics, mechanical oscillations, circuits containing nonlinear capacitors, predator-prey and competition interactions in ecology, and the kinetics of medical drugs.

Many applications from physics give rise to nonlinear second-order differential equations of the form $x'' = g(x, x')$. For example, in the analysis of free damped motion in Section 3.9, we assumed that the damping force was proportional to the velocity x'. Frequently the damping force is proportional to the square of the velocity, and the new differential equation becomes

$$x'' = -\frac{\beta}{m} x'|x'| - \frac{k}{m} x$$

The corresponding plane autonomous system is therefore

$$x' = y$$
$$y' = -\frac{\beta}{m} y|y| - \frac{k}{m} x$$

In this section we will use the results of Section 10.3 to analyze the nonlinear pendulum, the motion of a bead on a curve, the Lotka–Volterra predator-prey model, and the Lotka–Volterra competition model. Additional models will be presented in the exercises.

10.4 Applications of Autonomous Systems

Nonlinear Pendulum In Example 3 in Section 1.3 we showed that the displacement angle θ for a pendulum satisfies the second-order nonlinear differential equation

$$\frac{d^2\theta}{dt^2} + \frac{g}{l}\sin\theta = 0$$

Letting $x = \theta$ and $y = \theta'$, we can write this second-order differential equation as the plane autonomous system

$$x' = y$$
$$y' = -\frac{g}{l}\sin x$$

The critical points are $(\pm k\pi, 0)$, and the Jacobian matrix is easily shown to be

$$\mathbf{g}'((\pm k\pi, 0)) = \begin{bmatrix} 0 & 1 \\ (-1)^{k+1}\frac{g}{l} & 0 \end{bmatrix}$$

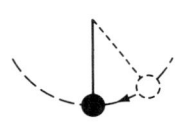

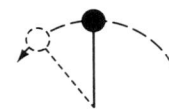

If $k = 2n + 1$, then $\Delta < 0$ and so all of critical points $(\pm(2n + 1)\pi, 0)$ are saddle points. In particular, the critical point at $(\pi, 0)$ is unstable as expected. See Figure 10.21. When $k = 2n$, the eigenvalues are pure imaginary and so the nature of these critical points remains in doubt. Since we have assumed that there are no damping forces acting on the pendulum, we expect that all the critical points $(\pm 2n\pi, 0)$ are centers. This can be verified using the phase-plane method.

From

$$\frac{dy}{dx} = \frac{dy/dt}{dx/dt} = -\frac{g}{l}\frac{\sin x}{y}$$

it follows that $y^2 = (2g/l)\cos x + c$. If $\mathbf{X}(0) = (x_0, 0)$, then

$$y^2 = \frac{2g}{l}(\cos x - \cos x_0)$$

Note that $y = 0$ when $x = -x_0$, and $(2g/l)(\cos x - \cos x_0) > 0$ for $|x| < |x_0| < \pi$. Thus, each such x has two corresponding values of y and so the solution $\mathbf{X} = \mathbf{X}(t)$ which satisfies $\mathbf{X}(0) = (x_0, 0)$ is periodic. We may conclude that $(0, 0)$ is a center. Figure 10.22 shows a family of solution curves.

(a) $\theta = 0, \theta' = 0$ (b) $\theta = \pi, \theta' = 0$

FIGURE 10.21

FIGURE 10.22

EXAMPLE 1 A pendulum in an equilibrium position with $\theta = 0$ is given an initial angular velocity of ω_0 radians/s. Determine under what conditions the resulting motion is periodic.

Solution We are asked to examine the solution of the plane autonomous system that satisfies $\mathbf{X}(0) = (0, \omega_0)$. From $y^2 = (2g/l)\cos x + c$, it follows that

$$y^2 = \frac{2g}{l}\left(\cos x - 1 + \frac{l}{2g}\omega_0^2\right)$$

To establish that the solution $\mathbf{X}(t)$ is periodic, it is sufficient to show that there are two x-intercepts, $x = \pm x_0$, between $-\pi$ and π and that the right-hand side is positive for $|x| < |x_0|$. Each such x will then have two corresponding values of y.

If $y = 0$, $\cos x = 1 - (l/2g)\omega_0^2$, and this equation will have two solutions $x = \pm x_0$ between $-\pi$ and π provided $1 - (l/2g)\omega_0^2 > -1$. Note that $(2g/l)(\cos x - \cos x_0)$ will then be positive for $|x| < |x_0|$. This restriction on the initial angular velocity may be written as

$$|\omega_0| < 2\sqrt{\frac{g}{l}}$$ □

Nonlinear Oscillations: Sliding Bead Suppose, as shown in Figure 10.23, a bead with mass m slides along a thin wire whose shape is described by the function $z = f(x)$. A wide variety of nonlinear oscillations can be obtained by changing the shape of the wire and by making different assumptions about the forces acting on the bead.

The tangential force $\mathbf{F}$ due to the weight $W = mg$ has magnitude $mg \sin \theta$, and therefore the x-component of $\mathbf{F}$ is $F_x = -mg \sin \theta \cos \theta$. Since $\tan \theta = f'(x)$, we may use the identities $1 + \tan^2 \theta = \sec^2 \theta$ and $\sin^2 \theta = 1 - \cos^2 \theta$ to conclude that

$$F_x = -mg \sin \theta \cos \theta = -mg \frac{f'(x)}{1 + [f'(x)]^2}$$

We will assume (as in Section 3.9) that a damping force $\mathbf{D}$ acting in the direction opposite to the motion is a constant multiple of the velocity of the bead. The x-component of $\mathbf{D}$ is therefore

$$D_x = -\beta \frac{dx}{dt}$$

If we ignore the frictional force between the wire and the bead and assume that no other external forces are impressed on the system, it follows from Newton's second law that

$$mx'' = -mg \frac{f'(x)}{1 + [f'(x)]^2} - \beta x'$$

and the corresponding plane autonomous system is

$$x' = y$$
$$y' = -g \frac{f'(x)}{1 + [f'(x)]^2} - \frac{\beta}{m} y$$

If $\mathbf{X}_1 = (x_1, y_1)$ is a critical point of the system, $y_1 = 0$ and therefore $f'(x_1) = 0$. The bead must therefore be at rest at a point on the wire where the tangent line is horizontal. When f is twice differentiable, the Jacobian matrix at $\mathbf{X}_1$ is

$$\mathbf{g}'(\mathbf{X}_1) = \begin{bmatrix} 0 & 1 \\ -gf''(x_1) & -\beta/m \end{bmatrix}$$

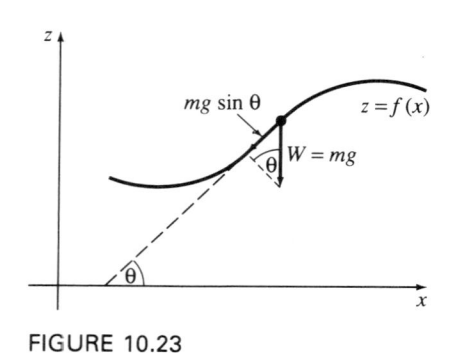

FIGURE 10.23

10.4 Applications of Autonomous Systems

and so $\tau = -\beta/m$, $\Delta = gf''(x_1)$, and $\tau^2 - 4\Delta = \beta^2/m^2 - 4gf''(x_1)$. Using the results of Section 10.3, we can make the following conclusions:

(i) $f''(x_1) < 0$: A relative maximum therefore occurs at $x = x_1$ and since $\Delta < 0$, an *unstable saddle point* occurs at $\mathbf{X}_1 = (x_1, 0)$.

(ii) $f''(x_1) > 0$ and $\beta > 0$: A relative minimum therefore occurs at $x = x_1$, and since $\tau < 0$ and $\Delta > 0$, $\mathbf{X}_1 = (x_1, 0)$ is a *stable critical point*. If $\beta^2 > 4gm^2 f''(x_1)$, the system is **overdamped** and the critical point is a *stable node*. If $\beta^2 < 4gm^2 f''(x_1)$, the system is **underdamped** and the critical point is a *stable spiral point*. The exact nature of the stable critical point is still in doubt if $\beta^2 = 4gm^2 f''(x_1)$.

(iii) $f''(x_1) > 0$ and the system is undamped ($\beta = 0$): In this case the eigenvalues are pure imaginary, but the phase-plane method can be used to show that the critical point is a *center*. Therefore, solutions with $\mathbf{X}(0) = (x(0), x'(0))$ near $\mathbf{X}_1 = (x_1, 0)$ are periodic.

EXAMPLE 2 A 10-g bead slides along the graph of $z = \sin x$. According to (ii), the relative minima at $x_1 = -\pi/2$ and $3\pi/2$ give rise to stable critical points (see Figure 10.24). Since $f''(-\pi/2) = f''(3\pi/2) = 1$, the system will be underdamped provided $\beta^2 < 4gm^2$. If we use SI units of measurement, $m = 0.01$ kg, $g = 9.8$ m/s^2, and so the condition for an underdamped system becomes $\beta^2 < 3.92 \times 10^{-3}$.

If $\beta = 0.01$ is the damping constant, both of these critical points will be *stable spiral points*. The two solutions corresponding to initial conditions $\mathbf{X}(0) = ((x(0), x'(0)) = (-2\pi, 10)$ and $\mathbf{X}(0) = (-2\pi, 15)$, respectively, were obtained numerically and are shown in Figure 10.25. When $x'(0) = 10$, the bead has enough momentum to make it over the hill at $x = -3\pi/2$ but not over the hill at $x = \pi/2$. The bead then approaches the relative minimum based at $x = -\pi/2$. If $x'(0) = 15$, the bead has the momentum to make it over both hills but then rocks back and forth in the valley based at $x = 3\pi/2$ and approaches the point $(3\pi/2, -1)$ on the wire.

Figure 10.26 shows a collection of solution curves for the undamped case. Since $\beta = 0$, the critical points corresponding to $x_1 = -\pi/2$ and $3\pi/2$ are now

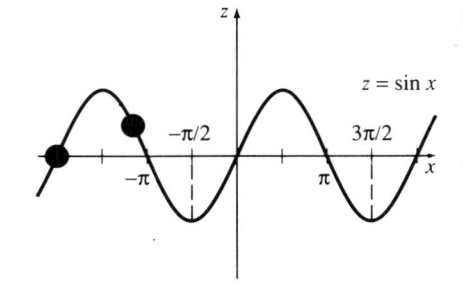

FIGURE 10.24

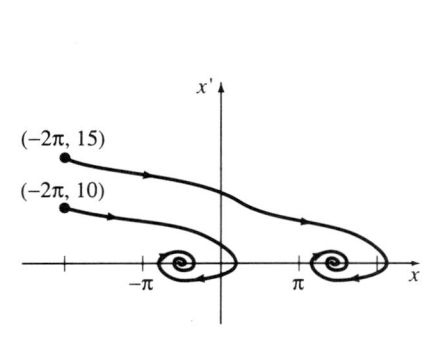

FIGURE 10.25

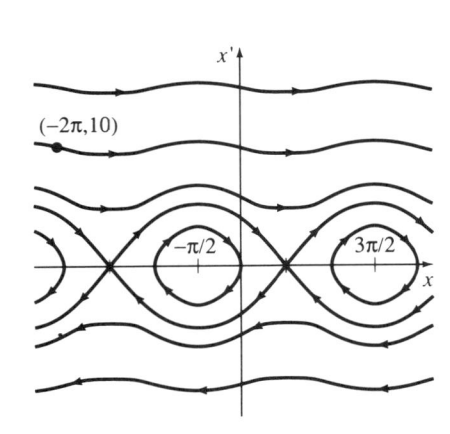

FIGURE 10.26

centers. When $\mathbf{X}(0) = (-2\pi, 10)$, the bead has sufficient momentum to move over *all* hills. The figure also indicates that when the bead is released from rest at a position on the wire between $x = -3\pi/2$ and $x = \pi/2$, the resulting motion is periodic. □

Lotka–Volterra Predator-Prey Model A **predator-prey interaction** between two species occurs when one species (the predator) feeds on the second species (the prey). For example, the snowy owl feeds almost exclusively on a common arctic rodent called a lemming, while the lemming uses arctic tundra plants as its food supply. Interest in using mathematics to help explain predator-prey interactions has been stimulated by the observation of population cycles in many arctic mammals. In the MacKenzie River district of Canada, for example, the principal prey of the lynx is the snowshoe hare, and both populations cycle with a period of about 10 years.

There are many predator-prey models that lead to plane autonomous systems with at least one periodic solution. The first such model was constructed independently by pioneer biomathematicians A. Lotka (1925) and V. Volterra (1926). If x denotes the number of predators and y denotes the number of prey, then the Lotka–Volterra model takes the form

$$x' = -ax + bxy = x(-a + by)$$
$$y' = -cxy + dy = y(-cx + d)$$

where a, b, c, and d are positive constants.

Note that in the absence of predators ($x = 0$), $y' = dy$ and so the number of prey grows exponentially. In the absence of prey, $x' = -ax$ and so the predator population becomes extinct. The term $-cxy$ represents the death rate due to predation. The model therefore assumes that this death rate is directly proportional to the number of possible encounters xy between predator and prey at a particular time t, and the term bxy represents the resulting positive contribution to the predator population.

The critical points of this plane autonomous system are $(0, 0)$ and $(d/c, a/b)$, and the corresponding Jacobian matrices are

$$\mathbf{A}_1 = \mathbf{g}'((0,0)) = \begin{bmatrix} -a & 0 \\ 0 & d \end{bmatrix} \quad \text{and} \quad \mathbf{A}_2 = \mathbf{g}'\left(\left(\frac{d}{c}, \frac{a}{b}\right)\right) = \begin{bmatrix} 0 & bd/c \\ -ac/b & 0 \end{bmatrix}$$

The critical point at $(0, 0)$ is a saddle point, and Figure 10.27 shows a typical profile of solutions that are in the first quadrant near $(0, 0)$.

Since the matrix $\mathbf{A}_2$ has pure imaginary eigenvalues $\lambda = \pm\sqrt{ad}\, i$, the critical point $(d/c, a/b)$ *may* be a center. This possibility can be investigated using the phase-plane method. Since

$$\frac{dy}{dx} = \frac{y(-cx + d)}{x(-a + by)}$$

we may separate variables and obtain

$$\int \frac{-a + by}{y}\, dy = \int \frac{-cx + d}{x}\, dx$$

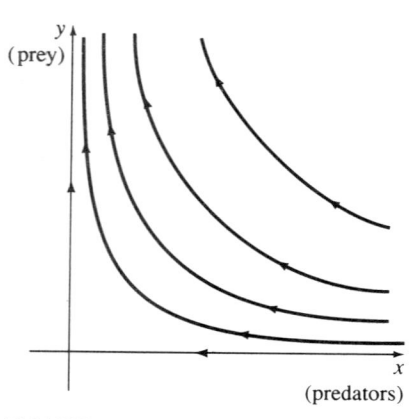

FIGURE 10.27

10.4 Applications of Autonomous Systems

so that
$$-a \ln y + by = -cx + d \ln x + c_1$$
or
$$(x^d e^{-cx})(y^a e^{-by}) = c_0$$

The following argument will establish that all solution curves that originate in the first quadrant are periodic.

Typical graphs of the nonnegative functions $F(x) = x^d e^{-cx}$ and $G(y) = y^a e^{-by}$ are shown in Figure 10.28. It is not hard to show that $F(x)$ has an absolute maximum at $x = d/c$, while $G(y)$ has an absolute maximum at $y = a/b$. Note that, with the exception of zero and the absolute maximum, F and G both take on all values in their range precisely twice.

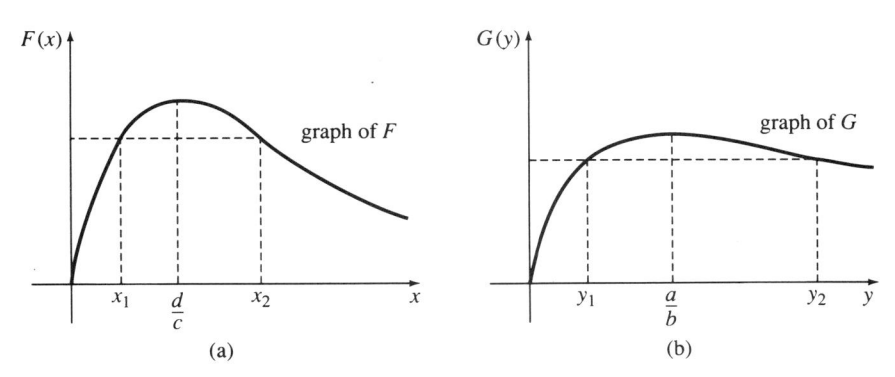

FIGURE 10.28

These graphs can be used to establish the following properties of a solution curve which originates at a noncritical point (x_0, y_0) in the first quadrant:

1. If $y = a/b$, the equation $F(x)G(y) = c_0$ has exactly two solutions x_m and x_M that satisfy $x_m < d/c < x_M$.
2. If $x_m < x_1 < x_M$ and $x = x_1$, then $F(x)G(y) = c_0$ has exactly two solutions y_1 and y_2 that satisfy $y_1 < a/b < y_2$.
3. If x is outside the interval $[x_m, x_M]$, then $F(x)G(y) = c_0$ has no solutions.

We will demonstrate property (1) and outline properties (2) and (3) in the the exercises. Since $(x_0, y_0) \neq (d/c, a/b)$, $F(x_0)G(y_0) < F(d/c)G(a/b)$. If $y = a/b$, then

$$0 < \frac{c_0}{G(a/b)} = \frac{F(x_0)G(y_0)}{G(a/b)} < \frac{F(d/c)G(a/b)}{G(a/b)} = F\left(\frac{d}{c}\right)$$

Therefore $F(x) = c_0/G(a/b)$ has precisely two solutions x_m and x_M that satisfy $x_m < d/c < x_M$. The graph of a typical periodic solution is shown in Figure 10.29.

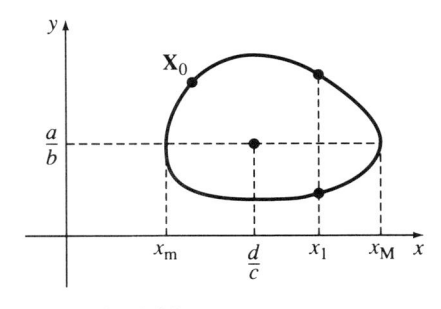

FIGURE 10.29

EXAMPLE 3 If we let $a = 0.1$, $b = 0.002$, $c = 0.0025$, and $d = 0.2$ in the Lotka–Volterra predator-prey model, the critical point in the first quadrant is $(d/c, a/b) = (80, 50)$, and we know that this critical point is a center. See

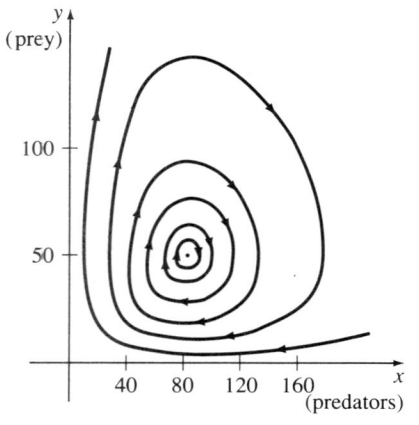

Figure 10.30. The closer the initial condition $\mathbf{X}_0$ is to (80, 50), the more the periodic solutions resemble the elliptical solutions to the corresponding linear system. The eigenvalues of $\mathbf{g}'((80, 50))$ are $\pm\sqrt{ad}\,i = \pm\sqrt{2}/10\,i$, and so the solutions near the critical point have period $p \approx 10\sqrt{2}\pi$, or about 44.4. □

Lotka–Volterra Competition Model A **competitive interaction** occurs when two or more species compete for the food, water, light, and space resources of an ecosystem. The use of one of these resources by one population therefore inhibits the ability of another population to survive and grow. Under what conditions can two competing species coexist? A number of mathematical models have been constructed that offer insights into conditions that permit coexistence. If x denotes the number in species I and y denotes the number in species II, then the Lotka–Volterra model takes the form

$$x' = \frac{r_1}{K_1} x(K_1 - x - \alpha_{12} y)$$

$$y' = \frac{r_2}{K_2} y(K_2 - y - \alpha_{21} x)$$

Note that in the absence of species II ($y = 0$), $x' = (r_1/K_1)x(K_1 - x)$, and so the first population grows logistically and approaches the steady-state population K_1 (see Section 2.11 and Example 3 in Section 10.3). A similar statement holds for species II growing in the absence of species I. The term $-\alpha_{21}xy$ in the second equation stems from the competitive effect of species I on species II. The model therefore assumes that this rate of inhibition is directly proportional to the number of possible competitive pairs xy at a particular time t.

This plane autonomous system has critical points at (0, 0), (K_1, 0), and (0, K_2). When $\alpha_{12}\alpha_{21} \neq 0$, the lines $K_1 - x - \alpha_{12}y = 0$ and $K_2 - y - \alpha_{21}x = 0$ intersect to produce a fourth critical point $\hat{\mathbf{X}} = (\hat{x}, \hat{y})$. Figure 10.31 shows the two conditions under which $(\hat{x}, \hat{y})$ is in the first quadrant. The trace and determinant of the Jacobian matrix at $(\hat{x}, \hat{y})$ are

$$\tau = -\hat{x}\frac{r_1}{K_1} - \hat{y}\frac{r_2}{K_2}$$

$$\Delta = (1 - \alpha_{12}\alpha_{21})\hat{x}\hat{y}\frac{r_1 r_2}{K_1 K_2}$$

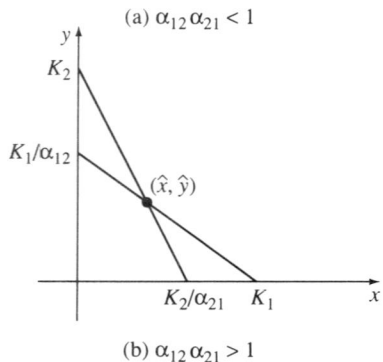

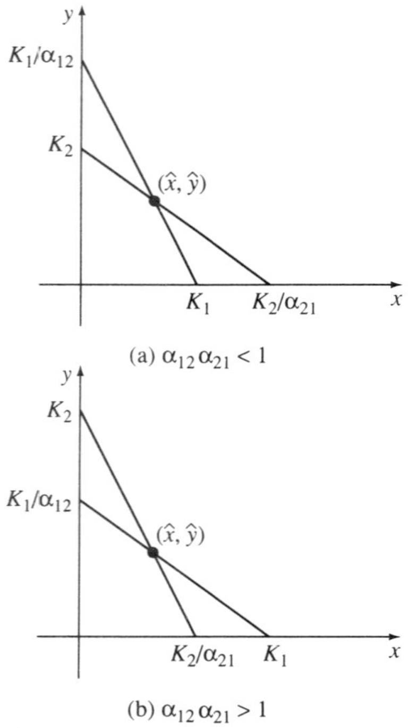

In case (a), $K_1/\alpha_{12} > K_2$ and $K_2/\alpha_{21} > K_1$. It follows that $\alpha_{12}\alpha_{21} < 1$, $\tau < 0$, and $\Delta > 0$. Since

$$\tau^2 - 4\Delta = \left(\hat{x}\frac{r_1}{K_1} + \hat{y}\frac{r_2}{K_2}\right)^2 + 4(\alpha_{12}\alpha_{21} - 1)\hat{x}\hat{y}\frac{r_1 r_2}{K_1 K_2}$$

$$= \left(\hat{x}\frac{r_1}{K_1} - \hat{y}\frac{r_2}{K_2}\right)^2 + 4\alpha_{12}\alpha_{21}\hat{x}\hat{y}\frac{r_1 r_2}{K_1 K_2}$$

$\tau^2 - 4\Delta > 0$, and so $(\hat{x}, \hat{y})$ is a stable node. Therefore, if $\mathbf{X}(0) = \mathbf{X}_0$ is sufficiently close to $\hat{\mathbf{X}} = (\hat{x}, \hat{y})$,

$$\lim_{t \to \infty} \mathbf{X}(t) = \hat{\mathbf{X}}$$

FIGURE 10.30

FIGURE 10.31
(a) $\alpha_{12}\alpha_{21} < 1$
(b) $\alpha_{12}\alpha_{21} > 1$

and we may conclude that coexistence is possible. The demonstration that case (b) leads to a saddle point and the investigation of the nature of critical points at $(0, 0)$, $(K_1, 0)$, and $(0, K_2)$ are left as exercises.

When the competitive interactions between two species are weak, both of the coefficients α_{12} and α_{21} will be small and so the conditions $K_1/\alpha_{12} > K_2$ and $K_2/\alpha_{21} > K_1$ may be satisfied. This might occur when there is a small overlap in the ranges of two predator species that hunt for a common prey.

EXAMPLE 4 A competitive interaction is described by the Lotka–Volterra competition model

$$x' = 0.004x(50 - x - 0.75y)$$
$$y' = 0.001y(100 - y - 3.0x)$$

Find and classify all critical points of the system.

Solution Critical points occur at $(0, 0)$, $(50, 0)$, $(0, 100)$, and the solution $(20, 40)$ of the system

$$x + 0.75y = 50$$
$$3.0x + y = 100$$

Since $\alpha_{12}\alpha_{21} = 2.25 > 1$, we have case (b), and the critical point at $(20, 40)$ is a saddle point. The Jacobian matrix is

$$\mathbf{g}'(\mathbf{X}) = \begin{bmatrix} 0.2 - 0.008x - 0.003y & -0.003x \\ -0.003y & 0.1 - 0.002y - 0.003x \end{bmatrix}$$

and we obtain

$$\mathbf{g}'((0, 0)) = \begin{bmatrix} 0.2 & 0 \\ 0 & 0.1 \end{bmatrix} \quad \mathbf{g}'((50, 0)) = \begin{bmatrix} -0.2 & -0.15 \\ 0 & -0.05 \end{bmatrix} \quad \mathbf{g}'((0, 100)) = \begin{bmatrix} -0.1 & 0 \\ -0.3 & -0.1 \end{bmatrix}$$

Therefore, $(0, 0)$ is an unstable node, whereas both $(50, 0)$ and $(0, 100)$ are stable nodes. □

Coexistence can also occur in the Lotka–Volterra competition model if there is at least one periodic solution lying entirely in the first quadrant. In Section 10.5 we will show that this model has no periodic solutions.

EXERCISES 10.4 *Answers to odd-numbered problems begin on page A-59.*

Nonlinear Pendulum

1. A pendulum is released at $\theta = \pi/3$ and is given an initial angular velocity of ω_0 radians/s. Determine under what conditions the resulting motion is periodic.

2. (a) If a pendulum is released from rest at $\theta = \theta_0$, show that the angular velocity is again 0 when $\theta = -\theta_0$.

(b) The period T of the pendulum is the amount of time needed for θ to change from θ_0 to $-\theta_0$ and back to θ_0. Show that

$$T = \sqrt{\frac{2l}{g}} \int_{-\theta_0}^{\theta_0} \frac{1}{\sqrt{\cos\theta - \cos\theta_0}} d\theta$$

Nonlinear Oscillations—The Sliding Bead

3. A bead with mass m slides along a thin wire whose shape is described by the function $z = f(x)$. If $\mathbf{X}_1 = (x_1, y_1)$ is a critical point of the plane autonomous system associated with the sliding bead, verify that the Jacobian matrix at $\mathbf{X}_1$ is

$$\mathbf{g}'(\mathbf{X}_1) = \begin{bmatrix} 0 & 1 \\ -gf''(x_1) & -\beta/m \end{bmatrix}$$

4. A bead with mass m slides along a thin wire whose shape is described by the function $z = f(x)$. When $f'(x_1) = 0$, $f''(x_1) > 0$, and the system is undamped, the critical point $\mathbf{X}_1 = (x_1, 0)$ is a center. Estimate the period of the bead when $x(0)$ is near x_1 and $x'(0) = 0$.

5. A bead is released from the position $x(0) = x_0$ on the curve $z = x^2/2$ with initial velocity $x'(0) = v_0$ cm/s.

   (a) Use the phase-plane method to show that the resulting solution is periodic when the system is undamped.

   (b) Show that the maximum height z_{max} to which the bead rises is given by

$$z_{max} = \frac{1}{2}[e^{v_0^2/g}(1 + x_0^2) - 1]$$

6. Rework Problem 5 with $z = \cosh x$.

Lotka–Volterra Predator-Prey Model

7. Refer to Figure 10.29. If $x_m < x_1 < x_M$ and $x = x_1$, show that $F(x)G(y) = c_0$ has exactly two solutions y_1 and y_2 that satisfy $y_1 < a/b < y_2$. [*Hint:* First show that $G(y) = c_0/F(x_1) < G(a/b)$.]

8. From (1) and (3) on page 703, conclude that the maximum number of predators occurs when $y = a/b$.

9. In many fishery science models, the rate at which a species is caught is assumed to be directly proportional to its abundance. If both predator and prey are being exploited in this manner, the Lotka–Volterra differential equations take the form

$$x' = -ax + bxy - \varepsilon_1 x$$
$$y' = -cxy + dy - \varepsilon_2 y$$

where ε_1 and ε_2 are positive constants.

   (a) When $\varepsilon_2 < d$, show that there is a new critical point in the first quadrant that is a center.

   (b) **Volterra's principle** states that a moderate amount of exploitation increases the average number of prey and decreases the average number of predators. Is the fishery science model consistent with Volterra's principle?

Lotka–Volterra Competition Model

10. Show that $(0, 0)$ is always an unstable node.

11. Show that $(K_1, 0)$ is a stable node when $K_1 > K_2/\alpha_{21}$ and a saddle point when $K_1 < K_2/\alpha_{21}$.

12. Use Problems 10 and 11 to establish that $(0, 0)$, $(K_1, 0)$, and $(0, K_2)$ are unstable when $\hat{\mathbf{X}} = (\hat{x}, \hat{y})$ is a stable node.

13. Show that $\hat{\mathbf{X}} = (\hat{x}, \hat{y})$ is a saddle point when $K_1/\alpha_{12} < K_2$ and $K_2/\alpha_{21} < K_1$.

Additional Applications

14. If we assume that a damping force acts in a direction opposite to the motion of a pendulum and with a magnitude directly proportional to the angular velocity $d\theta/dt$, then the displacement angle θ for the pendulum satisfies the second-order nonlinear differential equation

$$ml\frac{d^2\theta}{dt^2} = -mg\sin\theta - \beta\frac{d\theta}{dt}$$

   (a) Write the second-order differential equation as a plane autonomous system and find all critical points.

   (b) Find a condition on m, l, and β that will make $(0, 0)$ a stable spiral point.

15. In the analysis of free damped motion in Section 3.9 we assumed that the damping force was proportional to the velocity x'. Frequently this damping force is proportional to the square of the velocity, and the new differential equation becomes

$$x'' = -\frac{\beta}{m}x'|x'| - \frac{k}{m}x$$

   (a) Write the second-order differential equation as a plane autonomous system and find all critical points.

   (b) The system is called *overdamped* when $(0, 0)$ is a stable node and is called *underdamped* when $(0, 0)$ is a stable spiral point. Physical considerations suggest that $(0, 0)$ must be an asymptotically stable critical point. Show that the system is necessarily underdamped. [*Hint:* $\dfrac{d}{dy}(y|y|) = 2|y|$]

16. A bead with mass m slides along a thin wire whose shape may be described by the function $z = f(x)$. Small stretches of the wire act like an inclined plane, and in mechanics it is assumed that the magnitude of the frictional force between the bead and the wire is directly proportional to $mg\cos\theta$ (see Figure 10.23).

(a) Show the new differential equation for the x-coordinate of the bead is

$$x'' = g\frac{\mu - f'(x)}{1 + [f'(x)]^2} - \frac{\beta}{m}x'$$

for some positive constant μ.

(b) Characterize the critical points of the corresponding plane autonomous system. Under what conditions is a critical point a saddle point? a stable spiral point?

17. An undamped oscillation satisfies a second-order nonlinear differential equation of the form $x'' + f(x) = 0$, where $f(0) = 0$ and $xf(x) > 0$ for $x \neq 0$ and $-d < x < d$. Use the phase-plane method to show that it is not possible for the critical point $(0, 0)$ to be a stable spiral point. [*Hint:* Let $F(x) = \int_0^x f(u)\,du$ and show that $y^2 + 2F(x) = c$.]

18. The Lotka–Volterra predator-prey model assumes that in the absence of predators, the number of prey grows exponentially. If we make the alternative assumption that the prey population grows logistically, the new plane autonomous system is

$$x' = -ax + bxy$$

$$y' = -cxy + \frac{r}{K}y(K - y)$$

where a, b, c, r, and K are positive and $K > a/b$.

(a) Show that the system has critical points at $(0, 0)$, $(0, K)$, and $(\hat{x}, \hat{y})$, where

$$\hat{y} = a/b \quad \text{and} \quad c\hat{x} = \left(\frac{r}{K}\right)(K - \hat{y})$$

(b) Show that the critical points at $(0, 0)$ and $(0, K)$ are saddle points, whereas the critical point at $(\hat{x}, \hat{y})$ is either a stable node or a stable spiral point.

(c) Show that $(\hat{x}, \hat{y})$ is a stable spiral point if

$$\hat{y} < \frac{4bK^2}{r + 4bK}.$$ Explain why this case will occur when the carrying capacity K of the prey is large.

19. The plane autonomous system

$$x' = \alpha\frac{y}{1+y}x - x$$

$$y' = -\frac{y}{1+y}x - y + \beta$$

arises in a model for the growth of microorganisms in a chemostat, a simple laboratory device in which a nutrient from a supply source flows into a growth chamber. In the system, x denotes the concentration of the microorganisms in the growth chamber, y denotes the concentration of nutrients, and α and β are positive constants that can be adjusted by the experimenter. Show that when $\alpha > 1$ and $\beta(\alpha - 1) > 1$, the system has a unique critical point $(\hat{x}, \hat{y})$ in the first quadrant, and demonstrate that this critical point is a stable node.

[O] 10.5 PERIODIC SOLUTIONS, LIMIT CYCLES, AND GLOBAL STABILITY

In this section we will investigate the existence of periodic solutions of nonlinear plane autonomous systems and introduce special periodic solutions called limit cycles.

We saw in Sections 10.2 and 10.3 that an analysis of critical points using linearization can provide valuable information on the behavior of solutions near critical points and insight into a variety of biological and physical phenomena. There are, however, some inherent limitations to this approach. When the eigenvalues of the Jacobian matrix are pure imaginary, we cannot conclude that there are periodic solutions near the critical point. In some cases we were able to solve $dy/dx = Q(x, y)/P(x, y)$, obtain an implicit representation $f(x, y) = c$ of the solution curves, and investigate whether any of these solutions formed closed curves. More often than not, this differential

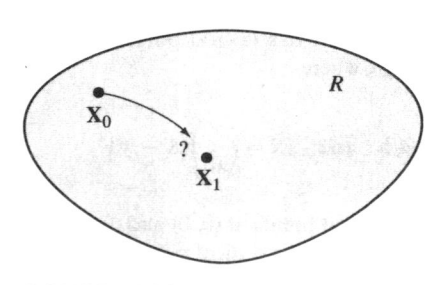

FIGURE 10.32

equation will not possess closed form solutions. For example, the Lotka-Volterra competition model cannot be handled by this procedure. The first goal of this section is therefore:

Goal 1: To determine conditions under which we can either exclude the possibility of periodic solutions or assert their existence.

We encountered an additional problem in studying the models in Section 10.4. Figure 10.32 illustrates the common situation in which a region R contains a *single* asymptotically stable critical point X_1. We can assert that $\lim_{t \to \infty} X(t) = X_1$ when the initial position $X(0) = X_0$ is "near" X_1, but under what conditions is $\lim_{t \to \infty} X(t) = X_1$ for *all initial positions* in R? Such a critical point is called **globally stable** in R. A second goal is therefore:

Goal 2: To determine conditions under which an asymptotically stable critical point is globally stable.

In motivating and discussing the methods in this section, we will use the fact that the vector field $V(x, y) = (P(x, y), Q(x, y))$ can be interpreted as defining a fluid flow in a region of the plane, and a solution to the autonomous system may be interpreted as the resulting path of a particle as it moves through the region.

Negative Criteria A number of results can sometimes be used to establish that there are no periodic solutions in a given region R of the plane. We will assume that $P(x, y)$ and $Q(x, y)$ have continuous first partial derivatives in R and that R is simply connected. Recall that in a simply connected region, any simple closed curve C in R encloses only points in R. Therefore, if there is a periodic solution $X = X(t)$ in R, then R will contain all points in the interior of the resulting curve.

> **THEOREM 10.4   Cycles and Critical Points**
>
> If a plane autonomous system has a periodic solution $X = X(t)$ in a simply connected region R, then the system has at least one critical point inside the corresponding simple closed curve C. If there is a single critical point inside C, then that critical point cannot be a saddle point.

> **COROLLARY**
>
> If a simply connected region R either contains no critical points of a plane autonomous system or contains a single saddle point, then there are no periodic solutions in R.

10.5 Periodic Solutions, Limit Cycles, and Global Stability

EXAMPLE 1 Show that the plane autonomous system

$$x' = xy$$
$$y' = -1 - x^2 - y^2$$

has no periodic solutions.

Solution If (x, y) is a critical point, then, from the first equation, either $x = 0$ or $y = 0$. If $x = 0$, then $-1 - y^2 = 0$ or $y^2 = -1$. Likewise, $y = 0$ implies $x^2 = -1$. Therefore, this plane autonomous system has no critical points, and by the corollary possesses no periodic solutions in the plane. □

EXAMPLE 2 Show that the Lotka–Volterra competition model

$$x' = 0.004x(50 - x - 0.75y)$$
$$y' = 0.001y(100 - y - 3.0x)$$

has no periodic solutions in the first quadrant.

Solution In Example 4 in Section 10.4 we showed that this system has critical points at $(0, 0)$, $(50, 0)$, $(0, 100)$, and $(20, 40)$, and that $(20, 40)$ is a saddle point. Since only $(20, 40)$ lies in the first quadrant, by the corollary there are no periodic solutions in the first quadrant. □

Another sometimes useful result can be formulated in terms of the divergence of the vector field $\mathbf{V}(x, y) = (P(x, y), Q(x, y))$:

THEOREM 10.5 Bendixson Negative Criterion

If $\operatorname{div} \mathbf{V} = \partial P/\partial x + \partial Q/\partial y$ does not change sign in a simply connected region R, then the plane autonomous system has no periodic solutions in R.

Proof Suppose, to the contrary, that there is a periodic solution $\mathbf{X} = \mathbf{X}(t)$ lying in R, and let C be the resulting simple closed curve and R_1 the region bounded by C. Green's theorem states that

$$\int_C M(x, y)\, dx + N(x, y)\, dy = \iint_{R_1} \left(\frac{\partial N}{\partial x} - \frac{\partial M}{\partial y} \right) dx\, dy$$

whenever $M(x, y)$ and $N(x, y)$ have continuous first partials in R. If we let $N = P$ and $M = -Q$, we obtain

$$\int_C -Q(x, y)\, dx + P(x, y)\, dy = \iint_{R_1} \left(\frac{\partial P}{\partial x} + \frac{\partial Q}{\partial y} \right) dx\, dy$$

Since $\mathbf{X} = \mathbf{X}(t)$ is a solution with period p, we have $x'(t) = P(x(t), y(t))$ and $y'(t) = Q(x(t), y(t))$, and so

$$\int_C -Q(x, y)\, dx + P(x, y)\, dy = \int_0^p [-Q(x(t), y(t))x'(t) + P(x(t), y(t))y'(t)]\, dt$$

$$= \int_0^p [-QP + PQ]\, dt = 0$$

Since div $\mathbf{V} = \partial P/\partial x + \partial Q/\partial y$ is continuous and does not change sign in R, it follows that either div $\mathbf{V} \geq 0$ in R or div $\mathbf{V} \leq 0$ in R, and so

$$\iint_{R_1} \left(\frac{\partial P}{\partial x} + \frac{\partial Q}{\partial y} \right) dx\, dy \neq 0$$

This contradiction establishes that there are no periodic solutions in R. ■

EXAMPLE 3 Use the Bendixson negative criterion to investigate possible periodic solutions of each plane autonomous system.

(a) $x' = x + 2y + 4x^3 - y^2$
$\ y' = -x + 2y + yx^2 + y^3$

(b) $x' = y + x(2 - x^2 - y^2)$
$\ y' = -x + y(2 - x^2 - y^2)$

Solution (a) We have div $\mathbf{V} = \partial P/\partial x + \partial Q/\partial y = 1 + 12x^2 + 2 + x^2 + 3y^2 \geq 3$, and so there are no periodic solutions in the plane.

(b) For this system div $\mathbf{V} = (2 - 3x^2 - y^2) + (2 - x^2 - 3y^2) = 4 - 4(x^2 + y^2)$. Therefore, if R is the interior of the circle $x^2 + y^2 = 1$, div $\mathbf{V} > 0$ and so there are no periodic solutions inside this disk. Note that div $\mathbf{V} < 0$ on the exterior of the circle. If R is any simply connected subset of the exterior, then there are no periodic solutions in R. It follows that if there is a periodic solution in the exterior, it must enclose the circle $x^2 + y^2 = 1$. In fact, the reader can verify that $\mathbf{X}(t) = (\sqrt{2} \sin t, \sqrt{2} \cos t)$ is a periodic solution which generates the circle $x^2 + y^2 = 2$. □

EXAMPLE 4 The sliding bead discussed in Section 10.4 satisfies the second-order nonlinear equation

$$mx'' = -mg \frac{f'(x)}{1 + [f'(x)]^2} - \beta x'$$

Show that there are no periodic solutions.

Solution The corresponding plane autonomous system is

$$x' = y$$

$$y' = -g \frac{f'(x)}{1 + [f'(x)]^2} - \frac{\beta}{m} y$$

and so div $\mathbf{V} = \dfrac{\partial P}{\partial x} + \dfrac{\partial Q}{\partial y} = -\dfrac{\beta}{m} < 0.$ □

10.5 Periodic Solutions, Limit Cycles, and Global Stability

The following theorem is a generalization of the Bendixson negative criterion, which leaves it to the reader to construct an appropriate function $\delta(x, y)$:

THEOREM 10.6 Dulac Negative Criterion

If $\delta(x, y)$ has continuous first partial derivatives in a simply connected region R and $\dfrac{\partial(\delta P)}{\partial x} + \dfrac{\partial(\delta Q)}{\partial y}$ does not change sign in R, then the plane autonomous system has no periodic solutions in R.

There are no general techniques for constructing an appropriate function $\delta(x, y)$. Instead, we experiment with simple functions of the form $ax^2 + by^2$, e^{ax+by}, $x^a y^b$, and so on, and try to determine constants for which $\partial(\delta P)/\partial x + \partial(\delta Q)/\partial y$ is nonzero in a given region.

EXAMPLE 5 Use the Dulac negative criterion to show that the second-order nonlinear differential equation

$$x'' = x^2 + (x')^2 - x - x'$$

has no periodic solutions.

Solution The corresponding plane autonomous system is

$$x' = y$$
$$y' = x^2 + y^2 - x - y$$

If we let $\delta(x, y) = e^{ax+by}$, then

$$\frac{\partial(\delta P)}{\partial x} + \frac{\partial(\delta Q)}{\partial y} = e^{ax+by}(ay + 2y - 1) + e^{ax+by}b(x^2 + y^2 - x - y)$$

If we set $a = -2$ and $b = 0$, then $\partial(\delta P)/\partial x + \partial(\delta Q)/\partial y = -e^{ax+by}$, which is always negative. Therefore, by the Dulac negative criterion, the second-order differential equation has no periodic solutions. □

EXAMPLE 6 Use $\delta(x, y) = 1/(xy)$ in the Dulac negative criterion to show that the Lotka–Volterra competition equations

$$x' = \frac{r_1}{K_1} x(K_1 - x - \alpha_{12} y)$$

$$y' = \frac{r_2}{K_2} y(K_2 - y - \alpha_{21} x)$$

have no periodic solutions in the first quadrant.

Solution If $\delta(x, y) = 1/(xy)$, then

$$\delta P = \frac{r_1}{K_1}\left[\frac{K_1}{y} - \frac{x}{y} - \alpha_{12}\right]$$

$$\delta Q = \frac{r_2}{K_2}\left[\frac{K_2}{x} - \frac{y}{x} - \alpha_{21}\right]$$

and so

$$\frac{\partial(\delta P)}{\partial x} + \frac{\partial(\delta Q)}{\partial y} = \frac{r_1}{K_1}\left(-\frac{1}{y}\right) + \frac{r_2}{K_2}\left(-\frac{1}{x}\right)$$

For (x, y) in the first quadrant, the latter expression is always negative. Therefore, there are no periodic solutions. □

Positive Criteria: Poincaré–Bendixson Theory The Poincaré–Bendixson theorem is an advanced result that describes the long-range behavior of a *bounded* solution to a plane autonomous system. Rather than present the result in its full generality, we will concentrate on a number of special cases that occur frequently in applications. One of these cases will lead to a new type of periodic solution called a *limit cycle*.

DEFINITION 10.3 Invariant Region

A region R is called an **invariant region** for a plane autonomous system if, whenever $\mathbf{X}_0$ is in R, the solution $\mathbf{X} = \mathbf{X}(t)$ satisfying $\mathbf{X}(0) = \mathbf{X}_0$ remains in R.

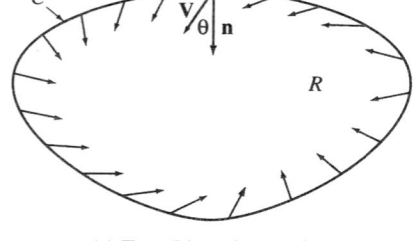

Figure 10.33 shows two standard types of invariant regions. A **Type I invariant region** R is bounded by a simple closed curve C, and the flow at the boundary defined by the vector field $\mathbf{V}(x, y) = (P(x, y), Q(x, y))$ is always directed into the region. This prevents a particle from crossing the boundary. A **Type II invariant region** is an annular region bounded by simple closed curves C_1 and C_2, and the flow at the boundary is again directed toward the interior of R. The following theorem provides a method for verifying that a given region is invariant:

(a) Type I invariant region

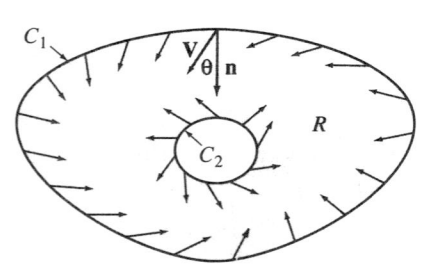

(b) Type II invariant region

FIGURE 10.33

THEOREM 10.7 Normal Vectors and Invariant Regions

If $\mathbf{n}(x, y)$ denotes a normal vector on the boundary that points inside the region, then R will be an invariant region for the plane autonomous system provided $\mathbf{V}(x, y) \cdot \mathbf{n}(x, y) \geq 0$ for all points (x, y) on the boundary.

10.5 Periodic Solutions, Limit Cycles, and Global Stability

Proof If θ is the angle between $\mathbf{V}(x, y)$ and $\mathbf{n}(x, y)$, then from $\mathbf{V} \cdot \mathbf{n} = \|\mathbf{V}\|\|\mathbf{n}\| \cos \theta$, we may conclude that $\cos \theta \geq 0$ and so θ is between $0°$ and $90°$. The flow is therefore directed into the region (or at worst along the boundary) for any boundary point (x, y). This prevents a solution that starts in R from leaving R. Therefore, R is an invariant region for the plane autonomous system. ∎

The problem of finding an invariant region for a given nonlinear system is an extremely difficult one. An excellent first step is to use software that plots the vector field $\mathbf{V}(x, y) = (P(x, y), Q(x, y))$ together with the curves $P(x, y) = 0$ (along which the vectors are vertical) and $Q(x, y) = 0$ (along which the vectors are horizontal). This can lead to choices for R. In the following examples, we will construct invariant regions bounded by lines and circles. In more complicated cases, we will be content to offer empirical evidence that an invariant region exists.

EXAMPLE 7 Find a circular region with center at $(0, 0)$ that serves as an invariant region for the plane autonomous system

$$x' = -y - x^3$$
$$y' = x - y^3$$

Solution For the circle $x^2 + y^2 = r^2$, $\mathbf{n} = (-2x, -2y)$ is a normal vector that points toward the interior of the circle. Since

$$\mathbf{V} \cdot \mathbf{n} = (-y - x^3, x - y^3) \cdot (-2x, -2y) = 2(x^4 + y^4)$$

we may conclude that $\mathbf{V} \cdot \mathbf{n} \geq 0$ on the circle $x^2 + y^2 = r^2$. Therefore, by Theorem 10.7 the circular region defined by $x^2 + y^2 \leq r^2$ serves as an invariant region for the system for any $r > 0$. □

EXAMPLE 8 Find an annular region bounded by circles that serves as an invariant region for the plane autonomous system

$$x' = x - y - 5x(x^2 + y^2) + x^5$$
$$y' = x + y - 5y(x^2 + y^2) + y^5$$

Solution As in Example 7, the normal vector $\mathbf{n}_1 = (-2x, -2y)$ points inside the circle $x^2 + y^2 = r^2$, while the normal vector $\mathbf{n}_2 = -\mathbf{n}_1$ is directed toward the exterior. Computing $\mathbf{V} \cdot \mathbf{n}_1$ and simplifying, we obtain

$$\mathbf{V} \cdot \mathbf{n}_1 = -2(r^2 - 5r^4 + x^6 + y^6)$$

Note that $r^2 - 5r^4 = r^2(1 - 5r^2)$ takes on both positive and negative values.

If $r = 1$, $\mathbf{V} \cdot \mathbf{n}_1 = 8 - 2(x^6 + y^6) \geq 0$, since the maximum value of $x^6 + y^6$ on the circle $x^2 + y^2 = 1$ is 1. The flow is therefore directed toward the interior of the circular region $x^2 + y^2 \leq 1$.

If $r = \frac{1}{4}$, $\mathbf{V} \cdot \mathbf{n}_1 \leq -2(r^2 - 5r^4) < 0$, and so $\mathbf{V} \cdot \mathbf{n}_2 = -\mathbf{V} \cdot \mathbf{n}_1 > 0$. The flow is therefore directed toward the exterior of the circle $x^2 + y^2 = \frac{1}{16}$, and so the annular region R defined by $\frac{1}{16} \leq x^2 + y^2 \leq 1$ is an invariant region for the system. □

EXAMPLE 9 The Van der Pol equation is a nonlinear second-order differential equation that arises in electronics, and as a plane autonomous system it takes the form

$$x' = y$$
$$y' = -\mu(x^2 - 1)y - x$$

Figure 10.34 shows the corresponding vector field for $\mu = 1$, together with the curves $y = 0$ and $(x^2 - 1)y = -x$ along which the vectors are vertical and horizontal, respectively. (For convenience we have sketched the normalized vector field $\mathbf{V}/\|\mathbf{V}\|$). It is not possible to find a simple invariant region whose boundary consists of lines or circles. The figure does offer empirical evidence that an invariant region R, with $(0, 0)$ in its interior, does exist. Advanced methods are required to demonstrate this mathematically*. □

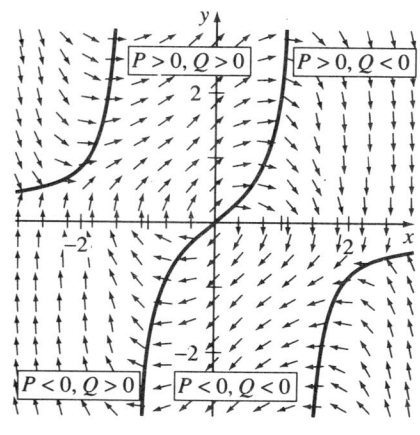

FIGURE 10.34

We next present two important special cases of the Poincaré–Bendixson theorem that guarantee the existence of periodic solutions.

THEOREM 10.8 Poincaré–Bendixson I

Let R be an invariant region for a plane autonomous system and suppose that R has no critical points on its boundary.

(i) If R is a Type I region that has a single unstable node or an unstable spiral point in its interior, then there is at least one periodic solution in R.

(ii) If R is a Type II region that contains no critical points of the system, then there is at least one periodic solution in R.

In either of the two cases, if $\mathbf{X} = \mathbf{X}(t)$ is a nonperiodic solution in R, then $\mathbf{X}(t)$ spirals toward a cycle which is a solution to the system. This periodic solution is called a **limit cycle**.

The flow interpretation portrayed in Figure 10.33 can be used to make the result plausible. If a particle is released at a point $\mathbf{X}_0$ in a Type II invariant region R, then, with no escape from the region and with no resting points,

---

*See M. Hirsch and S. Smale, *Differential Equations, Dynamical Systems, and Linear Algebra* (New York: Academic Press, 1974).

10.5 Periodic Solutions, Limit Cycles, and Global Stability

the particle will begin to rotate around the boundary C_2 and settle into a periodic orbit. It is not possible for the particle to return to an earlier position unless the solution is itself periodic.

EXAMPLE 10 Use Theorem 10.8 to show that the system

$$x' = -y + x(1 - x^2 - y^2) - y(x^2 + y^2)$$
$$y' = x + y(1 - x^2 - y^2) + x(x^2 + y^2)$$

has at least one periodic solution.

Solution We first construct an invariant region which is bounded by circles. If $\mathbf{n}_1 = (-2x, -2y)$, then $\mathbf{V} \cdot \mathbf{n}_1 = -2r^2(1 - r^2)$. If we let $r = 2$ and then $r = \frac{1}{2}$, we may conclude that the annular region R defined by $\frac{1}{4} \leq x^2 + y^2 \leq 4$ is invariant. If (x_1, y_1) is a critical point of the system, then $\mathbf{V} \cdot \mathbf{n}_1 = (0, 0) \cdot \mathbf{n}_1 = 0$. Therefore, $r = 0$ or $r = 1$. If $r = 0$, then $(x_1, y_1) = (0, 0)$ is a critical point. If $r = 1$, the system reduces to

$$-2y = 0$$
$$2x = 0$$

and we have reached a contradiction. Therefore, $(0, 0)$ is the only critical point, and this critical point is not in R. By part (*ii*) of Theorem 10.8, the system has at least one periodic solution in R.

The reader can verify that $\mathbf{X}(t) = (\cos 2t, \sin 2t)$ is a periodic solution. □

EXAMPLE 11 Show that the Van der Pol differential equation

$$x'' + \mu(x^2 - 1)x' + x = 0$$

has a periodic solution when $\mu > 0$.

Solution We will assume that there is a Type I invariant region R for the corresponding plane autonomous system and that this region contains $(0, 0)$ in its interior (see Example 9 and Figure 10.35). The only critical point is $(0, 0)$, and the Jacobian matrix is given by

$$\mathbf{g}'((0, 0)) = \begin{bmatrix} 0 & 1 \\ -1 & \mu \end{bmatrix}$$

Therefore, $\tau = \mu$, $\Delta = 1$, and $\tau^2 - 4\Delta = \mu^2 - 4$. Since $\mu > 0$, the critical point is either an unstable spiral point or an unstable node. By part (*i*) of Theorem 10.8, the system has at least one periodic solution in R. Figure 10.35 shows solutions corresponding to $\mathbf{X}(0) = (0.5, 0.5)$ and $\mathbf{X}(0) = (3, 3)$ for $\mu = 1$. Each of these solutions spirals around the origin and approaches a limit cycle. It can be shown that the Van der Pol differential equation has a unique limit cycle for all values of the parameter μ. □

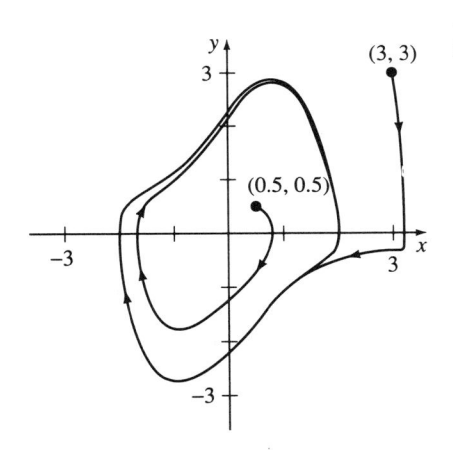

FIGURE 10.35

Global Stability Another version of the Poincaré–Bendixson theorem can be used to show that a locally stable critical point is globally stable:

THEOREM 10.9 Poincaré–Bendixson II

Let R be a Type I invariant region for a plane autonomous system which has *no* periodic solutions in R.

(i) If R has a finite number of nodes or spiral points, then given any initial position $\mathbf{X}_0$ in R, $\lim_{t \to \infty} \mathbf{X}(t) = \mathbf{X}_1$ for some critical point $\mathbf{X}_1$.

(ii) If R has a *single* stable node or stable spiral point $\mathbf{X}_1$ in its interior and no critical points on its boundary, then $\lim_{t \to \infty} \mathbf{X}(t) = \mathbf{X}_1$ for all initial positions $\mathbf{X}_0$ in R.

In Theorem 10.9 the particle cannot escape from R, cannot return to any of its prior positions, and therefore, in the absence of cycles, must be attracted to some stable critical point $\mathbf{X}_1$.

EXAMPLE 12 Investigate global stability for the system from Example 7:

$$x' = -y - x^3$$
$$y' = x - y^3$$

Solution In Example 7 we showed that the circular region defined by $x^2 + y^2 \leq r^2$ serves as an invariant region for the system for any $r > 0$. Since $\partial P/\partial x + \partial Q/\partial y = -3x^2 - 3y^2$ does not change sign, there are no periodic

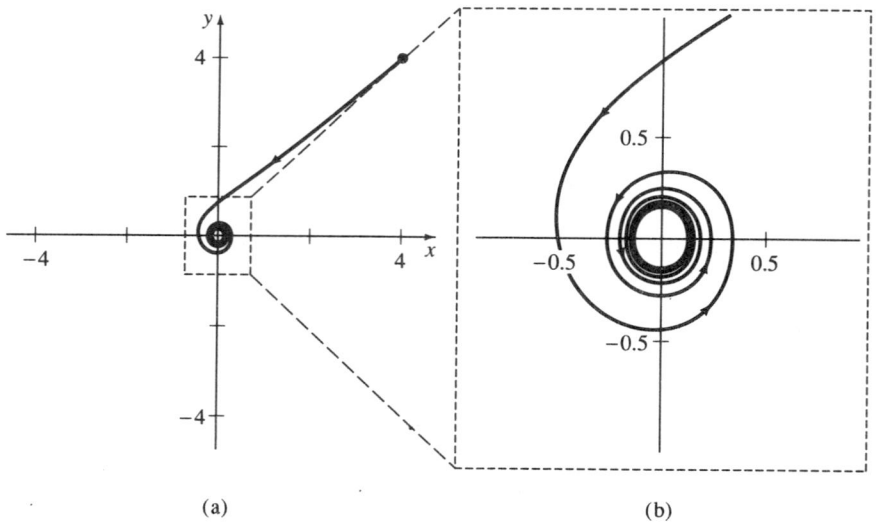

FIGURE 10.36

solutions by the Bendixson negative criterion. It is not hard to show that (0, 0) is the only critical point and that the Jacobian matrix is

$$\mathbf{g}'((0, 0)) = \begin{bmatrix} 0 & -1 \\ 1 & 0 \end{bmatrix}$$

Since $\tau = 0$ and $\Delta = 1$, (0, 0) may be either a stable or an unstable spiral (it cannot be a center). Theorem 10.9, however, guarantees that $\lim_{t \to \infty} \mathbf{X}(t) = \mathbf{X}_1$ for some critical point $\mathbf{X}_1$. Since (0, 0) is the only critical point, we must have $\lim_{t \to \infty} \mathbf{X}(t) = (0, 0)$ for *any* initial position $\mathbf{X}_0$ in the plane. The critical point is therefore a globally stable spiral point. Figure 10.36 shows two views of the solution satisfying $\mathbf{X}(0) = (4, 4)$. Part (b) is an enlarged view of the curve around (0, 0). Notice how slowly the solution spirals toward (0, 0). □

EXERCISES 10.5 Answers to odd-numbered problems begin on page A-60.

In Problems 1–8 show that the given plane autonomous system (or second-order differential equation) has no periodic solutions.

1. $x' = 2 + xy$
   $y' = x - y$

2. $x' = 2x - xy$
   $y' = -1 - x^2 + 2x - y^2$

3. $x' = -x + y^2$
   $y' = x - y$

4. $x' = xy^2 - x^2 y$
   $y' = x^2 y - 1$

5. $x' = -\mu x - y$
   $y' = x + y^3$ for $\mu < 0$

6. $x' = 2x + y^2$
   $y' = xy - y$

7. $x'' - 2x + (x')^4 = 0$

8. $x'' + x = [\frac{1}{2} + 3(x')^2]x' - x^2$

In Problems 9 and 10 use the Dulac negative criterion to show that the given plane autonomous system has no periodic solutions. Experiment with simple functions of the form $\delta(x, y) = ax^2 + by^2$, e^{ax+by}, or $x^a y^b$.

9. $x' = -2x + xy$
   $y' = 2y - x^2$

10. $x' = -x^3 + 4xy$
    $y' = -5x^2 - y^2$

11. Show that the plane autonomous system

    $$x' = x(1 - x^2 - 3y^2)$$
    $$y' = y(3 - x^2 - 3y^2)$$

    has no periodic solutions in an elliptical region about the origin.

12. If $\partial g/\partial x' \neq 0$ in a region R, prove that $x'' = g(x, x')$ has no periodic solutions in R.

13. Show that the predator-prey model

    $$x' = -ax + bxy$$
    $$y' = -cxy + \frac{r}{K} y(K - y)$$

    presented in Problem 18 of Exercises 10.4 has no periodic solutions in the first quadrant.

In Problems 14 and 15 find a circular invariant region for the given plane autonomous system.

14. $x' = -y - xe^{x+y}$
    $y' = x - ye^{x+y}$

15. $x' = -x + y + xy$
    $y' = x - y - x^2 - y^3$

16. Verify that the region bounded by the closed curve $x^6 + 3y^2 = 1$ is an invariant region for the second-order nonlinear differential equation $x'' + x' = -(x')^3 - x^5$. See Figure 10.37 on page 718.

17. The plane autonomous system in Example 8 has only one critical point. Can we conclude that this system has at least one periodic solution?

18. Use the Poincaré–Bendixson theorem to show that the second-order nonlinear differential equation $x'' = x'[1 - 3x^2 - 2(x')^2] - x$ has at least one periodic solution. [*Hint*: Find an invariant annular region for the corresponding plane autonomous system.]

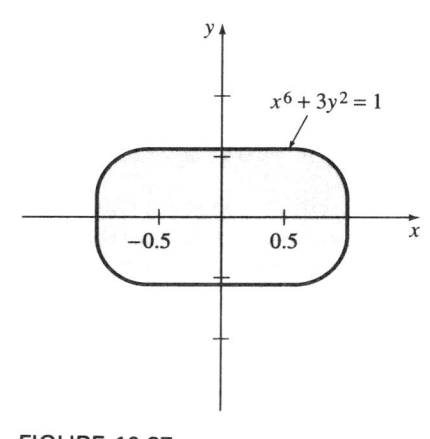

FIGURE 10.37

$x^6 + 3y^2 = 1$

-0.5 0.5

19. Let $\mathbf{X} = \mathbf{X}(t)$ be the solution of the plane autonomous system

$$x' = y$$
$$y' = -x - (1-x^2)y$$

that satisfies $\mathbf{X}(0) = (x_0, y_0)$. Show that if $x_0^2 + y_0^2 < 1$, then $\lim_{t \to \infty} \mathbf{X}(t) = (0, 0)$. [*Hint*: Select $r < 1$ with $x_0^2 + y_0^2 < r^2$ and first show that the circular region R defined by $x^2 + y^2 \leq r^2$ is an invariant region.]

20. Investigate global stability for the system

$$x' = y - x$$
$$y' = -x - y^3$$

21. Empirical evidence suggests that the plane autonomous system

$$x' = x^2 y - x + 1$$
$$y' = -x^2 y + \tfrac{1}{2}$$

has a Type I invariant region R which lies inside the rectangle $0 \leq x \leq 2, 0 \leq y \leq 1$.

(a) Use the Bendixson negative criterion to show that there are no periodic solutions in R.

(b) If $\mathbf{X}_0$ is in R and $\mathbf{X} = \mathbf{X}(t)$ is the solution satisfying $\mathbf{X}(t) = \mathbf{X}_0$, use Theorem 10.9 to find $\lim_{t \to \infty} \mathbf{X}(t)$.

22. (a) Find and classify all critical points of the plane autonomous system

$$x' = x\left(\frac{2y}{y+2} - 1\right)$$
$$y' = y\left(1 - \frac{2x}{y+2} - \frac{y}{8}\right)$$

(b) Figure 10.38 shows the vector field $\mathbf{V}/\|\mathbf{V}\|$ and offers empirical evidence that there is an invariant region R in the first quadrant with a critical point in its interior. Assuming that such a region exists, prove that there is at least one periodic solution.

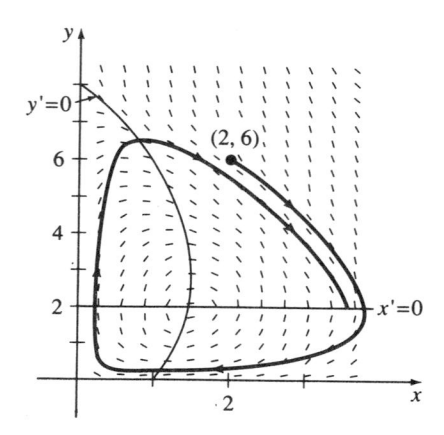

FIGURE 10.38

SUMMARY

The solutions $\mathbf{X}(t) = (x(t), y(t))$ of a **plane autonomous system**

$$\frac{dx}{dt} = P(x, y)$$
$$\frac{dy}{dt} = Q(x, y)$$

are either **critical points** (corresponding to constant solutions), **arcs** (in which the solution curve does not cross itself), or **cycles** (which correspond to periodic solutions). Critical points may be found by solving the system of alge-

10.5 Periodic Solutions, Limit Cycles, and Global Stability

braic equations

$$P(x, y) = 0$$
$$Q(x, y) = 0$$

whereas periodic solutions can sometimes be found by changing to polar coordinates.

When the plane autonomous system is the linear system $\mathbf{X}' = \mathbf{A}\mathbf{X}$ with $\det \mathbf{A} \neq 0$, a precise geometric description of solutions near the critical point $(0, 0)$ can be given once the eigenvalues and eigenvectors are known. The critical point is classified as a **stable or unstable node, stable or unstable spiral point, center**, or **saddle point**.

When the plane autonomous system is nonlinear, an isolated critical point $\mathbf{X}_1 = (x_1, y_1)$ can be classified by finding the eigenvalues of the Jacobian matrix

$$\mathbf{A} = \begin{bmatrix} \dfrac{\partial P}{\partial x}\bigg|_{(x_1, y_1)} & \dfrac{\partial P}{\partial y}\bigg|_{(x_1, y_1)} \\ \dfrac{\partial Q}{\partial x}\bigg|_{(x_1, y_1)} & \dfrac{\partial Q}{\partial y}\bigg|_{(x_1, y_1)} \end{bmatrix}$$

If the eigenvalues of $\mathbf{A}$ have negative real parts, then $\mathbf{X}_1$ is an **asymptotically stable** critical point. If $\mathbf{A}$ has an eigenvalue with a positive real part, then $\mathbf{X}_1$ is an **unstable** critical point.

In five separate cases (stable node, stable spiral point, unstable spiral point, unstable node, and saddle), the critical point of the nonlinear system can be categorized like the critical point in the corresponding linear system $\mathbf{X}' = \mathbf{A}\mathbf{X}$ and solutions close to the critical point have the same general geometric features as the solutions to the linear system. When the eigenvalues of $\mathbf{A}$ are pure imaginary, no conclusions can be made and further analysis is necessary.

The **phase-plane method** is designed to find y as a function of x by solving the first-order differential equation

$$\frac{dy}{dx} = \frac{Q(x, y)}{P(x, y)}$$

The method can sometimes be used to sketch solution curves and to decide whether a critical point is a stable spiral, an unstable spiral, or a center.

Plane autonomous systems can serve as mathematical models for such diverse physical phenomena as mechanical oscillations and predator-prey or competition interactions in ecology.

A periodic solution in a simply connected region must have at least one critical point inside the cycle it generates. The nonexistence of periodic solutions in a simply connected region can be investigated using the **Bendixson** and **Dulac negative criterion**.

A region R is called an **invariant region** for a plane autonomous system if, whenever $\mathbf{X}_0$ is in R, the solution $\mathbf{X} = \mathbf{X}(t)$ satisfying $\mathbf{X}(0) = \mathbf{X}_0$ remains in R. If $\mathbf{n}(x, y)$ denotes a normal vector on the boundary of R that points inside the region, then R will be an invariant region for the plane autonomous system provided $\mathbf{V}(x, y) \cdot \mathbf{n}(x, y) \geq 0$ for all points (x, y) on the boundary.

CHAPTER 10 REVIEW EXERCISES

Answers to odd-numbered problems begin on page A-60.

Answer Problems 1–10 without referring back to the text. Fill in the blank or answer true/false.

1. The second-order differential equation $x'' + f(x') + g(x) = 0$ can be written as a plane autonomous system. _____

2. If $\mathbf{X} = \mathbf{X}(t)$ is a solution of a plane autonomous system and $\mathbf{X}(t_1) = \mathbf{X}(t_2)$ for $t_1 \neq t_2$, then $\mathbf{X}(t)$ is a periodic solution. _____

3. If the trace of the matrix $\mathbf{A}$ is 0 and $\det \mathbf{A} \neq 0$, then the critical point $(0, 0)$ of the linear system $\mathbf{X}' = \mathbf{A}\mathbf{X}$ may be classified as _____.

4. If the critical point $(0, 0)$ of the linear system $\mathbf{X}' = \mathbf{A}\mathbf{X}$ is a stable spiral point, then the eigenvalues of $\mathbf{A}$ are _____.

5. If the critical point $(0, 0)$ of the linear system $\mathbf{X}' = \mathbf{A}\mathbf{X}$ is a saddle point and $\mathbf{X} = \mathbf{X}(t)$ is a solution, then $\lim_{t \to \infty} \mathbf{X}(t)$ does not exist. _____

6. If the Jacobian matrix $\mathbf{A} = \mathbf{g}'(\mathbf{X}_1)$ at a critical point of a plane autonomous system has positive trace and determinant, then the critical point $\mathbf{X}_1$ is unstable. _____

7. It is possible to show that a nonlinear plane autonomous system has periodic solutions using linearization. _____

8. All solutions to the pendulum equation
$$\frac{d^2\theta}{dt^2} + \frac{g}{l}\sin\theta = 0$$
are periodic. _____

9. If a simply connected region R contains no critical points of a plane autonomous system, then there are no periodic solutions in R. _____

10. If a plane autonomous system has no critical points in an annular invariant region R, then there is at least one periodic solution in R. _____

11. Solve the following nonlinear plane autonomous system by switching to polar coordinates, and describe the geometric behavior of the solution that satisfies the given initial condition:
$$\begin{aligned} x' &= -y - x(\sqrt{x^2+y^2})^3 \\ y' &= x - y(\sqrt{x^2+y^2})^3 \end{aligned}, \quad \mathbf{X}(0) = (1, 0)$$

12. Discuss the geometric nature of the solutions to the linear system $\mathbf{X}' = \mathbf{A}\mathbf{X}$ given the general solution:

   (a) $\mathbf{X}(t) = c_1 \begin{bmatrix} 1 \\ 1 \end{bmatrix} e^{-t} + c_2 \begin{bmatrix} 1 \\ -2 \end{bmatrix} e^{-2t}$

   (b) $\mathbf{X}(t) = c_1 \begin{bmatrix} 1 \\ -1 \end{bmatrix} e^{-t} + c_2 \begin{bmatrix} 1 \\ 2 \end{bmatrix} e^{2t}$

13. Classify the critical point $(0, 0)$ of the given linear system by computing the trace τ and determinant Δ:

   (a) $\begin{aligned} x' &= -3x + 4y \\ y' &= -5x + 3y \end{aligned}$ (b) $\begin{aligned} x' &= -3x + 2y \\ y' &= -2x + y \end{aligned}$

14. Find and classify (if possible) the critical points of the plane autonomous system
$$x' = x + xy - 3x^2$$
$$y' = 4y - 2xy - y^2$$
Does this system have any periodic solutions in the first quadrant?

15. Classify the critical point $(0, 0)$ of the plane autonomous system corresponding to the second-order nonlinear differential equation
$$x'' + \mu(x^2 - 1)x' + x = 0$$
where μ is a real constant.

16. Without solving explicitly, classify (if possible) the critical points of the first-order autonomous differential equation $x' = (x^2 - 1)e^{-x/2}$ as asymptotically stable or unstable.

17. Use the phase-plane method to show that the solutions of the second-order nonlinear differential equation $x'' = -2x\sqrt{(x')^2 + 1}$ that satisfy $x(0) = x_0$ and $x'(0) = 0$ are periodic.

18. (*Hard spring*) In Section 3.8 we assumed that the restoring force F of the spring satisfied Hookes' law $F = ks$, where s is the elongation of the spring and k is a positive constant of proportionality. If we replace this assumption with the nonlinear law $F = ks^3$, then the new differential equation for damped motion becomes $mx'' = -\beta x' - k(s + x)^3 + mg$, where $ks^3 = mg$. The system is called overdamped when $(0, 0)$ is a stable node and

is called underdamped when (0, 0) is a stable spiral point. Find new conditions on m, k, and β that will lead to overdamping and underdamping.

19. Show that the plane autonomous system

$$x' = 4x + 2y - 2x^2$$
$$y' = 4x - 3y + 4xy$$

has no periodic solutions.

20. Use the Poincaré–Bendixson theorem to show that the plane autonomous system

$$x' = \varepsilon x + y - x(x^2 + y^2)$$
$$y' = -x + \varepsilon y - y(x^2 + y^2)$$

has at least one periodic solution when $\varepsilon > 0$. What occurs when $\varepsilon < 0$?

21. (*The rotating pendulum*) The rod of a pendulum is attached to a movable joint at point P and rotates at an angular speed of ω (radians/s) in the plane perpendicular to the rod (see Figure 10.39). As a result, the bob of the pendulum experiences an additional centripetal force and the new differential equation for θ becomes

$$ml\frac{d^2\theta}{dt^2} = \omega^2 ml \sin\theta \cos\theta - mg \sin\theta - \beta\frac{d\theta}{dt}$$

(a) Establish that there are no periodic solutions.

(b) If $\omega^2 < g/l$, show that (0, 0) is a stable critical point and is the only critical point in the domain $-\pi < \theta < \pi$. Describe what occurs physically when $\theta(0) = \theta_0$, $\theta'(0) = 0$, and θ_0 is small.

(c) If $\omega^2 > g/l$, show that (0, 0) is unstable and there are two additional stable critical points $(\pm\hat{\theta}, 0)$ in the domain $-\pi < \theta < \pi$. Describe what occurs physically when $\theta(0) = \theta_0$, $\theta'(0) = 0$, and θ_0 is small.

(d) Determine under what conditions the critical points in parts (a) and (b) are stable spiral points.

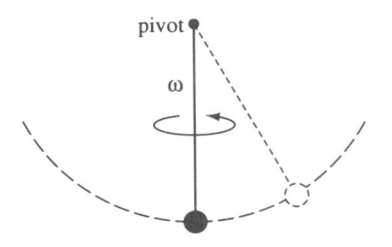

FIGURE 10.39

22. The nonlinear second-order differential equation $x'' - 2kx' + c(x')^3 + \omega^2 x = 0$ arises in modeling the motion of an electrically driven tuning fork. See Figure 10.40, where $k = c = 0.1$ and $\omega = 1$. Assume that this differential equation possesses a Type I invariant region that contains (0, 0). Show that there is at least one periodic solution.

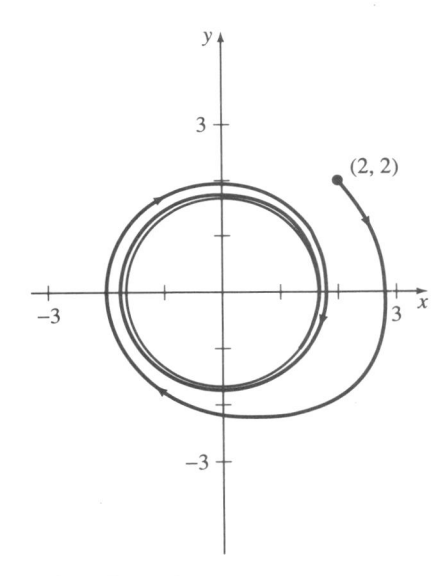

FIGURE 10.40

PART IV

Fourier Series and Boundary-Value Problems

11 ORTHOGONAL FUNCTIONS AND FOURIER SERIES

12 BOUNDARY-VALUE PROBLEMS IN RECTANGULAR COORDINATES

13 BOUNDARY-VALUE PROBLEMS IN OTHER COORDINATE SYSTEMS

14 INTEGRAL TRANSFORM METHOD

ORTHOGONAL FUNCTIONS AND FOURIER SERIES

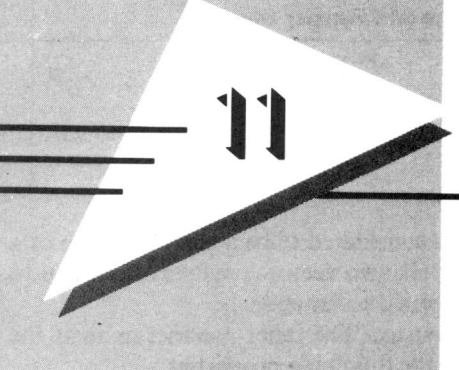

TOPICS TO REVIEW

integration by parts
general solution of $y'' + \lambda y = 0$ and $y'' - \lambda y = 0$ (3.3)
Bessel's differential equation (5.4)
Bessel functions (5.4)
Legendre's differential equation (5.4)
Legendre polynomials (5.4)

IMPORTANT CONCEPTS

orthogonal functions
orthogonal set
norm
square norm
orthogonality with respect to a weight function
generalized Fourier series
Fourier series
even function
odd function
cosine series
sine series
half-range expansions
eigenvalues
eigenfunctions
Sturm–Liouville problem
Fourier–Bessel series
Fourier–Legendre series

11.1 Orthogonal Functions
11.2 Fourier Series
11.3 Fourier Cosine and Sine Series
11.4 Sturm–Liouville Problem
11.5 Bessel and Legendre Series
Summary
Chapter 11 Review Exercises

INTRODUCTION

In this chapter we shall set the stage for the material in Chapters 12 and 13. Fundamental to the entire discussion are the two concepts of orthogonal functions and the expansion of a function in a series of orthogonal functions.

11.1 ORTHOGONAL FUNCTIONS

In advanced mathematics a function is considered to be a generalization of a vector. In this section we shall see how the two vector concepts of inner (dot) product and orthogonality can be extended to functions.

Suppose **u** and **v** are vectors in 3-space. The inner product (**u**, **v**) of the vectors, also written as **u · v**, possesses the following properties:

(i) (**u**, **v**) = (**v**, **u**)

(ii) (k**u**, **v**) = k(**u**, **v**),   k a scalar

(iii) (**u**, **u**) = 0 if **u** = **0**   and   (**u**, **u**) > 0 if **u** ≠ **0**

(iv) (**u** + **v**, **w**) = (**u**, **w**) + (**v**, **w**)

We expect that a generalization of the inner product concept should have these same properties.

Inner Product   Now suppose f_1 and f_2 are functions defined on an interval $[a, b]$.* Since a definite integral on the interval of the product $f_1(x)f_2(x)$ also possesses properties (i)–(iv) whenever the integrals exist, we are prompted to make the following definition.

DEFINITION 11.1   Inner Product

The **inner product** of two functions f_1 and f_2 on an interval $[a, b]$ is the number

$$(f_1, f_2) = \int_a^b f_1(x)f_2(x)\,dx$$

Orthogonal Functions   Motivated by the fact that two vectors **u** and **v** are orthogonal whenever their inner product is zero, we define **orthogonal functions** in a similar manner:

DEFINITION 11.2   Orthogonal Functions

Two functions f_1 and f_2 are said to be **orthogonal** on an interval $[a, b]$ if

$$(f_1, f_2) = \int_a^b f_1(x)f_2(x)\,dx = 0 \tag{1}$$

---

*The interval could also be $(-\infty, \infty)$, $[0, \infty)$, and so on.

11.1 Orthogonal Functions

Unlike vector analysis, where the word *orthogonal* is a synonym for *perpendicular*, in this context the term *orthogonal* and condition (1) have no geometric significance.

EXAMPLE 1 The functions $f_1(x) = x^2$ and $f_2(x) = x^3$ are orthogonal on the interval $[-1, 1]$ since

$$(f_1, f_2) = \int_{-1}^{1} f_1(x) f_2(x)\, dx$$

$$= \int_{-1}^{1} x^2 \cdot x^3\, dx = \frac{1}{6} x^6 \Big|_{-1}^{1}$$

$$= \frac{1}{6}[1 - (-1)^6] = 0 \qquad \square$$

Orthogonal Sets We are primarily interested in infinite sets of orthogonal functions.

DEFINITION 11.3 Orthogonal Set

A set of real-valued functions

$$\{\phi_0(x), \phi_1(x), \phi_2(x), \ldots\}$$

is said to be **orthogonal** on an interval $[a, b]$ if

$$(\phi_m, \phi_n) = \int_a^b \phi_m(x)\phi_n(x)\, dx = 0, \qquad m \ne n \qquad (2)$$

The norm, or length $\|\mathbf{u}\|$, of a vector $\mathbf{u}$ can be expressed in terms of the inner product—namely, $(\mathbf{u}, \mathbf{u}) = \|\mathbf{u}\|^2$ or $\|\mathbf{u}\| = \sqrt{(\mathbf{u}, \mathbf{u})}$. The **norm**, or generalized length, of a function ϕ_n is $\|\phi_n(x)\| = \sqrt{(\phi_n, \phi_n)}$; that is,

$$\|\phi_n(x)\| = \sqrt{\int_a^b \phi_n^2(x)\, dx}$$

The number
$$\|\phi_n(x)\|^2 = \int_a^b \phi_n^2(x)\, dx \qquad (3)$$

is called the **square norm** of ϕ_n. If $\{\phi_n(x)\}$ is an orthogonal set of functions on the interval $[a, b]$ with the property that $\|\phi_n(x)\| = 1$ for $n = 0, 1, 2, \ldots,$ then $\{\phi_n(x)\}$ is said to be an **orthonormal set** on the interval.

EXAMPLE 2 Show that the set $\{1, \cos x, \cos 2x, \ldots\}$ is orthogonal on the interval $[-\pi, \pi]$.

Solution If we make the identification $\phi_0(x) = 1$ and $\phi_n(x) = \cos nx$, we must then show that $\int_{-\pi}^{\pi} \phi_0(x)\phi_n(x)\, dx = 0$, $n \ne 0$, and $\int_{-\pi}^{\pi} \phi_m(x)\phi_n(x)\, dx = 0$,

$m \neq n$. We have, in the first case,

$$(\phi_0, \phi_n) = \int_{-\pi}^{\pi} \phi_0(x)\phi_n(x)\, dx$$

$$= \int_{-\pi}^{\pi} \cos nx\, dx$$

$$= \frac{1}{n} \sin nx \Big|_{-\pi}^{\pi}$$

$$= \frac{1}{n}[\sin n\pi - \sin(-n\pi)] = 0, \quad n \neq 0$$

and in the second,

$$(\phi_m, \phi_n) = \int_{-\pi}^{\pi} \phi_m(x)\phi_n(x)\, dx$$

$$= \int_{-\pi}^{\pi} \cos mx \cos nx\, dx \quad \text{trig identity}$$

$$= \frac{1}{2} \int_{-\pi}^{\pi} [\cos(m+n)x + \cos(m-n)x]\, dx$$

$$= \frac{1}{2}\left[\frac{\sin(m+n)x}{m+n} + \frac{\sin(m-n)x}{m-n}\right]_{-\pi}^{\pi} = 0, \quad m \neq n \quad \square$$

EXAMPLE 3 Find the norms of each function in the orthogonal set given in Example 2.

Solution For $\phi_0(x) = 1$ we have from (3),

$$\|\phi_0(x)\|^2 = \int_{-\pi}^{\pi} dx = 2\pi$$

so that $\|\phi_0(x)\| = \sqrt{2\pi}$. For $\phi_n(x) = \cos nx$, $n > 0$, it follows that

$$\|\phi_n(x)\|^2 = \int_{-\pi}^{\pi} \cos^2 nx\, dx$$

$$= \frac{1}{2} \int_{-\pi}^{\pi} [1 + \cos 2nx]\, dx = \pi$$

Thus, for $n > 0$, $\|\phi_n(x)\| = \sqrt{\pi}$. $\quad\square$

Any orthogonal set of nonzero functions $\{\phi_n(x)\}$, $n = 0, 1, 2, \ldots$, can be *normalized*—that is, made into an orthogonal set—by dividing each function by its norm.

EXAMPLE 4 It follows from Examples 2 and 3 that the set

$$\left\{\frac{1}{\sqrt{2\pi}}, \frac{\cos x}{\sqrt{\pi}}, \frac{\cos 2x}{\sqrt{\pi}}, \ldots\right\}$$

is orthonormal on $[-\pi, \pi]$. □

We shall make one more analogy between vectors and functions. Suppose $\mathbf{v}_1, \mathbf{v}_2$, and $\mathbf{v}_3$ are three mutually orthogonal nonzero vectors in 3-space. Such an orthogonal set can be used as a basis for 3-space; that is, any three-dimensional vector can be written as a linear combination

$$\mathbf{u} = c_1 \mathbf{v}_1 + c_2 \mathbf{v}_2 + c_3 \mathbf{v}_3 \tag{4}$$

where the c_i, $i = 1, 2, 3$, are scalars called the components of the vector. Each component c_i can be expressed in terms of $\mathbf{u}$ and the corresponding vector $\mathbf{v}_i$. To see this, we take the inner product of (4) with $\mathbf{v}_1$:

$$(\mathbf{u}, \mathbf{v}_1) = c_1(\mathbf{v}_1, \mathbf{v}_1) + c_2(\mathbf{v}_2, \mathbf{v}_1) + c_3(\mathbf{v}_3, \mathbf{v}_1)$$
$$= c_1 \|\mathbf{v}_1\|^2 + c_2 \cdot 0 + c_3 \cdot 0$$

Hence, $$c_1 = \frac{(\mathbf{u}, \mathbf{v}_1)}{\|\mathbf{v}_1\|^2}$$

In like manner we find that the components c_2 and c_3 are given by

$$c_2 = \frac{(\mathbf{u}, \mathbf{v}_2)}{\|\mathbf{v}_2\|^2} \quad \text{and} \quad c_3 = \frac{(\mathbf{u}, \mathbf{v}_3)}{\|\mathbf{v}_3\|^2}$$

Hence, (4) can be expressed as

$$\mathbf{u} = \frac{(\mathbf{u}, \mathbf{v}_1)}{\|\mathbf{v}_1\|^2} \mathbf{v}_1 + \frac{(\mathbf{u}, \mathbf{v}_2)}{\|\mathbf{v}_2\|^2} \mathbf{v}_2 + \frac{(\mathbf{u}, \mathbf{v}_3)}{\|\mathbf{v}_3\|^2} \mathbf{v}_3$$

or $$\mathbf{u} = \sum_{n=1}^{3} \frac{(\mathbf{u}, \mathbf{v}_n)}{\|\mathbf{v}_n\|^2} \mathbf{v}_n \tag{5}$$

Generalized Fourier Series Suppose $\{\phi_n(x)\}$ is an infinite orthogonal set of functions on an interval $[a, b]$. We ask: If $y = f(x)$ is a function defined on the interval $[a, b]$, is it possible to determine a set of coefficients c_n, $n = 0, 1, 2, \ldots$, for which

$$f(x) = c_0 \phi_0(x) + c_1 \phi_1(x) + \cdots + c_n \phi_n(x) + \cdots \tag{6}$$

As in the foregoing discussion on finding components of a vector, we can find the coefficients c_n by utilizing the inner product. Multiplying (6) by $\phi_m(x)$ and integrating over the interval $[a, b]$ give

$$\int_a^b f(x)\phi_m(x)\,dx = c_0 \int_a^b \phi_0(x)\phi_m(x)\,dx + c_1 \int_a^b \phi_1(x)\phi_m(x)\,dx + \cdots + c_n \int_a^b \phi_n(x)\phi_m(x)\,dx + \cdots$$
$$= c_0(\phi_0, \phi_m) + c_1(\phi_1, \phi_m) + \cdots + c_n(\phi_n, \phi_m) + \cdots$$

By orthogonality each term on the right-hand side of the last equation is zero *except* when $m = n$. In this case we have

$$\int_a^b f(x)\phi_n(x)\,dx = c_n \int_a^b \phi_n^2(x)\,dx$$

It follows that the required coefficients are

$$c_n = \frac{\int_a^b f(x)\phi_n(x)\,dx}{\int_a^b \phi_n^2(x)\,dx}, \qquad n = 0, 1, 2, \ldots$$

In other words,
$$f(x) = \sum_{n=0}^{\infty} c_n \phi_n(x) \tag{7}$$

where
$$c_n = \frac{\int_a^b f(x)\phi_n(x)\,dx}{\|\phi_n(x)\|^2} \tag{8}$$

With inner product notation, (7) becomes

$$f(x) = \sum_{n=0}^{\infty} \frac{(f, \phi_n)}{\|\phi_n(x)\|^2} \phi_n(x) \tag{9}$$

Thus, (9) is seen to be the functional analogue of the vector result given in (5).

DEFINITION 11.4 Orthogonal Set/Weight Function

A set of functions $\{\phi_n(x)\}$, $n = 0, 1, 2, \ldots$, is said to be **orthogonal with respect to a weight function** $w(x)$ on an interval $[a, b]$ if

$$\int_a^b w(x)\phi_m(x)\phi_n(x)\,dx = 0, \qquad m \neq n$$

The usual assumption is that $w(x) > 0$ on the interval of orthogonality $[a, b]$.

EXAMPLE 5 The set $\{1, \cos x, \cos 2x, \ldots\}$ is orthogonal with respect to the constant weight function $w(x) = 1$ on the interval $[-\pi, \pi]$. □

If $\{\phi_n(x)\}$ is orthogonal with respect to a weight function $w(x)$ on $[a, b]$, then multiplying (6) by $w(x)\phi_m(x)$ and integrating yield

$$c_n = \frac{\int_a^b f(x)w(x)\phi_n(x)\,dx}{\|\phi_n(x)\|^2} \tag{10}$$

where
$$\|\phi_n(x)\|^2 = \int_a^b w(x)\phi_n^2(x)\,dx \tag{11}$$

The series (7) with coefficients given by either (8) or (10) is called a **generalized Fourier series**.

Complete Sets We note that the procedure outlined for determining the c_n was *formal*; that is, basic questions on whether or not a series expansion such as (7) is actually possible were ignored. Also, to expand f in a series of orthogonal functions, it is certainly necessary that f not be orthogonal to each ϕ_n of the orthogonal set $\{\phi_n(x)\}$. (If f were orthogonal to every ϕ_n, then $c_n = 0$, $n = 0, 1, 2, \ldots$.) To avoid the latter problem we shall assume, for the remainder of the chapter, that an orthogonal set is **complete**. This means that the only function orthogonal to each member of the set is the zero function.

EXERCISES 11.1   Answers to odd-numbered problems begin on page A-60.

In Problems 1–6 show that the given functions are orthogonal on the indicated interval.

1. $f_1(x) = x$, $f_2(x) = x^2$; $[-2, 2]$
2. $f_1(x) = x^3$, $f_2(x) = x^2 + 1$; $[-1, 1]$
3. $f_1(x) = e^x$, $f_2(x) = xe^{-x} - e^{-x}$; $[0, 2]$
4. $f_1(x) = \cos x$, $f_2(x) = \sin^2 x$; $[0, \pi]$
5. $f_1(x) = x$, $f_2(x) = \cos 2x$; $[-\pi/2, \pi/2]$
6. $f_1(x) = e^x$, $f_2(x) = \sin x$; $[\pi/4, 5\pi/4]$

In Problems 7–12 show that the given set of functions is orthogonal on the indicated interval. Find the norm of each function in the set.

7. $\{\sin x, \sin 3x, \sin 5x, \ldots\}$; $[0, \pi/2]$
8. $\{\cos x, \cos 3x, \cos 5x, \ldots\}$; $[0, \pi/2]$
9. $\{\sin nx\}$, $n = 1, 2, 3, \ldots$; $[0, \pi]$
10. $\left\{\sin \dfrac{n\pi}{p} x\right\}$, $n = 1, 2, 3, \ldots$; $[0, p]$
11. $\left\{1, \cos \dfrac{n\pi}{p} x\right\}$, $n = 1, 2, 3, \ldots$; $[0, p]$
12. $\left\{1, \cos \dfrac{n\pi}{p} x, \sin \dfrac{m\pi}{p} x\right\}$, $n = 1, 2, 3, \ldots, m = 1, 2, 3, \ldots$; $[-p, p]$

In Problems 13 and 14 verify by direct integration that the functions are orthogonal with respect to the indicated weight function on the given interval.

13. $H_0(x) = 1$, $H_1(x) = 2x$, $H_2(x) = 4x^2 - 2$; $w(x) = e^{-x^2}$, $(-\infty, \infty)$
14. $L_0(x) = 1$, $L_1(x) = -x + 1$, $L_2(x) = \tfrac{1}{2}x^2 - 2x + 1$; $w(x) = e^{-x}$, $[0, \infty)$
15. Let $\{\phi_n(x)\}$ be an orthogonal set of functions on $[a, b]$ such that $\phi_0(x) = 1$. Show that $\int_a^b \phi_n(x)\,dx = 0$ for $n = 1, 2, \ldots$.
16. Let $\{\phi_n(x)\}$ be an orthogonal set of functions on $[a, b]$ such that $\phi_0(x) = 1$ and $\phi_1(x) = x$. Show that $\int_a^b (\alpha x + \beta)\phi_n(x)\,dx = 0$ for $n = 2, 3, \ldots$ and any constants α and β.
17. Let $\{\phi_n(x)\}$ be an orthogonal set of functions on $[a, b]$. Show that $\|\phi_m(x) + \phi_n(x)\|^2 = \|\phi_m(x)\|^2 + \|\phi_n(x)\|^2$, $m \neq n$.
18. From Problem 1 we know that $f_1(x) = x$ and $f_2(x) = x^2$ are orthogonal on $[-2, 2]$. Find constants c_1 and c_2 such that $f_3(x) = x + c_1 x^2 + c_2 x^3$ is orthogonal to both f_1 and f_2 on the same interval.
19. The set of functions $\{\sin nx\}$, $n = 1, 2, 3, \ldots$, is orthogonal on the interval $[-\pi, \pi]$. Show that the set is not complete.
20. Suppose f_1, f_2, and f_3 are functions continuous on the interval $[a, b]$. Show that $(f_1 + f_2, f_3) = (f_1, f_3) + (f_2, f_3)$.

11.2 FOURIER SERIES

The set of functions

$$\left\{1, \cos\frac{\pi}{p}x, \cos\frac{2\pi}{p}x, \ldots, \sin\frac{\pi}{p}x, \sin\frac{2\pi}{p}x, \sin\frac{3\pi}{p}x, \ldots\right\} \quad (1)$$

is orthogonal on the interval $[-p, p]$ (see Problem 12 in Exercises 11.1.) Suppose f is a function defined on the interval $(-p, p)$ that can be expanded in the trigonometric series

$$f(x) = \frac{a_0}{2} + \sum_{n=1}^{\infty}\left(a_n \cos\frac{n\pi}{p}x + b_n \sin\frac{n\pi}{p}x\right) \quad (2)$$

Then the coefficients $a_0, a_1, a_2, \ldots, b_1, b_2, \ldots$ can be determined in exactly the same manner as in the discussion of generalized Fourier series in Section 11.1.*

Integrating both sides of (2) from $-p$ to p gives

$$\int_{-p}^{p} f(x)\, dx = \frac{a_0}{2}\int_{-p}^{p} dx + \sum_{n=1}^{\infty}\left(a_n \int_{-p}^{p}\cos\frac{n\pi}{p}x\, dx + b_n \int_{-p}^{p}\sin\frac{n\pi}{p}x\, dx\right) \quad (3)$$

Since each function $\cos(n\pi x/p)$, $\sin(n\pi x/p)$, $n > 1$, is orthogonal to 1 on the interval, the right side of (3) reduces to a single term and, consequently,

$$\int_{-p}^{p} f(x)\, dx = \frac{a_0}{2}\int_{-p}^{p} dx = \frac{a_0}{2}x\bigg|_{-p}^{p} = pa_0$$

Solving for a_0 yields

$$\boxed{a_0 = \frac{1}{p}\int_{-p}^{p} f(x)\, dx} \quad (4)$$

Now multiply (2) by $\cos(m\pi x/p)$ and integrate:

$$\int_{-p}^{p} f(x)\cos\frac{m\pi}{p}x\, dx = \frac{a_0}{2}\int_{-p}^{p}\cos\frac{m\pi}{p}x\, dx$$
$$+ \sum_{n=1}^{\infty}\left(a_n\int_{-p}^{p}\cos\frac{m\pi}{p}x\cos\frac{n\pi}{p}x\, dx + b_n\int_{-p}^{p}\cos\frac{m\pi}{p}x\sin\frac{n\pi}{p}x\, dx\right) \quad (5)$$

---

*We have chosen to write the coefficient of 1 in the series (2) as $a_0/2$ rather than as a_0. This is for convenience only; the formula for a_n then reduces to a_0 when $n = 0$.

By orthogonality we have

$$\int_{-p}^{p} \cos \frac{m\pi}{p} x \, dx = 0, \qquad m > 0$$

$$\int_{-p}^{p} \cos \frac{m\pi}{p} x \cos \frac{n\pi}{p} x \, dx \begin{cases} = 0, & m \neq n \\ = p, & m = n \end{cases}$$

and

$$\int_{-p}^{p} \cos \frac{m\pi}{p} x \sin \frac{n\pi}{p} x \, dx = 0$$

Thus, (5) reduces to

$$\int_{-p}^{p} f(x) \cos \frac{n\pi}{p} x \, dx = a_n p$$

and so

$$\boxed{a_n = \frac{1}{p} \int_{-p}^{p} f(x) \cos \frac{n\pi}{p} x \, dx} \tag{6}$$

Finally, if we multiply (2) by $\sin(m\pi x/p)$, integrate, and make use of the results

$$\int_{-p}^{p} \sin \frac{m\pi}{p} x \, dx = 0, \qquad m > 0$$

$$\int_{-p}^{p} \sin \frac{m\pi}{p} x \cos \frac{n\pi}{p} x \, dx = 0$$

$$\int_{-p}^{p} \sin \frac{m\pi}{p} x \sin \frac{n\pi}{p} x \, dx \begin{cases} = 0, & m \neq n \\ = p, & m = n \end{cases}$$

we find that

$$\boxed{b_n = \frac{1}{p} \int_{-p}^{p} f(x) \sin \frac{n\pi}{p} x \, dx} \tag{7}$$

The trigonometric series (2) with coefficients a_0, a_n, and b_n defined by (4), (6), and (7), respectively, is said to be the **Fourier series** of the function f.* The coefficients obtained from (4), (6), and (7) are referred to as **Fourier coefficients** of f.

***Jean-Baptiste Joseph Fourier (1766–1830)** A French mathematical physicist, Fourier used such trigonometric series in his investigations into the theory of heat, and they appear throughout his 1822 treatise *Théorie analytique de la chaleur*. However, Fourier did not "invent" Fourier series. The development of the theory of expanding functions in trigonometric series was due principally to Daniel Bernoulli and Leonhard Euler. The integral formulas that define the coefficients a_0, a_n, and b_n were discovered by Euler in 1777. Today, Fourier series, the Fourier integral, and the Fourier transform constitute a branch of mathematical analysis that is invaluable in the study of wave phenomena.

A friend and confidant of Napoleon, Fourier served in the emperor's retinue during Napoleon's 1798 campaign to "civilize" Egypt. Fourier is also remembered for his patronage of the young Jean François Champollion, who was the first to decipher Egyptian hieroglyphics through his work on the Rosetta stone.

In finding the coefficients a_0, a_n, and b_n, we assumed that f was integrable on the interval and that (2), as well as the series obtained by multiplying (2) by $\cos(m\pi x/p)$, converged in such a manner as to permit term-by-term integration. Until (2) is shown to be convergent for a given function f, the equality sign is not to be taken in a strict or literal sense.* We summarize the results:

DEFINITION 11.5 Fourier Series

The **Fourier series** of a function f defined on the interval $(-p, p)$ is given by

$$f(x) = \frac{a_0}{2} + \sum_{n=1}^{\infty}\left(a_n \cos\frac{n\pi}{p}x + b_n \sin\frac{n\pi}{p}x\right) \qquad (8)$$

where
$$a_0 = \frac{1}{p}\int_{-p}^{p} f(x)\,dx \qquad (9)$$

$$a_n = \frac{1}{p}\int_{-p}^{p} f(x)\cos\frac{n\pi}{p}x\,dx \qquad (10)$$

$$b_n = \frac{1}{p}\int_{-p}^{p} f(x)\sin\frac{n\pi}{p}x\,dx \qquad (11)$$

EXAMPLE 1 Expand $f(x) = \begin{cases} 0, & -\pi < x < 0 \\ \pi - x, & 0 < x < \pi \end{cases}$ (12)

in a Fourier series.

Solution The graph of f is given in Figure 11.1. With $p = \pi$, we have from (9) and (10) that

$$a_0 = \frac{1}{\pi}\int_{-\pi}^{\pi} f(x)\,dx = \frac{1}{\pi}\left[\int_{-\pi}^{0} 0\,dx + \int_{0}^{\pi}(\pi - x)\,dx\right] = \frac{1}{\pi}\left[\pi x - \frac{x^2}{2}\right]_0^{\pi} = \frac{\pi}{2}$$

$$a_n = \frac{1}{\pi}\int_{-\pi}^{\pi} f(x)\cos nx\,dx = \frac{1}{\pi}\left[\int_{-\pi}^{0} 0\,dx + \int_{0}^{\pi}(\pi - x)\cos nx\,dx\right]$$

$$= \frac{1}{\pi}\left[(\pi - x)\frac{\sin nx}{n}\bigg|_0^{\pi} + \frac{1}{n}\int_0^{\pi}\sin nx\,dx\right]$$

$$= -\frac{1}{n\pi}\frac{\cos nx}{n}\bigg|_0^{\pi} = \frac{-\cos n\pi + 1}{n^2\pi} = \frac{1 - (-1)^n}{n^2\pi} \qquad \boxed{\cos n\pi = (-1)^n}$$

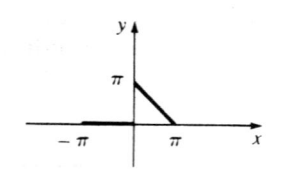

FIGURE 11.1

---

*Some texts use the symbol $\sim$ in place of $=$. In view of the fact that most functions in applications are of a type that guarantee convergence of the series, we shall use the equality symbol.

11.2 Fourier Series

In like manner we find from (11) that

$$b_n = \frac{1}{\pi}\int_0^\pi (\pi - x)\sin nx\, dx = \frac{1}{n}$$

Therefore,
$$f(x) = \frac{\pi}{4} + \sum_{n=1}^{\infty}\left\{\frac{1-(-1)^n}{n^2\pi}\cos nx + \frac{1}{n}\sin nx\right\} \qquad (13)$$

Note that a_n defined by (10) reduces to a_0 given by (9) when we set $n = 0$. But as Example 1 shows, this may not be the case *after* the integral for a_n is evaluated.

Convergence of a Fourier Series The following theorem gives sufficient conditions for the convergence of a Fourier series at a point.

THEOREM 11.1 Conditions for Convergence

Let f and f' be piecewise continuous on the interval $(-p, p)$; that is, let f and f' be continuous except at a finite number of points in the interval and have only finite discontinuities at these points. Then the Fourier series of f on the interval converges to $f(x)$ at a point of continuity. At a point of discontinuity, the Fourier series will converge to the average

$$\frac{f(x+) + f(x-)}{2}$$

where $f(x+)$ and $f(x-)$ denote the limit of f at x from the right and from the left, respectively.*

EXAMPLE 2 The function (12) given in Example 1 satisfies the conditions of Theorem 11.1. Thus, for every x in the interval $(-\pi, \pi)$, except at $x = 0$, the series (13) converges to $f(x)$. At $x = 0$ the function is discontinuous and so the series (13) converges to

$$\frac{f(0+) + f(0-)}{2} = \frac{\pi + 0}{2} = \frac{\pi}{2}$$

Periodic Extension Observe that the functions in the basic set (1) have a common period $2p$. Hence, the right side of (2) is periodic. We conclude that a

---

*In other words, for x a point in the interval and $h > 0$,

$$f(x+) = \lim_{h \to 0} f(x + h) \qquad f(x-) = \lim_{h \to 0} f(x - h)$$

Fourier series not only represents the function on the interval $(-p, p)$ but also gives the **periodic extension** of f outside this interval. We can now apply Theorem 11.1 to the periodic extension of f, or we may assume from the outset that the given function is periodic with period $2p$ (that is, $f(x + 2p) = f(x)$). When f is piecewise continuous and the right- and left-hand derivatives exist at $x = -p$ and $x = p$, respectively, then the series (8) will converge to the average $[f(p-) + f(-p+)]/2$ at these endpoints and to this value extended periodically to $\pm 3p$, $\pm 5p$, $\pm 7p$, and so on.

EXAMPLE 3 The Fourier series (13) converges to the periodic extension of (12) onto the entire x-axis. The dots in Figure 11.2 represent the value

$$\frac{f(0+) + f(0-)}{2} = \frac{\pi}{2}$$

at 0, $\pm 2\pi$, $\pm 4\pi$, ... At $\pm \pi$, $\pm 3\pi$, $\pm 5\pi$, . . . the series converges to the value

$$\frac{f(\pi-) + f(-\pi+)}{2} = 0$$

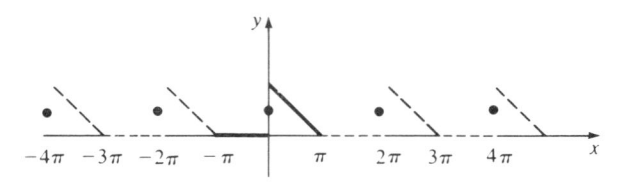

FIGURE 11.2

EXERCISES 11.2 Answers to odd-numbered problems begin on page A-61.

In Problems 1–16 find the Fourier series of f on the given interval.

1. $f(x) = \begin{cases} 0, & -\pi < x < 0 \\ 1, & 0 \leq x < \pi \end{cases}$

2. $f(x) = \begin{cases} -1, & -\pi < x < 0 \\ 2, & 0 \leq x < \pi \end{cases}$

3. $f(x) = \begin{cases} 1, & -1 < x < 0 \\ x, & 0 \leq x < 1 \end{cases}$

4. $f(x) = \begin{cases} 0, & -1 < x < 0 \\ x, & 0 \leq x < 1 \end{cases}$

5. $f(x) = \begin{cases} 0, & -\pi < x < 0 \\ x^2, & 0 \leq x < \pi \end{cases}$

6. $f(x) = \begin{cases} \pi^2, & -\pi < x < 0 \\ \pi^2 - x^2, & 0 \leq x < \pi \end{cases}$

7. $f(x) = x + \pi$, $-\pi < x < \pi$

8. $f(x) = 3 - 2x$, $-\pi < x < \pi$

9. $f(x) = \begin{cases} 0, & -\pi < x < 0 \\ \sin x, & 0 \leq x < \pi \end{cases}$

10. $f(x) = \begin{cases} 0, & -\pi/2 < x < 0 \\ \cos x, & 0 \leq x < \pi/2 \end{cases}$

11. $f(x) = \begin{cases} 0, & -2 < x < -1 \\ -2, & -1 \leq x < 0 \\ 1, & 0 \leq x < 1 \\ 0, & 1 \leq x < 2 \end{cases}$

12. $f(x) = \begin{cases} 0, & -2 < x < 0 \\ x, & 0 \leq x < 1 \\ 1, & 1 \leq x < 2 \end{cases}$

13. $f(x) = \begin{cases} 1, & -5 < x < 0 \\ 1 + x, & 0 \le x < 5 \end{cases}$

14. $f(x) = \begin{cases} 2 + x, & -2 < x < 0 \\ 2, & 0 \le x < 2 \end{cases}$

15. $f(x) = e^x, \quad -\pi < x < \pi$

16. $f(x) = \begin{cases} 0, & -\pi < x < 0 \\ e^x - 1, & 0 \le x < \pi \end{cases}$

17. Use the result of Problem 5 to show that

$$\frac{\pi^2}{6} = 1 + \frac{1}{2^2} + \frac{1}{3^2} + \frac{1}{4^2} + \cdots \quad \text{and} \quad \frac{\pi^2}{12} = 1 - \frac{1}{2^2} + \frac{1}{3^2} - \frac{1}{4^2} + \cdots$$

18. Use Problem 17 to find a series giving the numerical value of $\dfrac{\pi^2}{8}$.

19. Use the result of Problem 7 to show that

$$\frac{\pi}{4} = 1 - \frac{1}{3} + \frac{1}{5} - \frac{1}{7} + \cdots$$

20. Use the result of Problem 9 to show that

$$\frac{\pi}{4} = \frac{1}{2} + \frac{1}{1 \cdot 3} - \frac{1}{3 \cdot 5} + \frac{1}{5 \cdot 7} - \frac{1}{7 \cdot 9} + \cdots$$

21. (a) Use the complex exponential form of the cosine and sine:

$$\cos\frac{n\pi}{p}x = \frac{e^{in\pi x/p} + e^{-in\pi x/p}}{2} \qquad \sin\frac{n\pi}{p}x = \frac{e^{in\pi x/p} - e^{-in\pi x/p}}{2i}$$

to show that (8) can be written in the **complex form**

$$f(x) = \sum_{n=-\infty}^{\infty} c_n e^{in\pi x/p}$$

where $c_0 = a_0/2$, $c_n = (a_n - ib_n)/2$, and $c_{-n} = (a_n + ib_n)/2$, $n = 1, 2, 3, \ldots$, and $i^2 = -1$.

(b) Show that c_0, c_n, and c_{-n} of part (a) can be written as one integral

$$c_n = \frac{1}{2p}\int_{-p}^{p} f(x)e^{-in\pi x/p}\,dx, \qquad n = 0, \pm 1, \pm 2, \ldots$$

22. Use the results of Problem 21 to find the complex form of the Fourier series of $f(x) = e^{-x}$ on the interval $-\pi < x < \pi$.

11.3 FOURIER COSINE AND SINE SERIES

Even and Odd Functions   The reader may recall that a function f is said to be **even** if

$$\boxed{f(-x) = f(x)} \quad \text{whereas if} \quad \boxed{f(-x) = -f(x)}$$

then f is said to be an **odd** function.

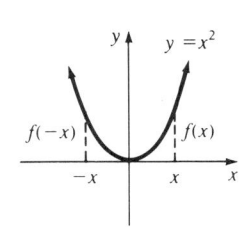

FIGURE 11.3

EXAMPLE 1   (a) $f(x) = x^2$ is even since

$$f(-x) = (-x)^2 = x^2 = f(x)$$

See Figure 11.3.

(b) $f(x) = x^3$ is odd since

$$f(-x) = (-x)^3 = -x^3 = -f(x)$$

See Figure 11.4.

FIGURE 11.4

As illustrated in Figures 11.3 and 11.4, the graph of an even function is symmetric with respect to the y-axis, and the graph of an odd function is symmetric with respect to the origin.

EXAMPLE 2 Since $\cos(-x) = \cos x$ and $\sin(-x) = -\sin x$, the cosine and sine are even and odd functions, respectively. □

Properties of Even and Odd Functions The following is a list of some properties of even and odd functions:

(*i*) The product of two even functions is even.

(*ii*) The product of two odd functions is even.

(*iii*) The product of an even function and an odd function is odd.

(*iv*) The sum (difference) of two even functions is even.

(*v*) The sum (difference) of two odd functions is odd.

(*vi*) If f is even, then $\int_{-a}^{a} f(x)\, dx = 2\int_{0}^{a} f(x)\, dx$.

vii) If f is odd, then $\int_{-a}^{a} f(x)\, dx = 0$.

To prove (*ii*) let us suppose that f and g are odd functions. Then we have $f(-x) = -f(x)$ and $g(-x) = -g(x)$. If we define the product of f and g as $F(x) = f(x)g(x)$, then

$$F(-x) = f(-x)g(-x) = (-f(x))(-g(x)) = f(x)g(x) = F(x)$$

This shows that the product F of two odd functions is an even function. The proofs of the remaining properties are left as exercises.

Cosine and Sine Series If f is an even function on $(-p, p)$, then in view of the foregoing properties, the coefficients (9), (10), and (11) of Section 11.2 become

$$a_0 = \frac{1}{p}\int_{-p}^{p} f(x)\, dx = \frac{2}{p}\int_{0}^{p} f(x)\, dx$$

$$a_n = \frac{1}{p}\underbrace{\int_{-p}^{p} f(x)\cos\frac{n\pi}{p}x\, dx}_{\text{even}} = \frac{2}{p}\int_{0}^{p} f(x)\cos\frac{n\pi}{p}x\, dx$$

$$b_n = \frac{1}{p}\underbrace{\int_{-p}^{p} f(x)\sin\frac{n\pi}{p}x\, dx}_{\text{odd}} = 0$$

Similarly, when f is odd on the interval $(-p, p)$,

$$a_n = 0, \quad n = 0, 1, 2, \ldots \qquad b_n = \frac{2}{p}\int_{0}^{p} f(x)\sin\frac{n\pi}{p}x\, dx$$

We summarize the results:

DEFINITION 11.6 Fourier Cosine and Sine Series

(*i*) The Fourier series of an even function on the interval $(-p, p)$ is the **cosine series**

$$f(x) = \frac{a_0}{2} + \sum_{n=1}^{\infty} a_n \cos \frac{n\pi}{p} x \qquad (1)$$

where

$$a_0 = \frac{2}{p} \int_0^p f(x)\, dx \qquad (2)$$

$$a_n = \frac{2}{p} \int_0^p f(x) \cos \frac{n\pi}{p} x\, dx \qquad (3)$$

(*ii*) The Fourier series of an odd function on the interval $(-p, p)$ is the **sine series**

$$f(x) = \sum_{n=1}^{\infty} b_n \sin \frac{n\pi}{p} x \qquad (4)$$

where

$$b_n = \frac{2}{p} \int_0^p f(x) \sin \frac{n\pi}{p} x\, dx \qquad (5)$$

EXAMPLE 3   Expand $f(x) = x$, $-2 < x < 2$, in a Fourier series.

Solution   We expand f in a sine series, since inspection of Figure 11.5 shows that the function is odd on the interval $(-2, 2)$. With the identification $2p = 4$, or $p = 2$, we can write (5) as

$$b_n = \int_0^2 x \sin \frac{n\pi}{2} x\, dx$$

Integration by parts then yields

$$b_n = \frac{4(-1)^{n+1}}{n\pi}$$

Therefore,

$$f(x) = \frac{4}{\pi} \sum_{n=1}^{\infty} \frac{(-1)^{n+1}}{n} \sin \frac{n\pi}{2} x \qquad (6)$$

FIGURE 11.5 $y = x$, $-2 < x < 2$

EXAMPLE 4 The function in Example 3 satisfies the conditions of Theorem 11.1. Hence the series (6) converges to the function on $(-2, 2)$ and the periodic extension (of period 4) given in Figure 11.6.

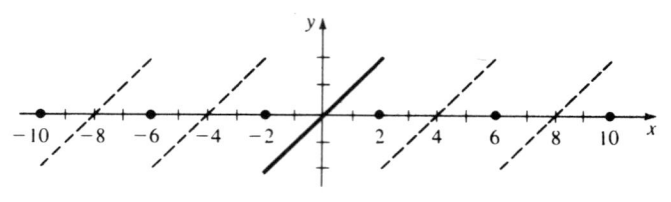

FIGURE 11.6

EXAMPLE 5 The function $f(x) = \begin{cases} -1, & -\pi < x < 0 \\ 1, & 0 \leq x < \pi \end{cases}$

shown in Figure 11.7 is odd on the interval $(-\pi, \pi)$. With $p = \pi$ we have from (5)

$$b_n = \frac{2}{\pi} \int_0^\pi (1) \sin nx \, dx = \frac{2}{\pi} \frac{1 - (-1)^n}{n}$$

and so

$$f(x) = \frac{2}{\pi} \sum_{n=1}^{\infty} \frac{1 - (-1)^n}{n} \sin nx \qquad (7)$$

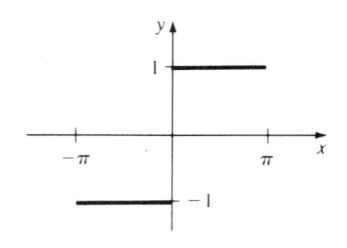

FIGURE 11.7

Sequence of Partial Sums It is interesting to see how the sequence of partial sums of a Fourier series approximates a function. In Figure 11.8 the graph of the function f in Example 5 is compared with the graphs of the first three partial sums of (7):

$$S_1 = \frac{4}{\pi} \sin x, \quad S_2 = \frac{4}{\pi}\left(\sin x + \frac{\sin 3x}{3}\right), \quad S_3 = \frac{4}{\pi}\left(\sin x + \frac{\sin 3x}{3} + \frac{\sin 5x}{5}\right)$$

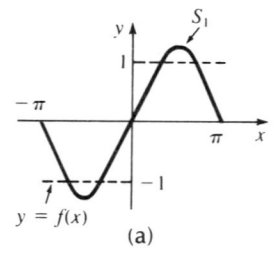

FIGURE 11.8

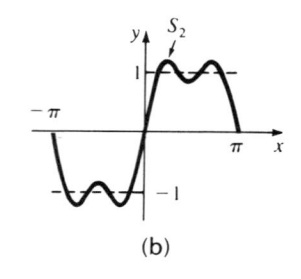

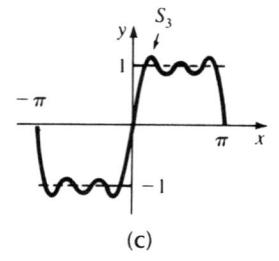

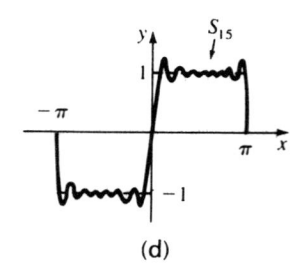

As seen in Figure 11.8(d), the graph of the partial sum S_{15} has pronounced spikes near the discontinuities at $x = 0$, $x = \pi$, $x = -\pi$, and so on. This "overshooting" of the partial sums S_N from the functional values near a point

11.3 Fourier Cosine and Sine Series

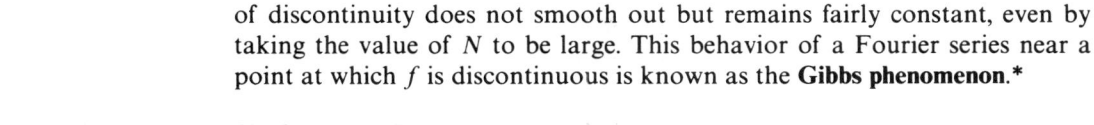

of discontinuity does not smooth out but remains fairly constant, even by taking the value of N to be large. This behavior of a Fourier series near a point at which f is discontinuous is known as the **Gibbs phenomenon.***

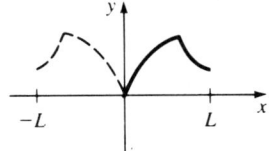

Half-Range Expansions Throughout the preceding discussion it was understood that a function f was defined on an interval with the origin as midpoint—that is, $-p < x < p$. However, in many instances we are interested in representing a function that is defined for only $0 < x < L$ by a trigonometric series. This can be done in many different ways by supplying an arbitrary *definition* of the function to the interval $-L < x < 0$. For brevity we consider the three most important cases.

If $y = f(x)$ is defined on the interval $0 < x < L$,

FIGURE 11.9

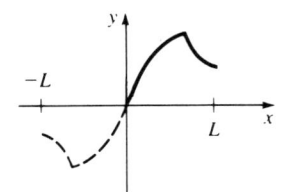

FIGURE 11.10

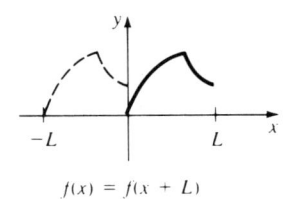

FIGURE 11.11
$f(x) = f(x + L)$

(i) reflect the graph of the function about the y-axis onto $-L < x < 0$; the function is now even on $-L < x < L$ (see Figure 11.9); or

(ii) reflect the graph of the function through the origin onto $-L < x < 0$; the function is now odd on $-L < x < L$ (see Figure 11.10); or

(iii) define f on $-L < x < 0$ by $f(x) = f(x + L)$ (see Figure 11.11).

Note that the coefficients of the series (1) and (4) utilize only the definition of the function on $0 < x < p$ (that is, half of the interval $-p < x < p$). Hence in practice there is no actual need to make the reflections described in (i) and (ii). If f is defined on $0 < x < L$, we simply identify the half-period as the length of the interval $p = L$. The coefficient formulas (2), (3), and (5) and the corresponding series yield either an even or an odd periodic extension of period $2L$ of the original function. The cosine and sine series obtained in this manner are known as **half-range expansions**. Last, in case (iii) we are defining the functional values on the interval $-L < x < 0$ to be the same as the values on $0 < x < L$. As in the previous two cases, there is no real need to do this. It can be shown that the set of functions in (1) of Section 11.2 is orthogonal on $a \leq x \leq a + 2p$ for any real number a. By choosing $a = -p$, we obtain the limits of integration in (9), (10), and (11) of that section. But for $a = 0$ the limits of integration are from $x = 0$ to $x = 2p$. Thus if f is defined over the interval $0 < x < L$, we identify $2p = L$ or $p = L/2$. The resulting Fourier series will give the periodic extension of f with period L. In this manner the values to which the series converges on $-L < x < 0$ will be the same as the values on $0 < x < L$.

***Josiah Willard Gibbs (1839–1903)** This "overshooting" phenomenon was discovered by the American physicist and physical chemist Josiah Gibbs in 1899. He is also remembered for his pioneering work in three-dimensional vector analysis. In 1881 Gibbs, for the benefit of his students at Yale, published a pamphlet entitled *Elements of Vector Analysis*. E. B. Wilson incorporated this pamphlet and Gibbs's lectures in the text *Vector Analysis*, which was published in 1901. In these works Gibbs introduced much of the notation that is still used in modern vector analysis.

EXAMPLE 6 Expand $f(x) = x^2$, $0 < x < L$, (a) in a cosine series, (b) in a sine series, and (c) in a Fourier series.

Solution The graph of the function is given in Figure 11.12.

(a) We have
$$a_0 = \frac{2}{L}\int_0^L x^2 \, dx = \frac{2}{3}L^2$$

and, integrating by parts, we get

$$a_n = \frac{2}{L}\int_0^L x^2 \cos\frac{n\pi}{L} x \, dx$$

$$= \frac{2}{L}\left[\frac{Lx^2 \sin\frac{n\pi}{L} x}{n\pi}\bigg|_0^L - \frac{2L}{n\pi}\int_0^L x \sin\frac{n\pi}{L} x \, dx\right]$$

$$= -\frac{4}{n\pi}\left[-\frac{Lx \cos\frac{n\pi}{L} x}{n\pi}\bigg|_0^L + \frac{L}{n\pi}\int_0^L \cos\frac{n\pi}{L} x \, dx\right]$$

$$= \frac{4L^2(-1)^n}{n^2\pi^2}$$

Thus,
$$f(x) = \frac{L^2}{3} + \frac{4L^2}{\pi^2}\sum_{n=1}^\infty \frac{(-1)^n}{n^2}\cos\frac{n\pi}{L}x \qquad (8)$$

(b) In this case
$$b_n = \frac{2}{L}\int_0^L x^2 \sin\frac{n\pi}{L} x \, dx$$

After integrating by parts, we find

$$b_n = \frac{2L^2(-1)^{n+1}}{n\pi} + \frac{4L^2}{n^3\pi^3}[(-1)^n - 1]$$

Thus,
$$f(x) = \frac{2L^2}{\pi}\sum_{n=1}^\infty \left\{\frac{(-1)^{n+1}}{n} + \frac{2}{n^3\pi^2}[(-1)^n - 1]\right\}\sin\frac{n\pi}{L}x \qquad (9)$$

(c) With $p = L/2$, $1/p = 2/L$, and $n\pi/p = 2n\pi/L$, we have

$$a_0 = \frac{2}{L}\int_0^L x^2 \, dx = \frac{2}{3}L^2$$

$$a_n = \frac{2}{L}\int_0^L x^2 \cos\frac{2n\pi}{L} x \, dx = \frac{L^2}{n^2\pi^2}$$

$$b_n = \frac{2}{L}\int_0^L x^2 \sin\frac{2n\pi}{L} x \, dx = -\frac{L^2}{n\pi}$$

Therefore,

$$f(x) = \frac{L^2}{3} + \frac{L^2}{\pi}\sum_{n=1}^\infty\left\{\frac{1}{n^2\pi}\cos\frac{2n\pi}{L}x - \frac{1}{n}\sin\frac{2n\pi}{L}x\right\} \qquad (10)$$

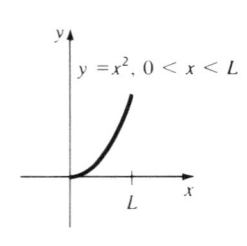

FIGURE 11.12

11.3 Fourier Cosine and Sine Series

The series (8), (9), and (10) converge to the $2L$-periodic even extension of f, to the $2L$-periodic odd extension of f, and to the L-periodic extension of f, respectively. The graphs of these periodic extensions are shown in Figure 11.13.

(a) cosine series

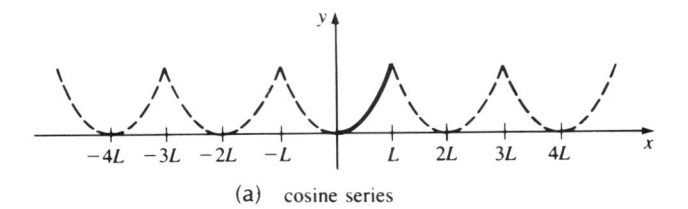

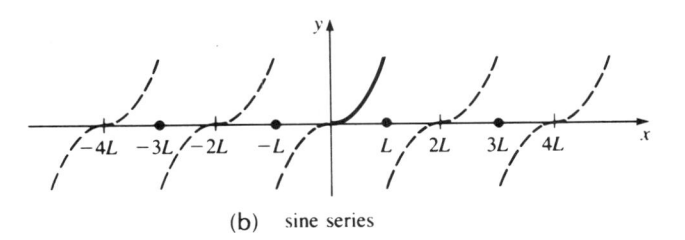

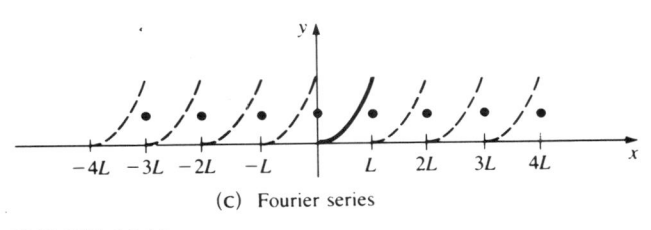

(b) sine series

(c) Fourier series

FIGURE 11.13

Periodic Driving Force Fourier series are sometimes useful in determining a particular solution of a differential equation describing a physical system in which the input or driving force $f(t)$ is periodic. In the next example we shall find a particular solution of the differential equation

$$m\frac{d^2x}{dt^2} + kx = f(t) \tag{11}$$

by first representing f by a half-range sine expansion

$$f(t) = \sum_{n=1}^{\infty} b_n \sin\frac{n\pi}{p}t$$

and then assuming a particular solution of the form

$$x_p(t) = \sum_{n=1}^{\infty} B_n \sin\frac{n\pi}{p}t \tag{12}$$

EXAMPLE 7 An undamped spring–mass system in which the mass $m = \frac{1}{16}$ slug and the spring constant $k = 4$ lb/ft is driven by the 2-periodic external force $f(t)$ shown in Figure 11.14. Although the force $f(t)$ acts on the system for $t > 0$, note that if we extend the graph of the function in a 2-periodic manner to the negative t-axis, we obtain an odd function. In practical terms this means that we need only find the half-range sine expansion of $f(t) = \pi t$, $0 < t < 1$. With $p = 1$ it follows from (5) and integration by parts that

$$b_n = 2\int_0^1 \pi t \sin n\pi t\, dt = \frac{2(-1)^{n+1}}{n}$$

From (11) the differential equation of motion is seen to be

$$\frac{1}{16}\frac{d^2x}{dt^2} + 4x = \sum_{n=1}^{\infty} \frac{2(-1)^{n+1}}{n} \sin n\pi t \tag{13}$$

To find a particular solution $x_p(t)$ of (13) we substitute (12) into the equation and equate coefficients of $\sin n\pi t$. This yields

$$\left(-\frac{1}{16}n^2\pi^2 + 4\right)B_n = \frac{2(-1)^{n+1}}{n} \quad \text{or} \quad B_n = \frac{32(-1)^{n+1}}{n(64 - n^2\pi^2)}$$

Thus, $$x_p(t) = \sum_{n=1}^{\infty} \frac{32(-1)^{n+1}}{n(64 - n^2\pi^2)} \sin n\pi t \tag{14}$$ □

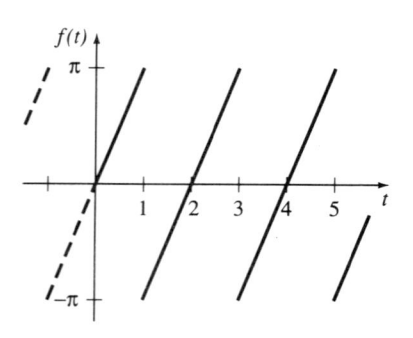

FIGURE 11.14

Observe that in the solution (14) there is no integer $n \geq 1$ for which the denominator $64 - n^2\pi^2$ of B_n is zero. In general, if there *is* a value of n, say N, for which $N\pi/p = \omega$, where $\omega = \sqrt{k/m}$, then the system described by (11) would be in a state of pure resonance. In other words, we have pure resonance if the Fourier series expansion of the driving force $f(t)$ contains a term $\sin(N\pi/L)t$ (or $\cos(N\pi/L)t$) that has the same frequency as the frequency of free vibrations.

Of course, if the $2p$-periodic extension of the driving force f onto the negative t-axis yields an even function, then we expand f in a cosine series.

EXERCISES 11.3   *Answers to odd-numbered problems begin on page A-61.*

In Problems 1–10 determine whether the function is even, odd, or neither.

1. $f(x) = \sin 3x$
2. $f(x) = x \cos x$
3. $f(x) = x^2 + x$
4. $f(x) = x^3 - 4x$
5. $f(x) = e^{|x|}$
6. $f(x) = |x^5|$
7. $f(x) = \begin{cases} x^2, & -1 < x < 0 \\ -x^2, & 0 \leq x < 1 \end{cases}$
8. $f(x) = \begin{cases} x + 5, & -2 < x < 0 \\ -x + 5, & 0 \leq x < 2 \end{cases}$
9. $f(x) = x^3$, $0 \leq x \leq 2$
10. $f(x) = 2|x| - 1$

In Problems 11–24 expand the given function in an appropriate cosine or sine series.

11. $f(x) = \begin{cases} -1, & -\pi < x < 0 \\ 1, & 0 \leq x < \pi \end{cases}$
12. $f(x) = \begin{cases} 1, & -2 < x < -1 \\ 0, & -1 \leq x < 1 \\ 1, & 1 \leq x < 2 \end{cases}$

13. $f(x) = |x|$, $-\pi < x < \pi$
14. $f(x) = x$, $-\pi < x < \pi$
15. $f(x) = x^2$, $-1 < x < 1$
16. $f(x) = x|x|$, $-1 < x < 1$
17. $f(x) = \pi^2 - x^2$, $-\pi < x < \pi$
18. $f(x) = x^3$, $-\pi < x < \pi$
19. $f(x) = \begin{cases} x - 1, & -\pi < x < 0 \\ x + 1, & 0 \le x < \pi \end{cases}$
20. $f(x) = \begin{cases} x + 1, & -1 < x < 0 \\ x - 1, & 0 \le x < 1 \end{cases}$
21. $f(x) = \begin{cases} 1, & -2 < x < -1 \\ -x, & -1 \le x < 0 \\ x, & 0 \le x < 1 \\ 1, & 1 \le x < 2 \end{cases}$
22. $f(x) = \begin{cases} -\pi, & -2\pi < x < \pi \\ x, & -\pi \le x < \pi \\ \pi, & \pi \le x < 2\pi \end{cases}$
23. $f(x) = |\sin x|$, $-\pi < x < \pi$
24. $f(x) = \cos x$, $-\frac{\pi}{2} < x < \frac{\pi}{2}$

In Problems 25–34 find the half-range cosine and sine expansions of the given function.

25. $f(x) = \begin{cases} 1, & 0 < x < \frac{1}{2} \\ 0, & \frac{1}{2} \le x < 1 \end{cases}$
26. $f(x) = \begin{cases} 0, & 0 < x < \frac{1}{2} \\ 1, & \frac{1}{2} \le x < 1 \end{cases}$
27. $f(x) = \cos x$, $0 < x < \frac{\pi}{2}$
28. $f(x) = \sin x$, $0 < x < \pi$
29. $f(x) = \begin{cases} x, & 0 < x < \frac{\pi}{2} \\ \pi - x, & \frac{\pi}{2} \le x < \pi \end{cases}$
30. $f(x) = \begin{cases} 0, & 0 < x < \pi \\ x - \pi, & \pi \le x < 2\pi \end{cases}$
31. $f(x) = \begin{cases} x, & 0 < x < 1 \\ 1, & 1 \le x < 2 \end{cases}$
32. $f(x) = \begin{cases} 1, & 0 < x < 1 \\ 2 - x, & 1 \le x < 2 \end{cases}$
33. $f(x) = x^2 + x$, $0 < x < 1$
34. $f(x) = x(2 - x)$, $0 < x < 2$

In Problems 35–38 expand the given function in a Fourier series.

35. $f(x) = x^2$, $0 < x < 2\pi$
36. $f(x) = x$, $0 < x < \pi$
37. $f(x) = x + 1$, $0 < x < 1$
38. $f(x) = 2 - x$, $0 < x < 2$

In Problems 39 and 40 proceed as in Example 7 to find a particular solution of (11) when $m = 1$, $k = 10$, and the driving force $f(t)$ is as given. Assume that when $f(t)$ is extended to the negative t-axis in a periodic manner, the resulting function is odd.

39. $f(t) = \begin{cases} 5, & 0 < t < \pi \\ -5, & \pi \le t < 2\pi \end{cases}$, $f(t + 2\pi) = f(t)$
40. $f(t) = 1 - t$, $0 < t < 2$, $f(t + 2) = f(t)$

In Problems 41 and 42 find a particular solution of (11) when $m = \frac{1}{4}$, $k = 12$, and the driving force $f(t)$ is as given. Assume that when $f(t)$ is extended to the negative t-axis in a periodic manner, the resulting function is even.

41. $f(t) = 2\pi t - t^2$, $0 < t < 2\pi$, $f(t + 2\pi) = f(t)$
42. $f(t) = \begin{cases} t, & 0 < t < \frac{1}{2} \\ 1 - t, & \frac{1}{2} \le t < 1 \end{cases}$, $f(t + 1) = f(t)$

43. Suppose a uniform beam of length L is simply supported at $x = 0$ and $x = L$. If the load per unit length is given by $w(x) = w_0 x/L$, $0 < x < L$, then the differential equation for the deflection $y(x)$ is

$$EI \frac{d^4 y}{dx^4} = \frac{w_0 x}{L}$$

where E, I, and w_0 are constants. (See (14) of Section 1.3.)

(a) Expand $w(x)$ in a half-range sine series.
(b) Use the method of Example 7 to find a particular solution $y(x)$ of the differential equation.

44. Proceed as in Problem 43 to find the deflection $y(x)$ when $w(x)$ is as given in Figure 11.15.

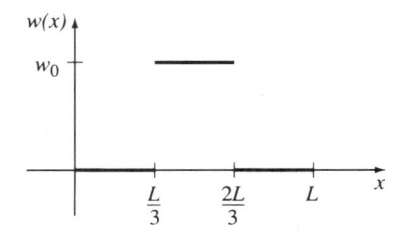

FIGURE 11.15

Miscellaneous Problems

45. Prove property (*i*). **46.** Prove property (*iii*).

47. Prove property (*iv*). **48.** Prove property (*vi*).

49. Prove property (*vii*).

50. Prove that any function f can be written as a sum of an even and an odd function. [*Hint:* Use the identity

$$f(x) = \frac{f(x) + f(-x)}{2} + \frac{f(x) - f(-x)}{2}$$

51. Find the Fourier series of

$$f(x) = \begin{cases} 0, & -\pi < x < 0 \\ x, & 0 \leq x < \pi \end{cases}$$

using the identity $f(x) = (|x| + x)/2$, $-\pi < x < \pi$, and the results of Problems 13 and 14. Observe that $|x|/2$ and $x/2$ are even and odd, respectively, on the interval (see Problem 50).

52. The **double sine series** for a function $f(x, y)$ defined over a rectangular region $0 \leq x \leq b$, $0 \leq y \leq c$ is given by

$$f(x, y) = \sum_{m=1}^{\infty} \sum_{n=1}^{\infty} A_{mn} \sin \frac{m\pi}{b} x \sin \frac{n\pi}{c} y$$

where $A_{mn} = \dfrac{4}{bc} \int_0^c \int_0^b f(x, y) \sin \dfrac{m\pi}{b} x \sin \dfrac{n\pi}{c} y \, dx \, dy$

Find the double sine series of $f(x, y) = 1$, $0 \leq x \leq \pi$, $0 \leq y \leq \pi$.

53. The **double cosine series** of a function $f(x, y)$ defined over a rectangular region $0 \leq x \leq b$, $0 \leq y \leq c$ is given by

$$f(x, y) = A_{00} + \sum_{m=1}^{\infty} A_{m0} \cos \frac{m\pi}{b} x + \sum_{n=1}^{\infty} A_{0n} \cos \frac{n\pi}{c} y$$

$$+ \sum_{m=1}^{\infty} \sum_{n=1}^{\infty} A_{mn} \cos \frac{m\pi}{b} x \cos \frac{n\pi}{c} y$$

where $A_{00} = \dfrac{1}{bc} \int_0^c \int_0^b f(x, y) \, dx \, dy$

$A_{m0} = \dfrac{2}{bc} \int_0^c \int_0^b f(x, y) \cos \dfrac{m\pi}{b} x \, dx \, dy$

$A_{0n} = \dfrac{2}{bc} \int_0^c \int_0^b f(x, y) \cos \dfrac{n\pi}{c} y \, dx \, dy$

$A_{mn} = \dfrac{4}{bc} \int_0^c \int_0^b f(x, y) \cos \dfrac{m\pi}{b} x \cos \dfrac{n\pi}{c} y \, dx \, dy$

Find the double cosine series of $f(x, y) = xy$, $0 \leq x \leq 1$, $0 \leq y \leq 1$.

11.4 STURM–LIOUVILLE PROBLEM

REVIEW

For convenience we present here a brief review of some of the differential equations and their general solutions that will be of importance in the sections and chapters that follow.

(i) **Linear first-order equation:** $y' + ky = 0$, k a constant

General solution: $y = c_1 e^{-kx}$

(ii) **Linear second-order equation:** $y'' + \lambda y = 0$, $\lambda > 0$

General solution: $y = c_1 \cos \sqrt{\lambda} x + c_2 \sin \sqrt{\lambda} x$

(iii) **Linear second-order equation:** $y'' - \lambda y = 0$, $\lambda > 0$

The general solution of this differential equation has two real forms.

General solution: $y = c_1 \cosh \sqrt{\lambda} x + c_2 \sinh \sqrt{\lambda} x$

$y = c_1 e^{-\sqrt{\lambda}x} + c_2 e^{\sqrt{\lambda}x}$

It should be noted that often in practice the exponential form is used when the domain of x is an infinite or semi-infinite interval* and the hyperbolic form is used when the domain of x is a finite interval.

(iv) **Cauchy–Euler equation:** $x^2 y'' + xy' - \lambda^2 y = 0$

General solution: $\lambda \neq 0$: $y = c_1 x^\lambda + c_2 x^{-\lambda}$

$\lambda = 0$: $y = c_1 + c_2 \ln x$

(v) **Parametric Bessel equation:** $x^2 y'' + xy' + (\lambda^2 x^2 - n^2) y = 0$, $n = 0, 1, 2, \ldots$

General solution: $y = c_1 J_n(\lambda x) + c_2 Y_n(\lambda x)$

It is important that you recognize Bessel's differential equation when $n = 0$:

$$xy'' + y' + \lambda^2 x y = 0$$

General solution: $y = c_1 J_0(\lambda x) + c_2 Y_0(\lambda x)$

Recall that $Y_n(\lambda x) \to -\infty$ as $x \to 0^+$.

(vi) **Legendre's differential equation:**
$(1 - x^2) y'' - 2xy' + n(n+1) y = 0$, $n = 0, 1, 2, \ldots$
Particular solutions are the Legendre polynomials $y = P_n(x)$, where

$$y = P_0(x) = 1, \quad y = P_1(x) = x, \quad y = P_2(x) = \frac{1}{2}(3x^2 - 1), \quad \ldots$$

Orthogonal functions often arise in the solution of differential equations. More to the point, an orthogonal set of functions can be generated by solving a two-point boundary-value problem involving a linear second-order differential equation containing a parameter λ.

EXAMPLE 1 Solve $y'' + \lambda y = 0$ subject to $y(0) = 0$, $y(L) = 0$.

Solution We consider three cases: $\lambda = 0$, $\lambda < 0$, and $\lambda > 0$.

CASE I For $\lambda = 0$ the solution of $y'' = 0$ is $y = c_1 x + c_2$. Applying the conditions $y(0) = 0$ and $y(L) = 0$ implies, in turn, $c_2 = 0$ and $c_1 = 0$. Hence for $\lambda = 0$ the only solution of the boundary-value problem is the trivial solution $y = 0$.

---

*Infinite: $(-\infty, \infty)$; semi-infinite: $(-\infty, 0]$, $(1, \infty)$, and so on.

CASE II For $\lambda < 0$ we have $y = c_1 \cosh \sqrt{-\lambda}\,x + c_2 \sinh \sqrt{-\lambda}\,x$.*
Again, $y(0) = 0$ gives $c_1 = 0$, and so $y = c_2 \sinh \sqrt{-\lambda}\,x$. The second condition $y(L) = 0$ dictates that $c_2 \sinh \sqrt{-\lambda}\,L = 0$. Since $\sinh \sqrt{-\lambda}\,L \neq 0$, we must have $c_2 = 0$. Thus, $y = 0$.

CASE III For $\lambda > 0$ the general solution of $y'' + \lambda y = 0$ is $y = c_1 \cos \sqrt{\lambda}\,x + c_2 \sin \sqrt{\lambda}\,x$. As before, $y(0) = 0$ yields $c_1 = 0$, but $y(L) = 0$ implies

$$c_2 \sin \sqrt{\lambda}\,L = 0 \tag{1}$$

If $c_2 = 0$, then necessarily $y = 0$. However, if $c_2 \neq 0$, then (1) is satisfied if $\sin \sqrt{\lambda}\,L = 0$. The last condition implies that the argument of the sine function must be an integer multiple of π:

$$\sqrt{\lambda}\,L = n\pi \quad \text{or} \quad \lambda = \frac{n^2\pi^2}{L^2}, \quad n = 1, 2, 3, \ldots \tag{2}$$

Therefore for any real nonzero c_2, $y = c_2 \sin(n\pi x/L)$ is a solution of the problem for each n. Since the differential equation is homogeneous, we may, if desired, not write c_2. In other words, for a given number in the sequence

$$\frac{\pi^2}{L^2}, \frac{4\pi^2}{L^2}, \frac{9\pi^2}{L^2}, \ldots$$

the *corresponding* function in the sequence

$$\sin \frac{\pi}{L}x, \sin \frac{2\pi}{L}x, \sin \frac{3\pi}{L}x, \ldots$$

is a nontrivial solution of the original problem. The reader should recognize that $\{\sin(n\pi x/L)\}$, $n = 1, 2, 3, \ldots$, is the orthogonal set of functions on $[0, L]$ used as the basis for the Fourier sine series. □

Eigenvalues and Eigenfunctions The numbers $\lambda = n^2\pi^2/L^2$, $n = 1, 2, 3, \ldots$, for which the boundary-value problem in Example 1 has a nontrivial solution are known as **characteristic values** or, more commonly, **eigenvalues**. The solutions depending on these values of λ, $y = c_2 \sin \frac{n\pi}{L}x$ (or simply $y = \sin \frac{n\pi}{L}x$), are called **characteristic functions**, or **eigenfunc-**

---

*$\sqrt{-\lambda}$ looks a little strange, but bear in mind that $\lambda < 0$ is equivalent to $-\lambda > 0$.

tions.* For example,

$$y'' + \left(\frac{9\pi^2}{L^2}\right)y = 0$$
$$y(0) = 0, \quad y(L) = 0$$

(eigenvalue: $\frac{9\pi^2}{L^2}$) solution is $y = \sin\frac{3\pi}{L}x$ (eigenfunction)

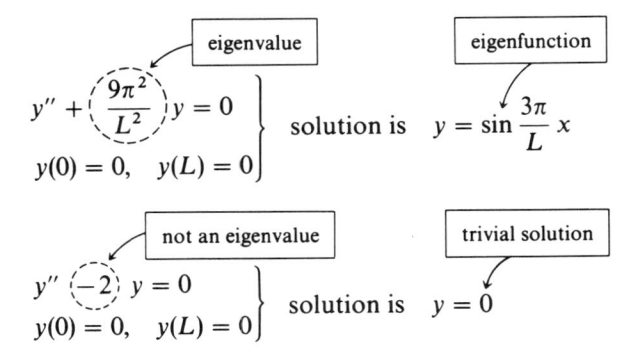

$$y'' + (-2)y = 0$$
$$y(0) = 0, \quad y(L) = 0$$

(not an eigenvalue: -2) solution is $y = 0$ (trivial solution)

EXAMPLE 2 The eigenvalues and eigenfunctions of the boundary-value problem

$$y'' + \lambda y = 0$$
$$y'(0) = 0, \quad y'(L) = 0$$

are, respectively, $\lambda = n^2\pi^2/L^2$, $n = 0, 1, 2, \ldots$, and $y = c_1 \cos(n\pi x/L)$, $c_1 \neq 0$. In contrast to Example 1, $\lambda = 0$ is an eigenvalue of this problem and $y = 1$ is the corresponding eigenfunction. The latter comes from solving $y'' = 0$, $y'(0) = 0$, $y'(L) = 0$ but can be incorporated in $y = \cos(n\pi x/L)$ by permitting $n = 0$ and $c_1 = 1$. The set $\{\cos(n\pi x/L)\}$, $n = 0, 1, 2, \ldots$, is orthogonal on the interval $[0, L]$ and is used in Fourier cosine expansions. □

Regular Sturm–Liouville Problem The boundary-value problem in Examples 1 and 2 are special cases of the following general problem:

Let p, q, r, and r' be real-valued functions continuous on an interval $[a, b]$, and let $r(x) > 0$ and $p(x) > 0$ for every x in the interval. The two-point boundary-value problem:

Solve $\quad\displaystyle\frac{d}{dx}[r(x)y'] + (q(x) + \lambda p(x))y = 0 \quad$ (3)

subject to the homogeneous boundary conditions:

$$\alpha_1 y(a) + \beta_1 y'(a) = 0 \quad (4)$$
$$\alpha_2 y(b) + \beta_2 y'(b) = 0 \quad (5)$$

---

*Notice in (2) that we need only take n to be a positive integer. If n is a negative integer, $n = -k$, $k = 1, 2, 3, \ldots$, we can use the trigonometric identity $\sin(-\theta) = -\sin\theta$ to rewrite $\sin(-k\pi x/L)$ as $-\sin(k\pi x/L)$. The factor -1 can be absorbed in the coefficient c_2. If $n = 0$, then $\lambda = 0$ and, from case I, $y = 0$.

is said to be a *regular Sturm–Liouville problem*.* *The coefficients in* (4) *and* (5) *are assumed to be real and independent of* λ. *In addition,* α_1 *and* β_1 *are not both zero, and* α_2 *and* β_2 *are not both zero.*

Since the differential equation (3) is also homogeneous, the Sturm–Liouville problem always possesses the trivial solution $y = 0$. However, this solution is of no interest to us. As in Examples 1 and 2, in solving such a problem we seek to find numbers λ (eigenvalues) and nontrivial solutions y that depend on λ (eigenfunctions).

Properties The following theorem is a list of some of the more important of the many properties of the regular Sturm–Liouville problem. We shall prove only the last property.

THEOREM 11.2 Properties of the Regular Sturm–Liouville Problem

(i) There exists an infinite number of real eigenvalues that can be arranged in increasing order $\lambda_1 < \lambda_2 < \lambda_3 < \cdots < \lambda_n < \cdots$ such that $\lambda_n \to \infty$ as $n \to \infty$.

(ii) For each eigenvalue there is only one eigenfunction (except for nonzero multiples).

(iii) Eigenfunctions corresponding to different eigenvalues are linearly independent.

(iv) The set of eigenfunctions corresponding to the set of eigenvalues is orthogonal with respect to the weight function $p(x)$ on the interval $[a, b]$.

---

*__Charles Jacques François Sturm (1803–1855)__ Born in Switzerland, Sturm spent his adult life in Paris. His primary interests were fluid mechanics and differential equations. Sturm was the first to accurately determine the speed of sound in water. He also won a prize for an article on compressible fluids. In mathematics he won the coveted Grand Prix des Sciences Mathématiques for his work in differential equations. Sturm held the chair of mechanics at the Sorbonne and was elected a member of the French Academy of Sciences. The work done with Liouville on differential equations and boundary-value problems appeared in the *Journal des mathématiques pures et appliquées* during the period 1836 to 1838.

Joseph Liouville (1809–1882) Besides his work in differential equations, the Frenchman Joseph Liouville is remembered for his work in number theory and complex variables. His research in number theory led to the discovery of transcendental numbers—that is, numbers that are not roots of any polynomial equation with integer coefficients. The numbers e and π are examples of transcendental numbers, although Liouville was not able to show this. In complex analysis his notable contribution was Liouville's theorem, which states that every bounded entire function must be a constant. Liouville, like Sturm, was a member of the French Academy of Sciences and a professor at the Sorbonne. He was the founder and editor of the *Journal des mathématiques pures et appliquées*.

11.4 Sturm–Liouville Problem

Proof of Orthogonality Let y_m and y_n be eigenfunctions corresponding to eigenvalues λ_m and λ_n, respectively. Then

$$\frac{d}{dx}[r(x)y'_m] + (q(x) + \lambda_m p(x))y_m = 0 \qquad (6)$$

$$\frac{d}{dx}[r(x)y'_n] + (q(x) + \lambda_n p(x))y_n = 0 \qquad (7)$$

Multiplying (6) by y_n and (7) by y_m and subtracting the two equations give

$$(\lambda_m - \lambda_n)p(x)y_m y_n = y_m \frac{d}{dx}[r(x)y'_n] - y_n \frac{d}{dx}[r(x)y'_m]$$

Integrating this last result by parts from $x = a$ to $x = b$ then yields

$$(\lambda_m - \lambda_n) \int_a^b p(x)y_m y_n \, dx = r(b)[y_m(b)y'_n(b) - y_n(b)y'_m(b)]$$
$$- r(a)[y_m(a)y'_n(a) - y_n(a)y'_m(a)] \qquad (8)$$

Now the eigenfunctions y_m and y_n must both satisfy the boundary conditions (4) and (5). In particular, from (4) we have

$$\alpha_1 y_m(a) + \beta_1 y'_m(a) = 0$$
$$\alpha_1 y_n(a) + \beta_1 y'_n(a) = 0$$

In order that this system be satisfied by α_1 and β_1 not both zero, the determinant of the coefficients must be zero:

$$y_m(a)y'_n(a) - y_n(a)y'_m(a) = 0$$

A similar argument applied to (5) also gives

$$y_m(b)y'_n(b) - y_n(b)y'_m(b) = 0$$

Since both members of the right side of (8) are zero, we have established the orthogonality relation

$$\int_a^b p(x)y_m y_n \, dx = 0, \qquad \lambda_m \neq \lambda_n \qquad (9)$$

∎

EXAMPLE 3 Solve
$$y'' + \lambda y = 0$$
subject to $\quad y(0) = 0, \quad y(1) + y'(1) = 0.$ $\qquad (10)$

Solution (10) is a regular Sturm–Liouville problem with $r(x) = 1$, $q(x) = 0$, $p(x) = 1$, $\alpha_1 = 1$, $\beta_1 = 0$, $\alpha_2 = 1$, $\beta_2 = 1$. Now for $\lambda > 0$ the condition $y(0) = 0$ implies $c_1 = 0$ in the general solution $y = c_1 \cos \sqrt{\lambda} x + c_2 \sin \sqrt{\lambda} x$. With

$y = c_2 \sin\sqrt{\lambda}\,x$, the boundary condition at $x = 1$ is satisfied if

$$c_2 \sin\sqrt{\lambda} + c_2\sqrt{\lambda}\cos\sqrt{\lambda} = 0$$

Choosing $c_2 \neq 0$, we see that the last equation is equivalent to

$$\tan\sqrt{\lambda} = -\sqrt{\lambda} \qquad (11)$$

Equation (11) can be solved only by a numerical procedure such as Newton's method (see Problem 21). Nonetheless, if we let $x = \sqrt{\lambda}$, then Figure 11.16 shows the plausibility that there exists an infinite number of roots of $\tan x = -x$. The eigenvalues of the problem are then $\lambda_n = x_n^2$, where the x_n, $n = 1, 2, 3, \ldots$, are the consecutive positive roots. The corresponding eigenfunctions are $y = \sin\sqrt{\lambda_n}\,x$. Furthermore, $\{\sin\sqrt{\lambda_n}\,x\}$, $n = 1, 2, 3, \ldots$, is an orthogonal set with respect to the weight function $p(x) = 1$ on the interval $[0, 1]$.

Observe that even though $\lambda = 0$ satisfies (11), $\lambda = 0$ is not an eigenvalue, since the solution of the problem $y'' = 0$, $y(0) = 0$, $y(1) + y'(1) = 0$ is $y = 0$. Similarly, for $\lambda < 0$ the boundary-value problem has only the trivial solution. □

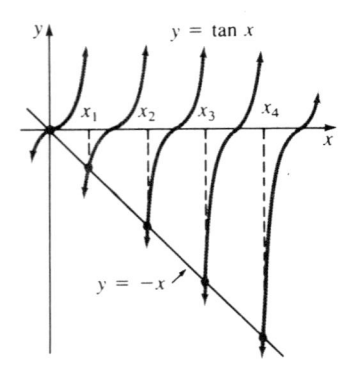

FIGURE 11.16

In some circumstances we can prove the orthogonality of the solutions of (3) without the necessity of specifying a boundary condition at $x = a$ and at $x = b$.

Singular Sturm–Liouville Problem If $r(a) = 0$, or $r(b) = 0$, or $r(a) = r(b) = 0$, the Sturm–Liouville problem is said to be **singular**. In the first case, $x = a$ may be a singular point of the differential equation, and consequently a solution of (3) may become unbounded as $x \to a$. However, we see from (8) that if $r(a) = 0$, then no boundary condition is required at $x = a$ to prove orthogonality of the eigenfunctions provided these solutions are bounded at that point. This latter requirement guarantees the existence of the integrals involved. By assuming the solutions of (3) are bounded on the closed interval $[a, b]$, we can say:

(*i*) if $r(a) = 0$, then the orthogonality relation (9) holds with no boundary condition at $x = a$;

(*ii*) if $r(b) = 0$, then the orthogonality relation (9) holds with no boundary condition at $x = b$;* and

(*iii*) if $r(a) = r(b) = 0$, then the orthogonality relation (9) holds with no boundary condition at either $x = a$ or $x = b$.

Self-Adjoint Form If the coefficients are continuous and $a(x) \neq 0$ for all x in some interval, then any second-order differential equation

$$a(x)y'' + b(x)y' + (c(x) + \lambda d(x))y = 0$$

---

*Conditions (*i*) and (*ii*) are equivalent to choosing $\alpha_1 = 0$, $\beta_1 = 0$ and $\alpha_2 = 0$, $\beta_2 = 0$, respectively.

can be put into the so-called **self-adjoint form** (3) by multiplying the equation by the integrating factor

$$\frac{1}{a(x)} e^{\int (b/a)\,dx} \tag{12}$$

EXAMPLE 4 In Section 5.4 we saw that the general solution of the parametric Bessel equation $x^2 y'' + xy' + (\lambda^2 x^2 - n^2)y = 0$, $n = 0, 1, 2, \ldots$, was $y = c_1 J_n(\lambda x) + c_2 Y_n(\lambda x)$. From (12) we find that $1/x$ is an integrating factor. Multiplying Bessel's equation by this factor yields the self-adjoint form

$$\frac{d}{dx}[xy'] + \left(\lambda^2 x - \frac{n^2}{x}\right) y = 0$$

where we identify $r(x) = x$, $q(x) = -n^2/x$, $p(x) = x$, and λ is replaced with λ^2. Now $r(0) = 0$, and of the two solutions $J_n(\lambda x)$ and $Y_n(\lambda x)$, only $J_n(\lambda x)$ is bounded at $x = 0$. Thus, in view of (i), the set $\{J_n(\lambda_i x)\}$, $i = 1, 2, 3, \ldots$, is orthogonal with respect to the weight function $p(x) = x$ on an interval $[0, b]$.

$$\int_0^b x J_n(\lambda_i x) J_n(\lambda_j x)\, dx = 0, \qquad \lambda_i \neq \lambda_j$$

provided the eigenvalues λ_i, $i = 1, 2, 3, \ldots$, are defined by means of a boundary condition at $x = b$ of the type

$$\alpha_2 J_n(\lambda b) + \beta_2 \lambda J_n'(\lambda b) = 0^* \tag{13}$$

For any choice of α_2 and β_2 not both zero, it is known that (13) has an infinite number of roots $x_i = \lambda_i b$ from which we get $\lambda_i = x_i/b$. More will be said about the eigenvalues in the next chapter. □

EXAMPLE 5 The Legendre polynomials $P_n(x)$ are bounded solutions of Legendre's differential equation $(1 - x^2)y'' - 2xy' + n(n + 1)y = 0$, $n = 0, 1, 2, \ldots$, on the interval $[-1, 1]$. From the self-adjoint form

$$\frac{d}{dx}[(1 - x^2)y'] + n(n + 1)y = 0$$

we read off $r(x) = 1 - x^2$, $q(x) = 0$, $p(x) = 1$, and $\lambda = n(n + 1)$. Observing that $r(-1) = r(1) = 0$, we find from (iii) that the set $\{P_n(x)\}$, $n = 0, 1, 2, \ldots$, is orthogonal with respect to the weight function $p(x) = 1$ on $[-1, 1]$:

$$\int_{-1}^1 P_m(x) P_n(x)\, dx = 0, \qquad m \neq n \qquad \square$$

---

*The extra factor of λ comes from the chain rule: $\dfrac{d}{dx} J_n(\lambda x) = \lambda J_n'(\lambda x)$.

EXERCISES 11.4 Answers to odd-numbered problems begin on page A-62.

In Problems 1–12 find the eigenvalues and eigenfunctions of the given boundary-value problem.

1. $y'' + \lambda y = 0$
   $y(0) = 0, y(\pi) = 0$

2. $y'' + \lambda y = 0$
   $y(0) = 0, y(\pi/4) = 0$

3. $y'' + \lambda y = 0$
   $y'(0) = 0, y(L) = 0$

4. $y'' + \lambda y = 0$
   $y(0) = 0, y'(\pi/2) = 0$

5. $y'' + \lambda y = 0$
   $y'(0) = 0, y'(\pi) = 0$

6. $y'' + \lambda y = 0$
   $y(-\pi) = 0, y(\pi) = 0$

7. $y'' + \lambda y = 0$
   $y'(0) = 0, y(1) + y'(1) = 0$

8. $y'' + \lambda y = 0$
   $y(0) + y'(0) = 0, y(1) = 0$

9. $y'' + 2y' + (\lambda + 1)y = 0$
   $y(0) = 0, y(5) = 0$

10. $y'' + (\lambda + 1)y = 0$
    $y'(0) = 0, y'(1) = 0$

11. $y'' + \lambda^2 y = 0$
    $y(0) = 0, y(L) = 0$

12. $y'' + \lambda^2 y = 0$
    $y(0) = 0, y'(3\pi) = 0$

13. When a load P is applied to a long slender column of uniform cross-section and length L, the y-coordinate of a point on its deflection curve is given by

    $$EI \frac{d^2y}{dx^2} + Py = 0$$

    where the constants E and I are, respectively, Young's modulus of elasticity and the moment of inertia of the cross-section. See Figure 11.17. If the column is constrained from rotating at its top and bottom, the boundary conditions are $y(0) = 0$ and $y(L) = 0$.

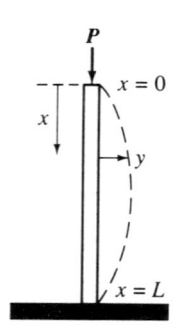

    (a) Show that this boundary-value problem indicates that the column does not bend at all unless the load P has one of the values $P_n = (n^2\pi^2/L^2)EI$, $n = 1, 2, 3, \ldots$. These different loads are called **critical loads**.

    (b) Show that the deflection curve corresponding to the lowest critical load $P_1 = (\pi^2/L^2)EI$, called the **Euler load**, is $y_1(x) = c_2 \sin(\pi x/L)$. Here $y_1(x)$ is called the **first buckling mode**.

14. In (9) of Section 1.3 we saw that the differential equation for the displacement $y(x)$ of a rotating string in which the tension T is not constant is given by

    $$\frac{d}{dx}\left[T(x)\frac{dy}{dx}\right] + \rho\omega^2 y = 0$$

    Suppose that $1 < x < e$ and that the tension is given by $T(x) = x^2$. If $y(1) = 0$, $y(e) = 0$, and $\rho\omega^2 > 0.25$, show that the **critical speeds** of angular rotation are $\omega_n = \frac{1}{2}\sqrt{(4n^2\pi^2 + 1)/\rho}$ and that the corresponding deflection curves are $y_n(x) = c_2 x^{-1/2} \sin(n\pi \ln x)$, $n = 1, 2, 3, \ldots$.

In Problems 15 and 16 find the eigenvalues and eigenfunctions of the given boundary-value problem. Put the Cauchy–Euler equation in self-adjoint form. Give an orthogonality relation.

15. $x^2 y'' + xy' + \lambda y = 0$
    $y(1) = 0, y(e^\pi) = 0$

16. $x^2 y'' + 2xy' + \lambda y = 0$
    $y(1) = 0, y(e^2) = 0$

17. Find the square norm of each eigenfunction in Problem 7.

18. Show that for the eigenfunctions in Example 3,

    $$\|\sin \sqrt{\lambda_n} x\|^2 = \frac{1}{2}[1 + \cos^2 \sqrt{\lambda_n}]$$

19. **Laguerre's differential equation*** $xy'' + (1 - x)y' + ny = 0$, $n = 0, 1, 2, \ldots$, has polynomial solutions $L_n(x)$. Put the equation in self-adjoint form and give an orthogonality relation.

---

*Named after the French mathematician **Edmond Laguerre** (1834–1886).

20. **Hermite's differential equation*** $y'' - 2xy' + 2ny = 0$, $n = 0, 1, 2, \ldots$, has polynomial solutions $H_n(x)$. Put the equation in self-adjoint form and give an orthogonality relation.

21. Find the first four eigenvalues in Example 3 to four decimal places. [*Hint:* See Section 15.1]

22. Find the first four eigenvalues in Problem 7 to four decimal places.

23. Show that the orthogonality relation (9) holds if $r(a) = r(b)$ and the eigenfunctions satisfy the **periodic boundary conditions** $y(a) = y(b)$, $y'(a) = y'(b)$.

24. Consider $y'' + \lambda y = 0$ subject to the periodic boundary conditions $y(-L) = y(L)$, $y'(-L) = y'(L)$. Show that the eigenfunctions are $\left\{ 1, \cos\dfrac{\pi}{L} x, \cos\dfrac{2\pi}{L} x, \ldots, \sin\dfrac{\pi}{L} x, \sin\dfrac{2\pi}{L} x, \sin\dfrac{3\pi}{L} x, \ldots \right\}$. This set, which is orthogonal on $[-L, L]$, is the basis for the Fourier series.

---

*Named after the French mathematician **Charles Hermite** (1822–1901). Hermite polynomials are important in some aspects of quantum mechanics.

11.5 BESSEL AND LEGENDRE SERIES

Fourier series, Fourier cosine series, and Fourier sine series are three ways of expanding a function in terms of an orthogonal set of functions. But such expansions are by no means limited to orthogonal sets of trigonometric functions. We saw in Section 11.1 that a function f defined on an interval (a, b) could be expanded at least formally in terms of any set of functions $\{\phi_n(x)\}$ that is orthogonal with respect to a weight function on $[a, b]$. Many of these so-called generalized Fourier series come from Sturm–Liouville problems arising in physical applications of linear partial differential equations. Fourier series and generalized Fourier series, as well as the two series considered in this section, will appear again in the subsequent consideration of these applications.

11.5.1 Fourier–Bessel Series

In Example 4 of the preceding section it was seen that the set of Bessel functions $J_n(\lambda_i x)$ is orthogonal with respect to the weight function $p(x) = x$ on an interval $[0, b]$ when the eigenvalues are defined by means of a boundary condition

$$\alpha_2 J_n(\lambda b) + \beta_2 \lambda J_n'(\lambda b) = 0 \tag{1}$$

From (7) and (8) of Section 11.1 the generalized Fourier expansion of a function f defined on $(0, b)$ in terms of this orthogonal set is

$$f(x) = \sum_{n=1}^{\infty} c_i J_n(\lambda_i x) \tag{2}$$

where

$$c_i = \frac{\int_0^b x J_n(\lambda_i x) f(x)\, dx}{\int_0^b x J_n^2(\lambda_i x)\, dx} \tag{3}$$

The series (2) with coefficients (3) is called a **Fourier–Bessel series**.

Differential Recurrence Relations The differential recurrence relations that were discussed in Section 5.4:

$$\frac{d}{dx}[x^n J_n(x)] = x^n J_{n-1}(x) \tag{4}$$

$$\frac{d}{dx}[x^{-n} J_n(x)] = -x^{-n} J_{n+1}(x) \tag{5}$$

are often useful in the evaluation of the coefficients in (3).

Square Norm The value of the square norm $\|J_n(\lambda_i x)\|^2 = \int_0^b x J_n^2(\lambda_i x)\, dx$ depends on how the eigenvalues λ_i are defined. If $y = J_n(\lambda x)$, then we know from Example 4 of Section 11.4 that

$$\frac{d}{dx}[xy'] + \left(\lambda^2 x - \frac{n^2}{x}\right) y = 0$$

After we multiply by $2xy'$, this equation can be written as

$$\frac{d}{dx}[xy']^2 + (\lambda^2 x^2 - n^2)\frac{d}{dx}[y]^2 = 0$$

Integrating by parts on $[0, b]$ then gives

$$2\lambda^2 \int_0^b xy^2\, dx = \left([xy']^2 + (\lambda^2 x^2 - n^2) y^2\right)\Big|_0^b$$

Since $y = J_n(\lambda x)$, the lower limit is zero for $n > 0$ since $J_n(0) = 0$. For $n = 0$ the quantity $[xy']^2 + \lambda^2 x^2 y^2$ is zero at $x = 0$. Thus,

$$2\lambda^2 \int_0^b x J_n^2(\lambda x)\, dx = \lambda^2 b^2 [J_n'(\lambda b)]^2 + (\lambda^2 b^2 - n^2)[J_n(\lambda b)]^2 \tag{6}$$

where we have used $y' = \lambda J_n'(\lambda x)$.

We now consider three cases of (1):

CASE I If we choose $\alpha_2 = 1$ and $\beta_2 = 0$, then (1) is

$$J_n(\lambda b) = 0 \tag{7}$$

There is an infinite number of positive roots x_i of (7) (see Figure 5.4), and so the eigenvalues are *positive** and are given by $\lambda_i = x_i/b$. The number

---

*No new eigenvalues result from the negative roots of (7) since $J_n(-x) = (-1)^n J_n(x)$ (see Problem 38 of Exercises 5.4).

11.5 Bessel and Legendre Series

0 is not an eigenvalue for any n, since $J_n(0) = 0$ for $n = 1, 2, 3, \ldots$ and $J_0(0) = 1$. In other words, if $\lambda = 0$, we get the trivial function (which is never an eigenfunction) for $n = 1, 2, 3, \ldots$, and for $n = 0$, $\lambda = 0$ does not satisfy (7). Using (5) in the form $xJ'_n(x) = nJ_n(x) - xJ_{n+1}(x)$, we find from (6) and (7) that the square norm of $J_n(\lambda_i x)$ is

$$\|J_n(\lambda_i x)\|^2 = \frac{b^2}{2} J^2_{n+1}(\lambda_i b) \tag{8}$$

CASE II If we choose $\alpha_2 = h \geq 0$, $\beta_2 = b$, then (1) is

$$hJ_n(\lambda b) + \lambda b J'_n(\lambda b) = 0 \tag{9}$$

Equation (9) has an infinite number of positive roots x_i for each positive integer $n = 1, 2, 3, \ldots$. As before, the eigenvalues are obtained from $\lambda_i = x_i/b$. $\lambda = 0$ is not eigenvalue for $n = 1, 2, 3, \ldots$. Substituting $\lambda_i b J'_n(\lambda_i b) = -hJ_n(\lambda_i b)$ in (6), we find that the square norm of $J_n(\lambda_i x)$ is now

$$\|J_n(\lambda_i x)\|^2 = \frac{\lambda_i^2 b^2 - n^2 + h^2}{2\lambda_i^2} J_n^2(\lambda_i b) \tag{10}$$

CASE III If $h = 0$ and $n = 0$ in (9), the eigenvalues λ_i are defined from the roots of

$$J'_0(\lambda b) = 0 \tag{11}$$

Even though (11) is just a special case of (9), it is the only situation for which $\lambda = 0$ is an eigenvalue. To see this observe that for $n = 0$ the result in (5) implies that

$$J'_0(\lambda b) = 0 \quad \text{is equivalent to} \quad J_1(\lambda b) = 0 \tag{12}$$

Since $x_1 = 0$ is a root of this equation and since $J_0(0) = 1$ is nontrivial, we conclude that $\lambda_1 = 0$ is an eigenvalue. But obviously we cannot use (10) when $h = 0$, $n = 0$, and $\lambda_1 = 0$. However, by the definition of the square norm, we have

$$\|1\|^2 = \int_0^b x\, dx = \frac{b^2}{2} \tag{13}$$

For $\lambda_i > 0$ we can use (10) with $h = 0$ and $n = 0$:

$$\|J_0(\lambda_i x)\|^2 = \frac{b^2}{2} J_0^2(\lambda_i b) \tag{14}$$

The following summary gives the corresponding three forms of the series (2):

DEFINITION 11.7 **Fourier–Bessel Series**

The **Fourier–Bessel series** of a function f defined on the interval $(0, b)$ is given by

(i)
$$f(x) = \sum_{i=1}^{\infty} c_i J_n(\lambda_i x) \tag{15}$$

$$c_i = \frac{2}{b^2 J_{n+1}^2(\lambda_i b)} \int_0^b x J_n(\lambda_i x) f(x)\, dx \tag{16}$$

where the λ_i are defined by (7), or

(ii)
$$f(x) = \sum_{i=1}^{\infty} c_i J_n(\lambda_i x) \tag{17}$$

$$c_i = \frac{2\lambda_i^2}{(\lambda_i^2 b^2 - n^2 + h^2) J_n^2(\lambda_i b)} \int_0^b x J_n(\lambda_i x) f(x)\, dx \tag{18}$$

where the λ_i are defined by (9), or

(iii)
$$f(x) = c_1 + \sum_{i=2}^{\infty} c_i J_0(\lambda_i x) \tag{19}$$

$$c_1 = \frac{2}{b^2} \int_0^b x f(x)\, dx$$

$$c_i = \frac{2}{b^2 J_0^2(\lambda_i b)} \int_0^b x J_0(\lambda_i x) f(x)\, dx \tag{20}$$

where the λ_i are defined by (11).

Convergence of a Fourier–Bessel Series Sufficient conditions for the convergence of a Fourier–Bessel series are not particularly restrictive.

THEOREM 11.3 **Conditions for Convergence**

If f and f' are piecewise continuous on the open interval $(0, b)$, then a Fourier–Bessel expansion of f converges to $f(x)$ at any point where f is continuous and to the average $[f(x-) + f(x+)]/2$ at a point where f is discontinuous.

EXAMPLE 1 Expand $f(x) = x$, $0 < x < 3$, in a Fourier–Bessel series using Bessel functions of order 1 that satisfy the boundary condition $J_1(3\lambda) = 0$.

Solution We use (15) where the coefficients c_i are given by (16):

$$c_i = \frac{2}{3^2 J_2^2(3\lambda_i)} \int_0^3 x^2 J_1(\lambda_i x)\, dx$$

To evaluate this integral we let $t = \lambda_i x$ and use (4) in the form $(d/dt)[t^2 J_2(t)] = t^2 J_1(t)$:

$$c_i = \frac{2}{9\lambda_i^3 J_2^2(3\lambda_i)} \int_0^{3\lambda_i} \frac{d}{dt}[t^2 J_2(t)]\, dt$$

$$= \frac{2}{\lambda_i J_2(3\lambda_i)}$$

Therefore the desired expansion is

$$f(x) = 2 \sum_{i=1}^\infty \frac{1}{\lambda_i J_2(3\lambda_i)} J_1(\lambda_i x) \qquad \square$$

EXAMPLE 2 If the eigenvalues λ_i in Example 1 are defined by $J_1(3\lambda) + \lambda J_1'(3\lambda) = 0$, then the only thing that changes in the expansion is the value of the square norm. Multiplying the boundary condition by 3 gives $3J_1(3\lambda) + 3\lambda J_1'(3\lambda) = 0$, which now matches (9) when $h = 3$, $b = 3$, and $n = 1$. Thus, (18) and (17) yield, in turn,

$$c_i = \frac{18\lambda_i J_2(3\lambda_i)}{(9\lambda_i^2 + 8)J_1^2(3\lambda_i)}$$

and $$f(x) = 18 \sum_{i=1}^\infty \frac{\lambda_i J_2(3\lambda_i)}{(9\lambda_i^2 + 8)J_1^2(3\lambda_i)} J_1(\lambda_i x) \qquad \square$$

11.5.2 Fourier–Legendre Series

From Example 5 of Section 11.4 we know that the set of Legendre polynomials $\{P_n(x)\}$, $n = 0, 1, 2, \ldots$, is orthogonal with respect to the weight function $p(x) = 1$ on $[-1, 1]$. Furthermore, it can be proved that

$$\|P_n(x)\|^2 = \int_{-1}^1 P_n^2(x)\, dx$$

$$= \frac{2}{2n + 1}$$

The generalized Fourier series of Legendre polynomials is summarized as follows:

> **DEFINITION 11.8 Fourier–Legendre Series**
>
> The **Fourier–Legendre series** of a function f defined on the interval $(-1, 1)$ is given by
>
> $$f(x) = \sum_{n=0}^{\infty} c_n P_n(x) \qquad (21)$$
>
> where
> $$c_n = \frac{2n+1}{2} \int_{-1}^{1} f(x) P_n(x)\, dx \qquad (22)$$

Convergence of a Fourier–Legendre Series

> **THEOREM 11.4 Conditions for Convergence**
>
> If f and f' are piecewise continuous on $(-1, 1)$, then the Fourier–Legendre series (21) converges to $f(x)$ at a point of continuity and to $[f(x+) + f(x-)]/2$ at a point of discontinuity.

EXAMPLE 3 Write out the first four nonzero terms in the Fourier–Legendre expansion of

$$f(x) = \begin{cases} 0, & -1 < x < 0, \\ 1, & 0 < x < 1 \end{cases}$$

Solution The first several Legendre polynomials are listed on page 315. From these and (22) we find

$$c_0 = \frac{1}{2}\int_{-1}^{1} f(x)P_0(x)\,dx = \frac{1}{2}\int_{0}^{1} 1 \cdot 1\,dx = \frac{1}{2}$$

$$c_1 = \frac{3}{2}\int_{-1}^{1} f(x)P_1(x)\,dx = \frac{3}{2}\int_{0}^{1} 1 \cdot x\,dx = \frac{3}{4}$$

$$c_2 = \frac{5}{2}\int_{-1}^{1} f(x)P_2(x)\,dx = \frac{5}{2}\int_{0}^{1} 1 \cdot \frac{1}{2}(3x^2 - 1)\,dx = 0$$

$$c_3 = \frac{7}{2}\int_{-1}^{1} f(x)P_3(x)\,dx = \frac{7}{2}\int_{0}^{1} 1 \cdot \frac{1}{2}(5x^3 - 3x)\,dx = -\frac{7}{16}$$

$$c_4 = \frac{9}{2}\int_{-1}^{1} f(x)P_4(x)\,dx = \frac{9}{2}\int_{0}^{1} 1\cdot\frac{1}{8}(35x^4 - 30x^2 + 3)\,dx = 0$$

$$c_5 = \frac{11}{2}\int_{-1}^{1} f(x)P_5(x)\,dx = \frac{11}{2}\int_{0}^{1} 1\cdot\frac{1}{8}(63x^5 - 70x^3 + 15x)\,dx = \frac{11}{32}$$

Hence, $\quad f(x) = \frac{1}{2}P_0(x) + \frac{3}{4}P_1(x) - \frac{7}{16}P_3(x) + \frac{11}{32}P_5(x) + \cdots$ □

Alternative Form In applications the Fourier–Legendre series appears in an alternative form. If we let $x = \cos\theta$, then $dx = -\sin\theta\,d\theta$ and (21) and (22) become

$$F(\theta) = \sum_{n=0}^{\infty} c_n P_n(\cos\theta) \qquad (23)$$

$$c_n = \frac{2n+1}{2}\int_0^{\pi} F(\theta)P_n(\cos\theta)\sin\theta\,d\theta \qquad (24)$$

where $f(\cos\theta)$ has been replaced by $F(\theta)$.

EXERCISES 11.5 Answers to odd-numbered problems begin on page A-62.

[11.5.1]

1. Find the first four eigenvalues λ_k defined by $J_1(3\lambda) = 0$ if the first four positive roots of $J_1(x) = 0$ are $x_1 = 3.8$, $x_2 = 7.0$, $x_3 = 10.2$, and $x_4 = 13.3$.

2. Find the first four eigenvalues λ_k defined by $J_0'(2\lambda) = 0$.

In Problems 3–6 expand $f(x) = 1$, $0 < x < 2$, in a Fourier–Bessel series using Bessel functions of order 0 that satisfy the given boundary condition.

3. $J_0(2\lambda) = 0$ **4.** $J_0'(2\lambda) = 0$

5. $J_0(2\lambda) + 2\lambda J_0'(2\lambda) = 0$ **6.** $J_0(2\lambda) + \lambda J_0'(2\lambda) = 0$

In Problems 7–10 expand the given function in a Fourier–Bessel series using Bessel functions of the same order as in the indicated boundary condition.

7. $f(x) = 5x$, $0 < x < 4$;
$3J_1(4\lambda) + 4\lambda J_1'(4\lambda) = 0$

8. $f(x) = x^2$, $0 < x < 1$;
$J_2(\lambda) = 0$

9. $f(x) = x^2$, $0 < x < 3$;
$J_0'(3\lambda) = 0$ [Hint: $t^3 = t^2 \cdot t$]

10. $f(x) = 1 - x^2$, $0 < x < 1$;
$J_0(\lambda) = 0$

[11.5.2]

A Fourier–Legendre expansion of a polynomial function defined on the interval $(-1, 1)$ is necessarily a finite series. (Why?) In Problems 11 and 12 find the Fourier–Legendre expansion of the given function.

11. $f(x) = x^2$ **12.** $f(x) = x^3$

In Problems 13 and 14 write out the first four nonzero terms in the Fourier–Legendre expansion of the given function.

13. $f(x) = \begin{cases} 0, & -1 < x < 0 \\ x, & 0 \leq x < 1 \end{cases}$

14. $f(x) = e^x$, $-1 < x < 1$

15. The first three Legendre polynomials are $P_0(x) = 1$, $P_1(x) = x$, and $P_2(x) = \frac{1}{2}(3x^2 - 1)$. If $x = \cos\theta$, then $P_0(\cos\theta) = 1$ and $P_1(\cos\theta) = \cos\theta$. Show that $P_2(\cos\theta) = \frac{1}{4}(3\cos 2\theta + 1)$.

16. Use the results given in Problem 15 to find a Fourier–Legendre expansion (23) of $F(\theta) = 1 - \cos 2\theta$.

17. A Legendre polynomial $P_n(x)$ is an even or odd function, depending on whether n is even or odd. Show that if f is an even function on $(-1, 1)$, then (21) and (22) become, respectively,

$$f(x) = \sum_{n=0}^{\infty} c_{2n} P_{2n}(x) \qquad (25)$$

$$c_{2n} = (4n+1) \int_0^1 f(x) P_{2n}(x)\, dx \qquad (26)$$

18. Show that if f is an odd function on the interval $(-1, 1)$, then (21) and (22) become, respectively,

$$f(x) = \sum_{n=0}^{\infty} c_{2n+1} P_{2n+1}(x) \qquad (27)$$

$$c_{2n+1} = (4n+3) \int_0^1 f(x) P_{2n+1}(x)\, dx \qquad (28)$$

The series (25) and (27) can also be used when f is defined on only $(0, 1)$. Both series represent f on $(0, 1)$, but on $(-1, 0)$, (25) represents its even extension, whereas (27) represents its odd extension. In Problems 19 and 20 write out the first three nonzero terms in the indicated expansion of the given function. What function does the series represent on $(-1, 1)$?

19. $f(x) = x,\ 0 < x < 1;\ (25)$

20. $f(x) = 1,\ 0 < x < 1;\ (27)$

SUMMARY

A set of functions $\{\phi_n(x)\}$, $n = 0, 1, 2, \ldots$, is said to be **orthogonal** on an interval $[a, b]$ if

$$\int_a^b \phi_m(x)\phi_n(x)\, dx \begin{cases} =0, & m \neq n \\ \neq 0, & m = n \end{cases}$$

A function f defined on the interval $(-p, p)$ can be expanded in terms of the trigonometric functions in the orthogonal set

$$\left\{ 1, \cos\frac{\pi}{p}x, \cos\frac{2\pi}{p}x, \ldots, \sin\frac{\pi}{p}x, \sin\frac{2\pi}{p}x, \ldots \right\}$$

We say

$$f(x) = \frac{a_0}{2} + \sum_{n=1}^{\infty}\left(a_n \cos\frac{n\pi}{p}x + b_n \sin\frac{n\pi}{p}x \right)$$

where

$$a_0 = \frac{1}{p}\int_{-p}^{p} f(x)\, dx \qquad a_n = \frac{1}{p}\int_{-p}^{p} f(x)\cos\frac{n\pi}{p}x\, dx$$

$$b_n = \frac{1}{p}\int_{-p}^{p} f(x)\sin\frac{n\pi}{p}x\, dx$$

is the **Fourier series** corresponding to f. The coefficients are obtained using the concept of orthogonality. If f is an even function on the interval, then $b_n = 0$, $n = 1, 2, 3, \ldots$, and

$$a_0 = \frac{2}{p}\int_0^p f(x)\, dx \quad \text{and} \quad a_n = \frac{2}{p}\int_0^p f(x)\cos\frac{n\pi}{p}x\, dx \qquad (1)$$

Similarly, if f is odd on the interval, then $a_n = 0$, $n = 0, 1, 2, \ldots$, and

$$b_n = \frac{2}{p}\int_0^p f(x)\sin\frac{n\pi}{p}x\, dx \qquad (2)$$

When f is defined on an interval $(0, L)$, it can be neither even nor odd. Nonetheless we can expand f in a cosine or a sine series. By defining $p = L$, we obtain a cosine series by using the coefficients (1) and a sine series by using (2). Such series are known as **half-range expansions**.

The **Fourier–Bessel series** of a function f defined on the interval $(0, b)$ is

$$f(x) = \sum_{i=1}^{\infty} c_i J_n(\lambda_i x)$$

where

$$c_i = \frac{2}{b^2 J_{n+1}^2(\lambda_i b)} \int_0^b x J_n(\lambda_i x) f(x)\, dx, \quad \lambda_i \text{ defined by } J_n(\lambda b) = 0$$

$$c_i = \frac{2\lambda_i^2}{(\lambda_i^2 b^2 - n^2 + h^2) J_n^2(\lambda_i b)} \int_0^b x J_n(\lambda_i x) f(x)\, dx, \quad \lambda_i \text{ defined by } h J_n(\lambda b) + \lambda b J_n'(\lambda b) = 0$$

$$c_1 = \frac{2}{b^2} \int_0^b x f(x)\, dx, \quad c_i = \frac{2}{b^2 J_0^2(\lambda_i b)} \int_0^b x J_0(\lambda_i x) f(x)\, dx, \quad \lambda_i \text{ defined by } J_0'(\lambda b) = 0$$

The **Fourier–Legendre series** of a function f defined on $(-1, 1)$ is

$$f(x) = \sum_{n=0}^{\infty} c_n P_n(x) \quad \text{where} \quad c_n = \frac{2n+1}{2} \int_{-1}^{1} f(x) P_n(x)\, dx$$

CHAPTER 11 REVIEW EXERCISES *Answers to odd-numbered problems begin on page A-63.*

Answer Problems 1–10 without referring back to the text. Fill in the blank or answer true/false.

1. The functions $f(x) = x^2 - 1$ and $f(x) = x^5$ are orthogonal on $[-\pi, \pi]$. _____

2. The product of an odd function and an odd function is _____.

3. To expand $f(x) = |x| + 1$, $-\pi < x < \pi$, in an appropriate Fourier series we would use a _____ series.

4. Since $f(x) = x^2$, $0 < x < 2$, is not an even function, it cannot be expanded in a Fourier cosine series. _____

5. The Fourier series of $f(x) = \begin{cases} 3, & -\pi < x < 0 \\ 0, & 0 \le x < \pi \end{cases}$ will converge to _____ at $x = 0$.

6. $y = 0$ is never an eigenfunction of a Sturm–Liouville problem. _____

7. $\lambda = 0$ is never an eigenvalue of a Sturm–Liouville problem. _____

8. For $\lambda = 25$ the corresponding eigenfunction for the boundary-value problem $y'' + \lambda y = 0$, $y'(0) = 0$, $y(\pi/2) = 0$ is _____.

9. **Chebyshev's differential equation** $(1 - x^2)y'' - xy' + n^2 y = 0$ has a polynomial solution $y = T_n(x)$ for $n = 0, 1, 2, \ldots$. The set of Chebyshev polynomials $\{T_n(x)\}$ is orthogonal with respect to the weight function _____ on the interval _____.

10. The set $\{P_n(x)\}$ of Legendre polynomials is orthogonal with respect to the weight function $p(x) = 1$ on $[-1, 1]$ and $P_0(x) = 1$. Hence, $\int_{-1}^{1} P_n(x)\, dx =$ _____ for $n > 0$.

11. Show that the set

$$\left\{ \sin \frac{\pi}{2L} x, \sin \frac{3\pi}{2L} x, \sin \frac{5\pi}{2L} x, \ldots \right\}$$

is orthogonal on the interval $0 \le x \le L$.

12. Find the norm of each function in Problem 11. Construct an orthonormal set.

13. Expand $f(x) = |x| - x$, $-1 < x < 1$, in a Fourier series.

14. Expand $f(x) = 2x^2 - 1$, $-1 < x < 1$, in a Fourier series.

15. Expand $f(x) = e^{-x}$, $0 < x < 1$, in a cosine series.

16. Expand the function given in Problem 15 in a sine series.

17. Find the eigenvalues and eigenfunctions of the boundary-value problem

$$x^2 y'' + xy' + 9\lambda y = 0$$
$$y'(1) = 0, \quad y(e) = 0$$

18. State an orthogonality relation for the eigenfunctions in Problem 17.

19. Expand $f(x) = \begin{cases} 1, & 0 < x < 2 \\ 0, & 2 \leq x < 4 \end{cases}$ in a Fourier–Bessel series using Bessel functions of order 0 that satisfy the boundary condition $J_0(4\lambda) = 0$.

20. Expand $f(x) = x^4$, $-1 < x < 1$, in a Fourier–Legendre series.

12

BOUNDARY-VALUE PROBLEMS IN RECTANGULAR COORDINATES

TOPICS TO REVIEW

linear first-order differential equations (2.5)
linear second-order differential equations with constant coefficients (3.3)
Fourier cosine and sine series (11.3)
generalized Fourier series (11.1)
boundary-value problem (11.4)

IMPORTANT CONCEPTS

linear partial differential equation
heat equation
wave equation
Laplace's equation
separation of variables
boundary-value problem
eigenvalues
eigenfunctions
homogeneous boundary conditions
nonhomogeneous boundary conditions
initial conditions

12.1 Separable Partial Differential Equations
12.2 Classical Equations and Boundary-Value Problems
12.3 Heat Equation
12.4 Wave Equation
12.5 Laplace's Equation
12.6 Nonhomogeneous Equations and Boundary Conditions
12.7 Use of Generalized Fourier Series
[O] 12.8 Boundary-Value Problems Involving Fourier Series in Two Variables
Summary
Chapter 12 Review Exercises

INTRODUCTION

In this and the next two chapters, the emphasis will be on two procedures used in solving partial differential equations that occur frequently in problems involving temperature distributions, vibrations, and potentials. These problems, called **boundary-value problems**, are described by relatively simple linear second-order partial differential equations. The thrust of both these procedures is to find particular solutions of a partial differential equation by reducing the latter to one or more ordinary differential equations.

We shall begin with the method of separation of variables. The application of this method leads back to the notions of eigenvalues, eigenfunctions, and the expansion of a function in a series of orthogonal functions.

12.1 SEPARABLE PARTIAL DIFFERENTIAL EQUATIONS

Linear Equations The general form of a **linear** second-order **partial differential equation** in two independent variables x and y is given by

$$A\frac{\partial^2 u}{\partial x^2} + B\frac{\partial^2 u}{\partial x \partial y} + C\frac{\partial^2 u}{\partial y^2} + D\frac{\partial u}{\partial x} + E\frac{\partial u}{\partial y} + Fu = G$$

where $A, B, C, \ldots, G$ are functions of x and y. When $G(x, y) = 0$, the equation is said to be **homogeneous**; otherwise it is **nonhomogeneous**.

EXAMPLE 1 The equation $\dfrac{\partial^2 u}{\partial x^2} + \dfrac{\partial^2 u}{\partial y^2} - u = 0$ is homogeneous, whereas the equation $\dfrac{\partial^2 u}{\partial x^2} - \dfrac{\partial u}{\partial y} = x^2$ is nonhomogeneous. □

Solutions A **solution** of a partial differential equation in two independent variables x and y is a function $u(x, y)$ that possesses all of the partial derivatives occurring in the equation and that satisfies the equation in some region of the xy-plane.

It is not our intention to focus on procedures for finding general solutions of partial differential equations. Unfortunately, for most second-order linear equations, even those with constant coefficients, a general solution cannot readily be obtained. But this is not as bad as it may sound since it is usually possible, and indeed easy, to find **particular solutions** of the important linear equations that appear in many applications.

Separation of Variables Although several methods may be tried to find particular solutions (see Problems 28 and 29), we shall be interested in just one: the so-called **method of separation of variables**. By seeking a particular solution in the form of a product of a function of x and a function of y,

$$u(x, y) = X(x)Y(y)$$

we can *sometimes* reduce a linear partial differential equation in two variables to two ordinary differential equations. To this end we note

$$\frac{\partial u}{\partial x} = X'Y \qquad \frac{\partial u}{\partial y} = XY'$$

and

$$\frac{\partial^2 u}{\partial x^2} = X''Y \qquad \frac{\partial^2 u}{\partial y^2} = XY''$$

where the primes denote ordinary differentiation.

12.1 Separable Partial Differential Equations

EXAMPLE 2 Find product solutions of $\dfrac{\partial^2 u}{\partial x^2} = 4 \dfrac{\partial u}{\partial y}$. (1)

Solution If $u(x, y) = X(x)Y(y)$, the equation becomes

$$X''Y = 4XY'$$

After dividing both sides by $4XY$, we have separated the variables:

$$\frac{X''}{4X} = \frac{Y'}{Y}$$

Since the left-hand side of the last equation is independent of y and is equal to the right-hand side, which is independent of x, we conclude that both sides of the equation are independent of *x and y*. In other words, each side of the equation must be a constant. In practice it is convenient to write this real *separation constant* as either λ^2 or $-\lambda^2$. We distinguish the following three cases:

CASE I When $\lambda^2 > 0$, the two equalities

$$\frac{X''}{4X} = \frac{Y'}{Y} = \lambda^2$$

lead to $\quad X'' - 4\lambda^2 X = 0 \quad$ and $\quad Y' - \lambda^2 Y = 0$

These latter equations have the solutions

$$X = c_1 \cosh 2\lambda x + c_2 \sinh 2\lambda x \quad \text{and} \quad Y = c_3 e^{\lambda^2 y}$$

respectively. Thus a particular solution of the partial differential equation is

$$\begin{aligned} u &= XY \\ &= (c_1 \cosh 2\lambda x + c_2 \sinh 2\lambda x)(c_3 e^{\lambda^2 y}) \\ &= A_1 e^{\lambda^2 y} \cosh 2\lambda x + B_1 e^{\lambda^2 y} \sinh 2\lambda x \end{aligned} \quad (2)$$

where $A_1 = c_1 c_3$ and $B_1 = c_2 c_3$.

CASE II When $-\lambda^2 < 0$, the two equalities

$$\frac{X''}{4X} = \frac{Y'}{Y} = -\lambda^2$$

give $\quad X'' + 4\lambda^2 X = 0 \quad$ and $\quad Y' + \lambda^2 Y = 0$

Since the solutions of these equations are

$$X = c_4 \cos 2\lambda x + c_5 \sin 2\lambda x \quad \text{and} \quad Y = c_6 e^{-\lambda^2 y}$$

respectively, another particular solution is

$$u = A_2 e^{-\lambda^2 y} \cos 2\lambda x + B_2 e^{-\lambda^2 y} \sin 2\lambda x \quad (3)$$

where $A_2 = c_4 c_6$ and $B_2 = c_5 c_6$.

CASE III When $\lambda^2 = 0$, it follows that

$$X'' = 0 \quad \text{and} \quad Y' = 0$$

In this case $\quad X = c_7 x + c_8 \quad \text{and} \quad Y = c_9$

so that $\quad u = A_3 x + B_3 \quad$ (4)

where $A_3 = c_7 c_9$ and $B_3 = c_8 c_9$. □

It is left as an exercise to verify that (2), (3), and (4) satisfy the given equation.

Separation of variables is not a general method for finding particular solutions; some linear partial differential equations are simply not separable. The reader should verify that the assumption $u = XY$ does not lead to a solution for $\partial^2 u/\partial x^2 - \partial u/\partial y = x$.

Superposition Principle The following theorem is analogous to Theorem 3.3 and is known as the **superposition principle**:

THEOREM 12.1 Superposition Principle

If $u_1, u_2, \ldots, u_k$ are solutions of a homogeneous linear partial differential equation, then the linear combination

$$u = c_1 u_1 + c_2 u_2 + \cdots + c_k u_k$$

where the c_i, $i = 1, 2, \ldots, k$, are constants, is also a solution.

Throughout the remainder of the chapter, we shall assume that, whenever we have an infinite set

$$u_1, u_2, u_3, \ldots$$

of solutions of a homogeneous linear equation, we can construct yet another solution u by forming the infinite series

$$u = \sum_{k=1}^{\infty} u_k$$

Classification of Equations A linear second-order partial differential equation in two independent variables with constant coefficients can be classified as one of three types. This classification depends only on the coefficients of the second-order derivatives. Of course, we assume that at least one of the coefficients A, B, and C is not zero.

> **DEFINITION 12.1 Classification of Equations**
>
> The linear second-order partial differential equation
>
> $$A \frac{\partial^2 u}{\partial x^2} + B \frac{\partial^2 u}{\partial x\, \partial y} + C \frac{\partial^2 u}{\partial y^2} + D \frac{\partial u}{\partial x} + E \frac{\partial u}{\partial y} + Fu = 0$$
>
> where A, B, C, D, E, and F are real constants, is said to be
>
> **hyperbolic** if $B^2 - 4AC > 0$
> **parabolic** if $B^2 - 4AC = 0$
> **elliptic** if $B^2 - 4AC < 0$

EXAMPLE 3 Classify each equation.

(a) $3 \dfrac{\partial^2 u}{\partial x^2} = \dfrac{\partial u}{\partial y}$ (b) $\dfrac{\partial^2 u}{\partial x^2} = \dfrac{\partial^2 u}{\partial y^2}$ (c) $\dfrac{\partial^2 u}{\partial x^2} + \dfrac{\partial^2 u}{\partial y^2} = 0$

Solution (a) By rewriting the given equation as

$$3 \frac{\partial^2 u}{\partial x^2} - \frac{\partial u}{\partial y} = 0$$

we can make the identifications $A = 3$, $B = 0$, and $C = 0$. Since $B^2 - 4AC = 0$, the equation is parabolic.

(b) By rewriting the equation as

$$\frac{\partial^2 u}{\partial x^2} - \frac{\partial^2 u}{\partial y^2} = 0$$

we see that $A = 1$, $B = 0$, and $C = -1$, so $B^2 - 4AC = -4(1)(-1) > 0$. The equation is hyperbolic.

(c) With $A = 1$, $B = 0$, and $C = 1$, $B^2 - 4AC = -4(1)(1) < 0$. The equation is elliptic. □

A detailed explanation of why we would want to classify a second-order partial differential equation is beyond the scope of this text. But the answer lies in the fact that we wish to solve partial differential equations subject to certain side conditions known as boundary and initial conditions. The kinds of side conditions appropriate for a given equation depend on whether the equation is hyperbolic, parabolic, or elliptic.

EXERCISES 12.1 Answers to odd-numbered problems begin on page A-63.

In Problems 1–16 use separation of variables to find, if possible, product solutions for the given partial differential equation.

1. $\dfrac{\partial u}{\partial x} = \dfrac{\partial u}{\partial y}$

2. $\dfrac{\partial u}{\partial x} + 3\dfrac{\partial u}{\partial y} = 0$

3. $u_x + u_y = u$

4. $u_x = u_y + u$

5. $x\dfrac{\partial u}{\partial x} = y\dfrac{\partial u}{\partial y}$

6. $y\dfrac{\partial u}{\partial x} + x\dfrac{\partial u}{\partial y} = 0$

7. $\dfrac{\partial^2 u}{\partial x^2} + \dfrac{\partial^2 u}{\partial x\,\partial y} + \dfrac{\partial^2 u}{\partial y^2} = 0$

8. $y\dfrac{\partial^2 u}{\partial x\,\partial y} + u = 0$

9. $k\dfrac{\partial^2 u}{\partial x^2} - u = \dfrac{\partial u}{\partial t}$, $k > 0$

10. $k\dfrac{\partial^2 u}{\partial x^2} = \dfrac{\partial u}{\partial t}$, $k > 0$

11. $a^2\dfrac{\partial^2 u}{\partial x^2} = \dfrac{\partial^2 u}{\partial t^2}$

12. $a^2\dfrac{\partial^2 u}{\partial x^2} = \dfrac{\partial^2 u}{\partial t^2} + 2k\dfrac{\partial u}{\partial t}$, $k > 0$

13. $\dfrac{\partial^2 u}{\partial x^2} + \dfrac{\partial^2 u}{\partial y^2} = 0$

14. $x^2\dfrac{\partial^2 u}{\partial x^2} + \dfrac{\partial^2 u}{\partial y^2} = 0$

15. $u_{xx} + u_{yy} = u$

16. $a^2 u_{xx} - g = u_{tt}$, g a constant

In Problems 17–26 classify the given partial differential equation as hyperbolic, parabolic, or elliptic.

17. $\dfrac{\partial^2 u}{\partial x^2} + \dfrac{\partial^2 u}{\partial x\,\partial y} + \dfrac{\partial^2 u}{\partial y^2} = 0$

18. $3\dfrac{\partial^2 u}{\partial x^2} + 5\dfrac{\partial^2 u}{\partial x\,\partial y} + \dfrac{\partial^2 u}{\partial y^2} = 0$

19. $\dfrac{\partial^2 u}{\partial x^2} + 6\dfrac{\partial^2 u}{\partial x\,\partial y} + 9\dfrac{\partial^2 u}{\partial y^2} = 0$

20. $\dfrac{\partial^2 u}{\partial x^2} - \dfrac{\partial^2 u}{\partial x\,\partial y} - 3\dfrac{\partial^2 u}{\partial y^2} = 0$

21. $\dfrac{\partial^2 u}{\partial x^2} = 9\dfrac{\partial^2 u}{\partial x\,\partial y}$

22. $\dfrac{\partial^2 u}{\partial x\,\partial y} - \dfrac{\partial^2 u}{\partial y^2} + 2\dfrac{\partial u}{\partial x} = 0$

23. $\dfrac{\partial^2 u}{\partial x^2} + 2\dfrac{\partial^2 u}{\partial x\,\partial y} + \dfrac{\partial^2 u}{\partial y^2} + \dfrac{\partial u}{\partial x} - 6\dfrac{\partial u}{\partial y} = 0$

24. $\dfrac{\partial^2 u}{\partial x^2} + \dfrac{\partial^2 u}{\partial y^2} = u$

25. $a^2\dfrac{\partial^2 u}{\partial x^2} = \dfrac{\partial^2 u}{\partial t^2}$

26. $k\dfrac{\partial^2 u}{\partial x^2} = \dfrac{\partial u}{\partial t}$, $k > 0$

27. Show that the equation
$$k\left(\dfrac{\partial^2 u}{\partial r^2} + \dfrac{1}{r}\dfrac{\partial u}{\partial r}\right) = \dfrac{\partial u}{\partial t}$$
possesses the product solution
$$u = e^{-k\lambda^2 t}(AJ_0(\lambda r) + BY_0(\lambda r))$$

28. (a) Show that the equation
$$a^2\dfrac{\partial^2 u}{\partial x^2} = \dfrac{\partial^2 u}{\partial t^2}$$
can be put into the form $\partial^2 u/\partial \eta\,\partial \xi = 0$ by means of the substitutions $\xi = x + at$, $\eta = x - at$.

(b) Show that a solution of the equation is
$$u = F(x + at) + G(x - at)$$
where F and G are arbitrary, twice-differentiable functions.

29. Find solutions of the form
$$u = e^{mx + ny} \text{ for } \dfrac{\partial^2 u}{\partial x^2} + \dfrac{\partial^2 u}{\partial x\,\partial y} - 6\dfrac{\partial^2 u}{\partial y^2} = 0.$$

30. Verify that the products (2), (3), and (4) satisfy (1).

31. Definition 12.1 generalizes to linear equations with coefficients that are functions of x and y. Determine the regions in the xy-plane for which the equation
$$(xy + 1)\dfrac{\partial^2 u}{\partial x^2} + (x + 2y)\dfrac{\partial^2 u}{\partial x\,\partial y} + \dfrac{\partial^2 u}{\partial y^2} + xy^2 u = 0$$
is hyperbolic, parabolic, or elliptic.

12.2 CLASSICAL EQUATIONS AND BOUNDARY-VALUE PROBLEMS

For the remainder of this and the next chapter, we shall be concerned principally with finding product solutions of the partial differential equations

$$k \frac{\partial^2 u}{\partial x^2} = \frac{\partial u}{\partial t}, \quad k > 0 \tag{1}$$

$$a^2 \frac{\partial^2 u}{\partial x^2} = \frac{\partial^2 u}{\partial t^2} \tag{2}$$

$$\frac{\partial^2 u}{\partial x^2} + \frac{\partial^2 u}{\partial y^2} = 0 \tag{3}$$

or slight variations of these equations. These classical equations of mathematical physics are known, respectively, as the **one-dimensional heat equation**, the **one-dimensional wave equation**, and **Laplace's equation in two dimensions**. "One-dimensional" refers to the fact that x denotes a spatial dimension, whereas t represents time. Laplace's equation is abbreviated as $\nabla^2 u = 0$, where

$$\nabla^2 u = \frac{\partial^2 u}{\partial x^2} + \frac{\partial^2 u}{\partial y^2}$$

is called the **two-dimensional Laplacian**. In three dimensions the Laplacian is

$$\nabla^2 u = \frac{\partial^2 u}{\partial x^2} + \frac{\partial^2 u}{\partial y^2} + \frac{\partial^2 u}{\partial z^2}$$

Note that the heat equation (1) is parabolic, the wave equation (2) is hyperbolic, and Laplace's equation (3) is an elliptic equation.

Heat Equation Equation (1) occurs in the theory of heat flow—that is, heat transferred by conduction in a rod or thin wire. The function $u(x, t)$ is temperature. Problems in mechanical vibrations often lead to the wave equation (2). For purposes of discussion, a solution $u(x, t)$ of (2) will represent the displacement of an idealized string. Last, a solution $u(x, y)$ of Laplace's equation (3) can be interpreted as the steady-state (that is, time-independent) temperature distribution throughout a thin, two-dimensional plate.

Even though we have to make many simplifying assumptions, it is worthwhile to see how equations such as (1) and (2) arise. Suppose a thin circular rod of length L has a cross-sectional area A and coincides with the x-axis on the interval $[0, L]$ (see Figure 12.1). Let us suppose:

FIGURE 12.1

- The flow of heat within the rod takes place only in the x-direction.
- The lateral, or curved, surface of the rod is insulated; that is, no heat escapes from this surface.

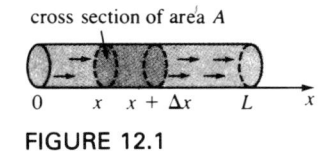

- No heat is being generated within the rod.
- The rod is homogeneous; that is, its mass per unit volume ρ is a constant.
- The specific heat γ and thermal conductivity K of the material of the rod are constants.

To derive the partial differential equation satisfied by the temperature $u(x, t)$ we need two empirical laws of heat conduction:

(i) *The quantity of heat Q in an element of mass m is*

$$Q = \gamma m u \qquad (4)$$

where u is the temperature of the element.

(ii) *The rate of heat flow Q_t through the cross-section indicated in Figure 12.1 is proportional to the area A of the cross-section and the partial derivative with respect to x of the temperature*

$$Q_t = -KAu_x \qquad (5)$$

Since heat flows in the direction of decreasing temperature, the minus sign in (5) is used to ensure that Q_t is positive for $u_x < 0$ (heat flow to the right) and negative for $u_x > 0$ (heat flow to the left). If the circular slice of the rod shown in Figure 12.1 between x and $x + \Delta x$ is very thin, then $u(x, t)$ can be taken as the approximate temperature at each point in the interval. Now the mass of the slice is $m = \rho(A \Delta x)$, and so it follows from (4) that the quantity of heat in it is

$$Q = \gamma \rho A \Delta x\, u \qquad (6)$$

Furthermore, when heat flows in the positive x-direction, we see from (5) that heat builds up in the slice at the net rate

$$-KAu_x(x, t) - (-KAu_x(x + \Delta x, t)) = KA[u_x(x + \Delta x, t) - u_x(x, t)] \qquad (7)$$

By differentiating (6) with respect to t, we see that this net rate is also given by

$$Q_t = \gamma \rho A \Delta x\, u_t \qquad (8)$$

Equations (7) and (8) give

$$\frac{K}{\gamma \rho} \frac{u_x(x + \Delta x, t) - u_x(x, t)}{\Delta x} = u_t \qquad (9)$$

Taking the limit of (9) as $\Delta x \to 0$ finally yields* (1) in the form

$$\frac{K}{\gamma \rho} u_{xx} = u_t$$

It is customary to let $k = K/\gamma \rho$ and call this positive constant the **thermal diffusivity**.

---

*Recall from calculus that $u_{xx} = \lim\limits_{\Delta x \to 0} \dfrac{u_x(x + \Delta x, t) - u_x(x, t)}{\Delta x}$.

Wave Equation Consider a string of length L, such as a guitar string, stretched taut between two points on the x-axis—say $x = 0$ and $x = L$. When the string starts to vibrate, assume that the motion takes place in the xy-plane in such a manner that each point on the string moves in a direction perpendicular to the x-axis (transverse vibrations). As shown in Figure 12.2(a), let $u(x, t)$ denote the vertical displacement of any point on the string measured from the x-axis for $t > 0$. We further assume:

- The string is perfectly flexible.
- The string is homogeneous; that is, its mass per unit length ρ is a constant.
- The displacements u are small compared to the length of the string.
- The slope of the curve is small at all points.
- The tension **T** of the string is constant.
- The tension is large compared with the force of gravity.
- No other external forces act on the string.

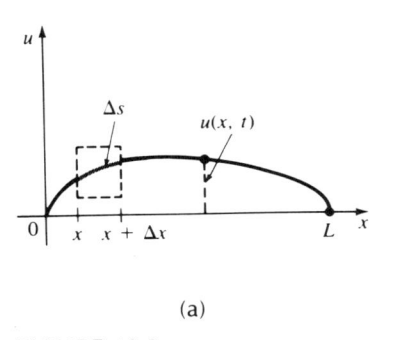

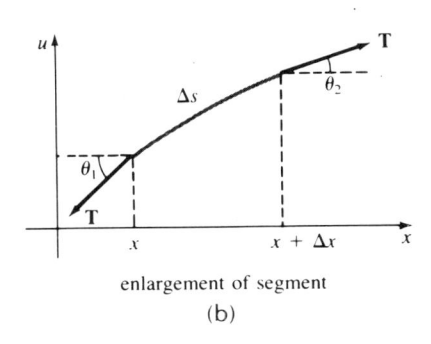

FIGURE 12.2

In Figure 12.2(b) the tension **T** is tangent to the ends of the curve on the interval $[x, x + \Delta x]$. For small θ_1 and θ_2, the net vertical force acting on the corresponding element Δs of the string is then

$$T \sin \theta_2 - T \sin \theta_1 \approx T \tan \theta_2 - T \tan \theta_1$$
$$= T[u_x(x + \Delta x, t) - u_x(x, t)]^*$$

where $T = \|\mathbf{T}\|$. Now $\rho \Delta s \approx \rho \Delta x$ is the mass of the string on $[x, x + \Delta x]$, and so Newton's second law gives

$$T[u_x(x + \Delta x, t) - u_x(x, t)] = \rho \Delta x\, u_{tt}$$

or

$$\frac{u_x(x + \Delta x, t) - u_x(x, t)}{\Delta x} = \frac{\rho}{T} u_{tt}$$

By taking the limit as $\Delta x \to 0$, we get for the last equation $u_{xx} = (\rho/T)u_{tt}$. This of course is (2) with $a^2 = T/\rho$.

---

*$\tan \theta_2 = u_x(x + \Delta x, t)$ and $\tan \theta_1 = u_x(x, t)$ are equivalent expressions for slope.

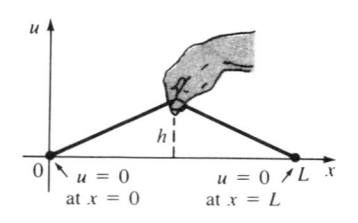

Laplace's Equation Although we shall not present its derivation, Laplace's equation in two and three dimensions occurs in time-independent problems involving potentials such as electrostatic, gravitational, and velocity potential in fluid mechanics. Moreover, a solution of Laplace's equation can also be interpreted as a steady-state temperature distribution. As illustrated in Figure 12.3, a solution $u(x, y)$ of (3) could represent the temperature of a rectangular plate that varies from point to point but not with time.

Initial Conditions We often wish to find solutions of (1), (2), and (3) that satisfy certain side conditions. Since solutions of (1) and (2) depend on time t, we can prescribe what happens at $t = 0$; that is, we can give **initial conditions**. If $f(x)$ denotes the initial temperature distribution throughout the rod in Figure 12.1, then a solution $u(x, t)$ of (1) must satisfy the single initial condition $u(x, 0) = f(x)$, $0 < x < L$. On the other hand, for a vibrating string, we can specify its initial displacement (or shape) $f(x)$ as well as its initial velocity $g(x)$. In mathematical terms we seek a function $u(x, t)$ satisfying (2) and the two initial conditions:

$$u(x, 0) = f(x), \quad \left.\frac{\partial u}{\partial t}\right|_{t=0} = g(x), \quad 0 < x < L \qquad (10)$$

For example, the string could be plucked, as shown in Figure 12.4, and be released from rest ($g(x) = 0$).

Boundary Conditions The string in Figure 12.4 is secured to the x-axis at $x = 0$ and $x = L$ for all time. We interpret this by the two **boundary conditions**:

$$u(0, t) = 0, \quad u(L, t) = 0, \quad t > 0$$

Note that in this context the function f in (10) is continuous and, consequently, $f(0) = 0$ and $f(L) = 0$. In general, there are three types of boundary conditions associated with (1), (2), and (3). On a boundary we can specify the values of *one* of the following:

$$(i) \ u, \quad (ii) \ \frac{\partial u}{\partial n}, \quad \text{or} \quad (iii) \ \frac{\partial u}{\partial n} + hu, \quad h \text{ a constant}$$

Here $\partial u/\partial n$ denotes the normal derivative of u (the directional derivative of u in the direction perpendicular to the boundary). A boundary condition of the first type (*i*) is called a **Dirichlet condition**; the second type (*ii*) is called a **Neumann condition**; and a boundary condition of the third type (*iii*) is known as a **Robin condition**.* For example, for $t > 0$, a typical condition at the right-

---

*Named after the French mathematician **Victor G. Robin** (1855–1897) and the German mathematicians **Peter G. L. Dirichlet** (1805–1859) and **Carl G. Neumann** (1832–1925).

hand end of the rod in Figure 12.1 can be

$(i)'$ $u(L, t) = u_0$, u_0 a constant

$(ii)'$ $\left.\dfrac{\partial u}{\partial x}\right|_{x=L} = 0$, or

$(iii)'$ $\left.\dfrac{\partial u}{\partial x}\right|_{x=L} = -h(u(L, t) - u_0)$, $h > 0$ and u_0 constants

---

Condition $(i)'$ simply states that the boundary $x = L$ is held by some means at a constant *temperature* u_0 for all time $t > 0$. Condition $(ii)'$ indicates that the boundary $x = L$ is *insulated*. From the empirical law of heat transfer, the flux of heat across a boundary (that is, the amount of heat per unit area per unit time conducted across the boundary) is proportional to the value of the normal derivative $\partial u/\partial n$ of the temperature u. Thus, when the boundary $x = L$ is thermally insulated, no heat flows into or out of the rod and so

$$\left.\frac{\partial u}{\partial x}\right|_{x=L} = 0$$

We can interpret $(iii)'$ to mean that *heat is lost* from the right-hand end of the rod by being in contact with a medium, such as air or water, that is held at a constant temperature. From Newton's law of cooling, the outward flux of heat from the rod is proportional to the difference between the temperature $u(L, t)$ at the boundary and the temperature u_0 of the surrounding medium. We note that if heat is lost from the left-hand end of the rod, the boundary condition is

$$\left.\frac{\partial u}{\partial x}\right|_{x=0} = h(u(0, t) - u_0)*$$

Of course, at the ends of the rod we can specify different conditions at the same time. For example, we could have

$$\left.\frac{\partial u}{\partial x}\right|_{x=0} = 0 \quad \text{and} \quad u(L, t) = u_0, \qquad t > 0$$

---

*The change in algebraic sign is consistent with the assumption that the rod is at a higher temperature than the medium surrounding the ends so that $u(0, t) > u_0$ and $u(L, t) > u_0$. At $x = 0$ and $x = L$, the slopes $u_x(0, t)$ and $u_x(L, t)$ must then be positive and negative, respectively.

Boundary-Value Problems Problems such as

$$\text{Solve: } a^2 \frac{\partial^2 u}{\partial x^2} = \frac{\partial^2 u}{\partial t^2}, \quad 0 < x < L, \quad t > 0$$

$$\text{Subject to: } u(0, t) = 0, \quad u(L, t) = 0, \quad t > 0 \quad (11)$$

$$u(x, 0) = f(x), \quad \left.\frac{\partial u}{\partial t}\right|_{t=0} = g(x), \quad 0 < x < L$$

and

$$\text{Solve: } \frac{\partial^2 u}{\partial x^2} + \frac{\partial^2 u}{\partial y^2} = 0, \quad 0 < x < a, \quad 0 < y < b$$

$$\text{Subject to: } \left.\frac{\partial u}{\partial x}\right|_{x=0} = 0, \quad \left.\frac{\partial u}{\partial x}\right|_{x=a} = 0, \quad 0 < y < b \quad (12)$$

$$u(x, 0) = 0, \quad u(x, b) = f(x), \quad 0 < x < a$$

are called **boundary-value problems**.

Modifications The partial differential equations (1), (2), and (3) must be modified to take into consideration internal or external influences acting on the physical system. More general forms of the one-dimensional heat and wave equations are, respectively,

$$k \frac{\partial^2 u}{\partial x^2} + G(x, t, u) = \frac{\partial u}{\partial t} \quad (13)$$

and

$$a^2 \frac{\partial^2 u}{\partial x^2} + F(x, t, u, u_t) = \frac{\partial^2 u}{\partial t^2} \quad (14)$$

For example, if there is heat transfer from the lateral surface of a rod into a surrounding medium that is held at a constant temperature u_0, then the heat equation (13) is

$$k \frac{\partial^2 u}{\partial x^2} - h(u - u_0) = \frac{\partial u}{\partial t}$$

In (14) the function F could represent the various forces acting on the string. For example, when external, damping, and elastic restoring forces are taken into account, (14) assumes the form

$$a^2 \frac{\partial^2 u}{\partial x^2} + \underbrace{f(x, t)}_{\text{external force}} = \frac{\partial^2 u}{\partial t^2} + \underbrace{c \frac{\partial u}{\partial t}}_{\text{damping force}} + \underbrace{ku}_{\text{restoring force}} \quad (15)$$

▲ **Remark** The analysis of a wide variety of diverse phenomena yields (1), (2), or (3), or their generalizations involving a greater number of spatial variables. For example, (1) is sometimes called the **diffusion equation**, since the diffusion

EXERCISES 12.2   Answers to odd-numbered problems begin on page A-63.

In Problems 1–4 a rod of length L coincides with the interval $[0, L]$ on the x-axis. Set up the boundary-value problem for the temperature $u(x, t)$.

1. The left end is held at temperature 0 and the right end is insulated. The initial temperature is $f(x)$ throughout.

2. The left end is held at temperature u_0 and the right end is held at temperature u_1. The initial temperature is 0 throughout.

3. The left end is held at temperature 100 and there is heat transfer from the right end into the surrounding medium at temperature 0. The initial temperature is $f(x)$ throughout.

4. The ends are insulated and there is heat transfer from the lateral surface into the surrounding medium at temperature 50. The initial temperature is 100 throughout.

In Problems 5–8 a string of length L coincides with the interval $[0, L]$ on the x-axis. Set up the boundary-value problem for the displacement $u(x, t)$.

5. The ends are secured to the x-axis. The string is released from rest from the initial displacement $x(L - x)$.

6. The ends are secured to the x-axis. Initially the string is undisplaced but has the initial velocity $\sin(\pi x/L)$.

7. The left end is secured to the x-axis but the right end moves in a transverse manner according to $\sin \pi t$. The string is released from rest from the initial displacement $f(x)$. For $t > 0$ the transverse vibrations are damped with a force proportional to the instantaneous velocity.

8. The ends are secured to the x-axis and the string is initially at rest on that axis. An external vertical force proportional to the horizontal distance from the left end acts on the string for $t > 0$.

In Problems 9 and 10 set up the boundary-value problem for the steady-state temperature $u(x, y)$.

9. A thin rectangular plate coincides with the region defined by $0 \leq x \leq 4$, $0 \leq y \leq 2$. The left end and the bottom of the plate are insulated. The top of the plate is held at temperature 0 and the right end of the plate is held at temperature $f(y)$.

10. A semi-infinite plate coincides with the region defined by $0 \leq x \leq \pi$, $y \geq 0$. The left end is held at temperature e^{-y} and the right end is held at temperature 100 for $0 < y \leq 1$ and 0 for $y > 1$. The bottom of the plate is held at temperature $f(x)$.

12.3 HEAT EQUATION

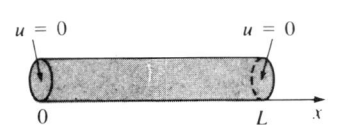

FIGURE 12.5

Consider a thin rod of length L with an initial temperature $f(x)$ throughout and with ends held at temperature 0 for all time $t > 0$. If the rod shown in Figure 12.5 satisfies the assumptions given on pages 771–772, the temperature $u(x, t)$ in the rod is determined from the boundary-value problem:

$$k\frac{\partial^2 u}{\partial x^2} = \frac{\partial u}{\partial t}, \quad 0 < x < L, \quad t > 0 \tag{1}$$

$$u(0, t) = 0, \quad u(L, t) = 0, \quad t > 0 \tag{2}$$

$$u(x, 0) = f(x), \quad 0 < x < L \tag{3}$$

Solution Using the product $u = X(x)T(t)$ and $-\lambda^2$ as a separation constant leads to

$$\frac{X''}{X} = \frac{T'}{kT} = -\lambda^2 \tag{4}$$

and
$$X'' + \lambda^2 X = 0 \tag{5}$$

$$T' + k\lambda^2 T = 0$$

$$X = c_1 \cos \lambda x + c_2 \sin \lambda x \tag{6}$$

$$T = c_3 e^{-k\lambda^2 t} \tag{7}$$

Now since
$$u(0, t) = X(0)T(t) = 0$$
$$u(L, t) = X(L)T(t) = 0$$

we must have $X(0) = 0$ and $X(L) = 0$. These boundary conditions together with (5) constitute a Sturm–Liouville problem. Applying the first of these conditions in (6) immediately gives $c_1 = 0$. Therefore,

$$X = c_2 \sin \lambda x$$

The second boundary condition now implies

$$X(L) = c_2 \sin \lambda L = 0$$

If $c_2 = 0$, then $X = 0$ so that $u = 0$. To obtain a nontrivial solution u, we must have $c_2 \neq 0$, and so the last equation is satisfied when

$$\sin \lambda L = 0$$

This implies that $\lambda L = n\pi$ or $\lambda = n\pi/L$, $n = 1, 2, 3, \ldots$. The values

$$\lambda = \frac{n\pi}{L}, \quad n = 1, 2, 3, \ldots \tag{8}$$

and the corresponding solutions

$$X = c_2 \sin \frac{n\pi}{L} x, \quad n = 1, 2, 3, \ldots \tag{9}$$

are the **eigenvalues** and **eigenfunctions**, respectively, of the problem.
From (7) we have $T = c_3 e^{-k(n^2\pi^2/L^2)t}$ and so

$$u_n = XT = A_n e^{-k(n^2\pi^2/L^2)t} \sin \frac{n\pi}{L} x \tag{10}$$

where we have replaced the constant $c_2 c_3$ by A_n. The products $u_n(x, t)$ satisfy the partial differential equation (1) and the boundary conditions (2) for each value of the positive integer n.* However, in order that the functions given in (10) satisfy the initial condition (3), we would have to choose the coefficient A_n in such a manner that

$$u_n(x, 0) = f(x) = A_n \sin \frac{n\pi}{L} x \tag{11}$$

In general, we would not expect condition (11) to be satisfied for an arbitrary, but reasonable, choice of f. Therefore we are forced to admit that $u_n(x, t)$ *is not a solution of the given problem*. Now by the superposition principle the function

$$u(x, t) = \sum_{n=1}^{\infty} u_n = \sum_{n=1}^{\infty} A_n e^{-k(n^2\pi^2/L^2)t} \sin \frac{n\pi}{L} x \tag{12}$$

must also, though formally, satisfy (1) and (2). Substituting $t = 0$ in (12) implies

$$u(x, 0) = f(x) = \sum_{n=1}^{\infty} A_n \sin \frac{n\pi}{L} x$$

This last expression is recognized as the half-range expansion of f in a sine series. Thus if we make the identification $A_n = b_n$, $n = 1, 2, 3, \ldots$, it follows from (5) of Section 11.3 that

$$A_n = \frac{2}{L} \int_0^L f(x) \sin \frac{n\pi}{L} x \, dx$$

We conclude that a solution of the boundary-value problem described in (1), (2), and (3) is given by the infinite series

$$u(x, t) = \frac{2}{L} \sum_{n=1}^{\infty} \left(\int_0^L f(x) \sin \frac{n\pi}{L} x \, dx \right) e^{-k(n^2\pi^2/L^2)t} \sin \frac{n\pi}{L} x$$

EXERCISES 12.3
Answers to odd-numbered problems begin on page A-64.

In Problems 1 and 2 solve the heat equation (1) subject to the given conditions. Assume a rod of length L.

1. $u(0, t) = 0$, $u(L, t) = 0$

$$u(x, 0) = \begin{cases} 1, & 0 < x < \dfrac{L}{2} \\ 0, & \dfrac{L}{2} \le x < L \end{cases}$$

2. $u(0, t) = 0$, $u(L, t) = 0$
$u(x, 0) = x(L - x)$

3. Find the temperature $u(x, t)$ in a rod of length L if the initial temperature is $f(x)$ throughout and if the ends $x = 0$ and $x = L$ are insulated.

---

*The reader is urged to verify that if the separation constant is chosen as either $\lambda^2 = 0$ or $\lambda^2 > 0$, then the only solution of (1) satisfying (2) is $u = 0$.

4. Solve Problem 3 if $L = 2$ and $f(x) = \begin{cases} x, & 0 < x < 1 \\ 0, & 1 \leq x < 2 \end{cases}$.

5. Suppose heat is lost from the lateral surface of a thin rod of length L into a surrounding medium at temperature 0. If the linear law of heat transfer applies, then the heat equation takes on the form $k \dfrac{\partial^2 u}{\partial x^2} - hu = \dfrac{\partial u}{\partial t}$, $0 < x < L$, $t > 0$, h a constant. Find the temperature $u(x, t)$ if the initial temperature is $f(x)$ throughout and the ends $x = 0$ and $x = L$ are insulated (see Figure 12.6).

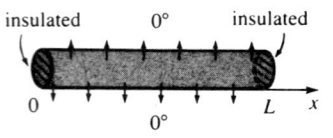

FIGURE 12.6

6. Solve Problem 5 if the ends $x = 0$ and $x = L$ are held at temperature 0.

12.4 WAVE EQUATION

We are now in a position to solve the boundary-value problem (11) of Section 12.2. The vertical displacement $u(x, t)$ of the vibrating string of length L shown in Figure 12.2(a) is determined from

$$a^2 \frac{\partial^2 u}{\partial x^2} = \frac{\partial^2 u}{\partial t^2}, \quad 0 < x < L, \quad t > 0 \tag{1}$$

$$u(0, t) = 0, \quad u(L, t) = 0, \quad t > 0 \tag{2}$$

$$u(x, 0) = f(x), \quad \left.\frac{\partial u}{\partial t}\right|_{t=0} = g(x), \quad 0 < x < L \tag{3}$$

Solution Separating variables in (1) gives

$$\frac{X''}{X} = \frac{T''}{a^2 T} = -\lambda^2$$

so that

$$X'' + \lambda^2 X = 0$$
$$T'' + \lambda^2 a^2 T = 0$$

and therefore

$$X = c_1 \cos \lambda x + c_2 \sin \lambda x$$
$$T = c_3 \cos \lambda a t + c_4 \sin \lambda a t$$

As before, the boundary conditions (2) translate into $X(0) = 0$ and $X(L) = 0$. In turn we find

$$c_1 = 0 \quad \text{and} \quad c_2 \sin \lambda L = 0$$

This last equation yields the eigenvalues $\lambda = n\pi/L$, $n = 1, 2, 3, \ldots$. The corresponding eigenfunctions are

$$X = c_2 \sin \frac{n\pi}{L} x, \quad n = 1, 2, 3, \ldots$$

Thus, solutions of (1) satisfying the boundary conditions (2) are

$$u_n = \left(A_n \cos \frac{n\pi a}{L} t + B_n \sin \frac{n\pi a}{L} t \right) \sin \frac{n\pi}{L} x \tag{4}$$

12.4 Wave Equation

and
$$u(x, t) = \sum_{n=1}^{\infty} \left(A_n \cos \frac{n\pi a}{L} t + B_n \sin \frac{n\pi a}{L} t \right) \sin \frac{n\pi}{L} x \quad (5)$$

Setting $t = 0$ in (5) gives
$$u(x, 0) = f(x) = \sum_{n=1}^{\infty} A_n \sin \frac{n\pi}{L} x$$

which is a half-range expansion for f in a sine series. As in the discussion of the heat equation, we can write $A_n = b_n$:
$$A_n = \frac{2}{L} \int_0^L f(x) \sin \frac{n\pi}{L} x \, dx \quad (6)$$

To determine B_n we differentiate (5) with respect to t and then set $t = 0$:
$$\frac{\partial u}{\partial t} = \sum_{n=1}^{\infty} \left(-A_n \frac{n\pi a}{L} \sin \frac{n\pi a}{L} t + B_n \frac{n\pi a}{L} \cos \frac{n\pi a}{L} t \right) \sin \frac{n\pi}{L} x$$

$$\left. \frac{\partial u}{\partial t} \right|_{t=0} = g(x) = \sum_{n=1}^{\infty} \left(B_n \frac{n\pi a}{L} \right) \sin \frac{n\pi}{L} x$$

In order for this last series to be the half-range sine expansion of g on the interval, the *total* coefficient $B_n n\pi a/L$ must be given by the form of (5) in Section 11.3—that is,
$$B_n \frac{n\pi a}{L} = \frac{2}{L} \int_0^L g(x) \sin \frac{n\pi}{L} x \, dx$$

from which we obtain
$$B_n = \frac{2}{n\pi a} \int_0^L g(x) \sin \frac{n\pi}{L} x \, dx \quad (7)$$

The solution of the problem consists of the series (5) with A_n and B_n defined by (6) and (7), respectively.

We note that when the string is released from *rest*, $g(x) = 0$ for every x in $0 \le x \le L$ and, consequently, $B_n = 0$.

Standing Waves Recall from the derivation of the wave equation in Section 12.2 that the constant a appearing in the solution of the boundary-value problem in (1), (2), and (3) is given by $\sqrt{T/\rho}$, where ρ is mass per unit length and T is the magnitude of the tension in the string. When T is large enough, the vibrating string will produce a musical sound. This sound is the result of standing waves. The solution (5) is a superposition of product solutions called **standing waves** or **normal modes**:
$$u(x, t) = u_1(x, t) + u_2(x, t) + u_3(x, t) + \cdots$$

In view of (8) and (9) of Section 3.8, the product solutions (4) can be written as
$$u_n(x, t) = C_n \sin\left(\frac{n\pi a}{L} t + \phi_n \right) \sin \frac{n\pi}{L} x \quad (8)$$

where $C_n = \sqrt{A_n^2 + B_n^2}$ and ϕ_n is defined by $\sin \phi_n = A_n/C_n$ and $\cos \phi_n = B_n/C_n$. For $n = 1, 2, 3, \ldots$, the standing waves are essentially the graphs of $\sin(n\pi x/L)$, with a time-varying amplitude given by

$$C_n \sin\left(\frac{n\pi a}{L} t + \phi_n\right)$$

Alternatively, we see from (8) that at a fixed value of x, each product function $u_n(x, t)$ represents simple harmonic motion with amplitude $C_n|\sin(n\pi x/L)|$ and frequency $f_n = na/2L$. In other words, each point on a standing wave vibrates with a different amplitude but with the same frequency. When $n = 1$,

$$u_1(x, t) = C_1 \sin\left(\frac{\pi a}{L} t + \phi_1\right) \sin \frac{\pi}{L} x$$

is called the **first standing wave**, the **first normal mode**, or the **fundamental mode of vibration**. The first three standing waves, or normal modes, are shown in Figure 12.7. The dashed graphs represent the standing waves at various values of time. The points in the interval $(0, L)$ for which $\sin(n\pi/L)x = 0$ correspond to points on a standing wave where there is no motion. These points are called **nodes**. For example, in Figures 12.7(b) and (c) we see that the second standing wave has one node at $L/2$ and the third standing wave has two nodes at $L/3$ and $2L/3$. In general, the nth normal mode of vibration has $n - 1$ nodes.

The frequency

$$f_1 = \frac{a}{2L} = \frac{1}{2L} \sqrt{\frac{T}{\rho}}$$

of the first normal mode is called the **fundamental frequency**, or **first harmonic**, and is directly related to the pitch produced by a stringed instrument. It is apparent that the greater the tension in the string, the greater the pitch of the sound. The frequencies f_n of the other normal modes, which are integer multiples of the fundamental frequency, are called **overtones**. The second harmonic is the first overtone, and so on.

(a) first standing wave

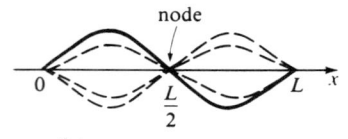

(b) second standing wave

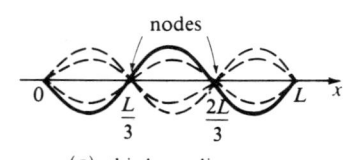

(c) third standing wave

FIGURE 12.7

EXERCISES 12.4   Answers to odd-numbered problems begin on page A-64.

In Problems 1–8 solve the wave equation (1) subject to the given conditions.

1. $u(0, t) = 0, u(L, t) = 0$

$u(x, 0) = \frac{1}{4}x(L - x), \quad \left.\frac{\partial u}{\partial t}\right|_{t=0} = 0$

2. $u(0, t) = 0, u(L, t) = 0$

$u(x, 0) = 0, \quad \left.\frac{\partial u}{\partial t}\right|_{t=0} = x(L - x)$

3. $u(0, t) = 0, u(L, t) = 0$

$u(x, 0)$, as specified in Figure 12.8

$\left.\dfrac{\partial u}{\partial t}\right|_{t=0} = 0$

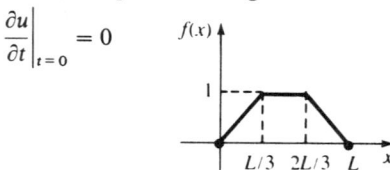

FIGURE 12.8

4. $u(0, t) = 0$, $u(\pi, t) = 0$
$u(x, 0) = \frac{1}{6}x(\pi^2 - x^2)$
$\left.\frac{\partial u}{\partial t}\right|_{t=0} = 0$

5. $u(0, t) = 0$, $u(\pi, t) = 0$
$u(x, 0) = 0$
$\left.\frac{\partial u}{\partial t}\right|_{t=0} = \sin x$

6. $u(0, t) = 0$, $u(1, t) = 0$
$u(x, 0) = 0.01 \sin 3\pi x$
$\left.\frac{\partial u}{\partial t}\right|_{t=0} = 0$

7. $u(0, t) = 0$, $u(L, t) = 0$, $u(x, 0) = \begin{cases} \dfrac{2hx}{L}, & 0 < x < \dfrac{L}{2} \\ 2h\left(1 - \dfrac{x}{L}\right), & \dfrac{L}{2} \leq x < L \end{cases}$

$\left.\frac{\partial u}{\partial t}\right|_{t=0} = 0$

The constant h is positive but small compared to L. This is referred to as the "plucked string" problem.

8. $\left.\dfrac{\partial u}{\partial x}\right|_{x=0} = 0$, $\left.\dfrac{\partial u}{\partial x}\right|_{x=L} = 0$, $u(x, 0) = x$, $\left.\dfrac{\partial u}{\partial t}\right|_{t=0} = 0$

This problem could describe the longitudinal displacement $u(x, t)$ of a vibrating elastic bar. The boundary conditions at $x = 0$ and $x = L$ are called **free-end conditions** (see Figure 12.9).

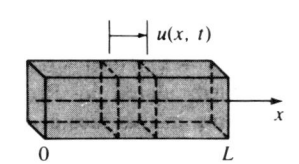

FIGURE 12.9

9. A string is stretched and secured on the x-axis at $x = 0$ and $x = \pi$ for $t > 0$. If the transverse vibrations take place in a medium that imparts a resistance proportional to the instantaneous velocity, then the wave equation takes on the form

$$\frac{\partial^2 u}{\partial x^2} = \frac{\partial^2 u}{\partial t^2} + 2\beta \frac{\partial u}{\partial t}, \quad 0 < \beta < 1, \quad t > 0$$

Find the displacement $u(x, t)$ if the string starts from rest from the initial displacement $f(x)$.

10. Show that a solution of the boundary-value problem

$$\frac{\partial^2 u}{\partial x^2} = \frac{\partial^2 u}{\partial t^2} + u, \quad 0 < x < \pi, \quad t > 0$$

$u(0, t) = 0$, $u(\pi, t) = 0$, $t > 0$

$u(x, 0) = \begin{cases} x, & 0 < x < \pi/2 \\ \pi - x, & \pi/2 \leq x < \pi \end{cases}$

$\left.\dfrac{\partial u}{\partial t}\right|_{t=0} = 0$, $0 < x < \pi$

is

$$u(x, t) = \frac{4}{\pi} \sum_{k=1}^{\infty} \frac{(-1)^{k+1}}{(2k-1)^2} \cos \sqrt{(2k-1)^2 + 1}\, t \sin(2k - 1)x$$

11. The transverse displacement $u(x, t)$ of a vibrating beam of length L is determined from a fourth-order partial differential equation:

$$a^2 \frac{\partial^4 u}{\partial x^4} + \frac{\partial^2 u}{\partial t^2} = 0, \quad 0 < x < L, \quad t > 0$$

If the beam is **simply supported**, as shown in Figure 12.10, boundary and initial conditions are

$\left. \begin{array}{l} u(0, t) = 0, \quad u(L, t) = 0 \\ \left.\dfrac{\partial^2 u}{\partial x^2}\right|_{x=0} = 0, \quad \left.\dfrac{\partial^2 u}{\partial x^2}\right|_{x=L} = 0 \end{array} \right\} t > 0$

$u(x, 0) = f(x)$, $\left.\dfrac{\partial u}{\partial t}\right|_{t=0} = g(x)$, $0 < x < L$

Solve for $u(x, t)$. [*Hint:* For convenience use λ^4 instead of λ^2 when separating variables.]

simply supported beam

FIGURE 12.10

12. If the ends of the beam in Problem 11 are **embedded** at $x = 0$ and $x = L$, then the boundary conditions become, for $t > 0$,

$u(0, t) = 0$, $u(L, t) = 0$

$\left.\dfrac{\partial u}{\partial x}\right|_{x=0} = 0$, $\left.\dfrac{\partial u}{\partial x}\right|_{x=L} = 0$

(a) Show that the eigenvalues of the problem are $\lambda = x_n/L$, where x_n, $n = 1, 2, 3, \ldots$, are the positive roots of the equation $\cosh x \cos x = 1$.

(b) Show graphically that the equation in part (a) has an infinite number of roots.

(c) Find the first four eigenvalues to four decimal places. [*Hint:* If Newton's method is used (Section 15.1), use the graphs in part (b) to pick an initial guess that is close to the root.]

13. Consider the boundary-value problem given in (1), (2), and (3) of this section. If $g(x) = 0$ on $0 < x < L$, show that the solution of the problem can be written as

$$u(x, t) = \frac{1}{2}[f(x + at) + f(x - at)]$$

[*Hint:* Use the identity $2 \sin \theta_1 \cos \theta_2 = \sin(\theta_1 + \theta_2) + \sin(\theta_1 - \theta_2)$.]

14. The vertical displacement $u(x, t)$ of an infinitely long string is determined from the initial-value problem

$$a^2 \frac{\partial^2 u}{\partial x^2} = \frac{\partial^2 u}{\partial t^2}, \quad -\infty < x < \infty, \quad t > 0$$

$$u(x, 0) = f(x), \quad \left.\frac{\partial u}{\partial t}\right|_{t=0} = g(x)$$

This problem can be solved without separating variables.

(a) Recall from Problem 28 of Exercises 12.1 that the wave equation can be put into the form $\partial^2 u/\partial \eta \, \partial \xi = 0$ by means of the substitutions $\xi = x + at$ and $\eta = x - at$. Integrating the last partial differential equation with respect to η and then with respect to ξ shows that $u(x, t) = F(x + at) + G(x - at)$, where F and G are arbitrary twice-differentiable functions, is a solution of the wave equation. Use this solution and the given initial conditions to show that

$$F(x) = \frac{1}{2}f(x) + \frac{1}{2a}\int_{x_0}^{x} g(s) \, ds + c$$

and $$G(x) = \frac{1}{2}f(x) - \frac{1}{2a}\int_{x_0}^{x} g(s) \, ds - c$$

where x_0 is arbitrary and c is a constant of integration.

(b) Use the results in part (a) to show that

$$u(x, t) = \frac{1}{2}[f(x + at) + f(x - at)] + \frac{1}{2a}\int_{x-at}^{x+at} g(s) \, ds \quad (9)$$

Note that when the initial velocity $g(x) = 0$, we obtain

$$u(x, t) = \frac{1}{2}[f(x + at) + f(x - at)], \quad -\infty < x < \infty$$

The last solution can be interpreted as a superposition of two **traveling waves**, one moving to the right (that is, $\frac{1}{2}f(x - at)$) and one moving to the left ($\frac{1}{2}f(x + at)$). Both waves travel with speed a and have the same basic shape as the initial displacement $f(x)$. The form of $u(x, t)$ given in (9) is called **d'Alembert's solution.**[*]

In Problems 15–17 use d'Alembert's solution (9) to solve the initial-value problem in Problem 14 subject to the given initial conditions.

15. $f(x) = \sin x, g(x) = 1$

16. $f(x) = \sin x, g(x) = \cos x$

17. $f(x) = 0, g(x) = \sin 2x$

18. Suppose $f(x) = 1/(1 + x^2)$, $g(x) = 0$, and $a = 1$ for the initial-value problem stated in Problem 14. Graph d'Alembert's solution in this case at times $t = 0$, $t = 1$, and $t = 3$.

12.5 LAPLACE'S EQUATION

Suppose we wish to find the steady-state temperature $u(x, y)$ in a rectangular plate with insulated boundaries, as shown in Figure 12.11. When no heat

---

*__Jean le Rond d'Alembert (1717–1783)__ Educated in law and medicine, the Frenchman d'Alembert devoted his life to the study of physics and mathematics. D'Alembert was also a collaborator with the famous philosopher Denis Diderot on the latter's *Encyclopédie*.

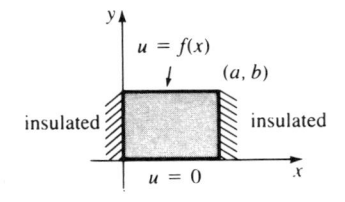

escapes from the lateral faces of the plate, we solve Laplace's equation

$$\frac{\partial^2 u}{\partial x^2} + \frac{\partial^2 u}{\partial y^2} = 0, \quad 0 < x < a, \quad 0 < y < b$$

subject to
$$\left.\frac{\partial u}{\partial x}\right|_{x=0} = 0, \quad \left.\frac{\partial u}{\partial x}\right|_{x=a} = 0, \quad 0 < y < b$$

$$u(x, 0) = 0, \quad u(x, b) = f(x), \quad 0 < x < a$$

Solution Separation of variables leads to

$$\frac{X''}{X} = -\frac{Y''}{Y} = -\lambda^2$$

$$X'' + \lambda^2 X = 0 \tag{1}$$

$$Y'' - \lambda^2 Y = 0 \tag{2}$$

$$X = c_1 \cos \lambda x + c_2 \sin \lambda x \tag{3}$$

and, since $0 < y < b$ is a finite interval,

$$Y = c_3 \cosh \lambda y + c_4 \sinh \lambda y \tag{4}$$

Now the first three boundary conditions translate into $X'(0) = 0$, $X'(a) = 0$, and $Y(0) = 0$. Differentiating X and setting $x = 0$ imply $c_2 = 0$ and, therefore, $X = c_1 \cos \lambda x$. Differentiating again and then setting $x = a$ give $-c_1 \lambda \sin \lambda a = 0$. This last condition is satisfied when $\lambda = 0$ or when $\lambda a = n\pi$, or $\lambda = n\pi/a$, $n = 1, 2, \ldots$. Observe that $\lambda = 0$ implies (1) is $X'' = 0$. The general solution of this equation is given by the linear function $X = c_1 + c_2 x$ and *not* by (3). In this case the boundary conditions $X'(0) = 0$, $X'(a) = 0$ demand that $X = c_1$. In this example, unlike the previous two examples, we are forced to conclude that $\lambda = 0$ is an eigenvalue. Corresponding $\lambda = 0$ with $n = 0$, the eigenfunctions are

$$X = c_1, \quad n = 0, \quad \text{and} \quad X = c_1 \cos \frac{n\pi}{a} x, \quad n = 1, 2, \ldots$$

Finally, the condition that $Y(0) = 0$ dictates that $c_3 = 0$ in (4) when $\lambda > 0$. However, when $\lambda = 0$, (2) becomes $Y'' = 0$ and thus the solution is given by $Y = c_3 + c_4 y$ rather than by (4). But $Y(0) = 0$ implies again that $c_3 = 0$ and so $Y = c_4 y$. Thus product solutions of the equation satisfying the first three boundary conditions are

$$A_0 y, \quad n = 0, \quad \text{and} \quad A_n \sinh \frac{n\pi}{a} y \cos \frac{n\pi}{a} x, \quad n = 1, 2, \ldots$$

The superposition principle yields another solution

$$u(x, y) = A_0 y + \sum_{n=1}^{\infty} A_n \sinh \frac{n\pi}{a} y \cos \frac{n\pi}{a} x \tag{5}$$

Substituting $y = b$ in (5) gives

$$u(x, b) = f(x) = A_0 b + \sum_{n=1}^{\infty} \left(A_n \sinh \frac{n\pi}{a} b \right) \cos \frac{n\pi}{a} x$$

which, in this case, is a half-range expansion of f in a cosine series. If we make the identifications $A_0 b = a_0/2$ and $A_n \sinh(n\pi b/a) = a_n$, $n = 1, 2, 3, \ldots$, it follows from (2) and (3) of Section 11.3 that

$$2A_0 b = \frac{2}{a} \int_0^a f(x)\, dx$$

$$A_0 = \frac{1}{ab} \int_0^a f(x)\, dx \qquad (6)$$

and

$$A_n \sinh \frac{n\pi}{a} b = \frac{2}{a} \int_0^a f(x) \cos \frac{n\pi}{a} x\, dx$$

$$A_n = \frac{2}{a \sinh \dfrac{n\pi}{a} b} \int_0^a f(x) \cos \frac{n\pi}{a} x\, dx \qquad (7)$$

The solution of this problem consists of the series given in (5), where A_0 and A_n are defined by (6) and (7), respectively.

Dirichlet Problem A boundary-value problem in which we seek a solution to an elliptic partial differential equation, such as Laplace's equation $\nabla^2 u = 0$, within a region such that u takes on prescribed values on the entire boundary of the region is called a **Dirichlet problem**.

Superposition Principle A Dirichlet problem for a rectangle can be solved readily by separation of variables when homogeneous boundary conditions are specified on two *parallel* boundaries. However, the method of separation of variables is not applicable to a Dirichlet problem when the boundary conditions on all four sides of the rectangle are nonhomogeneous. To get around this difficulty we break the problem

$$\frac{\partial^2 u}{\partial x^2} + \frac{\partial^2 u}{\partial y^2} = 0, \quad 0 < x < a, \quad 0 < y < b$$

$$u(0, y) = F(y), \quad u(a, y) = G(y), \quad 0 < y < b \qquad (8)$$

$$u(x, 0) = f(x), \quad u(x, b) = g(x), \quad 0 < x < a$$

into two problems, each of which has homogeneous boundary conditions on parallel boundaries, as shown:

Problem 1	Problem 2

$\dfrac{\partial^2 u_1}{\partial x^2} + \dfrac{\partial^2 u_1}{\partial y^2} = 0, \quad 0 < x < a, \quad 0 < y < b$ $\dfrac{\partial^2 u_2}{\partial x^2} + \dfrac{\partial^2 u_2}{\partial y^2} = 0, \quad 0 < x < a, \quad 0 < y < b$

$u_1(0, y) = 0, \quad u_1(a, y) = 0, \quad 0 < y < b$ $u_2(0, y) = F(y), \quad u_2(a, y) = G(y), \quad 0 < y < b$

$u_1(x, 0) = f(x), \quad u_1(x, b) = g(x), \quad 0 < x < a$ $u_2(x, 0) = 0, \quad u_2(x, b) = 0, \quad 0 < x < a$

12.5 Laplace's Equation

Suppose u_1 and u_2 are the solutions of Problems 1 and 2, respectively. If we define $u(x, y) = u_1(x, y) + u_2(x, y)$, it is seen that u satisfies all boundary conditions in the original problem (8). For example,

$$u(0, y) = u_1(0, y) + u_2(0, y) = 0 + F(y) = F(y)$$
$$u(x, b) = u_1(x, b) + u_2(x, b) = g(x) + 0 = g(x)$$

and so on. Furthermore, u is a solution of Laplace's equation by Theorem 12.1. In other words, by solving Problems 1 and 2 and adding their solutions, we have solved the original problem. This additive property of solutions is known as the **superposition principle** (see Figure 12.12).

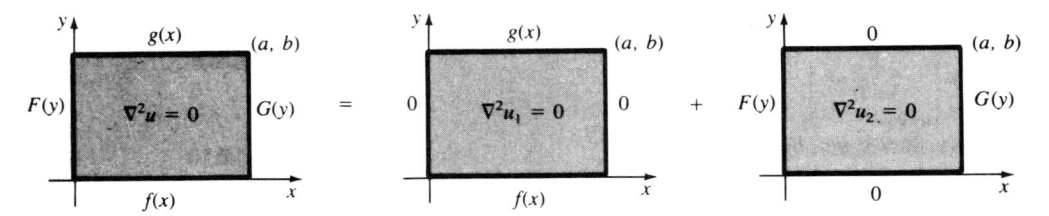

Solution u = Solution u_1 of Problem 1 + Solution u_2 of Problem 2

FIGURE 12.12

We leave it as exercises (see Problems 11 and 12) to show that a solution of Problem 1 is

$$u_1(x, y) = \sum_{n=1}^{\infty} \left\{ A_n \cosh \frac{n\pi}{a} y + B_n \sinh \frac{n\pi}{a} y \right\} \sin \frac{n\pi}{a} x$$

where $\quad A_n = \dfrac{2}{a} \displaystyle\int_0^a f(x) \sin \dfrac{n\pi}{a} x \, dx$

$$B_n = \frac{1}{\sinh \dfrac{n\pi}{a} b} \left(\frac{2}{a} \int_0^a g(x) \sin \frac{n\pi}{a} x \, dx - A_n \cosh \frac{n\pi}{a} b \right)$$

and that a solution of Problem 2 is

$$u_2(x, y) = \sum_{n=1}^{\infty} \left\{ A_n \cosh \frac{n\pi}{b} x + B_n \sinh \frac{n\pi}{b} x \right\} \sin \frac{n\pi}{b} y$$

where $\quad A_n = \dfrac{2}{b} \displaystyle\int_0^b F(y) \sin \dfrac{n\pi}{b} y \, dy$

$$B_n = \frac{1}{\sinh \dfrac{n\pi}{b} a} \left(\frac{2}{b} \int_0^b G(y) \sin \frac{n\pi}{b} y \, dy - A_n \cosh \frac{n\pi}{b} a \right)$$

EXERCISES 12.5 Answers to odd-numbered problems begin on page A-64.

In Problems 1–8 find the steady-state temperature for a rectangular plate with boundary conditions as given.

1. $u(0, y) = 0, u(a, y) = 0$
   $u(x, 0) = 0, u(x, b) = f(x)$

2. $u(0, y) = 0, u(a, y) = 0$
   $\left.\dfrac{\partial u}{\partial y}\right|_{y=0} = 0, u(x, b) = f(x)$

3. $u(0, y) = 0, u(a, y) = 0$
   $u(x, 0) = f(x), u(x, b) = 0$

4. $\left.\dfrac{\partial u}{\partial x}\right|_{x=0} = 0, \left.\dfrac{\partial u}{\partial x}\right|_{x=a} = 0$
   $u(x, 0) = x, u(x, b) = 0$

5. $u(0, y) = 0, u(1, y) = 1 - y$
   $\left.\dfrac{\partial u}{\partial y}\right|_{y=0} = 0, \left.\dfrac{\partial u}{\partial y}\right|_{y=1} = 0$

6. $u(0, y) = g(y), \left.\dfrac{\partial u}{\partial x}\right|_{x=1} = 0$
   $\left.\dfrac{\partial u}{\partial y}\right|_{y=0} = 0, \left.\dfrac{\partial u}{\partial y}\right|_{y=\pi} = 0$

7. $\left.\dfrac{\partial u}{\partial x}\right|_{x=0} = u(0, y), u(\pi, y) = 1$
   $u(x, 0) = 0, u(x, \pi) = 0$

8. $u(0, y) = 0, u(1, y) = 0$
   $\left.\dfrac{\partial u}{\partial y}\right|_{y=0} = u(x, 0), u(x, 1) = f(x)$

In Problems 9 and 10 find the steady-state temperature in the given semi-infinite plate extending in the positive y-direction. In each case assume $u(x, y)$ is bounded as $y \to \infty$.

9.

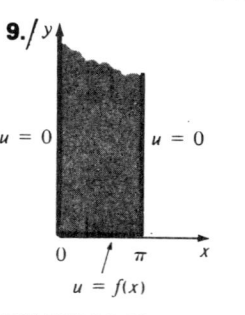

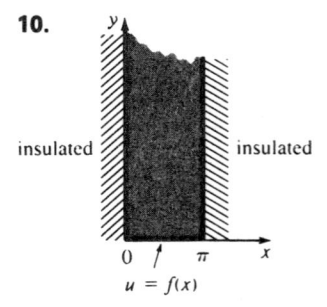

$u = 0$ $u = 0$

$u = f(x)$

FIGURE 12.13

10.

insulated   insulated

$u = f(x)$

FIGURE 12.14

In Problems 11 and 12 find the steady-state temperature for a rectangular plate with boundary conditions as given.

11. $u(0, y) = 0, u(a, y) = 0$
    $u(x, 0) = f(x), u(x, b) = g(x)$

12. $u(0, y) = F(y), u(a, y) = G(y)$
    $u(x, 0) = 0, u(x, b) = 0$

In Problems 13 and 14 use the superposition principle to find the steady-state temperature for a square plate with boundary conditions as given.

13. $u(0, y) = 1, u(\pi, y) = 1$
    $u(x, 0) = 0, u(x, \pi) = 1$

14. $u(0, y) = 0, u(2, y) = y(2 - y)$
    $u(x, 0) = 0, u(x, 2) = \begin{cases} x, & 0 < x < 1 \\ 2 - x, & 1 \leq x < 2 \end{cases}$

12.6 NONHOMOGENEOUS EQUATIONS AND BOUNDARY CONDITIONS

The method of separation of variables may not be applicable to a boundary-value problem when the partial differential equation or boundary conditions are nonhomogeneous. For example, when heat is generated at a constant rate r within a rod of finite length, the form of the heat equation is

$$k\dfrac{\partial^2 u}{\partial x^2} + r = \dfrac{\partial u}{\partial t} \qquad (1)$$

12.6 Nonhomogeneous Equations and Boundary Conditions

Equation (1) is nonhomogeneous and is readily shown to be not separable. On the other hand, suppose we wish to solve the usual heat equation $ku_{xx} = u_t$ when the boundaries $x = 0$ and $x = L$ are held at nonzero temperatures k_1 and k_2. Even though the assumption $u(x, t) = X(t)T(t)$ separates the partial differential equation, we quickly find ourselves at an impasse in determining eigenvalues and eigenfunctions, since no conclusions can be obtained from $u(0, t) = X(0)T(t) = k_1$ and $u(L, t) = X(L)T(t) = k_2$.

A few problems involving nonhomogeneous equations or nonhomogeneous boundary conditions can be solved by means of a change of dependent variable

$$u = v + \psi$$

The basic idea is to determine ψ, a function of *one* variable, in such a manner that v, a function of *two* variables, is made to satisfy a homogeneous partial differential equation and homogeneous boundary conditions. The following example will illustrate the procedure.

EXAMPLE 1 Solve (1) subject to

$$u(0, t) = 0, \quad u(1, t) = u_0, \quad t > 0$$
$$u(x, 0) = f(x), \quad 0 < x < 1$$

Solution Both the equation and the boundary condition at $x = 1$ are nonhomogeneous. If we let $u(x, t) = v(x, t) + \psi(x)$, then

$$\frac{\partial^2 u}{\partial x^2} = \frac{\partial^2 v}{\partial x^2} + \psi'' \quad \text{and} \quad \frac{\partial u}{\partial t} = \frac{\partial v}{\partial t}$$

Substituting these results into (1) gives

$$k\frac{\partial^2 v}{\partial x^2} + k\psi'' + r = \frac{\partial v}{\partial t} \qquad (2)$$

Equation (2) reduces to a homogeneous equation if we demand that ψ satisfy

$$k\psi'' + r = 0 \quad \text{or} \quad \psi'' = -\frac{r}{k}$$

Integrating the last equation twice reveals that

$$\psi(x) = -\frac{r}{2k}x^2 + c_1 x + c_2 \qquad (3)$$

Furthermore,
$$u(0, t) = v(0, t) + \psi(0) = 0$$
$$u(1, t) = v(1, t) + \psi(1) = u_0$$

We will have $v(0, t) = 0$ and $v(1, t) = 0$, provided

$$\psi(0) = 0 \quad \text{and} \quad \psi(1) = u_0$$

Applying the latter two conditions to (3) gives, in turn, $c_2 = 0$ and $c_1 = (r/2k) + u_0$. Consequently,

$$\psi(x) = -\frac{r}{2k}x^2 + \left(\frac{r}{2k} + u_0\right)x$$

Last, the initial condition $u(x, 0) = v(x, 0) + \psi(x)$ implies $v(x, 0) = u(x, 0) - \psi(x) = f(x) - \psi(x)$. Thus to determine $v(x, t)$ we solve the *new* boundary-value problem

$$k\frac{\partial^2 v}{\partial x^2} = \frac{\partial v}{\partial t}, \quad 0 < x < 1, \quad t > 0$$

$$v(0, t) = 0, \quad v(1, t) = 0, \quad t > 0$$

$$v(x, 0) = f(x) + \frac{r}{2k}x^2 - \left(\frac{r}{2k} + u_0\right)x$$

by separation of variables. In the usual manner we find

$$v(x, t) = \sum_{n=1}^{\infty} A_n e^{-kn^2\pi^2 t} \sin n\pi x$$

where

$$A_n = 2\int_0^1 \left[f(x) + \frac{r}{2k}x^2 - \left(\frac{r}{2k} + u_0\right)x\right] \sin n\pi x\, dx \quad (4)$$

Finally, a solution of the original problem is obtained by adding $\psi(x)$ and $v(x, t)$:

$$u(x, t) = -\frac{r}{2k}x^2 + \left(\frac{r}{2k} + u_0\right)x + \sum_{n=1}^{\infty} A_n e^{-kn^2\pi^2 t} \sin n\pi x \quad (5)$$

where the A_n are defined in (4). □

Observe in (5) that $u(x, t) \to \psi(x)$ as $t \to \infty$. In the context of solving forms of the heat equation, ψ is called a **steady-state solution**. Since $v(x, t) \to 0$ as $t \to \infty$, v is called a **transient solution**.

The substitution $u = v + \psi$ can also be used on problems involving forms of the wave equation as well as Laplace's equation.

EXERCISES 12.6   Answers to odd-numbered problems begin on page A-65.

In Problems 1 and 2 solve the heat equation $ku_{xx} = u_t$, $0 < x < 1$, $t > 0$, subject to the given conditions.

1. $u(0, t) = 100, u(1, t) = 100$
$u(x, 0) = 0$

2. $u(0, t) = u_0, u(1, t) = 0$
$u(x, 0) = f(x)$

In Problems 3 and 4 solve the partial differential equation (1) subject to the given conditions.

3. $u(0, t) = u_0, u(1, t) = u_0$
$u(x, 0) = 0$

4. $u(0, t) = u_0, u(1, t) = u_1$
$u(x, 0) = f(x)$

5. Solve the boundary-value problem:

$$k\frac{\partial^2 u}{\partial x^2} + Ae^{-\beta x} = \frac{\partial u}{\partial t}, \quad \beta > 0, \quad 0 < x < 1, \quad t > 0$$

$$u(0, t) = 0, \quad u(1, t) = 0, \quad t > 0$$

$$u(x, 0) = f(x), \quad 0 < x < 1$$

The partial differential equation is a form of the heat equation when heat is generated within a thin rod due to radioactive decay of the material.

6. Solve the boundary-value problem:

$$k\frac{\partial^2 u}{\partial x^2} - hu = \frac{\partial u}{\partial t}, \quad 0 < x < \pi, \quad t > 0$$

$$u(0, t) = 0, \quad u(\pi, t) = u_0, \quad t > 0$$

$$u(x, 0) = 0, \quad 0 < x < \pi$$

7. Find a steady-state solution $\psi(x)$ of the boundary-value problem:

$$k\frac{\partial^2 u}{\partial x^2} - h(u - u_0) = \frac{\partial u}{\partial t}, \quad 0 < x < 1, \quad t > 0$$

$$u(0, t) = u_0, \quad u(1, t) = 0, \quad t > 0$$

$$u(x, 0) = f(x), \quad 0 < x < 1$$

8. Find a steady-state solution $\psi(x)$ if the rod in Problem 7 is semi-infinite extending in the positive x-direction, radiates from its lateral surface into a medium at temperature 0, and

$$u(0, t) = u_0, \quad \lim_{x \to \infty} u(x, t) = 0, \quad t > 0$$

$$u(x, 0) = f(x), \quad x > 0$$

9. When a vibrating string is subjected to an external vertical force that varies with the horizontal distance from the left end, the wave equation takes on the form

$$a^2 \frac{\partial^2 u}{\partial x^2} + Ax = \frac{\partial^2 u}{\partial t^2}$$

Solve this partial differential equation subject to

$$u(0, t) = 0, \quad u(1, t) = 0, \quad t > 0$$

$$u(x, 0) = 0, \quad \left.\frac{\partial u}{\partial t}\right|_{t=0} = 0, \quad 0 < x < 1$$

10. A string initially at rest on the x-axis is secured on the x-axis at $x = 0$ and $x = 1$. If the string is allowed to fall under its own weight for $t > 0$, the displacement $u(x, t)$ satisfies

$$a^2 \frac{\partial^2 u}{\partial x^2} - g = \frac{\partial^2 u}{\partial t^2}, \quad 0 < x < 1, \quad t > 0$$

where g is the acceleration of gravity. Solve for $u(x, t)$.

11. Find the steady-state temperature $u(x, y)$ in the semi-infinite plate shown in Figure 12.15. Assume that the temperature is bounded as $x \to \infty$. [*Hint:* Try $u(x, y) = v(x, y) + \psi(y)$.]

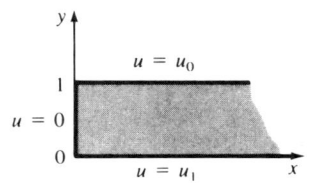

FIGURE 12.15

12. Poisson's equation*

$$\frac{\partial^2 u}{\partial x^2} + \frac{\partial^2 u}{\partial y^2} = -h, \quad h > 0$$

occurs in many problems involving electric potential. Solve the equation subject to the conditions

$$u(0, y) = 0, \quad u(\pi, y) = 1, \quad y > 0$$

$$u(x, 0) = 0, \quad 0 < x < \pi$$

---

*Simeon-Denis Poisson (1781–1840) was a French mathematician and physicist.

12.7 USE OF GENERALIZED FOURIER SERIES

For certain types of boundary conditions, the method of separation of variables and the superposition principle will lead to an expansion of a function in a trigonometric series that is not a Fourier series. To solve the problems in this section we shall utilize the concept of generalized Fourier series developed in Section 11.1.

EXAMPLE 1 The temperature in a rod of unit length in which there is heat transfer from its right boundary into a surrounding medium kept at a constant temperature 0 is determined from

$$k\frac{\partial^2 u}{\partial x^2} = \frac{\partial u}{\partial t}, \quad 0 < x < 1, \quad t > 0$$

$$u(0, t) = 0, \quad \left.\frac{\partial u}{\partial x}\right|_{x=1} = -hu(1, t), \quad h > 0, \quad t > 0$$

$$u(x, 0) = 1, \quad 0 < x < 1$$

Solve for $u(x, t)$.

Solution The method of separation of variables gives

$$X'' + \lambda^2 X = 0 \tag{1}$$

$$T' + k\lambda^2 T = 0$$

and

$$X(x) = c_1 \cos \lambda x + c_2 \sin \lambda x$$

$$T(t) = c_3 e^{-k\lambda^2 t}$$

Since $u = XT$, the boundary conditions become

$$X(0) = 0 \quad \text{and} \quad X'(1) = -hX(1) \tag{2}$$

The first condition in (2) immediately gives $c_1 = 0$. Applying the second condition in (2) to $X(x) = c_2 \sin \lambda x$ yields

$$\lambda \cos \lambda = -h \sin \lambda \quad \text{or} \quad \tan \lambda = -\frac{\lambda}{h} \tag{3}$$

From the analysis in Example 3 of Section 11.4, we know that the last equation in (3) has an infinite number of roots. The consecutive positive roots λ_n, $n = 1, 2, 3, \ldots$, are the eigenvalues of the problem, and the corresponding eigenfunctions are $X(x) = c_2 \sin \lambda_n x$, $n = 1, 2, 3, \ldots$. Thus,

$$u_n = XT = A_n e^{-k\lambda_n^2 t} \sin \lambda_n x$$

and

$$u(x, t) = \sum_{n=1}^{\infty} A_n e^{-k\lambda_n^2 t} \sin \lambda_n x$$

Now at $t = 0$, $u(x, 0) = 1$, $0 < x < 1$, so that

$$1 = \sum_{n=1}^{\infty} A_n \sin \lambda_n x \tag{4}$$

The series in (4) is not a Fourier sine series; it is an expansion in terms of the orthogonal functions arising from the Sturm–Liouville problem consisting of the differential equation (1) and boundary conditions (2). It follows that the set of eigenfunctions $\{\sin \lambda_n x\}$, $n = 1, 2, 3, \ldots$, where the λ's are defined by $\tan \lambda = -\lambda/h$, is orthogonal with respect to the weight function $p(x) = 1$ on the interval $[0, 1]$. With $f(x) = 1$ in (8) of Section 11.1, we can write

$$A_n = \frac{\int_0^1 \sin \lambda_n x \, dx}{\int_0^1 \sin^2 \lambda_n x \, dx} \tag{5}$$

To evaluate the square norm of each of the eigenfunctions we use a trigonometric identity:

$$\int_0^1 \sin^2 \lambda_n x \, dx = \frac{1}{2} \int_0^1 [1 - \cos 2\lambda_n x] \, dx = \frac{1}{2}\left[1 - \frac{1}{2\lambda_n} \sin 2\lambda_n\right] \tag{6}$$

Using $\sin 2\lambda_n = 2 \sin \lambda_n x \cos \lambda_n x$ and $\lambda \cos \lambda = -h \sin \lambda$, we simplify (6) to

$$\int_0^1 \sin^2 \lambda_n x \, dx = \frac{1}{2h}[h + \cos^2 \lambda_n]$$

Also,
$$\int_0^1 \sin \lambda_n x \, dx = -\frac{1}{\lambda_n} \cos \lambda_n x \Big|_0^1 = \frac{1}{\lambda_n}[1 - \cos \lambda_n].$$

Consequently (5) becomes

$$A_n = \frac{2h(1 - \cos \lambda_n)}{\lambda_n(h + \cos^2 \lambda_n)}$$

Finally, a solution of the boundary-value problem is

$$u(x, t) = 2h \sum_{n=1}^{\infty} \frac{1 - \cos \lambda_n}{\lambda_n(h + \cos^2 \lambda_n)} e^{-k\lambda_n^2 t} \sin \lambda_n x \qquad \square$$

EXAMPLE 2 The twist angle $\theta(x, t)$ of a torsionally vibrating shaft of unit length is determined from

$$a^2 \frac{\partial^2 \theta}{\partial x^2} = \frac{\partial^2 \theta}{\partial t^2}, \quad 0 < x < 1, \quad t > 0$$

$$\theta(0, t) = 0, \quad \frac{\partial \theta}{\partial x}\Big|_{x=1} = 0, \quad t > 0$$

$$\theta(x, 0) = x, \quad \frac{\partial \theta}{\partial t}\Big|_{t=0} = 0, \quad 0 < x < 1$$

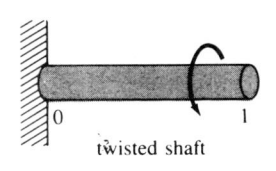

FIGURE 12.16

(see Figure 12.16). The boundary condition at $x = 1$ is called a free-end condition. Solve for $\theta(x, t)$.

Solution Using $\theta = XT$, we find

$$X'' + \lambda^2 X = 0$$
$$T'' + a^2 \lambda^2 T = 0$$

and
$$X(x) = c_1 \cos \lambda x + c_2 \sin \lambda x$$
$$T(t) = c_3 \cos a\lambda t + c_4 \sin a\lambda t$$

The boundary conditions $X(0) = 0$ and $X'(1) = 0$ give $c_1 = 0$ and

$$c_2 \cos \lambda = 0$$

respectively. Since the cosine function is 0 at odd multiples of $\pi/2$, the eigenvalues of the problem are $\lambda = (2n-1)(\pi/2)$, $n = 1, 2, 3, \ldots$. The initial condition $T'(0) = 0$ gives $c_4 = 0$ and so

$$\theta_n = XT = A_n \cos a\left(\frac{2n-1}{2}\right)\pi t \sin\left(\frac{2n-1}{2}\right)\pi x$$

In order to satisfy the remaining initial condition, we form

$$\theta(x, t) = \sum_{n=1}^{\infty} A_n \cos a\left(\frac{2n-1}{2}\right)\pi t \sin\left(\frac{2n-1}{2}\right)\pi x \quad (7)$$

When $t = 0$, we must have for $0 < x < 1$,

$$\theta(x, 0) = x = \sum_{n=1}^{\infty} A_n \sin\left(\frac{2n-1}{2}\right)\pi x \quad (8)$$

As in Example 1, the set of eigenfunctions $\left\{\sin\left(\frac{2n-1}{2}\right)\pi x\right\}$, $n = 1, 2, 3, \ldots$, is orthogonal with respect to the weight function $p(x) = 1$ on the interval $[0, 1]$. The series $\sum_{n=1}^{\infty} A_n \sin\left(\frac{2n-1}{2}\right)\pi x$ is not a Fourier sine series since the argument of the sine is not an integer multiple of $\pi x/L$ ($L = 1$ in this case). The series is again a generalized Fourier series. Hence, from (8) of Section 11.1 the coefficients in (7) are

$$A_n = \frac{\int_0^1 x \sin\left(\frac{2n-1}{2}\right)\pi x \, dx}{\int_0^1 \sin^2\left(\frac{2n-1}{2}\right)\pi x \, dx}$$

Carrying out the two integrations, we arrive at

$$A_n = \frac{8(-1)^{n+1}}{(2n-1)^2 \pi^2}$$

The twist angle is then

$$\theta(x, t) = \frac{8}{\pi^2} \sum_{n=1}^{\infty} \frac{(-1)^{n+1}}{(2n-1)^2} \cos a\left(\frac{2n-1}{2}\right)\pi t \sin\left(\frac{2n-1}{2}\right)\pi x \quad \square$$

EXERCISES 12.7   Answers to odd-numbered problems begin on page A-65.

1. In Example 1 find the temperature $u(x, t)$ when the left end of the rod is insulated.

2. Solve the boundary-value problem:
$$k\frac{\partial^2 u}{\partial x^2} = \frac{\partial u}{\partial t}, \quad 0 < x < 1, \quad t > 0$$
$$u(0, t) = 0, \quad \frac{\partial u}{\partial x}\bigg|_{x=1} = -h(u(1, t) - u_0), \quad h > 0, \quad t > 0$$
$$u(x, 0) = f(x), \quad 0 < x < 1$$

3. Find the steady-state temperature for a rectangular plate for which the boundary conditions are
$$u(0, y) = 0, \quad \frac{\partial u}{\partial x}\bigg|_{x=a} = -hu(a, y) \quad 0 < y < b$$
$$u(x, 0) = 0, \quad u(x, b) = f(x), \quad 0 < x < a$$

4. Solve the boundary-value problem:
$$\frac{\partial^2 u}{\partial x^2} + \frac{\partial^2 u}{\partial y^2} = 0, \quad 0 < y < 1, \quad x > 0$$
$$u(0, y) = u_0, \quad \lim_{x \to \infty} u(x, y) = 0, \quad 0 < y < 1$$
$$\frac{\partial u}{\partial y}\bigg|_{y=0} = 0, \quad \frac{\partial u}{\partial y}\bigg|_{y=1} = -hu(x, 1), \quad h > 0, \quad x > 0$$

5. Find the temperature $u(x, t)$ in a rod of length L if the initial temperature is $f(x)$ throughout and if the end $x = 0$ is kept at temperature 0 and the end $x = L$ is insulated.

6. Solve the boundary-value problem:
$$a^2 \frac{\partial^2 u}{\partial x^2} = \frac{\partial^2 u}{\partial t^2}, \quad 0 < x < L, \quad t > 0$$
$$u(0, t) = 0, \quad E\frac{\partial u}{\partial x}\bigg|_{x=L} = F_0, \quad t > 0$$
$$u(x, 0) = 0, \quad \frac{\partial u}{\partial t}\bigg|_{t=0} = 0, \quad 0 < x < L$$

The solution $u(x, t)$ represents the longitudinal displacement of a vibrating elastic bar that is anchored at its left end and is subjected to a constant force of magnitude F_0 at its right end (see Figure 12.9). E is a constant called the modulus of elasticity.

7. Solve the boundary-value problem:
$$\frac{\partial^2 u}{\partial x^2} + \frac{\partial^2 u}{\partial y^2} = 0, \quad 0 < x < 1, \quad 0 < y < 1$$
$$\frac{\partial u}{\partial x}\bigg|_{x=0} = 0, \quad u(1, y) = u_0, \quad 0 < y < 1$$
$$u(x, 0) = 0, \quad \frac{\partial u}{\partial y}\bigg|_{y=1} = 0, \quad 0 < x < 1$$

8. The initial temperature in a rod of unit length is $f(x)$ throughout. There is heat transfer from both ends, $x = 0$ and $x = 1$, into a surrounding medium kept at a constant temperature 0. Show that
$$u(x, t) = \sum_{n=1}^{\infty} A_n e^{-k\lambda_n^2 t}(\lambda_n \cos \lambda_n x + h \sin \lambda_n x)$$
where
$$A_n = \frac{2}{(\lambda_n^2 + 2h + h^2)} \int_0^1 f(x)(\lambda_n \cos \lambda_n x + h \sin \lambda_n x) \, dx$$
and the λ_n, $n = 1, 2, 3, \ldots$, are the consecutive positive roots of $\tan \lambda = 2\lambda h/(\lambda^2 - h^2)$.

[O] 12.8 BOUNDARY-VALUE PROBLEMS INVOLVING FOURIER SERIES IN TWO VARIABLES

Suppose the rectangular region in Figure 12.17 is a thin plate in which the temperature u is a function of time t and position (x, y). Then, under suitable conditions, $u(x, y, t)$ can be shown to satisfy the **two-dimensional heat equation**

$$k\left(\frac{\partial^2 u}{\partial x^2} + \frac{\partial^2 u}{\partial y^2}\right) = \frac{\partial u}{\partial t} \tag{1}$$

CHAPTER 12 Boundary-Value Problems in Rectangular Coordinates

On the other hand, suppose Figure 12.17 represents a rectangular frame over which a thin flexible membrane has been stretched. If the membrane is set in motion, then its displacement u, measured from the xy-plane (transverse vibrations), is also a function of time t and position (x, y). When the vibrations are small, free, and undamped, $u(x, y, t)$ satisfies the **two-dimensional wave equation**

$$a^2\left(\frac{\partial^2 u}{\partial x^2} + \frac{\partial^2 u}{\partial y^2}\right) = \frac{\partial^2 u}{\partial t^2} \tag{2}$$

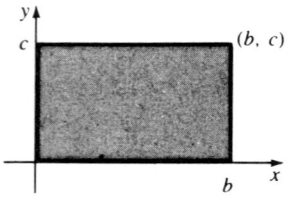

FIGURE 12.17

As the next example will show, solutions of boundary-value problems involving (1) and (2) by the method of separation of variables lead to the notion of Fourier series in two variables.

EXAMPLE 1 Find the temperature $u(x, y, t)$ in the plate shown in Figure 12.17 if the initial temperature is $f(x, y)$ throughout and if the boundaries are held at temperature 0.

Solution We must solve

$$k\left(\frac{\partial^2 u}{\partial x^2} + \frac{\partial^2 u}{\partial y^2}\right) = \frac{\partial u}{\partial t}, \quad 0 < x < b, \quad 0 < y < c, \quad t > 0$$

subject to
$$u(0, y, t) = 0, \quad u(b, y, t) = 0, \quad 0 < y < c, \quad t > 0$$
$$u(x, 0, t) = 0, \quad u(x, c, t) = 0, \quad 0 < x < b, \quad t > 0$$
$$u(x, y, 0) = f(x, y), \quad 0 < x < b, \quad 0 < y < c$$

To separate variables in the partial differential equation in three independent variables we try to find a product solution $u(x, y, t) = X(x)Y(y)T(t)$. Substituting, we get

$$k(X''YT + XY''T) = XYT'$$

or
$$\frac{X''}{X} = -\frac{Y''}{Y} + \frac{T'}{kT} \tag{3}$$

Since the left side of (3) depends only on x and the right side depends only on y and t, we must have both sides equal to a constant $-\lambda^2$:

$$\frac{X''}{X} = -\frac{Y''}{Y} + \frac{T'}{kT} = -\lambda^2$$

and so
$$X'' + \lambda^2 X = 0 \tag{4}$$
$$\frac{Y''}{Y} = \frac{T'}{kT} + \lambda^2 \tag{5}$$

By the same reasoning, if we introduce another separation constant $-\mu^2$ in (5), then

$$\frac{Y''}{Y} = -\mu^2 \quad \text{and} \quad \frac{T'}{kT} + \lambda^2 = -\mu^2$$
$$Y'' + \mu^2 Y = 0 \quad \text{and} \quad T' + k(\lambda^2 + \mu^2)T = 0 \tag{6}$$

12.8 Boundary-Value Problems Involving Fourier Series in Two Variables

Now the solutions of the equations in (4) and (6) are, respectively,

$$X(x) = c_1 \cos \lambda x + c_2 \sin \lambda x \qquad (7)$$

$$Y(y) = c_3 \cos \mu y + c_4 \sin \mu y \qquad (8)$$

$$T(t) = c_5 e^{-k(\lambda^2 + \mu^2)t} \qquad (9)$$

But the boundary conditions

$$\left.\begin{array}{l} u(0, y, t) = 0, \quad u(b, y, t) = 0 \\ u(x, 0, t) = 0, \quad u(x, c, t) = 0 \end{array}\right\} \quad \text{imply} \quad \begin{cases} X(0) = 0, \quad X(b) = 0 \\ Y(0) = 0, \quad Y(c) = 0 \end{cases}$$

Applying these conditions to (7) and (8) gives $c_1 = 0$, $c_3 = 0$, and $c_2 \sin \lambda b = 0$, $c_4 \sin \mu c = 0$. The latter equations in turn imply

$$\lambda = \frac{m\pi}{b}, \quad m = 1, 2, 3, \ldots \qquad \mu = \frac{n\pi}{c}, \quad n = 1, 2, 3, \ldots$$

Thus a product solution of the two-dimensional heat equation that satisfies the boundary conditions is

$$u_{mn}(x, y, t) = A_{mn} e^{-k[(m\pi/b)^2 + (n\pi/c)^2]t} \sin \frac{m\pi}{b} x \sin \frac{n\pi}{c} y$$

where A_{mn} is an arbitrary constant. Because we have two independent sets of eigenvalues, we are prompted to try the superposition principle in the form of a double sum

$$u(x, y, t) = \sum_{m=1}^{\infty} \sum_{n=1}^{\infty} A_{mn} e^{-k[(m\pi/b)^2 + (n\pi/c)^2]t} \sin \frac{m\pi}{b} x \sin \frac{n\pi}{c} y \qquad (10)$$

Now at $t = 0$ we must have

$$u(x, y, 0) = f(x, y) = \sum_{m=1}^{\infty} \sum_{n=1}^{\infty} A_{mn} \sin \frac{m\pi}{b} x \sin \frac{n\pi}{c} y \qquad (11)$$

We can find the coefficients A_{mn} by multiplying the double sum (11) by the product $\sin(m\pi x/b) \sin(n\pi y/c)$ and integrating over the rectangle $0 \leq x \leq b$, $0 \leq y \leq c$. It follows that

$$A_{mn} = \frac{4}{bc} \int_0^c \int_0^b f(x, y) \sin \frac{m\pi}{b} x \sin \frac{n\pi}{c} y \, dx \, dy \qquad (12)$$

Thus, the solution of the boundary-value problem consists of (1) with the A_{mn} defined by (12). □

The series (11) with coefficients (12) is called a **sine series in two variables**, or a **double sine series** (see Problem 52 in Exercises 11.3).

EXERCISES 12.8   *Answers to odd-numbered problems begin on page A-65.*

In Problems 1 and 2 solve the heat equation (1) subject to the given conditions.

1. $u(0, y, t) = 0$, $u(\pi, y, t) = 0$
$u(x, 0, t) = 0$, $u(x, \pi, t) = 0$
$u(x, y, 0) = u_0$

2. $\dfrac{\partial u}{\partial x}\bigg|_{x=0} = 0$, $\dfrac{\partial u}{\partial x}\bigg|_{x=1} = 0$
$\dfrac{\partial u}{\partial y}\bigg|_{y=0} = 0$, $\dfrac{\partial u}{\partial y}\bigg|_{y=1} = 0$
$u(x, y, 0) = xy$
[*Hint:* See Problem 53 in Exercises 11.3.]

In Problems 3 and 4 solve the wave equation (2) subject to the given conditions.

3. $u(0, y, t) = 0$, $u(\pi, y, t) = 0$
$u(x, 0, t) = 0$, $u(x, \pi, t) = 0$
$u(x, y, 0) = xy(x - \pi)(y - \pi)$
$\dfrac{\partial u}{\partial t}\bigg|_{t=0} = 0$

4. $u(0, y, t) = 0$, $u(b, y, t) = 0$
$u(x, 0, t) = 0$, $u(x, c, t) = 0$
$u(x, y, 0) = f(x, y)$
$\dfrac{\partial u}{\partial t}\bigg|_{t=0} = g(x, y)$

The steady-state temperature $u(x, y, z)$ in the rectangular parallelepiped shown in Figure 12.18 satisfies Laplace's equation in three dimensions

$$\frac{\partial^2 u}{\partial x^2} + \frac{\partial^2 u}{\partial y^2} + \frac{\partial^2 u}{\partial z^2} = 0 \qquad (13)$$

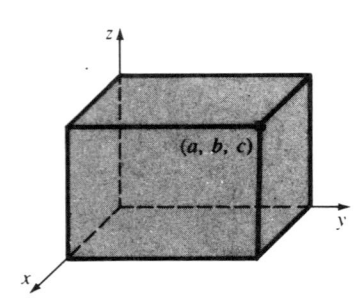

FIGURE 12.18

5. Solve Laplace's equation (13) if the top ($z = c$) of the parallelepiped is kept at temperature $f(x, y)$ and the remaining sides are kept at temperature 0.

6. Solve Laplace's equation (13) if the bottom ($z = 0$) of the parallelepiped is kept at temperature $f(x, y)$ and the remaining sides are kept at temperature 0.

SUMMARY

A particular solution of a linear partial differential equation in two variables can possibly be found by assuming a solution in the form of a product $u = XY$, where X is a function of x only and Y is a function of y only. If applicable, this **method of separation of variables** leads to two ordinary differential equations.

A **boundary-value problem** consists of finding a function that satisfies a partial differential equation as well as side conditions consisting of perhaps both boundary conditions and initial conditions. We applied the method of separation of variables to obtain solutions of certain boundary-value problems involving the heat equation, the wave equation, and Laplace's equation. The procedure consisted of five basic steps.

(*i*) Separate the variables.

(*ii*) Solve the separated ordinary differential equations and find the eigenvalues and eigenfunctions of the problem.

CHAPTER 12 REVIEW EXERCISES   *Answers to odd-numbered problems begin on page A-65.*

1. Use separation of variables to find product solutions of

$$\frac{\partial^2 u}{\partial x\, \partial y} = u$$

2. Use separation of variables to find product solutions of

$$\frac{\partial^2 u}{\partial x^2} + \frac{\partial^2 u}{\partial y^2} + 2\frac{\partial u}{\partial x} + 2\frac{\partial u}{\partial y} = 0$$

   Is it possible to choose a constant of separation so that both X and Y are oscillatory functions?

3. Find a steady-state solution $\psi(x)$ of the boundary-value problem:

$$k\frac{\partial^2 u}{\partial x^2} = \frac{\partial u}{\partial t}, \quad 0 < x < \pi, \quad t > 0$$

$$u(0, t) = u_0, \quad -\frac{\partial u}{\partial x}\bigg|_{x=\pi} = u(\pi, t) - u_1, \quad t > 0$$

$$u(x, 0) = 0, \quad 0 < x < \pi$$

4. Give a physical interpretation for the boundary conditions in Problem 3.

5. At $t = 0$ a string of unit length is stretched on the positive x-axis. The ends of the string $x = 0$ and $x = 1$ are secured on the x-axis for $t > 0$. Find the displacement $u(x, t)$ if the initial velocity $g(x)$ is as given in Figure 12.19.

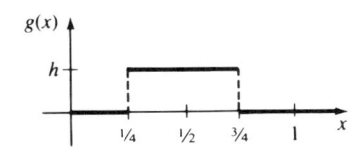

FIGURE 12.19

6. The partial differential equation

$$\frac{\partial^2 u}{\partial x^2} + x^2 = \frac{\partial^2 u}{\partial t^2}$$

   is a form of the wave equation when an external vertical force proportional to the square of the horizontal distance from the left end is applied to the string. The string is secured at $x = 0$ one unit above the x-axis and on the x-axis at $x = 1$ for $t > 0$. Find the displacement $u(x, t)$ if the string starts from rest from the initial displacement $f(x)$.

7. Find the steady-state temperature $u(x, y)$ in the square plate shown in Figure 12.20.

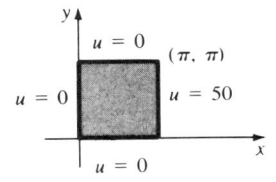

FIGURE 12.20

8. Find the steady-state temperature $u(x, y)$ in the semi-infinite plate shown in Figure 12.21.

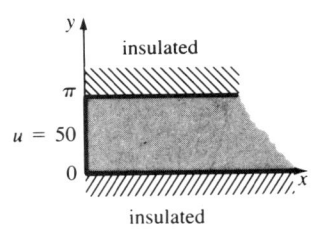

FIGURE 12.21

9. Solve Problem 8 if the boundaries $y = 0$ and $y = \pi$ are held at temperature 0 for all time.

10. Find the temperature $u(x, t)$ in the infinite plate of width $2L$ shown in Figure 12.22 if the initial temperature is u_0 throughout. [*Hint:* $u(x, 0) = u_0$, $-L < x < L$, is an even function of x.]

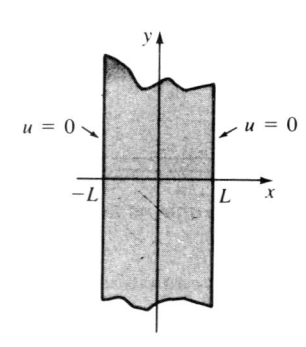

FIGURE 12.22

11. Solve the boundary-value problem:

$$\frac{\partial^2 u}{\partial x^2} = \frac{\partial u}{\partial t}, \quad 0 < x < \pi, \quad t > 0$$

$$u(0, t) = 0, \quad u(\pi, t) = 0, \quad t > 0$$

$$u(x, 0) = \sin x, \quad 0 < x < \pi$$

12. Find a formal series solution of the problem

$$\frac{\partial^2 u}{\partial x^2} + 2\frac{\partial u}{\partial x} = \frac{\partial^2 u}{\partial t^2} + 2\frac{\partial u}{\partial t} + u, \quad 0 < x < \pi, \quad t > 0$$

$$u(0, t) = 0, \quad u(\pi, t) = 0, \quad t > 0$$

$$\left.\frac{\partial u}{\partial t}\right|_{t=0} = 0, \quad 0 < x < \pi$$

Do not attempt to evaluate the coefficients in the series.

13

BOUNDARY-VALUE PROBLEMS IN OTHER COORDINATE SYSTEMS

TOPICS TO REVIEW

polar coordinates
cylindrical coordinates
spherical coordinates
Cauchy–Euler equations (5.1)
Bessel functions (5.4)
Fourier–Bessel series (11.5)
Legendre polynomials (5.4)
Fourier–Legendre series (11.5)

IMPORTANT CONCEPTS

Laplacian in
   polar coordinates
   cylindrical coordinates
   spherical coordinates
radial symmetry
standing waves
nodal lines

13.1 Problems Involving Laplace's Equation in Polar Coordinates
13.2 Problems in Polar and Cylindrical Coordinates: Bessel Functions
13.3 Problems in Spherical Coordinates: Legendre Polynomials
      Summary
      Chapter 13 Review Exercises

INTRODUCTION

All the boundary-value problems that have been considered so far have been expressed in terms of a rectangular coordinate system. If, however, we wish to find temperatures in a circular disk, in a circular cylinder, or in a sphere, we would naturally try to describe the problem in terms of polar coordinates, cylindrical coordinates, or spherical coordinates, respectively. It is important then that we express the Laplacian $\nabla^2 u$ in terms of these different coordinates. In Sections 13.2 and 13.3 we will put to practical use the theory of Fourier–Bessel series and Fourier–Legendre series.

13.1 PROBLEMS INVOLVING LAPLACE'S EQUATION IN POLAR COORDINATES

Laplacian in Polar Coordinates The relationship between polar coordinates in the plane and rectangular coordinates is given by

$$x = r \cos \theta \qquad y = r \sin \theta$$

and

$$r^2 = x^2 + y^2 \qquad \tan \theta = \frac{y}{x}$$

See Figure 13.1. The first pair of equations transforms polar coordinates (r, θ) into rectangular coordinates (x, y); the second pair of equations enables us to transform rectangular coordinates into polar coordinates. These equations also make it possible to convert the two-dimensional Laplacian $\nabla^2 u = \partial^2 u/\partial x^2 + \partial^2 u/\partial y^2$ into polar coordinates. The reader is encouraged to work through the details of the chain rule and show that

$$\frac{\partial u}{\partial x} = \frac{\partial u}{\partial r}\frac{\partial r}{\partial x} + \frac{\partial u}{\partial \theta}\frac{\partial \theta}{\partial x}$$

$$= \cos \theta \frac{\partial u}{\partial r} - \frac{\sin \theta}{r}\frac{\partial u}{\partial \theta}$$

$$\frac{\partial u}{\partial y} = \frac{\partial u}{\partial r}\frac{\partial r}{\partial y} + \frac{\partial u}{\partial \theta}\frac{\partial \theta}{\partial y}$$

$$= \sin \theta \frac{\partial u}{\partial r} + \frac{\cos \theta}{r}\frac{\partial u}{\partial \theta}$$

$$\frac{\partial^2 u}{\partial x^2} = \cos^2 \theta \frac{\partial^2 u}{\partial r^2} - \frac{2 \sin \theta \cos \theta}{r}\frac{\partial^2 u}{\partial r \partial \theta}$$

$$+ \frac{\sin^2 \theta}{r^2}\frac{\partial^2 u}{\partial \theta^2} + \frac{\sin^2 \theta}{r}\frac{\partial u}{\partial r} + \frac{2 \sin \theta \cos \theta}{r}\frac{\partial u}{\partial \theta} \qquad (1)$$

$$\frac{\partial^2 u}{\partial y^2} = \sin^2 \theta \frac{\partial^2 u}{\partial r^2} + \frac{2 \sin \theta \cos \theta}{r}\frac{\partial^2 u}{\partial r \partial \theta}$$

$$+ \frac{\cos^2 \theta}{r^2}\frac{\partial^2 u}{\partial \theta^2} + \frac{\cos^2 \theta}{r}\frac{\partial u}{\partial r} - \frac{2 \sin \theta \cos \theta}{r}\frac{\partial u}{\partial \theta} \qquad (2)$$

Adding (1) and (2) and simplifying yield

$$\nabla^2 u = \frac{\partial^2 u}{\partial x^2} + \frac{\partial^2 u}{\partial y^2} = \frac{\partial^2 u}{\partial r^2} + \frac{1}{r}\frac{\partial u}{\partial r} + \frac{1}{r^2}\frac{\partial^2 u}{\partial \theta^2}$$

In this section we shall focus only on problems involving Laplace's equation in polar coordinates:

$$\frac{\partial^2 u}{\partial r^2} + \frac{1}{r}\frac{\partial u}{\partial r} + \frac{1}{r^2}\frac{\partial^2 u}{\partial \theta^2} = 0$$

The first example is the Dirichlet problem for a circular disk.

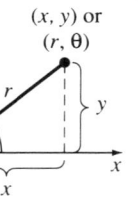

FIGURE 13.1

13.1 Problems Involving Laplace's Equation in Polar Coordinates

EXAMPLE 1 Find the steady-state temperature $u(r, \theta)$ in the circular plate of radius c if the temperature of the circumference is $u(c, \theta) = f(\theta)$, $0 < \theta < 2\pi$. See Figure 13.2.

Solution We must solve Laplace's equation

$$\frac{\partial^2 u}{\partial r^2} + \frac{1}{r}\frac{\partial u}{\partial r} + \frac{1}{r^2}\frac{\partial^2 u}{\partial \theta^2} = 0, \quad 0 < \theta < 2\pi, \quad 0 < r < c$$

subject to $\quad u(c, \theta) = f(\theta), \quad 0 < \theta < 2\pi$

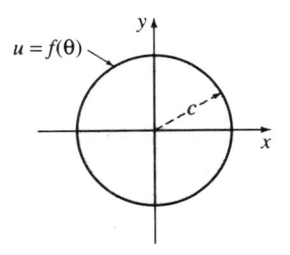

FIGURE 13.2

If we define $u = R(r)\Theta(\theta)$, then separation of variables gives

$$\frac{r^2 R'' + rR'}{R} = -\frac{\Theta''}{\Theta} = \lambda^2$$

and
$$r^2 R'' + rR' - \lambda^2 R = 0 \quad (3)$$
$$\Theta'' + \lambda^2 \Theta = 0 \quad (4)$$

The solution of (4) is

$$\Theta = c_1 \cos \lambda\theta + c_2 \sin \lambda\theta \quad (5)$$

Equation (3) is recognized as a Cauchy–Euler equation. Its general solution is

$$R = c_3 r^\lambda + c_4 r^{-\lambda} \quad (6)$$

There are no explicit conditions in the statement of this problem that enable us to determine any of the coefficients or the eigenvalues. However, there are several implicit conditions.

First, our physical intuition leads us to expect that the temperature $u(r, \theta)$ should be bounded inside the circle $r = c$. Moreover, the temperature u should be the same at a specified point in the circle regardless of the polar description of that point. Since $(r, \theta + 2\pi)$ is an equivalent description of the point (r, θ), we must have

$$u(r, \theta) = u(r, \theta + 2\pi)$$

In other words, the temperature $u(r, \theta)$ must be periodic with period 2π. But the only way the solutions of (4) can be 2π-periodic is to take $\lambda = n$, where $n = 0, 1, 2, \ldots$. Thus, for $n > 0$, (5) and (6) become

$$\Theta = c_1 \cos n\theta + c_2 \sin n\theta \quad (7)$$
$$R = c_3 r^n + c_4 r^{-n} \quad (8)$$

Now observe in (8) that $r^{-n} = 1/r^n$. In order that our solution be bounded at the center of the plate, that is, at $r = 0$, we must define $c_4 = 0$. Hence, for $\lambda = n > 0$,

$$u_n = R\Theta = r^n(A_n \cos n\theta + B_n \sin n\theta) \quad (9)$$

where $c_3 c_1$ and $c_3 c_2$ have been replaced by A_n and B_n, respectively.

There is one last item of consideration: namely, $n = 0$ is an eigenvalue. For this value the solutions of (3) and (4) are not given by (5) and (6). The solutions of the differential equations $\Theta'' = 0$ and $r^2R'' + rR' = 0$ are

$$\Theta = c_5\theta + c_6 \quad \text{and} \quad R = c_7 + c_8 \ln r$$

Periodicity demands that we take $c_5 = 0$, whereas the expectation that the temperature is finite at $r = 0$ demands that we take $c_8 = 0$. In this case a product solution of the partial differential equation is simply

$$u_0 = A_0 \tag{10}$$

where A_0 represents $c_6 c_7$.

The superposition principle then gives

$$u(r, \theta) = \sum_{n=0}^{\infty} u_n$$

or

$$u(r, \theta) = A_0 + \sum_{n=1}^{\infty} r^n (A_n \cos n\theta + B_n \sin n\theta) \tag{11}$$

Finally, by applying the boundary condition at $r = c$ to the result in (11), we recognize

$$f(\theta) = A_0 + \sum_{n=1}^{\infty} c^n (A_n \cos n\theta + B_n \sin n\theta)$$

as an expansion of f in a full Fourier series. Consequently, we can make the identifications

$$A_0 = \frac{a_0}{2}, \quad c^n A_n = a_n, \quad \text{and} \quad c^n B_n = b_n$$

that is,

$$A_0 = \frac{1}{2\pi} \int_0^{2\pi} f(\theta)\, d\theta \tag{12}$$

$$A_n = \frac{1}{c^n \pi} \int_0^{2\pi} f(\theta) \cos n\theta\, d\theta \tag{13}$$

$$B_n = \frac{1}{c^n \pi} \int_0^{2\pi} f(\theta) \sin n\theta\, d\theta \tag{14}$$

The solution of the problem consists of the series given in (11), where A_0, A_n, and B_n are defined in (12), (13), and (14), respectively. □

Observe in Example 1 that corresponding to each positive eigenvalue, $n = 1, 2, 3, \ldots$, there are two different eigenfunctions—namely, $\cos n\theta$ and $\sin n\theta$. In this situation the eigenvalues are sometimes called **double eigenvalues**.

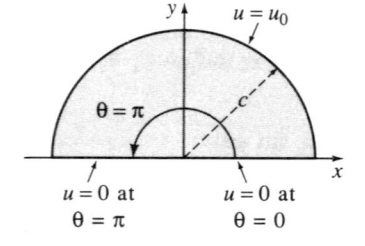

FIGURE 13.3

EXAMPLE 2 Find the steady-state temperature $u(r, \theta)$ in the semicircular plate shown in Figure 13.3.

Solution The boundary-value problem is

$$\frac{\partial^2 u}{\partial r^2} + \frac{1}{r}\frac{\partial u}{\partial r} + \frac{1}{r^2}\frac{\partial^2 u}{\partial \theta^2} = 0, \quad 0 < \theta < \pi, \quad 0 < r < c$$

$$u(c, \theta) = u_0, \quad 0 < \theta < \pi$$

$$u(r, 0) = 0, \quad u(r, \pi) = 0, \quad 0 < r < c$$

Defining $u = R(r)\Theta(\theta)$ and separating variables give

$$\frac{r^2 R'' + rR'}{R} = -\frac{\Theta''}{\Theta} = \lambda^2$$

and
$$r^2 R'' + rR' - \lambda^2 R = 0 \tag{15}$$
$$\Theta'' + \lambda^2 \Theta = 0 \tag{16}$$

Applying the boundary conditions $\Theta(0) = 0$ and $\Theta(\pi) = 0$ to the solution $\Theta = c_1 \cos \lambda\theta + c_2 \sin \lambda\theta$ of (16) gives, in turn, $c_1 = 0$ and $\lambda = n$, $n = 1, 2, 3, \ldots$. Hence, $\Theta = c_2 \sin n\theta$. Unlike the problem in Example 1, $n = 0$ is not an eigenvalue. With $\lambda = n$ the solution of (15) is

$$R = c_3 r^n + c_4 r^{-n}$$

But the assumption that $u(r, \theta)$ is bounded at $r = 0$ prompts us to define $c_4 = 0$. Therefore,

$$u_n = R(r)\Theta(\theta) = A_n r^n \sin n\theta$$

and
$$u(r, \theta) = \sum_{n=1}^{\infty} A_n r^n \sin n\theta$$

The remaining boundary condition at $r = c$ gives the sine series

$$u_0 = \sum_{n=1}^{\infty} A_n c^n \sin n\theta$$

Consequently,
$$A_n c^n = \frac{2}{\pi} \int_0^\pi u_0 \sin n\theta \, d\theta$$

and so
$$A_n = \frac{2u_0}{\pi c^n} \frac{1 - (-1)^n}{n}$$

Hence, the solution of the problem is given by

$$u(r, \theta) = \frac{2u_0}{\pi} \sum_{n=1}^{\infty} \frac{1 - (-1)^n}{n} \left(\frac{r}{c}\right)^n \sin n\theta \qquad \square$$

EXERCISES 13.1 Answers to odd-numbered problems begin on page A-65.

In Problems 1–4 find the steady-state temperature $u(r, \theta)$ in a circular plate of radius 1 if the temperature on the circumference is as given.

1. $u(1, \theta) = \begin{cases} u_0, & 0 < \theta < \pi \\ 0, & \pi \leq \theta < 2\pi \end{cases}$

2. $u(1, \theta) = \begin{cases} \theta, & 0 < \theta < \pi \\ \pi - \theta, & \pi \leq \theta < 2\pi \end{cases}$

3. $u(1, \theta) = 2\pi\theta - \theta^2, 0 < \theta < 2\pi$

4. $u(1, \theta) = \theta, 0 < \theta < 2\pi$

5. Solve the exterior Dirichlet problem for a circular disk of radius c if $u(c, \theta) = f(\theta), 0 < \theta < 2\pi$. In other words, find the steady-state temperature $u(r, \theta)$ in a plate that coincides with the entire xy-plane in which a circular hole of radius c has been cut out around the origin and the temperature on the circumference of the hole is $f(\theta)$. [*Hint:* Assume that the temperature is bounded as $r \to \infty$.]

6. Find the steady-state temperature in the quarter-circular plate shown in Figure 13.4.

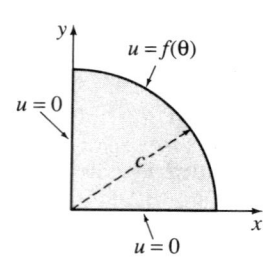

FIGURE 13.4

7. If the boundaries $\theta = 0$ and $\theta = \pi/2$ in Figure 13.4 are insulated, we then have, respectively,

$$\left.\frac{\partial u}{\partial \theta}\right|_{\theta=0} = 0 \qquad \left.\frac{\partial u}{\partial \theta}\right|_{\theta=\pi/2} = 0$$

Find the steady-state temperature if

$$u(c, \theta) = \begin{cases} 1, & 0 < \theta < \pi/4 \\ 0, & \pi/4 \leq \theta \leq \pi/2 \end{cases}$$

8. Find the steady-state temperature in the infinite wedge-shaped plate shown in Figure 13.5. [*Hint:* Assume that the temperature is bounded as $r \to 0$ and as $r \to \infty$.]

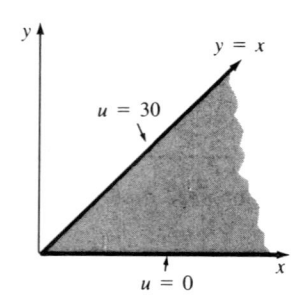

FIGURE 13.5

9. Find the steady-state temperature $u(r, \theta)$ in the circular ring shown in Figure 13.6. [*Hint:* Proceed as in Example 1.]

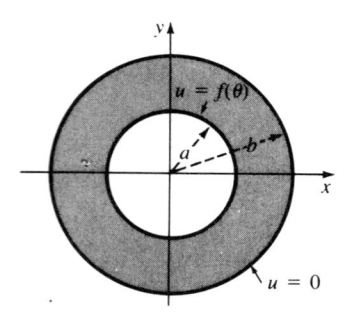

FIGURE 13.6

10. If the boundary conditions for the circular ring in Figure 13.6 are $u(a, \theta) = u_0, u(b, \theta) = u_1, 0 < \theta < 2\pi, u_0$ and u_1 constants, show that the steady-state temperature is given by $u(r, \theta) = [u_0 \ln(r/b) - u_1 \ln(r/a)]/\ln(a/b)$. [*Hint:* Try a solution of the form $u(r, \theta) = v(r, \theta) + \psi(r)$.]

11. Find the steady-state temperature $u(r, \theta)$ in a semicircular plate of radius 1 if

$$u(1, \theta) = u_0, \quad 0 < \theta < \pi$$
$$u(r, 0) = 0, \quad u(r, \pi) = u_0, \quad 0 < r < 1$$

12. Show that the steady-state temperature in Example 1 can be written as the integral

$$u(r, \theta) = \frac{1}{2\pi} \int_0^{2\pi} \frac{c^2 - r^2}{c^2 - 2cr\cos(t - \theta) + r^2} f(t) \, dt$$

This result is known as **Poisson's integral formula for a circle**. [*Hint:* First show that

$$u(r, \theta) = \frac{1}{2\pi} \int_0^{2\pi} f(t) \left[1 + 2 \sum_{n=1}^{\infty} \left(\frac{r}{c}\right)^n \cos n(t - \theta) \right] dt$$

Then use $\cos nv = \frac{1}{2}(e^{inv} + e^{-inv})$ and geometric series to show that

$$1 + 2 \sum_{n=1}^{\infty} u^n \cos nv = \frac{1 - u^2}{1 - 2u\cos v + u^2}, \quad |u| < 1.]$$

13.2 PROBLEMS IN POLAR AND CYLINDRICAL COORDINATES: BESSEL FUNCTIONS

The two-dimensional heat and wave equations,

$$k\left(\frac{\partial^2 u}{\partial x^2} + \frac{\partial^2 u}{\partial y^2}\right) = \frac{\partial u}{\partial t} \quad \text{and} \quad a^2\left(\frac{\partial^2 u}{\partial x^2} + \frac{\partial^2 u}{\partial y^2}\right) = \frac{\partial^2 u}{\partial t^2}$$

expressed in polar coordinates are, respectively,

$$k\left(\frac{\partial^2 u}{\partial r^2} + \frac{1}{r}\frac{\partial u}{\partial r} + \frac{1}{r^2}\frac{\partial^2 u}{\partial \theta^2}\right) = \frac{\partial u}{\partial t} \quad \text{and} \quad a^2\left(\frac{\partial^2 u}{\partial r^2} + \frac{1}{r}\frac{\partial u}{\partial r} + \frac{1}{r^2}\frac{\partial^2 u}{\partial \theta^2}\right) = \frac{\partial^2 u}{\partial t^2}$$

where $u = u(r, \theta, t)$. To solve a boundary-value problem involving either of these equations by separation of variables we would define $u = R(r)\Theta(\theta)T(t)$. As in Section 12.8, this assumption leads to multiple infinite series (see Problem 12). In the following discussion we shall consider the simpler, but still important, problems that possess **radial symmetry**—that is, problems that are independent of the angular coordinate θ. In this case the heat and wave equations take the forms

$$k\left(\frac{\partial^2 u}{\partial r^2} + \frac{1}{r}\frac{\partial u}{\partial r}\right) = \frac{\partial u}{\partial t} \quad \text{and} \quad a^2\left(\frac{\partial^2 u}{\partial r^2} + \frac{1}{r}\frac{\partial u}{\partial r}\right) = \frac{\partial^2 u}{\partial t^2} \quad (1)$$

where $u = u(r, t)$. Vibrations described by the second equation in (1) are said to be **radial vibrations**.

The first example will deal with the free radial vibrations of a thin circular membrane. We assume that the displacements are small and that the motion is such that each point on the membrane moves in a direction perpendicular to the xy-plane (transverse vibrations); that is, the u-axis is perpendicular to the xy-plane. A physical model to keep in mind while reading this example is a vibrating drumhead.

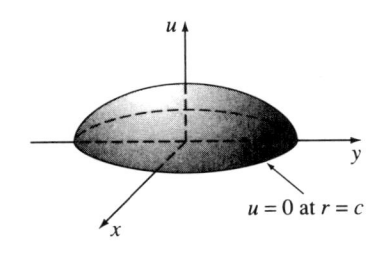

FIGURE 13.7

EXAMPLE 1 Find the displacement $u(r, t)$ of a circular membrane of radius c clamped along its circumference if its initial displacement is $f(r)$ and its initial velocity is $g(r)$. See Figure 13.7.

Solution The boundary-value problem to be solved is

$$a^2\left(\frac{\partial^2 u}{\partial r^2} + \frac{1}{r}\frac{\partial u}{\partial r}\right) = \frac{\partial^2 u}{\partial t^2}, \quad 0 < r < c, \quad t > 0$$

$$u(c, t) = 0, \quad t > 0$$

$$u(r, 0) = f(r), \quad \left.\frac{\partial u}{\partial t}\right|_{t=0} = g(r), \quad 0 < r < c$$

Substituting $u = R(r)T(t)$ into the partial differential equation and separating variables give

$$\frac{R'' + \frac{1}{r}R'}{R} = \frac{T''}{a^2 T} = -\lambda^2$$

and
$$rR'' + R' + \lambda^2 rR = 0 \qquad (2)$$
$$T'' + a^2\lambda^2 T = 0 \qquad (3)$$

Now (2) is not a Cauchy–Euler equation but is the parametric Bessel differential equation with order $v = 0$. Its general solution is

$$R = c_1 J_0(\lambda r) + c_2 Y_0(\lambda r)$$

The general solution of the familiar equation (3) is

$$T = c_3 \cos a\lambda t + c_4 \sin a\lambda t$$

Now recall that $Y_0(\lambda r) \to -\infty$ as $r \to 0^+$, and so the implicit assumption that the displacement $u(r, t)$ should be bounded at $r = 0$ forces us to define $c_2 = 0$. Thus, $R = c_1 J_0(\lambda r)$.

Since the boundary condition $u(c, t) = 0$ is equivalent to $R(c) = 0$, we must have $c_1 J_0(\lambda c) = 0$. We rule out $c_1 = 0$ (this would lead to a trivial solution of the partial differential equation), so consequently

$$J_0(\lambda c) = 0 \qquad (4)$$

This equation defines the positive eigenvalues λ_n of the problem: If x_n are the positive roots of (4), then $\lambda_n = x_n/c$. Product solutions that satisfy the partial differential equation and the boundary condition are

$$u_n = RT = (A_n \cos a\lambda_n t + B_n \sin a\lambda_n t) J_0(\lambda_n r) \qquad (5)$$

where we have done the usual relabeling of the constants. The superposition principle then gives

$$u(r, t) = \sum_{n=1}^{\infty} (A_n \cos a\lambda_n t + B_n \sin a\lambda_n t) J_0(\lambda_n r) \qquad (6)$$

The given initial conditions will determine the coefficients A_n and B_n.

Setting $t = 0$ in (6) and using $u(r, 0) = f(r)$ give

$$f(r) = \sum_{n=1}^{\infty} A_n J_0(\lambda_n r) \qquad (7)$$

This last result is recognized as the Fourier–Bessel expansion of the function f on the interval $(0, c)$. Hence, by a direct comparison of (4) and (7) with (7) and (15) of Section 11.5, we can identify the coefficients A_n with those given in (16) of Section 11.5:

$$A_n = \frac{2}{c^2 J_1^2(\lambda_n c)} \int_0^c r J_0(\lambda_n r) f(r) \, dr \qquad (8)$$

13.2 Problems in Polar and Cylindrical Coordinates: Bessel Functions

Next, we differentiate (6) with respect to t, set $t = 0$, and use $u_t(r, 0) = g(r)$:

$$g(r) = \sum_{n=1}^{\infty} a\lambda_n B_n J_0(\lambda_n r)$$

This is now a Fourier–Bessel expansion of the function g. By identifying the total coefficient $a\lambda_n B_n$ with (16) of Section 11.5, we can write

$$B_n = \frac{2}{a\lambda_n c^2 J_1^2(\lambda_n c)} \int_0^c r J_0(\lambda_n r) g(r)\, dr \qquad (9)$$

Finally, the solution of the given boundary-value problem is the series (6) with A_n and B_n defined in (8) and (9). □

Analogous to (8) of Section 12.4, the product solutions (5) are called **standing waves**. For $n = 1, 2, 3, \ldots$, the standing waves are basically the graph of $J_0(\lambda_n r)$ with the time-varying amplitude

$$A_n \cos a\lambda_n t + B_n \sin a\lambda_n t$$

The standing waves at different values of time are represented by the dashed lines in Figure 13.8. The zeros of each standing wave in the interval $(0, c)$ are the roots of $J_0(\lambda_n r) = 0$ and correspond to the set of points on a standing wave where there is no motion. This set of points is called a **nodal line**. If (as in Example 1) the positive roots of $J_0(\lambda_n c) = 0$ are denoted by x_n, then $\lambda_n c = x_n$ implies $\lambda_n = x_n/c$, and consequently the zeros of the standing waves are determined from

$$J_0(\lambda_n r) = J_0\left(\frac{x_n}{c} r\right) = 0$$

Now the first three positive zeros of J_0 are (approximately) $x_1 = 2.4$, $x_2 = 5.5$, and $x_3 = 8.7$. Thus for $n = 1$, the first positive root of

$$J_0\left(\frac{x_1}{c} r\right) = 0 \quad \text{is} \quad \frac{2.4}{c} r = 2.4 \quad \text{or} \quad r = c$$

Since we are seeking zeros of the standing waves in the open interval $(0, c)$, the last result means that the first standing wave has no nodal line. For $n = 2$, the first two positive roots of

$$J_0\left(\frac{x_2}{c} r\right) = 0 \quad \text{are determined from} \quad \frac{5.5}{c} r = 2.4 \quad \text{and} \quad \frac{5.5}{c} r = 5.5$$

Thus, the second standing wave has one nodal line defined by $r = x_1 c/x_2 = 2.4c/5.5$. Note that $r \approx 0.44c < c$. For $n = 3$, a similar analysis shows that there are two nodal lines defined by $r = x_1 c/x_3 \approx 2.4c/8.7$ and $r = x_2 c/x_3 \approx 5.5c/8.7$. In general, the nth standing wave has $n - 1$ nodal lines $r = x_1 c/x_n$, $r = x_2 c/x_n, \ldots, r = x_{n-1} c/x_n$. Since $r = $ *constant* is an equation of a circle in polar coordinates, we see in Figure 13.8 that nodal lines of a standing wave are concentric circles.

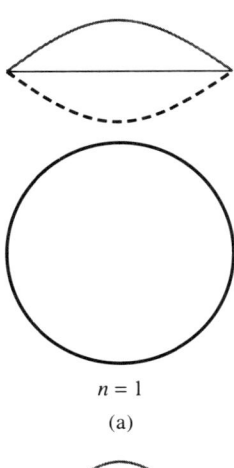

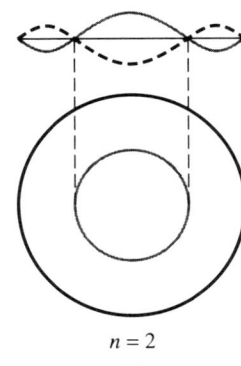

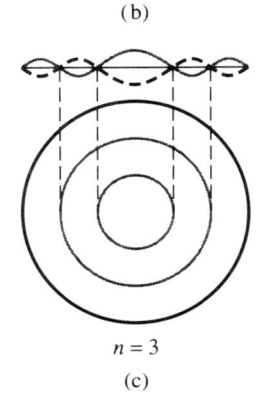

FIGURE 13.8

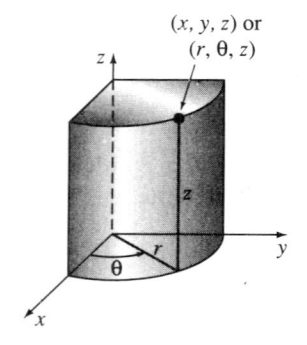

Laplacian in Cylindrical Coordinates From Figure 13.9 we see that the relationship between the cylindrical coordinates of a point in space and its rectangular coordinates is given by

$$x = r\cos\theta \qquad y = r\sin\theta \qquad z = z$$

It follows immediately that the Laplacian in cylindrical coordinates

$$\nabla^2 u = \frac{\partial^2 u}{\partial x^2} + \frac{\partial^2 u}{\partial y^2} + \frac{\partial^2 u}{\partial z^2} \quad \text{becomes} \quad \nabla^2 u = \frac{\partial^2 u}{\partial r^2} + \frac{1}{r}\frac{\partial u}{\partial r} + \frac{1}{r^2}\frac{\partial^2 u}{\partial \theta^2} + \frac{\partial^2 u}{\partial z^2}$$

EXAMPLE 2 Find the steady-state temperature in the circular cylinder shown in Figure 13.10.

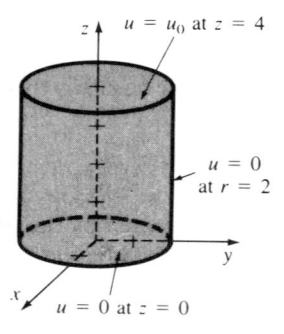

Solution The prescribed boundary conditions suggest that the temperature u has radial symmetry. Accordingly $u(r, z)$ is determined from:

$$\frac{\partial^2 u}{\partial r^2} + \frac{1}{r}\frac{\partial u}{\partial r} + \frac{\partial^2 u}{\partial z^2} = 0, \quad 0 < r < 2, \quad 0 < z < 4$$

$$u(2, z) = 0, \quad 0 < z < 4$$

$$u(r, 0) = 0, \quad u(r, 4) = u_0, \quad 0 < r < 2$$

Using $u = R(r)Z(z)$ and separating variables give

$$\frac{R'' + \frac{1}{r}R'}{R} = -\frac{Z''}{Z} = -\lambda^2$$

and
$$rR'' + R' + \lambda^2 rR = 0 \qquad (10)$$
$$Z'' - \lambda^2 Z = 0 \qquad (11)$$

A negative separation constant is used here since there is no reason to expect the solution $u(r, z)$ to be periodic in z. The solution of (10) is

$$R = c_1 J_0(\lambda r) + c_2 Y_0(\lambda r)$$

and since the solution of (11) is defined on the finite interval [0, 2], we write its general solution as

$$Z = c_3 \cosh \lambda z + c_4 \sinh \lambda z$$

As in Example 1, the assumption that the function u is bounded at $r = 0$ demands that $c_2 = 0$. The condition $u(2, z) = 0$ implies $R(2) = 0$. This equation,

$$J_0(2\lambda) = 0 \qquad (12)$$

defines the positive eigenvalues λ_n of the problem. Last, $Z(0) = 0$ implies $c_3 = 0$. Hence we have $R = c_1 J_0(\lambda_n r)$, $Z = c_4 \sinh \lambda_n z$, and

$$u_n = RZ = A_n \sinh \lambda_n z \, J_0(\lambda_n r)$$

$$u(r, z) = \sum_{n=1}^{\infty} A_n \sinh \lambda_n z \, J_0(\lambda_n r)$$

The remaining boundary condition at $z = 4$ then gives the Fourier–Bessel series

$$u_0 = \sum_{n=1}^{\infty} A_n \sinh 4\lambda_n J_0(\lambda_n r)$$

so that in view of (12) the coefficients are defined by (16) of Section 11.5:

$$A_n \sinh 4\lambda_n = \frac{2u_0}{2^2 J_1^2(2\lambda_n)} \int_0^2 r J_0(\lambda_n r)\, dr$$

To evaluate the last integral we first use the substitution $t = \lambda_n r$, followed by $(d/dt)[tJ_1(t)] = tJ_0(t)$:

$$A_n \sinh 4\lambda_n = \frac{u_0}{2\lambda_n^2 J_1^2(2\lambda_n)} \int_0^{2\lambda_n} \frac{d}{dt}[tJ_1(t)]\, dt$$

$$= \frac{u_0}{\lambda_n J_1(2\lambda_n)}$$

Finally we arrive at

$$A_n = \frac{u_0}{\lambda_n \sinh 4\lambda_n J_1(2\lambda_n)}$$

Thus, the temperature in the cylinder is given by

$$u(r, z) = u_0 \sum_{n=1}^{\infty} \frac{\sinh \lambda_n z J_0(\lambda_n r)}{\lambda_n \sinh 4\lambda_n J_1(2\lambda_n)} \qquad \square$$

EXERCISES 13.2

Answers to odd-numbered problems begin on page A-66.

1. Find the displacement $u(r, t)$ in Example 1 if $f(r) = 0$ and the circular membrane is given an initial unit velocity in the upward direction.

2. A circular membrane of radius 1 is clamped along its circumference. Find the displacement $u(r, t)$ if the membrane starts from rest from the initial displacement $f(r) = 1 - r^2$, $0 < r < 1$. [*Hint:* See Problem 10 in Exercises 11.5.]

3. Find the steady-state temperature $u(r, z)$ in the cylinder in Example 2 if the boundary conditions are $u(2, z) = 0$, $0 < z < 4$, $u(r, 0) = u_0$, $u(r, 4) = 0$, $0 < r < 2$.

4. If the lateral side of the cylinder in Example 2 is insulated, then $\left.\frac{\partial u}{\partial r}\right|_{r=2} = 0$, $0 < z < 4$.

   (a) Find the steady-state temperature $u(r, z)$ when $u(r, 4) = f(r)$, $0 < r < 2$.

   (b) Show that the steady-state temperature in part (a) reduces to $u(r, z) = u_0 z/4$ when $f(r) = u_0$. [*Hint:* Use (11) of Section 11.5.]

5. The temperature in a circular plate of radius c is determined from the boundary-value problem

$$k\left(\frac{\partial^2 u}{\partial r^2} + \frac{1}{r}\frac{\partial u}{\partial r}\right) = \frac{\partial u}{\partial t}, \quad 0 < r < c, \quad t > 0$$

$$u(c, t) = 0, \quad t > 0$$

$$u(r, 0) = f(r), \quad 0 < r < c$$

Solve for $u(r, t)$.

6. Solve Problem 5 if the edge $r = c$ of the plate is insulated.

7. When there is heat transfer from the lateral side of an infinite circular cylinder of unit radius (see Figure 13.11)

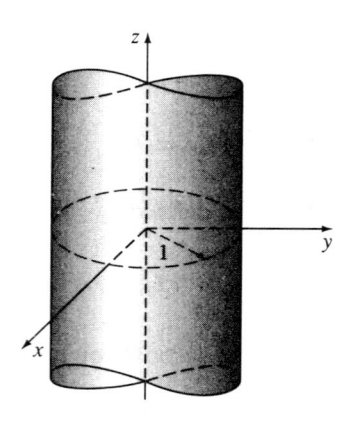

FIGURE 13.11

into a surrounding medium at temperature 0, the temperature inside the cylinder is determined from

$$k\left(\frac{\partial^2 u}{\partial r^2} + \frac{1}{r}\frac{\partial u}{\partial r}\right) = \frac{\partial u}{\partial t}, \quad 0 < r < 1, \quad t > 0$$

$$\left.\frac{\partial u}{\partial r}\right|_{r=1} = -hu(1, t), \quad h > 0, \quad t > 0$$

$$u(r, 0) = f(r), \quad 0 < r < 1.$$

Solve for $u(r, t)$.

8. Find the steady-state temperature $u(r, z)$ in a semi-infinite cylinder of unit radius ($z \geq 0$) if there is heat transfer from its lateral side into a surrounding medium at temperature 0 and if the temperature of the base $z = 0$ is held at a constant temperature u_0.

9. A circular plate is a composite of two different materials in the form of concentric circles (see Figure 13.12).

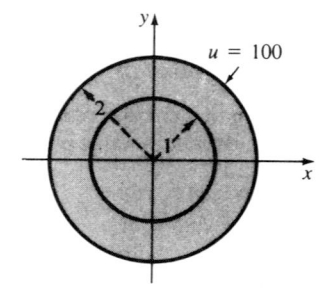

FIGURE 13.12

The temperature in the plate is determined from the boundary-value problem:

$$\frac{\partial^2 u}{\partial r^2} + \frac{1}{r}\frac{\partial u}{\partial r} = \frac{\partial u}{\partial t}, \quad 0 < r < 2, \quad t > 0$$

$$u(2, t) = 100, \quad t > 0$$

$$u(r, 0) = \begin{cases} 200, & 0 < r < 1 \\ 100, & 1 \leq r \leq 2 \end{cases}$$

Solve for $u(r, t)$. [*Hint:* Let $u(r, t) = v(r, t) + \psi(r)$.]

10. Solve the boundary-value problem:

$$\frac{\partial^2 u}{\partial r^2} + \frac{1}{r}\frac{\partial u}{\partial r} + \beta = \frac{\partial u}{\partial t}, \quad 0 < r < 1, \quad t > 0, \quad \beta \text{ a constant}$$

$$u(1, t) = 0, \quad t > 0$$

$$u(r, 0) = 0, \quad 0 < r < 1$$

11. The horizontal displacement $u(x, t)$ of a heavy chain of length L oscillating in a vertical plane satisfies

$$g\frac{\partial}{\partial x}\left(x\frac{\partial u}{\partial x}\right) = \frac{\partial^2 u}{\partial t^2}, \quad 0 < x < L, \quad t > 0$$

(see Figure 13.13).

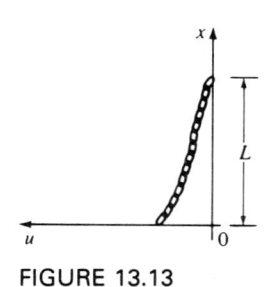

FIGURE 13.13

(a) Using $-\lambda^2$ as a separation constant, show that the ordinary differential equation in the spatial variable x is $xX'' + X' + \lambda^2 X = 0$. Solve this equation by means of the substitution $x = \tau^2/4$.

(b) Use the result of part (a) to solve the given partial differential equation subject to

$$u(L, t) = 0, \quad t > 0,$$

$$u(x, 0) = f(x), \quad \left.\frac{\partial u}{\partial t}\right|_{t=0} = 0, \quad 0 < x < L$$

[*Hint:* Assume the oscillations at the free end $x = 0$ are finite.]

12. In this problem we consider the general case of the vibrating circular membrane of radius c:

$$a^2\left(\frac{\partial^2 u}{\partial r^2} + \frac{1}{r}\frac{\partial u}{\partial r} + \frac{1}{r^2}\frac{\partial^2 u}{\partial \theta^2}\right) = \frac{\partial^2 u}{\partial t^2}, \quad 0 < r < c, \quad t > 0$$

$$u(c, \theta, t) = 0, \quad 0 < \theta < 2\pi, \quad t > 0$$

$u(r, \theta, 0) = f(r, \theta)$, $0 < r < c$, $0 < \theta < 2\pi$

$\left.\dfrac{\partial u}{\partial t}\right|_{t=0} = g(r, \theta)$, $0 < r < c$, $0 < \theta < 2\pi$

(a) Assume that $u = R(r)\Theta(\theta)T(t)$ and that the separation constants are $-\lambda^2$ and $-v^2$. Show that the separated differential equations are

$$T'' + a^2\lambda^2 T = 0$$

$$\Theta'' + v^2\Theta = 0$$

$$r^2R'' + rR' + (\lambda^2 r^2 - v^2)R = 0$$

(b) Solve the separated equations.

(c) Show that the eigenvalues and eigenfunctions of the problem are:

eigenvalues: $v = n$, $n = 0, 1, 2, \ldots$;

eigenfunctions: 1, $\cos n\theta$, $\sin n\theta$

eigenvalues: $\lambda_{ni} = \dfrac{x_{ni}}{c}$, $i = 1, 2, \ldots$, where for each n, x_{ni} are the positive roots of $J_n(\lambda c) = 0$;

eigenfunctions: $J_n(\lambda_{ni} r)$

(d) Use the superposition principle to determine a multiple series solution. Do not attempt to evaluate the coefficients.

13.3 PROBLEMS IN SPHERICAL COORDINATES: LEGENDRE POLYNOMIALS

In the spherical coordinate system shown in Figure 13.14, we have $x = r\sin\theta\cos\phi$, $y = r\sin\theta\sin\phi$, $z = r\cos\theta$. The Laplacian $\nabla^2 u = \dfrac{\partial^2 u}{\partial x^2} + \dfrac{\partial^2 u}{\partial y^2} + \dfrac{\partial^2 u}{\partial z^2}$ then becomes

$$\nabla^2 u = \frac{\partial^2 u}{\partial r^2} + \frac{2}{r}\frac{\partial u}{\partial r} + \frac{1}{r^2 \sin^2\theta}\frac{\partial^2 u}{\partial \phi^2} + \frac{1}{r^2}\frac{\partial^2 u}{\partial \theta^2} + \frac{\cot\theta}{r^2}\frac{\partial u}{\partial \theta} \tag{1}$$

As the reader might imagine, problems involving (1) can be quite formidable. Consequently we shall consider only a few of the simpler problems that are independent of the azimuthal angle ϕ.

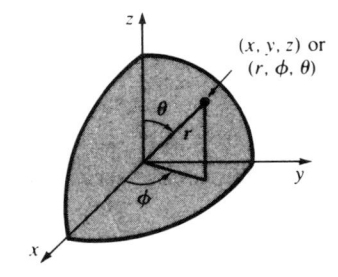

FIGURE 13.14

EXAMPLE 1 Find the steady-state temperature $u(r, \theta)$ in the sphere shown in Figure 13.15.

Solution The temperature is determined from

$$\frac{\partial^2 u}{\partial r^2} + \frac{2}{r}\frac{\partial u}{\partial r} + \frac{1}{r^2}\frac{\partial^2 u}{\partial \theta^2} + \frac{\cot\theta}{r^2}\frac{\partial u}{\partial \theta} = 0, \quad 0 < r < a, \quad 0 < \theta < \pi$$

$$u(a, \theta) = f(\theta), \quad 0 < \theta < \pi$$

If $u = R(r)\Theta(\theta)$, the partial differential equation separates as

$$\frac{r^2 R'' + 2rR'}{R} = -\frac{\Theta'' + \cot\theta\, \Theta'}{\Theta} = \lambda^2$$

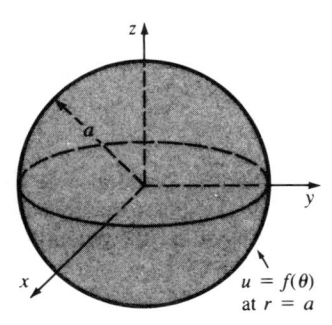

FIGURE 13.15

and so
$$r^2 R'' + 2rR' - \lambda^2 R = 0 \qquad (2)$$
$$\sin\theta\, \Theta'' + \cos\theta\, \Theta' + \lambda^2 \sin\theta\, \Theta = 0 \qquad (3)$$

After we substitute $x = \cos\theta$, $0 \leq \theta \leq \pi$, (3) becomes

$$(1 - x^2)\frac{d^2\Theta}{dx^2} - 2x\frac{d\Theta}{dx} + \lambda^2 \Theta = 0, \quad -1 \leq x \leq 1 \qquad (4)$$

The latter equation is a form of Legendre's equation (see Problem 41 in Exercises 5.4). Now the only solutions of (4) that are continuous and have continuous derivatives on the closed interval $[-1, 1]$ are the Legendre polynomials $P_n(x)$ corresponding to $\lambda^2 = n(n + 1)$, $n = 0, 1, 2, \ldots$. Thus we take the solutions of (3) to be

$$\Theta = P_n(\cos\theta)$$

Furthermore when $\lambda^2 = n(n + 1)$, the general solution of the Cauchy–Euler equation (2) is

$$R = c_1 r^n + c_2 r^{-(n+1)}$$

Since we again expect $u(r, \theta)$ to be bounded at $r = 0$, we define $c_2 = 0$. Hence,

$$u_n = A_n r^n P_n(\cos\theta)$$

and
$$u(r, \theta) = \sum_{n=0}^{\infty} A_n r^n P_n(\cos\theta)$$

At $r = a$,
$$f(\theta) = \sum_{n=0}^{\infty} A_n a^n P_n(\cos\theta)$$

Therefore, $A_n a^n$ are the coefficients of the Fourier–Legendre series (23) of Section 11.5.

$$A_n = \frac{2n + 1}{2a^n} \int_0^{\pi} f(\theta) P_n(\cos\theta) \sin\theta\, d\theta$$

It follows that the solution is

$$u(r, \theta) = \sum_{n=0}^{\infty} \left(\frac{2n + 1}{2} \int_0^{\pi} f(\theta) P_n(\cos\theta) \sin\theta\, d\theta\right) \left(\frac{r}{a}\right)^n P_n(\cos\theta) \qquad \square$$

EXERCISES 13.3

Answers to odd-numbered problems begin on page A-66.

1. Solve the problem in Example 1 if
$$f(\theta) = \begin{cases} 50, & 0 < \theta < \pi/2 \\ 0, & \pi/2 \leq \theta < \pi \end{cases}$$
Write out the first four nonzero terms of the series solution. [*Hint:* See Example 3 in Section 11.5.]

2. The solution $u(r, \theta)$ in Example 1 could also be interpreted as the potential inside the sphere due to a charge distribution $f(\theta)$ on its surface. Find the potential outside the sphere.

3. Find the solution of the problem in Example 1 if $f(\theta) = \cos\theta$, $0 < \theta < \pi$. [*Hint:* $P_1(\cos\theta) = \cos\theta$. Use orthogonality.]

4. Find the solution of the problem in Example 1 if $f(\theta) = 1 - \cos 2\theta$, $0 < \theta < \pi$. [*Hint:* See Problem 16 in Exercises 11.5.]

5. Find the steady-state temperature $u(r, \theta)$ within a hollow sphere $a < r < b$ if its inner surface $r = a$ is kept at temperature $f(\theta)$ and its outer surface $r = b$ is kept at temperature 0. The sphere in the first octant is shown in Figure 13.16.

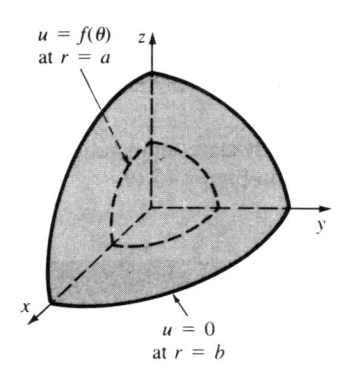

FIGURE 13.16

6. The time-dependent temperature within a sphere of unit radius is determined from

$$\frac{\partial^2 u}{\partial r^2} + \frac{2}{r}\frac{\partial u}{\partial r} = \frac{\partial u}{\partial t}, \quad 0 < r < 1, \quad t > 0$$

$$u(1, t) = 100, \quad t > 0,$$

$$u(r, 0) = 0, \quad 0 < r < 1$$

Solve for $u(r, t)$. [*Hint:* Verify that the left side of the partial differential equation can be written as $\dfrac{1}{r}\dfrac{\partial^2}{\partial r^2}(ru)$. Let $ru(r, t) = v(r, t) + \psi(r)$. Use only functions that are bounded as $r \to 0$.]

7. The steady-state temperature in a hemisphere of radius a and insulated base is determined from

$$\frac{\partial^2 u}{\partial r^2} + \frac{2}{r}\frac{\partial u}{\partial r} + \frac{1}{r^2}\frac{\partial^2 u}{\partial \theta^2} + \frac{\cot\theta}{r^2}\frac{\partial u}{\partial \theta} = 0, \quad 0 < r < a, \quad 0 < \theta < \frac{\pi}{2}$$

$$\left.\frac{\partial u}{\partial \theta}\right|_{\theta = \pi/2} = 0, \quad 0 < r < a$$

$$u(a, \theta) = f(\theta), \quad 0 < \theta < \frac{\pi}{2}$$

Solve for $u(r, \theta)$. [*Hint:* $P'_n(0) = 0$ only if n is even. Also see Problem 17 in Exercises 11.5.]

8. A uniform solid sphere of radius 1 at an initial constant temperature u_0 throughout is dropped into a large container of fluid that is kept at a constant temperature u_1 ($u_1 > u_0$) for all time (see Figure 13.17). Since there is heat transfer across the boundary $r = 1$, the temperature $u(r, t)$ in the sphere is determined from the boundary-value problem

$$\frac{\partial^2 u}{\partial r^2} + \frac{2}{r}\frac{\partial u}{\partial r} = \frac{\partial u}{\partial t}, \quad 0 < r < 1, \quad t > 0$$

$$\left.\frac{\partial u}{\partial r}\right|_{r=1} = -h(u(1, t) - u_1), \quad 0 < h < 1$$

$$u(r, 0) = u_0, \quad 0 < r < 1$$

Solve for $u(r, t)$. [*Hint:* Proceed as in Problem 6. Use only functions that are bounded as $r \to 0$. Review Section 12.7.]

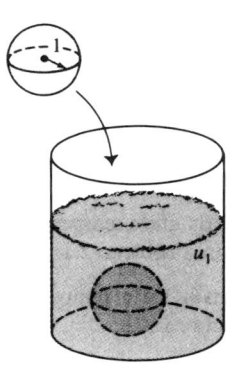

FIGURE 13.17

SUMMARY

Certain kinds of applied problems are best analyzed in a coordinate system other than rectangular coordinates. In this chapter, we considered boundary-value problems involving the heat equation, wave equation, and Laplace's equation in **polar, cylindrical**, and **spherical coordinates**. In each of these coordinate systems, the method of separation of variables leads to an ordinary differential equation with variable coefficients. We encountered the Cauchy–Euler equation in the course of solving problems involving Laplace's equation in polar coordinates, Bessel's equation arose in the solution of some radially symmetric problems in polar and cylindrical coordinates, and Legendre's equation occurred in special kinds of problems in spherical coordinates. The solution of those problems in polar or cylindrical coordinates that gave rise to Bessel's differential equation depended in the final analysis on our ability to expand functions in an infinite series of Bessel functions — that is, in a **Fourier–Bessel series**. Similarly, we used the concept of **Fourier–Legendre series** to solve problems in spherical coordinates.

CHAPTER 13 REVIEW EXERCISES *Answers to odd-numbered problems begin on page A-66.*

1. Find the steady-state temperature $u(r, \theta)$ in a circular plate of radius c if the temperature on the circumference is given by

$$u(c, \theta) = \begin{cases} u_0, & 0 < \theta < \pi \\ -u_0, & \pi \leq \theta < 2\pi \end{cases}$$

2. Find the steady-state temperature in the circular plate in Problem 1 if

$$u(c, \theta) = \begin{cases} 1, & 0 \leq \theta < \pi/2 \\ 0, & \pi/2 \leq \theta < 3\pi/2 \\ 1, & 3\pi/2 \leq \theta < 2\pi \end{cases}$$

3. Find the steady-state temperature $u(r, \theta)$ in a semicircular plate of radius 1 if

$$u(1, \theta) = u_0(\pi\theta - \theta^2), \quad 0 < \theta < \pi$$
$$u(r, 0) = 0, \quad u(r, \pi) = 0, \quad 0 < r < 1$$

4. Find the steady-state temperature $u(r, \theta)$ in the semicircular plate in Problem 3 if $u(1, \theta) = \sin \theta$, $0 < \theta < \pi$.

5. Find the steady-state temperature $u(r, \theta)$ in the plate shown in Figure 13.18.

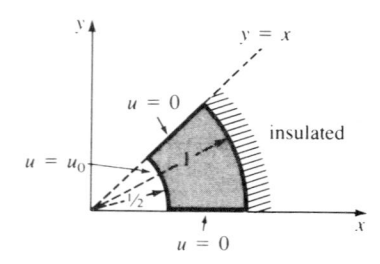

FIGURE 13.18

6. Find the steady-state temperature $u(r, \theta)$ in the infinite plate shown in Figure 13.19.

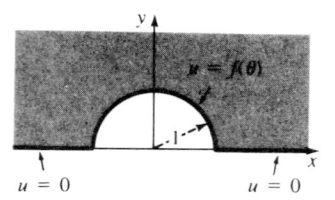

FIGURE 13.19

7. Suppose heat is lost from the flat surfaces of a very thin circular unit disk into a surrounding medium at temperature 0. If the linear law of heat transfer applies, the heat equation assumes the form

$$\frac{\partial^2 u}{\partial r^2} + \frac{1}{r}\frac{\partial u}{\partial r} - hu = \frac{\partial u}{\partial t}, \quad h > 0, \quad 0 < r < 1, \quad t > 0$$

(see Figure 13.20). Find the temperature $u(r, t)$ if the edge $r = 1$ is kept at temperature 0 and if initially the temperature of the plate is unity throughout.

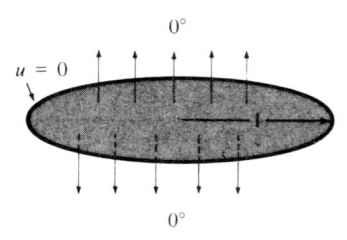

FIGURE 13.20

8. Suppose x_k is a positive zero of J_0. Show that a solution of the boundary-value problem

$$a^2\left(\frac{\partial^2 u}{\partial r^2} + \frac{1}{r}\frac{\partial u}{\partial r}\right) = \frac{\partial^2 u}{\partial t^2}, \quad 0 < r < 1, \quad t > 0$$

$$u(1, t) = 0, \quad t > 0$$

$$u(r, 0) = u_0 J_0(x_k r), \quad \left.\frac{\partial u}{\partial t}\right|_{t=0} = 0, \quad 0 < r < 1$$

is $u(r, t) = u_0 J_0(x_k r) \cos a x_k t$.

9. Find the steady-state temperature $u(r, z)$ in the cylinder in Figure 13.10 if the lateral side is kept at temperature 0, the top $z = 4$ is kept at temperature 50, and the base $z = 0$ is insulated.

10. Solve the boundary-value problem

$$\frac{\partial^2 u}{\partial r^2} + \frac{1}{r}\frac{\partial u}{\partial r} + \frac{\partial^2 u}{\partial z^2} = 0, \quad 0 < r < 1, \quad 0 < z < 1$$

$$\left.\frac{\partial u}{\partial r}\right|_{r=1} = 0, \quad 0 < z < 1$$

$$u(r, 0) = f(r), \quad u(r, 1) = g(r), \quad 0 < r < 1$$

11. Find the steady-state temperature $u(r, \theta)$ in a sphere of unit radius if the surface is kept at

$$u(1, \theta) = \begin{cases} 100, & 0 < \theta < \pi/2 \\ -100, & \pi/2 \le \theta < \pi \end{cases}$$

[*Hint:* See Problem 20 in Exercises 11.5.]

14

INTEGRAL TRANSFORM METHOD

TOPICS TO REVIEW

linear first-order differential equations (2.5)
linear second-order differential equations (3.3, 3.4)
properties of the Laplace transform (4.3, 4.4)

IMPORTANT CONCEPTS

error function
complementary error function
Fourier integral
cosine integral
sine integral
transform pairs
Fourier transform
inverse Fourier transform
Fourier sine transform
inverse Fourier sine transform
Fourier cosine transform
inverse Fourier cosine transform

14.1 Error Function
14.2 Applications of the Laplace Transform
14.3 Fourier Integral
14.4 Fourier Transforms
Summary
Chapter 14 Review Exercises

INTRODUCTION

The method of separation of variables is a powerful, but not universally applicable, method for solving boundary-value problems. If the partial differential equation is nonhomogeneous, if the boundary conditions are time-dependent, or if the domain of a spatial variable is infinite or semi-infinite, we may be able to use an integral transform. In Section 14.2 we shall solve problems involving the heat and wave equations by means of the familiar Laplace transform. In Section 14.4 three new integral transforms will be introduced.

14.1 ERROR FUNCTION

erf(x) and erfc(x) Before applying the Laplace transform to boundary-value problems, we need to consider a function that plays an important role in many applications. This function, like the natural logarithm, is defined by means of an integral.

DEFINITION 14.1 Error Function

The **error function**, denoted by erf(x), is

$$\operatorname{erf}(x) = \frac{2}{\sqrt{\pi}} \int_0^x e^{-u^2}\, du \qquad (1)$$

The **complementary error function** erfc(x) is defined in terms of the error function:

$$\operatorname{erfc}(x) = 1 - \operatorname{erf}(x) \qquad (2)$$

In calculus it is usually shown that $\int_0^\infty e^{-u^2}\, du = \sqrt{\pi}/2$ or $(2/\sqrt{\pi}) \int_0^\infty e^{-u^2}\, du = 1$. Hence, (2) can be written as

$$\operatorname{erfc}(x) = \frac{2}{\sqrt{\pi}} \left[\int_0^\infty e^{-u^2}\, du - \int_0^x e^{-u^2}\, du \right]$$

or

$$\operatorname{erfc}(x) = \frac{2}{\sqrt{\pi}} \int_x^\infty e^{-u^2}\, du$$

The graphs of erf(x) and erfc(x) are given in Figure 14.1. Note that erf(0) = 0, erfc(0) = 1, and erf(x) → 1 and erfc(x) → 0 as $x \to \infty$. Numerical values of erf(x) and erfc(x) have been extensively tabulated. In tables the error function is often referred to as the **probability integral**. The following table of Laplace

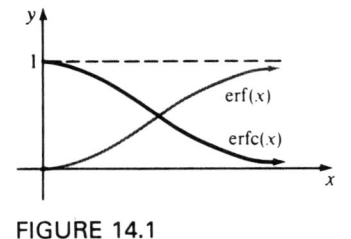

FIGURE 14.1

transforms will be useful in the exercises of the next section. The proofs of these results are complicated and will not be given.

$f(t)$, $a > 0$	$\mathcal{L}\{f(t)\} = F(s)$
1. $\dfrac{1}{\sqrt{\pi t}} e^{-a^2/4t}$	$\dfrac{e^{-a\sqrt{s}}}{\sqrt{s}}$
2. $\dfrac{a}{2\sqrt{\pi t^3}} e^{-a^2/4t}$	$e^{-a\sqrt{s}}$
3. $\mathrm{erfc}\left(\dfrac{a}{2\sqrt{t}}\right)$	$\dfrac{e^{-a\sqrt{s}}}{s}$
4. $2\sqrt{\dfrac{t}{\pi}}\, e^{-a^2/4t} - a\,\mathrm{erfc}\left(\dfrac{a}{2\sqrt{t}}\right)$	$\dfrac{e^{-a\sqrt{s}}}{s\sqrt{s}}$
5. $e^{ab}e^{b^2 t}\,\mathrm{erfc}\left(b\sqrt{t} + \dfrac{a}{2\sqrt{t}}\right)$	$\dfrac{e^{-a\sqrt{s}}}{\sqrt{s}(\sqrt{s} + b)}$
6. $-e^{ab}e^{b^2 t}\,\mathrm{erfc}\left(b\sqrt{t} + \dfrac{a}{2\sqrt{t}}\right) + \mathrm{erfc}\left(\dfrac{a}{2\sqrt{t}}\right)$	$\dfrac{b e^{-a\sqrt{s}}}{s(\sqrt{s} + b)}$

EXERCISES 14.1 Answers to odd-numbered problems begin on page A-66.

1. (a) Show that $\mathrm{erf}(\sqrt{t}) = \dfrac{1}{\sqrt{\pi}} \displaystyle\int_0^t \dfrac{e^{-\tau}}{\sqrt{\tau}}\, d\tau$.

   (b) Use the convolution theorem and the results of Problems 43 and 44 in Exercises 4.1 to show that
   $$\mathcal{L}\{\mathrm{erf}(\sqrt{t})\} = \dfrac{1}{s\sqrt{s+1}}.$$

2. Use the result of Problem 1 to show that
$$\mathcal{L}\{\mathrm{erfc}(\sqrt{t})\} = \dfrac{1}{s}\left[1 - \dfrac{1}{\sqrt{s+1}}\right].$$

3. Use the result of Problem 1 to show that
$$\mathcal{L}\{e^t\,\mathrm{erf}(\sqrt{t})\} = \dfrac{1}{\sqrt{s(s-1)}}.$$

4. Use the result of Problem 2 to show that
$$\mathcal{L}\{e^t\,\mathrm{erfc}(\sqrt{t})\} = \dfrac{1}{\sqrt{s}(\sqrt{s}+1)}.$$

5. Let C, G, R, and x be constants. Use the table in this section to show that
$$\mathcal{L}^{-1}\left\{\dfrac{C}{Cs + G}\left(1 - e^{-x\sqrt{RCs+RG}}\right)\right\} = e^{-Gt/C}\,\mathrm{erf}\left(\dfrac{x}{2}\sqrt{\dfrac{RC}{t}}\right)$$

6. Let a be a constant. Show that
$$\mathcal{L}^{-1}\left\{\dfrac{\sinh a\sqrt{s}}{s \sinh \sqrt{s}}\right\} = \sum_{n=0}^{\infty}\left[\mathrm{erf}\left(\dfrac{2n+1+a}{2\sqrt{t}}\right) - \mathrm{erf}\left(\dfrac{2n+1-a}{2\sqrt{t}}\right)\right]$$

[*Hint:* Use the exponential definition of the hyperbolic sine. Expand $1/(1 - e^{-2\sqrt{s}})$ in a geometric series.]

7. Use the Laplace transform and the table in this section to solve the integral equation
$$y(t) = 1 - \int_0^t \dfrac{y(\tau)}{\sqrt{t - \tau}}\, d\tau$$

8. Use the third and fifth entries in the table in this section to derive the sixth entry.

9. Show that $\displaystyle\int_a^b e^{-u^2}\, du = \dfrac{\sqrt{\pi}}{2}[\mathrm{erf}(b) - \mathrm{erf}(a)]$.

10. Show that $\displaystyle\int_{-a}^a e^{-u^2}\, du = \sqrt{\pi}\,\mathrm{erf}(a)$.

14.2 APPLICATIONS OF THE LAPLACE TRANSFORM

We define the Laplace transform of a function of two variables $u(x, t)$ with respect to the variable t by

$$\mathscr{L}\{u(x, t)\} = \int_0^\infty e^{-st} u(x, t) \, dt = U(x, s)$$

where x is treated as a parameter. Throughout this section we shall assume that all the operational properties of Section 4.3 apply to functions of two variables. For example, by Theorem 4.8 the transforms of the partial derivatives $\partial u/\partial t$ and $\partial^2 u/\partial t^2$ are, respectively,

$$\mathscr{L}\left\{\frac{\partial u}{\partial t}\right\} = s\mathscr{L}\{u(x, t)\} - u(x, 0)$$
$$= sU(x, s) - u(x, 0) \qquad (1)$$

$$\mathscr{L}\left\{\frac{\partial^2 u}{\partial t^2}\right\} = s^2 \mathscr{L}\{u(x, t)\} - su(x, 0) - \left.\frac{\partial u}{\partial t}\right|_{t=0}$$
$$= s^2 U(x, s) - su(x, 0) - u_t(x, 0) \qquad (2)$$

Since we are transforming with respect to t, we further suppose that it is legitimate to interchange integration and differentiation in the transform of $\partial^2 u/\partial x^2$:

$$\mathscr{L}\left\{\frac{\partial^2 u}{\partial x^2}\right\} = \int_0^\infty e^{-st} \frac{\partial^2 u}{\partial x^2} \, dt = \int_0^\infty \frac{\partial^2}{\partial x^2}[e^{-st} u(x, t)] \, dt$$
$$= \frac{d^2}{dx^2} \int_0^\infty e^{-st} u(x, t) \, dt = \frac{d^2}{dx^2} \mathscr{L}\{u(x, t)\}$$

that is,
$$\mathscr{L}\left\{\frac{\partial^2 u}{\partial x^2}\right\} = \frac{d^2 U}{dx^2} \qquad (3)$$

In view of properties (1) and (2), we see that the Laplace transform is suited to problems with initial conditions—namely, those problems associated with the heat equation or the wave equation.

EXAMPLE 1 Find the Laplace transform of the wave equation
$a^2 \dfrac{\partial^2 u}{\partial x^2} = \dfrac{\partial^2 u}{\partial t^2}, t > 0.$

Solution From (2) and (3),

$$\mathscr{L}\left\{a^2 \frac{\partial^2 u}{\partial x^2}\right\} = \mathscr{L}\left\{\frac{\partial^2 u}{\partial t^2}\right\}$$

becomes $$a^2 \frac{d^2}{dx^2} \mathscr{L}\{u(x, t)\} = s^2 \mathscr{L}\{u(x, t)\} - su(x, 0) - u_t(x, 0)$$

or $$a^2 \frac{d^2 U}{dx^2} - s^2 U = -su(x, 0) - u_t(x, 0) \tag{4}$$

The Laplace transform with respect to t of either the wave equation or the heat equation eliminates that variable and, for the one-dimensional equations, the transformed equations are then *ordinary differential equations* in the spatial variable x. In solving a transformed equation, we treat s as a parameter.

EXAMPLE 2 Solve $\quad \dfrac{\partial^2 u}{\partial x^2} = \dfrac{\partial^2 u}{\partial t^2}, \quad 0 < x < 1, \quad t > 0$

subject to $\quad u(0, t) = 0, \quad u(1, t) = 0, \quad t > 0$

$$u(x, 0) = 0, \quad \left.\frac{\partial u}{\partial t}\right|_{t=0} = \sin \pi x, \quad 0 < x < 1$$

Solution The partial differential equation is recognized as the wave equation with $a = 1$. From (4) and the given initial conditions, the transformed equation is

$$\frac{d^2 U}{dx^2} - s^2 U = -\sin \pi x \tag{5}$$

where $U(x, s) = \mathscr{L}\{u(x, t)\}$. Since the boundary conditions are functions of t, we must also find their Laplace transforms:

$$\mathscr{L}\{u(0, t)\} = U(0, s) = 0 \quad \text{and} \quad \mathscr{L}\{u(1, t)\} = U(1, s) = 0 \tag{6}$$

The results in (6) are boundary conditions for the ordinary differential equation (5). Now the complementary function of (5) is

$$U_c(x, s) = c_1 \cosh sx + c_2 \sinh sx$$

The method of undetermined coefficients yields a particular solution

$$U_p(x, s) = \frac{1}{s^2 + \pi^2} \sin \pi x$$

Hence,

$$U(x, s) = c_1 \cosh sx + c_2 \sinh sx + \frac{1}{s^2 + \pi^2} \sin \pi x$$

But the conditions $U(0, s) = 0$ and $U(1, s) = 0$ yield, in turn, $c_1 = 0$ and $c_2 = 0$. We conclude that

$$U(x, s) = \frac{1}{s^2 + \pi^2} \sin \pi x$$

$$u(x, t) = \mathscr{L}^{-1}\left\{\frac{1}{s^2 + \pi^2} \sin \pi x\right\}$$

$$= \frac{1}{\pi} \sin \pi x \, \mathscr{L}^{-1}\left\{\frac{\pi}{s^2 + \pi^2}\right\}$$

Therefore, $\quad u(x, t) = \dfrac{1}{\pi} \sin \pi x \sin \pi t \quad\square$

EXAMPLE 3 A very long string is initially at rest on the nonnegative x-axis. The string is secured at $x = 0$ and its distant right end slides down a frictionless vertical support. The string is set in motion by letting it fall under its own weight. Find the displacement $u(x, t)$.

Solution Since the force of gravity is taken into consideration, it can be shown that the wave equation has the form

$$a^2 \frac{\partial^2 u}{\partial x^2} - g = \frac{\partial^2 u}{\partial t^2}, \quad x > 0, \quad t > 0$$

The boundary and initial conditions are, respectively,

$$u(0, t) = 0, \quad \lim_{x \to \infty} \frac{\partial u}{\partial x} = 0, \quad t > 0$$

$$u(x, 0) = 0, \quad \left.\frac{\partial u}{\partial t}\right|_{t=0} = 0, \quad x > 0$$

The second boundary condition $\lim_{x \to \infty} \partial u/\partial x = 0$ indicates that the string is horizontal at a great distance from the left end. Now from (2) and (3),

$$\mathscr{L}\left\{a^2 \frac{\partial^2 u}{\partial x^2}\right\} - \mathscr{L}\{g\} = \mathscr{L}\left\{\frac{\partial^2 u}{\partial t^2}\right\}$$

becomes $\quad a^2 \dfrac{d^2 U}{dx^2} - \dfrac{g}{s} = s^2 U - s u(x, 0) - u_t(x, 0)$

or, in view of the initial conditions,

$$\frac{d^2 U}{dx^2} - \frac{s^2}{a^2} U = \frac{g}{a^2 s}$$

The transforms of the boundary conditions are

$$\mathscr{L}\{u(0, t)\} = U(0, s) = 0 \quad \text{and} \quad \mathscr{L}\left\{\lim_{x \to \infty} \frac{\partial u}{\partial x}\right\} = \lim_{x \to \infty} \frac{dU}{dx} = 0$$

With the aid of undetermined coefficients, the general solution of the transformed equation is found to be

$$U(x, s) = c_1 e^{-(x/a)s} + c_2 e^{(x/a)s} - \frac{g}{s^3}$$

The boundary condition $\lim_{x\to\infty} dU/dx = 0$ implies $c_2 = 0$ and $U(0, s) = 0$ gives $c_1 = g/s^3$. Therefore,

$$U(x, s) = \frac{g}{s^3} e^{-(x/a)s} - \frac{g}{s^3}$$

Now by the second translation theorem, we have

$$u(x, t) = \mathscr{L}^{-1}\left\{\frac{g}{s^3} e^{-(x/a)s} - \frac{g}{s^3}\right\}$$

$$= \frac{1}{2} g\left(t - \frac{x}{a}\right)^2 \mathscr{U}\left(t - \frac{x}{a}\right) - \frac{1}{2} gt^2$$

or

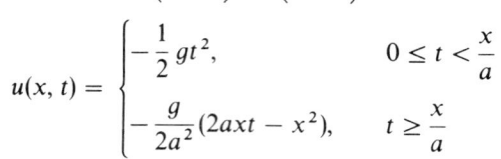

$$u(x, t) = \begin{cases} -\dfrac{1}{2} gt^2, & 0 \le t < \dfrac{x}{a} \\ -\dfrac{g}{2a^2}(2axt - x^2), & t \ge \dfrac{x}{a} \end{cases}$$

To interpret the solution, let us suppose $t > 0$ is fixed. For $0 \le x \le at$ the string is the shape of a parabola passing through $(0, 0)$ and $(at, -\frac{1}{2}gt^2)$. For $x > at$ the string is described by the horizontal line $u = -\frac{1}{2}gt^2$ (see Figure 14.2). □

FIGURE 14.2

The reader should observe that the problem in the next example could be solved by the procedure in Section 12.6. The Laplace transform provides an alternative solution.

EXAMPLE 4 Solve the heat equation

$$\frac{\partial^2 u}{\partial x^2} = \frac{\partial u}{\partial t}, \quad 0 < x < 1, \quad t > 0$$

subject to
$$u(0, t) = 0, \quad u(1, t) = u_0, \quad t > 0$$
$$u(x, 0) = 0, \quad 0 < x < 1$$

Solution From (1) and (3) and the given initial condition,

$$\mathscr{L}\left\{\frac{\partial^2 u}{\partial x^2}\right\} = \mathscr{L}\left\{\frac{\partial u}{\partial t}\right\}$$

becomes
$$\frac{d^2 U}{dx^2} - sU = 0 \tag{7}$$

The transforms of the boundary conditions are

$$U(0, s) = 0 \quad \text{and} \quad U(1, s) = \frac{u_0}{s} \tag{8}$$

Since we are concerned with a finite interval on the x-axis, we choose to write the general solution of (7) as

$$U(x, s) = c_1 \cosh(\sqrt{s}\,x) + c_2 \sinh(\sqrt{s}\,x)$$

Applying the two boundary conditions in (8) yields, in turn, $c_1 = 0$ and $c_2 = u_0/(s \sinh \sqrt{s})$. Thus,

$$U(x, s) = u_0 \frac{\sinh(\sqrt{s}\, x)}{s \sinh \sqrt{s}}$$

Now, the inverse transform of the latter function cannot be found in most tables. However, by writing

$$\frac{\sinh(\sqrt{s}\, x)}{s \sinh \sqrt{s}} = \frac{e^{\sqrt{s}\, x} - e^{-\sqrt{s}\, x}}{s(e^{\sqrt{s}} - e^{-\sqrt{s}})} = \frac{e^{(x-1)\sqrt{s}} - e^{-(x+1)\sqrt{s}}}{s(1 - e^{-2\sqrt{s}})}$$

and using the geometric series

$$\frac{1}{1 - e^{-2\sqrt{s}}} = \sum_{n=0}^{\infty} e^{-2n\sqrt{s}}$$

we find

$$\frac{\sinh(\sqrt{s}\, x)}{s \sinh \sqrt{s}} = \sum_{n=0}^{\infty} \left[\frac{e^{-(2n+1-x)\sqrt{s}}}{s} - \frac{e^{-(2n+1+x)\sqrt{s}}}{s} \right].$$

If we assume that the inverse Laplace transform can be done term by term, it follows from entry 3 of the table in Section 14.1 that

$$u(x, t) = u_0 \mathscr{L}^{-1} \left\{ \frac{\sinh(\sqrt{s}\, x)}{s \sinh \sqrt{s}} \right\}$$

$$= u_0 \sum_{n=0}^{\infty} \left[\mathscr{L}^{-1} \left\{ \frac{e^{-(2n+1-x)\sqrt{s}}}{s} \right\} - \mathscr{L}^{-1} \left\{ \frac{e^{-(2n+1+x)\sqrt{s}}}{s} \right\} \right]$$

$$= u_0 \sum_{n=0}^{\infty} \left[\operatorname{erfc}\left(\frac{2n+1-x}{2\sqrt{t}} \right) - \operatorname{erfc}\left(\frac{2n+1+x}{2\sqrt{t}} \right) \right] \quad (9)$$

The solution (9) can be rewritten in terms of the error function using $\operatorname{erfc}(x) = 1 - \operatorname{erf}(x)$:

$$u(x, t) = u_0 \sum_{n=0}^{\infty} \left[\operatorname{erf}\left(\frac{2n+1+x}{2\sqrt{t}} \right) - \operatorname{erf}\left(\frac{2n+1-x}{2\sqrt{t}} \right) \right] \quad \square$$

EXERCISES 14.2   Answers to odd-numbered problems begin on page A-66.

In the following problems use tables as necessary.

1. A string is stretched along the x-axis between $(0, 0)$ and $(L, 0)$. Find the displacement $u(x, t)$ if the string starts from rest in the initial position $A \sin(\pi x/L)$.

2. Solve the boundary-value problem:

$$\frac{\partial^2 u}{\partial x^2} = \frac{\partial^2 u}{\partial t^2}, \quad 0 < x < 1, \quad t > 0$$

$$u(0, t) = 0, \quad u(1, t) = 0$$

$$u(x, 0) = 0, \quad \left. \frac{\partial u}{\partial t} \right|_{t=0} = 2 \sin \pi x + 4 \sin 3\pi x$$

3. The displacement of a semi-infinite elastic string is determined from

$$a^2 \frac{\partial^2 u}{\partial x^2} = \frac{\partial^2 u}{\partial t^2}, \quad x > 0, \quad t > 0$$

$$u(0, t) = f(t), \quad \lim_{x \to \infty} u(x, t) = 0, \quad t > 0$$

$$u(x, 0) = 0, \quad \left.\frac{\partial u}{\partial t}\right|_{t=0} = 0, \quad x > 0$$

Solve for $u(x, t)$.

4. Solve the boundary-value problem in Problem 3 when

$$f(t) = \begin{cases} \sin \pi t, & 0 \le t \le 1 \\ 0, & t > 1 \end{cases}$$

Sketch the displacement $u(x, t)$ for $t > 1$.

5. In Example 3 find the displacement $u(x, t)$ when the left end of the string at $x = 0$ is given an oscillatory motion described by $f(t) = A \sin \omega t$.

6. The displacement $u(x, t)$ of a string that is driven by an external force is determined from

$$\frac{\partial^2 u}{\partial x^2} + \sin \pi x \sin \omega t = \frac{\partial^2 u}{\partial t^2}, \quad 0 < x < 1, \quad t > 0$$

$$u(0, t) = 0, \quad u(1, t) = 0, \quad t > 0$$

$$u(x, 0) = 0, \quad \left.\frac{\partial u}{\partial t}\right|_{t=0} = 0, \quad 0 < x < 1$$

Solve for $u(x, t)$.

7. A uniform bar is clamped at $x = 0$ and is initially at rest. If a constant force F_0 is applied to the free end at $x = L$, the longitudinal displacement $u(x, t)$ of a cross-section of the bar is determined from

$$a^2 \frac{\partial^2 u}{\partial x^2} = \frac{\partial^2 u}{\partial t^2}, \quad 0 < x < L, \quad t > 0$$

$$u(0, t) = 0, \quad \left.E \frac{\partial u}{\partial x}\right|_{x=L} = F_0, \quad E \text{ a constant}, \quad t > 0$$

$$u(x, 0) = 0, \quad \left.\frac{\partial u}{\partial t}\right|_{t=0} = 0, \quad 0 < x < L$$

Solve for $u(x, t)$. [*Hint:* Expand $1/(1 + e^{-2sL/a})$ in a geometric series.]

8. A uniform semi-infinite elastic beam moving along the x-axis with a constant velocity $-v_0$ is brought to a stop by hitting a wall at time $t = 0$ (see Figure 14.3). The

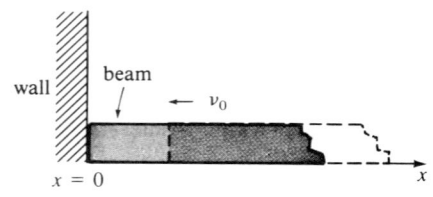

FIGURE 14.3

longitudinal displacement $u(x, t)$ is determined from

$$a^2 \frac{\partial^2 u}{\partial x^2} = \frac{\partial^2 u}{\partial t^2}, \quad x > 0, \quad t > 0$$

$$u(0, t) = 0, \quad \lim_{x \to \infty} \frac{\partial u}{\partial x} = 0, \quad t > 0$$

$$u(x, 0) = 0, \quad \left.\frac{\partial u}{\partial t}\right|_{t=0} = -v_0, \quad x > 0$$

Solve for $u(x, t)$.

9. Solve the boundary-value problem:

$$\frac{\partial^2 u}{\partial x^2} = \frac{\partial^2 u}{\partial t^2}, \quad x > 0, \quad t > 0$$

$$u(0, t) = 0, \quad \lim_{x \to \infty} u(x, t) = 0, \quad t > 0$$

$$u(x, 0) = xe^{-x}, \quad \left.\frac{\partial u}{\partial t}\right|_{t=0} = 0, \quad x > 0$$

10. Solve the boundary-value problem:

$$\frac{\partial^2 u}{\partial x^2} = \frac{\partial^2 u}{\partial t^2}, \quad x > 0, \quad t > 0$$

$$u(0, t) = 1, \quad \lim_{x \to \infty} u(x, t) = 0, \quad t > 0$$

$$u(x, 0) = e^{-x}, \quad \left.\frac{\partial u}{\partial t}\right|_{t=0} = 0, \quad x > 0$$

11. The temperature in a semi-infinite solid is determined from

$$k \frac{\partial^2 u}{\partial x^2} = \frac{\partial u}{\partial t}, \quad x > 0, \quad t > 0$$

$$u(0, t) = u_0, \quad \lim_{x \to \infty} u(x, t) = 0, \quad t > 0$$

$$u(x, 0) = 0, \quad x > 0$$

Solve for $u(x, t)$.

12. In Problem 11 if there is a constant flux of heat into the solid at its left-hand boundary, then the boundary condition is $\left.\frac{\partial u}{\partial x}\right|_{x=0} = -A, t > 0$. Solve for $u(x, t)$.

In Problems 13–20 use the Laplace transform to solve the heat equation $u_{xx} = u_t$, $x > 0$, $t > 0$, subject to the given conditions.

13. $u(0, t) = u_0, \quad \lim_{x \to \infty} u(x, t) = u_1$

$u(x, 0) = u_1$

14. $u(0, t) = u_0, \quad \lim_{x \to \infty} u(x, t)/x = u_1$

$u(x, 0) = u_1 x$

15. $\left.\dfrac{\partial u}{\partial x}\right|_{x=0} = u(0, t), \quad \lim_{x \to \infty} u(x, t) = u_0$

   $u(x, 0) = u_0$

16. $\left.\dfrac{\partial u}{\partial x}\right|_{x=0} = u(0, t) - 50, \quad \lim_{x \to \infty} u(x, t) = 0$

   $u(x, 0) = 0$

17. $u(0, t) = f(t), \quad \lim_{x \to \infty} u(x, t) = 0$

   $u(x, 0) = 0$

   [*Hint:* Use the convolution theorem.]

18. $\left.\dfrac{\partial u}{\partial x}\right|_{x=0} = -f(t), \quad \lim_{x \to \infty} u(x, t) = 0$

   $u(x, 0) = 0$

19. $u(0, t) = 60 + 40\mathcal{U}(t - 2), \quad \lim_{x \to \infty} u(x, t) = 60$

   $u(x, 0) = 60$

20. $u(0, t) = \begin{cases} 20, & 0 < t < 1, \\ 0, & t \geq 1, \end{cases} \quad \lim_{x \to \infty} u(x, t) = 100$

   $u(x, 0) = 100$

21. Solve the boundary-value problem

    $\dfrac{\partial^2 u}{\partial x^2} = \dfrac{\partial u}{\partial t}, \quad -\infty < x < 1, \quad t > 0$

    $\left.\dfrac{\partial u}{\partial x}\right|_{x=1} = 100 - u(1, t), \quad \lim_{x \to -\infty} u(x, t) = 0, \quad t > 0$

    $u(x, 0) = 0, \quad -\infty < x < 1$

22. Show that a solution of the boundary-value problem

    $k\dfrac{\partial^2 u}{\partial x^2} + r = \dfrac{\partial u}{\partial t}, \quad x > 0, \quad t > 0$

    $u(0, t) = 0, \quad \lim_{x \to \infty} \dfrac{\partial u}{\partial x} = 0, \quad t > 0$

    $u(x, 0) = 0, \quad x > 0$

    where r is a constant, is given by

    $u(x, t) = rt - r\int_0^t \mathrm{erfc}\left(\dfrac{x}{2\sqrt{k\tau}}\right) d\tau$

23. A rod of length L is held at a constant temperature u_0 at its ends $x = 0$ and $x = L$. If the initial temperature of the rod is $u_0 + u_0 \sin(x\pi/L)$, solve the heat equation $u_{xx} = u_t$, $0 < x < L$, $t > 0$, for the temperature $u(x, t)$.

24. If there is heat transfer from the lateral surface of a thin wire of length L into a medium at constant temperature u_1, then the heat equation takes on the form

    $k\dfrac{\partial^2 u}{\partial x^2} - h(u - u_1) = \dfrac{\partial u}{\partial t}, \quad 0 < x < L, \quad t > 0$

   h a constant. Find the temperature $u(x, t)$ if the initial temperature is a constant u_0 throughout and the ends $x = 0$ and $x = L$ are insulated.

25. A rod of unit length L is insulated at $x = 0$ and is kept at temperature 0 at $x = 1$. If the initial temperature of the rod is a constant u_0, solve $ku_{xx} = u_t$, $0 < x < 1$, $t > 0$, for the temperature $u(x, t)$. [*Hint:* Expand $1/(1 + e^{-2\sqrt{s/k}})$ in a geometric series.]

26. An infinite porous slab of unit width is immersed in a solution of constant concentration C_0. A dissolved substance in the solution diffuses into the slab. The concentration $C(x, t)$ in the slab is determined from

    $D\dfrac{\partial^2 C}{\partial x^2} = \dfrac{\partial C}{\partial t}, \quad 0 < x < 1, \quad t > 0$

    $C(0, t) = C_0, \quad C(1, t) = C_0, \quad t > 0$

    $C(x, 0) = 0, \quad 0 < x < 1$

    Solve for $C(x, t)$.

27. A very long telephone transmission line is initially at a constant potential u_0. If the line is grounded at $x = 0$ and insulated at the distant right end, then the potential $u(x, t)$ at a point x along the line at time t is determined from

    $\dfrac{\partial^2 u}{\partial x^2} - RC\dfrac{\partial u}{\partial t} - RGu = 0, \quad x > 0, \quad t > 0$

    $u(0, t) = 0, \quad \lim_{x \to \infty} \dfrac{\partial u}{\partial x} = 0, \quad t > 0$

    $u(x, 0) = u_0, \quad x > 0$

    where R, C, and G are constants known as resistance, capacitance, and conductance, respectively. Solve for $u(x, t)$. [*Hint:* See Problem 5 in Exercises 14.1.]

28. Show that a solution of the boundary-value problem

    $\dfrac{\partial^2 u}{\partial x^2} - hu = \dfrac{\partial u}{\partial t}, \quad x > 0, \quad t > 0$

    $u(0, t) = u_0, \quad \lim_{x \to \infty} u(x, t) = 0, \quad t > 0$

    $u(x, 0) = 0, \quad x > 0$

    is $u(x, t) = \dfrac{u_0 x}{2\sqrt{\pi}} \int_0^t \dfrac{e^{-h\tau - x^2/4\tau}}{\tau^{3/2}} d\tau$

29. Solve the boundary-value problem:

    $k\dfrac{\partial^2 u}{\partial x^2} = \dfrac{\partial u}{\partial t}, \quad x > 0, \quad t > 0$

    $\left.\dfrac{\partial u}{\partial x}\right|_{x=0} = -A\delta(t), \quad \lim_{x \to \infty} u(x, t) = 0, \quad t > 0$

    $u(x, 0) = 0, \quad x > 0,$

where $\delta(t)$ is the Dirac delta function (see Section 4.5). Interpret the boundary condition at $x = 0$.

30. Starting at $t = 0$, a concentrated load of magnitude F_0 moves with a constant velocity v_0 along a semi-infinite string. In this case the wave equation becomes

$$a^2 \frac{\partial^2 u}{\partial x^2} = \frac{\partial^2 u}{\partial t^2} + F_0 \delta\left(t - \frac{x}{v_0}\right)$$

Solve this partial differential equation subject to

$$u(0, t) = 0, \quad \lim_{x \to \infty} u(x, t) = 0, \quad t > 0$$

$$u(x, 0) = 0, \quad \left.\frac{\partial u}{\partial t}\right|_{t=0} = 0, \quad x > 0$$

(a) when $v_0 \neq a$ and **(b)** when $v_0 = a$.

14.3 FOURIER INTEGRAL

In Chapters 11 and 12 we used Fourier series to represent a function f defined on a *finite* interval such as $(-p, p)$ or $(0, L)$. When f and f' are piecewise continuous on such an interval, a Fourier series represents the function on the interval and converges to a periodic extension of f outside of the interval. In this way we are justified in saying that Fourier series are associated only with periodic functions. We shall now derive, in a nonrigorous fashion, a means of representing certain kinds of nonperiodic functions that are defined on either an *infinite* interval $(-\infty, \infty)$ or a *semi-infinite* interval $(0, \infty)$.

From Fourier Series to Fourier Integral   Suppose a function f is defined on $(-p, p)$ and that

$$\begin{aligned} f(x) &= \frac{a_0}{2} + \sum_{n=1}^{\infty} \left(a_n \cos \frac{n\pi}{p} x + b_n \sin \frac{n\pi}{p} x\right) \\ &= \frac{1}{2p} \int_{-p}^{p} f(t)\, dt + \frac{1}{p} \sum_{n=1}^{\infty} \left[\left(\int_{-p}^{p} f(t) \cos \frac{n\pi}{p} t\, dt\right) \cos \frac{n\pi}{p} x \right. \\ &\quad \left. + \left(\int_{-p}^{p} f(t) \sin \frac{n\pi}{p} t\, dt\right) \sin \frac{n\pi}{p} x\right] \end{aligned} \quad (1)$$

If we let $\alpha_n = n\pi/p$, $\Delta\alpha = \alpha_{n+1} - \alpha_n = \pi/p$, then (1) becomes

$$f(x) = \frac{1}{2\pi}\left(\int_{-p}^{p} f(t)\, dt\right) \Delta\alpha + \frac{1}{\pi} \sum_{n=1}^{\infty} \left[\left(\int_{-p}^{p} f(t) \cos \alpha_n t\, dt\right) \cos \alpha_n x \right. \\ \left. + \left(\int_{-p}^{p} f(t) \sin \alpha_n t\, dt\right) \sin \alpha_n x\right] \Delta\alpha \quad (2)$$

We now expand the interval $(-p, p)$ by letting $p \to \infty$. Since $p \to \infty$ implies $\Delta\alpha \to 0$, the limit of (2) has the form $\lim_{\Delta\alpha \to 0} \sum_{n=1}^{\infty} F(\alpha_n) \Delta\alpha$, which is suggestive of the definition of the integral $\int_0^{\infty} F(\alpha)\, d\alpha$. Thus, if $\int_{-\infty}^{\infty} f(t)\, dt$ exists, the limit of the first term in (2) is zero and the limit of the sum becomes

$$f(x) = \frac{1}{\pi} \int_0^{\infty} \left[\left(\int_{-\infty}^{\infty} f(t) \cos \alpha t\, dt\right) \cos \alpha x + \left(\int_{-\infty}^{\infty} f(t) \sin \alpha t\, dt\right) \sin \alpha x\right] d\alpha \quad (3)$$

14.3 Fourier Integral

The result given in (3) is called the **Fourier integral** of f on $(-\infty, \infty)$. As the following summary shows, the basic structure of the Fourier integral is reminiscent of that of a Fourier series.

DEFINITION 14.2 Fourier Integral

The **Fourier integral** of a function f defined on the interval $(-\infty, \infty)$ is given by

$$f(x) = \frac{1}{\pi} \int_0^\infty [A(\alpha) \cos \alpha x + B(\alpha) \sin \alpha x] \, d\alpha \qquad (4)$$

where

$$A(\alpha) = \int_{-\infty}^\infty f(x) \cos \alpha x \, dx \qquad (5)$$

$$B(\alpha) = \int_{-\infty}^\infty f(x) \sin \alpha x \, dx \qquad (6)$$

Convergence of a Fourier Integral Sufficient conditions under which a Fourier integral converges to $f(x)$ are similar to, but slightly more restrictive than, the conditions for a Fourier series.

THEOREM 14.1 Conditions for Convergence

Let f and f' be piecewise continuous on every finite interval and let f be absolutely integrable on $(-\infty, \infty)$.* Then the Fourier integral of f on the interval converges to $f(x)$ at a point of continuity. At a point of discontinuity the Fourier integral will converge to the average

$$\frac{f(x+) + f(x-)}{2}$$

where $f(x+)$ and $f(x-)$ denote the limit of f at x from the right and from the left, respectively.

---

*This means that the integral $\int_{-\infty}^\infty |f(x)| \, dx$ converges.

EXAMPLE 1 Find the Fourier integral representation of the function

$$f(x) = \begin{cases} 0, & x < 0 \\ 1, & 0 < x < 2 \\ 0, & x > 0 \end{cases}$$

Solution The function, whose graph is shown in Figure 14.4, satisfies the hypotheses of Theorem 14.1. Hence, from (5) and (6) we have at once

$$A(\alpha) = \int_{-\infty}^{\infty} f(x) \cos \alpha x \, dx$$

$$= \int_{-\infty}^{0} f(x) \cos \alpha x \, dx + \int_{0}^{2} f(x) \cos \alpha x \, dx + \int_{2}^{\infty} f(x) \cos \alpha x \, dx$$

$$= \int_{0}^{2} \cos \alpha x \, dx$$

$$= \frac{\sin 2\alpha}{\alpha}$$

and

$$B(\alpha) = \int_{-\infty}^{\infty} f(x) \sin \alpha x \, dx$$

$$= \int_{0}^{2} \sin \alpha x \, dx$$

$$= \frac{1 - \cos 2\alpha}{\alpha}$$

Substituting these coefficients in (4) then gives

$$f(x) = \frac{1}{\pi} \int_{0}^{\infty} \left[\left(\frac{\sin 2\alpha}{\alpha} \right) \cos \alpha x + \left(\frac{1 - \cos 2\alpha}{\alpha} \right) \sin \alpha x \right] d\alpha$$

When we use trigonometric identities, the last integral simplifies to

$$f(x) = \frac{2}{\pi} \int_{0}^{\infty} \frac{\sin \alpha \cos \alpha(x - 1)}{\alpha} d\alpha \qquad (7)$$

□

The Fourier integral can be used to evaluate integrals. For example, at $x = 1$ it follows from Theorem 14.1 that (7) converges to $f(1)$; that is,

$$\int_{0}^{\infty} \frac{\sin \alpha}{\alpha} d\alpha = \frac{\pi}{2}$$

The latter result is worthy of special note since it cannot be obtained in the "usual" manner; the integrand $(\sin x)/x$ does not possess an antiderivative that is an elementary function.

Cosine and Sine Integrals When f is an even function on the interval $(-\infty, \infty)$, then the product $f(x) \cos \alpha x$ is also an even function, whereas

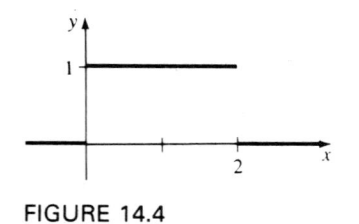

14.3 Fourier Integral

$f(x) \sin \alpha x$ is an odd function. As a consequence of property (*vii*) of Section 11.3, $B(\alpha) = 0$ and so (4) becomes

$$f(x) = \frac{2}{\pi} \int_0^\infty \left(\int_0^\infty f(t) \cos \alpha t \, dt \right) \cos \alpha x \, d\alpha$$

Here we have also used property (*vi*) of Section 11.3 to write

$$\int_{-\infty}^\infty f(t) \cos \alpha t \, dt = 2 \int_0^\infty f(t) \cos \alpha t \, dt$$

Similarly, when f is an odd function on $(-\infty, \infty)$, the products $f(x) \cos \alpha x$ and $f(x) \sin \alpha x$ are odd and even functions, respectively. Therefore $A(\alpha) = 0$ and

$$f(x) = \frac{2}{\pi} \int_0^\infty \left(\int_0^\infty f(t) \sin \alpha t \, dt \right) \sin \alpha x \, d\alpha$$

We summarize:

DEFINITION 14.3   Fourier Cosine and Sine Integrals

(i) The Fourier integral of an even function on the interval $(-\infty, \infty)$ is the **cosine integral**

$$f(x) = \frac{2}{\pi} \int_0^\infty A(\alpha) \cos \alpha x \, d\alpha \qquad (8)$$

where

$$A(\alpha) = \int_0^\infty f(x) \cos \alpha x \, dx \qquad (9)$$

(ii) The Fourier integral of an odd function on the interval $(-\infty, \infty)$ is the **sine integral**

$$f(x) = \frac{2}{\pi} \int_0^\infty B(\alpha) \sin \alpha x \, d\alpha \qquad (10)$$

where

$$B(\alpha) = \int_0^\infty f(x) \sin \alpha x \, dx \qquad (11)$$

EXAMPLE 2   Find the Fourier integral representation of the function

$$f(x) = \begin{cases} 1, & |x| < a \\ 0, & |x| > a \end{cases}$$

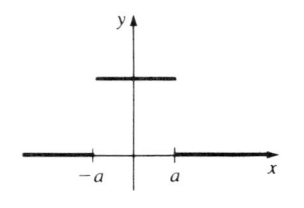

FIGURE 14.5

Solution It is apparent from Figure 14.5 that f is an even function. Hence, we represent f by the Fourier cosine integral (8). From (9) we obtain

$$A(\alpha) = \int_0^\infty f(x) \cos \alpha x \, dx$$

$$= \int_0^a f(x) \cos \alpha x \, dx + \int_a^\infty f(x) \cos \alpha x \, dx$$

$$= \int_0^a \cos \alpha x \, dx$$

$$= \frac{\sin a\alpha}{\alpha}$$

and so
$$f(x) = \frac{2}{\pi} \int_0^\infty \frac{\sin a\alpha \cos \alpha x}{\alpha} \, d\alpha \qquad \square$$

The integrals (8) and (10) can be used when f is neither odd nor even and defined only on the half-line $(0, \infty)$. In this case, (8) represents f on the interval $(0, \infty)$ and its even (but not periodic) extension to $(-\infty, 0)$, whereas (10) represents f on $(0, \infty)$ and its odd extension to the interval $(-\infty, 0)$. The next example will illustrate this concept.

EXAMPLE 3 Represent $f(x) = e^{-x}$, $x > 0$, (a) by a cosine integral and (b) by a sine integral.

Solution The graph of the function is given in Figure 14.6.

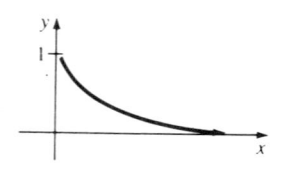

FIGURE 14.6

(a) Using integration by parts, we find

$$A(\alpha) = \int_0^\infty e^{-x} \cos \alpha x \, dx = \frac{1}{1 + \alpha^2}$$

Therefore the cosine integral of f is

$$f(x) = \frac{2}{\pi} \int_0^\infty \frac{\cos \alpha x}{1 + \alpha^2} \, d\alpha$$

(b) Similarly, we have

$$B(\alpha) = \int_0^\infty e^{-x} \sin \alpha x \, dx = \frac{\alpha}{1 + \alpha^2}$$

The sine integral of f is then

$$f(x) = \frac{2}{\pi} \int_0^\infty \frac{\alpha \sin \alpha x}{1 + \alpha^2} \, d\alpha$$

Figure 14.7 shows the graphs of the functions and their extensions represented by the two integrals.

14.3 Fourier Integral

(a) cosine integral   (b) sine integral

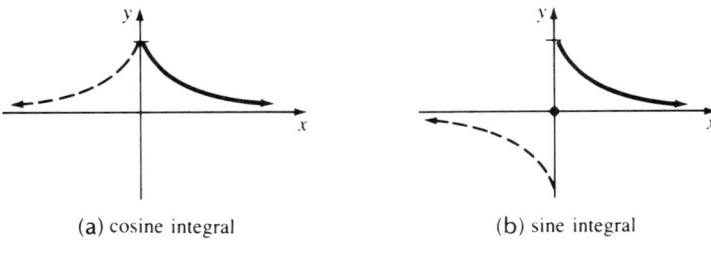

FIGURE 14.7

Complex Form The Fourier integral (4) also possesses an equivalent **complex form**, or **exponential form**, that is analogous to the complex form of a Fourier series (see Problem 21 in Exercises 11.2). If (5) and (6) are substituted in (4), then

$$f(x) = \frac{1}{\pi} \int_0^\infty \int_{-\infty}^\infty f(t)[\cos \alpha t \cos \alpha x + \sin \alpha t \sin \alpha x] \, dt \, d\alpha$$

$$= \frac{1}{\pi} \int_0^\infty \int_{-\infty}^\infty f(t) \cos \alpha(t - x) \, dt \, d\alpha$$

$$= \frac{1}{2\pi} \int_{-\infty}^\infty \int_{-\infty}^\infty f(t) \cos \alpha(t - x) \, dt \, d\alpha \tag{12}$$

$$= \frac{1}{2\pi} \int_{-\infty}^\infty \int_{-\infty}^\infty f(t)[\cos \alpha(t - x) + i \sin \alpha(t - x)] \, dt \, d\alpha \tag{13}$$

$$= \frac{1}{2\pi} \int_{-\infty}^\infty \int_{-\infty}^\infty f(t) e^{i\alpha(t - x)} \, dt \, d\alpha$$

$$= \frac{1}{2\pi} \int_{-\infty}^\infty \left(\int_{-\infty}^\infty f(t) e^{i\alpha t} \, dt \right) e^{-i\alpha x} \, d\alpha \tag{14}$$

We note that (12) follows from the fact that the integrand is an even function of α. In (13) we have simply added zero to the integrand:

$$i \int_{-\infty}^\infty \int_{-\infty}^\infty f(t) \sin \alpha(t - x) \, dt \, d\alpha = 0$$

because the integrand is an odd function of α. The integral in (14) can be expressed as

$$f(x) = \frac{1}{2\pi} \int_{-\infty}^\infty C(\alpha) e^{-i\alpha x} \, d\alpha \tag{15}$$

where

$$C(\alpha) = \int_{-\infty}^\infty f(x) e^{i\alpha x} \, dx \tag{16}$$

This latter form of the Fourier integral will be put to use in the next section when we return to the solution of boundary-value problems.

EXERCISES 14.3   Answers to odd-numbered problems begin on page A-67.

In Problems 1–6 find the Fourier integral representation of the given function.

1. $f(x) = \begin{cases} 0, & x < -1 \\ -1, & -1 < x < 0 \\ 2, & 0 < x < 1 \\ 0, & x > 1 \end{cases}$

2. $f(x) = \begin{cases} 0, & x < \pi \\ 4, & \pi < x < 2\pi \\ 0, & x > 2\pi \end{cases}$

3. $f(x) = \begin{cases} 0, & x < 0 \\ x, & 0 < x < 3 \\ 0, & x > 3 \end{cases}$

4. $f(x) = \begin{cases} 0, & x < 0 \\ \sin x, & 0 \leq x \leq \pi \\ 0, & x > \pi \end{cases}$

5. $f(x) = \begin{cases} 0, & x < 0 \\ e^{-x}, & x > 0 \end{cases}$

6. $f(x) = \begin{cases} e^x, & |x| < 1 \\ 0, & |x| > 1 \end{cases}$

In Problems 7–12 represent the given function by an appropriate cosine or sine integral.

7. $f(x) = \begin{cases} 0, & x < -1 \\ -5, & -1 < x < 0 \\ 5, & 0 < x < 1 \\ 0, & x > 1 \end{cases}$

8. $f(x) = \begin{cases} 0, & |x| < 1 \\ \pi, & 1 < |x| < 2 \\ 0, & |x| > 2 \end{cases}$

9. $f(x) = \begin{cases} |x|, & |x| < \pi \\ 0, & |x| > \pi \end{cases}$

10. $f(x) = \begin{cases} x, & |x| < \pi \\ 0, & |x| > \pi \end{cases}$

11. $f(x) = e^{-|x|} \sin x$*

12. $f(x) = xe^{-|x|}$

In Problems 13–16 find the cosine and sine integral representations of the given function.

13. $f(x) = e^{-kx}, k > 0, x > 0$

14. $f(x) = e^{-x} - e^{-3x}, x > 0$

15. $f(x) = xe^{-2x}, x > 0$

16. $f(x) = e^{-x} \cos x, x > 0$

In Problems 17 and 18 solve the given integral equation.

17. $\int_0^\infty f(x) \cos \alpha x \, dx = e^{-\alpha}$

18. $\int_0^\infty f(x) \sin \alpha x \, dx = \begin{cases} 1, & 0 < \alpha < 1 \\ 0, & \alpha > 1 \end{cases}$

19. (a) Use (7) to show that
$$\int_0^\infty \frac{\sin 2x}{x} dx = \frac{\pi}{2}$$
[Hint: α is a dummy variable of integration.]

(b) Show in general that for $k > 0$,
$$\int_0^\infty \frac{\sin kx}{x} dx = \frac{\pi}{2}$$

20. Use the complex form (15) to find the Fourier integral representation of $f(x) = e^{-|x|}$. Show that the result is the same as that obtained from (8).

*The integrals in Problems 11–16 will not be difficult for students who remember some Laplace transform results.

14.4 FOURIER TRANSFORMS

Transform Pairs   Up to now we have been using a symbolic representation for the inverse of a Laplace transform: $\mathscr{L}^{-1}\{F(s)\} = f(t)$. Actually, the inverse Laplace transform is itself an integral. If
$$\mathscr{L}\{f(t)\} = \int_0^\infty e^{-st} f(t) \, dt = F(s) \qquad (1)$$

then it can be proved that

$$\mathscr{L}^{-1}\{F(s)\} = \frac{1}{2\pi i}\int_{\gamma-i\infty}^{\gamma+i\infty} e^{st}F(s)\,ds = f(t) \tag{2}$$

The foregoing integral is called a **contour integral**; its evaluation requires the use of complex variables. The point is this: Integral transforms occur in **transform pairs**—that is, if

$$F(\alpha) = \int_a^b f(x)K(\alpha, x)\,dx \tag{3}$$

is an integral that transforms $f(x)$ into $F(\alpha)$, then the function f is recovered by means of another integral transform

$$f(x) = \int_c^d F(\alpha)H(\alpha, x)\,dx \tag{4}$$

called the **inversion integral** or **inverse transform**. The functions K and H are called the **kernels** of transforms (3) and (4), respectively. For example, in (1) we identify the kernel of the transform as $K(s, t) = e^{-st}$, whereas in (2) we can take the kernel of the inverse transform to be $H(s, t) = e^{st}/2\pi i$.

Fourier Transforms The Fourier integral is the source of three new integral transforms. From (16)–(15), (11)–(10), and (9)–(8) of the preceding section we are prompted to define the following transform pairs:

DEFINITION 14.4 Fourier Transform Pairs

(i) **Fourier transform:** $\quad\mathscr{F}\{f(x)\} = \displaystyle\int_{-\infty}^{\infty} f(x)e^{i\alpha x}\,dx = F(\alpha)\quad$ (5)

   Inverse Fourier transform: $\quad\mathscr{F}^{-1}\{F(\alpha)\} = \dfrac{1}{2\pi}\displaystyle\int_{-\infty}^{\infty} F(\alpha)e^{-i\alpha x}\,d\alpha = f(x)\quad$ (6)

(ii) **Fourier sine transform:** $\quad\mathscr{F}_s\{f(x)\} = \displaystyle\int_0^{\infty} f(x)\sin\alpha x\,dx = F(\alpha)\quad$ (7)

   Inverse Fourier sine transform: $\quad\mathscr{F}_s^{-1}\{F(\alpha)\} = \dfrac{2}{\pi}\displaystyle\int_0^{\infty} F(\alpha)\sin\alpha x\,d\alpha = f(x)\quad$ (8)

(iii) **Fourier cosine transform:** $\quad\mathscr{F}_c\{f(x)\} = \displaystyle\int_0^{\infty} f(x)\cos\alpha x\,dx = F(\alpha)\quad$ (9)

   Inverse Fourier cosine transform: $\quad\mathscr{F}_c^{-1}\{F(\alpha)\} = \dfrac{2}{\pi}\displaystyle\int_0^{\infty} F(\alpha)\cos\alpha x\,d\alpha = f(x)\quad$ (10)

Existence The conditions under which (5), (7), and (9) exist are more stringent than those for the Laplace transform. For example, the reader should verify that $\mathscr{F}\{1\}$, $\mathscr{F}_s\{1\}$, and $\mathscr{F}_c\{1\}$ do not exist. Sufficient conditions for existence are that f be absolutely integrable on the appropriate interval and that f and f' be piecewise continuous on every finite interval.

Operational Properties   Since our immediate goal is to apply these new transforms to boundary-value problems, we need to examine the transforms of derivatives.

Fourier Transform   Suppose that f is continuous and absolutely integrable on the interval $(-\infty, \infty)$ and f' is piecewise continuous on every finite interval. If $f(x) \to 0$ as $x \to \pm\infty$, then integration by parts gives

$$\mathscr{F}\{f'(x)\} = \int_{-\infty}^{\infty} f'(x) e^{i\alpha x} \, dx$$

$$= f(x) e^{i\alpha x} \Big|_{-\infty}^{\infty} - i\alpha \int_{-\infty}^{\infty} f(x) e^{i\alpha x} \, dx$$

$$= -i\alpha \int_{-\infty}^{\infty} f(x) e^{i\alpha x} \, dx$$

that is,

$$\boxed{\mathscr{F}\{f'(x)\} = -i\alpha F(\alpha)} \tag{11}$$

Similarly, under the added assumptions that f' is continuous on $(-\infty, \infty)$, $f''(x)$ is piecewise continuous on every finite interval, and $f'(x) \to 0$ as $x \to \pm\infty$, we have

$$\boxed{\mathscr{F}\{f''(x)\} = (-i\alpha)^2 \mathscr{F}\{f(x)\} = -\alpha^2 F(\alpha)} \tag{12}$$

It is important to be aware of the fact that the sine and cosine transforms are not suitable for transforming the first derivative (or for that matter, any derivative of *odd* order). It is left as an exercise to show that

$$\mathscr{F}_s\{f'(x)\} = -\alpha \mathscr{F}_c\{f(x)\} \tag{13}$$

and

$$\mathscr{F}_c\{f'(x)\} = \alpha \mathscr{F}_s\{f(x)\} - f(0) \tag{14}$$

The difficulty is apparent; the transform of $f'(x)$ is not expressed in terms of the original integral transform.

Fourier Sine Transform   Suppose that f and f' are continuous, f is absolutely integrable on the interval $[0, \infty)$, f'' is piecewise continuous on every finite interval. If $f \to 0$ and $f' \to 0$ as $x \to \infty$, then

$$\mathscr{F}_s\{f''(x)\} = \int_0^{\infty} f''(x) \sin \alpha x \, dx$$

$$= f'(x) \sin \alpha x \Big|_0^{\infty} - \alpha \int_0^{\infty} f'(x) \cos \alpha x \, dx$$

$$= -\alpha\left[f(x)\cos\alpha x\Big|_0^\infty + \alpha\int_0^\infty f(x)\sin\alpha x\,dx\right]$$

$$= \alpha f(0) - \alpha^2 \mathscr{F}_s\{f(x)\}$$

that is,
$$\boxed{\mathscr{F}_s\{f''(x)\} = -\alpha^2 F(\alpha) + \alpha f(0)} \tag{15}$$

Fourier Cosine Transform Under the same assumptions leading to (15), we find the Fourier cosine transform of $f''(x)$ to be

$$\boxed{\mathscr{F}_c\{f''(x)\} = -\alpha^2 F(\alpha) - f'(0)} \tag{16}$$

A natural question is: How do we know which transform to use on a given boundary-value problem? Clearly, to use a Fourier transform, the domain of the variable to be eliminated must be $(-\infty, \infty)$. To utilize a sine or cosine transform, the domain of at least one of the variables in the problem must be $[0, \infty)$. But the determining factor in choosing between the sine transform and the cosine transform is the type of boundary condition specified at zero.

In the problems that follow, we shall assume without further mention that both u and $\partial u/\partial x$ (or $\partial u/\partial y$) approach zero as $x \to \pm\infty$. This is not a major restriction since these conditions hold in most applications.

EXAMPLE 1 Solve the heat equation

$$k\frac{\partial^2 u}{\partial x^2} = \frac{\partial u}{\partial t}, \quad -\infty < x < \infty, \quad t > 0$$

subject to
$$u(x, 0) = f(x), \text{ where } f(x) = \begin{cases} 1, & |x| \leq 1 \\ 0 & |x| > 1. \end{cases}$$

Solution The problem can be interpreted as finding the temperature $u(x, t)$ in an infinite rod. Since the domain of x is the infinite interval $(-\infty, \infty)$, we use the Fourier transform (5) and define

$$\mathscr{F}\{u(x, t)\} = \int_{-\infty}^\infty u(x, t)e^{i\alpha x}\,dx = U(\alpha, t)$$

Transforming the partial differential equation and using (12) yield

$$\mathscr{F}\left\{k\frac{\partial^2 u}{\partial x^2}\right\} = \mathscr{F}\left\{\frac{\partial u}{\partial t}\right\}$$

$$-k\alpha^2 U(\alpha, t) = \frac{dU}{dt} \quad \text{or} \quad \frac{dU}{dt} + k\alpha^2 U = 0$$

Thus,
$$U(\alpha, t) = ce^{-k\alpha^2 t} \tag{17}$$

Now the transform of the initial condition is

$$\mathscr{F}\{u(x, 0)\} = \int_{-\infty}^{\infty} f(x)e^{i\alpha x}\, dx = \int_{-1}^{1} e^{i\alpha x}\, dx$$

$$= \frac{e^{i\alpha} - e^{-i\alpha}}{i\alpha}$$

or $$U(\alpha, 0) = 2\,\frac{\sin \alpha}{\alpha}$$

Using this last condition in (17) gives $U(\alpha, 0) = c = (2 \sin \alpha)/\alpha$, and so

$$U(\alpha, t) = 2\,\frac{\sin \alpha}{\alpha}\, e^{-k\alpha^2 t}$$

It then follows from the inversion integral (6) that

$$u(x, t) = \frac{1}{\pi} \int_{-\infty}^{\infty} \frac{\sin \alpha}{\alpha}\, e^{-k\alpha^2 t} e^{-i\alpha x}\, d\alpha$$

The last expression can be simplified somewhat by writing $e^{-i\alpha x} = \cos \alpha x - i \sin \alpha x$ and noting $\int_{-\infty}^{\infty} \frac{\sin \alpha}{\alpha}\, e^{-k\alpha^2 t} \sin \alpha x\, d\alpha = 0$, since the integrand is an odd function of α. Hence, we finally have

$$u(x, t) = \frac{1}{\pi} \int_{-\infty}^{\infty} \frac{\sin \alpha \cos \alpha x}{\alpha}\, e^{-k\alpha^2 t}\, d\alpha \qquad (18)$$

□

EXAMPLE 2 The steady-state temperature in a semi-infinite plate is determined from

$$\frac{\partial^2 u}{\partial x^2} + \frac{\partial^2 u}{\partial y^2} = 0, \quad 0 < x < \pi, \quad y > 0$$

$$u(0, y) = 0, \quad u(\pi, y) = e^{-y}, \quad y > 0$$

$$\left.\frac{\partial u}{\partial y}\right|_{y=0} = 0, \quad 0 < x < \pi$$

Solve for $u(x, y)$.

Solution The domain of the variable y and the prescribed condition at $y = 0$ indicate that the Fourier cosine transform is suitable for the problem. We define

$$\mathscr{F}_c\{u(x, y)\} = \int_0^{\infty} u(x, y) \cos \alpha y\, dy = U(x, \alpha)$$

In view of (16),

$$\mathscr{F}_c\left\{\frac{\partial^2 u}{\partial x^2}\right\} + \mathscr{F}_c\left\{\frac{\partial^2 u}{\partial y^2}\right\} = \mathscr{F}_c\{0\}$$

becomes

$$\frac{d^2U}{dx^2} - \alpha^2 U(x,\alpha) - u_y(x,0) = 0 \quad \text{or} \quad \frac{d^2U}{dx^2} - \alpha^2 U = 0$$

Since the domain of x is a finite interval, we choose to write the solution of the ordinary differential equation as

$$U(x,\alpha) = c_1 \cosh \alpha x + c_2 \sinh \alpha x \tag{19}$$

Now $\mathscr{F}_c\{u(0,y)\} = \mathscr{F}_c\{0\}$ and $\mathscr{F}_c\{u(\pi,y)\} = \mathscr{F}_c\{e^{-y}\}$ are in turn equivalent to

$$U(0,\alpha) = 0 \quad \text{and} \quad U(\pi,\alpha) = \frac{1}{1+\alpha^2}$$

Applying these latter conditions, the solution (19) gives $c_1 = 0$ and $c_2 = 1/[(1+\alpha^2)\sinh \alpha\pi]$. Therefore,

$$U(x,\alpha) = \frac{\sinh \alpha x}{(1+\alpha^2)\sinh \alpha\pi}$$

and so from (10) we arrive at

$$u(x,y) = \frac{2}{\pi}\int_0^\infty \frac{\sinh \alpha x}{(1+\alpha^2)\sinh \alpha\pi} \cos \alpha y \, d\alpha \tag{20}$$

□

Had $u(x,0)$ been given in Example 2 rather than $u_y(x,0)$, the sine transform would have been appropriate.

▲ **Remark** The reader may be wondering whether the integrals in (18) and (20) can be evaluated. In *most* cases the answer is no (see Problem 22 for an exception). One way of finding a specific value, say, of the temperature $u(1,1)$ in (20) is to resort to numerical integration.

EXERCISES 14.4 *Answers to odd-numbered problems begin on page A-67.*

In Problems 1–18 use the Fourier integral transforms of this section to solve the given boundary-value problem. Make assumptions about boundedness where necessary.

1. $k\dfrac{\partial^2 u}{\partial x^2} = \dfrac{\partial u}{\partial t}, \quad -\infty < x < \infty \quad t > 0$
   $u(x,0) = e^{-|x|}, \quad -\infty < x < \infty$

2. $k\dfrac{\partial^2 u}{\partial x^2} = \dfrac{\partial u}{\partial t}, \quad -\infty < x < \infty \quad t > 0$
   $u(x,0) = \begin{cases} 0, & x < -1 \\ -100, & -1 < x < 0 \\ 100, & 0 < x < 1 \\ 0, & x > 1 \end{cases}$

3. Use the result $\mathscr{F}\{e^{-x^2/4p^2}\} = 2\sqrt{\pi}p e^{-p^2\alpha^2}$ to solve the boundary-value problem

   $k\dfrac{\partial^2 u}{\partial x^2} = \dfrac{\partial u}{\partial t}, \quad -\infty < x < \infty, \quad t > 0$

   $u(x,0) = e^{-x^2}, \quad -\infty < x < \infty$

4. (a) If $\mathscr{F}\{f(x)\} = F(\alpha)$ and $\mathscr{F}\{g(x)\} = G(\alpha)$, then the **convolution theorem** for the Fourier transform is given by

   $$\int_{-\infty}^\infty f(\tau)g(x-\tau)\,d\tau = \mathscr{F}^{-1}\{F(\alpha)G(\alpha)\}$$

Use this result and $\mathscr{F}\{e^{-x^2/4p^2}\} = 2\sqrt{\pi}pe^{-p^2\alpha^2}$ to show that a solution of the boundary-value problem

$$k\frac{\partial^2 u}{\partial x^2} = \frac{\partial u}{\partial t}, \quad -\infty < x < \infty, \quad t > 0$$

$$u(x, 0) = f(x), \quad -\infty < x < \infty$$

is given by

$$u(x, t) = \frac{1}{2\sqrt{k\pi t}} \int_{-\infty}^{\infty} f(\tau)e^{-(x-\tau)^2/4kt}\, d\tau$$

(b) By using the change of variables $u = (x - \tau)/2\sqrt{kt}$ and Problem 9 of Exercises 14.1, show that the solution of part (a) when

$$f(x) = \begin{cases} u_0, & |x| < 1 \\ 0, & |x| > 1 \end{cases}$$

is $\displaystyle u(x, t) = \frac{u_0}{2}\left[\operatorname{erf}\left(\frac{x+1}{2\sqrt{kt}}\right) - \operatorname{erf}\left(\frac{x-1}{2\sqrt{kt}}\right)\right]$

5. Find the temperature $u(x, t)$ in a semi-infinite rod if $u(0, t) = u_0, t > 0$, and $u(x, 0) = 0, x > 0$.

6. Use the result $\displaystyle\int_0^\infty \frac{\sin \alpha x}{\alpha}\, d\alpha = \frac{\pi}{2}, x > 0$, to show that the solution of Problem 5 can be written as

$$u(x, t) = u_0 - \frac{2u_0}{\pi}\int_0^\infty \frac{\sin \alpha x}{\alpha} e^{-k\alpha^2 t}\, d\alpha$$

7. Find the temperature $u(x, t)$ in a semi-infinite rod if $u(0, t) = 0, t > 0$, and $u(x, 0) = \begin{cases} 1, & 0 < x < 1 \\ 0, & x > 1. \end{cases}$

8. Solve Problem 5 if the condition at the left boundary is

$$\left.\frac{\partial u}{\partial x}\right|_{x=0} = -A, \quad t > 0$$

9. Solve Problem 7 if the end $x = 0$ is insulated.

10. Find the temperature $u(x, t)$ in a semi-infinite rod if $u(0, t) = 1, t > 0$, and $u(x, 0) = e^{-x}, x > 0$.

11. (a) $a^2 \dfrac{\partial^2 u}{\partial x^2} = \dfrac{\partial^2 u}{\partial t^2}, \quad -\infty < x < \infty, t > 0$

$$u(x, 0) = f(x), \left.\frac{\partial u}{\partial t}\right|_{t=0} = g(x), \quad -\infty < x < \infty$$

(b) If $g(x) = 0$ show that the solution of part (a) can be written as $u(x, t) = \tfrac{1}{2}[f(x + at) + f(x - at)]$.

12. Find the displacement $u(x, t)$ of a semi-infinite string if

$$u(0, t) = 0, t > 0, u(x, 0) = xe^{-x}, \left.\frac{\partial u}{\partial t}\right|_{t=0} = 0, x > 0.$$

13. Solve the problem in Example 2 if the boundary conditions at $x = 0$ and $x = \pi$ are reversed: $u(0, y) = e^{-y}$, $u(\pi, y) = 0, y > 0$.

14. Solve the problem in Example 2 if the boundary condition at $y = 0$ is $u(x, 0) = 1, 0 < x < \pi$.

15. Find the steady-state temperature $u(x, y)$ in a plate defined by $x \geq 0, y \geq 0$ if the boundary $x = 0$ is insulated, and at $y = 0, u(x, 0) = \begin{cases} 50, & 0 < x < 1 \\ 0, & x > 1. \end{cases}$

16. Solve Problem 15 if the boundary condition at $x = 0$ is $u(0, y) = 0, y > 0$.

17. $\dfrac{\partial^2 u}{\partial x^2} + \dfrac{\partial^2 u}{\partial y^2} = 0, x > 0, 0 < y < 2$

$u(0, y) = 0, 0 < y < 2$
$u(x, 0) = f(x), u(x, 2) = 0, x > 0$

18. $\dfrac{\partial^2 u}{\partial x^2} + \dfrac{\partial^2 u}{\partial y^2} = 0, 0 < x < \pi, y > 0$

$u(0, y) = f(y), \left.\dfrac{\partial u}{\partial x}\right|_{x=\pi} = 0, y > 0$

$\left.\dfrac{\partial u}{\partial y}\right|_{y=0} = 0, 0 < x < \pi$

In Problems 19 and 20 find the steady-state temperature in the plate given in the figure. [*Hint:* One way of proceeding is to express Problems 19 and 20 as two- and three-boundary-value problems, respectively. Use the superposition principle (see Section 12.5).]

19.

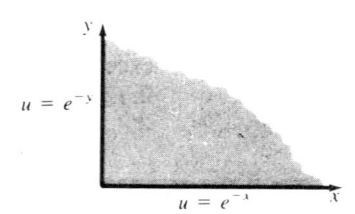

FIGURE 14.8

20.

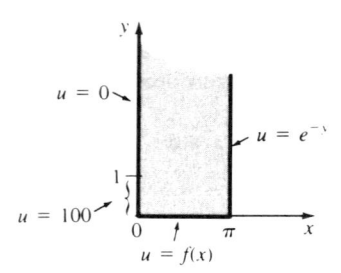

FIGURE 14.9

21. Use the transform $\mathscr{F}\{e^{-x^2/4p^2}\}$ given in Problem 3 to find the steady-state temperature in the infinite strip shown in Figure 14.10.

22. The solution of Problem 16 can be integrated. Use entries 44 and 45 of the table in Appendix II to show

$$u(x, y) = \frac{100}{\pi}\left[\arctan\frac{x}{y} - \frac{1}{2}\arctan\frac{x+1}{y} - \frac{1}{2}\arctan\frac{x-1}{y}\right]$$

23. Derive (13). **24.** Derive (14).

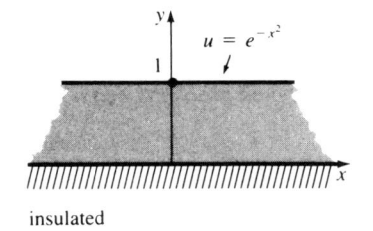

FIGURE 14.10

SUMMARY

The **error function** and **complementary error function**

$$\text{erf}(x) = \frac{2}{\sqrt{\pi}}\int_0^x e^{-u^2}\,du$$

$$\text{erfc}(x) = 1 - \text{erf}(x) = \frac{2}{\sqrt{\pi}}\int_x^\infty e^{-u^2}\,du$$

occur often in finding inverse Laplace transforms and in the solution of boundary-value problems.

The **Fourier integral**

$$f(x) = \frac{1}{\pi}\int_0^\infty [A(\alpha)\cos\alpha x + B(\alpha)\sin\alpha x]\,d\alpha \qquad (1)$$

where

$$A(\alpha) = \int_{-\infty}^\infty f(x)\cos\alpha x\,dx$$

$$B(\alpha) = \int_{-\infty}^\infty f(x)\sin\alpha x\,dx$$

is a way of representing a nonperiodic function that is defined on the infinite interval $(-\infty, \infty)$. When f is an even function, (1) becomes a **cosine integral**; when f is odd, (1) reduces to a **sine integral**.

Integral transforms occur in **transform pairs**: the transform itself and the inversion integral. To solve a boundary-value problem by an integral transform, these are the steps to take:

(*i*) Transform the partial differential equation into an ordinary differential equation.

(*ii*) Transform all boundary or initial conditions that depend on the transformed variable.

(*iii*) Solve the ordinary differential equation subject to the conditions found in step (*ii*).

(*iv*) Find the inverse transform of the function found in step (*iii*). This is the solution of the original boundary-value problem.

CHAPTER 14 REVIEW EXERCISES
Answers to odd-numbered problems begin on page A-67.

In Problems 1–11 solve the given boundary-value problem by an appropriate integral transform.

1. $\dfrac{\partial^2 u}{\partial x^2} + \dfrac{\partial^2 u}{\partial y^2} = 0,\ x > 0,\ 0 < y < \pi$

    $\left.\dfrac{\partial u}{\partial x}\right|_{x=0} = 0,\ 0 < y < \pi$

    $u(x, 0) = 0,\ \left.\dfrac{\partial u}{\partial y}\right|_{y=\pi} = e^{-x},\ x > 0$

2. $\dfrac{\partial^2 u}{\partial x^2} = \dfrac{\partial u}{\partial t},\ 0 < x < 1,\ t > 0$

    $u(0, t) = 0,\ u(1, t) = 0,\ t > 0$

    $u(x, 0) = 50 \sin 2\pi x,\ 0 < x < 1$

3. $\dfrac{\partial^2 u}{\partial x^2} - hu = \dfrac{\partial u}{\partial t},\ h > 0,\ x > 0,\ t > 0$

    $u(0, t) = 0,\ \lim\limits_{x \to \infty} \dfrac{\partial u}{\partial x} = 0,\ t > 0$

    $u(x, 0) = u_0,\ x > 0$

4. $\dfrac{\partial u}{\partial t} - \dfrac{\partial^2 u}{\partial x^2} = e^{-|x|},\ -\infty < x < \infty,\ t > 0$

    $u(x, 0) = 0,\ -\infty < x < \infty$

5. $\dfrac{\partial^2 u}{\partial x^2} = \dfrac{\partial u}{\partial t},\ x > 0,\ t > 0$

    $u(0, t) = t,\ \lim\limits_{x \to \infty} u(x, t) = 0$

    $u(x, 0) = 0,\ x > 0$ [*Hint:* Use Theorem 4.9.]

6. $\dfrac{\partial^2 u}{\partial x^2} = \dfrac{\partial^2 u}{\partial t^2},\ 0 < x < 1,\ t > 0$

    $u(0, t) = 0,\ u(1, t) = 0,\ t > 0$

    $u(x, 0) = \sin \pi x,\ \left.\dfrac{\partial u}{\partial t}\right|_{t=0} = -\sin \pi x,\ 0 < x < 1$

7. $k\dfrac{\partial^2 u}{\partial x^2} = \dfrac{\partial u}{\partial t},\ -\infty < x < \infty,\ t > 0$

    $u(x, 0) = \begin{cases} 0, & x < 0 \\ u_0, & 0 < x < \pi \\ 0, & x > \pi \end{cases}$

8. $\dfrac{\partial^2 u}{\partial x^2} + \dfrac{\partial^2 u}{\partial y^2} = 0,\ 0 < x < \pi,\ y > 0$

    $u(0, y) = 0,\ u(\pi, y) = \begin{cases} 0, & 0 < y < 1 \\ 1, & 1 < y < 2 \\ 0, & y > 2 \end{cases}$

    $\left.\dfrac{\partial u}{\partial y}\right|_{y=0} = 0,\ 0 < x < \pi$

9. $\dfrac{\partial^2 u}{\partial x^2} + \dfrac{\partial^2 u}{\partial y^2} = 0,\ x > 0,\ y > 0$

    $u(0, y) = \begin{cases} 50, & 0 < y < 1 \\ 0, & y > 1 \end{cases}$

    $u(x, 0) = \begin{cases} 100, & 0 < x < 1 \\ 0, & x > 1 \end{cases}$

10. $\dfrac{\partial^2 u}{\partial x^2} + r = \dfrac{\partial u}{\partial t},\ 0 < x < 1,\ t > 0$

    $\left.\dfrac{\partial u}{\partial t}\right|_{x=0} = 0,\ u(1, t) = 0,\ t > 0$

    $u(x, 0) = 0,\ 0 < x < 1$

11. $\dfrac{\partial^2 u}{\partial x^2} + \dfrac{\partial^2 u}{\partial y^2} = 0,\ x > 0,\ 0 < y < \pi$

    $u(0, y) = A,\ 0 < y < \pi$

    $\left.\dfrac{\partial u}{\partial y}\right|_{y=0} = 0,\ \left.\dfrac{\partial u}{\partial y}\right|_{y=\pi} = Be^{-x},\ x > 0$

PART V

Numerical Analysis

15 NUMERICAL METHODS

15

NUMERICAL METHODS

15.1 Newton's Method
15.2 Approximate Integration
15.3 Direction Fields
15.4 The Euler Methods
15.5 The Three-Term Taylor Method
15.6 The Runge–Kutta Method
15.7 Multistep Methods, Errors
15.8 Higher-Order Equations and Systems
15.9 Second-Order Boundary-Value Problems
15.10 Numerical Methods for Partial Differential Equations: Elliptic Equations
15.11 Numerical Methods for Partial Differential Equations: Parabolic Equations
15.12 Numerical Methods for Partial Differential Equations: Hyperbolic Equations
15.13 Approximation of Eigenvalues
Summary
Chapter 15 Review Exercises

TOPICS TO REVIEW

roots of equations
rational roots of polynomial equations
definition of the definite integral
area under a curve
solution of a differential equation
initial-value problem
boundary-value problem
Laplace's equation
heat equation
wave equation
eigenvalues of a matrix

IMPORTANT CONCEPTS

Newton's method
midpoint rule
trapezoidal rule
Simpson's rule
error bounds
lineal elements
slope field
Euler's methods
step size
three-term Taylor method
first and second-order Runge–Kutta method
fourth-order Runge–Kutta method
predictor and corrector
single-step methods
multistep methods
Adams–Bashforth/Adams–Moulton method
difference equation
mesh size
Gauss–Seidel iteration
Crank–Nicholson method
dominant eigenvalue
power method
scaling
method of deflation

INTRODUCTION

A differential equation does not have to possess a solution, and even if a solution exists, we cannot always find explicit or implicit solutions of the equation. In many instances, particularly in the study of nonlinear differential equations (see Chapter 10), we have to be content with an *approximation* to a solution. In this chapter we shall study methods for approximating roots of equations, definite integrals, solutions of differential equations, and eigenvalues and eigenvectors of a matrix.

15.1 NEWTON'S METHOD

There are few straightforward methods for finding the roots of an equation $f(x) = 0$. For polynomial equations of degree 4 or less, we can always solve the equation by means of an algebraic formula that expresses the unknown x in terms of the coefficients of $f(x)$. We know, of course, that $ax^2 + bx + c = 0$, $a \neq 0$, can be solved by the quadratic formula. One of the major achievements in mathematics was the proof in the nineteenth century that polynomial equations of degree greater than 4 cannot be solved by means of an algebraic formula (that is, in terms of radicals).* Thus, solving the algebraic equation

$$x^5 - 3x^2 + 4x - 6 = 0$$

poses a quandary unless the polynomial factors. Furthermore, in scientific analyses, one is often asked to find roots of transcendental equations such as

$$2x = \tan x$$

For equations such as these[†], it is common practice to use a technique that yields an *approximation* or *estimation* of the roots. One such procedure, known as **Newton's method**,[†] employs the derivative of a function.

An Iterative Technique Suppose f is differentiable and suppose c represents the unknown root of $f(x) = 0$; that is, $f(c) = 0$. Let x_0 denote a number that is chosen arbitrarily as a first guess to c. If $f(x_0) \neq 0$, compute $f'(x_0)$ and, as shown in Figure 15.1, construct a tangent to the graph of f at $(x_0, f(x_0))$. If we now let x_1 denote the x-intercept of this line, we must have

$$\text{slope of line} = f'(x_0) = \frac{f(x_0)}{x_0 - x_1}$$

Solving for x_1 then gives

$$x_1 = x_0 - \frac{f(x_0)}{f'(x_0)}$$

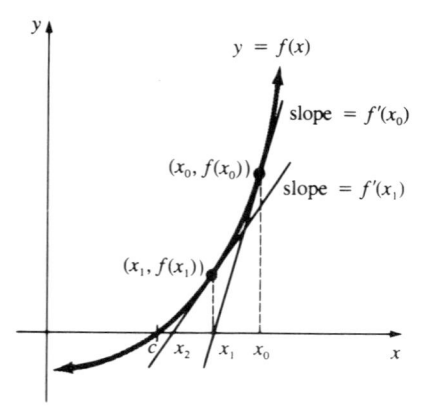

FIGURE 15.1

---

*Lodovico Farrari (1522–1565) In around 1540 this Italian mathematician discovered an algebraic formula for the roots of the general fourth-degree polynomial equation. For nearly the next 300 years mathematicians labored to find such formulas for polynomial equations of degree 5 or greater.

Neils Henrik Abel (1802–1829) In 1824 the Norwegian mathematician Abel was the first to prove that it is impossible to solve the general fifth-degree polynomial equation $ax^5 + bx^4 + cx^3 + dx^2 + ex + f = 0$ in terms of radicals.

---

*See Example 1 in Section 12.7.
[†]This is also called the **Newton–Raphson method**.

15.1 Newton's Method

Repeat the procedure at $(x_1, f(x_1))$ and let x_2 be the x-intercept of the second tangent line. From

$$f'(x_1) = \frac{f(x_1)}{x_1 - x_2} \quad \text{we find} \quad x_2 = x_1 - \frac{f(x_1)}{f'(x_1)}$$

Continuing in this fashion, we determine x_{n+1} from

$$\boxed{x_{n+1} = x_n - \frac{f(x_n)}{f'(x_n)}} \tag{1}$$

The repetitive use, or **iteration**, of (1) yields a sequence $x_1, x_2, x_3, \ldots$ of approximations that we expect *converges* to the root c; that is, $x_n \to c$ as n increases.

Graphical Analysis Before applying (1), let's try to determine the existence and number of real roots of $f(x) = 0$ through graphical means; for example, the irrational number $\sqrt{3}$ can be interpreted as either

(i) a root of the quadratic equation $x^2 - 3 = 0$ and, hence, a zero of the continuous function $f(x) = x^2 - 3$, or
(ii) the x-coordinate of a point of intersection of the graphs of $y = x^2$ and $y = 3$.

Both interpretations are illustrated in Figure 15.2. Of course, another reason for a graph is to enable us to choose the initial guess x_0 so that it is close to the root c.

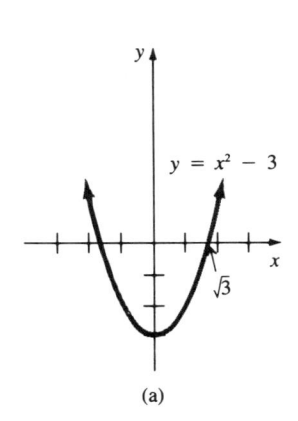

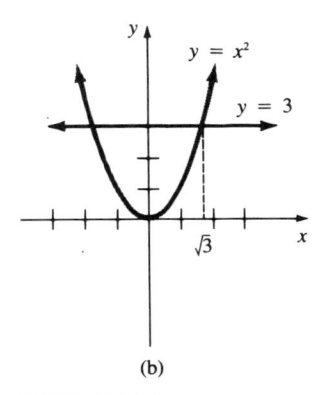

FIGURE 15.2

EXAMPLE 1 Determine the number of real roots of $x^3 - x + 1 = 0$.

Solution From Figure 15.3 we see that the graphs of the functions

$$y = x^3 \quad \text{and} \quad y = x - 1$$

intersect at one point. Hence, we conclude that the equation

$$x^3 = x - 1 \quad \text{or} \quad x^3 - x + 1 = 0$$

possesses only one real root. □

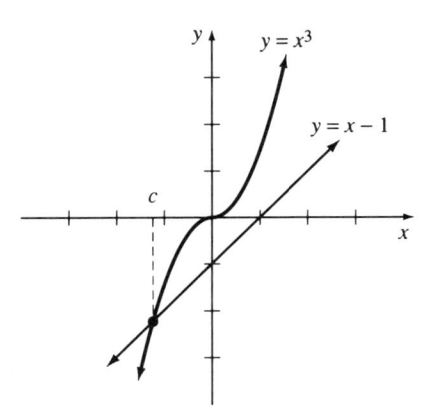

FIGURE 15.3

Although the actual calculation of the number $\sqrt{3}$ is trivial on a calculator, this calculation serves nicely as an introduction to the use of Newton's method.

EXAMPLE 2 Approximate $\sqrt{3}$ by Newton's method.

Solution If we define $f(x) = x^2 - 3$, then $f'(x) = 2x$ and (1) becomes

$$x_{n+1} = x_n - \frac{x_n^2 - 3}{2x_n} \quad \text{or} \quad x_{n+1} = \frac{1}{2}\left(x_n + \frac{3}{x_n}\right)$$

Since $1 < \sqrt{3} < 2$ it seems reasonable to choose $x_0 = 1$. Thus,

$$x_1 = \frac{1}{2}\left(x_0 + \frac{3}{x_0}\right) = \frac{1}{2}(1 + 3) = 2$$

$$x_2 = \frac{1}{2}\left(x_1 + \frac{3}{x_1}\right) = \frac{1}{2}\left(2 + \frac{3}{2}\right) = 1.75$$

$$x_3 = \frac{1}{2}\left(x_2 + \frac{3}{x_2}\right) = \frac{1}{2}\left(\frac{7}{4} + \frac{12}{7}\right) \approx 1.7321$$

$$x_4 = \frac{1}{2}\left(x_3 + \frac{3}{x_3}\right) \approx 1.7321$$

Since there is no significant difference between x_3 and x_4, it makes sense to stop the iteration. Indeed, $\sqrt{3} = 1.73205$ is accurate to five decimal places. □

EXAMPLE 3 Use Newton's method to find an approximation to the real root of the equation $x^3 - x + 1 = 0$.

Solution Let $f(x) = x^3 - x + 1$ so that $f'(x) = 3x^2 - 1$. Hence, (1) is

$$x_{n+1} = x_n - \frac{x_n^3 - x_n + 1}{3x_n^2 - 1} \quad \text{or} \quad x_{n+1} = \frac{2x_n^3 - 1}{3x_n^2 - 1}$$

If we are interested in three- and possibly four-decimal-place accuracy, we carry out the iteration until two successive iterants agree to four decimal places. Figure 15.3 prompts us to make $x_0 = -1.5$ the initial guess. Consequently,

$$x_1 = \frac{2x_0^3 - 1}{3x_0^2 - 1} = \frac{2(-1.5)^3 - 1}{3(-1.5)^2 - 1} \approx -1.3478$$

$$x_2 = \frac{2x_1^3 - 1}{3x_1^2 - 1} \approx -1.3252$$

$$x_3 = \frac{2x_2^3 - 1}{3x_2^2 - 1} \approx -1.3247$$

$$x_4 = \frac{2x_3^3 - 1}{3x_3^2 - 1} \approx -1.3247$$

Hence, the root of the given equation is approximately -1.3247. □

15.1 Newton's Method

A BASIC program for Newton's method is provided in Appendix IV.

Polynomial Equations In general an nth-degree polynomial equation with integer coefficients

$$a_n x^n + a_{n-1} x^{n-1} + \cdots + a_1 x + a_0 = 0 \tag{2}$$

has at most n real roots. An odd-degree polynomial equation, such as the equation in Example 3, always has at least one real root because complex roots must always appear in conjugate pairs $a + ib$ and $a - ib$, $i^2 = -1$, $b \neq 0$. Real roots can be rational or irrational numbers. Recall from algebra that if (2) has a rational root p/q (p and q integers, $q \neq 0$), then p is a factor of a_0 and q is a factor of the lead coefficient a_n.

The next example will require either a calculator with graphing capabilities or a computer with graphing software. In addition, we shall make use of a computer program to carry out the iteration of (1).

EXAMPLE 4 Use Newton's method to approximate the real roots of

$$3x^{11} + 5x^9 - 15x^5 - 8x^4 + 5x^2 - x + 6 = 0 \tag{3}$$

Solution Using a computer, we obtain the graph of the polynomial function $f(x) = 3x^{11} + 5x^9 - 15x^5 - 8x^4 + 5x^2 - x + 6$ shown in Figure 15.4. The figure suggests that (3) has three real roots in the neighborhood of the origin. Now, to check whether these roots are rational we list all the factors of a_0 and a_n, respectively:

$$p: \pm 1, \pm 2, \pm 3, \pm 6$$
$$q: \pm 1, \pm 3$$

Thus, the possible rational roots of (3) are

$$-1, 1, -2, 2, -3, 3, -6, 6, -\frac{1}{3}, \frac{1}{3}, -\frac{2}{3}, \frac{2}{3}$$

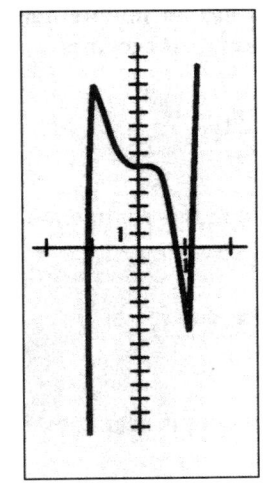

FIGURE 15.4

Since Figure 15.4 clearly indicates that the roots are in the intervals $(-2, -1)$, $(0, 1)$, and $(1, 2)$, the only candidates for rational roots in those intervals are $\frac{1}{3}$ and $\frac{2}{3}$. Although the graph indicates that these are unlikely roots, we can test the numbers either by direct substitution in (3), or, better, by synthetic division. In this manner we find that neither $\frac{1}{3}$ nor $\frac{2}{3}$ is a root. We conclude that the roots of (3) in the three intervals are irrational.

Now $f'(x) = 33x^{10} + 45x^8 - 75x^4 - 32x^3 + 10x - 1$, so that (1) becomes

$$x_{n+1} = x_n - \frac{3x_n^{11} + 5x_n^9 - 15x_n^5 - 8x_n^4 + 5x_n^2 - x_n + 6}{33x_n^{10} + 45x_n^8 - 75x_n^4 - 32x_n^3 + 10x_n - 1} \tag{4}$$

The last expression will not be simplified as in Example 3, in view of line 50 in the BASIC program found in Appendix IV.

We begin by approximating the root in the interval (0, 1). Figure 15.4 suggests that $x_0 = 0.9$ is a reasonable initial guess. Iteration of (4) then gives

$$x_1 \approx 0.8382729$$
$$x_2 \approx 0.8372948$$
$$x_3 \approx 0.8372939$$
$$x_4 \approx 0.8372939$$

We conclude that the root in (0, 1) is approximately 0.8372939. Using the initial values $x_0 = 1.1$ and $x_0 = -1.1$, we find in the same manner that approximations to the roots in (1, 2) and (-2, -1) are 1.160470 and -1.140433, respectively. □

EXAMPLE 5 Find the smallest positive root of $2x = \tan x$.

Solution Figure 15.5 shows that the equation has an infinite number of roots. With $f(x) = 2x - \tan x$ and $f'(x) = 2 - \sec^2 x$, (1) becomes

$$x_{n+1} = x_n - \frac{2x_n - \tan x_n}{2 - \sec^2 x_n}$$

Since calculators and computers do not possess a secant routine, we express the last equation in terms of $\sin x$ and $\cos x$:

$$x_{n+1} = x_n - \frac{2x_n \cos^2 x_n - \sin x_n \cos x_n}{2 \cos^2 x_n - 1} \tag{5}$$

It appears from Figure 15.5 that the first positive root is near $x_0 = 1$. Using the program, iteration of (5) yields

$$x_1 \approx 1.310478$$
$$x_2 \approx 1.223929$$
$$x_3 \approx 1.176051$$
$$x_4 \approx 1.165927$$
$$x_5 \approx 1.165562$$
$$x_6 \approx 1.165561$$
$$x_7 \approx 1.165561$$

We conclude that the first positive root is approximately 1.165561. □

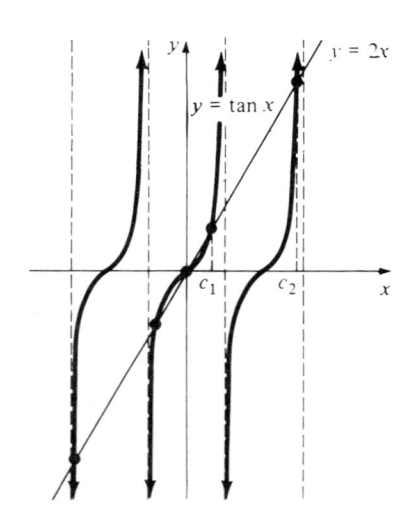

FIGURE 15.5

Example 5 illustrates the importance of the selection of the initial value x_0. The reader should verify that the choice $x_0 = \frac{1}{2}$ in (5) leads to a sequence of values $x_1, x_2, x_3, \ldots$ that converge to the one obvious root $c = 0$.

▲ **Remark** There are problems with Newton's method.

(*i*) We must compute $f'(x)$. Needless to say, the form of $f'(x)$ could be formidable when the equation $f(x) = 0$ is complicated.

(*ii*) If the root c of $f(x) = 0$ is near a value for which $f'(x) = 0$, then the denominator in (1) is approaching zero. This necessitates a computation of $f(x_n)$ and $f'(x_n)$ to a high degree of accuracy. A calculation of this kind usually requires a computer with a double precision routine.

(*iii*) It is necessary to find an approximate location of a root of $f(x) = 0$ before x_0 is chosen. Attendant to this are the usual difficulties in graphing. But, worse, the iteration of (1) *may not converge* for an imprudent or perhaps blindly chosen x_0. In Figure 15.6(a) we see that x_2 is undefined because $f'(x_1) = 0$. In Figure 15.6(b) we see what may happen to the tangent lines when x_0 is not close to c. In Figure 15.6(c) observe that when $f(x_0) = -f(x_1)$ and $f'(x_0) = f'(x_1)$, the tangent lines "bounce" back and forth between two points $(x_0, f(x_0))$ and $(x_1, f(x_1))$. (See Problems 27 and 28.)

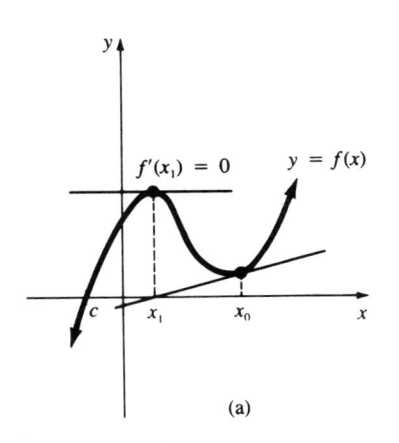

FIGURE 15.6

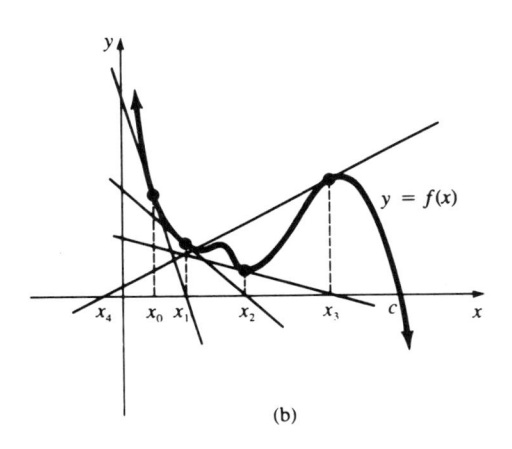

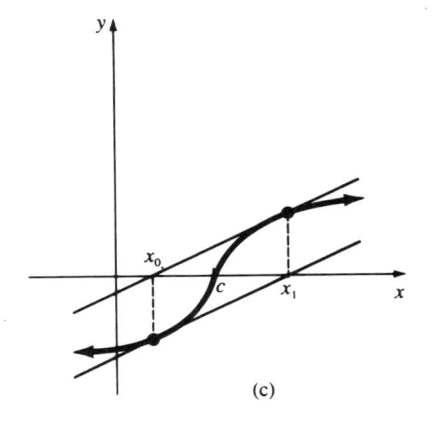

These three problems notwithstanding, the major advantage of Newton's method is that when it converges to a root, it usually does so rather rapidly. It can be shown that under certain conditions Newton's method converges *quadratically*. Very roughly, this means that the number of places of accuracy can, but will not necessarily, double with each iteration.

▶ **EXERCISES 15.1** *Answers to odd-numbered problems begin on page A-68.*

Where appropriate carry out the iteration of (1) until the successive iterants agree to four decimal places.

In Problems 1–4 determine graphically whether the given equation possesses any real roots.

1. $x^3 = -2 + \sin x$
2. $x^3 - 3x = x^2 - 1$
3. $x^4 + x^2 - 2x + 3 = 0$
4. $\tan x = \cos x$

In Problems 5–8 use Newton's method to find an approximation for the given number.

5. $\sqrt{10}$ **6.** $1 + \sqrt{5}$
7. $\sqrt[3]{4}$ **8.** $\sqrt[5]{2}$

In Problems 9–14 use Newton's method, if necessary, to find approximations to all real roots of the given equations.

9. $x^3 = -x + 1$ **10.** $x^3 - x^2 + 1 = 0$
11. $x^4 + x^2 - 3 = 0$ **12.** $x^4 = 2x + 1$
13. $x^2 = \sin x$ **14.** $x + \cos x = 0$

15. Find the smallest positive x-intercept of the graph of $f(x) = 3\cos x + 4\sin x$.

16. Consider the function $f(x) = x^5 + x^2$. Use Newton's method to approximate the smallest positive number for which $f(x) = 4$.

17. A cantilever beam 20 ft long with a load of 600 lb at its end is deflected by an amount $d = (60x^2 - x^3)/16{,}000$, where d is measured in inches and x in feet. See Figure 15.7. Use Newton's method to approximate the value of x that corresponds to a deflection of 0.01 in.

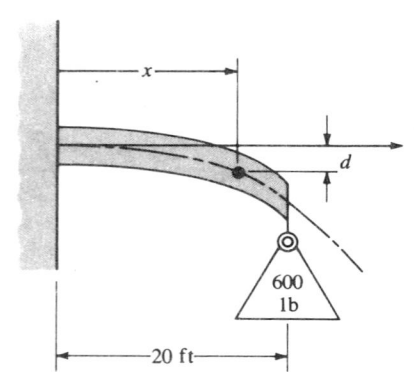

FIGURE 15.7

18. A vertical solid cylindrical column of fixed radius r that supports its own weight will eventually buckle when its height is increased. It can be proved that the maximum, or critical, height of such a column is $h_{cr} = kr^{2/3}$, where k is a constant and r is measured in meters. Use Newton's method to approximate the diameter of a column for which $h_{cr} = 10$ m and $k = 35$.

19. A beam of light originating at point P in medium A, whose index of refraction is n_1, strikes the surface of medium B, whose index of refraction is n_2. We can prove from Snell's law that the beam is refracted tangent to the surface for the critical angle determined from $\sin \theta_c = n_2/n_1$, $0 < \theta_c < 90°$. For angles of incidence greater than

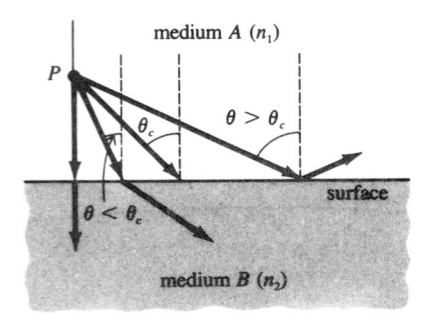

FIGURE 15.8

the critical angle, all light is reflected internally to medium A. See Figure 15.8. If $n_2 = 1$ for air and $n_1 = 1.5$ for glass, use Newton's method to approximate θ_c in radians.

20. For a suspension bridge, the length s of a cable between two vertical supports whose span is l (horizontal distance) is related to the sag d of the cable by

$$s = l + \frac{8d^2}{3l} - \frac{32d^4}{5l^3}$$

See Figure 15.9. If $s = 404$ ft and $l = 400$ ft, use Newton's method to approximate the sag. Round the answer to one decimal place.* [*Hint:* The root c satisfies $20 < c < 30$.]

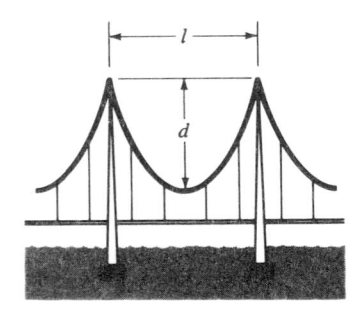

FIGURE 15.9

21. A rectangular block of steel is hollowed out, making a tub with a uniform thickness t. The dimensions of the tub are shown in Figure 15.10(a). For an object to float in water, as shown in Figure 15.10(b), the weight of the water displaced must equal the weight of the tub (Archimedes' principle). If the weight density of water is 62.4 lb/ft^3 and the weight density of the steel is 490 lb/ft^3,

---

*The formula for s is itself only an approximation.

then

$$\text{weight of water displaced} = 62.4 \times \begin{pmatrix} \text{volume of water} \\ \text{displaced} \end{pmatrix}$$

weight of tub = 490 × (volume of steel in tub)

(a) Show that t satisfies the equation

$$t^3 - 7t^2 + \frac{61}{4}t - \frac{1638}{1225} = 0$$

(b) Use Newton's method to approximate the largest positive root of the equation in part (a).

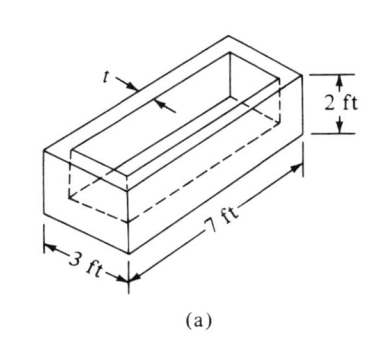

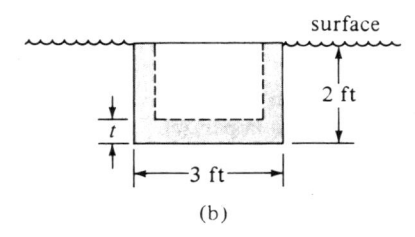

FIGURE 15.10

22. A flexible strip of metal 10 ft long is bent into the shape of a circular arc by the ends being secured together by means of a cable that is 8 ft long. See Figure 15.11. Use Newton's method to approximate the radius r of the circular arc.

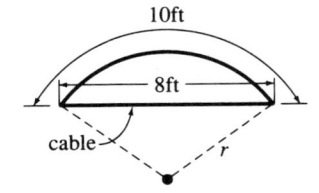

FIGURE 15.11

23. Two ends of a railroad track L feet long are pushed l feet closer together so that the track bows upward in

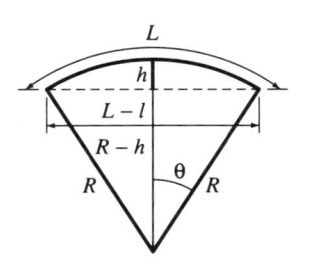

FIGURE 15.12

the arc of a circle of radius R. See Figure 15.12. What is the height h above ground of the highest point on the track?

(a) Use Figure 15.12 to show that

$$h = \frac{L(1 - l/L)^2 \theta}{2(1 + \sqrt{1 - (1 - l/L)^2 \theta^2})}$$

where $\theta > 0$ satisfies $\sin \theta = (1 - l/L)\theta$. [Hint: In a circular sector, how are the arc length, the radius, and the central angle related?]

(b) If $L = 5280$ ft and $l = 1$ ft, use Newton's method to approximate θ and then solve for the corresponding value of h.

(c) If l/L and θ are very small, then $h \approx L\theta/4$ and $\sin \theta \approx \theta - \theta^3/6$. Use these two approximations to show that $h \approx \sqrt{3lL/8}$. Use this formula with $L = 5280$ ft and $l = 1$ ft, and compare with the result in part (b).

24. At a foundry a metal sphere of radius 2 ft is recast in the form of a rod that is a right circular cylinder 15 ft long surmounted by a hemisphere at one end. The radius r of the hemisphere is the same as the base radius of the cylinder. Use Newton's method to approximate r.

25. A round but unbalanced wheel of mass M and radius r is connected by a rope and frictionless pulleys to a mass m as shown in Figure 15.13. O is the center of the wheel and P is its center of mass. If the wheel is released from rest, it can be shown that the angle θ at which the wheel

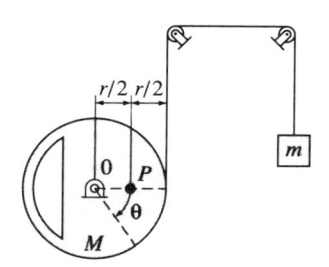

FIGURE 15.13

first stops satisfies the equation

$$Mg\frac{r}{2}\sin\theta - mgr\theta = 0$$

where g is the acceleration of gravity. Use Newton's method to approximate θ if the mass of the wheel is 4 times the mass m.

26. Two ladders of lengths $L_1 = 40$ ft and $L_2 = 30$ ft are placed against two vertical walls as shown in Figure 15.14. The height of the point where the ladders cross is $h = 10$ ft.

    (a) Show that the indicated height x in the figure can be determined from the equation

    $$x^4 - 2hx^3 + (L_1^2 - L_2^2)x^2 - 2h(L_1^2 - L_2^2)x + h^2(L_1^2 - L_2^2) = 0$$

    (b) Use Newton's method to approximate the solution of the equation in part (a). Why does it make sense to choose $x_0 \geq 10$?
    (c) Approximate the distance z between the two walls.

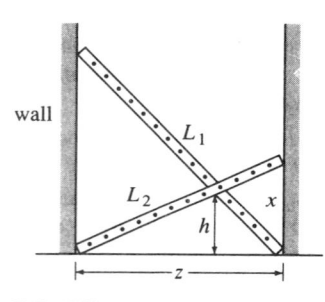

FIGURE 15.14

27. Let f be a differentiable function. Show that if $f(x_0) = -f(x_1)$ and $f'(x_0) = f'(x_1)$, then (1) implies $x_2 = x_0$.

28. Given

$$f(x) = \begin{cases} -\sqrt{4-x}, & x < 4 \\ \sqrt{x-4}, & x \geq 4 \end{cases}$$

observe that $f(4) = 0$. Show that for any choice of x_0, Newton's method will fail to converge to the root. [*Hint:* See Problem 27.]

In Problems 29 and 30 use a graphics calculator or computer to obtain the graph of the given function. Use Newton's method to approximate the roots of $f(x) = 0$ that are discovered from the graph.

29. $f(x) = 2x^5 + 3x^4 - 7x^3 + 2x^2 + 8x - 8$
30. $f(x) = 4x^{12} + x^{11} - 4x^8 + 3x^3 - 2x^2 + x - 10$
31. (a) Use a graphics calculator or computer to obtain the graphs of $f(x) = 0.5x^3 - x$ and $g(x) = \cos x$ on the same coordinate axes.
    (b) Use a graphics calculator or computer to obtain the graph of $y = f(x) - g(x)$, where f and g are given in part (a).
    (c) Use the graphs in part (a) or the graph in part (b) to determine the number of roots of the equation $0.5x^3 - x = \cos x$.
    (d) Use Newton's method to approximate the roots of the equation in part (c).

15.2 APPROXIMATE INTEGRATION

Recall from calculus that we may not be able to express the antiderivative of a continuous function in terms of elementary functions such as polynomial, rational, or trigonometric functions. Hence, the fundamental theorem of calculus, $\int_a^b f(x)\,dx = F(b) - F(a)$, where F is any antiderivative of f, cannot be used to evaluate every definite integral. Sometimes the best we can do is to *approximate* the value of $\int_a^b f(x)\,dx$. In this section we shall consider three numerical procedures for approximating definite integrals.

Throughout the discussion that follows it is useful to interpret the definite integral $\int_a^b f(x)\,dx$ as the area under the graph of the function f on the interval $[a, b]$. Although continuity of f is essential, there is no actual requirement that $f(x) \geq 0$ on the interval.

Midpoint Rule One way of approximating a definite integral is suggested in Figure 15.15(a)—namely, constructing rectangular elements under

15.2 Approximate Integration

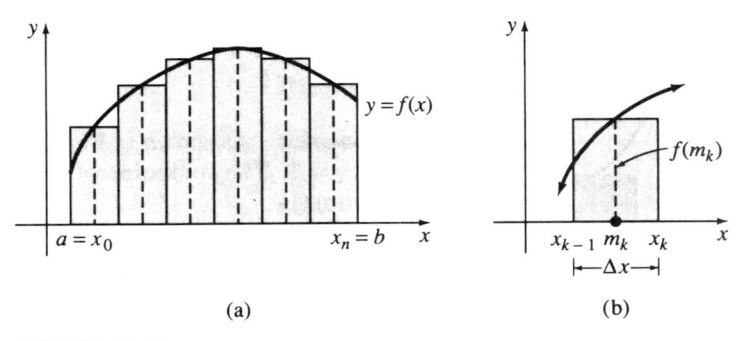

FIGURE 15.15

(a) (b)

the graph of f and adding their areas. In particular, let us suppose that $y = f(x)$ is continuous on $[a, b]$ and that this interval is partitioned into n subintervals of equal length $\Delta x = (b - a)/n$. (This is called a regular partition.) A simple but fairly accurate approximation rule consists of adding the areas of n rectangular elements whose lengths are calculated at the **midpoint** of each subinterval. Let m_k denote the midpoint of $[x_{k-1}, x_k]$ as in Figure 15.15(b). The dashed lines in Figure 15.15(a) represent the lengths of the elements and are $f(m_k)$, $k = 1, 2, \ldots, n$. Now since $m_k = (x_{k-1} + x_k)/2$, the area of the rectangular element shown in Figure 15.15(b) is

$$A_k = f(m_k) \Delta x_k = f\left(\frac{x_{k-1} + x_k}{2}\right) \Delta x_k$$

By identifying $a = x_0$ and $b = x_n$ and summing the n areas, we get

$$\int_a^b f(x)\, dx \approx f\left(\frac{x_0 + x_1}{2}\right) \Delta x + f\left(\frac{x_1 + x_2}{2}\right) \Delta x + \cdots + f\left(\frac{x_{n-1} + x_n}{2}\right) \Delta x$$

If we replace Δx by $(b - a)/n$, this midpoint approximation rule can be summarized as follows:

DEFINITION 15.1 Midpoint Rule

The **midpoint rule** is the approximation $\int_a^b f(x)\, dx \approx M_n$, where

$$M_n = \frac{b - a}{n}\left[f\left(\frac{x_0 + x_1}{2}\right) + f\left(\frac{x_1 + x_2}{2}\right) + \cdots + f\left(\frac{x_{n-1} + x_n}{2}\right) \right] \quad (1)$$

Since the function $f(x) = 1/x$ is continuous on any interval $[a, b]$ that does not include the origin, we know that $\int_a^b dx/x$ exists.

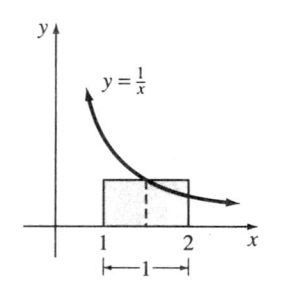

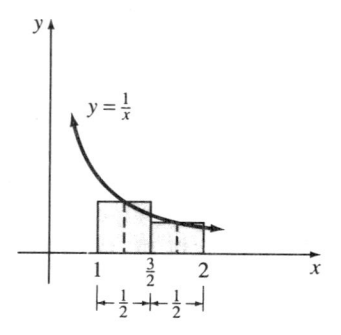

EXAMPLE 1 Approximate $\int_1^2 dx/x$ by the midpoint rule for $n = 1$, $n = 2$, and $n = 5$.

Solution As shown in Figure 15.16, the case $n = 1$ is one rectangle in which $\Delta x = 1$. The midpoint of the interval is $m_1 = \frac{3}{2}$ and $f(\frac{3}{2}) = \frac{2}{3}$. Therefore, from (1),

$$M_1 = 1 \cdot \frac{2}{3} \approx 0.6666$$

When $n = 2$, Figure 15.17 shows that $\Delta x = \frac{1}{2}$, $x_0 = 1$, $x_1 = 1 + \Delta x = \frac{3}{2}$, and $x_2 = 1 + 2\Delta x = 2$. The midpoints of $[1, \frac{3}{2}]$ and $[\frac{3}{2}, 2]$ are, respectively, $m_1 = \frac{5}{4}$ and $m_2 = \frac{7}{4}$, and so $f(\frac{5}{4}) = \frac{4}{5}$ and $f(\frac{7}{4}) = \frac{4}{7}$. Hence (1) gives

$$M_2 = \frac{1}{2}\left[\frac{4}{5} + \frac{4}{7}\right] \approx 0.6857$$

Finally, for $n = 5$, $\Delta x = \frac{1}{5}$, $x_0 = 1$, $x_1 = 1 + \Delta x = \frac{6}{5}$, $x_2 = 1 + 2\Delta x = \frac{7}{5}$, ..., $x_5 = 1 + 5\Delta x = 2$. The midpoints of the five subintervals $[1, \frac{6}{5}]$, $[\frac{6}{5}, \frac{7}{5}]$, $[\frac{7}{5}, \frac{8}{5}]$, $[\frac{8}{5}, \frac{9}{5}]$, and $[\frac{9}{5}, 2]$ and the corresponding functional values are shown

k	1	2	3	4	5
m_k	$\frac{11}{10}$	$\frac{13}{10}$	$\frac{15}{10}$	$\frac{17}{10}$	$\frac{19}{10}$
$f(m_k)$	$\frac{10}{11}$	$\frac{10}{13}$	$\frac{10}{15}$	$\frac{10}{17}$	$\frac{10}{19}$

in the accompanying table. Using the information in the table then gives

$$M_5 = \frac{1}{5}\left[\frac{10}{11} + \frac{10}{13} + \frac{10}{15} + \frac{10}{17} + \frac{10}{19}\right] \approx 0.6919$$

In other words, $\int_1^2 dx/x \approx M_5$ or $\int_1^2 dx/x \approx 0.6919$. □

Error in the Midpoint Rule Suppose $I = \int_a^b f(x)\,dx$ and M_n is an approximation to I using n rectangles. We define the error in the method to be

$$E_n = |I - M_n|$$

An upper bound for the error can be obtained by means of the next result. The proof is omitted.

THEOREM 15.1 Error Bound for Midpoint Rule

If there exists a number $M > 0$ such that $|f''(x)| \leq M$ for all x in $[a, b]$, then

$$E_n \leq \frac{M(b-a)^3}{24n^2} \quad (2)$$

15.2 Approximate Integration

Observe that this upper bound for the error E_n is inversely proportional to n^2. Hence, the accuracy in the method improves as we take more and more rectangles. For example, if the number of rectangles is doubled, the error E_{2n} is less than $\frac{1}{4}$ the error bound for E_n. Thus we see that $\lim_{n \to \infty} M_n = I$.

The next example will illustrate how the error bound (2) can be utilized to determine the number of rectangles that will yield a prescribed accuracy.

EXAMPLE 2 Determine a value of n so that (1) will give an approximation to $\int_1^2 dx/x$ that is accurate to two decimal places.

Solution The midpoint rule will be accurate to two decimal places for those values of n for which the upper bound $M(b-a)^3/24n^2$ for the error is strictly less than 0.005.*

For $f(x) = 1/x$ we have $f'''(x) = 2/x^3$. Since f'' decreases on $[1, 2]$, it follows that $f''(x) \le f''(1) = 2$ for all x in the interval. Thus, with $M = 2$, $b - a = 1$, we want

$$\frac{2(1)^3}{24n^2} < 0.005 \quad \text{or} \quad n^2 > \frac{50}{3} \approx 16.67$$

By taking $n \ge 5$ we obtain the desired accuracy. □

Example 2 indicates that the third approximation $\int_1^2 dx/x \approx 0.6919$ obtained in Example 1 is accurate to two decimal places. By way of comparison, it is known that the estimate $\int_1^2 dx/x \approx 0.6931$ is correct to four decimal places. Thus, for $n = 5$ the error in the method E_5 is approximately 0.0012.

Trapezoidal Rule A more popular method for approximating an integral is based on the plausibility that a better estimate of $\int_a^b f(x)\,dx$ can be obtained by adding the areas of trapezoids instead of the areas of rectangles. See Figure 15.18(a). The area of the trapezoid shown in Figure 15.18(b) is $h(l_1 + l_2)/2$. Thus, the area A_k of the trapezoidal element shown in Figure 15.18(c) is

$$A_k = \Delta x \, \frac{f(x_{k-1}) + f(x_k)}{2}$$

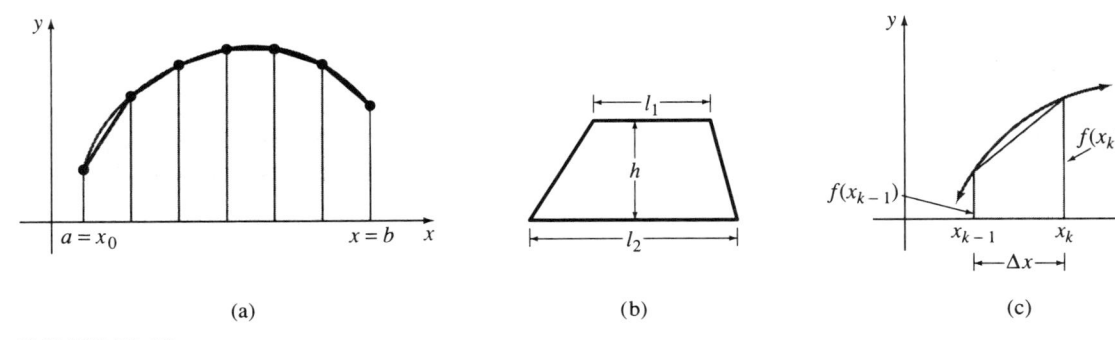

FIGURE 15.18

---

*If we want accuracy to three decimal places, we use 0.0005, and so on.

Thus, for a regular partition of the interval $[a, b]$ on which f is continuous we obtain

$$\int_a^b f(x)\,dx \approx \Delta x \frac{f(x_0) + f(x_1)}{2} + \Delta x \frac{f(x_1) + f(x_2)}{2} + \cdots + \Delta x \frac{f(x_{n-1}) + f(x_n)}{2}$$

We will summarize this new approximation rule in the next definition after we combine like terms and substitute $\Delta x = (b - a)/n$.

DEFINITION 15.2 Trapezoidal Rule

The **trapezoidal rule** is the approximation $\int_a^b f(x)\,dx \approx T_n$, where

$$T_n = \frac{b - a}{2n}[f(x_0) + 2f(x_1) + 2f(x_2) + \cdots + 2f(x_{n-1}) + f(x_n)] \quad (3)$$

Error in the Trapezoidal Rule The error in the method for the trapezoidal rule is given by $E_n = |I - T_n|$, where $I = \int_a^b f(x)\,dx$. As the next theorem will show, the error bound for the trapezoidal rule is almost the same as that for the midpoint rule.

THEOREM 15.2 Error Bound for Trapezoidal Rule

If there exists a number $M > 0$ such that $|f''(x)| \leq M$ for all x in $[a, b]$, then

$$E_n \leq \frac{M(b - a)^3}{12n^2} \quad (4)$$

EXAMPLE 3 Determine a value of n so that the trapezoidal rule will give an approximation to $\int_1^2 dx/x$ that is accurate to two decimal places. Approximate the integral.

Solution Using the information in Example 2, we have immediately that

$$\frac{2(1)^3}{12n^2} < 0.005 \quad \text{or} \quad n^2 > \frac{100}{3} \approx 33.33$$

15.2 Approximate Integration

k	0	1	2	3	4	5	6
x_k	1	$\frac{7}{6}$	$\frac{4}{3}$	$\frac{3}{2}$	$\frac{5}{3}$	$\frac{11}{6}$	2
$f(x_k)$	1	$\frac{6}{7}$	$\frac{3}{4}$	$\frac{2}{3}$	$\frac{3}{5}$	$\frac{6}{11}$	$\frac{1}{2}$

In this case we must take $n \geq 6$ to obtain the desired accuracy. Hence $\Delta x = \frac{1}{6}$, $x_0 = 1$, $x_1 = 1 + \Delta x = \frac{7}{6}$, and so on. From the information in the accompanying table, (3) gives

$$T_6 = \frac{1}{12}\left[1 + 2\left(\frac{6}{7}\right) + 2\left(\frac{3}{4}\right) + 2\left(\frac{2}{3}\right) + 2\left(\frac{3}{5}\right) + 2\left(\frac{6}{11}\right) + \frac{1}{2}\right] \approx 0.6949 \quad \square$$

EXAMPLE 4 Approximate $\int_{1/2}^{1} \cos \sqrt{x}\, dx$ by the trapezoidal rule so that the error is less than 0.001.

Solution The second derivative of $f(x) = \cos \sqrt{x}$ is

$$f''(x) = \frac{1}{4x}\left(\frac{\sin \sqrt{x}}{\sqrt{x}} - \cos \sqrt{x}\right)$$

For x in the interval $[\frac{1}{2}, 1]$ we have $0 < (\sin \sqrt{x})/\sqrt{x} \leq 1$ and $0 < \cos \sqrt{x} \leq 1$ and consequently $|f''(x)| \leq \frac{1}{4x}$. Therefore, on the interval, $|f''(x)| \leq \frac{1}{2}$. Thus, with $M = \frac{1}{2}$ and $b - a = \frac{1}{2}$, it follows from (4) that we want

$$\frac{\frac{1}{2}(\frac{1}{2})^3}{12n^2} < 0.001 \quad \text{or} \quad n^2 > \frac{125}{24} \approx 5.21$$

k	0	1	2	3
x_k	$\frac{1}{2}$	$\frac{2}{3}$	$\frac{5}{6}$	1
$f(x_k)$	0.7602	0.6848	0.6115	0.5403

Hence, to obtain the desired accuracy it suffices to choose $n = 3$ and $\Delta x = \frac{1}{6}$. With the aid of a calculator to obtain the information in the accompanying table, we find the following approximation for $\int_{1/2}^{1} \cos \sqrt{x}\, dx$ from (3):

$$T_3 = \frac{1}{12}\left[\cos \sqrt{\frac{1}{2}} + 2 \cos \sqrt{\frac{2}{3}} + 2 \cos \sqrt{\frac{5}{6}} + \cos 1\right] \approx 0.3244 \quad \square$$

Although not obvious from a figure, an improved method of approximating a definite integral $\int_a^b f(x)\, dx$ can be obtained by considering a series of parabolic arcs instead of a series of chords used in the trapezoidal rule. It can be proved that a parabolic arc that passes through *three* specified points will "fit" the graph of f better than a single straight line. See Figure 15.19. By adding the areas under the parabolic arcs, we obtain an approximation to the integral.

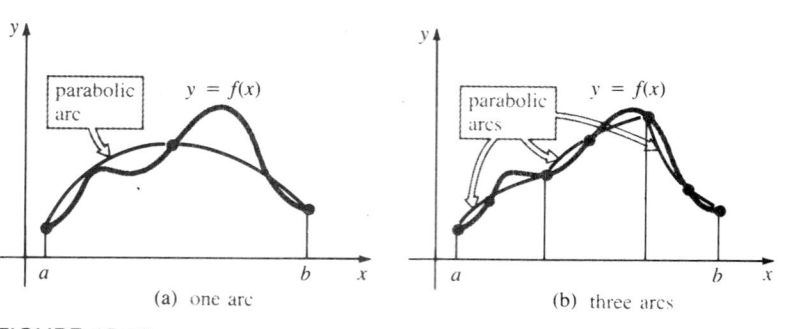

FIGURE 15.19

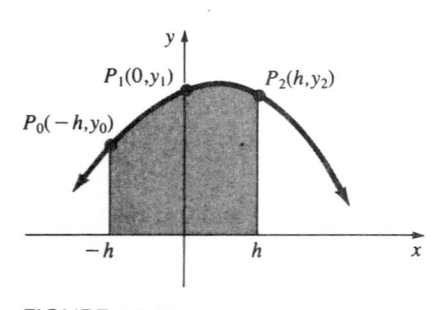

FIGURE 15.20

To begin, let us find the area under an arc of a parabola that passes through three points $P_0(x_0, y_0)$, $P_1(x_1, y_1)$, and $P_2(x_2, y_2)$, where $x_0 < x_1 < x_2$ and $x_1 - x_0 = x_2 - x_1 = h$. As shown in Figure 15.20, this can be done by finding the area under the graph of $y = Ax^2 + Bx + C$ on the interval $[-h, h]$ so that P_0, P_1, and P_2 have coordinates $(-h, y_0)$, $(0, y_1)$, and (h, y_2), respectively. The interval $[-h, h]$ is chosen for simplicity; the area in question does not depend on the location of the y-axis. The area is

$$\int_{-h}^{h} (Ax^2 + Bx + C)\, dx = \frac{h}{3}(2Ah^2 + 6C) \tag{5}$$

But, since the graph is to pass through $(-h, y_0)$, $(0, y_1)$, and (h, y_2), we must have

$$y_0 = Ah^2 - Bh + C \tag{6}$$
$$y_1 = C \tag{7}$$
$$y_2 = Ah^2 + Bh + C \tag{8}$$

By adding (6) and (8) and using (7), we find $2Ah^2 = y_0 + y_2 - 2y_1$. Thus, (5) can be expressed as

$$\text{area} = \frac{h}{3}(y_0 + 4y_1 + y_2) \tag{9}$$

Simpson's Rule Now suppose that the interval $[a, b]$ is partitioned into n subintervals of equal with $\Delta x = (b - a)/n$, where *n is an even integer*. As shown in Figure 15.21, on each subinterval $[x_{k-2}, x_k]$ of width $2\Delta x$ we approximate the graph of f by an arc of a parabola through points P_{k-2}, P_{k-1}, and P_k on the graph that corresponds to the endpoints and midpoint

15.2 Approximate Integration

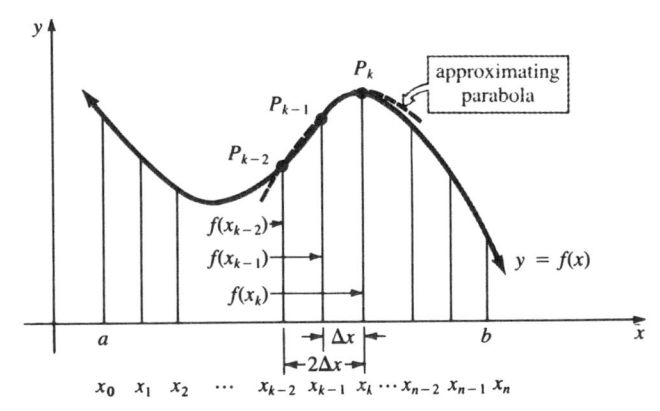

FIGURE 15.21

of the subinterval. If A_k denotes the area under the parabola on $[x_{k-2}, x_k]$, it follows from (9) that

$$A_k = \frac{\Delta x}{3}[f(x_{k-2}) + 4f(x_{k-1}) + f(x_k)]$$

Thus, summing all the A_k values gives

$$\int_a^b f(x)\,dx \approx \frac{\Delta x}{3}[f(x_0) + 4f(x_1) + f(x_2)]$$
$$+ \frac{\Delta x}{3}[f(x_2) + 4f(x_3) + f(x_4)] + \cdots + \frac{\Delta x}{3}[f(x_{n-2}) + 4f(x_{n-1}) + f(x_n)]$$

This approximation rule, named after the English mathematician Thomas Simpson (1710–1761), is summarized in the last definition.

DEFINITION 15.3 Simpson's Rule

Simpson's rule is the approximation $\int_a^b f(x)\,dx \approx S_n$, where

$$S_n = \frac{b-a}{3n}[f(x_0) + 4f(x_1) + 2f(x_2) + 4f(x_3) + \cdots + 2f(x_{n-2}) + 4f(x_{n-1}) + f(x_n)] \quad (10)$$

We note again that the integer n in (10) must be even, since each A_k represents the area under a parabolic arc on a subinterval of width $2\Delta x$.

Error in Simpson's Rule If $I = \int_a^b f(x)\,dx$, then the next theorem establishes an upper bound for the error in the method $E_n = |I - S_n|$ using an upper bound on the fourth derivative.

> **THEOREM 15.3 Error Bound for Simpson's Rule**
>
> If there exists a number $M > 0$ such that $|f^{(4)}(x)| \leq M$ for all x in $[a, b]$, then
>
> $$E_n \leq \frac{M(b-a)^5}{180n^4} \tag{11}$$

EXAMPLE 5 Determine a value of n so that (10) will give an approximation to $\int_1^2 dx/x$ that is accurate to two decimal places.

Solution For $f(x) = 1/x$, $f^{(4)}(x) = 24/x^5$ and $[1, 2]$, $f^{(4)}(x) \leq f^{(4)}(1) = 24$. Thus, with $M = 24$ it follows from (11) that

$$\frac{24(1)^6}{180n^4} < 0.005 \quad \text{or} \quad n^4 > \frac{80}{3} \approx 26.67$$

and so $n > 2.27$. Since n must be an even integer, it suffices to take $n \geq 4$. □

EXAMPLE 6 Approximate $\int_1^2 dx/x$ by Simpson's rule for $n = 4$.

Solution When $n = 4$ we have $\Delta x = \frac{1}{4}$. From (10) and the accompanying

k	0	1	2	3	4
x_k	1	$\frac{5}{4}$	$\frac{3}{2}$	$\frac{7}{4}$	2
$f(x_k)$	1	$\frac{4}{5}$	$\frac{2}{3}$	$\frac{4}{7}$	$\frac{1}{2}$

table we obtain

$$S_4 = \frac{1}{12}\left[1 + 4\left(\frac{4}{5}\right) + 2\left(\frac{2}{3}\right) + 4\left(\frac{4}{7}\right) + \frac{1}{2}\right] \approx 0.6933 \quad \square$$

In Example 6, keep in mind that even though we are using $n = 4$, the definite integral $\int_1^2 dx/x$ is being approximated by the area under only two parabolic arcs. Recall that the midpoint rule gave $\int_1^2 dx/x \approx 0.6919$ with $n = 5$, the trapezoidal rule gave $\int_1^2 dx/x \approx 0.6949$ with $n = 6$, and 0.6931 is an estimation of the integral correct to four decimal places.

In some applications it may be possible to obtain only numerical values of a quantity $Q(x)$—say, by measurements or by experiment—at specific points in some interval $[a, b]$, and yet it may be necessary to have some idea

15.2 Approximate Integration

of the value of the definite integral $\int_a^b Q(x)\,dx$. Even though Q is not defined by means of an explicit formula, we may still be able to apply the trapezoidal rule or Simpson's rule to approximate the integral.

EXAMPLE 7 Suppose we wish to find the area of an irregularly shaped piece of land that is bounded by a straight road and the shore of a lake. The side boundaries of the land are indicated by the dashed lines in Figure 15.22(a). Suppose we divide the indicated 1-mi boundary along the road into, say, $n = 8$ subintervals and then, as shown in Figure 15.22(b), measure the

(a)   (b)

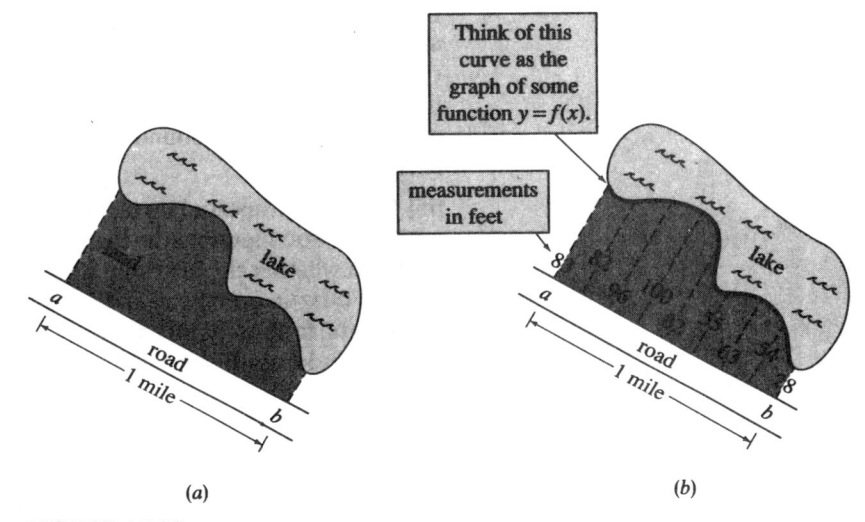

FIGURE 15.22

perpendicular distances from the road to the shore of the lake. We are now in a position to approximate the area of the land $A = \int_a^b f(x)\,dx$ by Simpson's rule. With $b - a = 1$ mi $= 5280$ ft, $\Delta x = 5280/8 = 660$ and the identifications $f(x_0) = 83, \ldots, f(x_8) = 28$, (10) gives the following approximation for A

$$S_8 = \frac{660}{3}[83 + 4(82) + 2(96) + 4(100) + 2(82) + 4(55) + 2(63) + 4(54) + 28] = 386{,}540 \text{ ft}^2$$

Using the fact that 1 acre $= 43{,}560$ ft^2, we see that the land area is approximately 8.9 acres. □

Listings of BASIC programs for the trapezoidal rule and Simpson's rule are found in Appendix IV. It is left as an exercise to write a program for the midpoint rule.

▲ **Remarks** (i) The popularity of the trapezoidal rule notwithstanding, a direct comparison of the error bounds (2) and (4) shows that the midpoint rule is actually more accurate than the trapezoidal rule. Specifically, (2) suggests that in some cases the error in the midpoint rule can be one-half the error in the trapezoidal rule. See Problem 33.

(ii) Under some circumstances the rules considered in the foregoing discussion will give the exact value of an integral $\int_a^b f(x)\,dx$. The error bounds (2) and (4) indicate that M_n and T_n will yield the precise value whenever f is a linear function. See Problems 31 and 32. Simpson's rule will give the exact value of $\int_a^b f(x)\,dx$ whenever f is a linear, quadratic, or a cubic polynomial function. See Problems 34 and 42.

(iii) In general, Simpson's rule will give greater accuracy than either the midpoint or the trapezoidal rule. So why should we even bother with these other rules? In some instances the slightly simpler midpoint and trapezoidal rules will yield accuracy that is sufficient for the purpose at hand. Furthermore, the requirement that n must be an even integer in Simpson's rule may prevent its application to a given problem. Also, to find an error bound for Simpson's rule we must compute and then find an upper bound for the fourth derivative. The expression for $f^{(4)}(x)$ can, of course, be very complicated or may not exist. The error bounds for the other two rules depend on the second derivative.

(iv) Since the trapezoidal rule and Simpson's rule are based on fitting linear and quadratic functions to the graph of a given function f, the reader may question whether the next step is to try to fit the graph of f with arcs of cubic or even quartic functions. Indeed, that can be done. A cubic approximation would utilize four points over three intervals of equal width Δx. A quartic arc would use five points, and so on. This method generates a sequence of approximation formulas known as **Newton–Cotes formulas**. Because of increasing complexity and other inherent problems, Newton–Cotes formulas of order higher than 2 are seldom used.

EXERCISES 15.2   Answers to odd-numbered problems begin on page A-68.

In Problems 1 and 2 compare the exact value of the integral with the approximation obtained from the midpoint rule for the indicated value of n.

1. $\int_1^4 (3x^2 + 2x)\,dx;\ n = 3$

2. $\int_0^{\pi/6} \cos x\,dx;\ n = 4$

In Problems 3 and 4 compare the exact value of the integral with the approximation obtained from the trapezoidal rule for the indicated value of n.

3. $\int_1^3 (x^3 + 1)\,dx;\ n = 4$

4. $\int_0^2 \sqrt{x + 1}\,dx;\ n = 6$

In Problems 5–12 use the midpoint rule and the trapezoidal rule to obtain an approximation to the given integral for the indicated value of n.

5. $\int_1^6 \dfrac{dx}{x};\ n = 5$

6. $\int_0^2 \dfrac{dx}{3x + 1};\ n = 4$

7. $\int_0^1 \sqrt{x^2 + 1}\,dx;\ n = 10$

8. $\int_1^2 \dfrac{dx}{\sqrt{x^3 + 1}};\ n = 5$

9. $\int_0^\pi \dfrac{\sin x}{x + \pi}\,dx;\ n = 6$

10. $\int_0^{\pi/4} \tan x\,dx;\ n = 3$

11. $\int_0^2 \cos x^2\,dx;\ n = 6$

12. $\int_0^1 \dfrac{\sin x}{x}\,dx;\ n = 5$ [*Hint*: Define $f(0) = 1$.]

In Problems 13 and 14 compare the exact value of the integral with the approximation obtained from Simpson's rule for the indicated value of n.

13. $\int_0^4 \sqrt{2x+1}\, dx;\ n=4$

14. $\int_0^{\pi/2} \sin^2 x\, dx;\ n=2$

In Problems 15–22 use Simpson's rule to obtain an approximation to the given integral for the indicated value of n.

15. $\int_{1/2}^{5/2} \dfrac{dx}{x};\ n=4$
16. $\int_0^5 \dfrac{dx}{x+2};\ n=6$
17. $\int_0^1 \dfrac{dx}{1+x^2};\ n=4$
18. $\int_{-1}^1 \sqrt{x^2+1}\, dx;\ n=2$
19. $\int_0^\pi \dfrac{\sin x}{x+\pi}\, dx;\ n=6$
20. $\int_0^1 \cos\sqrt{x}\, dx;\ n=4$
21. $\int_2^4 \sqrt{x^3+x}\, dx;\ n=4$
22. $\int_{\pi/4}^{\pi/2} \dfrac{dx}{2+\sin x};\ n=2$

23. Determine the number of rectangles needed so that an approximation to $\int_{-1}^2 dx/(x+3)$ is accurate to two decimal places.

24. Determine the number of trapezoids needed so that the error in an approximation to $\int_0^{1.5} \sin^2 x\, dx$ is less than 0.0001.

25. Use the trapezoidal rule so that an approximation to the area under the graph of $f(x)=1/(1+x^2)$ on $[0,2]$ is accurate to two decimal places. [*Hint:* Examine $f'''(x)$.]

26. The domain of $f(x)=10^x$ is the set of real numbers and $f(x)>0$ for all x. Use the trapezoidal rule to approximate the area under the graph of f on $[-2, 2]$ with $n=4$.

27. Using Simpson's rule, determine n so that the error in approximating $\int_1^3 dx/x$ is less than 10^{-5}. Compare with the n needed in the trapezoidal rule to give the same accuracy.

28. Find an upper bound for the error in approximating $\int_0^3 dx/(2x+1)$ by Simpson's rule with $n=6$.

In Problems 29 and 30 use the data given in the table and an appropriate rule to approximate the indicated definite integral.

29.

x	$f(x)$
2.05	4.91
2.10	4.80
2.15	4.66
2.20	4.41
2.25	3.93
2.30	3.58

; $\int_{2.05}^{2.30} f(x)\, dx$

30.

x	$f(x)$
0.0	−0.72
0.1	−0.55
0.2	−0.16
0.4	0.62
0.6	0.78
0.8	1.34
0.9	1.47
1.00	1.61
1.20	1.51

; $\int_0^{1.20} f(x)\, dx$

31. Compare the exact value of the integral $\int_0^4 (2x+5)\, dx$ with the approximation obtained from the midpoint rule with $n=2$ and $n=4$.

32. Repeat Problem 31 using the trapezoidal rule.

33. (a) Find the exact value of the integral
$I=\int_{-1}^1 (x^3+x^2)\, dx$.
(b) Use the midpoint rule with $n=8$ to find an approximation to I.
(c) Use the trapezoidal rule with $n=8$ to find an approximation to I.
(d) Compare the errors $E_8=|I-M_8|$ and $E_8=|I-T_8|$.

34. Compare the exact value of the integral $\int_{-1}^3 (x^3-x^2)\, dx$ with the approximations obtained from Simpson's rule with $n=2$ and $n=4$.

35. Use the data given in Figure 15.23 and Simpson's rule to find an approximation to the area under the graph of the continuous function f on the interval $[1, 4]$.

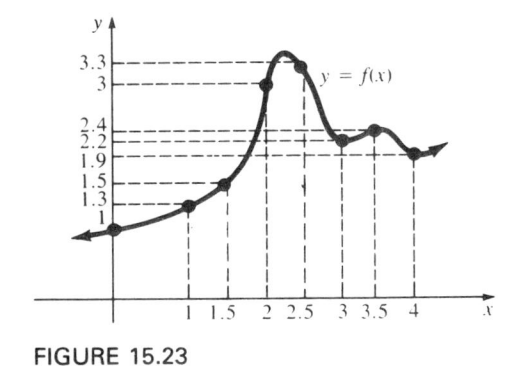

FIGURE 15.23

36. Use the trapezoidal rule with $n=9$ to find an approximation to the area of the shaded region in Figure 15.24.

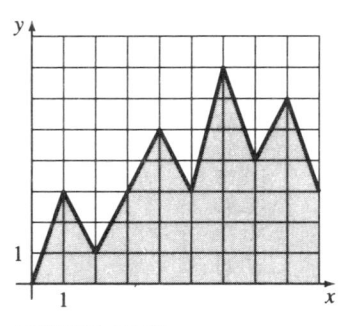

FIGURE 15.24

Does the trapezoidal rule give the exact value of the area?

37. The large irregularly shaped fish pond shown in Figure 15.25 is filled with water to a uniform depth of 4 ft. Use Simpson's rule to find an approximation to the number of gallons of water in the pond. Measurements are in feet. There are 7.48 gallons in a cubic foot of water.

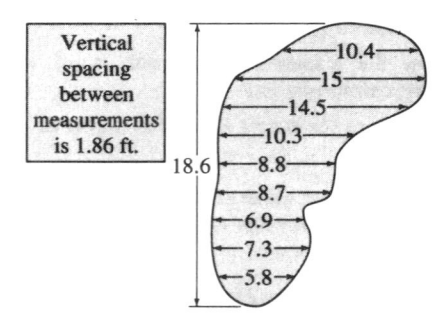

FIGURE 15.25

38. The moment of inertia I of a three-bladed ship's propeller whose dimensions are shown in Figure 15.26(a) is given by

$$I = \frac{3\rho\pi}{2g} + \frac{3\rho}{g}\int_1^{4.5} r^2 A\, dr$$

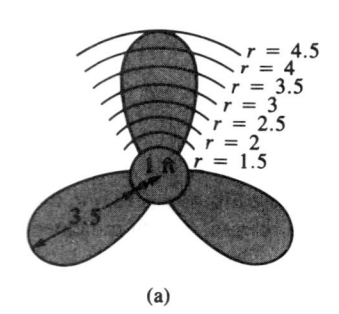

(a)

r (ft)	1	1.5	2	2.5	3	3.5	4	4.5
A (ft)	0.3	0.50	0.62	0.70	0.60	0.50	0.27	0

(b)

FIGURE 15.26

where ρ is the density of the metal, g is the acceleration of gravity, and A is the area of a cross-section of the propeller at a distance r ft from the center of the hub. If $\rho = 570$ lb/ft^3 for bronze, use the data in Figure 15.26(b) and the trapezoidal rule to find an approximation to I.

In Problems 39 and 40 use a graphics calculator or computer to obtain the graph of the given function. Use Simpson's rule to approximate the area bounded by the graph of f and the x-axis on the indicated interval. Use $n = 10$.

39. $f(x) = \sqrt[5]{(5^{2.5} - |x|^{2.5})^2}$; $[-5, 5]$

40. $f(x) = 1 + |\sin x|^x$; $[0, 2\pi]$ [*Hint:* Use the graph to discern $f(0)$.]

41. Write a computer program for the midpoint rule.

42. Prove that Simpson's rule will give the exact value of $\int_a^b f(x)\, dx$ when $f(x) = c_3 x^3 + c_2 x^2 + c_1 x + c_0$.

15.3 DIRECTION FIELDS

Introduction If a solution of a differential equation exists, it represents a locus of points (points connected by a smooth curve) in the Cartesian plane. Beginning in Section 15.4 we shall consider numerical procedures that utilize the differential equation to obtain a sequence of distinct points whose coordinates approximate the coordinates of the points on the actual solution curve (see Figure 15.27).

15.3 Direction Fields

For the next four sections of this chapter we shall confine our attention to first-order differential equations $dy/dx = f(x, y)$. As we saw in Chapter 9, higher-order equations can always be reduced to a system of first-order equations. In Section 15.8 we shall see that the numerical methods developed for first-order equations are easily adapted to systems of first-order equations.

We begin with the study of direction fields. Although it is not a numerical method, the concept of a direction field enables us to obtain a rough sketch of a solution of a first-order differential equation without solving it.

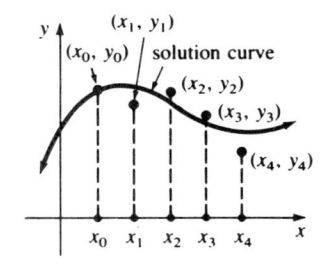

FIGURE 15.27

Lineal Elements Suppose for the moment that we do not know the general solution of the simple equation $y' = y$. Specifically, the differential equation implies that slopes of tangent lines to a solution curve are given by the function $f(x, y) = y$. When $f(x, y)$ is held constant—that is, when

$$y = c \tag{1}$$

where c is any constant—we are in effect stating that the slope of the tangents to the solution curves is the same constant value along a horizontal line. For example, for $y = 2$ let us draw a sequence of short line segments, or **lineal elements**, each having slope 2 and its midpoint on the line. As shown in Figure 15.28, the solution curves pass through this horizontal line at every point tangent to the lineal elements.

FIGURE 15.28

Isoclines and Direction Fields Equation (1) represents a one-parameter family of horizontal lines. In general, any member of the family $f(x, y) = c$ is called an **isocline**, which literally means a curve along which the inclination (of the tangents) is the same. As the parameter c is varied, we obtain a collection of isoclines on which the lineal elements are judiciously constructed. The totality of these lineal elements is called a **direction field**, **slope field**, or **lineal element field** of the differential equation $y' = f(x, y)$. As we see in Figure 15.29(a), the direction field suggests the "flow pattern" for the family of solution curves of the differential equation $y' = y$. In particular, if we want the one solution passing through the point (0, 1), then, as indicated in Figure 15.29(b), we construct a curve through this point and passing through the isoclines with the appropriate slopes.

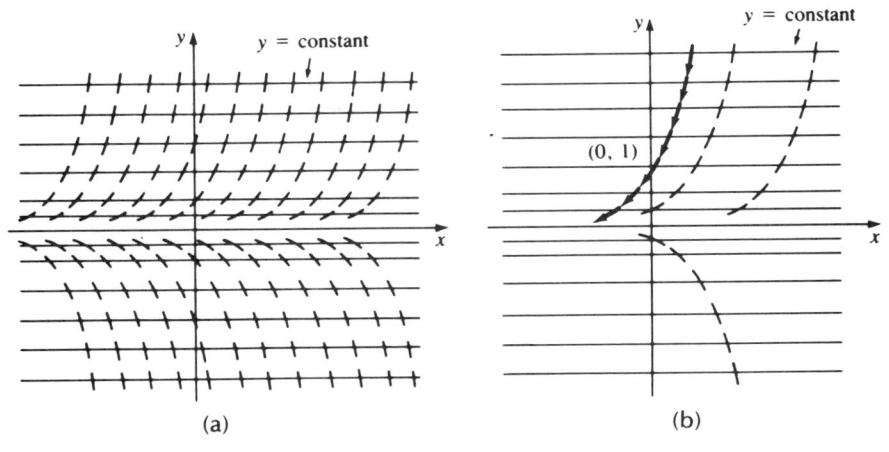

FIGURE 15.29

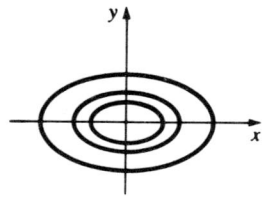

FIGURE 15.30

EXAMPLE 1 Determine the isoclines for the differential equation

$$\frac{dy}{dx} = 4x^2 + 9y^2$$

Solution For $c > 0$ the isoclines are the curves

$$4x^2 + 9y^2 = c$$

As Figure 15.30 shows, the curves are a concentric family of ellipses with major axis along the x-axis. □

EXAMPLE 2 Sketch the direction field and indicate several possible members of the family of solution curves for

$$\frac{dy}{dx} = \frac{x}{y}$$

Solution Before sketching the direction field corresponding to the isoclines $x/y = c$ or $y = x/c$, we note that the differential equation gives the following information:

(a) If a solution curve crosses the x-axis ($y = 0$), it does so tangent to a vertical lineal element at every point except possibly $(0, 0)$.

(b) If a solution curve crosses the y-axis ($x = 0$), it does so tangent to a horizontal lineal element at every point except possibly $(0, 0)$.

(c) The lineal elements corresponding to the isoclines $c = 1$ and $c = -1$ are collinear with the lines $y = x$ and $y = -x$, respectively. Indeed, it is easily verified that these isoclines are both particular solutions of the given dif-

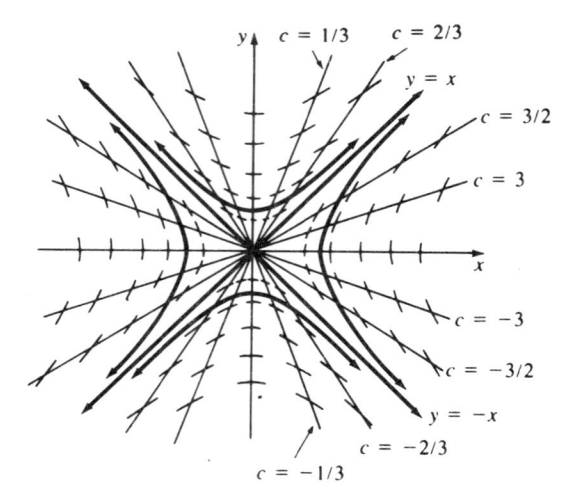

FIGURE 15.31

15.3 Direction Fields

ferential equation. However, it should be noted that *in general*, isoclines are themselves not solutions to a differential equation.*

Figure 15.31 shows the direction field and several possible solution curves. □

EXAMPLE 3 In Section 2.1 we indicated that the differential equation

$$\frac{dy}{dx} = x^2 + y^2$$

cannot be solved in terms of elementary functions. Use a direction field to locate an appropriate solution satisfying $y(0) = 1$.

Solution The isoclines are concentric circles defined by

$$x^2 + y^2 = c, \quad c > 0$$

By choosing $c = \frac{1}{4}$, $c = 1$, $c = \frac{9}{4}$ and $c = 4$, we obtain the circles with radii $\frac{1}{2}$, 1, $\frac{3}{2}$, and 2 shown in Figure 15.32(a). The lineal elements superimposed on each circle have slope corresponding to the particular value of c. It seems plausible from inspection of Figure 15.32(a) that a solution curve of the given initial-value problem might have the shape given in Figure 15.32(b). Unfortunately, we are not able to obtain any formula that describes this curve.

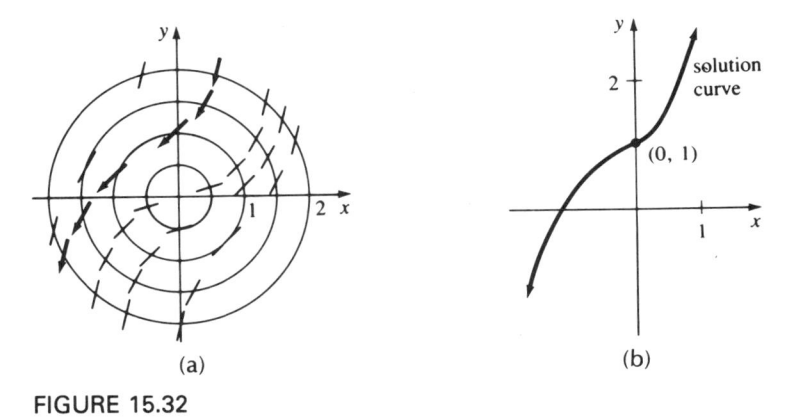

FIGURE 15.32 □

The concept of the direction field is used primarily to establish the existence of, and possibly to locate an approximate solution curve for, a first-order differential equation that cannot be solved by the usual standard techniques. However, the preceding discussion is of little value in determining specific values of a solution $y(x)$ at given points. For example, if we want to

---

*When the isoclines are straight lines it is easy to determine which, if any, of these isoclines are also particular solutions of the differential equation (see Problems 21–26).

know the approximate value of $y(0.5)$ for the solution of

$$\frac{dy}{dx} = x^2 + y^2, \quad y(0) = 1$$

then Figure 15.32(b) can do nothing more for us than indicate that $y(0.5)$ may be in the same "ball park" as $y = 2$.

▲ **Remark** Sketching a direction field by hand is a straightforward but time-consuming task. If it is available, we recommend that computer software be used.

EXERCISES 15.3   Answers to odd-numbered problems begin on page A-68.

In Problems 1–10 identify the isoclines for the given differential equation.

1. $\dfrac{dy}{dx} = x + 4$

2. $\dfrac{dy}{dx} = 2x + y$

3. $\dfrac{dy}{dx} = x^2 - y^2$

4. $\dfrac{dy}{dx} = y - x^2$

5. $y' = \sqrt{x^2 + y^2 + 2y + 1}$

6. $y' = (x^2 + y^2)^{-1}$

7. $\dfrac{dy}{dx} = y(x + y)$

8. $\dfrac{dy}{dx} = y + e^x$

9. $\dfrac{dy}{dx} = \dfrac{y-1}{x-2}$

10. $\dfrac{dy}{dx} = \dfrac{x-y}{x+y}$

In Problems 11–18 sketch the direction field for the given differential equation and indicate several possible solution curves.

11. $y' = x$

12. $y' = x + y$

13. $y\dfrac{dy}{dx} = -x$

14. $\dfrac{dy}{dx} = \dfrac{1}{y}$

15. $\dfrac{dy}{dx} = xy$

16. $\dfrac{dy}{dx} = 1 - xy$

17. $y' = y - \cos\dfrac{\pi}{2}x$

18. $y' = 1 - \dfrac{y}{x}$

Miscellaneous Problems

19. Formally show that the isoclines for the differential equation

$$\frac{dy}{dx} = \frac{\alpha x + \beta y}{\gamma x + \delta y}$$

are straight lines through the origin.

20. Show that $y = cx$ is a solution of the differential equation in Problem 19 if and only if $(\beta - \gamma)^2 + 4\alpha\delta \geq 0$.

EXAMPLE 4 The isoclines of the differential equation

$$y' = 2x + y \quad (2)$$

are the straight lines

$$2x + y = c \quad (3)$$

A line in this latter family will be a solution of the differential equation whenever its slope is the same as c. In other words, both the original equation and the line will satisfy $y' = c$. Since the slope of (3) is -2 if we choose $c = -2$, $2x + y = -2$ is a solution of (2).   □

In Problems 21–26 find those isoclines that are also solutions of the given differential equation (see Problems 19 and 20).

21. $y' = 3x + 2y$

22. $y' = 6x - 2y$

23. $y' = \dfrac{2x}{y}$

24. $y' = \dfrac{2y}{x+y}$

25. $\dfrac{dy}{dx} = \dfrac{4x + 3y}{y}$

26. $\dfrac{dy}{dx} = \dfrac{5x + 10y}{-4x + 3y}$

15.4 THE EULER METHODS

15.4.1 Euler's Method

One of the simplest techniques for approximating solutions of differential equations is known as **Euler's method**, or the method of **tangent lines**. Suppose we wish to approximate the solution of the initial-value problem

$$y' = f(x, y), \qquad y(x_0) = y_0$$

If h is a positive increment on the x-axis, then as Figure 15.33 shows, we can find a point $(x_1, y_1) = (x_0 + h, y_1)$ on the line tangent to the unknown solution curve at (x_0, y_0).

By the point–slope form of the equation of a line, we have

$$\frac{y_1 - y_0}{(x_0 + h) - x_0} = y'_0 \quad \text{or} \quad y_1 = y_0 + hy'_0$$

where $y'_0 = f(x_0, y_0)$. If we label $x_0 + h$ by x_1, then the point (x_1, y_1) on the tangent line is an approximation to the point $(x_1, y(x_1))$ on the solution curve—that is, $y_1 \approx y(x_1)$. Of course, the accuracy of the approximation depends heavily on the size of the increment h. Usually we must choose this **step size** to be "reasonably small."

Assuming a uniform (constant) value of h, we can obtain a succession of points $(x_1, y_1), (x_2, y_2), \ldots, (x_n, y_n)$, which we hope are close to the points $(x_1, y(x_1)), (x_2, y(x_2)), \ldots, (x_n, y(x_n))$ (see Figure 15.34). Now using (x_1, y_1), we can obtain the value of y_2, which is the ordinate of a point on a new "tangent" line. We have

$$\frac{y_2 - y_1}{h} = y'_1 \quad \text{or} \quad y_2 = y_1 + hy'_1 = y_1 + hf(x_1, y_1)$$

In general, it follows that

$$\boxed{\begin{aligned} y_{n+1} &= y_n + hy'_n \\ &= y_n + hf(x_n, y_n) \end{aligned}} \qquad (1)$$

FIGURE 15.33

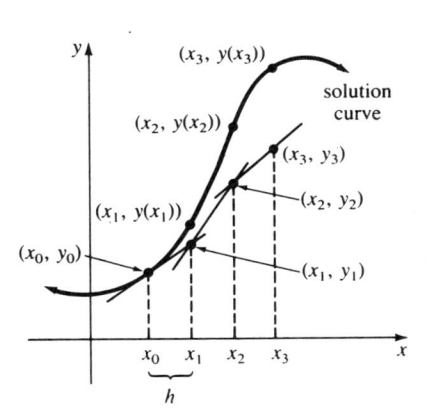

FIGURE 15.34

where $x_n = x_0 + nh$.

As an example, suppose we try the iteration scheme (1) on a differential equation for which we know the explicit solution; in this way we can compare the estimated values y_n and the true values $y(x_n)$.

EXAMPLE 1 Consider the initial-value problem

$$y' = 2xy, \qquad y(1) = 1$$

Use the Euler method to obtain an approximation to $y(1.5)$ using first $h = 0.1$ and then $h = 0.05$.

Solution We first identify $f(x, y) = 2xy$ so that (1) becomes

$$y_{n+1} = y_n + h(2x_n y_n)$$

Then for $h = 0.1$ we find

$$y_1 = y_0 + (0.1)(2x_0 y_0)$$
$$= 1 + (0.1)[2(1)(1)] = 1.2$$

which is an estimate to the value of $y(1.1)$. However, if we use $h = 0.05$, it takes *two* iterations to reach $x = 1.1$. We have

$$y_1 = 1 + (0.05)[2(1)(1)] = 1.1$$
$$y_2 = 1.1 + (0.05)[2(1.05)(1.1)] = 1.2155$$

Here we note that $y_1 \approx y(1.05)$ and $y_2 \approx y(1.1)$. The remainder of the calculations are summarized in Tables 15.1 and 15.2. Each entry is rounded to four decimal places.

TABLE 15.1 Euler's Method with $h = 0.1$

x_n	y_n	True Value	Abs. Error	% Rel. Error
1.00	1.0000	1.0000	0.0000	0.00
1.10	1.2000	1.2337	0.0337	2.73
1.20	1.4640	1.5527	0.0887	5.71
1.30	1.8154	1.9937	0.1784	8.95
1.40	2.2874	2.6117	0.3244	12.42
1.50	2.9278	3.4904	0.5625	16.12

TABLE 15.2 Euler's Method with $h = 0.05$

x_n	y_n	True Value	Abs. Error	% Rel. Error
1.00	1.0000	1.0000	0.0000	0.00
1.05	1.1000	1.1079	0.0079	0.72
1.10	1.2155	1.2337	0.0182	1.47
1.15	1.3492	1.3806	0.0314	2.27
1.20	1.5044	1.5527	0.0483	3.11
1.25	1.6849	1.7551	0.0702	4.00
1.30	1.8955	1.9937	0.0982	4.93
1.35	2.1419	2.2762	0.1343	5.90
1.40	2.4311	2.6117	0.1806	6.92
1.45	2.7714	3.0117	0.2403	7.98
1.50	3.1733	3.4904	0.3171	9.08

In Example 1 the true values were calculated from the known solution $y = e^{x^2 - 1}$. Also, the **absolute error** is defined to be

$$|\text{true value} - \text{approximation}|$$

The **relative error** and the **percentage relative error** are, respectively,

$$\frac{|\text{true value} - \text{approximation}|}{|\text{true value}|}$$

and

$$\frac{|\text{true value} - \text{approximation}|}{|\text{true value}|} \times 100 = \frac{\text{absolute error}}{|\text{true value}|} \times 100$$

It should be apparent that in the case of the step size $h = 0.1$, a 16% relative error in the calculation of the approximation to $y(1.5)$ is totally unacceptable. At the expense of doubling the number of calculations, a slight improvement in accuracy is obtained by halving the step size to $h = 0.05$.

Of course in many instances we may not know the solution of a particular differential equation, or for that matter whether a solution of an initial-value problem actually exists. The nonlinear equation in Example 2 does possess a solution in closed form, but we leave it as an exercise for the reader to find it (see Problem 1).

EXAMPLE 2 Use the Euler method to obtain the approximate value of $y(0.5)$ for the solution of

$$y' = (x + y - 1)^2, \qquad y(0) = 2$$

Solution For $n = 0$ and $h = 0.1$, we have

$$y_1 = y_0 + (0.1)(x_0 + y_0 - 1)^2$$
$$= 2 + (0.1)(1)^2 = 2.1$$

The remaining calculations are summarized in Tables 15.3 and 15.4 for $h = 0.1$ and $h = 0.05$, respectively.

TABLE 15.3 Euler's Method with $h = 0.1$

x_n	y_n
0.00	2.0000
0.10	2.1000
0.20	2.2440
0.30	2.4525
0.40	2.7596
0.50	3.2261

TABLE 15.4 Euler's Method with $h = 0.05$

x_n	y_n
0.00	2.0000
0.05	2.0500
0.10	2.1105
0.15	2.1838
0.20	2.2727
0.25	2.3812
0.30	2.5142
0.35	2.6788
0.40	2.8845
0.45	3.1455
0.50	3.4823

We may want greater accuracy than that displayed, say, in Table 15.2, and so we could try a step size even smaller than $h = 0.05$. However, rather than resorting to this extra labor, it probably would be more advantageous to employ an alternative numerical procedure. The Euler formula by itself, though attractive in its simplicity, is seldom used in serious calculations.

15.4.2 Improved Euler's Method

The formula

$$y_{n+1} = y_n + h \frac{f(x_n, y_n) + f(x_{n+1}, y^*_{n+1})}{2}$$

where

$$y^*_{n+1} = y_n + hf(x_n, y_n)$$

(2)

is known as the **improved Euler formula**, or **Heun's formula**. The values $f(x_n, y_n)$ and $f(x_{n+1}, y^*_{n+1})$ are approximations to the slope of the curve at $(x_n, y(x_n))$ and $(x_{n+1}, y(x_{n+1}))$ and, consequently, the quotient

$$\frac{f(x_n, y_n) + f(x_{n+1}, y^*_{n+1})}{2}$$

can be interpreted as an average slope on the interval between x_n and x_{n+1}.

The equations in (2) can be readily visualized. In Figure 15.35 we show the case in which $n = 0$. Note that

$$f(x_0, y_0) \quad \text{and} \quad f(x_1, y^*_1)$$

are slopes of the indicated straight lines passing through the points (x_0, y_0) and (x_1, y^*_1), respectively. By taking an average of these slopes, we obtain the slope of the dashed skew lines. Rather than advancing along the line with slope $m = f(x_0, y_0)$ to the point with ordinate y^*_1 obtained by the usual Euler method, we advance instead along the line through (x_0, y_0) with slope m_{ave} until we reach x_1. It seems plausible from inspection of the figure that y_1 is an improvement over y^*_1.

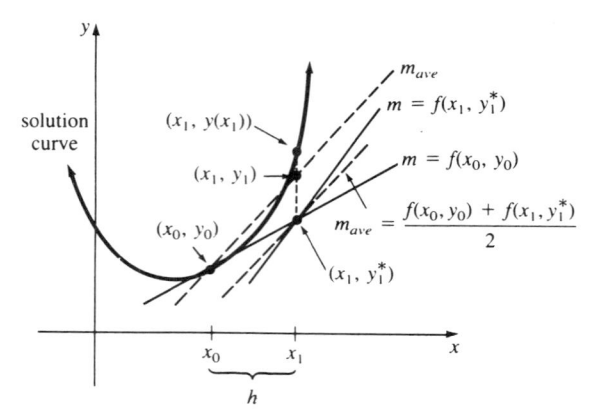

FIGURE 15.35

We might also say that the value of

$$y^*_1 = y_0 + hf(x_0, y_0)$$

15.4 The Euler Methods

predicts a value of $y(x_1)$, whereas

$$y_1 = y_0 + h \frac{f(x_0, y_0) + f(x_1, y_1^*)}{2}$$

corrects this estimate.

EXAMPLE 3 Use the improved Euler formula to obtain the approximate value of $y(1.5)$ for the solution of the initial-value problem in Example 1. Compare the results for $h = 0.1$ and $h = 0.05$.

Solution For $n = 0$ and $h = 0.1$, we first compute

$$y_1^* = y_0 + (0.1)(2x_0 y_0) = 1.2$$

Then from (2),

$$y_1 = y_0 + (0.1)\frac{2x_0 y_0 + 2x_1 y_1^*}{2}$$

$$= 1 + (0.1)\frac{2(1)(1) + 2(1.1)(1.2)}{2} = 1.232$$

The comparative values of the calculations for $h = 0.1$ and $h = 0.5$ are given in Tables 15.5 and 15.6, respectively.

TABLE 15.5 Improved Euler's Method with $h = 0.1$

x_n	y_n	True Value	Abs. Error	% Rel. Error
1.00	1.0000	1.0000	0.0000	0.00
1.10	1.2320	1.2337	0.0017	0.14
1.20	1.5479	1.5527	0.0048	0.31
1.30	1.9832	1.9937	0.0106	0.53
1.40	2.5908	2.6117	0.0209	0.80
1.50	3.4509	3.4904	0.0394	1.13

TABLE 15.6 Improved Euler's Method with $h = 0.05$

x_n	y_n	True Value	Abs. Error	% Rel. Error
1.00	1.0000	1.0000	0.0000	0.00
1.05	1.1077	1.1079	0.0002	0.02
1.10	1.2332	1.2337	0.0004	0.04
1.15	1.3798	1.3806	0.0008	0.06
1.20	1.5514	1.5527	0.0013	0.08
1.25	1.7531	1.7551	0.0020	0.11
1.30	1.9909	1.9937	0.0029	0.14
1.35	2.2721	2.2762	0.0041	0.18
1.40	2.6060	2.6117	0.0057	0.22
1.45	3.0038	3.0117	0.0079	0.26
1.50	3.4795	3.4904	0.0108	0.31

□

A brief word of caution is in order here. We cannot compute all the values of y_n^* first and then substitute these values in the first formula of (2). In other words, we cannot use the data in Table 15.1 to help construct the values in Table 15.5. Why not?

EXAMPLE 4 Use the improved Euler formula to obtain the approximate value of $y(0.5)$ for the solution of the initial-value problem in Example 2.

Solution For $n = 0$ and $h = 0.1$, we have

$$y_1^* = y_0 + (0.1)(x_0 + y_0 - 1)^2 = 2.1$$

and so
$$y_1 = y_0 + (0.1)\frac{(x_0 + y_0 - 1)^2 + (x_1 + y_1^* - 1)^2}{2}$$

$$= 2 + (0.1)\frac{1 + 1.44}{2} = 2.122$$

The remaining calculations are summarized in Tables 15.7 and 15.8 for $h = 0.1$ and $h = 0.05$, respectively.

TABLE 15.7 Improved Euler's Method with $h = 0.1$

x_n	y_n
0.00	2.0000
0.10	2.1220
0.20	2.3049
0.30	2.5858
0.40	3.0378
0.50	3.8254

TABLE 15.8 Improved Euler's Method with $h = 0.05$

x_n	y_n
0.00	2.0000
0.05	2.0553
0.10	2.1228
0.15	2.2056
0.20	2.3075
0.25	2.4342
0.30	2.5931
0.35	2.7953
0.40	3.0574
0.45	3.4057
0.50	3.8840

BASIC Programs BASIC programs for Euler methods are given in Appendix IV.

EXERCISES 15.4   *Answers to odd-numbered problems begin on page A-69.*

1. Solve the initial-value problem

    $$y' = (x + y - 1)^2, \quad y(0) = 2$$

    in terms of elementary functions.

2. Let $y(x)$ be the solution of the initial-value problem given in Problem 1. Round to four decimal places, and compute the exact values of $y(0.1)$, $y(0.2)$, $y(0.3)$, $y(0.4)$, and $y(0.5)$. Compare these values with the entries in Tables 15.3, 15.4, 15.7, and 15.8.

Given the initial-value problems in Problems 3–12, use the Euler formula to obtain a four-decimal approximation to the indicated value. First use **(a)** $h = 0.1$ and then **(b)** $h = 0.05$.

3. $y' = 2x - 3y + 1$, $y(1) = 5$; $y(1.5)$

4. $y' = 4x - 2y$, $y(0) = 2$; $y(0.5)$

5. $y' = 1 + y^2$, $y(0) = 0$; $y(0.5)$

6. $y' = x^2 + y^2$, $y(0) = 1$; $y(0.5)$

7. $y' = e^{-y}$, $y(0) = 0$; $y(0.5)$

8. $y' = x + y^2$, $y(0) = 0$; $y(0.5)$

9. $y' = (x - y)^2$, $y(0) = 0.5$; $y(0.5)$

10. $y' = xy + \sqrt{y}$, $y(0) = 1$; $y(0.5)$

11. $y' = xy^2 - \dfrac{y}{x}$, $y(1) = 1$; $y(1.5)$

12. $y' = y - y^2$, $y(0) = 0.5$; $y(0.5)$

13. As parts (a)–(e) of this problem, repeat the calculations of Problems 3, 5, 7, 9, and 11 using the improved Euler formula.

14. As parts (a)–(e) of this problem, repeat the calculations of Problems 4, 6, 8, 10, and 12 using the improved Euler formula.

15. Although it may not be obvious from the differential equation, its solution could "behave badly" near a point x at which we wish to approximate $y(x)$. Numerical procedures may then give widely differing results near this point. Let $y(x)$ be the solution of the initial-value problem

$$y' = x^2 + y^3, \quad y(1) = 1$$

Using the step size $h = 0.1$, compare the results obtained from the Euler formula with the results from the improved Euler formula in the approximation of $y(1.4)$.

15.5 THE THREE-TERM TAYLOR METHOD

The numerical method considered in this section, the **three-term Taylor method**, is more of theoretical interest than of practical importance, since the results obtained using the following formula (5) will not differ substantially from those obtained using the improved Euler method.

In the study of numerical solutions of differential equations, many computational algorithms can be derived from a Taylor series expansion. Recall from calculus that the form of this expansion centered at a point a is

$$y(x) = y(a) + y'(a)\dfrac{(x - a)}{1!} + y''(a)\dfrac{(x - a)^2}{2!} + \cdots \quad (1)$$

It is understood that the function $y(x)$ possesses derivatives of all orders and that the series (1) converges in some interval defined by $|x - a| < R$. Notice, in particular, that if we set $a = x_n$ and $x = x_n + h$, then (1) becomes

$$y(x_n + h) = y(x_n) + y'(x_n)h + y''(x_n)\dfrac{h^2}{2} + \cdots \quad (2)$$

Euler's Method Revisited Let us now assume that the function $y(x)$ is a solution of the first-order differential equation

$$y' = f(x, y)$$

If we then truncate the series (2) after, say, two terms, we obtain the approximation

$$y(x_n + h) \approx y(x_n) + y'(x_n)h$$

or

$$y(x_n + h) \approx y(x_n) + f(x_n, y(x_n))h \quad (3)$$

Observe that we can obtain the Euler formula

$$y_{n+1} = y_n + hf(x_n, y_n) \qquad (4)$$

of the preceding section by replacing $y(x_n + h)$ and $y(x_n)$ in (3) by their approximations y_{n+1} and y_n, respectively. The approximation symbol $\approx$ is replaced by an equality, since we are defining the left side of (4) by the numbers obtained from the right-hand member.

Taylor's Method By retaining three terms in the series (2), we can write

$$y(x_n + h) \approx y(x_n) + y'(x_n)h + y''(x_n)\frac{h^2}{2}$$

After using the replacements noted in the preceding material, we get

$$\boxed{y_{n+1} = y_n + y'_n h + y''_n \frac{h^2}{2}} \qquad (5)$$

The second derivative y'' can be obtained by differentiating $y' = f(x, y)$.

At this point let us reexamine the two initial-value problems of the preceding section.

EXAMPLE 1 Use the three-term Taylor formula to obtain the approximate value of $y(1.5)$ for the solution of

$$y' = 2xy, \qquad y(1) = 1$$

Compare the results for $h = 0.1$ and $h = 0.05$.

Solution Since $y' = 2xy$, it follows by the product rule that $y'' = 2xy' + 2y$. Thus, for example, when $h = 0.1$, $n = 0$, we can first calculate

$$y'_0 = 2x_0 y_0 = 2(1)(1) = 2$$

and then
$$y''_0 = 2x_0 y'_0 + 2y_0$$
$$= 2(1)(2) + 2(1) = 6$$

Hence (5) becomes

$$y_1 = y_0 + y'_0(0.1) + y''_0 \frac{(0.1)^2}{2}$$
$$= 1 + 2(0.1) + 6(0.005)$$
$$= 1.23$$

The results of the iteration, along with the comparative exact values, are summarized in Tables 15.9 and 15.10.

15.5 The Three-Term Taylor Method

TABLE 15.9 Three-Term Taylor Method with $h = 0.1$

x_n	y_n	True Value	Abs. Error	% Rel. Error
1.00	1.0000	1.0000	0.0000	0.00
1.10	1.2300	1.2337	0.0037	0.30
1.20	1.5427	1.5527	0.0100	0.65
1.30	1.9728	1.9937	0.0210	1.05
1.40	2.5721	2.6117	0.0396	1.52
1.50	3.4188	3.4904	0.0715	2.05

TABLE 15.10 Three-Term Taylor Method with $h = 0.05$

x_n	y_n	True Value	Abs. Error	% Rel. Error
1.00	1.0000	1.0000	0.0000	0.00
1.05	1.1075	1.1079	0.0004	0.04
1.10	1.2327	1.2337	0.0010	0.08
1.15	1.3788	1.3806	0.0018	0.13
1.20	1.5499	1.5527	0.0028	0.18
1.25	1.7509	1.7551	0.0041	0.23
1.30	1.9879	1.9937	0.0059	0.29
1.35	2.2681	2.2762	0.0081	0.36
1.40	2.6006	2.6117	0.0111	0.43
1.45	2.9967	3.0117	0.0150	0.50
1.50	3.4702	3.4904	0.0202	0.58

□

EXAMPLE 2 Use the three-term Taylor formula to obtain the approximate value of $y(0.5)$ for the solution of

$$y' = (x + y - 1)^2, \quad y(0) = 2$$

Solution In this case we compute y'' by the power rule. We have

$$y'' = 2(x + y - 1)(1 + y')$$

The results are summarized in Tables 15.11 and 15.12 for $h = 0.1$ and $h = 0.05$, respectively.

TABLE 15.11 Three-Term Taylor's Method with $h = 0.1$

x_n	y_n
0.00	2.0000
0.10	2.1200
0.20	2.2992
0.30	2.5726
0.40	3.0077
0.50	3.7511

TABLE 15.12 Three-Term Taylor's Method with $h = 0.05$

x_n	y_n
0.00	2.0000
0.05	2.0550
0.10	2.1222
0.15	2.2045
0.20	2.3058
0.25	2.4315
0.30	2.5890
0.35	2.7889
0.40	3.0475
0.45	3.3898
0.50	3.8574

□

A comparison of the last two examples with the corresponding results obtained from the improved Euler method shows no startling dissimilarities. In fact, when $f(x, y)$ is linear in both variables x and y, the Taylor method gives the *same* values of y_n as the improved Euler method for a given value of h. See Problems 12 and 13.

A BASIC program for the three-term Taylor method is given in Appendix IV.

EXERCISES 15.5   *Answers to odd-numbered problems begin on page A-72.*

Given the initial-value problems in Problems 1–10, use the three-term Taylor formula to obtain a four-decimal approximation to the indicated value. First use (a) $h = 0.1$ and then (b) $h = 0.05$.

1. $y' = 2x - 3y + 1$, $y(1) = 5$; $y(1.5)$
2. $y' = 4x - 2y$, $y(0) = 2$; $y(0.5)$
3. $y' = 1 + y^2$, $y(0) = 0$; $y(0.5)$
4. $y' = x^2 + y^2$, $y(0) = 1$; $y(0.5)$
5. $y' = e^{-y}$, $y(0) = 0$; $y(0.5)$
6. $y' = x + y^2$, $y(0) = 0$; $y(0.5)$
7. $y' = (x - y)^2$, $y(0) = 0.5$; $y(0.5)$
8. $y' = xy + \sqrt{y}$, $y(0) = 1$; $y(0.5)$
9. $y' = xy^2 - \dfrac{y}{x}$, $y(1) = 1$; $y(1.5)$
10. $y' = y - y^2$, $y(0) = 0.5$; $y(0.5)$

11. Let $y(x)$ be the solution of the initial-value problem
$$y' = x^2 + y^3, \quad y(1) = 1$$

Use $h = 0.1$ and the three-term Taylor formula to obtain an approximation to $y(1.4)$. Compare your answer with the results obtained in Problem 15 in Exercises 15.4

12. Consider the differential equation $y' = f(x, y)$, where f is linear in x and y. In this case prove that the improved Euler formula is the same as the three-term Taylor formula. [*Hint:* Recall from calculus that a Taylor series for a function g of two variables is

$$g(a + h, b + k) = g(a, b) + g_x(a, b)h + g_y(a, b)k$$
$$+ \frac{1}{2}(h^2 g_{xx} + 2hk g_{xy} + k^2 g_{yy})\bigg|_{(a,b)}$$

+ terms involving higher-order derivatives

Apply this result to $f(x_n + h, y_n + hf(x_n, y_n))$ in the improved Euler formula. Also use the fact that $y''(x) = (d/dx)y'(x) = f_x + f_y y'$.]

13. Compare the approximate values of $y(1.5)$ for
$$y' = x + y - 1, \quad y(1) = 5$$
using the three-term Taylor method and the improved Euler method with $h = 0.1$. Solve the initial-value problem and compute the true values $y(x_n)$, $n = 0, 1, \ldots, 5$.

15.6 THE RUNGE–KUTTA METHOD

Probably one of most popular as well as accurate numerical procedures used in obtaining approximate solutions to differential equations is the **fourth-order Runge–Kutta method**.* As the name suggests, there are Runge–Kutta's methods of different orders.

---

*Carl D. T. Runge (1856–1927) A German professor of applied mathematics at the University of Göttingen, Runge devised this numerical method around 1895. (M. W. Kutta expanded on this work in 1901.) In mathematics Runge also did notable work in the field of Diophantine equations. In physics Runge is remembered for his work on the Zeeman effect (the spectral lines in the emission spectrum of an element are affected by the presence of a magnetic field).

Martin W. Kutta (1867–1944) Also a German applied mathematician, Kutta made significant contributions to the field of aerodynamics.

15.6 The Runge–Kutta Method

For the moment let us consider a **second-order** procedure. This consists of finding constants a, b, α, and β such that the formula

$$y_{n+1} = y_n + ak_1 + bk_2 \tag{1}$$

where
$$k_1 = hf(x_n, y_n)$$
$$k_2 = hf(x_n + \alpha h, y_n + \beta k_1) \tag{2}$$

agrees with a Taylor series expansion to as many terms as possible. The obvious purpose is to achieve the accuracy of the Taylor method without having to compute higher-order derivatives. Now it can be shown that whenever the constants satisfy

$$a + b = 1 \qquad b\alpha = \frac{1}{2} \qquad b\beta = \frac{1}{2}$$

then (1) agrees with a Taylor expansion out to h^2 or the third term. It should be of interest to observe that when $a = \frac{1}{2}$, $b = \frac{1}{2}$, $\alpha = 1$, $\beta = 1$, then (1) reduces to the improved Euler method. Thus we can conclude that the three-term Taylor formula is essentially equivalent to the improved Euler formula. Also, the basic Euler method is a **first-order** Runge–Kutta procedure.

Notice too that the sum $ak_1 + bk_2$, $a + b = 1$ in (1) is simply a *weighted average* of k_1 and k_2. The numbers k_1 and k_2 are multiples of approximations to the slope at two different points.

Fourth-Order Runge–Kutta Formula The **fourth-order** Runge–Kutta method consists of determining appropriate constants so that a formula such as

$$y_{n+1} = y_n + ak_1 + bk_2 + ck_3 + dk_4$$

agrees with a Taylor expansion out to h^4 or the fifth term. As in (2) the k_i are constant multiples of $f(x, y)$ evaluated at select points. The derivation of the actual method is tedious, to say the least, so we state the results:

$$\boxed{\begin{aligned} y_{n+1} &= y_n + \frac{1}{6}(k_1 + 2k_2 + 2k_3 + k_4) \\ k_1 &= hf(x_n, y_n) \\ k_2 &= hf\left(x_n + \frac{1}{2}h, y_n + \frac{1}{2}k_1\right) \\ k_3 &= hf\left(x_n + \frac{1}{2}h, y_n + \frac{1}{2}k_2\right) \\ k_4 &= hf(x_n + h, y_n + k_3) \end{aligned}} \tag{3}$$

The reader is advised to look carefully at the formulas in (3); note that k_2 depends on k_1, k_3 depends on k_2, and so on. Also, k_2 and k_3 are approximations to the slope at the midpoint of the interval between x_n and $x_{n+1} = x_n + h$.

EXAMPLE 1 Use the Runge–Kutta method with $h = 0.1$ to obtain an approximation to $y(1.5)$ for the solution of

$$y' = 2xy, \quad y(1) = 1$$

Solution For the sake of illustration, let us compute the case when $n = 0$. From (3) we find

$$k_1 = (0.1)f(x_0, y_0)$$
$$= (0.1)(2x_0 y_0) = 0.2$$
$$k_2 = (0.1)f\left(x_0 + \frac{1}{2}(0.1), y_0 + \frac{1}{2}(0.2)\right)$$
$$= (0.1)2\left(x_0 + \frac{1}{2}(0.1)\right)\left(y_0 + \frac{1}{2}(0.2)\right) = 0.231$$
$$k_3 = (0.1)f\left(x_0 + \frac{1}{2}(0.1), y_0 + \frac{1}{2}(0.231)\right)$$
$$= (0.1)2\left(x_0 + \frac{1}{2}(0.1)\right)\left(y_0 + \frac{1}{2}(0.231)\right) = 0.234255$$
$$k_4 = (0.1)f(x_0 + 0.1, y_0 + 0.234255)$$
$$= (0.1)2(x_0 + 0.1)(y_0 + 0.234255) = 0.2715361$$

and, therefore,

$$y_1 = y_0 + \frac{1}{6}(k_1 + 2k_2 + 2k_3 + k_4)$$
$$= 1 + \frac{1}{6}(0.2 + 2(0.231) + 2(0.234255) + 0.2715361) = 1.23367435$$

When we round to the usual four decimal places, we obtain

$$y_1 = 1.2337$$

Table 15.13 should convince the student why the Runge–Kutta method is so popular. Of course, there is no need to use any smaller step size.

TABLE 15.13 Runge–Kutta's Method with $h = 0.1$

x_n	y_n	True Value	Abs. Error	% Rel. Error
1.00	1.0000	1.0000	0.0000	0.00
1.10	1.2337	1.2337	0.0000	0.00
1.20	1.5527	1.5527	0.0000	0.00
1.30	1.9937	1.9937	0.0000	0.00
1.40	2.6116	2.6117	0.0001	0.00
1.50	3.4902	3.4904	0.0001	0.00

The reader might be interested in inspecting Tables 15.14 and 15.15 at this point. These tables compare the results obtaining from the various for-

15.6 The Runge–Kutta Method

mulas that we have examined applied to the two specific problems

$$y' = 2xy, \quad y(1) = 1$$
$$y' = (x + y - 1)^2, \quad y(0) = 2$$

that we have considered throughout the last three sections.

TABLE 15.14 $y' = 2xy$, $y(1) = 1$

x_n	Euler	Improved Euler	Three-Term Taylor	Runge–Kutta	True Value
\multicolumn{6}{c}{Comparison of Numerical Methods with $h = 0.1$}
1.00	1.0000	1.0000	1.0000	1.0000	1.0000
1.10	1.2000	1.2320	1.2300	1.2337	1.2337
1.20	1.4640	1.5479	1.5427	1.5527	1.5527
1.30	1.8154	1.9832	1.9728	1.9937	1.9937
1.40	2.2874	2.5908	2.5721	2.6116	2.6117
1.50	2.9278	3.4509	3.4188	3.4902	3.4904
\multicolumn{6}{c}{Comparison of Numerical Methods with $h = 0.05$}
1.00	1.0000	1.0000	1.0000	1.0000	1.0000
1.05	1.1000	1.1077	1.1075	1.1079	1.1079
1.10	1.2155	1.2332	1.2327	1.2337	1.2337
1.15	1.3492	1.3798	1.3788	1.3806	1.3806
1.20	1.5044	1.5514	1.5499	1.5527	1.5527
1.25	1.6849	1.7531	1.7509	1.7551	1.7551
1.30	1.8955	1.9909	1.9879	1.9937	1.9937
1.35	2.1419	2.2721	2.2681	2.2762	2.2762
1.40	2.4311	2.6060	2.6006	2.6117	2.6117
1.45	2.7714	3.0038	2.9967	3.0117	3.0117
1.50	3.1733	3.4795	3.4702	3.4903	3.4904

TABLE 15.15 $y' = (x + y - 1)^2$, $y(0) = 2$

x_n	Euler	Improved Euler	Three-Term Taylor	Runge–Kutta	True Value
\multicolumn{6}{c}{Comparison of Numerical Methods with $h = 0.1$}
0.00	2.0000	2.0000	2.0000	2.0000	2.0000
0.10	2.1000	2.1220	2.1200	2.1230	2.1230
0.20	2.2440	2.3049	2.2992	2.3085	2.3085
0.30	2.4525	2.5858	2.5726	2.5958	2.5958
0.40	2.7596	3.0378	3.0077	3.0649	3.0650
0.50	3.2261	3.8254	3.7511	3.9078	3.9082

(*continued*)

TABLE 15.15 (*continued*)

	Comparison of Numerical Methods with $h = 0.05$				
x_n	Euler	Improved Euler	Three-Term Taylor	Runge–Kutta	True Value
0.00	2.0000	2.0000	2.0000	2.0000	2.0000
0.05	2.0500	2.0553	2.0550	2.0554	2.0554
0.10	2.1105	2.1228	2.1222	2.1230	2.1230
0.15	2.1838	2.2056	2.2045	2.2061	2.2061
0.20	2.2727	2.3075	2.3058	2.3085	2.3085
0.25	2.3812	2.4342	2.4315	2.4358	2.4358
0.30	2.5142	2.5931	2.5890	2.5958	2.5958
0.35	2.6788	2.7953	2.7889	2.7998	2.7997
0.40	2.8845	3.0574	3.0475	3.0650	3.0650
0.45	3.1455	3.4057	3.3898	3.4189	3.4189
0.50	3.4823	3.8840	3.8574	3.9082	3.9082

BASIC Program   A listing of a BASIC program for the Runge–Kutta method is given in Appendix IV.

EXERCISES 15.6   *Answers to odd-numbered problems begin on page A-73.*

Given the initial-value problems in Problems 1–10, use the Runge–Kutta method with $h = 0.1$ to obtain a four-decimal approximation to the indicated value.

1. $y' = 2x - 3y + 1$, $y(1) = 5$; $y(1.5)$
2. $y' = 4x - 2y$, $y(0) = 2$; $y(0.5)$
3. $y' = 1 + y^2$, $y(0) = 0$; $y(0.5)$
4. $y' = x^2 + y^2$, $y(0) = 1$; $y(0.5)$
5. $y' = e^{-y}$, $y(0) = 0$; $y(0.5)$
6. $y' = x + y^2$, $y(0) = 0$; $y(0.5)$
7. $y' = (x - y)^2$, $y(0) = 0.5$; $y(0.5)$
8. $y' = xy + \sqrt{y}$, $y(0) = 1$; $y(0.5)$
9. $y' = xy^2 - \dfrac{y}{x}$, $y(1) = 1$; $y(1.5)$
10. $y' = y - y^2$, $y(0) = 0.5$; $y(0.5)$

11. If air resistance is proportional to the square of the instantaneous velocity, the velocity v of mass m dropped from a height h is determined from

$$m\frac{dv}{dt} = mg - kv^2, \quad k > 0$$

If $v(0) = 0$, $k = 0.125$, $m = 5$ slugs, and $g = 32$ ft/s^2, use the Runge–Kutta method to find an approximation to the velocity of the falling mass at $t = 5$ seconds. Use $h = 1$.

12. Solve the initial-value problem in Problem 11 by one of the methods of Chapter 2. Find the true value $v(5)$.

13. A mathematical model for the area A (in cm^2) that a colony of bacteria (*B. dendroides*) occupies is given by*

$$\frac{dA}{dt} = A(2.128 - 0.0432A)$$

If $A(0) = 0.24$ cm^2, use the Runge–Kutta method to complete the following table. Use $h = 0.5$.

t (days)	1	2	3	4	5
A (observed)	2.78	13.53	36.30	47.50	49.40
A (approximated)					

14. Solve the initial-value problem in Problem 13. Compute the values $A(1)$, $A(2)$, $A(3)$, $A(4)$, and $A(5)$. [*Hint:* See Section 2.11.]

---

*See V. A. Kostitzin, *Mathematical Biology* (London: Harrap, 1939).

15. Let $y(x)$ be the solution of the initial-value problem

$$y' = x^2 + y^3, \qquad y(1) = 1$$

Determine whether the Runge–Kutta formula can be used to obtain an approximation for $y(1.4)$. Use $h = 0.1$

16. Consider the differential equation $y' = f(x)$. In this case show that the fourth-order Runge–Kutta method reduces to Simpson's rule for the integral of $f(x)$ on the interval $x_n \leq x \leq x_{n+1}$.

15.7 MULTISTEP METHODS, ERRORS

Adams–Bashforth/Adams–Moulton Method Many additional formulas can be applied to obtain approximations to solutions of differential equations. Although it is not our intention to survey the vast field of numerical methods, several additional formulas deserve mention. The **Adams–Bashforth/Adams–Moulton method**, like the improved Euler formula, is a predictor–corrector method. By first using the Adams–Bashforth formula

$$y_{n+1}^* = y_n + \frac{h}{24}(55y_n' - 59y_{n-1}' + 37y_{n-2}' - 9y_{n-3}') \qquad (1)$$

where
$$y_n' = f(x_n, y_n)$$
$$y_{n-1}' = f(x_{n-1}, y_{n-1})$$
$$y_{n-2}' = f(x_{n-2}, y_{n-2})$$
$$y_{n-3}' = f(x_{n-3}, y_{n-3})$$

$n \geq 3$, as a predictor, we are able to substitute the value of y_{n+1}^* into the Adams–Moulton corrector:

$$y_{n+1} = y_n + \frac{h}{24}(9y_{n+1}' + 19y_n' - 5y_{n-1}' + y_{n-2}') \qquad (2)$$

where
$$y_{n+1}' = f(x_{n+1}, y_{n+1}^*)$$

Notice (1) requires that we know the values of $y_0, y_1, y_2,$ and y_3 in order to obtain y_4. The value of y_0 is, of course, the given initial condition; the values of $y_1, y_2,$ and y_3 are computed by an accurate method such as the Runge–Kutta formula.

Since the Adams–Bashforth/Adams–Moulton formulas demand that we know more than just y_n to compute y_{n+1}, the procedure is called a **multistep**, or **continuing**, method. The Euler formulas, the three-term Taylor formula, and the Runge–Kutta formulas are examples of **single-step**, or **starting**, methods.

EXAMPLE 1 Use the Adams–Bashforth/Adams–Moulton method with $h = 0.2$ to obtain an approximation to $y(0.8)$ for the solution of

$$y' = x + y - 1, \qquad y(0) = 1$$

Solution With a step size of $h = 0.2$, $y(0.8)$ will be approximated by y_4. To get started we use the Runge–Kutta method with $x_0 = 0$, $y_0 = 1$, and $h = 0.2$ to obtain

$$y_1 = 1.02140000$$
$$y_2 = 1.09181796$$
$$y_3 = 1.22210646$$

Now with the identifications $x_0 = 0$, $x_1 = 0.2$, $x_2 = 0.4$, $x_3 = 0.6$, and $f(x, y) = x + y - 1$, we find

$$y'_0 = f(x_0, y_0) = (0) + (1) - 1 = 0$$
$$y'_1 = f(x_1, y_1) = (0.2) + (1.02140000) - 1 = 0.22140000$$
$$y'_2 = f(x_2, y_2) = (0.4) + (1.09181796) - 1 = 0.49181796$$
$$y'_3 = f(x_3, y_3) = (0.6) + (1.22210646) - 1 = 0.82210646$$

With the foregoing values, the predictor (1) gives

$$y_4^* = y_3 + \frac{0.2}{24}(55y'_3 - 59y'_2 + 37y'_1 - 9y'_0) = 1.42535975$$

To use the corrector (2) we first need

$$y'_4 = f(x_4, y_4^*) = (0.8) + (1.42535975) - 1 = 1.22535975$$

Finally, (2) yields

$$y_4 = y_3 + \frac{0.2}{24}(9y'_4 + 19y'_3 - 5y'_2 + y'_1) = 1.42552788 \qquad \square$$

The reader should verify that the exact value of $y(0.8)$ in Example 1 is 1.42554093.

Milne's Method Another multistep method, which admittedly has limited use, is called **Milne's method**. In this method the predictor is

$$y_{n+1}^* = y_{n-3} + \frac{4h}{3}(2y'_n - y'_{n-1} + 2y'_{n-2}) \tag{3}$$

where
$$y'_n = f(x_n, y_n)$$
$$y'_{n-1} = f(x_{n-1}, y_{n-1})$$
$$y'_{n-2} = f(x_{n-2}, y_{n-2})$$

for $n \geq 3$. The corrector for this formula, based on Simpson's rule of integration, is given by

$$y_{n+1} = y_{n-1} + \frac{h}{3}(y'_{n+1} + 4y'_n + y'_{n-1}) \tag{4}$$

where
$$y'_{n+1} = f(x_{n+1}, y_{n+1}^*)$$

As in the Adams–Bashforth/Adams–Moulton procedure, we generally use the Runge–Kutta method to compute y_1, y_2, and y_3.

Errors In a serious and detailed study of numerical solutions of differential equations, we would have to pay close attention to the various sources of errors. For some kinds of computation, accumulation of errors might reduce the accuracy of an approximation to the point of being useless.

By using only three terms of a Taylor series to approximate the value of a function, the method itself naturally will be a source of error. As we have seen, the Euler formula is essentially two terms of a Taylor series expansion; by advancing along a tangent line, we do not necessarily get to a point on or even near the solution curve. The errors inherent to these methods are known as **truncation errors, formula errors**, or **discretization errors**.

Any calculator or computer can compute to only at most a finite number of decimal places. Suppose for the sake of illustration that we have a calculator that can display six digits while carrying eight digits internally. If we multiply two numbers, each having six decimals, then the product actually contains 12 decimal places. But the number that we see is rounded to six decimal places, while the machine has stored a number rounded to eight decimal places. In one calculation such as this, the **round-off error** may not be significant, but a problem could arise if many calculations are performed with rounded numbers. The effects of round-off can be minimized on a computer if it has double-precision capabilities.

When iterating a formula such as

$$y_{n+1} = y_n + hf(x_n, y_n)$$

we obtain a sequence of values

$$y_1, y_2, y_3, \ldots$$

The value of y_1 is, of course, in error, and unfortunately y_2 depends on y_1. Thus, y_2 must also be in error. In turn, y_3 inherits an error from y_2. The error resulting from the inheritance of errors in preceding calculations is known as **propagation error**. To make matters worse, formulas can be **unstable**. This means that errors occurring in the early stages of calculation are not only propagated but also *compounded* at each step of the iteration. The error may grow so fast that the subsequent approximations are completely overwhelmed. Under certain circumstances the corrector formula in Milne's method is unstable.

EXERCISES 15.7 *Answers to odd-numbered problems begin on page A-74.*

1. Find the exact solution of the initial-value problem in Example 1. Compare the exact values of $y(0.2)$, $y(0.4)$, $y(0.6)$, and $y(0.8)$ with the approximations y_1, y_2, y_3, and y_4.

2. Write a computer program for the Adams–Bashforth/Adams–Moulton method.

In Problems 3 and 4 use the Adams–Bashforth/Adams–Moulton method to approximate $y(0.8)$, where $y(x)$ is the solution of the given initial-value problem. Use $h = 0.2$ and the Runge–Kutta method to compute y_1, y_2, and y_3.

3. $y' = 2x - 3y + 1$, $y(0) = 1$

4. $y' = 4x - 2y$, $y(0) = 2$

In Problems 5–8 use the Adams–Bashforth/Adams–Moulton method to approximate $y(1.0)$, where $y(x)$ is the solution of the given initial-value problem. Use $h = 0.2$ and $h = 0.1$ and the Runge–Kutta method to compute y_1, y_2, and y_3.

5. $y' = 1 + y^2$, $y(0) = 0$
6. $y' = y + \cos x$, $y(0) = 1$
7. $y' = (x - y)^2$, $y(0) = 0$
8. $y' = xy + \sqrt{y}$, $y(0) = 1$

9. Use Milne's method to approximate the value of $y(0.4)$, where $y(x)$ is the solution of the initial-value problem in Example 1. Use $h = 0.1$ and the Runge–Kutta method to compute y_1, y_2, and y_3.

10. Consider the recurrence formula
$$y_{n+1} = k(1 - y_n)$$
where $n = 0, 1, 2, \ldots$ and k is a constant. Suppose that the initial value y_0 has an absolute error $\varepsilon = y_0 - y$, where y is the true value. Show that the formula is unstable for increasing n when $|k| > 1$ and stable when $|k| < 1$.

15.8 HIGHER-ORDER EQUATIONS AND SYSTEMS

Second-Order Initial-Value Problem   The numerical procedures discussed so far in this chapter have been applied to only a first-order equation $dy/dx = f(x, y)$ subject to an initial condition $y(x_0) = y_0$. To approximate the solution of, say, a second-order initial-value problem

$$\frac{d^2 y}{dx^2} = f(x, y, y'), \qquad y(x_0) = y_0, \quad y'(x_0) = y'_0 \qquad (1)$$

we begin by reducing the second-order differential equation to a second-order system. If we let $y' = u$, then the equation in (1) becomes

$$\begin{aligned} y' &= u \\ u' &= f(x, y, u) \end{aligned} \qquad (2)$$

We now apply a particular method to each equation in the resulting system. For example, the basic **Euler method** for the system (2) is

$$\begin{aligned} y_{n+1} &= y_n + h u_n \\ u_{n+1} &= u_n + h f(x_n, y_n, u_n) \end{aligned} \qquad (3)$$

EXAMPLE 1   Use the Euler method to obtain the approximate value of $y(0.2)$, where $y(x)$ is the solution of the initial-value problem

$$y'' + xy' + y = 0, \qquad y(0) = 1, \quad y'(0) = 2$$

Solution   In terms of the substitution $y' = u$, the equation is equivalent to the system

$$\begin{aligned} y' &= u \\ u' &= -xu - y \end{aligned}$$

Thus, from (3) we obtain

$$\begin{aligned} y_{n+1} &= y_n + h u_n \\ u_{n+1} &= u_n + h[-x_n u_n - y_n] \end{aligned}$$

15.8 Higher-Order Equations and Systems

Using the step size $h = 0.1$ and $y_0 = 1$, $u_0 = 2$, we find

$$y_1 = y_0 + (0.1)u_0$$
$$= 1 + (0.1)2 = 1.2$$
$$u_1 = u_0 + (0.1)[-x_0 u_0 - y_0]$$
$$= 2 + (0.1)[-(0)(2) - 1] = 1.9$$
$$y_2 = y_1 + (0.1)u_1$$
$$= 1.2 + (0.1)(1.9) = 1.39$$
$$u_2 = u_1 + (0.1)[-x_1 u_1 - y_1]$$
$$= 1.9 + (0.1)[-(0.1)(1.9) - 1.2] = 1.761$$

In other words, $y(0.2) \approx 1.39$ and $y'(0.2) \approx 1.761$. □

Systems An initial-value problem for a second-order system—that is, a system of two first-order differential equations—is

$$x' = f(t, x, y)$$
$$y' = g(t, x, y) \qquad (4)$$
$$x(t_0) = x_0, \quad y(t_0) = y_0$$

As we did in (3), to approximate a solution of this problem we apply a numerical method to each equation in (4). The **fourth-order Runge–Kutta method** looks like this:

$$x_{n+1} = x_n + \frac{1}{6}(m_1 + 2m_2 + 2m_3 + m_4)$$
$$y_{n+1} = y_n + \frac{1}{6}(k_1 + 2k_2 + 2k_3 + k_4) \qquad (5)$$

where

$$m_1 = hf(t_n, x_n, y_n) \qquad k_1 = hg(t_n, x_n, y_n)$$
$$m_2 = hf\left(t_n + \frac{1}{2}h, x_n + \frac{1}{2}m_1, y_n + \frac{1}{2}k_1\right) \qquad k_2 = hg\left(t_n + \frac{1}{2}h, x_n + \frac{1}{2}m_1, y_n + \frac{1}{2}k_1\right)$$
$$m_3 = hf\left(t_n + \frac{1}{2}h, x_n + \frac{1}{2}m_2, y_n + \frac{1}{2}k_2\right) \qquad k_3 = hg\left(t_n + \frac{1}{2}h, x_n + \frac{1}{2}m_2, y_n + \frac{1}{2}k_2\right) \qquad (6)$$
$$m_4 = hf(t_n + h, x_n + m_3, y_n + k_3) \qquad k_4 = hg(t_n + h, x_n + m_3, y_n + k_3)$$

EXAMPLE 2 Consider the initial-value problem

$$x' = 2x + 4y$$
$$y' = -x + 6y$$
$$x(0) = -1, \quad y(0) = 6$$

Use the fourth-order Runge–Kutta method to approximate $x(0.6)$ and $y(0.6)$. Compare the results for $h = 0.2$ and $h = 0.1$.

Solution We shall illustrate the computations of x_1 and y_1 with the step size of $h = 0.2$. With the identifications $f(t, x, y) = 2x + 4y$, $g(t, x, y) = -x + 6y$, $t_0 = 0$, $x_0 = -1$, and $y_0 = 6$, we see from (6) that

$$m_1 = hf(t_0, x_0, y_0) = 0.2f(0, -1, 6) = 0.2[2(-1) + 4(6)] = 4.4000$$

$$k_1 = hg(t_0, x_0, y_0) = 0.2g(0, -1, 6) = 0.2[-1(-1) + 6(6)] = 7.4000$$

$$m_2 = hf\left(t_0 + \frac{1}{2}h, x_0 + \frac{1}{2}m_1, y_0 + \frac{1}{2}k_1\right) = 0.2f(0.1, 1.2, 9.7) = 8.2400$$

$$k_2 = hg\left(t_0 + \frac{1}{2}h, x_0 + \frac{1}{2}m_1, y_0 + \frac{1}{2}k_1\right) = 0.2g(0.1, 1.2, 9.7) = 11.4000$$

$$m_3 = hf\left(t_0 + \frac{1}{2}h, x_0 + \frac{1}{2}m_2, y_0 + \frac{1}{2}k_2\right) = 0.2f(0.1, 3.12, 11.7) = 10.6080$$

$$k_3 = hg\left(t_0 + \frac{1}{2}h, x_0 + \frac{1}{2}m_2, y_0 + \frac{1}{2}k_2\right) = 0.2g(0.1, 3.12, 11.7) = 13.4160$$

$$m_4 = hf(t_0 + h, x_0 + m_3, y_0 + k_3) = 0.2f(0.2, 8, 20.216) = 19.3760$$

$$k_4 = hg(t_0 + h, x_0 + m_3, y_0 + k_3) = 0.2g(0.2, 8, 20.216) = 21.3776$$

Therefore, from (5) we get

$$x_1 = x_0 + \frac{1}{6}(m_1 + 2m_2 + 2m_3 + m_4)$$

$$= -1 + \frac{1}{6}(4.4 + 2(8.24) + 2(10.608) + 19.3760) = 9.2453$$

$$y_1 = y_0 + \frac{1}{6}(k_1 + 2k_2 + 2k_3 + k_4)$$

$$= 6 + \frac{1}{6}(7.4 + 2(11.4) + 2(13.416) + 21.3776) = 19.0683$$

where, as usual, the computed values are rounded to four decimal places. These numbers give the approximations $x_1 \approx x(0.2)$ and $y_1 \approx y(0.2)$. The subsequent values, obtained with the aid of a computer, are listed in Tables 15.16 and 15.17.

TABLE 15.16 Runge–Kutta Method with $h = 0.2$

m_1	m_2	m_3	m_4	k_1	k_2	k_3	k_4	t_n	x_n	y_n
								0.00	−1.0000	6.000
4.4000	8.2400	10.6080	19.3760	7.4000	11.4000	13.4160	21.3776	0.20	9.2453	19.0683
18.9527	31.1564	37.8870	63.6848	21.0329	31.7573	36.9716	57.8214	0.40	46.0327	55.1203
62.5093	97.7863	116.0063	187.3669	56.9378	84.8495	98.0688	151.4191	0.60	158.9430	150.8192

TABLE 15.17 Runge–Kutta Method with $h = 0.1$

m_1	m_2	m_3	m_4	k_1	k_2	k_3	k_4	t_n	x_n	y_n
								0.00	−1.0000	6.0000
2.2000	3.1600	3.4560	4.8720	3.7000	4.7000	4.9520	6.3256	0.10	2.3840	10.8883
4.8321	6.5742	7.0778	9.5870	6.2946	7.9413	8.3482	10.5957	0.20	9.3379	19.1332
9.5208	12.5821	13.4258	17.7609	10.5461	13.2339	13.8872	17.5358	0.30	22.5541	32.8539
17.6524	22.9090	24.3055	31.6554	17.4569	21.8114	22.8549	28.7393	0.40	46.5103	55.4420
31.4788	40.3496	42.6387	54.9202	28.6141	35.6245	37.2840	46.7207	0.50	88.5729	92.3006
54.6348	69.4029	73.1247	93.4107	46.5231	57.7482	60.3774	75.4370	0.60	160.7563	152.0025

□

The reader should verify that the solution of the initial-value problem in Example 2 is given by $x(t) = (26t - 1)e^{4t}$, $y(t) = (13t + 6)e^{4t}$. From these equations we see that the exact values are $x(0.6) = 160.9384$ and $y(0.6) = 152.1198$.

In conclusion, we state **Euler's method** for the general system (4):

$$x_{n+1} = x_n + f(t_n, x_n, y_n)$$
$$y_{n+1} = y_n + g(t_n, x_n, y_n)$$

EXERCISES 15.8 *Answers to odd-numbered problems begin on page A-74.*

1. Use the Euler method to approximate $y(0.2)$, where $y(x)$ is the solution of the initial-value problem

$$y'' - 4y' + 4y = 0$$
$$y(0) = -2, \quad y'(0) = 1$$

Use $h = 0.1$. Find the exact solution of the problem and compare the exact value of $y(0.2)$ with y_2.

2. Use the Euler method to approximate $y(1.2)$, where $y(x)$ is the solution of the initial-value problem

$$x^2 y'' - 2xy' + 2y = 0, \quad x > 0$$
$$y(1) = 4, \quad y'(1) = 9$$

Use $h = 0.1$. Find the exact solution of the problem and compare the exact value of $y(1.2)$ with y_2.

3. Repeat Problem 1 using the Runge–Kutta method with $h = 0.2$. with $h = 0.1$.

4. Repeat Problem 2 using the Runge–Kutta method with $h = 0.2$. with $h = 0.1$.

5. Use the Runge–Kutta method to obtain the approximate value of $y(0.2)$, where $y(x)$ is a solution of the initial-value problem

$$y'' - 2y' + 2y = e^t \cos t$$
$$y(0) = 1, \quad y'(0) = 2$$

Use $h = 0.2$ and $h = 0.1$.

6. When $E = 100$ volts, $R = 10$ ohms, and $L = 1$ henry, the system of differential equations for the currents $i_1(t)$ and $i_3(t)$ in an electrical network is given by

$$\frac{di_1}{dt} = -20 i_1 + 10 i_3 + 100$$

$$\frac{di_3}{dt} = 10 i_1 - 20 i_3$$

where $i_1(0) = 0$ and $i_3(0) = 0$. Use the Runge–Kutta method to approximate $i_1(t)$ and $i_3(t)$ at $t = 0.1, 0.2, 0.3, 0.4$, and 0.5. Use $h = 0.1$.

In Problems 7–10 use the Runge–Kutta method to approximate $x(0.2)$ and $y(0.2)$. Compare the results for $h = 0.2$ and $h = 0.1$.

7. $x' = 2x - y$
$y' = x$
$x(0) = 6, y(0) = 2$

8. $x' = x + 2y$
$y' = 4x + 3y$
$x(0) = 1, y(0) = 1$

9. $x' = -y + t$
$y' = x - t$
$x(0) = -3, y(0) = 5$

10. $x' = 6x + y + 6t$
$y' = 4x + 3y - 10t + 4$
$x(0) = 0.5, y(0) = 0.2$

15.9 SECOND-ORDER BOUNDARY-VALUE PROBLEMS

In the last four sections we have focused on techniques that yield an approximation to a solution of a first-order initial-value problem $y' = f(x, y)$, $y(x_0) = y_0$. In addition, we saw in the last section that we can adapt the approximation techniques to a second-order initial-value problem $y'' = f(x, y, y')$, y_0, $y'(x_0) = y_1$ by reducing the second-order differential equation to a system of first-order equations. In this section we shall examine a method for approximating a solution of a second-order boundary-value problem $y'' = f(x, y, y')$, $y(a) = \alpha$, $y(b) = \beta$. We note at that outset that this method does not require reducing the second-order differential equation to a system of equations.

Finite Difference Approximations Recall from (2) of Section 15.5 that a Taylor's series expansion of a function $y(x)$ can be written as

$$y(x + h) = y(x) + y'(x)h + y''(x)\frac{h^2}{2} + y'''(x)\frac{h^3}{6} + \cdots \quad (1)$$

and so

$$y(x - h) = y(x) - y'(x)h + y''(x)\frac{h^2}{2} - y'''(x)\frac{h^3}{6} + \cdots \quad (2)$$

If h is small, we can ignore terms involving $h^4, h^5, \ldots$, since these values are negligible. Indeed, if we ignore all terms involving h^2 and higher, then (1) and (2) yield, in turn, the following approximations for $y'(x)$:

$$y'(x) \approx \frac{1}{h}[y(x + h) - y(x)] \quad (3)$$

$$y'(x) \approx \frac{1}{h}[y(x) - y(x - h)] \quad (4)$$

Subtracting (1) and (2) also gives

$$y'(x) \approx \frac{1}{2h}[y(x + h) - y(x - h)] \quad (5)$$

On the other hand, if we add (1) and (2) and solve for $y''(x)$, we get

$$y''(x) \approx \frac{1}{h^2}[y(x + h) - 2y(x) + y(x - h)] \quad (6)$$

The right sides of (3), (4), (5), and (6) are called **difference quotients**. The expressions

$$y(x + h) - y(x) \qquad y(x) - y(x - h)$$
$$y(x + h) - y(x - h) \qquad y(x + h) - 2y(x) + y(x - h)$$

15.9 Second-Order Boundary-Value Problems

are called **finite differences**. Specifically, $y(x + h) - y(x)$ is called a **forward difference**, $y(x) - y(x - h)$ is a **backward difference**, and both $y(x + h) - y(x - h)$ and $y(x + h) - 2y(x) + y(x - h)$ are called **central differences**. The results given in (5) and (6) are referred to as **central difference approximations** for the derivatives y' and y'', respectively.

Difference Equation   Consider now a linear second-order boundary-value problem

$$y'' + P(x)y' + Q(x)y = f(x)$$
$$y(a) = \alpha, \quad y(b) = \beta \tag{7}$$

Suppose $a = x_0 < x_1 < x_2 < \cdots < x_{n-1} < x_n = b$ represents a regular partition of the interval $[a, b]$; that is, $x_i = a + ih$, where $i = 0, 1, 2, \ldots, n$ and $h = (b - a)/n$. The points

$$x_1 = a + h, \quad x_2 = a + 2h, \quad \ldots, \quad x_{n-1} = a + (n - 1)h$$

are called **interior mesh points** of the interval $[a, b]$. If we let

$$y_i = y(x_i), \quad P_i = P(x_i), \quad Q_i = Q(x_i), \quad \text{and} \quad f_i = f(x_i)$$

and if y'' and y' in (7) are replaced by the central difference approximations (5) and (6), we get

$$\frac{y_{i+1} - 2y_i + y_{i-1}}{h^2} + P_i \frac{y_{i+1} - y_{i-1}}{2h} + Q_i y_i = f_i$$

or, after simplifying,

$$\boxed{\left(1 + \frac{h}{2} P_i\right) y_{i+1} + (-2 + h^2 Q_i) y_i + \left(1 - \frac{h}{2} P_i\right) y_{i-1} = h^2 f_i} \tag{8}$$

The last equation, known as a **finite difference equation**, is an approximation to the differential equation. It enables us to approximate the solution $y(x)$ of (7) at the interior mesh points $x_1, x_2, \ldots, x_{n-1}$ of the interval $[a, b]$. By letting i take on the values $1, 2, \ldots, n - 1$ in (8), we obtain $n - 1$ equations in the $n - 1$ unknowns $y_1, y_2, \ldots, y_{n-1}$. Bear in mind that we know y_0 and y_n since these are the prescribed boundary conditions $y_0 = y(x_0) = y(a) = \alpha$ and $y_n = y(x_n) = y(b) = \beta$.

In the first example we shall consider a boundary-value problem for which we can compare the approximate values found with the exact values of an explicit solution.

EXAMPLE 1   Use the difference equation (8) with $n = 4$ to approximate the solution of

$$y'' - 4y = 0, \quad y(0) = 0, \quad y(1) = 5$$

Solution To use (8) we identify $P(x) = 0$, $Q(x) = -4$, $f(x) = 0$, $h = (1-0)/4 = \frac{1}{4}$. Hence the difference equation is

$$y_{i+1} - 2.25y_i + y_{i-1} = 0 \tag{9}$$

Now, the interior points are $x_1 = 0 + \frac{1}{4}$, $x_2 = 0 + \frac{2}{4}$, $x_3 = 0 + \frac{3}{4}$, and so for $i = 1, 2$, and 3, (9) yields the following system for the corresponding y_1, y_2, and y_3:

$$y_2 - 2.25y_1 + y_0 = 0$$
$$y_3 - 2.25y_2 + y_1 = 0$$
$$y_4 - 2.25y_3 + y_2 = 0$$

Using the boundary conditions $y_0 = 0$ and $y_4 = 5$, the foregoing system becomes

$$-2.25y_1 + y_2 = 0$$
$$y_1 - 2.25y_2 + y_3 = 0$$
$$y_2 - 2.25y_3 = -5$$

Solving the system gives,

$$y_1 = 0.7256, \quad y_2 = 1.6327 \quad \text{and} \quad y_3 = 2.9479$$

Now the general solution to the given differential equation is $y = c_1 \cosh 2x + c_2 \sinh 2x$. The condition $y(0) = 0$ implies $c_1 = 0$. The other boundary condition gives c_2. In this way we see that an explicit solution of the boundary-value problem is $y(x) = (5 \sinh 2x)/\sinh 2$. Thus, the exact values (rounded to four decimal places) of this solution at the interior points are

$$y(0.25) = 0.7184, \, y(0.5) = 1.6201 \quad \text{and} \quad y(0.75) = 2.9354 \quad \square$$

The accuracy of the approximations in Example 1 can be improved by using a smaller value of h. Of course, the trade-off is that a smaller value of h necessitates solving a larger system of equations. It is left as an exercise to show that with $h = \frac{1}{8}$, approximations to $y(0.25)$, $y(0.5)$, and $y(0.75)$ are 0.7202, 1.6233, and 2.9386, respectively. See Problem 11.

EXAMPLE 2 Use the difference equation (8) with $n = 10$ to approximate the solution of

$$y'' + 3y' + 2y = 4x^2, \quad y(1) = 1, \quad y(2) = 6$$

Solution In this case we identify $P(x) = 3$, $Q(x) = 2$, $f(x) = 4x^2$, $h = (2-1)/10 = 0.1$, and so (8) becomes

$$1.15y_{i+1} - 1.98y_i + 0.85y_{i-1} = 0.04x_i^2 \tag{10}$$

Now, the interior points are $x_1 = 1.1$, $x_2 = 1.2$, $x_3 = 1.3$, $x_4 = 1.4$, $x_5 = 1.5$, $x_6 = 1.6$, $x_7 = 1.7$, $x_8 = 1.8$, and $x_9 = 1.9$. Thus, for $i = 1, 2, \ldots, 9$ and $y_0 = 1$, $y_{10} = 6$, (10) gives a system of nine equations and nine unknowns:

$$1.15y_2 - 1.98y_1 \qquad\qquad = -0.8016$$
$$1.15y_3 - 1.98y_2 + 0.85y_1 = 0.0576$$
$$1.15y_4 - 1.98y_3 + 0.85y_2 = 0.0676$$
$$1.15y_5 - 1.98y_4 + 0.85y_3 = 0.0784$$
$$1.15y_6 - 1.98y_5 + 0.85y_4 = 0.0900$$
$$1.15y_7 - 1.98y_6 + 0.85y_5 = 0.1024$$
$$1.15y_8 - 1.98y_7 + 0.85y_6 = 0.1156$$
$$1.15y_9 - 1.98y_8 + 0.85y_7 = 0.1296$$
$$\qquad\quad - 1.98y_9 + 0.85y_8 = -6.7556$$

We can solve this large system using Gaussian elimination or, with relative ease, by means of a computer algebra system such as Mathematica. The result is found to be $y_1 = 2.4047$, $y_2 = 3.4432$, $y_3 = 4.2010$, $y_4 = 4.7469$, $y_5 = 5.1359$, $y_6 = 5.4124$, $y_7 = 5.6117$, $y_8 = 5.7620$, and $y_9 = 5.8855$. □

▲ **Remark** The approximation method that we have just considered can be extended to boundary-value problems in which the first derivative is specified at a boundary—for example, a problem such as $y'' = f(x, y, y')$, $y'(a) = \alpha$, $y(b) = \beta$. See Problem 13. Moreover, in the two sections that follow we shall see that the finite difference method can also be used to approximate the solutions of elliptic, parabolic, and hyperbolic partial differential equations.

EXERCISES 15.9   Answers to odd-numbered problems begin on page A-74.

In Problems 1–10 use the finite difference method and the indicated value of n to approximate the solution of the given boundary-value problem.

1. $y'' + 9y = 0$, $y(0) = 4$, $y(2) = 1$; $n = 4$
2. $y'' - y = x^2$, $y(0) = 0$, $y(1) = 0$; $n = 4$
3. $y'' + 2y' + y = 5x$, $y(0) = 0$, $y(1) = 0$; $n = 5$
4. $y'' - 10y' + 25y = 1$, $y(0) = 1$, $y(1) = 0$; $n = 5$
5. $y'' - 4y' + 4y = (x + 1)e^{2x}$, $y(0) = 3$, $y(1) = 0$; $n = 6$
6. $y'' + 5y' = 4\sqrt{x}$, $y(1) = 1$, $y(2) = -1$; $n = 6$
7. $x^2y'' + 3xy' + 3y = 0$, $y(1) = 5$, $y(2) = 0$; $n = 8$
8. $x^2y'' - xy' + y = \ln x$, $y(1) = 0$, $y(2) = -2$; $n = 8$
9. $y'' + (1 - x)y' + xy = x$, $y(0) = 0$, $y(1) = 2$; $n = 10$
10. $y'' + xy' + y = x$, $y(0) = 1$, $y(1) = 0$; $n = 10$
11. Rework Example 1 using $n = 8$.
12. The electrostatic potential u between two concentric spheres of radius $r = 1$ and $r = 4$ is determined from

$$\frac{d^2u}{dr^2} + \frac{2}{r}\frac{du}{dr} = 0, \quad u(1) = 50, \quad u(4) = 100$$

Use the method of this section with $n = 6$ to approximate the solution of this boundary-value problem.

13. Consider the boundary-value problem $y'' + xy = 0$, $y'(0) = 1$, $y(1) = -1$.
    (a) Find the difference equation corresponding to the differential equation. Show that for $i = 0, 1, 2, \ldots,$ $n - 1$, the difference equation yields n equations in $n + 1$ unknowns $y_{-1}, y_0, y_1, y_2, \ldots, y_{n-1}$. Here y_{-1} and y_0 are unknowns, since y_{-1} represents an approximation to y at the exterior point $x = -h$ and y_0 is not specified at $x = 0$.
    (b) Use the central difference approximation (5) to show that $y_1 - y_{-1} = 2h$. Use this equation to eliminate y_{-1} from the system in part (a).
    (c) Use $n = 5$ and the system of equations found in parts (a) and (b) to approximate the solution of the original boundary-value problem.

15.10 NUMERICAL METHODS FOR PARTIAL DIFFERENTIAL EQUATIONS: ELLIPTIC EQUATIONS

Using the difference quotients derived in the preceding section, we can easily replace a linear second-order partial differential equation by a difference equation. In the discussion that follows we shall confine our attention to elliptic partial differential equations such as Laplace's equation.

A Difference Equation Replacement for Laplace's Equation If u is a function of two variables x and y, we can form two central differences

$$u(x + h, y) - 2u(x, y) + u(x - h, y) \quad \text{and} \quad u(x, y + h) - 2u(x, y) + u(x, y - h)$$

It then follows from (6) of Section 15.9 that finite difference approximations for the second partial derivatives u_{xx} and u_{yy} are given by

$$\frac{\partial^2 u}{\partial x^2} \approx \frac{1}{h^2}[u(x + h, y) - 2u(x, y) + u(x - h, y)] \tag{1}$$

$$\frac{\partial^2 u}{\partial y^2} \approx \frac{1}{h^2}[u(x, y + h) - 2u(x, y) + u(x, y - h)] \tag{2}$$

Now by adding (1) and (2) we obtain a **five-point approximation** to the Laplacian:

$$\frac{\partial^2 u}{\partial x^2} + \frac{\partial^2 u}{\partial y^2} = \frac{1}{h^2}[u(x + h, y) + u(x, y + h) + u(x - h, y) + u(x, y - h) - 4u(x, y)]$$

Hence, we can replace Laplace's equation $\partial^2 u/\partial x^2 + \partial^2 u/\partial y^2 = 0$ by the **difference equation**:

$$u(x + h, y) + u(x, y + h) + u(x - h, y) + u(x, y - h) - 4u(x, y) = 0 \tag{3}$$

If we adopt the notation

$$u(x, y) = u_{ij}$$
$$u(x + h, y) = u_{i+1, j}$$
$$u(x, y + h) = u_{i, j+1}$$
$$u(x - h, y) = u_{i-1, j}$$
$$u(x, y - h) = u_{i, j-1}$$

then (3) becomes

$$\boxed{u_{i+1, j} + u_{i, j+1} + u_{i-1, j} + u_{i, j-1} - 4u_{ij} = 0} \tag{4}$$

To understand (4) a little better, suppose a rectangular grid consisting of horizontal lines spaced h units apart and vertical lines spaced h units apart

15.10 Elliptic Equations

is placed over a region R bounded by a curve C in which we are seeking a solution of Laplace's equation. The number h is called the **mesh size**. See Figure 15.36(a). The points of intersection $P_{ij} = P(ih, jh)$, i and j integers, of the lines are called **mesh points** or **lattice points**. A mesh point is an **interior point** if its four nearest neighboring mesh points are points of R. Points in R or on C that are not interior points are called **boundary points**. For example, in Figure 15.36(a) we have

$$P_{20} = P(2h, 0), \quad P_{11} = P(h, h), \quad P_{21} = P(2h, h), \quad P_{22} = P(2h, 2h)$$

and so on. Of the points listed, P_{21} and P_{22} are interior points, P_{20} and P_{11} are boundary points. Now from (4) we see that

$$u_{ij} = \frac{1}{4}[u_{i+1,j} + u_{i,j+1} + u_{i-1,j} + u_{i,j-1}] \quad (5)$$

and so, as shown in Figure 15.36(b), the value of u_{ij} at an interior mesh point of R is the average of the values of u at four neighboring mesh points. The neighboring points $P_{i+1,j}$, $P_{i,j+1}$, $P_{i-1,j}$, and $P_{i,j-1}$ correspond, respectively, to the four points on a compass E, N, W, and S.

Dirichlet Problem Recall that in the Dirichlet problem for Laplace's equation $\nabla^2 u = 0$, the values of $u(x, y)$ are prescribed on the boundary C of a region R. The basic idea is to find an approximate solution to Laplace's equation at interior mesh points by replacing the partial differential equation at these points by the difference equation (4). Hence, the approximate values of u at the mesh points—namely, the u_{ij}—are related to each other and, possibly, to known values of u if a mesh point lies on the boundary C. In this manner we obtain a system of linear algebraic equations which we solve for the unknown u_{ij}. The following example will illustrate the method for a square region.

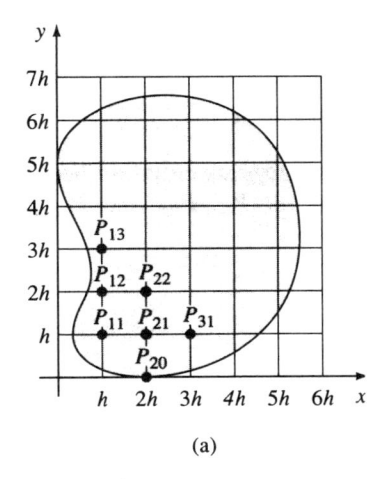

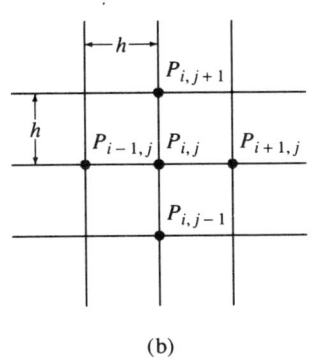

FIGURE 15.36

EXAMPLE 1 In Problem 14 of Exercises 12.5 you were asked to solve the boundary-value problem

$$\frac{\partial^2 u}{\partial x^2} + \frac{\partial^2 u}{\partial y^2} = 0, \quad 0 < x < 2, \quad 0 < y < 2$$

$$u(0, y) = 0, \quad u(2, y) = y(2 - y), \quad 0 < y < 2$$

$$u(x, 0) = 0, \quad u(x, 2) = \begin{cases} x, & 0 < x < 1 \\ 2 - x, & 1 \leq x < 2 \end{cases}$$

utilizing the superposition principle. To apply the present numerical method let us start with a mesh size of $h = \frac{2}{3}$. As we see in Figure 15.37, that choice yields four interior points and eight boundary points. The numbers listed next to the boundary points are the exact values of u obtained from the specified condition along that boundary. For example, at $P_{31} = P(3h, h) = P(2, \frac{2}{3})$, we have $x = 2$ and $y = \frac{2}{3}$ and so the condition $u(2, y)$ gives $u(2, \frac{2}{3}) = \frac{2}{3}(2 - \frac{2}{3}) = \frac{8}{9}$. Similarly, at $P_{13} = P(\frac{2}{3}, 2)$, the condition $u(x, 2)$ gives $u(\frac{2}{3}, 2) = \frac{2}{3}$. We now

FIGURE 15.37

apply (4) at each interior point. For example, at P_{11} we have $i = 1$ and $j = 1$, so (4) becomes

$$u_{21} + u_{12} + u_{01} + u_{10} - 4u_{11} = 0$$

Since $u_{01} = u(0, \frac{2}{3}) = 0$ and $u_{10} = u(\frac{2}{3}, 0) = 0$, the foregoing equation becomes $-4u_{11} + u_{21} + u_{12} = 0$. Repeating this, in turn, at P_{21}, P_{12}, and P_{22} produces three additional equations:

$$\begin{aligned} -4u_{11} + u_{21} + u_{12} &= 0 \\ u_{11} - 4u_{21} + u_{22} &= -\frac{8}{9} \\ u_{11} - 4u_{12} + u_{22} &= -\frac{2}{3} \\ u_{21} + u_{12} - 4u_{22} &= -\frac{14}{9} \end{aligned} \quad (6)$$

Using Mathematica to solve the system, we find the approximate temperatures at the four interior points to be:

$$u_{11} = \frac{7}{36} = 0.1944 \qquad u_{21} = \frac{5}{12} = 0.4167$$

$$u_{12} = \frac{13}{36} = 0.3611 \qquad u_{22} = \frac{7}{12} = 0.5833 \qquad \square$$

As in the discussion of ordinary differential equations, we expect that a smaller value of h will improve the accuracy of the approximation. However, using a smaller mesh size means, of course, that there are more interior mesh points, and correspondingly there is a much larger system of equations to be solved. For a *square* region with length of side L, a mesh size of $h = L/n$ will yield a total of $(n - 1)^2$ interior mesh points. In Example 1, for $n = 8$ the mesh size is a reasonable $h = \frac{2}{8} = \frac{1}{4}$, but the number of interior points is $(8 - 1)^2 = 49$. Thus, we have 49 equations in 49 unknowns. In the next example, we shall use a mesh size of $h = \frac{1}{2}$.

EXAMPLE 2 As we see in Figure 15.38, with $n = 4$, a mesh size $h = \frac{2}{4} = \frac{1}{2}$ for the square in Example 1 gives $3^2 = 9$ interior mesh points. Applying (4) at

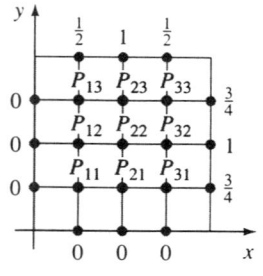

FIGURE 15.38

these points and using the indicated boundary conditions give nine equations in nine unknowns. So that the reader can verify the results, we give the system in an unsimplified form:

$$\begin{aligned} u_{21} + u_{12} + 0 + 0 - 4u_{11} &= 0 \\ u_{31} + u_{22} + u_{11} + 0 - 4u_{21} &= 0 \\ \tfrac{3}{4} + u_{32} + u_{21} + 0 - 4u_{31} &= 0 \\ u_{22} + u_{13} + u_{11} + 0 - 4u_{12} &= 0 \\ u_{32} + u_{23} + u_{12} + u_{21} - 4u_{22} &= 0 \\ 1 + u_{33} + u_{22} + u_{31} - 4u_{32} &= 0 \\ u_{23} + \tfrac{1}{2} + 0 + u_{12} - 4u_{13} &= 0 \\ u_{33} + 1 + u_{13} + u_{22} - 4u_{23} &= 0 \\ \tfrac{3}{4} + \tfrac{1}{2} + u_{23} + u_{32} - 4u_{33} &= 0 \end{aligned} \qquad (7)$$

In this case, Mathematica yields

$$u_{11} = \frac{7}{64} = 0.1094 \qquad u_{21} = \frac{51}{224} = 0.2277 \qquad u_{31} = \frac{177}{448} = 0.3951$$

$$u_{12} = \frac{47}{224} = 0.2098 \qquad u_{22} = \frac{13}{32} = 0.4063 \qquad u_{32} = \frac{135}{224} = 0.6027$$

$$u_{13} = \frac{145}{448} = 0.3237 \qquad u_{23} = \frac{131}{224} = 0.5848 \qquad u_{33} = \frac{39}{64} = 0.6094 \qquad \square$$

After we simplify (7), it is interesting to note that the 9×9 matrix of coefficients is

$$\begin{bmatrix} -4 & 1 & 0 & 1 & 0 & 0 & 0 & 0 & 0 \\ 1 & -4 & 1 & 0 & 1 & 0 & 0 & 0 & 0 \\ 0 & 1 & -4 & 0 & 0 & 1 & 0 & 0 & 0 \\ 1 & 0 & 0 & -4 & 1 & 0 & 1 & 0 & 0 \\ 0 & 1 & 0 & 1 & -4 & 1 & 0 & 1 & 0 \\ 0 & 0 & 1 & 0 & 1 & -4 & 0 & 0 & 1 \\ 0 & 0 & 0 & 1 & 0 & 0 & -4 & 1 & 0 \\ 0 & 0 & 0 & 0 & 1 & 0 & 1 & -4 & 1 \\ 0 & 0 & 0 & 0 & 0 & 1 & 0 & 1 & -4 \end{bmatrix} \qquad (8)$$

This is an example of a **sparse matrix**, in that a large percentage of the entries are zeros. The matrix (8) is also an example of a **banded matrix**. These kinds of matrices are characterized by the property that the entries on the main diagonal and on diagonals (or bands) parallel to the main diagonal are all nonzero.

Gauss–Siedel Iteration Problems requiring approximations of solutions of partial differential equations invariably lead to large systems of linear algebraic equations. It is not uncommon to have to solve systems involving hundreds of equations. Although a direct method of solution such as Gaussian elimination leaves unchanged the zero entries outside the bands in a matrix such as (8), it does fill in the positions between the bands with non-zeros. Since storing very large matrices uses up a large portion of computer memory, it is usual practice to solve a large system in an indirect manner. One popular indirect method is called **Gauss–Seidel iteration**.

We shall illustrate this method for the system in (6).* For the sake of simplicity, let us replace the double-subscripted variables u_{11}, u_{21}, u_{12}, and u_{22} by x_1, x_2, x_3, and x_4, respectively.

EXAMPLE 3 **Step 1** *Solve each equation for the variables on the main diagonal of the system.* That is, in (6) solve the first equation for x_1, the second equation for x_2, and so on:

$$\begin{aligned} x_1 &= \phantom{0.25x_1 +{}} 0.25x_2 + 0.25x_3 \\ x_2 &= 0.25x_1 + 0.25x_4 + 0.2222 \\ x_3 &= 0.25x_1 + 0.25x_4 + 0.1667 \\ x_4 &= \phantom{0.25x_1 +{}} 0.25x_2 + 0.25x_3 + 0.3889 \end{aligned} \qquad (9)$$

These equations can be obtained directly by using (5) rather than (4) at the interior points.

Step 2 *Iterate.* We start by making an initial guess for the values of x_1, x_2, x_3, and x_4. If this were simply a system of linear equations and we knew nothing about the solution, we could start with $x_1 = 0$, $x_2 = 0$, $x_3 = 0$, and $x_4 = 0$. But since the solution of (9) represents approximations to a solution of a boundary-value problem, it seems reasonable to use as the initial guess for the values of $x_1 = u_{11}$, $x_2 = u_{21}$, $x_3 = u_{12}$, and $x_4 = u_{22}$ the average of all the boundary conditions. In this case the average of the numbers at the eight boundary points shown in Figure 15.37 is approximately 0.4. Thus, our initial guess will be $x_1 = 0.4$, $x_2 = 0.4$, $x_3 = 0.4$, and $x_4 = 0.4$. Iterations of the Gauss–Seidel method use the x-values as soon as they are computed. Note that the first equation in (9) depends only on x_2 and x_3; thus, substituting $x_2 = 0.4$ and $x_3 = 0.4$ gives $x_1 = 0.2$. Since the second and third equations depend on x_1 and x_4, we use the newly calculated values $x_1 = 0.2$ and $x_4 = 0.4$ to obtain $x_2 = 0.3722$ and $x_3 = 0.3167$. The fourth equation depends on x_2 and x_3, so we use the new values $x_2 = 0.3722$ and $x_3 = 0.3167$ to get $x_4 = 0.5611$. In summary, the first iteration has given the values

$$x_1 = 0.2 \qquad x_2 = 0.3722 \qquad x_3 = 0.3167 \qquad x_4 = 0.5611$$

Note how close these numbers are already to the actual values given at the end of Example 1.

---

*In the context of approximating a solution to Laplace's equation, this iteration technique is often referred to as **Liebman's method**.

The second iteration starts with substituting $x_2 = 0.3722$ and $x_3 = 0.3167$ into the first equation. This gives $x_1 = 0.1722$. From $x_1 = 0.1722$ and the last computed value of x_4, namely, $x_4 = 0.5611$, the second and third equations give, in turn, $x_2 = 0.4055$ and $x_3 = 0.3500$. Using these two values, we find from the fourth equation that $x_4 = 0.5678$. At the end of the second iteration we have

$$x_1 = 0.1722 \qquad x_2 = 0.4055 \qquad x_3 = 0.3500 \qquad x_4 = 0.5678$$

The third through seventh iterations are summarized in the accompanying table.

Iteration	3rd	4th	5th	6th	7th
x_1	0.1889	0.1931	0.1941	0.1944	0.1944
x_2	0.4139	0.4160	0.4165	0.4166	0.4166
x_3	0.3584	0.3605	0.3610	0.3611	0.3611
x_4	0.5820	0.5830	0.5833	0.5833	0.5833

As the BASIC computer program given in Appendix IV indicates, we continue the iterations until the difference between successive iterants is less than a preassigned tolerance (TOL).

Note that to apply Gauss–Seidel iteration to a general system of n linear equations in n unknowns, the variable x_i must actually appear in the ith equation of the system. Moreover, after each equation is solved for x_i, $i = 1, 2, \ldots, n$, the resulting system has the form $\mathbf{X} = \mathbf{AX} + \mathbf{B}$, where all the entries on the main diagonal of $\mathbf{A}$ are zero.

▲ **Remarks** (*i*) In the examples given in this section, the values of u_{ij} were determined using known values of u at boundary points. But what do we do if the region is such that boundary points do not coincide with the actual boundary C of the region R? In that case, the required values can be obtained by interpolation.

(*ii*) It is sometimes possible to reduce the number of equations to solve by employing symmetry. Consider the rectangular region shown in Figure 15.39. The boundary conditions are $u = 0$ along $x = 0$, $x = 2$, $y = 1$ and $u = 100$ along $y = 0$. The region is symmetric about the lines $x = 1$ and $y = \frac{1}{2}$, and the interior points P_{11} and P_{31} are equidistant from the neighboring boundary points at which the specified values of u are the same. Consequently, we assume that $u_{11} = u_{31}$ and so the system of three equations in three unknowns reduces to two equations in two unknowns. See Problem 2.

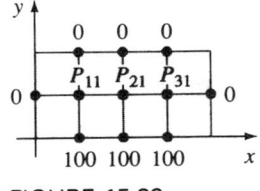

FIGURE 15.39

(*iii*) On a computer this may not be noticeable, but convergence of Gauss–Seidel iteration may not be particularly fast. Also, in a more general setting, Gauss–Seidel iteration may not converge at all. If a system of n linear equations in n unknowns is written in the standard form $\mathbf{AX} = \mathbf{B}$, then a sufficient condition for convergence is that the coefficient matrix $\mathbf{A}$ be **diagonally dominant**. This means that the absolute value of each entry on the main diagonal of $\mathbf{A}$ is greater than the sum of the absolute values of the remaining entries in the same row. For example, the coefficient matrix (8) for the system (7) is clearly diagonally dominant, since in each row $|-4|$ is greater than the sum of the absolute values of the other entries. You should verify that the coefficient matrix for the system (6) has the same property. Diagonal dominance guarantees that Gauss–Seidel iteration converges for any initial values.

EXERCISES 15.10 *Answers to odd-numbered problems begin on page A-75.*

In Problems 1–4 use (4) to approximate the solution of Laplace's equation at the interior points of the given region. Use symmetry when possible.

1. $u(0, y) = 0$, $u(3, y) = y(2 - y)$, $0 < y < 2$
   $u(x, 0) = 0$, $u(x, 2) = x(3 - x)$, $0 < x < 3$
   mesh size: $h = 1$

2. $u(0, y) = 0$, $u(2, y) = 0$, $0 < y < 1$
   $u(x, 0) = 100$, $u(x, 1) = 0$, $0 < x < 2$,
   mesh size: $h = \frac{1}{2}$

3. $u(0, y) = 0$, $u(1, y) = 0$, $0 < y < 1$
   $u(x, 0) = 0$, $u(x, 1) = \sin \pi x$, $0 < x < 1$
   mesh size: $h = \frac{1}{3}$

4. $u(0, y) = 108y^2(1 - y)$, $u(1, y) = 0$, $0 < y < 1$
   $u(x, 0) = 0$, $u(x, 1) = 0$, $0 < x < 1$
   mesh size: $h = \frac{1}{3}$

In Problems 5 and 6 use (5) and Gauss–Seidel iteration to approximate the solution of Laplace's equation at the interior points of a unit square. Use mesh size $h = \frac{1}{4}$. In Problem 5 the boundary conditions are given; in Problem 6 the values of u at boundary points are given in Figure 15.40.

5. $u(0, y) = 0$, $u(1, y) = 100y$, $0 < y < 1$
   $u(x, 0) = 0$, $u(x, 1) = 100x$, $0 < x < 1$

6.

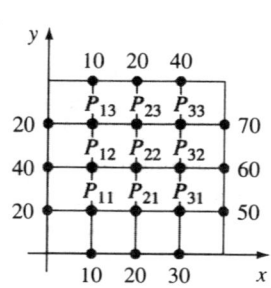

FIGURE 15.40

7. (a) In Problem 12 of Exercises 12.6 we solved a potential problem using a special form of Poisson's equation $\partial^2 u/\partial x^2 + \partial^2 u/\partial y^2 = f(x, y)$. Show that the difference equation replacement for Poisson's equation is

$$u_{i+1,j} + u_{i,j+1} + u_{i-1,j} + u_{i,j-1} - 4u_{ij} = h^2 f(x, y).$$

(b) Use the result in part (a) to approximate the solution of the Poisson equation $\partial^2 u/\partial x^2 + \partial^2 u/\partial y^2 = -2$ at the interior points of the region in Figure 15.41. The mesh size is $h = \frac{1}{2}$, $u = 1$ at every point along $ABCD$, and $u = 0$ at every point along $DEFGA$. Use symmetry and, if necessary, Gauss–Seidel iteration.

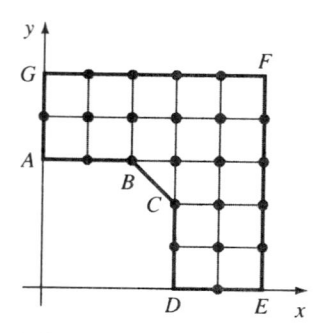

FIGURE 15.41

8. Use the result in part (a) of Problem 7 to approximate the solution of the Poisson equation

$$\frac{\partial^2 u}{\partial x^2} + \frac{\partial^2 u}{\partial y^2} = -64$$

at the interior points of the region in Figure 15.42. The mesh size is $h = \frac{1}{8}$ and $u = 0$ at every point on the boundary of the region. If necessary use Gauss–Seidel iteration.

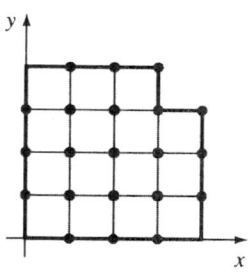

FIGURE 15.42

9. Use the approximation

$$\frac{\partial u}{\partial t} \approx \frac{u(x, t + k) - u(x, t)}{k}$$

to find a difference equation replacement for the heat equation $\partial^2 u/\partial x^2 = \partial u/\partial t$. Note that since x and t usually represent, respectively, spatial and time variables, we do not assume that the mesh size h in the x-direction is the same as the mesh size k in the t-direction.

10. Find a difference equation replacement for the wave equation $\partial^2 u/\partial x^2 = \partial^2 u/\partial t^2$.

15.11 NUMERICAL METHODS FOR PARTIAL DIFFERENTIAL EQUATIONS: PARABOLIC EQUATIONS

In this section we shall examine two methods for approximating the solution of a boundary-value problem of the type

$$c\frac{\partial^2 u}{\partial x^2} = \frac{\partial u}{\partial t}, \quad 0 < x < a, \quad t > 0$$
$$u(0, t) = T_1, \quad u(a, t) = T_2, \quad t > 0 \quad (1)$$
$$u(x, 0) = f(x), \quad 0 < x < a$$

The equation in this problem is the one-dimensional heat equation and is, as we know from Section 12.1, a parabolic partial differential equation. The constants T_1 and T_2 represent temperatures at the endpoints of the rod. Recall that the function f can be interpreted as the initial temperature distribution in a homogeneous rod extending from $x = 0$ to $x = a$. Although we shall not prove it, the boundary-value problem has a unique solution when f is continuous on the closed interval $[0, a]$. This latter condition will be assumed, and so we shall replace the initial condition in (1) by $u(x, 0) = f(x), 0 \leq x \leq a$.

A Difference Equation Replacement for the Heat Equation To approximate the solution of (1) we first replace the heat equation by a difference equation. Using the central difference approximation (1) of Section 15.10

$$\frac{\partial^2 u}{\partial x^2} \approx \frac{1}{h^2}[u(x + h, t) - 2u(x, t) + u(x - h, t)]$$

and the forward difference approximation (3) of Section 15.9

$$\frac{\partial u}{\partial t} \approx \frac{1}{k}[u(x, t + k) - u(x, t)]$$

we can replace the heat equation $c\partial^2 u/\partial x^2 = \partial u/\partial t$ by the following difference equation:

$$\frac{c}{h^2}[u(x + h, t) - 2u(x, t) + u(x - h, t)] = \frac{1}{k}[u(x, t + k) - u(x, t)] \quad (2)$$

If we let $\lambda = ck/h^2$ and

$$u(x, t) = u_{ij}$$
$$u(x + h, t) = u_{i+1, j}$$
$$u(x - h, t) = u_{i-1, j}$$
$$u(x, t + k) = u_{i, j+1}$$

then (2) becomes, after simplifying,

$$u_{i, j+1} = \lambda u_{i+1, j} + (1 - 2\lambda)u_{ij} + \lambda u_{i-1, j} \qquad (3)$$

We wish to approximate a solution of a problem such as (1) on a rectangular region in the xt-plane defined by the inequalities $0 \leq x \leq a, 0 \leq t \leq T$ for some value of time T. Over this region we place a rectangular grid consisting of vertical lines h units apart and horizontal lines k units apart. If we choose two positive integers n and m and define

$$h = \frac{a}{n} \quad \text{and} \quad k = \frac{T}{m}$$

then the vertical and horizontal grid lines are defined by

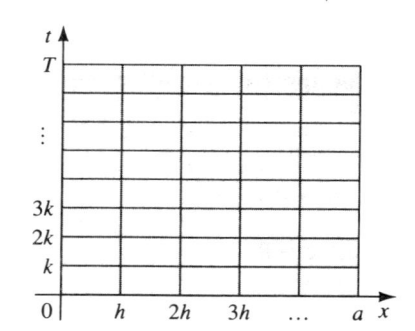

$$x_i = ih, \quad i = 0, 1, 2, \ldots, n$$
$$t_j = jk, \quad j = 0, 1, 2, \ldots, m$$

FIGURE 15.43

See Figure 15.43.
As illustrated in Figure 15.44, the basic idea here is to use (3) to estimate the values of the solution $u(x, t)$ at the points on the $(j + 1)$st time line using only values from the jth time line. For example, the values on the first time line ($j = 1$) depend on the initial condition $u_{i,0} = u(x_i, 0) = f(x_i)$ given on the zeroth time line ($j = 0$). This kind of numerical procedure is called an **explicit finite difference method**.

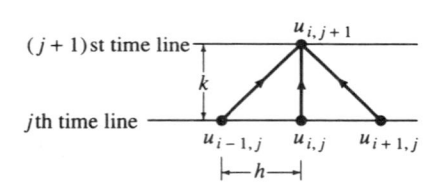

FIGURE 15.44

EXAMPLE 1 Consider the boundary-value problem

$$\frac{\partial^2 u}{\partial x^2} = \frac{\partial u}{\partial t}, \quad 0 < x < 1, \quad 0 < t < 0.5$$
$$u(0, t) = 0, \quad u(1, t) = 0, \quad 0 \leq t \leq 0.5$$
$$u(x, 0) = \sin \pi x, \quad 0 \leq x \leq 1$$

First we identify $c = 1$, $a = 1$, and $T = 0.5$. If we choose, say, $n = 5$ and $m = 50$, then $h = \frac{1}{5} = 0.2$, $k = \frac{0.5}{50} = 0.01$, $\lambda = 0.25$, and

$$x_i = i\frac{1}{5}, \quad i = 0, 1, 2, 3, 4, 5,$$
$$t_j = j\frac{1}{100}, \quad j = 0, 1, 2, \ldots, 50$$

15.11 Parabolic Equations

Thus (3) becomes

$$u_{i,j+1} = 0.25(u_{i+1,j} + 2u_{ij} + u_{i-1,j})$$

By setting $j = 0$ in this formula, we get a formula for the approximations to the temperature u on the first time line:

$$u_{i,1} = 0.25(u_{i+1,0} + 2u_{i,0} + u_{i-1,0})$$

If we then let $i = 1, \ldots, 4$ in the last equation, we obtain, in turn,

$$u_{11} = 0.25(u_{20} + 2u_{10} + u_{00})$$
$$u_{21} = 0.25(u_{30} + 2u_{20} + u_{10})$$
$$u_{31} = 0.25(u_{40} + 2u_{30} + u_{20})$$
$$u_{41} = 0.25(u_{50} + 2u_{40} + u_{40})$$

The first equation in this list is interpreted as

$$u_{11} = 0.25(u(x_2, 0) + 2u(x_1, 0) + u(0, 0))$$
$$= 0.25(u(0.4, 0) + 2u(0.2, 0) + u(0, 0))$$

From the initial condition $u(x, 0) = \sin \pi x$, the last line becomes

$$u_{11} = 0.25(0.951056516 + 2(0.587785252) + 0) = 0.531656755$$

This number represents an approximation to the temperature $u(0.2, 0.01)$. The remaining calculations are summarized in Table 15.18.

TABLE 15.18 Explicit Difference Equation Approximation with $h = 0.2$, $k = 0.01$, $\lambda = 0.25$

Time	$x = 0.20$	$x = 0.40$	$x = 0.60$	$x = 0.80$
0.00	0.5878	0.9511	0.9511	0.5878
0.01	0.5317	0.8602	0.8602	0.5317
0.02	0.4809	0.7781	0.7781	0.4809
0.03	0.4350	0.7038	0.7038	0.4350
0.04	0.3934	0.6366	0.6366	0.3934
0.05	0.3559	0.5758	0.5758	0.3559
0.06	0.3219	0.5208	0.5208	0.3219
0.07	0.2911	0.4711	0.4711	0.2911
0.08	0.2633	0.4261	0.4261	0.2633
0.09	0.2382	0.3854	0.3854	0.2382
0.10	0.2154	0.3486	0.3486	0.2154
0.11	0.1949	0.3153	0.3153	0.1949
0.12	0.1763	0.2852	0.2852	0.1763
0.13	0.1594	0.2580	0.2580	0.1594
0.14	0.1442	0.2333	0.2333	0.1442
0.15	0.1304	0.2111	0.2111	0.1304
0.16	0.1180	0.1909	0.1909	0.1180
0.17	0.1067	0.1727	0.1727	0.1067

(continued)

TABLE 15.18 (*continued*)

Time	$x = 0.20$	$x = 0.40$	$x = 0.60$	$x = 0.80$
0.18	0.0965	0.1562	0.1562	0.0965
0.19	0.0873	0.1413	0.1413	0.0873
0.20	0.0790	0.1278	0.1278	0.0790
0.21	0.0714	0.1156	0.1156	0.0714
0.22	0.0646	0.1045	0.1045	0.0646
0.23	0.0584	0.0946	0.0946	0.0584
0.24	0.0529	0.0855	0.0855	0.0529
0.25	0.0478	0.0774	0.0774	0.0478
0.26	0.0432	0.0700	0.0700	0.0432
0.27	0.0391	0.0633	0.0633	0.0391
0.28	0.0354	0.0572	0.0572	0.0354
0.29	0.0320	0.0518	0.0518	0.0320
0.30	0.0289	0.0468	0.0468	0.0289
0.31	0.0262	0.0424	0.0424	0.0262
0.32	0.0237	0.0383	0.0383	0.0237
0.33	0.0214	0.0347	0.0347	0.0214
0.34	0.0194	0.0313	0.0313	0.0194
0.35	0.0175	0.0284	0.0284	0.0175
0.36	0.0159	0.0256	0.0256	0.0159
0.37	0.0143	0.0232	0.0232	0.0143
0.38	0.0130	0.0210	0.0210	0.0130
0.39	0.0117	0.0190	0.0190	0.0117
0.40	0.0106	0.0172	0.0172	0.0106
0.41	0.0096	0.0155	0.0155	0.0096
0.42	0.0087	0.0140	0.0140	0.0087
0.43	0.0079	0.0127	0.0127	0.0079
0.44	0.0071	0.0115	0.0115	0.0071
0.45	0.0064	0.0104	0.0104	0.0064
0.46	0.0058	0.0094	0.0094	0.0058
0.47	0.0053	0.0085	0.0085	0.0053
0.48	0.0048	0.0077	0.0077	0.0048
0.49	0.0043	0.0070	0.0070	0.0043
0.50	0.0039	0.0063	0.0063	0.0039

□

A BASIC program for the method illustrated in Example 1 is found in Appendix IV.

The reader should use the methods of Chapter 12 to verify that the exact solution of the boundary-value problem in Example 1 is given by $u(x, t) = e^{-\pi^2 t} \sin \pi x$. The following are some sample values:

Exact	Approximation
$u(0.4, 0.05) = 0.5806$	$u_{25} = 0.5758$
$u(0.6, 0.06) = 0.5261$	$u_{36} = 0.5208$
$u(0.2, 0.10) = 0.2191$	$u_{1,10} = 0.2154$
$u(0.8, 0.14) = 0.1476$	$u_{4,14} = 0.1442$

These approximations are comparable to the exact values and are accurate enough for some purposes. But there is a difficulty with the foregoing

15.11 Parabolic Equations

method. Recall that a numerical method is **unstable** if round-off errors or any other errors grow too rapidly as the computations proceed. The numerical procedure illustrated in Example 1 can exhibit this kind of behavior. It can be proved that the procedure is stable if λ is less than or equal to 0.5 but unstable otherwise. Note that in Example 1 we chose the values of h and k so that $\lambda = 0.25$.

The next example will illustrate what happens when a larger step size k is chosen so that λ is greater than 0.5.

EXAMPLE 2 Approximate the solution of the boundary-value problem

$$\frac{\partial^2 u}{\partial x^2} = \frac{\partial u}{\partial t}, \quad 0 < x < 1, \quad 0 < t < 1$$

$$u(0, t) = 0, \quad u(1, t) = 0, \quad 0 \leq t \leq 1$$

$$u(x, 0) = \sin \pi x, \quad 0 \leq x \leq 1$$

using (3) with $n = 5$ and $m = 25$.

Solution In this case $c = 1$, $a = 1$, $T = 1$, $h = \frac{1}{5} = 0.2$, $k = \frac{1}{25} = 0.04$, and $\lambda = 1$. With the aid of the program given in Appendix IV, we obtain the results in Table 15.19.

TABLE 15.19 Explicit Difference Equation Approximation with $h = 0.2$, $k = 0.04$, $\lambda = 1$

Time	$x = 0.20$	$x = 0.40$	$x = 0.60$	$x = 0.80$
0.00	0.5878	0.9511	0.9511	0.5878
0.04	0.3633	0.5878	0.5878	0.3633
0.08	0.2245	0.3633	0.3633	0.2245
0.12	0.1388	0.2245	0.2245	0.1388
0.16	0.0858	0.1388	0.1388	0.0858
0.20	0.0530	0.0857	0.0858	0.0530
0.24	0.0327	0.0530	0.0530	0.0328
0.28	0.0203	0.0327	0.0328	0.0202
0.32	0.0124	0.0204	0.0201	0.0126
0.36	0.0080	0.0121	0.0129	0.0075
0.40	0.0041	0.0088	0.0067	0.0054
0.44	0.0047	0.0020	0.0075	0.0012
0.48	−0.0026	0.0102	−0.0043	0.0063
0.52	0.0128	−0.0171	0.0208	−0.0106
0.56	−0.0299	0.0507	−0.0485	0.0314
0.60	0.0806	−0.1291	0.1306	−0.0799
0.64	−0.2097	0.3403	−0.3396	0.2105
0.68	0.5500	−0.8895	0.8903	−0.5500
0.72	−1.4395	2.3298	−2.3299	1.4403
0.76	3.7693	−6.0992	6.1000	−3.7702
0.80	−9.8685	15.9685	−15.9693	9.8702
0.84	25.8370	−41.8063	41.8080	−25.8395
0.88	−67.6432	109.4512	−109.4538	67.6475
0.92	177.0945	−286.5482	286.5525	−177.1013
0.96	−463.6426	750.1950	−750.2019	463.6537
1.00	1213.8374	−1964.0393	1964.0503	−1213.8555

□

For values of t from 0 to 0.4 the results given in Table 15.19 are believable, although the accuracy, upon closer examination, is found to be poor. The results for values of t greater than 0.5 are not believable. Because the values of $u(x, 0) = \sin \pi x$ on the interval $0 \leq x \leq 1$ are never negative, we expect that the solution $u(x, t)$ should never be negative in this example. Moreover, the large values near the end of the output clearly reveal the large errors produced by the instability of the procedure for this choice of h and k. Of course, $\lambda = 1$ exceeds 0.5 and so the instability was predictable. For $\lambda = 0.5$, the numerical results are satisfactory, but some increase in accuracy results from using smaller values of k with consequent smaller values of λ. The necessity of using very small step sizes in the time direction is the principal fault of this method.

Crank–Nicholson Method   There are **implicit finite difference methods** for solving parabolic partial differential equations. These methods require that we solve a system of equations to determine the approximate values of u on the $(j + 1)$st time line. However, implicit methods do not suffer from instability problems.

The algorithm introduced by J. Crank and P. Nicholson in 1947 is used mostly for solving the heat equation and consists of replacing the second partial derivative in $c\dfrac{\partial^2 u}{\partial x^2} = \dfrac{\partial u}{\partial t}$ by an average of two central difference quotients, one evaluated at t and the other at $t + k$:

$$\frac{c}{2}\left[\frac{u(x + h, t) - 2u(x, t) + u(x - h, t)}{h^2} + \frac{u(x + h, t + k) - 2u(x, t + k) + u(x - h, t + k)}{h^2}\right]$$
$$= \frac{1}{k}[u(x, t + k) - u(x, t)] \tag{4}$$

If we again define $\lambda = ck/h^2$ and rearrange terms, (4) can be written as

$$\boxed{-u_{i-1, j+1} + \alpha u_{i, j+1} - u_{i+1, j+1} = u_{i+1, j} - \beta u_{ij} + u_{i-1, j}} \tag{5}$$

where $\alpha = 2(1 + 1/\lambda)$ and $\beta = 2(1 - 1/\lambda)$, $j = 0, 1, \ldots, m - 1$ and $i = 1, 2, \ldots, n - 1$.

For each choice of j the difference equation (5) for $i = 1, 2, \ldots, n - 1$ gives $n - 1$ equations in $n - 1$ unknowns $u_{i, j+1}$. Because of the prescribed boundary conditions, the values of $u_{i, j+1}$ are known for $i = 0$ and for $i = n$. For example, in the case $n = 4$ the system of equations for determining the approximate values of u on the $(j + 1)$st time line is

$$-u_{0, j+1} + \alpha u_{1, j+1} - u_{2, j+1} = u_{2, j} - \beta u_{1, j} + u_{0, j}$$
$$-u_{1, j+1} + \alpha u_{2, j+1} - u_{3, j+1} = u_{3, j} - \beta u_{2, j} + u_{1, j}$$
$$-u_{2, j+1} + \alpha u_{3, j+1} - u_{4, j+1} = u_{4, j} - \beta u_{3, j} + u_{2, j}$$

or

$$\begin{aligned} \alpha u_{1, j+1} - u_{2, j+1} &= b_1 \\ -u_{1, j+1} + \alpha u_{2, j+1} - u_{3, j+1} &= b_2 \\ -u_{2, j+1} + \alpha u_{3, j+1} &= b_3 \end{aligned} \tag{6}$$

where

$$b_1 = u_{2, j} - \beta u_{1, j} + u_{0, j} + u_{0, j+1}$$
$$b_2 = u_{3, j} - \beta u_{2, j} + u_{1, j}$$
$$b_3 = u_{4, j} - \beta u_{3, j} + u_{2, j} + u_{4, j+1}$$

15.11 Parabolic Equations

In general, if we use the difference equation (5) to determine values of u on the $(j+1)$st time line, we need to solve a linear system $\mathbf{AX} = \mathbf{B}$, where the coefficient matrix $\mathbf{A}$ is a **tridiagonal matrix**:

$$\mathbf{A} = \begin{bmatrix} \alpha & -1 & 0 & 0 & 0 & \cdots & & 0 \\ -1 & \alpha & -1 & 0 & 0 & & & 0 \\ 0 & -1 & \alpha & -1 & 0 & & & 0 \\ 0 & 0 & -1 & \alpha & -1 & & & 0 \\ \vdots & & & & & \ddots & & \vdots \\ 0 & 0 & 0 & 0 & 0 & & \alpha & -1 \\ 0 & 0 & 0 & 0 & 0 & \cdots & -1 & \alpha \end{bmatrix}$$

and the entries of the column matrix $\mathbf{B}$ are

$$\begin{aligned} b_1 &= u_{2,j} - \beta u_{1,j} + u_{0,j} + u_{0,j+1} \\ b_2 &= u_{3,j} - \beta u_{2,j} + u_{1,j} \\ b_3 &= u_{4,j} - \beta u_{3,j} + u_{2,j} \\ &\vdots \\ b_{n-1} &= u_{n,j} - \beta u_{n-1,j} + u_{n-2,j} + u_{n,j+1} \end{aligned}$$

A BASIC program for the Crank–Nicholson method is given in Appendix IV.

EXAMPLE 3 Use the Crank–Nicholson method to approximate the solution of the boundary-value problem

$$0.25 \frac{\partial^2 u}{\partial x^2} = \frac{\partial u}{\partial t}, \quad 0 < x < 2, \quad 0 < t < 0.3$$

$$u(0, t) = 0, \quad u(2, t) = 0, \quad 0 \le t \le 0.3$$

$$u(x, 0) = \sin \pi x, \quad 0 \le x \le 2$$

using $n = 8$ and $m = 30$.

Solution From the identifications $a = 2$, $T = 0.3$, $h = \frac{1}{4} = 0.25$, $k = \frac{1}{100} = 0.01$, and $c = 0.25$, we get $\lambda = 0.04$. With the aid of the computer program in Appendix IV we get the results in Table 15.20.

TABLE 15.20 Crank–Nicholson Method with $h = 0.25$, $k = 0.01$, $\lambda = 0.25$

Time	$x = 0.25$	$x = 0.50$	$x = 0.75$	$x = 1.00$	$x = 1.25$	$x = 1.50$	$x = 1.75$
0.00	0.7071	1.0000	0.7071	0.0000	−0.7071	−1.0000	−0.7071
0.01	0.6907	0.9768	0.6907	0.0000	−0.6907	−0.9768	−0.6907
0.02	0.6747	0.9542	0.6747	0.0000	−0.6747	−0.9542	−0.6747
0.03	0.6591	0.9321	0.6591	0.0000	−0.6591	−0.9321	−0.6591
0.04	0.6438	0.9105	0.6438	0.0000	−0.6438	−0.9105	−0.6438
0.05	0.6289	0.8894	0.6289	0.0000	−0.6289	−0.8894	−0.6289
0.06	0.6144	0.8688	0.6144	0.0000	−0.6144	−0.8688	−0.6144
0.07	0.6001	0.8487	0.6001	0.0000	−0.6001	−0.8487	−0.6001
0.08	0.5862	0.8291	0.5862	0.0000	−0.5862	−0.8291	−0.5862
0.09	0.5727	0.8099	0.5727	0.0000	−0.5727	−0.8099	−0.5727
0.10	0.5594	0.7911	0.5594	0.0000	−0.5594	−0.7911	−0.5594
0.11	0.5464	0.7728	0.5464	0.0000	−0.5464	−0.7728	−0.5464

(continued)

Table 15.20 (Continued)

Time	x = 0.25	x = 0.50	x = 0.75	x = 1.00	x = 1.25	x = 1.50	x = 1.75
0.00	0.7071	1.0000	0.7071	0.0000	−0.7071	−1.0000	−0.7071
0.13	0.5214	0.7374	0.5214	0.0000	−0.5214	−0.7374	−0.5214
0.14	0.5093	0.7203	0.5093	0.0000	−0.5093	−0.7203	−0.5093
0.15	0.4975	0.7036	0.4975	0.0000	−0.4975	−0.7036	−0.4975
0.16	0.4860	0.6873	0.4860	0.0000	−0.4860	−0.6873	−0.4860
0.17	0.4748	0.6714	0.4748	0.0000	−0.4748	−0.6714	−0.4748
0.18	0.4638	0.6559	0.4638	0.0000	−0.4638	−0.6559	−0.4638
0.19	0.4530	0.6407	0.4530	0.0000	−0.4530	−0.6407	−0.4530
0.20	0.4425	0.6258	0.4425	0.0000	−0.4425	−0.6258	−0.4425
0.21	0.4323	0.6114	0.4323	0.0000	−0.4323	−0.6114	−0.4323
0.22	0.4223	0.5972	0.4223	0.0000	−0.4223	−0.5972	−0.4223
0.23	0.4125	0.5834	0.4125	0.0000	−0.4125	−0.5834	−0.4125
0.24	0.4029	0.5699	0.4029	0.0000	−0.4029	−0.5699	−0.4029
0.25	0.3936	0.5567	0.3936	0.0000	−0.3936	−0.5567	−0.3936
0.26	0.3845	0.5438	0.3845	0.0000	−0.3845	−0.5438	−0.3845
0.27	0.3756	0.5312	0.3756	0.0000	−0.3756	−0.5312	−0.3756
0.28	0.3669	0.5189	0.3669	0.0000	−0.3669	−0.5189	−0.3669
0.29	0.3584	0.5068	0.3584	0.0000	−0.3584	−0.5068	−0.3584
0.30	0.3501	0.4951	0.3501	0.0000	−0.3501	−0.4951	−0.3501

The reader should verify that the function satisfying all the conditions in Example 3 is $u(x, t) = e^{-\pi^2 t/4} \sin \pi x$. Sample comparisons are given in the following table:

Exact	Approximation
$u(0.75, 0.05) = 0.6250$	$u_{35} = 0.6289$
$u(0.50, 0.20) = 0.6105$	$u_{2, 20} = 0.6259$
$u(0.25, 0.10) = 0.5525$	$u_{1, 10} = 0.5594$

These comparisons show that the absolute errors are of the order 10^{-2} or 10^{-3}. Smaller errors can be obtained by decreasing *either* h or k.

EXERCISES 15.11

Answers to odd-numbered problems begin on page A-75.

1. Use the difference equation (3) to approximate the solution of the boundary-value problem

$$\frac{\partial^2 u}{\partial x^2} = \frac{\partial u}{\partial t}, \quad 0 < x < 2, \quad 0 < t < 1$$

$$u(0, t) = 0, \quad u(2, t) = 0, \quad 0 \le t \le 1$$

$$u(x, 0) = \begin{cases} 1, & 0 \le x \le 1 \\ 0, & 1 < x \le 2 \end{cases}$$

Use $n = 8$ and $m = 40$.

2. Using the Fourier series solution obtained in Problem 1 of Exercises 12.3 with $L = 2$, one can sum the first 20 terms to estimate the values for $u(0.25, 0.1)$, $u(1, 0.5)$, and $u(1.5, 0.8)$ for the solution $u(x, t)$ of Problem 1. A student wrote a computer program to do this and obtained the results $u(0.25, 0.1) = 0.3794$, $u(1, 0.5) = 0.1854$, and $u(1.5, 0.8) = 0.0623$. Assume these results are accurate for all digits given. Compare these values with the approximations obtained in Problem 1. Find the absolute errors in each case.

3. Solve Problem 1 by the Crank–Nicholson method with $n = 8$ and $m = 40$. Use the values for $u(0.25, 0.1)$, $u(1, 0.5)$, and $u(1.5, 0.8)$ given in Problem 2 to compute the absolute errors.

4. Repeat Problem 1 using $n = 8$ and $m = 20$. Use the values for $u(0.25, 0.1)$, $u(1, 0.5)$, and $u(1.5, 0.8)$ given in Problem 2 to compute the absolute errors. Why are the approximations so inaccurate in this case?

5. Solve Problem 1 by the Crank–Nicholson method with $n = 8$ and $m = 20$. Use the values for $u(0.25, 0.1)$, $u(1, 0.5)$, and $u(1.5, 0.8)$ given in Problem 2 to compute the absolute errors. Compare the absolute errors with those obtained in Problem 4.

6. It was shown in Section 12.2 that if a rod of length L is made of a material with thermal conductivity K, specific heat γ, and density ρ, then the temperature $u(x, t)$ satisfies the partial differential equation

$$\frac{K}{\gamma\rho}\frac{\partial^2 u}{\partial x^2} = \frac{\partial u}{\partial t}, \quad 0 < x < L$$

Consider the boundary-value problem consisting of the foregoing equation and conditions

$$u(0, t) = 0, \quad u(L, t) = 0, \quad 0 \le t \le 10$$

$$u(x, 0) = f(x), \quad 0 \le x \le L$$

Use the difference equation (3) in this section with $n = 10$ and $m = 10$ to approximate the solution of the boundary-value problem when:

(a) $L = 20$, $K = 0.15$, $\rho = 8.0$, $\gamma = 0.11$, $f(x) = 30$
(b) $L = 50$, $K = 0.15$, $\rho = 8.0$, $\gamma = 0.11$, $f(x) = 30$
(c) $L = 20$, $K = 1.10$, $\rho = 2.7$, $\gamma = 0.22$, $f(x) = 0.5x(20 - x)$
(d) $L = 100$, $K = 1.04$, $\rho = 10.6$, $\gamma = 0.06$,
$$f(x) = \begin{cases} 0.8x, & 0 \le x \le 50 \\ 40 - 0.8(x - 50), & 50 < x \le 100 \end{cases}$$

7. Solve Problem 6 by the Crank–Nicholson method with $n = 10$ and $m = 10$.

8. Repeat Problem 6 if the endpoint temperatures are $u(0, t) = 0$ and $u(L, t) = 20$, $0 \le t \le 10$.

9. Solve Problem 8 by the Crank–Nicholson method.

10. Consider the boundary-value problem in Example 3. Assume that $n = 4$.
    (a) Find the new value of λ.
    (b) Use the Crank–Nicholson difference equation (5) to find the system of equations for u_{11}, u_{21}, and u_{31}, that is, the approximate values of u on the first time line. [*Hint:* Set $j = 0$ in (5) and let i take on the values 1, 2, 3.]
    (c) Solve the system of three equations without the aid of the BASIC program in Appendix IV. Compare your results with the corresponding entries in Table 15.20.

11. Consider a rod whose length is $L = 20$ for which $K = 1.05$, $\rho = 10.6$, and $\gamma = 0.056$. Suppose

$$u(0, t) = 20, \quad u(20, t) = 30$$

$$u(x, 0) = 50$$

(a) Use the method outlined in Section 12.6 to find the steady-state solution $\psi(x)$.
(b) Use the Crank–Nicholson method to approximate the temperatures $u(x, t)$ for $0 \le t \le T_{max}$. Select T_{max} large enough to allow the temperatures to approach the steady-state values. Compare the approximations for $t = T_{max}$ with the values of $\psi(x)$ found in part (a).

15.12 NUMERICAL METHOD FOR PARTIAL DIFFERENTIAL EQUATIONS: HYPERBOLIC EQUATIONS

We turn now to a numerical method for approximating a solution of a boundary-value problem involving the one-dimensional wave equation

$$c^2 \frac{\partial^2 u}{\partial x^2} = \frac{\partial^2 u}{\partial t^2}, \quad 0 < x < a, \quad t > 0$$

$$u(0, t) = 0, \quad u(a, t) = 0, \quad t > 0 \qquad (1)$$

$$u(x, 0) = f(x), \quad \left.\frac{\partial u}{\partial t}\right|_{t=0} = g(x), \quad 0 \le x \le a$$

The wave equation is an example of a hyperbolic partial differential equation. This problem has a unique solution if the functions f and g have continuous second-order derivatives on the interval $(0, a)$ and $f(a) = f(0) = 0$.

A Difference Equation Replacement for the Wave Equation As in the preceding two sections, our first objective is to replace the partial differential equation (1) by a difference equation. In this case we replace the two second-order partial derivatives by the central difference approximations

$$\frac{\partial^2 u}{\partial x^2} \approx \frac{1}{h^2}[u(x + h, t) - 2u(x, t) + u(x - h, t)]$$

and
$$\frac{\partial^2 u}{\partial t^2} \approx \frac{1}{k^2}[u(x, t + k) - 2u(x, t) + u(x, t - k)]$$

Thus, we replace $c^2(\partial^2 u/\partial x^2) = \partial^2 u/\partial t^2$ by

$$\frac{c^2}{h^2}[u(x + h, t) - 2u(x, t) + u(x - h, t)] = \frac{1}{k^2}[u(x, t + k) - 2u(x, t) + u(x, t - k)] \quad (2)$$

If we let $\lambda = ck/h$, then (2) can be expressed as

$$u_{i, j+1} = \lambda^2 u_{i+1, j} + 2(1 - \lambda^2)u_{ij} + \lambda^2 u_{i-1, j} - u_{i, j-1} \quad (3)$$

for $i = 1, 2, \ldots, n - 1$ and $j = 1, 2, \ldots, m - 1$.

The numerical method based on (3), like the first method considered in Section 15.11, is an explicit finite difference method. As before, we use the difference equation to approximate the solution $u(x, t)$ of (1) over a rectangular region in the xt-plane defined by the inequalities $0 \le x \le a$, $0 \le t \le T$. If n and m are positive integers and

$$h = \frac{a}{n} \qquad k = \frac{T}{m}$$

the vertical and horizontal grid lines on this region are defined by

$$x_i = ih, \qquad i = 0, 1, 2, \ldots, n$$
$$t_j = jk, \qquad j = 0, 1, 2, \ldots, n$$

As shown in Figure 15.45, (3) enables us to obtain the approximation $u_{i, j+1}$ on the $(j + 1)$st time line from the values indicated on the jth and the $(j - 1)$st time lines. Moreover, we use

$u_{0, j} = u(0, jk) = 0, \qquad u_{n, j} = u(a, 0) = 0$ boundary conditions

and $u_{i, 0} = u(x_i, 0) = f(x_i)$ initial condition

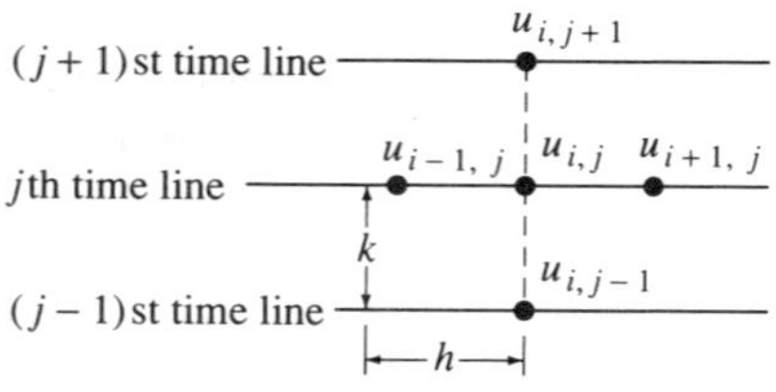

There is one minor problem in getting started. One can see from (3) that for $j = 1$ we need to know the values of $u_{i,1}$—that is, the estimates of u on the first time line—in order to find $u_{i,2}$. But from Figure 15.45, with $j = 0$, we see that the values of $u_{i,1}$ on the first time line depend on the values of $u_{i,0}$

FIGURE 15.45

on the zeroth time line and on the values of $u_{i,-1}$. To compute these latter values we make use of the initial-velocity condition $u_t(x, 0) = g(x)$. At $t = 0$ it follows from (5) of Section 15.9 that

$$g(x_i) = u_t(x_i, 0) \approx \frac{u(x_i, k) - u(x_i, -k)}{2k} \tag{4}$$

In order to make sense of the term $u(x_i, -k) = u_{i,-1}$ in (4), we have to imagine $u(x, t)$ extended backward in time. It follows from (4) that

$$u(x_i, -k) \approx u(x_i, k) - 2kg(x_i)$$

This last result suggests that we define

$$u_{i,-1} = u_{i,1} - 2kg(x_i) \tag{5}$$

in the iteration of (3). By substituting (5) into (3) when $j = 0$, we get the special case

$$u_{i,1} = \frac{\lambda^2}{2}(u_{i+1,0} + u_{i-1,0}) + (1 - \lambda^2)u_{i,0} + kg(x_i) \tag{6}$$

A BASIC program for the preceding method is given in Appendix IV.

EXAMPLE 1 Approximate the solution of the boundary-value problem

$$4\frac{\partial^2 u}{\partial x^2} = \frac{\partial^2 u}{\partial t^2}, \quad 0 < x < 1, \quad 0 < t < 1,$$

$$u(0, t) = 0, \quad u(1, t) = 0, \quad 0 \leq t \leq 1$$

$$u(x, 0) = \sin \pi x, \quad \left.\frac{\partial u}{\partial t}\right|_{t=0} = 0, \quad 0 \leq x \leq 1$$

using (3) with $n = 5$ and $m = 20$.

Solution We make the identifications $c = 2$, $a = 1$, and $T = 1$. With $n = 5$ and $m = 20$, we get $h = \frac{1}{5} = 0.2$, $k = \frac{1}{20} = 0.05$, and $\lambda = 0.5$. Thus, with $g(x) = 0$, (6) and (3) become, respectively,

$$u_{i,1} = 0.125(u_{i+1,0} + u_{i-1,0}) + 0.75u_{i,0} \tag{7}$$

$$u_{i,j+1} = 0.25u_{i+1,j} + 1.5u_{ij} + 0.25u_{i-1,j} - u_{i,j-1} \tag{8}$$

For $i = 1, 2, 3, 4$, (7) yields the following values for $u_{i,1}$ on the first time line:

$$\begin{aligned} u_{11} &= 0.125(u_{20} + u_{00}) + 0.75u_{10} = 0.55972100 \\ u_{21} &= 0.125(u_{30} + u_{10}) + 0.75u_{20} = 0.90564761 \\ u_{31} &= 0.125(u_{40} + u_{20}) + 0.75u_{30} = 0.90564761 \\ u_{41} &= 0.125(u_{50} + u_{30}) + 0.75u_{40} = 0.55972100 \end{aligned} \tag{9}$$

Note that the results in (9) were obtained from the initial condition $u(x, 0) = \sin \pi x$. For example, $u_{20} = \sin(0.2\pi)$, and so on. Now $j = 1$ in (8) gives

$$u_{i,2} = 0.25u_{i+1,1} + 1.5u_{i,1} + 0.25u_{i-1,1} - u_{i,0}$$

TABLE 15.21 Explicit Difference Equation Approximation with $h = 0.2$, $k = 0.05$, $\lambda = 0.5$

Time	$x = 0.20$	$x = 0.40$	$x = 0.60$	$x = 0.80$
0.00	0.5878	0.9511	0.9511	0.5878
0.05	0.5597	0.9056	0.9056	0.5597
0.10	0.4782	0.7738	0.7738	0.4782
0.15	0.3510	0.5680	0.5680	0.3510
0.20	0.1903	0.3080	0.3080	0.1903
0.25	0.0115	0.0185	0.0185	0.0115
0.30	−0.1685	−0.2727	−0.2727	−0.1685
0.35	−0.3324	−0.5378	−0.5378	−0.3324
0.40	−0.4645	−0.7516	−0.7516	−0.4645
0.45	−0.5523	−0.8936	−0.8936	−0.5523
0.50	−0.5873	−0.9503	−0.9503	−0.5873
0.55	−0.5663	−0.9163	−0.9163	−0.5663
0.60	−0.4912	−0.7947	−0.7947	−0.4912
0.65	−0.3691	−0.5973	−0.5973	−0.3691
0.70	−0.2119	−0.3428	−0.3428	−0.2119
0.75	−0.0344	−0.0556	−0.0556	−0.0344
0.80	0.1464	0.2369	0.2369	0.1464
0.85	0.3132	0.5068	0.5068	0.3132
0.90	0.4501	0.7283	0.7283	0.4501
0.95	0.5440	0.8803	0.8803	0.5440
1.00	0.5860	0.9482	0.9482	0.5860

and so for $i = 1, 2, 3, 4$, we get

$$u_{12} = 0.25u_{21} + 1.5u_{11} + 0.25u_{01} - u_{10}$$
$$u_{22} = 0.25u_{31} + 1.5u_{21} + 0.25u_{11} - u_{20}$$
$$u_{32} = 0.25u_{41} + 1.5u_{31} + 0.25u_{21} - u_{30}$$
$$u_{42} = 0.25u_{51} + 1.5u_{41} + 0.25u_{31} - u_{40}$$

From the boundary conditions, the initial conditions, and the data obtained in (9), these equations give us the approximations for u on the second time line. These results and the remaining calculations are listed in Table 15.21.

It is readily verified that the exact solution of the problem in Example 1 is $u(x, t) = \sin \pi x \cos 2\pi t$. With this function we can compare the exact results with the approximations given in Table 15.21. For example, some sample values are listed in the following table:

Exact	Approximation
$u(0.4, 0.25) = 0$	$u_{25} = 0.0185$
$u(0.6, 0.3) = -0.2939$	$u_{36} = -0.2727$
$u(0.2, 0.5) = -0.5878$	$u_{1,10} = -0.5873$
$u(0.8, 0.7) = -0.1816$	$u_{4,14} = -0.2119$

15.12 Hyperbolic Equations

Note that the approximations are in the same "ballpark" as the exact values, but the accuracy is not particularly impressive. We can, however, obtain more accurate results. The accuracy of this algorithm varies with the choice of λ. Of course, λ is determined by the choice of the integers n and m, which in turn determine the values of the step sizes h and k. It can be proved that the best accuracy is always obtained from this method when the ratio $\lambda = kc/h$ is equal to one—in other words, when the step in the time direction is $k = h/c$. For example, the choice $n = 8$ and $m = 16$ yields $h = \frac{1}{8}$, $k = \frac{1}{16}$, and $\lambda = 1$. The sample values are given in the table:

Exact	Approximation
$u(0.25, 0.3125) = -0.2706$	$u_{25} = -0.2706$
$u(0.375, 0.375) = -0.6533$	$u_{36} = -0.6533$
$u(0.125, 0.625) = -0.2706$	$u_{1,10} = -0.2706$

These values clearly show the improved accuracy.

We note in conclusion that this explicit finite difference method for the wave equation is stable when $\lambda \leq 1$ and unstable when $\lambda > 1$.

▲ **Remark** It is worthwhile to spend a few moments considering the sources of error in the finite difference methods discussed in the last several sections. The error in the numerical solution (that is, in the approximations) arises from two sources:

(*i*) replacement of the partial differential equation by a difference equation, and

(*ii*) round-off error in computer arithmetic.

The error in the values of the solution u which arises from approximation of the partial differential equation by a difference equation is an example of **discretization error**. If $u(x_i, t_j)$ denotes the exact value of the solution at the ijth grid point, and if u_{ij} is the numerical estimate arising from a finite difference method, then the total error at the ijth point is

$$u(x_i, t_j) - u_{ij} = \text{discretization error} + \text{round-off error}$$

One might expect the numerical estimates to improve as h and k are decreased. Eventually, however, the benefit from a decrease in the step size is offset by the increase in round-off error which results from increasing the amount of computer arithmetic required. Theoretical bounds from the discretization error can be established for some difference methods. Such bounds can sometimes be used to prove that the discretization error does decrease in magnitude as h and k are decreased. Unfortunately, such bounds are not ordinarily useful in practice for the actual estimation of the total error in the numerical solution. Such an analysis of the difference method presented in this section reveals that if $\lambda = 1$, then the only errors involved are round-off errors in the computer arithmetic. This explains why the accuracy in Example 1 can be increased by taking $\lambda = 1$.

EXERCISES 15.12   Answers to odd-numbered problems begin on page A-81.

1. Use the difference equation (3) to approximate the solution of the boundary-value problem

$$c^2 \frac{\partial^2 u}{\partial x^2} = \frac{\partial^2 u}{\partial t^2}, \quad 0 < x < a, \quad 0 < t < T$$

$$u(0, t) = 0, \quad u(a, t) = 0, \quad 0 \le t \le T$$

$$u(x, 0) = f(x), \quad \left.\frac{\partial u}{\partial t}\right|_{t=0} = 0, \quad 0 \le x \le a$$

when

(a) $c = 1, a = 1, T = 1, f(x) = x(1 - x); n = 4$ and $m = 10$
(b) $c = 1, a = 2, T = 1, f(x) = e^{-16(x-1)^2}; n = 5$ and $m = 10$
(c) $c = \sqrt{2}, a = 1, T = 1, f(x) = \begin{cases} 0, & 0 \le x \le 0.5 \\ 0.5, & 0.5 < x \le 1 \end{cases}$; $n = 10$ and $m = 25$

2. Given the boundary-value problem

$$\frac{\partial^2 u}{\partial x^2} = \frac{\partial^2 u}{\partial t^2}, \quad 0 < x < 1, \quad 0 < t < 0.5$$

$$u(0, t) = 0, \quad u(1, t) = 0, \quad 0 \le t \le 0.5$$

$$u(x, 0) = \sin \pi x, \quad \left.\frac{\partial u}{\partial t}\right|_{t=0} = 0, \quad 0 \le x \le 1$$

(a) Use the methods of Chapter 12 to verify that the solution of the problem is $u(x, t) = \sin \pi x \cos \pi t$.
(b) Use the method of this section to approximate the solution of the problem without the aid of the BASIC program in Appendix IV. Use $n = 4$ and $m = 5$.
(c) Compute the absolute error at each interior grid point.

3. Approximate the solution of the boundary-value problem in Problem 2 using a computer program with (a) $n = 5$, $m = 10$; (b) $n = 5$, $m = 20$; and (c) $n = 10$, $m = 50$.

4. Given the boundary-value problem

$$\frac{\partial^2 u}{\partial x^2} = \frac{\partial^2 u}{\partial t^2}, \quad 0 < x < 1, \quad 0 < t < 1$$

$$u(0, t) = 0, \quad u(1, t) = 0, \quad 0 \le t \le 1$$

$$u(x, 0) = x(1 - x), \quad \left.\frac{\partial u}{\partial t}\right|_{t=0} = 0, \quad 0 \le x \le 1$$

Use $h = k = \frac{1}{5}$ in (6) to compute the values of $u_{i,1}$ by hand.

5. It was shown in Section 12.2 that the equation of a vibrating string is

$$\frac{T}{\rho} \frac{\partial^2 u}{\partial x^2} = \frac{\partial^2 u}{\partial t^2}$$

where T is the magnitude of the tension in the string and ρ is its mass per unit length. Suppose a string of length 60 cm is secured to the x-axis at its ends and is released from rest from the initial displacement

$$f(x) = \begin{cases} 0.01x, & 0 \le x \le 30 \\ 0.30 - 0.01(x - 30), & 30 < x \le 60 \end{cases}$$

Use the difference equation (3) in this section to approximate the solution of the boundary-value problem when $h = 10$ and $k = 5\sqrt{\rho/T}$ and where $\rho = 0.0225$ g/cm, $T = 1.4 \times 10^7$ dynes. Use $m = 50$.

6. Repeat Problem 5 using

$$f(x) = \begin{cases} 0.02x, & 0 \le x \le 15 \\ 0.30 - \dfrac{x - 15}{150}, & 15 < x \le 60 \end{cases}$$

and $h = 10$, $k = 2.5\sqrt{\rho/T}$. Use $m = 50$.

15.13 APPROXIMATION OF EIGENVALUES

> **REVIEW**
>
> The **inner product** or **dot product** of two $n \times 1$ column vectors
>
> $$\mathbf{X} = \begin{bmatrix} x_1 \\ x_2 \\ \vdots \\ x_n \end{bmatrix} \quad \text{and} \quad \mathbf{Y} = \begin{bmatrix} y_1 \\ y_2 \\ \vdots \\ y_n \end{bmatrix}$$
>
> is the number $\mathbf{X} \cdot \mathbf{Y} = \mathbf{X}^T\mathbf{Y} = x_1 y_1 + x_2 y_2 + \cdots + x_n y_n$. The **norm** of a vector $\mathbf{X}$ is $\|\mathbf{X}\| = \sqrt{\mathbf{X}^T\mathbf{X}}$. See Section 7.8.

Recall that to find the eigenvalues for a matrix $\mathbf{A}$ we must find the roots of the polynomial equation $\det(\mathbf{A} - \lambda \mathbf{I}) = 0$. If $\mathbf{A}$ is a large matrix, the computations involved in obtaining this characteristic equation can be very formidable. Moreover, even if we could find the exact characteristic equation, it is likely that we would have to use a numerical procedure to approximate its roots. There are alternative numerical procedures for approximating eigenvalues and the corresponding eigenvectors. The procedure that we shall consider in this section deals with matrices that possess a **dominant eigenvalue**.

> **DEFINITION 15.4** **Dominant Eigenvalue**
>
> Let $\lambda_1, \lambda_2, \ldots, \lambda_n$ denote the eigenvalues of an $n \times n$ matrix $\mathbf{A}$. The eigenvalue λ_1 is said to be the **dominant eigenvalue** of $\mathbf{A}$ if
>
> $$|\lambda_1| > \lambda_i, \quad i = 2, 3, \ldots, n$$
>
> An eigenvector corresponding to λ_1 is called a **dominant eigenvector** of $\mathbf{A}$.

In other words, a dominant eigenvalue is one whose absolute value is greater than the absolute value of each of the remaining eigenvalues.

EXAMPLE 1 In Example 2 of Section 7.7 we saw that the eigenvalues of the matrix

$$\mathbf{A} = \begin{bmatrix} 1 & 2 & 1 \\ 6 & -1 & 0 \\ -1 & -2 & -1 \end{bmatrix} \quad \text{are} \quad \lambda_1 = 0, \lambda_2 = -4, \text{ and } \lambda_3 = 3$$

Since $|-4| > 0$ and $|-4| > 3$, we see that $\lambda_2 = -4$ is the dominant eigenvalue of **A**. □

A matrix **A** may not have a dominant eigenvalue.

EXAMPLE 2 (a) The matrix $\mathbf{A} = \begin{bmatrix} 4 & 0 \\ 0 & -4 \end{bmatrix}$ has eigenvalues $\lambda_1 = -2$ and $\lambda_2 = 2$. Since $|\lambda_1| = |\lambda_2| = 2$, it follows that there is no dominant eigenvalue.

(b) The eigenvalues of the matrix

$$\mathbf{A} = \begin{bmatrix} 2 & 0 & 0 \\ 0 & 5 & 1 \\ 0 & 0 & 5 \end{bmatrix} \quad \text{are} \quad \lambda_1 = 2, \lambda_2 = \lambda_3 = 5$$

Again, the matrix has no dominant eigenvalue. □

Power Method Suppose the $n \times n$ matrix **A** has a dominant eigenvalue λ_1. The iterative technique for approximating a corresponding dominant eigenvector is due to the German mathematician Richard Von Mises (1883–1953) and is called the **power method**. The basic idea of this procedure is first to compute an approximation to a dominant eigenvector by means of the sequence

$$\mathbf{X}_i = \mathbf{A}\mathbf{X}_{i-1}, \quad i = 1, 2, 3, \ldots \tag{1}$$

where $\mathbf{X}_0$ represents a nonzero $n \times 1$ vector that is an initial guess or approximation for the eigenvector we seek. Iterating (1) gives

$$\begin{aligned} \mathbf{X}_1 &= \mathbf{A}\mathbf{X}_0 \\ \mathbf{X}_2 &= \mathbf{A}\mathbf{X}_1 = \mathbf{A}^2\mathbf{X}_0 \\ &\vdots \\ \mathbf{X}_m &= \mathbf{A}\mathbf{X}_{m-1} = \mathbf{A}^m\mathbf{X}_0 \end{aligned} \tag{2}$$

Under certain circumstances for large values of m the vector defined by $\mathbf{X}_m = \mathbf{A}^m\mathbf{X}_0$ is an approximation to a dominant eigenvector. To see this, let us make some further assumptions about the matrix **A**. Let us assume that the eigenvalues of **A** are such that

$$|\lambda_1| > |\lambda_2| \geq |\lambda_3| \geq \cdots \geq |\lambda_n|$$

and that the corresponding n eigenvectors $\mathbf{K}_1, \mathbf{K}_2, \ldots, \mathbf{K}_n$ are linearly independent. Because of this last assumption, $\mathbf{K}_1, \mathbf{K}_2, \ldots, \mathbf{K}_n$ can serve as a basis for R^n (see Section 6.6). Thus, for any nonzero $n \times 1$ vector $\mathbf{X}_0$, constants $c_1, c_2, \ldots, c_n$ can be found such that

$$\mathbf{X}_0 = c_1\mathbf{K}_1 + c_2\mathbf{K}_2 + \cdots + c_n\mathbf{K}_n \tag{3}$$

We shall also assume X_0 is chosen so that $c_1 \neq 0$. Multiplying (3) by A then gives

$$AX_0 = c_1 AK_1 + c_2 AK_2 + \cdots + c_n AK_n$$

Since $AK_1 = \lambda_1 K_1$, $AK_2 = \lambda_2 K_2, \ldots, AK_n = \lambda_n K_n$, the last line becomes

$$AX_0 = c_1 \lambda_1 K_1 + c_2 \lambda_2 K_2 + \cdots + c_n \lambda_n AK_n \tag{4}$$

Multiplying (4) by A gives

$$A^2 X_0 = c_1 \lambda_1 AK_1 + c_2 \lambda_2 AK_2 + \cdots + c_n \lambda_n AK_n$$
$$= c_1 \lambda_1^2 K_1 + c_2 \lambda_2^2 K_2 + \cdots + c_n \lambda_n^2 K_n$$

Continuing in this manner, we find that

$$A^m X_0 = c_1 \lambda_1^m K_1 + c_2 \lambda_2^m K_2 + \cdots + c_n \lambda_n^m K_n \tag{5}$$

$$= \lambda_1^m \left[c_1 K_1 + c_2 \left(\frac{\lambda_2}{\lambda_1}\right)^m K_2 + \cdots + c_n \left(\frac{\lambda_n}{\lambda_1}\right)^m K_n \right] \tag{6}$$

Since $|\lambda_1| > |\lambda_i|$ for $i = 2, 3, \ldots, n$, we have $|\lambda_i/\lambda_1| < 1$ and consequently $\lim_{m \to \infty} (\lambda_i/\lambda_1)^m = 0$. Therefore, as $m \to \infty$, we see from (6) that

$$A^m X_0 \approx \lambda_1^m c_1 K_1 \tag{7}$$

Since a nonzero constant multiple of an eigenvector is an eigenvector, it follows from (7) that for large values of m and under all the assumptions that were made, the $n \times 1$ matrix $X_m = A^m X_0$ is an approximation to a dominant eigenvector associated with the dominant eigenvalue λ_1. The rate at which this method converges depends on the quotient λ_2/λ_1: If $|\lambda_2/\lambda_1|$ is very small, then convergence is fast, whereas if $|\lambda_2/\lambda_1|$ is close to one, the convergence is slow. Of course, this information is not as useful as it seems because we generally do not know the eigenvalues in advance.

It remains, then, to approximate the dominant eigenvalue itself. This can be done by means of the inner product. If K is an eigenvector of a matrix A corresponding to the eigenvalue λ, we have $AK = \lambda K$, and so we have $AK \cdot K = \lambda K \cdot K$. Since $AK \cdot K$ and $K \cdot K$ are scalars, we can solve this last equation for λ:

$$\lambda = \frac{AK \cdot K}{K \cdot K}$$

Hence, if $X_m = A^m X_0$ is an approximation to a dominant eigenvector obtained by the iteration of (1), then the dominant eigenvalue λ_1 can be approximated by the quotient

$$\lambda_1 \approx \frac{AX_m \cdot X_m}{X_m \cdot X_m} \tag{8}$$

The quotient in (8) is sometimes referred to as the **Rayleigh quotient**.

EXAMPLE 3 Use the power method to approximate the dominant eigenvalue and a corresponding dominant eigenvector of $\mathbf{A} = \begin{bmatrix} 4 & 2 \\ 3 & -1 \end{bmatrix}$.

Solution Since we do not know the eigenvalues and eigenvectors, we may as well take $\mathbf{X}_0 = \begin{bmatrix} 1 \\ 1 \end{bmatrix}$. Now the first two terms of the sequence of vectors defined by (1) are

$$\mathbf{X}_1 = \mathbf{A}\mathbf{X}_0 = \begin{bmatrix} 4 & 2 \\ 3 & -1 \end{bmatrix}\begin{bmatrix} 1 \\ 1 \end{bmatrix} = \begin{bmatrix} 6 \\ 2 \end{bmatrix}$$

$$\mathbf{X}_2 = \mathbf{A}\mathbf{X}_1 = \begin{bmatrix} 4 & 2 \\ 3 & -1 \end{bmatrix}\begin{bmatrix} 6 \\ 2 \end{bmatrix} = \begin{bmatrix} 28 \\ 16 \end{bmatrix}$$

The next five vectors obtained in this manner are given in the following table:

i	3	4	5	6	7
$\mathbf{X}_i$	$\begin{bmatrix} 144 \\ 68 \end{bmatrix}$	$\begin{bmatrix} 712 \\ 364 \end{bmatrix}$	$\begin{bmatrix} 3576 \\ 1772 \end{bmatrix}$	$\begin{bmatrix} 17{,}848 \\ 8956 \end{bmatrix}$	$\begin{bmatrix} 89{,}304 \\ 44{,}588 \end{bmatrix}$

At this point it may not appear that we are getting anywhere, since the entries of the vectors in the table appear to be growing without bound. But bear in mind that (7) indicates we are obtaining a constant multiple of a vector. If the power method converges, then by factoring the entry with the largest absolute value from $\mathbf{X}_m$ (for a large value of m), we should obtain a reasonable approximation to a dominant eigenvector. From the table,

$$\mathbf{X}_7 = 89{,}304 \begin{bmatrix} 1 \\ 0.4993 \end{bmatrix} \qquad (9)$$

It appears then that the vectors are approaching scalar multiples of $\begin{bmatrix} 1 \\ 0.5 \end{bmatrix}$.

We now use (8) to approximate the dominant eigenvalue λ_1. First we have

$$\mathbf{A}\mathbf{X}_7 = \begin{bmatrix} 4 & 2 \\ 3 & -1 \end{bmatrix}\begin{bmatrix} 1 \\ 0.4993 \end{bmatrix} = \begin{bmatrix} 4.9986 \\ 2.5007 \end{bmatrix}$$

$$\mathbf{A}\mathbf{X}_7 \cdot \mathbf{X}_7 = \begin{bmatrix} 4.9986 \\ 2.5007 \end{bmatrix}^T \begin{bmatrix} 1 \\ 0.4993 \end{bmatrix} = 6.2472$$

$$\mathbf{X}_7 \cdot \mathbf{X}_7 = \begin{bmatrix} 1 \\ 0.4993 \end{bmatrix}^T \begin{bmatrix} 1 \\ 0.4993 \end{bmatrix} = 1.2493$$

Finally, we have

$$\lambda_1 \approx \frac{\mathbf{A}\mathbf{X}_7 \cdot \mathbf{X}_7}{\mathbf{X}_7 \cdot \mathbf{X}_7} = \frac{6.2472}{1.2493} = 5.0006$$

15.13 Approximation of Eigenvalues

The reader should use the procedure of Section 7.7 to verify that the eigenvalues and corresponding eigenvectors of **A** are $\lambda_1 = 5$, $\lambda_2 = -2$, $\mathbf{K}_1 = \begin{bmatrix} 1 \\ 0.5 \end{bmatrix}$, and $\mathbf{K}_2 = \begin{bmatrix} 1 \\ -3 \end{bmatrix}$. □

Scaling As we have just seen, iteration of (1) often results in vectors whose entries become very large in absolute value. Large numbers can, of course, cause a problem if a computer is used to carry out a great number of iterations. The result in (9) suggests that one way around this difficulty is to use a **scaled-down** vector at each step of the iteration. To do this we simply multiply the vector $\mathbf{AX}_0$ by the reciprocal of the entry with the largest absolute value. That is, we multiply

$$\mathbf{AX}_0 = \begin{bmatrix} x_1 \\ \vdots \\ x_n \end{bmatrix} \quad \text{by} \quad \frac{1}{\max\{|x_1|, |x_2|, \ldots, |x_n|\}}$$

This resulting matrix, whose entries are now less than or equal to one, we label $\mathbf{X}_1$. We repeat the process with the vector $\mathbf{AX}_1$ to obtain the scaled-down vector $\mathbf{X}_2$, and so on.

EXAMPLE 4 Repeat the iterations of Example 3 using scaled-down vectors.

Solution From $\mathbf{AX}_0 = \begin{bmatrix} 4 & 2 \\ 3 & -1 \end{bmatrix} \begin{bmatrix} 1 \\ 1 \end{bmatrix} = \begin{bmatrix} 6 \\ 2 \end{bmatrix}$ we define

$$\mathbf{X}_1 = \frac{1}{6} \begin{bmatrix} 6 \\ 2 \end{bmatrix} = \begin{bmatrix} 1 \\ 0.3333 \end{bmatrix}$$

From $\mathbf{AX}_1 = \begin{bmatrix} 4 & 2 \\ 3 & -1 \end{bmatrix} \begin{bmatrix} 1 \\ 0.3333 \end{bmatrix} = \begin{bmatrix} 4.6666 \\ 2.6667 \end{bmatrix}$ we define

$$\mathbf{X}_2 = \frac{1}{4.6666} \begin{bmatrix} 4.6666 \\ 2.6667 \end{bmatrix} = \begin{bmatrix} 1 \\ 0.5714 \end{bmatrix}$$

We continue in this manner to construct the following table:

i	3	4	5	6	7
$\mathbf{X}_i$	$\begin{bmatrix} 1 \\ 0.4722 \end{bmatrix}$	$\begin{bmatrix} 1 \\ 0.5112 \end{bmatrix}$	$\begin{bmatrix} 1 \\ 0.4955 \end{bmatrix}$	$\begin{bmatrix} 1 \\ 0.5018 \end{bmatrix}$	$\begin{bmatrix} 1 \\ 0.4993 \end{bmatrix}$

In contrast to the table in Example 3, it is apparent from this table that the vectors are approaching $\begin{bmatrix} 1 \\ 0.5 \end{bmatrix}$. □

Method of Deflation

After we have found the dominant eigenvalue λ_1 of a matrix **A** it may still be necessary to find nondominant eigenvalues. The procedure we shall consider next is a modification of the power method and is called the **method of deflation**. We will limit the discussion to the case where **A** is a *symmetric* matrix.

Suppose λ_1 and $\mathbf{K}_1$ are, respectively, the dominant eigenvalue and a corresponding *normalized* eigenvector* (that is $\|\mathbf{K}_1\| = 1$) of a symmetric matrix **A**. Furthermore, suppose the eigenvalues of **A** are such that

$$|\lambda_1| > |\lambda_2| > |\lambda_3| \geq \cdots \geq |\lambda_n|$$

It can be proved that the matrix

$$\mathbf{B} = \mathbf{A} - \lambda_1 \mathbf{K}_1 \mathbf{K}_1^T \tag{10}$$

has eigenvalues $0, \lambda_2, \lambda_3, \ldots, \lambda_n$ and that eigenvectors of **B** are also eigenvectors of **A**. Note that λ_2 is now the dominant eigenvalue of **B**. We apply the power method to **B** to approximate λ_2 and a corresponding eigenvector.

EXAMPLE 5 Use the method of deflation to approximate the eigenvalues of

$$\mathbf{A} = \begin{bmatrix} 1 & 2 & -1 \\ 2 & 1 & 1 \\ -1 & 1 & 0 \end{bmatrix}$$

Solution We begin by using the power method with scaling to find the dominant eigenvalue and a corresponding eigenvector of **A**. Choosing $\mathbf{X}_0 = \begin{bmatrix} 1 \\ 1 \\ 1 \end{bmatrix}$, we see that

$$\mathbf{AX}_0 = \begin{bmatrix} 1 & 2 & -1 \\ 2 & 1 & 1 \\ -1 & 1 & 0 \end{bmatrix} \begin{bmatrix} 1 \\ 1 \\ 1 \end{bmatrix} = \begin{bmatrix} 2 \\ 4 \\ 0 \end{bmatrix} \quad \text{so} \quad \mathbf{X}_1 = \frac{1}{4}\begin{bmatrix} 2 \\ 4 \\ 0 \end{bmatrix} = \begin{bmatrix} 0.5 \\ 1 \\ 0 \end{bmatrix}$$

$$\mathbf{AX}_1 = \begin{bmatrix} 1 & 2 & -1 \\ 2 & 1 & 1 \\ -1 & 1 & 0 \end{bmatrix} \begin{bmatrix} 0.5 \\ 1 \\ 0 \end{bmatrix} = \begin{bmatrix} 2.5 \\ 2 \\ 0.5 \end{bmatrix} \quad \text{so} \quad \mathbf{X}_2 = \frac{1}{2.5}\begin{bmatrix} 2.5 \\ 2 \\ 0.5 \end{bmatrix} = \begin{bmatrix} 1 \\ 0.8 \\ 0.2 \end{bmatrix}$$

The scaled vectors $\mathbf{X}_3$ to $\mathbf{X}_{10}$ are given in the following table:

i	3	4	5	6	7	8	9	10
$\mathbf{X}_i$	$\begin{bmatrix} 0.8 \\ 1 \\ -0.0667 \end{bmatrix}$	$\begin{bmatrix} 1 \\ 0.8837 \\ 0.0698 \end{bmatrix}$	$\begin{bmatrix} 0.9134 \\ 1 \\ -0.0394 \end{bmatrix}$	$\begin{bmatrix} 1 \\ 0.9440 \\ 0.0293 \end{bmatrix}$	$\begin{bmatrix} 0.9614 \\ 1 \\ -0.0188 \end{bmatrix}$	$\begin{bmatrix} 1 \\ 0.9744 \\ 0.0129 \end{bmatrix}$	$\begin{bmatrix} 0.9828 \\ 1 \\ -0.0086 \end{bmatrix}$	$\begin{bmatrix} 1 \\ 0.9885 \\ 0.0058 \end{bmatrix}$

---

*See Example 3 in Section 7.8.

Utilizing $\mathbf{X}_{10}$ and (8), we find

$$\lambda_1 \approx \frac{\mathbf{A}\mathbf{X}_{10} \cdot \mathbf{X}_{10}}{\mathbf{X}_{10} \cdot \mathbf{X}_{10}} = 2.9997$$

It appears that the dominant eigenvalue and a corresponding eigenvector are, respectively, $\lambda_1 = 3$ and $\mathbf{K} = \begin{bmatrix} 1 \\ 1 \\ 0 \end{bmatrix}$.

Our next task is to construct the matrix $\mathbf{B}$ defined by (10). With $\|\mathbf{K}\| = \sqrt{2}$, the normalized eigenvector is $\mathbf{K}_1 = \begin{bmatrix} 1/\sqrt{2} \\ 1/\sqrt{2} \\ 0 \end{bmatrix}$. Thus,

$$\mathbf{B} = \begin{bmatrix} 1 & 2 & -1 \\ 2 & 1 & 1 \\ -1 & 1 & 0 \end{bmatrix} - 3 \begin{bmatrix} 1/\sqrt{2} \\ 1/\sqrt{2} \\ 0 \end{bmatrix} [1/\sqrt{2} \quad 1/\sqrt{2} \quad 0]$$

$$= \begin{bmatrix} 1 & 2 & -1 \\ 2 & 1 & 1 \\ -1 & 1 & 0 \end{bmatrix} - \begin{bmatrix} \frac{3}{2} & \frac{3}{2} & 0 \\ \frac{3}{2} & \frac{3}{2} & 0 \\ 0 & 0 & 0 \end{bmatrix} = \begin{bmatrix} -0.5 & 0.5 & -1 \\ 0.5 & -0.5 & 1 \\ -1 & 1 & 0 \end{bmatrix}$$

We now use the power method with scaling to find the dominant eigenvalue of $\mathbf{B}$. With $\mathbf{X}_0 = \begin{bmatrix} 1 \\ 1 \\ 1 \end{bmatrix}$ again, the results are summarized in the following table:

i	1	2	3	4	5	6	7
$\mathbf{X}_i$	$\begin{bmatrix} -1 \\ 1 \\ 0 \end{bmatrix}$	$\begin{bmatrix} 0.5 \\ -0.5 \\ 1 \end{bmatrix}$	$\begin{bmatrix} -1 \\ 1 \\ -0.6667 \end{bmatrix}$	$\begin{bmatrix} 0.8333 \\ -0.8333 \\ 1 \end{bmatrix}$	$\begin{bmatrix} -1 \\ 1 \\ -0.9091 \end{bmatrix}$	$\begin{bmatrix} 0.9545 \\ -0.9545 \\ 1 \end{bmatrix}$	$\begin{bmatrix} -1 \\ 1 \\ -0.9767 \end{bmatrix}$

Utilizing $\mathbf{X}_7$ and (8), we find

$$\lambda_2 \approx \frac{\mathbf{A}\mathbf{X}_7 \cdot \mathbf{X}_7}{\mathbf{X}_7 \cdot \mathbf{X}_7} = -1.9996$$

From these computations it seems apparent that the dominant eigenvalue of $\mathbf{B}$ and a corresponding eigenvector are $\lambda_2 = -2$ and $\mathbf{K} = \begin{bmatrix} -1 \\ 1 \\ -1 \end{bmatrix}$.

To find the last eigenvalue of $\mathbf{A}$, we repeat the deflation process to find the dominant eigenvalue and a corresponding eigenvector of the matrix

$$\mathbf{C} = \mathbf{B} - \lambda_2 \mathbf{K}_2 \mathbf{K}_2^T = \begin{bmatrix} 0.1667 & -0.1667 & -0.3333 \\ -0.1667 & 0.1667 & 0.3333 \\ -0.3333 & 0.3333 & 0.6667 \end{bmatrix}$$

where we have used $\mathbf{K}_2 = \begin{bmatrix} -1/\sqrt{3} \\ 1/\sqrt{3} \\ -1/\sqrt{3} \end{bmatrix}$. The student is encouraged to verify that $\lambda_3 = 1$. □

Example 5 is somewhat artificial since eigenvalues of a matrix need not be "nice" numbers such as 3, −2, and 1. Moreover, we used the exact values of the dominant eigenvalues λ_1 and λ_2 in the formation of the matrices **B** and **C**. In practice, of course, we may have to be content with working with approximations to the dominant eigenvalue λ_1 and a corresponding normalized dominant eigenvector $\mathbf{K}_1$ of **A**. If these approximations are used in (10), an error is introduced in the computation of the matrix **B**, and so more errors may be introduced in the computation of its dominant eigenvalue λ_2 and dominant eigenvector $\mathbf{K}_2$. If λ_2 and $\mathbf{K}_2$ are now used to construct the matrix **C**, it seems reasonable to conclude that errors are being compounded. In other words, the method of deflation can become increasingly inaccurate as more eigenvalues are computed.

Inverse Power Method In some applied problems we are more interested in approximating the eigenvalue of a matrix **A** of smallest absolute value than the dominant eigenvalue. If **A** is nonsingular, then the eigenvalues of **A** are nonzero (prove this), and if $\lambda_1, \lambda_2, \ldots, \lambda_n$ are the eigenvalues of **A**, then $1/\lambda_1, 1/\lambda_2, \ldots, 1/\lambda_n$ are the eigenvalues of $\mathbf{A}^{-1}$. The last fact can be seen by multiplying the equation $\mathbf{AK} = \lambda \mathbf{K}$, $\lambda \neq 0$, by $\mathbf{A}^{-1}$ and $1/\lambda$ to get $\mathbf{A}^{-1}\mathbf{K} = (1/\lambda)\mathbf{K}$. Now if the eigenvalues of **A** can be arranged in the order

$$|\lambda_1| \geq |\lambda_2| \geq |\lambda_3| \geq \cdots \geq |\lambda_{n-1}| > |\lambda_n|$$

then we see that $1/\lambda_n$ is the dominant eigenvalue of $\mathbf{A}^{-1}$. By applying the power method to $\mathbf{A}^{-1}$, we approximate the eigenvalue of largest magnitude and, by taking its reciprocal, we find the eigenvalue of **A** of least magnitude. This is called the **inverse power method**. See Problems 11–13.

▲ **Remark** It is recommended that the reader write a computer program, or use a computer algebra system such as Mathematica, to carry out the iterations required in the methods discussed in this section.

EXERCISES 15.13   *Answers to odd-numbered problems begin on page A-85.*

Each matrix in Problems 1–10 has a dominant eigenvalue.

In Problems 1 and 2 use the power method as illustrated in Example 3 to find the dominant eigenvalue and a corresponding dominant eigenvector of the given matrix.

1. $\begin{bmatrix} 1 & 1 \\ 2 & 0 \end{bmatrix}$
2. $\begin{bmatrix} -7 & 2 \\ 8 & -1 \end{bmatrix}$

In Problems 2–6 use the power method with scaling to find the dominant eigenvalue and a corresponding eigenvector of the given matrix.

3. $\begin{bmatrix} 2 & 4 \\ 3 & 13 \end{bmatrix}$
4. $\begin{bmatrix} -1 & 2 \\ -2 & 7 \end{bmatrix}$

5. $\begin{bmatrix} 5 & 4 & 2 \\ 4 & 5 & 2 \\ 2 & 2 & 2 \end{bmatrix}$ 6. $\begin{bmatrix} 3 & 1 & 1 \\ 0 & 1 & 1 \\ 0 & 0 & 2 \end{bmatrix}$

In Problems 7–10 use the method of deflation to find the eigenvalues of the given matrix.

7. $\begin{bmatrix} 3 & 2 \\ 2 & 6 \end{bmatrix}$ 8. $\begin{bmatrix} 1 & 3 \\ 3 & 9 \end{bmatrix}$

9. $\begin{bmatrix} 3 & -1 & 0 \\ -1 & 2 & -1 \\ 0 & -1 & 3 \end{bmatrix}$ 10. $\begin{bmatrix} 0 & 0 & -4 \\ 0 & -4 & 0 \\ -4 & 0 & 15 \end{bmatrix}$

In Problems 11 and 12 use the inverse power method to find the eigenvalue of least magnitude for the given matrix.

11. $\begin{bmatrix} 1 & 1 \\ 3 & 4 \end{bmatrix}$ 12. $\begin{bmatrix} -0.2 & 0.3 \\ 0.4 & -0.1 \end{bmatrix}$

13. In Problem 13 of Exercises 11.4 we saw that the deflection curve of a thin column under an applied load P was defined by the boundary-value problem

$$EI \frac{d^2 y}{dx^2} + Py = 0, \quad y(0) = 0, \quad y(L) = 0$$

In this problem we show how to apply matrix techniques to compute the lowest critical load.

Let the interval $[0, L]$ be divided into n subintervals of length $h = L/n$, and let $x_i = ih$, $i = 0, 1, \ldots, n$. For small values of h it follows from (6) of Section 15.9 that

$$\frac{d^2 y}{dx^2} \approx \frac{y_{i+1} - 2y_i + y_{i-1}}{h^2}$$

where $y_i = y(x_i)$.

(a) Show that the differential equation can be replaced by the difference equation

$$EI(y_{i+1} - 2y_i + y_{i-1}) + Ph^2 y_i = 0, \quad i = 1, 2, \ldots, n - 1$$

(b) Show that for $n = 4$ the difference equation in part (a) yields the system of linear equations

$$\begin{bmatrix} 2 & -1 & 0 \\ -1 & 2 & -1 \\ 0 & -1 & 2 \end{bmatrix} \begin{bmatrix} y_1 \\ y_2 \\ y_3 \end{bmatrix} = \frac{PL^2}{16EI} \begin{bmatrix} y_1 \\ y_2 \\ y_3 \end{bmatrix}$$

Note that this system has the form of the eigenvalue problem $\mathbf{AY} = \lambda \mathbf{Y}$, where $\lambda = PL^2/16EI$.

(c) Find $\mathbf{A}^{-1}$.

(d) Use the inverse power method to find, to two decimal places, the eigenvalue of $\mathbf{A}$ of least magnitude.

(e) Use the result of part (d) to compute the approximate lowest critical load. Compare your answer with that given in Problem 13 in Exercises 11.4.

14. Suppose the column in Problem 13 is tapered so that the moment of inertia of a cross-section I varies linearly from $I(0) = I_0 = 0.002$ to $I(L) = I_L = 0.001$.

(a) Use the difference equation in part (a) of Problem 13 with $n = 4$ to set up a system of equations analogous to that given in part (b).

(b) Proceed as in Problem 13 to find an approximation to the lowest critical load.

SUMMARY

In this chapter we considered methods for approximating roots of equations, definite integrals, solutions of ordinary and partial differential equations, and eigenvalues and eigenvectors of a matrix.

Newton's method utilizes a sequence of tangent lines to approximate a root of an equation. The **midpoint rule, trapezoidal rule**, and **Simpson's rule** utilize, respectively, rectangles, trapezoids, and parabolas to approximate a definite integral.

A solution of a differential equation may exist and yet we may not be able to determine it in terms of the familiar elementary functions. A way of convincing oneself that a first-order differential equation $y' = f(x, y)$ possesses a solution passing through a specific point (x_0, y_0) is to sketch the **direction field** associated with the equation. The equation $f(x, y) = c$ determines the **isoclines**, or curves of constant inclination. This means that every solution curve passing through a particular isocline does so with the same slope. The direction field is the totality of short line segments throughout two-dimensional space that have midpoints on the isoclines and that possess slope equal to the value of the parameter c. A sequence of **lineal elements** can suggest the shape of a solution curve passing through the given point.

At best, a direction field can give only the crudest approximation to a numerical value of the solution $y(x)$ of the initial-value problem $y' = f(x, y)$, $y(x_0) = y_0$ when x is close to x_0. To approximate values of $y(x)$ we used **Euler's formula**, the **improved Euler formula**, the **three-term Taylor formula**, the **Runge–Kutta formula**, and the **Adams–Bashforth/Adams–Moulton formulas**. Euler's method consists of approximating the solution curve by a sequence of straight lines. The improved Euler and Runge–Kutta methods use the idea of averaging slopes.

The first four formulas are known as **single-step**, or **starting, methods**. The Adams–Bashforth/Adams–Moulton method is an example of a **multistep**, or **continuing, method**. The improved Euler and Adams–Bashforth/Adams–Moulton formulas are also examples of a class of approximating formulas known as **predictor–corrector** formulas.

To obtain numerical approximations of solutions of higher-order ordinary differential equations, we can reduce the equation to a system of first-order equations. We then apply a particular numerical technique to each equation of the system.

An approximation to a solution of a boundary-value problem involving an elliptic, parabolic, or hyperbolic partial differential equation can be found by replacing the differential equation by a **finite difference equation**.

If an $n \times n$ matrix **A** has a dominant eigenvalue, the corresponding dominant eigenvector can be approximated by an iterative scheme known as the **power method**.

CHAPTER 15 REVIEW EXERCISES *Answers to odd-numbered problems begin on page A-85.*

In Problems 1 and 2 use Newton's method to find the indicated root. Carry out the method until two successive iterants agree to four decimal places.

1. $x^3 - 4x + 2 = 0$, largest positive root

2. $\left(\dfrac{\sin x}{x}\right)^2 = \dfrac{1}{2}$, smallest positive root

In Problems 3 and 4 use **(a)** the midpoint rule, **(b)** the trapezoidal rule, and **(c)** Simpson's rule to obtain an approximation to the given integral for the indicated value of n.

3. $\displaystyle\int_0^1 \sin x^2 \, dx$, $n = 4$

4. $\displaystyle\int_1^7 x^2\sqrt{2x - 1} \, dx$, $n = 6$

5. Use Simpson's rule to find an approximation to the area of the archway shown in Figure 15.46. [*Hint:* The shape of the archway above the 9-ft level is not parabolic.]

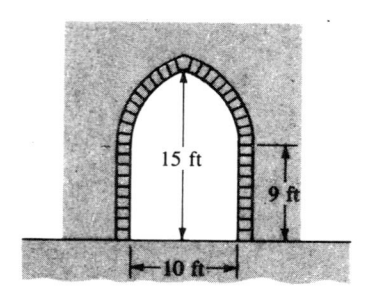

FIGURE 15.46

6. Suppose it is desired to approximate

$$\int_{1/2}^{3} \frac{dx}{x}$$

to an accuracy of three decimal places. Compare the value of n needed for the midpoint rule, the trapezoidal rule, and Simpson's rule.

In Problems 7 and 8 sketch the direction field for the given differential equation. Indicate several possible solution curves.

7. $y\,dx - x\,dy = 0$ **8.** $y' = 2x - y$

In Problems 9–12 construct a table comparing the indicated values of $y(x)$ using the Euler, improved Euler, three-term Taylor, and Runge–Kutta methods. Compute to four rounded decimal places. Use $h = 0.1$ and $h = 0.05$.

9. $y' = 2 \ln xy$, $y(1) = 2$
$y(1.1), y(1.2), y(1.3), y(1.4), y(1.5)$

10. $y' = \sin x^2 + \cos y^2$, $y(0) = 0$
$y(0.1), y(0.2), y(0.3), y(0.4), y(0.5)$

11. $y' = \sqrt{x + y}$, $y(0.5) = 0.5$
$y(0.6), y(0.7), y(0.8), y(0.9), y(1.0)$

12. $y' = xy + y^2$, $y(1) = 1$
$y(1.1), y(1.2), y(1.3), y(1.4), y(1.5)$

13. Use the Euler method to obtain the approximate value of $y(0.2)$, where $y(x)$ is the solution of the initial-value problem

$$y'' - (2x + 1)y = 0, \quad y(0) = 3, \quad y'(0) = 1$$

First use one step with $h = 0.2$ and then repeat the calculations using $h = 0.1$.

14. Use the Adams–Bashforth/Adams–Moulton method to approximate the value of $y(0.4)$, where $y(x)$ is the solution of

$$y' = 4x - 2y, \quad y(0) = 2$$

Use the Runge–Kutta formula and $h = 0.1$ to obtain the values of y_1, y_2, and y_3.

15. Consider the boundary-value problem

$$\frac{\partial^2 u}{\partial x^2} + \frac{\partial^2 u}{\partial y^2} = 0, \quad 0 < x < 2, \quad 0 < y < 1$$

$u(0, y) = 0, \quad u(2, y) = 50, \quad 0 < y < 1$

$u(x, 0) = 0, \quad u(x, 1) = 0, \quad 0 < x < 2$

Approximate the solution of the differential equation at the interior points of the region with mesh size $h = \frac{1}{2}$. Use Gaussian elimination or Gauss–Seidel iteration.

16. Solve Problem 15 with mesh size $h = \frac{1}{4}$. Use Gauss–Seidel iteration.

17. Light from a source at a point $S(a, b)$ is reflected by a spherical mirror of radius 1, centered at the origin, to an observer located at point $O(c, d)$ as shown in Figure 15.47. The point of reflection $P(x, y)$ from a spherical mirror lies in the plane determined by the source, the observer, and the center of the sphere. (The analysis of spherical mirrors occurs, among other places, in the study of radar design.)

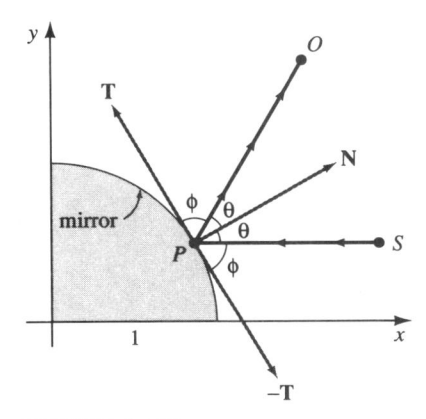

FIGURE 15.47

(a) Use Definition 6.3 twice, once with the angle θ and once with the angle ϕ, to show that the coordinates of the point of reflection $P(x, y)$ satisfy the equation

$$\frac{ax + by - 1}{ay - bx} = \frac{cx + dy - 1}{dx - cy}$$

[*Hint:* As shown in the figure, let **N** and **T** denote, respectively, a unit normal vector and a unit tangent to the circle at $P(x, y)$. If $\mathbf{N} = x\mathbf{i} + y\mathbf{j}$, what is **T** in terms of x and y?]

(b) Let $a = 2, b = 0, c = 0$, and $d = 3$. Use the relation $x^2 + y^2 = 1$ to show that the x-coordinate of the point of reflection is a root of a fourth-degree polynomial equation.

(c) Use Newton's method to find the point of reflection in part (b). You may have to consider all four roots of the equation in part (b) to find the one that corresponds to a solution of the equation in part (a).

PART VI

Complex Analysis

16 FUNCTIONS OF A COMPLEX VARIABLE

17 INTEGRATION IN THE COMPLEX PLANE

18 SERIES AND RESIDUES

19 CONFORMAL MAPPINGS AND APPLICATIONS

16

FUNCTIONS OF A COMPLEX VARIABLE

TOPICS TO REVIEW
complex numbers
factor theorem
definition of the derivative
rules of differentiation
partial differentiation

IMPORTANT CONCEPTS
complex number
real and imaginary parts
complex plane
conjugate of a complex number
modulus or absolute value of a complex number
polar form
argument
principal argument
DeMoivre's formula
roots of a complex number
principal *n*th root
neighborhood
open set
domain
complex function
mapping
derivative of a complex function
analyticity at a point
Cauchy–Riemann equations
harmonic functions
exponential and logarithmic functions
principal branch of the logarithm
trigonometric and hyperbolic functions

16.1 Complex Numbers
16.2 Polar Form of Complex Numbers; Powers and Roots
16.3 Sets of Points in the Complex Plane
16.4 Functions of a Complex Variable; Analyticity
16.5 Cauchy–Riemann Equations
16.6 Exponential and Logarithmic Functions
16.7 Trigonometric and Hyperbolic Functions
16.8 Inverse Trigonometric and Hyperbolic Functions
     Summary
     Chapter 16 Review Exercises

INTRODUCTION

Elementary algebra courses deal with the existence and some of the properties of complex numbers. But in courses such as calculus, students most likely do not even see a complex number. Introductory calculus is basically the study of functions of real variables. In advanced courses students use complex numbers only occasionally (see Sections 3.3, 7.7, and 9.4). In the next four chapters, however, we are going to introduce complex analysis. We shall examine limits, derivatives, integrals, sequences, series, and applications of functions of complex variables. Although there are many similarities between this and real analysis, there are many interesting differences and some surprises as well.

16.1 COMPLEX NUMBERS

Students have undoubtedly encountered complex numbers in earlier mathematics classes. From the quadratic formula we know that the zeros of the quadratic function $f(x) = ax^2 + bx + c$ are complex whenever the discriminant $b^2 - 4ac$ is negative. Simple equations such as

$$x^2 = -5 \quad \text{and} \quad x^2 + x + 1 = 0$$

have no real-number solutions.

DEFINITION 16.1 Complex Number

A number of the form $z = x + iy$, where x and y are real numbers and i is a number such that $i^2 = -1$, is called a **complex number**.

Terminology The number i in Definition 16.1 is called the **imaginary unit**. The real number x in $z = x + iy$ is called the **real part** of z; the real number y is called the **imaginary part** of z. The real and imaginary parts of a complex number z are abbreviated $\text{Re}(z)$ and $\text{Im}(z)$, respectively. For example, if $z = 4 - 9i$, then $\text{Re}(z) = 4$ and $\text{Im}(z) = -9$. A real constant multiple of the imaginary unit is called a **pure imaginary number**. For example, $z = 6i$ is a pure imaginary number. Two complex numbers are **equal** if their real and imaginary parts are equal. Since this simple concept is sometimes useful, we shall formalize the last statement in the next definition.

DEFINITION 16.2 Equality

Complex numbers $z_1 = x_1 + iy_1$ and $z_2 = x_2 + iy_2$ are **equal**, $z_1 = z_2$, if $\text{Re}(z_1) = \text{Re}(z_2)$ and $\text{Im}(z_1) = \text{Im}(z_2)$.

A complex number $x + iy = 0$ if $x = 0$ and $y = 0$.

Arithmetic Operations Complex numbers can be added, subtracted, multiplied, and divided. If $z_1 = x_1 + iy_1$ and $z_2 = x_2 + iy_2$, these operations are defined as follows:

Addition: $z_1 + z_2 = (x_1 + iy_1) + (x_2 + iy_2) = (x_1 + x_2) + i(y_1 + y_2)$

Subtraction: $z_1 - z_2 = (x_1 + iy_1) - (x_2 + iy_2) = (x_1 - x_2) + i(y_1 - y_2)$

Multiplication: $z_1 \cdot z_2 = (x_1 + iy_1)(x_2 + iy_2)$
$= x_1 x_2 - y_1 y_2 + i(y_1 x_2 + x_1 y_2)$

16.1 Complex Numbers

Division: $\dfrac{z_1}{z_2} = \dfrac{x_1 + iy_1}{x_2 + iy_2}$

$$= \frac{x_1 x_2 + y_1 y_2}{x_2^2 + y_2^2} + i \frac{y_1 x_2 - x_1 y_2}{x_2^2 + y_2^2}$$

The familiar commutative, associative, and distributive laws hold for complex numbers:

Commutative laws: $\begin{cases} z_1 + z_2 = z_2 + z_1 \\ z_1 z_2 = z_2 z_1 \end{cases}$

Associative laws: $\begin{cases} z_1 + (z_2 + z_3) = (z_1 + z_2) + z_3 \\ z_1(z_2 z_3) = (z_1 z_2) z_3 \end{cases}$

Distributive law: $z_1(z_2 + z_3) = z_1 z_2 + z_1 z_3$

In view of these laws, there is no need to memorize the definitions of addition, subtraction, and multiplication. To add (subtract) two complex numbers, we simply add (subtract) the corresponding real and imaginary parts. To multiply two complex numbers, we use the distributive law and the fact that $i^2 = -1$.

EXAMPLE 1 If $z_1 = 2 + 4i$ and $z_2 = -3 + 8i$, find (a) $z_1 + z_2$ and (b) $z_1 z_2$.

Solution (a) By adding the real and imaginary parts of the two numbers, we get

$$(2 + 4i) + (-3 + 8i) = (2 - 3) + (4 + 8)i = -1 + 12i$$

(b) Using the distributive law, we have

$$(2 + 4i)(-3 + 8i) = (2 + 4i)(-3) + (2 + 4i)(8i)$$
$$= -6 - 12i + 16i + 32i^2$$
$$= (-6 - 32) + (16 - 12)i = -38 + 4i \quad \square$$

There is also no need to memorize the definition of division, but before discussing that we need to introduce another concept.

Conjugate If z is a complex number, then the number obtained by changing the sign of its imaginary part is called the **complex conjugate** or, simply, the **conjugate** of z. If $z = x + iy$, then its conjugate is

$$\bar{z} = x - iy$$

For example, if $z = 6 + 3i$, then $\bar{z} = 6 - 3i$; if $z = -5 - i$, then $\bar{z} = -5 + i$. If z is a real number, say $z = 7$, then $\bar{z} = 7$. From the definition of addition it can be readily shown that the conjugate of a sum of two complex numbers is the sum of the conjugates:

$$\overline{z_1 + z_2} = \bar{z}_1 + \bar{z}_2$$

Moreover, we have the additional three properties

$$\overline{z_1 - z_2} = \bar{z}_1 - \bar{z}_2 \qquad \overline{z_1 z_2} = \bar{z}_1 \bar{z}_2 \qquad \overline{\left(\frac{z_1}{z_2}\right)} = \frac{\bar{z}_1}{\bar{z}_2}$$

The definitions of addition and multiplication show that the sum and product of a complex number z and its conjugate $\bar{z}$ are also real numbers:

$$z + \bar{z} = (x + iy) + (x - iy) = 2x \tag{1}$$
$$z\bar{z} = (x + iy)(x - iy) = x^2 - i^2 y^2 = x^2 + y^2 \tag{2}$$

The difference between a complex number z and its conjugate $\bar{z}$ is a pure imaginary number:

$$z - \bar{z} = (x + iy) - (x - iy) = 2iy \tag{3}$$

Since $x = \text{Re}(z)$ and $y = \text{Im}(z)$, (1) and (3) yield two useful formulas:

$$\text{Re}(z) = \frac{z + \bar{z}}{2} \quad \text{and} \quad \text{Im}(z) = \frac{z - \bar{z}}{2i}$$

However, (2) is the important relationship that enables us to approach division in a more practical manner: To divide z_1 by z_2, we multiply both numerator and denominator of z_1/z_2 by the conjugate of z_2. This will be illustrated in the next example.

EXAMPLE 2 If $z_1 = 2 - 3i$ and $z_2 = 4 + 6i$, find (a) $\dfrac{z_1}{z_2}$ and (b) $\dfrac{1}{z_1}$.

Solution In both parts of this example we shall multiply both numerator and denominator by the conjugate of the denominator and then use (2).

(a) $\dfrac{2 - 3i}{4 + 6i} = \dfrac{2 - 3i}{4 + 6i} \dfrac{4 - 6i}{4 - 6i} = \dfrac{8 - 12i - 12i + 18i^2}{16 + 36}$

$= \dfrac{-10 - 24i}{52} = -\dfrac{5}{26} - \dfrac{6}{13}i$

(b) $\dfrac{1}{2 - 3i} = \dfrac{1}{2 - 3i} \dfrac{2 + 3i}{2 + 3i} = \dfrac{2 + 3i}{4 + 9} = \dfrac{2}{13} + \dfrac{3}{13}i$ □

Geometric Interpretation A complex number $z = x + iy$ is uniquely determined by an *ordered pair* of real numbers (x, y). The first and second entries of the ordered pairs correspond, in turn, with the real and imaginary parts of the complex number. For example, the ordered pair $(2, -3)$ corresponds to the complex number $z = 2 - 3i$. Conversely, $z = 2 - 3i$ determines the ordered pair $(2, -3)$. In this manner we are able to associate a complex number $z = x + iy$ with a point (x, y) in a coordinate plane. But, as we saw in Section 6.1, an ordered pair of real numbers can be interpreted as the components of a vector. Thus, a complex number $z = x + iy$ can also be viewed

16.1 Complex Numbers

as a *vector* whose initial point is the origin and whose terminal point is (x, y). The coordinate plane illustrated in Figure 16.1 is called the **complex plane** or simply the **z-plane**. The horizontal or *x*-axis is called the **real axis** and the vertical or *y*-axis is called the **imaginary axis**. The length of a vector z, or the distance from the origin to the point (x, y), is clearly $\sqrt{x^2 + y^2}$. This real number is given a special name.

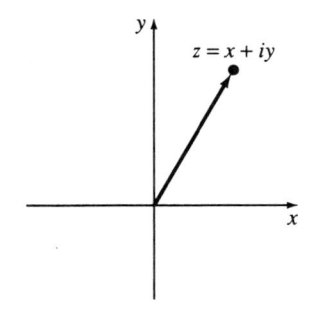

FIGURE 16.1

> **DEFINITION 16.3   Modulus or Absolute Value**
>
> The **modulus** or **absolute value** of $z = x + iy$, denoted by $|z|$, is the real number
>
> $$|z| = \sqrt{x^2 + y^2} = \sqrt{z\bar{z}} \tag{4}$$

EXAMPLE 3   If $z = 2 - 3i$, then $|z| = \sqrt{2^2 + (-3)^2} = \sqrt{13}$.   □

As Figure 16.2 shows, the sum of the vectors z_1 and z_2 is the vector $z_1 + z_2$. For the triangle given in the figure we know that the length of the side of the triangle corresponding to the vector $z_1 + z_2$ cannot be longer than the sum of the remaining two sides. In symbols this is

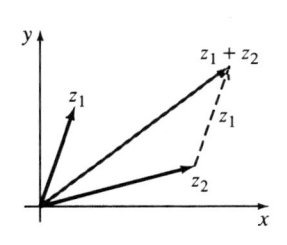

$$|z_1 + z_2| \leq |z_1| + |z_2| \tag{5}$$

FIGURE 16.2

The result in (5) is known as the **triangle inequality** and extends to any finite sum:

$$|z_1 + z_2 + z_3 + \cdots + z_n| \leq |z_1| + |z_2| + |z_3| + \cdots + |z_n| \tag{6}$$

Using (5) on $z_1 + z_2 + (-z_2)$, we obtain another important inequality:

$$|z_1 + z_2| \geq |z_1| - |z_2| \tag{7}$$

▲ **Remark**   Many of the properties of the real system hold in the complex number system, but there are some remarkable differences as well. For example, we cannot compare two complex numbers $z_1 = x_1 + iy_1$, $y_1 \neq 0$, and $z_2 = x_2 + iy_2$, $y_2 \neq 0$, by means of inequalities. In other words, statements such as $z_1 < z_2$ and $z_2 \geq z_1$ have no meaning except in the case when the two numbers z_1 and z_2 are real. We can, however, compare the absolute values of two complex numbers. Thus, if $z_1 = 3 + 4i$ and $z_2 = 5 - i$, then $|z_1| = 5$ and $|z_2| = \sqrt{26}$, and consequently $|z_1| < |z_2|$. This last inequality means that the point $(3, 4)$ is closer to the origin than is the point $(5, -1)$.

EXERCISES 16.1 *Answers to odd-numbered problems begin on page A-86.*

In Problems 1–26 write the given number in the form $a + ib$.

1. $2i^3 - 3i^2 + 5i$
2. $3i^5 - i^4 + 7i^3 - 10i^2 - 9$
3. i^8
4. i^{11}
5. $(5 - 9i) + (2 - 4i)$
6. $3(4 - i) - 3(5 + 2i)$
7. $i(5 + 7i)$
8. $i(4 - i) + 4i(1 + 2i)$
9. $(2 - 3i)(4 + i)$
10. $(\frac{1}{2} - \frac{1}{4}i)(\frac{2}{3} + \frac{5}{3}i)$
11. $(2 + 3i)^2$
12. $(1 - i)^3$
13. $\dfrac{2}{i}$
14. $\dfrac{i}{1 + i}$
15. $\dfrac{2 - 4i}{3 + 5i}$
16. $\dfrac{10 - 5i}{6 + 2i}$
17. $\dfrac{(3 - i)(2 + 3i)}{1 + i}$
18. $\dfrac{(1 + i)(1 - 2i)}{(2 + i)(4 - 3i)}$
19. $\dfrac{(5 - 4i) - (3 + 7i)}{(4 + 2i) + (2 - 3i)}$
20. $\dfrac{(4 + 5i) + 2i^3}{(2 + i)^2}$
21. $i(1 - i)(2 - i)(2 + 6i)$
22. $(1 + i)^2(1 - i)^3$
23. $(3 + 6i) + (4 - i)(3 + 5i) + \dfrac{1}{2 - i}$
24. $(2 + 3i)\left(\dfrac{2 - i}{1 + 2i}\right)^2$
25. $\left(\dfrac{i}{3 - i}\right)\left(\dfrac{1}{2 + 3i}\right)$
26. $\dfrac{1}{(1 + i)(1 - 2i)(1 + 3i)}$

In Problems 27–32 let $z = x + iy$. Find the indicated expression.

27. $\text{Re}(1/z)$
28. $\text{Re}(z^2)$
29. $\text{Im}(2z + 4\bar{z} - 4i)$
30. $\text{Im}(\bar{z}^2 + z^2)$
31. $|z - 1 - 3i|$
32. $|z + 5\bar{z}|$

In Problems 33–36 use Definition 16.2 to find a complex number z satisfying the given equation.

33. $2z = i(2 + 9i)$
34. $z - 2\bar{z} + 7 - 6i = 0$
35. $z^2 = i$
36. $\bar{z}^2 = 4z$

In Problems 37 and 38 determine which complex number is closer to the origin.

37. $10 + 8i, 11 - 6i$
38. $\frac{1}{2} - \frac{1}{4}i, \frac{2}{3} + \frac{1}{6}i$

39. Prove that $|z_1 - z_2|$ is the distance between the points z_1 and z_2 in the complex plane.

40. Show for all complex numbers z on the circle $x^2 + y^2 = 4$ that $|z + 6 + 8i| \le 12$.

16.2 POLAR FORM OF COMPLEX NUMBERS; POWERS AND ROOTS

Polar Form Recall from calculus that the polar form of the point (x, y) is (r, θ), where $x, y, r,$ and θ are related by $x = r\cos\theta$ and $y = r\sin\theta$. Thus, a nonzero complex number $z = x + iy$ can be written as $z = (r\cos\theta) + i(r\sin\theta)$ or

$$z = r(\cos\theta + i\sin\theta) \qquad (1)$$

We say that (1) is the **polar form** of the complex number. We see from Figure 16.3 that the coordinate r can be interpreted as the distance from the origin to the point (x, y). In other words, we adopt the convention that r is never negative so that we can take r to be the modulus of z: $r = |z|$. The angle θ of inclination of the vector z measured in radians from the positive real axis is positive when measured counterclockwise and negative when measured

FIGURE 16.3

16.2 Polar Form of Complex Numbers; Powers and Roots

clockwise. The angle θ is called an **argument** of z and is written $\theta = \arg z$. From Figure 16.3 we see that an argument of a complex number must satisfy the equation $\tan \theta = y/x$. The solutions of this equation are not unique, since if θ_0 is an argument of z, then necessarily the angles $\theta_0 \pm 2\pi$, $\theta_0 \pm 4\pi, \ldots$ are also arguments. The argument of a complex number in the interval $-\pi < \theta \leq \pi$ is called the **principal argument** of z and is denoted by $\text{Arg } z$. For example, $\text{Arg}(i) = \pi/2$.

EXAMPLE 1 Express $1 - \sqrt{3}i$ in polar form.

Solution With $x = 1$ and $y = -\sqrt{3}$, we obtain $r = |z| = \sqrt{(1)^2 + (-\sqrt{3})^2} = 2$. Now since the point $(1, -\sqrt{3})$ lies in the fourth quadrant, we can take the solution of $\tan \theta = -\sqrt{3}/1 = -\sqrt{3}$ to be $\theta = \arg z = 5\pi/3$. It follows from (1) that a polar form of the number is

$$z = 2\left(\cos \frac{5\pi}{3} + i \sin \frac{5\pi}{3}\right)$$

As we see in Figure 16.4, the argument of $1 - \sqrt{3}i$ that lies in the interval $(-\pi, \pi]$, the principal argument of z, is $\text{Arg } z = -\pi/3$. Thus, an alternative polar form of the complex number is

$$z = 2\left[\cos\left(-\frac{\pi}{3}\right) + i \sin\left(-\frac{\pi}{3}\right)\right] \qquad \square$$

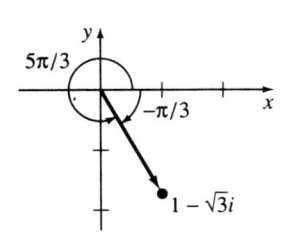

FIGURE 16.4

Multiplication and Division The polar form of a complex number is especially convenient to use when multiplying or dividing two complex numbers. Suppose

$$z_1 = r_1(\cos \theta_1 + i \sin \theta_1) \quad \text{and} \quad z_2 = r_2(\cos \theta_2 + i \sin \theta_2)$$

where θ_1 and θ_2 are any arguments of z_1 and z_2, respectively. Then

$$z_1 z_2 = r_1 r_2 [(\cos \theta_1 \cos \theta_2 - \sin \theta_1 \sin \theta_2) + i(\sin \theta_1 \cos \theta_2 + \cos \theta_1 \sin \theta_2)] \qquad (2)$$

and for $z_2 \neq 0$,

$$\frac{z_1}{z_2} = \frac{r_1}{r_2}[(\cos \theta_1 \cos \theta_2 + \sin \theta_1 \sin \theta_2) + i(\sin \theta_1 \cos \theta_2 - \cos \theta_1 \sin \theta_2)] \qquad (3)$$

From the addition formulas from trigonometry, (2) and (3) can be rewritten, in turn, as

$$\boxed{\begin{array}{l} z_1 z_2 = r_1 r_2 [\cos(\theta_1 + \theta_2) + i \sin(\theta_1 + \theta_2)] \qquad (4) \\[6pt] \dfrac{z_1}{z_2} = \dfrac{r_1}{r_2}[\cos(\theta_1 - \theta_2) + i \sin(\theta_1 - \theta_2)] \qquad (5) \end{array}}$$

and

Inspection of (4) and (5) shows that

$$|z_1 z_2| = |z_1||z_2| \qquad \left|\frac{z_1}{z_2}\right| = \frac{|z_1|}{|z_2|} \qquad (6)$$

and
$$\arg(z_1 z_2) = \arg z_1 + \arg z_2 \qquad \arg\left(\frac{z_1}{z_2}\right) = \arg z_1 - \arg z_2 \qquad (7)$$

EXAMPLE 2 We have seen that Arg $z_1 = \pi/2$ for $z_1 = i$. In Example 1 we saw that Arg $z_2 = -\pi/3$ for $z_2 = 1 - \sqrt{3}i$. Thus, for

$$z_1 z_2 = i(1 - \sqrt{3}i) = \sqrt{3} + i \quad \text{and} \quad \frac{z_1}{z_2} = \frac{1 - \sqrt{3}i}{i} = -\sqrt{3} - i$$

it follows from (7) that

$$\arg(z_1 z_2) = \frac{\pi}{2} - \frac{\pi}{3} = \frac{\pi}{6} \quad \text{and} \quad \arg\left(\frac{z_1}{z_2}\right) = \frac{\pi}{2} - \left(-\frac{\pi}{3}\right) = \frac{5\pi}{6} \qquad \square$$

In Example 2 we used the principal arguments of z_1 and z_2 and obtained $\arg(z_1 z_2) = \text{Arg}(z_1 z_2)$ and $\arg(z_1/z_2) = \text{Arg}(z_1/z_2)$. It should be observed, however, that this was a coincidence. Although (7) is true for any arguments of z_1 and z_2, *it is not true*, in general, that $\text{Arg}(z_1 z_2) = \text{Arg } z_1 + \text{Arg } z_2$ and $\text{Arg}(z_1/z_2) = \text{Arg } z_1 - \text{Arg } z_2$. See Problem 39.

Powers of z We can find integer powers of the complex number z from the results in (4) and (5). For example, if $z = r(\cos \theta + i \sin \theta)$, then with $z_1 = z$ and $z_2 = z$, (4) gives

$$z^2 = r^2[\cos(\theta + \theta) + i \sin(\theta + \theta)] = r^2(\cos 2\theta + i \sin 2\theta)$$

Since $z^3 = z^2 z$, it follows that

$$z^3 = r^3(\cos 3\theta + i \sin 3\theta)$$

Moreover, since $\arg(1) = 0$, it follows from (5) that

$$\frac{1}{z^2} = z^{-2} = r^{-2}[\cos(-2\theta) + i \sin(-2\theta)]$$

Continuing in this manner, we obtain a formula for the nth power of z for any integer n:

$$z^n = r^n(\cos n\theta + i \sin n\theta) \qquad (8)$$

EXAMPLE 3 Compute z^3 for $z = 1 - \sqrt{3}i$.

Solution In Example 1 we saw that

$$z = 2\left[\cos\left(-\frac{\pi}{3}\right) + i\sin\left(-\frac{\pi}{3}\right)\right]$$

Hence from (8) with $r = 2$, $\theta = -\pi/3$, and $n = 3$, we get

$$(1 - \sqrt{3}i)^3 = 2^3\left[\cos\left(3\left(-\frac{\pi}{3}\right)\right) + i\sin\left(3\left(-\frac{\pi}{3}\right)\right)\right]$$
$$= 8[\cos(-\pi) + i\sin(-\pi)] = -8 \qquad \square$$

DeMoivre's Formula   When $z = \cos\theta + i\sin\theta$, we have $|z| = r = 1$ and so (8) yields

$$\boxed{(\cos\theta + i\sin\theta)^n = \cos n\theta + i\sin n\theta} \qquad (9)$$

This last result is known as **DeMoivre's formula*** and is useful in deriving certain trigonometric identities.

Roots   A number w is said to be an nth **root** of a nonzero complex number z if $w^n = z$. If we let $w = \rho(\cos\phi + i\sin\phi)$ and $z = r(\cos\theta + i\sin\theta)$ be the polar forms of w and z, then in view of (8) $w^n = z$ becomes

$$\rho^n(\cos n\phi + i\sin n\phi) = r(\cos\theta + i\sin\theta)$$

From this we conclude that

$$\rho^n = r \quad \text{or} \quad \rho = r^{1/n}$$

and

$$\cos n\phi + i\sin n\phi = \cos\theta + i\sin\theta$$

By equating the real and imaginary parts, we get from this equation

$$\cos n\phi = \cos\theta \quad \text{and} \quad \sin n\phi = \sin\theta$$

These equalities imply that $n\phi = \theta + 2k\pi$, where k is an integer. Thus,

$$\phi = \frac{\theta + 2k\pi}{n}$$

As k takes on the successive integer values $k = 0, 1, 2, \ldots, n - 1$, we obtain n *distinct* roots with the same modulus but different arguments. But for $k \geq n$ we obtain the same roots because the sine and cosine are 2π-periodic. To see

---

*Named after the French mathematician Abraham DeMoivre (1667–1754).

this, suppose $k = n + m$, where $m = 0, 1, 2, \ldots$. Then

$$\phi = \frac{\theta + 2(n+m)\pi}{n} = \frac{\theta + 2m\pi}{n} + 2\pi$$

and so $\quad \sin\phi = \sin\left(\dfrac{\theta + 2m\pi}{n}\right) \quad \cos\phi = \cos\left(\dfrac{\theta + 2m\pi}{n}\right)$

We summarize this result. The nth roots of a nonzero complex number $z = r(\cos\theta + i\sin\theta)$ are given by

$$w_k = r^{1/n}\left[\cos\left(\frac{\theta + 2k\pi}{n}\right) + i\sin\left(\frac{\theta + 2k\pi}{n}\right)\right] \tag{10}$$

where $k = 0, 1, 2, \ldots, n-1$.

EXAMPLE 4 Find the three cube roots of $z = i$.

Solution With $r = 1$, $\theta = \arg z = \pi/2$, the polar form of the given number is $z = \cos(\pi/2) + i\sin(\pi/2)$. From (10) with $n = 3$ we obtain

$$w_k = (1)^{1/3}\left[\cos\left(\frac{\pi/2 + 2k\pi}{3}\right) + i\sin\left(\frac{\pi/2 + 2k\pi}{3}\right)\right], \quad k = 0, 1, 2$$

Hence, the three roots are:

$$k = 0, \quad w_0 = \cos\frac{\pi}{6} + i\sin\frac{\pi}{6} = \frac{\sqrt{3}}{2} + \frac{1}{2}i$$

$$k = 1, \quad w_1 = \cos\frac{5\pi}{6} + i\sin\frac{5\pi}{6} = -\frac{\sqrt{3}}{2} + \frac{1}{2}i$$

$$k = 2, \quad w_2 = \cos\frac{3\pi}{2} + i\sin\frac{3\pi}{2} = -i \quad \square$$

The root w of a complex number z obtained by using the principal argument of z with $k = 0$ is sometimes called the **principal nth root** of z. In Example 4, since $\text{Arg}(i) = \pi/2$, $w_0 = (\sqrt{3}/2) + (1/2)i$ is the principal third root of i.

Since the roots given by (8) have the same modulus, the n roots of a nonzero complex number z lie on a circle of radius $r^{1/n}$ centered at the origin in the complex plane. Moreover, since the difference between the arguments of any two successive roots is $2\pi/n$, the nth roots of z are equally spaced on this circle. Figure 16.5 shows the three roots of i equally spaced on a unit circle; the angle between roots (vectors) w_k and w_{k+1} is $2\pi/3$.

As the next example will show, the roots of a complex number do not have to be "nice" numbers as in Example 3.

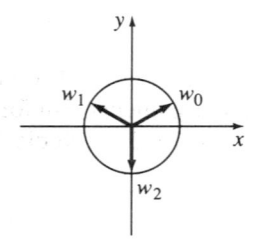

FIGURE 16.5

16.2 Polar Form of Complex Numbers; Powers and Roots

EXAMPLE 5 Find the four fourth roots of $z = 1 + i$.

Solution In this case, $r = \sqrt{2}$ and $\theta = \arg z = \pi/4$. From (10) with $n = 4$ we obtain

$$w_k = (2)^{1/4}\left[\cos\left(\frac{\pi/4 + 2k\pi}{4}\right) + i\sin\left(\frac{\pi/4 + 2k\pi}{4}\right)\right], \quad k = 0, 1, 2, 3$$

Thus,

$$k = 0, \quad w_0 = (2)^{1/4}\left[\cos\frac{\pi}{16} + i\sin\frac{\pi}{16}\right] = 1.1664 + 0.2320i$$

$$k = 1, \quad w_1 = (2)^{1/4}\left[\cos\frac{9\pi}{16} + i\sin\frac{9\pi}{16}\right] = -0.2320 + 1.1664i$$

$$k = 2, \quad w_2 = (2)^{1/4}\left[\cos\frac{17\pi}{16} + i\sin\frac{17\pi}{16}\right] = -1.1664 - 0.2320i$$

$$k = 3, \quad w_3 = (2)^{1/4}\left[\cos\frac{25\pi}{16} + i\sin\frac{25\pi}{16}\right] = 0.2320 - 1.1664i \quad \square$$

EXERCISES 16.2   Answers to odd-numbered problems begin on page A-86.

In Problems 1–10 write the given complex number in polar form.

1. 2
2. -10
3. $-3i$
4. $6i$
5. $1 + i$
6. $5 - 5i$
7. $-\sqrt{3} + i$
8. $-2 - 2\sqrt{3}i$
9. $\dfrac{3}{-1 + i}$
10. $\dfrac{12}{\sqrt{3} + i}$

In Problems 11–14 write the number given in polar form in the form $a + ib$.

11. $z = 5\left(\cos\dfrac{7\pi}{6} + i\sin\dfrac{7\pi}{6}\right)$
12. $z = 8\sqrt{2}\left(\cos\dfrac{11\pi}{4} + i\sin\dfrac{11\pi}{4}\right)$
13. $z = 6\left(\cos\dfrac{\pi}{8} + i\sin\dfrac{\pi}{8}\right)$
14. $z = 10\left(\cos\dfrac{\pi}{5} + i\sin\dfrac{\pi}{5}\right)$

In Problems 15 and 16 find $z_1 z_2$ and z_1/z_2. Write the number in the form $a + ib$.

15. $z_1 = 2\left(\cos\dfrac{\pi}{8} + i\sin\dfrac{\pi}{8}\right)$, $z_2 = 4\left(\cos\dfrac{3\pi}{8} + i\sin\dfrac{3\pi}{8}\right)$
16. $z_1 = \sqrt{2}\left(\cos\dfrac{\pi}{4} + i\sin\dfrac{\pi}{4}\right)$, $z_2 = \sqrt{3}\left(\cos\dfrac{\pi}{12} + i\sin\dfrac{\pi}{12}\right)$

In Problems 17–20 write each complex number in polar form. Then use either (4) or (5) to obtain a polar form of the given number. Write the polar form in the form $a + ib$.

17. $(3 - 3i)(5 + 5\sqrt{3}i)$
18. $(4 + 4i)(-1 + i)$
19. $\dfrac{-i}{2 - 2i}$
20. $\dfrac{\sqrt{2} + \sqrt{6}i}{-1 + \sqrt{3}i}$

In Problems 21–26 use (8) to compute the indicated power.

21. $(1 + \sqrt{3}i)^9$
22. $(2 - 2i)^5$
23. $(\tfrac{1}{2} + \tfrac{1}{2}i)^{10}$
24. $(-\sqrt{2} + \sqrt{6}i)^4$
25. $\left(\cos\dfrac{\pi}{8} + i\sin\dfrac{\pi}{8}\right)^{12}$

26. $\left[\sqrt{3}\left(\cos\dfrac{2\pi}{9} + i\sin\dfrac{2\pi}{9}\right)\right]^6$

In Problems 27–32 use (10) to compute all roots. Sketch these roots on an appropriate circle centered at the origin.

27. $(8)^{1/3}$
28. $(1)^{1/8}$
29. $(i)^{1/2}$
30. $(-1+i)^{1/3}$
31. $(-1+\sqrt{3}i)^{1/2}$
32. $(-1-\sqrt{3}i)^{1/4}$

In Problems 33 and 34 find all solutions of the given equation.

33. $z^4 + 1 = 0$
34. $z^8 - 2z^4 + 1 = 0$

In Problems 35 and 36 express the given complex number first in polar form and then in the form $a + ib$.

35. $\left(\cos\dfrac{\pi}{9} + i\sin\dfrac{\pi}{9}\right)^{12}\left[2\left(\cos\dfrac{\pi}{6} + i\sin\dfrac{\pi}{6}\right)\right]^5$

36. $\dfrac{\left[8\left(\cos\dfrac{3\pi}{8} + i\sin\dfrac{3\pi}{8}\right)\right]^3}{\left[2\left(\cos\dfrac{\pi}{16} + i\sin\dfrac{\pi}{16}\right)\right]^{10}}$

37. Use the result $(\cos\theta + i\sin\theta)^2 = \cos 2\theta + i\sin 2\theta$ to find trigonometric identities for $\cos 2\theta$ and $\sin 2\theta$.

38. Use the result $(\cos\theta + i\sin\theta)^3 = \cos 3\theta + i\sin 3\theta$ to find trigonometric identities for $\cos 3\theta$ and $\sin 3\theta$.

39. (a) If $z_1 = -1$ and $z_2 = 5i$, verify that
$\text{Arg}(z_1 z_2) \neq \text{Arg}(z_1) + \text{Arg}(z_2)$.

(b) If $z_1 = -1$ and $z_2 = -5i$, verify that
$\text{Arg}(z_1/z_2) \neq \text{Arg}(z_1) - \text{Arg}(z_2)$.

40. For the complex numbers given in Problem 39, verify in both parts (a) and (b) that
$\arg(z_1 z_2) = \arg(z_1) + \arg(z_2)$ and $\arg\left(\dfrac{z_1}{z_2}\right) = \arg(z_1) - \arg(z_2)$

16.3 SETS OF POINTS IN THE COMPLEX PLANE

Terminology Before discussing the concept of functions of a complex variable we need to introduce some essential terminology about sets in the complex plane.

Suppose $z_0 = x_0 + iy_0$. Since $|z - z_0| = \sqrt{(x-x_0)^2 + (y-y_0)^2}$ is the distance between the points $z = x + iy$ and $z_0 = x_0 + iy_0$, the points $z = x + iy$ that satisfy the equation

$$|z - z_0| = \rho$$

$\rho > 0$, lie on a **circle** of radius ρ centered at the point z_0. See Figure 16.6.

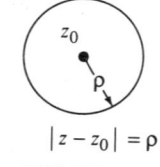

FIGURE 16.6 $|z - z_0| = \rho$

EXAMPLE 1 (a) $|z| = 1$ is the equation of a unit circle centered at the origin.
(b) $|z - 1 - 2i| = 5$ is the equation of a circle of radius 5 centered at $1 + 2i$. □

The points z satisfying the inequality

$$|z - z_0| < \rho$$

$\rho > 0$, lie within, but not on, a circle of radius ρ centered at the point z_0. This set is called a **neighborhood** of z_0 or an **open disk**. A point z_0 is said to be an **interior point** of a set S of the complex plane if there exists some neighboroood of z_0 that lies entirely within S. If every point z of a set S is an interior point, then S is said to be an **open set**. See Figure 16.7. For example, the inequality $\text{Re}(z) > 1$ defines a *right half-plane*, which is an open set. All complex numbers $z = x + iy$ for which $x > 1$ are in this set. If we choose,

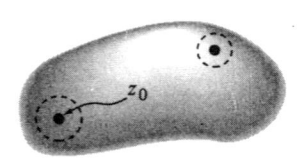

FIGURE 16.7

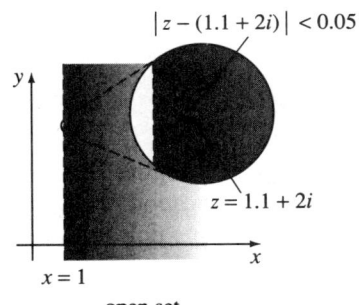

16.3 Sets of Points in the Complex Plane

FIGURE 16.8 open set — magnified view of a point near $x = 1$ ($|z - (1.1 + 2i)| < 0.05$, $z = 1.1 + 2i$)

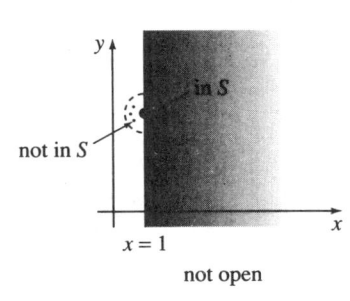

FIGURE 16.9 not open ($x = 1$)

for example, $z_0 = 1.1 + 2i$, then a neighborhood of z_0 lying entirely in the set is defined by $|z - (1.1 + 2i)| < 0.05$. See Figure 16.8. On the other hand, the set S of points in the complex plane defined by $\text{Re}(z) \geq 1$ is *not* open, since every neighborhood of a point on the line $x = 1$ must contain points in S and points not in S. See Figure 16.9.

EXAMPLE 2 Figure 16.10 illustrates some additional open sets.

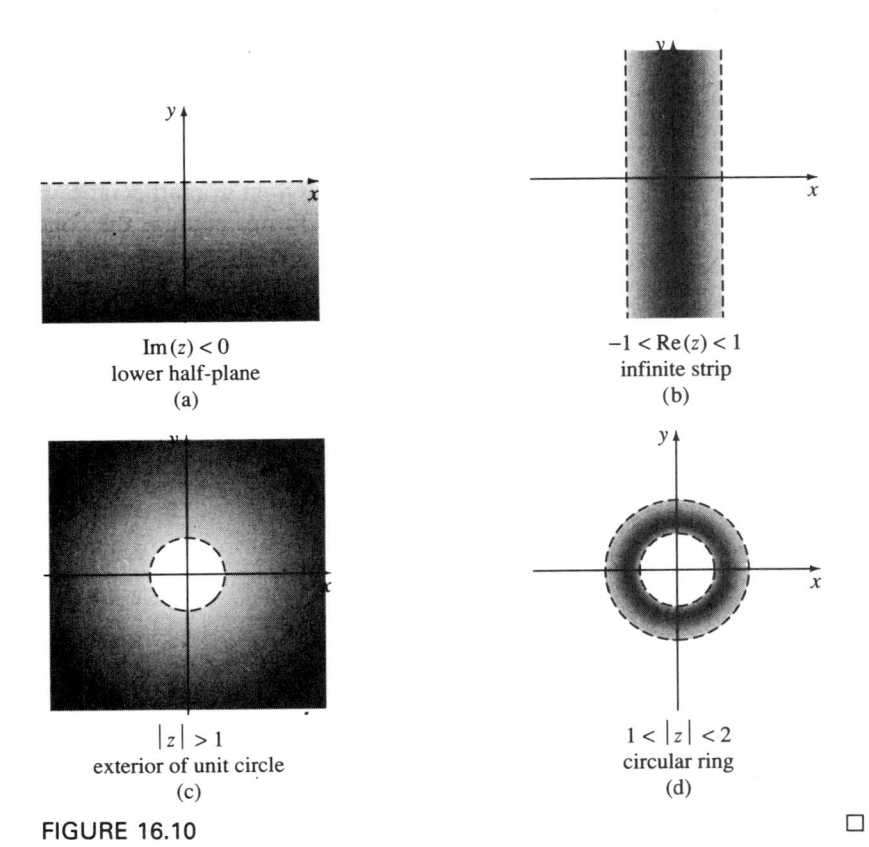

(a) $\text{Im}(z) < 0$ lower half-plane

(b) $-1 < \text{Re}(z) < 1$ infinite strip

(c) $|z| > 1$ exterior of unit circle

(d) $1 < |z| < 2$ circular ring

FIGURE 16.10

The set of numbers satisfying the inequality

$$\rho_1 < |z - z_0| < \rho_2$$

such as illustrated in Figure 16.10(d), is also called an open **annulus**.

If every neighborhood of a point z_0 contains at least one point that is in a set S and at least one point that is not in S, then z_0 is said to be a **boundary point** of S. The **boundary** of a set S is the set of all boundary points of S. For the set of points defined by $\text{Re}(z) \geq 1$, the points on the line $x = 1$ are boundary points. The points on the circle $|z - i| = 2$ are boundary points for the disk $|z - i| \leq 2$.

If any pair of points z_1 and z_2 in an open set S can be connected by a polygonal line that lies entirely in the set, then the open set S is said to be **connected**. See Figure 16.11. An open connected set is called a **domain**. All the open sets in Figure 16.10 are connected and so are domains. The set of numbers satisfying $\text{Re}(z) \neq 4$ is an open set but is not connected, since it is not possible to join points on either side of the vertical line $x = 4$ by a polygonal line without leaving the set (bear in mind that the points on $x = 4$ are not in the set).

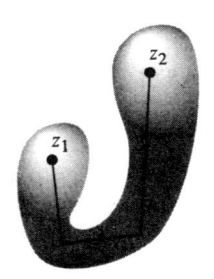

FIGURE 16.11

A **region** is a domain in the complex plane with all, some, or none of its boundary points. Since an open connected set does not contain any boundary points, it is automatically a region. A region containing all its boundary points is said to be **closed**. The disk defined by $|z - i| \leq 2$ is an example of a closed region and is referred to as a closed disk. A region may be neither open nor closed; the annular region defined by $1 \leq |z - 5| < 3$ contains only some of its boundary points and so is neither open nor closed.

▲ **Remark** Often in mathematics the same word is used in entirely different contexts. Do not confuse the concept of "domain" defined above with the concept of the "domain of a function."

EXERCISES 16.3 Answers to odd-numbered problems begin on page A-86.

In Problems 1–8 sketch the graph of the given equation.

1. $\text{Re}(z) = 5$
2. $\text{Im}(z) = -2$
3. $\text{Im}(\bar{z} + 3i) = 6$
4. $\text{Im}(z - i) = \text{Re}(z + 4 - 3i)$
5. $|z - 3i| = 2$
6. $|2z + 1| = 4$
7. $|z - 4 + 3i| = 5$
8. $|z + 2 + 2i| = 2$

In Problems 9–22 sketch the set of points in the complex plane satisfying the given inequality. Determine whether the set is a domain.

9. $\text{Re}(z) < -1$
10. $|\text{Re}(z)| > 2$
11. $\text{Im}(z) > 3$
12. $\text{Im}(z - i) < 5$
13. $2 < \text{Re}(z - 1) < 4$
14. $-1 \leq \text{Im}(z) < 4$
15. $\text{Re}(z^2) > 0$
16. $\text{Im}(1/z) < 1/2$
17. $0 \leq \arg(z) \leq 2\pi/3$
18. $|\arg(z)| < \pi/4$
19. $|z - i| > 1$
20. $|z - i| > 0$
21. $2 < |z - i| < 3$
22. $1 \leq |z - 1 - i| < 2$

23. Describe the set of points in the complex plane that satisfies $|z + 1| = |z - i|$.

24. Describe the set of points in the complex plane that satisfies $|\text{Re}(z)| \leq |z|$.

25. Describe the set of points in the complex plane that satisfies $z^2 + \bar{z}^2 = 2$.

26. Describe the set of points in the complex plane that satisfies $|z - i| + |z + i| = 1$.

16.4 FUNCTIONS OF A COMPLEX VARIABLE; ANALYTICITY

Function Recall that a function f is a rule of correspondence that assigns to each element of a set A one and only one element of a set B. If b is the element in B assigned to the element a in A by f, we say that b is the **image** of a and write $b = f(a)$. The set A is called the **domain** of the function f but is not necessarily a domain in the sense defined in Section 16.3. The set of all images in B is called the **range** of the function. For example, suppose the set A is a set of real numbers defined by $3 \leq x < \infty$ and the function is given by $f(x) = \sqrt{x - 3}$; then $f(3) = 0$, $f(4) = 1$, $f(8) = \sqrt{5}$, and so on. In other words, the range of f is the set given by $0 \leq y < \infty$. Since A is a set of real numbers, we say f is a function of a real variable x. When the domain A is the set of complex numbers z, we naturally say that f is a function of a complex variable z or a **complex function**. The image w of a complex number z will be some complex number $u + iv$, that is,

$$w = f(z) = u(x, y) + iv(x, y) \qquad (1)$$

where u and v are the real and imaginary parts of w and are real-valued functions. Inherent in the mathematical statement (1) is the fact:

We cannot draw a graph of a complex function $w = f(z)$

since a graph would require four axes in a four-dimensional coordinate system.

Some examples of functions of a complex variable are

$$f(z) = z^2 - 4z, \quad z \text{ any complex number}$$

$$f(z) = \frac{z}{z^2 + 1}, \quad z \neq i \text{ and } z \neq -i$$

$$f(z) = z + \text{Re}(z); \quad z \text{ any complex number}$$

Each of these functions can be expressed in form (1). For example,

$$f(z) = z^2 - 4z = (x + iy)^2 - 4(x + iy) = (x^2 - y^2 - 4x) + i(2xy - 4y)$$

Thus, $u(x, y) = x^2 - y^2 - 4x$ and $v(x, y) = 2xy - 4y$.

Although we cannot draw a graph, a complex function $w = f(z)$ can be interpreted as a **mapping** or **transformation** from the z-plane to the w-plane. See Figure 16.12.

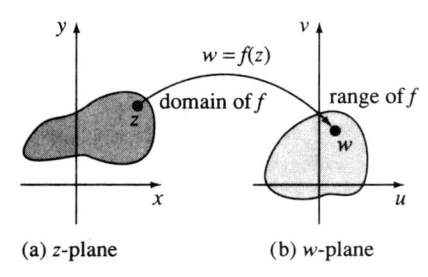

(a) z-plane
(b) w-plane

FIGURE 16.12

(a) z-plane

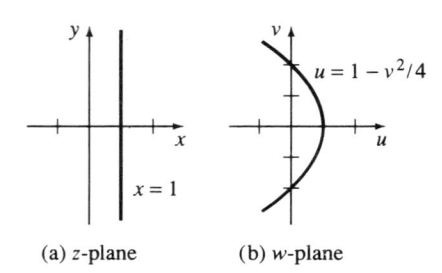

(b) w-plane

FIGURE 16.13

EXAMPLE 1 Find the image of the line $\text{Re}(z) = 1$ under the mapping $f(z) = z^2$.

Solution For the function $f(z) = z^2$ we have $u(x, y) = x^2 - y^2$ and $v(x, y) = 2xy$. Now, $\text{Re}(z) = x$ and so by substituting $x = 1$ into the functions u and v, we obtain

$$u = 1 - y^2 \quad \text{and} \quad v = 2y$$

These are parametric equations of a curve in the w-plane. Substituting $y = v/2$ into the first equation eliminates the parameter y to give $u = 1 - v^2/4$. In other words, the image of the line in Figure 16.13(a) is the parabola shown in Figure 16.13(b). □

We shall pursue the idea of $f(z)$ as a mapping in greater detail in Chapter 19.

It should be noted that a complex function is completely determined by the real-valued functions u and v. This means a complex function $w = f(z)$ can be defined by arbitrarily specifying $u(x, y)$ and $v(x, y)$, even though $u + iv$ may not be obtainable through the familiar operations on the symbol z alone. For example, if $u(x, y) = xy^2$ and $v(x, y) = x^2 - 4y^3$, then $f(z) = xy^2 + i(x^2 - 4y^3)$ is a function of a complex variable. To compute, say, $f(3 + 2i)$ we substitute $x = 3$ and $y = 2$ into u and v to obtain $f(3 + 2i) = 12 - 23i$.

Complex Functions As Flows We also may interpret a complex function $w = f(z)$ as a **two-dimensional fluid flow** by considering the complex number $f(z)$ as a vector based at the point z. The vector $f(z)$ specifies the speed and direction of the flow at a given point z. Figures 16.14 and 16.15 show the flows corresponding to the complex functions $f_1(z) = \bar{z}$ and $f_2(z) = z^2$, respectively.

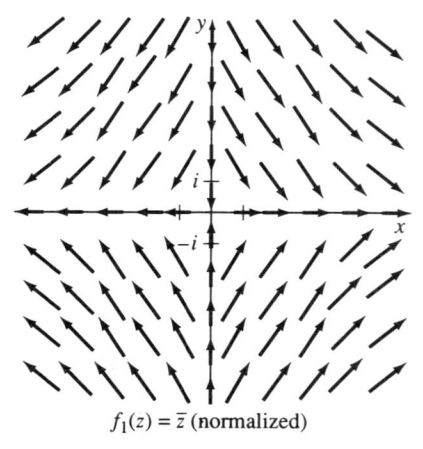

$f_1(z) = \bar{z}$ (normalized)

FIGURE 16.14

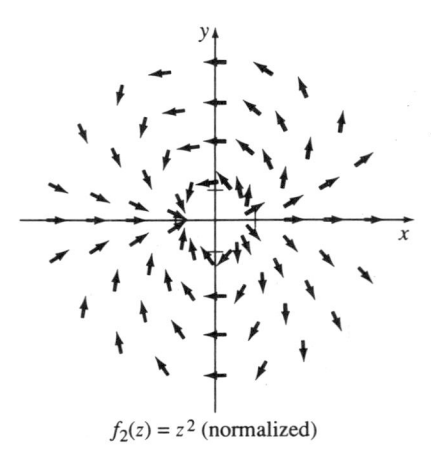

$f_2(z) = z^2$ (normalized)

FIGURE 16.15

16.4 Functions of a Complex Variable; Analyticity

If $x(t) + iy(t)$ is a parametric representation for the path of a particle in the flow, the tangent vector $\mathbf{T} = x'(t) + iy'(t)$ must coincide with $f(x(t) + iy(t))$. When $f(z) = u(x, y) + iv(x, y)$, it follows that the path of the particle must satisfy the system of differential equations

$$\frac{dx}{dt} = u(x, y)$$

$$\frac{dy}{dt} = v(x, y)$$

We call the family of solutions of this system the *streamlines* of the flow associated with $f(z)$.

EXAMPLE 2 Find the streamlines of the flows associated with the complex functions (a) $f_1(z) = \bar{z}$ and (b) $f_2(z) = z^2$.

Solution (a) The streamlines corresponding to $f_1(z) = x - iy$ satisfy the system

$$\frac{dx}{dt} = x$$

$$\frac{dy}{dt} = -y$$

and so $x(t) = c_1 e^t$ and $y(t) = c_2 e^{-t}$. By multiplying these two parametric equations, we see that the point $x(t) + iy(t)$ lies on the hyperbola $xy = c_1 c_2$.
(b) To find the streamlines corresponding to $f_2(z) = (x^2 - y^2) + i2xy$, note that $dx/dt = x^2 - y^2$, $dy/dt = 2xy$, and so

$$\frac{dy}{dx} = \frac{2xy}{x^2 - y^2}$$

This homogeneous differential equation has the solution $x^2 + y^2 = c_2 y$, a family of circles that have centers on the y-axis and pass through the origin. □

Limits and Continuity The definition of a limit of a complex function $f(z)$ as $z \to z_0$ has the same outward appearance as the limit in real variables.

DEFINITION 16.4 Limit of a Function

Suppose the function f is defined in some neighborhood of z_0, except possibly at z_0 itself. Then f is said to possess a **limit** at z_0, written

$$\lim_{z \to z_0} f(z) = L$$

if, for each $\varepsilon > 0$, there exists a $\delta > 0$ such that $|f(z) - L| < \varepsilon$ whenever $0 < |z - z_0| < \delta$.

948 CHAPTER 16 Functions of a Complex Variable

(a) δ-neighborhood (a) ε-neighborhood

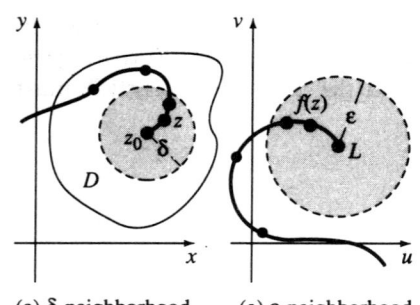

FIGURE 16.16

In words, $\lim_{z \to z_0} f(z) = L$ means that the points $f(z)$ can be made arbitrarily close to the point L if we choose the point z sufficiently close to, but not equal to, the point z_0. As shown in Figure 16.16, for each ε-neighborhood of L (defined by $|f(z) - L| < \varepsilon$) there is a δ-neighborhood of z_0 (defined by $|z - z_0| < \delta$) so that the images of all points $z \neq z_0$ in this neighborhood lie in the ε-neighborhood of L.

The fundamental difference between this definition and the limit concept in real variables lies in the understanding of $z \to z_0$. For a function f of a single real variable x, $\lim_{x \to x_0} f(x) = L$ means $f(x)$ approaches L as x approaches x_0 either from the right of x_0 or from the left of x_0 on the real number line. But since z and z_0 are points in the complex plane, when we say that $\lim_{z \to z_0} f(z)$ exists, we mean that $f(z)$ approaches L as the point z approaches z_0 from *any* direction.

The following theorem will summarize some properties of limits:

THEOREM 16.1 Limit of Sum, Product, Quotient

Suppose $\lim_{z \to z_0} f(z) = L_1$ and $\lim_{z \to z_0} g(z) = L_2$. Then:

(i) $\displaystyle\lim_{z \to z_0} [f(z) + g(z)] = L_1 + L_2$

(ii) $\displaystyle\lim_{z \to z_0} f(z)g(z) = L_1 L_2$

(iii) $\displaystyle\lim_{z \to z_0} \frac{f(z)}{g(z)} = \frac{L_1}{L_2}, \quad L_2 \neq 0$

DEFINITION 16.5 Continuity at a Point

A function f is **continuous** at a point z_0 if

$$\lim_{z \to z_0} f(z) = f(z_0)$$

As a consequence of Theorem 16.1, it follows that if two functions f and g are continuous at point z_0, then their sum and product are continuous at z_0. The quotient of the two functions is continuous at z_0 provided $g(z_0) \neq 0$.

A function f defined by

$$f(z) = a_n z^n + a_{n-1} z^{n-1} + \cdots + a_2 z^2 + a_1 z + a_0, \qquad a_n \neq 0 \qquad (2)$$

where n is a nonnegative integer and the coefficients a_i, $i = 0, \ldots, n$, are complex constants, is called a **polynomial** of degree n. Although we shall not prove it, the following limit result

$$\lim_{z \to z_0} z = z_0$$

16.4 Functions of a Complex Variable; Analyticity

indicates that the simple polynomial function $f(z) = z$ is continuous everywhere—that is, on the entire z-plane. With this result in mind and with repeated applications of Theorems 16.1 (*ii*) and (*i*), it follows that a polynomial function (2) is continuous everywhere. A **rational function**

$$f(z) = \frac{g(z)}{h(z)}$$

where g and h are polynomial functions, is continuous except at those points at which $h(z)$ is zero.

Derivative The derivative of a complex function is defined in terms of a limit. The symbol Δz used in the following definition is the complex number $\Delta x + i\,\Delta y$.

DEFINITION 16.6 Derivative

Suppose the complex function f is defined in a neighborhood of a point z_0. The **derivative** of f at z_0 is

$$f'(z_0) = \lim_{\Delta z \to 0} \frac{f(z_0 + \Delta z) - f(z_0)}{\Delta z} \tag{3}$$

provided this limit exists.

If the limit in (3) exists, the function f is said to be **differentiable** at z_0. The derivative of a function $w = f(z)$ is also written dw/dz.

As in real variables, differentiability implies continuity:

If f is differentiable at z_0, then f is continuous at z_0.

Moreover, the **rules of differentiation** are the same as in the calculus of real variables. If f and g are differentiable at a point z and c is a complex constant, then:

Constant rules: $\quad \dfrac{d}{dz} c = 0, \quad \dfrac{d}{dz} cf(z) = cf'(z) \tag{4}$

Sum rule: $\quad \dfrac{d}{dz}[f(z) + g(z)] = f'(z) + g'(z) \tag{5}$

Product rule: $\quad \dfrac{d}{dz}[f(z)g(z)] = f(z)g'(z) + g(z)f'(z) \tag{6}$

Quotient rule: $\quad \dfrac{d}{dz}\left[\dfrac{f(z)}{g(z)}\right] = \dfrac{g(z)f'(z) - f(z)g'(z)}{[g(z)]^2} \tag{7}$

Chain rule: $\quad \dfrac{d}{dz} f(g(z)) = f'(g(z))g'(z) \tag{8}$

The usual rule for differentiation of powers of z is also valid:

$$\frac{d}{dz} z^n = nz^{n-1}, \quad n \text{ an integer} \tag{9}$$

EXAMPLE 3 Differentiate (a) $f(z) = 3z^4 - 5z^3 + 2z$ and (b) $f(z) = \dfrac{z^2}{4z+1}$.

Solution (a) Using the power rule (9) along with the sum rule (5), we obtain

$$f'(z) = 3 \cdot 4z^3 - 5 \cdot 3z^2 + 2 = 12z^3 - 15z^2 + 2$$

(b) From the quotient rule (7),

$$f'(z) = \frac{(4z+1) \cdot 2z - z^2 \cdot 4}{(4z+1)^2} = \frac{4z^2 + 2z}{(4z+1)^2} \qquad \square$$

In order for a complex function f to be differentiable at a point z_0, $\lim\limits_{\Delta z \to 0} \dfrac{f(z_0 + \Delta z) - f(z_0)}{\Delta z}$ must approach the same complex number from any direction. Thus in the study of complex variables, to require the differentiability of a function is a greater demand than in real variables. If a complex function is made up, such as $f(z) = x + 4iy$, there is a good chance that it is not differentiable.

EXAMPLE 4 Show that the function $f(z) = x + 4iy$ is nowhere differentiable.

Solution With $\Delta z = \Delta x + i\,\Delta y$, we have

$$f(z + \Delta z) - f(z) = (x + \Delta x) + 4i(y + \Delta y) - x - 4iy = \Delta x + 4i\,\Delta y$$

and so
$$\lim_{\Delta z \to 0} \frac{f(z + \Delta z) - f(z)}{\Delta z} = \lim_{\Delta z \to 0} \frac{\Delta x + 4i\,\Delta y}{\Delta x + i\,\Delta y} \tag{10}$$

Now, if we let $\Delta z \to 0$ along a line parallel to the x-axis, then $\Delta y = 0$ and the value of (10) is 1. On the other hand, if we let $\Delta z \to 0$ along a line parallel to the y-axis, then $\Delta x = 0$ and the value of (1) is seen to be 4. Therefore, $f(z) = x + 4iy$ is not differentiable at any point z. $\qquad \square$

Analytic Functions While the requirement of differentiability is a stringent demand, there is a class of functions that is of great importance whose members satisfy even more severe requirements. These functions are called **analytic functions**.

> **DEFINITION 16.7 Analyticity at a Point**
>
> A complex function $w = f(z)$ is said to be **analytic at a point** z_0 if f is differentiable at z_0 and at every point in some neighborhood of z_0.

A function f is analytic in a domain D if it is analytic at every point in D.

The student should read Definition 16.7 carefully. Analyticity *at a point* is a neighborhood property. Analyticity at a point is, therefore, not the same as differentiability at a point. It is left as an exercise to show that the function $f(z) = |z|^2$ is differentiable at $z = 0$ but it differentiable nowhere else. Hence, $f(z) = |z|^2$ is nowhere analytic. In contrast, the simple polynomial $f(z) = z^2$ is differentiable at every point z in the complex plane. Hence, $f(z) = z^2$ is analytic everywhere. A function that is analytic at every point z is said to be an **entire function**. Polynomial functions are differentiable at every point z and so are entire functions.

▲ **Remark** Recall from algebra that a number c is a zero of a polynomial function if and only if $x - c$ is a factor of $f(x)$. The same result holds in complex analysis. For example, since $f(z) = z^4 + 5z^2 + 4 = (z^2 + 1)(z^2 + 4)$, the zeros of f are $-i, i, -2i$, and $2i$. Hence, $f(z) = (z + i)(z - i)(z + 2i)(z - 2i)$. Moreover, the quadratic formula is also valid. For example, using this formula, we can write

$$f(z) = z^2 - 2z + 2$$
$$= (z - (1 + i))(z - (1 - i))$$
$$= (z - 1 - i)(z - 1 + i)$$

See Problems 21 and 22.

EXERCISES 16.4 *Answers to odd-numbered problems begin on page A-87.*

In Problems 1–6 find the image of the given line under the mapping $f(z) = z^2$.

1. $y = 2$ **2.** $x = -3$
3. $x = 0$ **4.** $y = 0$
5. $y = x$ **6.** $y = -x$

In Problems 7–14 express the given function in the form $f(z) = u + iv$.

7. $f(z) = 6z - 5 + 9i$
8. $f(z) = 7z - 9i\bar{z} - 3 + 2i$
9. $f(z) = z^2 - 3z + 4i$ **10.** $f(z) = 3\bar{z}^2 + 2z$
11. $f(z) = z^3 - 4z$ **12.** $f(z) = z^4$
13. $f(z) = z + 1/z$ **14.** $f(z) = \dfrac{z}{z + 1}$

In Problems 15–18 evaluate the given function at the indicated points.

15. $f(z) = 2x - y^2 + i(xy^3 - 2x^2 + 1)$
  (a) $2i$ (b) $2 - i$ (c) $5 + 3i$

16. $f(z) = (x + 1 + 1/x) + i(4x^2 - 2y^2 - 4)$
  (a) $1 + i$ (b) $2 - i$ (c) $1 + 4i$

17. $f(z) = 4z + i\bar{z} + \text{Re}(z)$
  (a) $4 - 6i$ (b) $-5 + 12i$ (c) $2 - 7i$

18. $f(z) = e^x \cos y + ie^x \sin y$
  (a) $\pi i/4$ (b) $-1 - \pi i$ (c) $3 + \pi i/3$

In Problems 19–22 the given limit exists. Find its value.

19. $\lim\limits_{z \to i}(4z^3 - 5z^2 + 4z + 1 - 5i)$

20. $\lim\limits_{z \to 1-i} \dfrac{5z^2 - 2z + 2}{z + 1}$

21. $\lim\limits_{z \to i} \dfrac{z^4 - 1}{z - i}$

22. $\lim\limits_{z \to 1+i} \dfrac{z^2 - 2z + 2}{z^2 - 2i}$

In Problems 23 and 24 show that the given limit does not exist.

23. $\lim\limits_{z \to 0} \dfrac{z}{\bar{z}}$

24. $\lim\limits_{z \to 1} \dfrac{x + y - 1}{z - 1}$

In Problems 25 and 26 use (3) to obtain the indicated derivative of the given function.

25. $f(z) = z^2$, $f'(z) = 2z$

26. $f(z) = 1/z$, $f'(z) = -1/z^2$

In Problems 27–34 use (4)–(8) to find the derivative $f'(z)$ for the given function.

27. $f(z) = 4z^3 - (3 + i)z^2 - 5z + 4$

28. $f(z) = 5z^4 - iz^3 + (8 - i)z^2 - 6i$

29. $f(z) = (2z + 1)(z^2 - 4z + 8i)$

30. $f(z) = (z^5 + 3iz^3)(z^4 + iz^3 + 2z^2 - 6iz)$

31. $f(z) = (z^2 - 4i)^3$

32. $f(z) = (2z - 1/z)^6$

33. $f(z) = \dfrac{3z - 4 + 8i}{2z + i}$

34. $f(z) = \dfrac{5z^2 - z}{z^3 + 1}$

In Problems 35–38 give the points at which the given function will not be analytic.

35. $f(z) = \dfrac{z}{z - 3i}$

36. $f(z) = \dfrac{2i}{z^2 - 2z + 5iz}$

37. $f(z) = \dfrac{z^3 + z}{z^2 + 4}$

38. $f(z) = \dfrac{z - 4 + 3i}{z^2 - 6z + 25}$

39. Show that the function $f(z) = \bar{z}$ is nowhere differentiable.

40. The function $f(z) = |z|^2$ is continuous throughout the entire complex plane. Show, however, that f is differentiable only at the point $z = 0$. [*Hint*: Use (3) and consider two cases, $z = 0$ and $z \neq 0$. In the second case let Δz approach zero along a line parallel to the x-axis and then let Δz approach zero along a line parallel to the y-axis.]

In Problems 41–44 find the streamlines of the flow associated with the given complex function.

41. $f(z) = 2z$

42. $f(z) = iz$

43. $f(z) = 1/\bar{z}$

44. $f(z) = x^2 - iy^2$

In Problems 45 and 46 use a graphics calculator or computer to obtain the image of the given parabola under the mapping $f(z) = z^2$.

45. $y = \tfrac{1}{2}x^2$

46. $y = (x - 1)^2$

16.5 CAUCHY–RIEMANN EQUATIONS

In the preceding section we saw that a function f of a complex variable z is analytic in a domain D if f is differentiable at all points in D. We shall now develop a simple test for the analyticity of a complex-valued function f based on the partial derivatives of its real and imaginary parts. We will first prove a necessary condition for analyticity.

THEOREM 16.2 **Cauchy–Riemann Equations**

Suppose $f(z) = u(x, y) + iv(x, y)$ is continuous in some neighborhood of the point $z = x + iy$ and is differentiable at z. Then at the point z the first-order partial derivatives of u and v exist and satisfy the **Cauchy–Riemann equations**

$$\dfrac{\partial u}{\partial x} = \dfrac{\partial v}{\partial y} \quad \text{and} \quad \dfrac{\partial u}{\partial y} = -\dfrac{\partial v}{\partial x} \qquad (1)$$

16.5 Cauchy–Riemann Equation

Proof Since $f'(z)$ exists, we know that

$$f'(z) = \lim_{\Delta z \to 0} \frac{f(z + \Delta z) - f(z)}{\Delta z} \qquad (2)$$

By writing $f(z) = u(x, y) + iv(x, y)$ and $\Delta z = \Delta x + i\,\Delta y$, we get from (2)

$$f'(z) = \lim_{\Delta z \to 0} \frac{u(x + \Delta x, y + \Delta y) + iv(x + \Delta x, y + \Delta y) - u(x, y) - iv(x, y)}{\Delta x + i\,\Delta y} \qquad (3)$$

Since this limit exists, Δz can approach zero from any convenient direction. In particular, if $\Delta z \to 0$ horizontally, then $\Delta z = \Delta x$ and so (3) becomes

$$f'(z) = \lim_{\Delta x \to 0} \frac{u(x + \Delta x, y) - u(x, y)}{\Delta x} + i \lim_{\Delta x \to 0} \frac{v(x + \Delta x, y) - v(x, y)}{\Delta x} \qquad (4)$$

Since $f'(z)$ exists, the two limits in (4) exist. But by definition the limits in (4) are the first partial derivatives of u and v with respect to x. Thus, we have shown that

$$f'(z) = \frac{\partial u}{\partial x} + i\frac{\partial v}{\partial x} \qquad (5)$$

Now if we let $\Delta z \to 0$ vertically, then $\Delta z = i\,\Delta y$ and (3) becomes

$$f'(z) = \lim_{\Delta y \to 0} \frac{u(x, y + \Delta y) - u(x, y)}{i\,\Delta y} + i \lim_{\Delta y \to 0} \frac{v(x, y + \Delta y) - v(x, y)}{i\,\Delta y} \qquad (6)$$

which is the same as

$$f'(z) = -i\frac{\partial u}{\partial y} + \frac{\partial v}{\partial y} \qquad (7)$$

Equating the real and imaginary parts of (5) and (7) yields the pair of equations in (1). ■

If a complex function $f(z) = u(x, y) + iv(x, y)$ is analytic throughout a domain D, then the real functions u and v must satisfy the Cauchy–Riemann equations (1) at every point in D.

EXAMPLE 1 The polynomial $f(z) = z^2 + z$ is analytic for all z and $f(z) = x^2 - y^2 + x + i(2xy + y)$. Thus, $u(x, y) = x^2 - y^2 + x$ and $v(x, y) = 2xy + y$. For any point (x, y) we see that the Cauchy–Riemann equations are satisfied:

$$\frac{\partial u}{\partial x} = 2x + 1 = \frac{\partial v}{\partial y} \quad \text{and} \quad \frac{\partial u}{\partial y} = -2y = -\frac{\partial v}{\partial x} \qquad \square$$

EXAMPLE 2 Show that the function $f(z) = (2x^2 + y) + i(y^2 - x)$ is not analytic at any point.

Solution We identify $u(x, y) = 2x^2 + y$ and $v(x, y) = y^2 - x$. Now from

$$\frac{\partial u}{\partial x} = 4x \quad \text{and} \quad \frac{\partial v}{\partial y} = 2y$$

$$\frac{\partial u}{\partial y} = 1 \quad \text{and} \quad \frac{\partial v}{\partial x} = -1$$

we see that $\partial u/\partial y = -\partial v/\partial x$ but that the equality $\partial u/\partial x = \partial v/\partial y$ is satisfied only on the line $y = 2x$. However, for any point z on the line there is no neighborhood or open disk about z in which f is differentiable. We conclude that f is nowhere analytic. □

By themselves, the Cauchy–Riemann equations are not sufficient to ensure analyticity. However, when we add the condition of continuity to u and v *and* the four partial derivatives, the Cauchy–Riemann equations can be shown to imply analyticity. The proof is long and complicated and so we state only the result.

THEOREM 16.3 **Criterion for Analyticity**

Suppose the real-valued functions $u(x, y)$ and $v(x, y)$ are continuous and have continuous first-order partial derivatives in a domain D. If u and v satisfy the Cauchy–Riemann equations at all points of D, then the complex function $f(z) = u(x, y) + iv(x, y)$ is analytic in D.

EXAMPLE 3 For the function $f(z) = \dfrac{x}{x^2 + y^2} - i\dfrac{y}{x^2 + y^2}$ we have

$$\frac{\partial u}{\partial x} = \frac{y^2 - x^2}{(x^2 + y^2)^2} = \frac{\partial v}{\partial y} \quad \text{and} \quad \frac{\partial u}{\partial y} = -\frac{2xy}{(x^2 + y^2)^2} = -\frac{\partial v}{\partial x}$$

In other words, the Cauchy–Riemann equations are satisfied except at the point where $x^2 + y^2 = 0$, that is, at $z = 0$. We conclude from Theorem 16.3 that f is analytic in any domain not containing the point $z = 0$. □

The results in (5) and (7) were obtained under the basic assumption that f was differentiable at the point z. In other words, (5) and (7) give us a formula for computing $f'(z)$:

$$\boxed{f'(z) = \frac{\partial u}{\partial x} + i\frac{\partial v}{\partial x} = \frac{\partial v}{\partial y} - i\frac{\partial u}{\partial y}} \qquad (8)$$

For example, we know that $f(z) = z^2$ is differentiable for all z. With $u(x, y) = x^2 - y^2$, $\partial u/\partial x = 2x$, $v(x, y) = 2xy$, and $\partial v/\partial x = 2y$, we see that

$$f'(z) = 2x + i2y = 2(x + iy) = 2z$$

Recall that analyticity implies differentiability but not vice versa. Theorem 16.3 has an analogue that gives sufficient conditions for differentiability:

If the real-valued functions $u(x, y)$ and $v(x, y)$ are continuous and have continuous first partial derivatives in a neighborhood of z, and if u and v satisfy the Cauchy–Riemann equations at the point z, then the complex function $f(z) = u(x, y) + iv(x, y)$ is differentiable at z and $f'(z)$ is given by (8).

The function $f(z) = x^2 - y^2 i$ is nowhere analytic. With the identifications $u(x, y) = x^2$ and $v(x, y) = -y^2$, we see from

$$\frac{\partial u}{\partial x} = 2x, \frac{\partial v}{\partial y} = -2y \quad \text{and} \quad \frac{\partial u}{\partial y} = 0, \frac{\partial v}{\partial x} = 0$$

that the Cauchy–Riemann equations are satisfied only when $y = -x$. But since the functions u, $\partial u/\partial x$, $\partial u/\partial y$, v, $\partial v/\partial x$, and $\partial v/\partial y$ are continuous at every point, it follows that f is differentiable on the line $y = -x$ and on that line (8) gives the derivative $f'(z) = 2x = -2y$.

Harmonic Functions We saw in Chapter 12 that Laplace's equation $\partial^2 u/\partial x^2 + \partial^2 u/\partial y^2 = 0$ occurs in certain problems involving steady-state temperatures. This partial differential equation also plays an important role in many areas of applied mathematics. Indeed, as we now see, the real and imaginary parts of an analytic function cannot be chosen arbitrarily, since both u and v must satisfy Laplace's equation. It is this link between analytic functions and Laplace's equation that makes complex variables so essential in the serious study of applied mathematics.

DEFINITION 16.3 Harmonic Functions

A real-valued function $\phi(x, y)$ that has continuous second-order partial derivatives in a domain D and satisfies Laplace's equation is said to be **harmonic** in D.

THEOREM 16.4

Suppose $f(z) = u(x, y) + iv(x, y)$ is analytic in a domain D. Then the functions $u(x, y)$ and $v(x, y)$ are harmonic functions.

Proof In this proof we shall assume that u and v have continuous second-order partial derivatives. Since f is analytic, the Cauchy–Riemann equations are satisfied. Differentiating both sides of $\partial u/\partial x = \partial v/\partial y$ with respect to x and differentiating both sides of $\partial u/\partial y = -\partial v/\partial x$ with respect to y then give

$$\frac{\partial^2 u}{\partial x^2} = \frac{\partial^2 v}{\partial x\, \partial y} \quad \text{and} \quad \frac{\partial^2 u}{\partial y^2} = -\frac{\partial^2 v}{\partial y\, \partial x}$$

With the assumption of continuity, the mixed partials are equal. Hence, adding these two equations gives

$$\frac{\partial^2 u}{\partial x^2} + \frac{\partial^2 u}{\partial y^2} = 0$$

This shows that $u(x, y)$ is harmonic.

Now differentiating both sides of $\partial u/\partial x = \partial v/\partial y$ with respect to y and differentiating both sides of $\partial u/\partial y = -\partial v/\partial x$ with respect to x and subtracting yield

$$\frac{\partial^2 v}{\partial x^2} + \frac{\partial^2 v}{\partial y^2} = 0 \qquad \blacksquare$$

Conjugate Harmonic Functions If $f(z) = u(x, y) + iv(x, y)$ is analytic in a domain D, then u and v are harmonic in D. Now suppose $u(x, y)$ is a given function that is harmonic in D. It is then sometimes possible to find another function $v(x, y)$ that is harmonic in D so that $u(x, y) + iv(x, y)$ is an analytic function in D. The function v is called a **conjugate harmonic function** of u.

EXAMPLE 4 (a) Verify that the function $u(x, y) = x^3 - 3xy^2 - 5y$ is harmonic in the entire complex plane.

(b) Find the conjugate harmonic function of u.

Solution (a) From the partial derivatives

$$\frac{\partial u}{\partial x} = 3x^2 - 3y^2 \qquad \frac{\partial^2 u}{\partial x^2} = 6x \qquad \frac{\partial u}{\partial y} = -6xy - 5 \qquad \frac{\partial^2 u}{\partial y^2} = -6x$$

we see that u satisfies Laplace's equation:

$$\frac{\partial^2 u}{\partial x^2} + \frac{\partial^2 u}{\partial y^2} = 6x - 6x = 0$$

(b) Since the conjugate harmonic function v must satisfy the Cauchy–Riemann equations, we must have

$$\frac{\partial v}{\partial y} = \frac{\partial u}{\partial x} = 3x^2 - 3y^2 \quad \text{and} \quad \frac{\partial v}{\partial x} = -\frac{\partial u}{\partial y} = 6xy + 5 \qquad (9)$$

Partial integration of the first equation in (9) with respect to y gives $v(x, y) = 3x^2y - y^3 + h(x)$. From this we get

$$\frac{\partial v}{\partial x} = 6xy + h'(x)$$

Substituting this result into the second equation in (9) gives $h'(x) = 5$ and so $h(x) = 5x + C$. Therefore, the conjugate harmonic function of u is $v(x, y) = 3x^2y - y^3 + 5x + C$. The analytic function is

$$f(z) = x^3 - 3xy^2 - 5y + i(3x^2y - y^3 + 5x + C) \qquad \square$$

▲ **Remark** Suppose u and v are the harmonic functions that comprise the real and imaginary parts of an analytic function $f(z)$. The level curves $u(x, y) = c_1$ and $v(x, y) = c_2$ defined by these functions form two *orthogonal* families of curves. (See Problem 32.) For example, the level curves generated by the simple analytic function $f(z) = z = x + iy$ are $x = c_1$ and $y = c_2$. The family of vertical lines defined by $x = c_1$ is clearly orthogonal to the family of horizontal lines defined by $y = c_2$. In electrostatics, if $u(x, y) = c_1$ defines the equipotential curves, then the other, and orthogonal, family $v(x, y) = c_2$ defines the lines of force.

EXERCISES 16.5 Answers to odd-numbered problems begin on page A-87.

In Problems 1 and 2 the given function is analytic for all z. Show that the Cauchy–Riemann equations are satisfied at every point.

1. $f(z) = z^3$
2. $f(z) = 3z^2 + 5z - 6i$

In Problems 3–8 show that the given function is not analytic at any point.

3. $f(z) = \text{Re}(z)$
4. $f(z) = y + ix$
5. $f(z) = 4z - 6\bar{z} + 3$
6. $f(z) = \bar{z}^2$
7. $f(z) = x^2 + y^2$
8. $f(z) = \dfrac{y}{x^2 + y^2} + i\dfrac{x}{x^2 + y^2}$

In Problems 9–14 use Theorem 16.3 to show that the given function is analytic in an appropriate domain.

9. $f(z) = e^x \cos y + ie^x \sin y$
10. $f(z) = x + \sin x \cosh y + i(y + \cos x \sinh y)$
11. $f(z) = e^{x^2 - y^2} \cos 2xy + ie^{x^2 - y^2} \sin 2xy$
12. $f(z) = 4x^2 + 5x - 4y^2 + 9 + i(8xy + 5y - 1)$

13. $f(z) = \dfrac{x - 1}{(x - 1)^2 + y^2} - i\dfrac{y}{(x - 1)^2 + y^2}$

14. $f(z) = \dfrac{x^3 + xy^2 + x}{x^2 + y^2} + i\dfrac{x^2y + y^3 - y}{x^2 + y^2}$

In Problems 15 and 16 find real constants a, b, c, and d so that the given function is analytic.

15. $f(z) = 3x - y + 5 + i(ax + by - 3)$
16. $f(z) = x^2 + axy + by^2 + i(cx^2 + dxy + y^2)$

In Problems 17–20 show that the given function is not analytic at any point, but is differentiable along the indicated curve(s).

17. $f(z) = x^2 + y^2 + 2xyi$; x-axis
18. $f(z) = 3x^2y^2 - 6x^2y^2i$; coordinate axes
19. $f(z) = x^3 + 3xy^2 - x + i(y^3 + 3x^2y - y)$; coordinates axes
20. $f(z) = x^2 - x + y + i(y^2 - 5y - x)$; $y = x + 2$

21. Use (8) to find the derivative of the function in Problem 9.

22. Use (8) to find the derivative of the function in Problem 11.

In Problems 23–28 verify that the given function u is harmonic. Find v, the conjugate harmonic function of u. Form the corresponding analytic function $f(z) = u + iv$.

23. $u(x, y) = x$

24. $u(x, y) = 2x - 2xy$

25. $u(x, y) = x^2 - y^2$

26. $u(x, y) = 4xy^3 - 4x^3y + x$

27. $u(x, y) = \log_e(x^2 + y^2)$

28. $u(x, y) = e^x(x \cos y - y \sin y)$

29. Sketch the level curves $u(x, y) = c_1$ and $v(x, y) = c_2$ of the analytic function $f(z) = z^2$.

30. Consider the function $f(z) = 1/z$. Describe the level curves.

31. Consider the function $f(z) = z + 1/z$. Describe the level curve $v(x, y) = 0$.

32. Suppose u and v are the harmonic functions forming the real and imaginary parts of an analytic function. Show that the level curves $u(x, y) = c_1$ and $v(x, y) = c_2$ are orthogonal. [*Hint:* Consider the gradient of u and the gradient of v. Ignore the case where a gradient vector is the zero vector.]

16.6 EXPONENTIAL AND LOGARITHMIC FUNCTIONS

In this and the next section, we shall examine the exponential, logarithmic, trigonometric, and hyperbolic functions of a complex variable z. Although the definitions of these complex functions are motivated by their real-variable analogues, the properties of these complex functions will yield some surprises.

16.6.1 Exponential Function

Recall that in real variables the exponential function $f(x) = e^x$ has the properties:

$$f'(x) = f(x) \quad \text{and} \quad f(x_1 + x_2) = f(x_1)f(x_2) \qquad (1)$$

We certainly want the definition of the complex function $f(z) = e^z$, where $z = x + iy$, to reduce to e^x for $y = 0$ and to possess the properties in (1).

We have already used an exponential function with a pure imaginary exponent. Euler's formula:

$$e^{iy} = \cos y + i \sin y, \quad y \text{ a real number} \qquad (2)$$

played an important role in Section 3.3. We can formally establish this result by using the Maclaurin series for e^x and replacing x by iy and rearranging terms:

$$e^{iy} = \sum_{k=0}^{\infty} \frac{(iy)^k}{k!} = 1 + iy + \frac{(iy)^2}{2!} + \frac{(iy)^3}{3!} + \frac{(iy)^4}{4!} + \cdots$$

$$= \left(1 - \frac{y^2}{2!} + \frac{y^4}{4!} - \frac{y^6}{6!} + \cdots\right) + i\left(y - \frac{y^3}{3!} + \frac{y^5}{5!} - \frac{y^7}{7!} + \cdots\right)$$

$$= \cos y + i \sin y$$

For $z = x + iy$ it is natural to expect that

16.6 Exponential and Logarithmic Functions

$$e^{x+iy} = e^x e^{iy}$$

and so by (2),

$$e^{x+iy} = e^x(\cos y + i \sin y)$$

Inspired by this formal result, we make the following definition:

DEFINITION 16.9 Exponential Function

$$e^z = e^{x+iy} = e^x(\cos y + i \sin y) \qquad (3)$$

The exponential function e^z is also denoted by the symbol $\exp z$. Note that (3) reduces to e^x when $y = 0$.

EXAMPLE 1 Evaluate $e^{1.7+4.2i}$.

Solution With the identifications $x = 1.7$ and $y = 4.2$ and the aid of a calculator, we have, to four rounded decimal places:

$$e^{1.7} \cos 4.2 = -2.6837 \quad \text{and} \quad e^{1.7} \sin 4.2 = -4.7710$$

It follows from (3) that

$$e^{1.7+4.2i} = -2.6837 - 4.7710i \qquad \square$$

The real and imaginary parts of e^z, $u(x, y) = e^x \cos y$ and $v(x, y) = e^x \sin y$, are continuous and have continuous first partial derivatives at every point z of the complex plane. Moreover, the Cauchy–Riemann equations are satisfied at all points of the complex plane:

$$\frac{\partial u}{\partial x} = e^x \cos y = \frac{\partial v}{\partial y} \quad \text{and} \quad \frac{\partial u}{\partial y} = -e^x \sin y = -\frac{\partial v}{\partial x}$$

It follows from Theorem 16.3 that $f(z) = e^z$ is analytic for all z; in other words, f is an entire function.

We shall now demonstrate that e^z possesses the two desired properties given in (1). First, the derivative of f is given by (5) of Section 16.5:

$$f'(z) = e^x \cos y + i(e^x \sin y) = e^x(\cos y + i \sin y) = f(z)$$

As desired, we have established that

$$\boxed{\frac{d}{dz} e^z = e^z}$$

Second, if $z_1 = x_1 + iy_1$ and $z_2 = x_2 + iy_2$, then by multiplication of complex numbers and the addition formulas of trigonometry, we obtain

$$f(z_1)f(z_2) = e^{x_1}(\cos y_1 + i \sin y_1)e^{x_2}(\cos y_2 + i \sin y_2)$$
$$= e^{x_1+x_2}[(\cos y_1 \cos y_2 - \sin y_1 \sin y_2) + i(\sin y_1 \cos y_2 + \cos y_1 \sin y_2)]$$
$$= e^{x_1+x_2}[\cos(y_1 + y_2) + i \sin(y_1 + y_2)] = f(z_1 + z_2)$$

In other words,

$$e^{z_1} e^{z_2} = e^{z_1 + z_2} \tag{4}$$

It is left as an exercise to prove that

$$\frac{e^{z_1}}{e^{z_2}} = e^{z_1 - z_2}$$

Periodicity Unlike the real function e^x, the complex function $f(z) = e^z$ is **periodic** with the complex period $2\pi i$. Since $e^{2\pi i} = \cos 2\pi + i \sin 2\pi = 1$ and, in view of (4), $e^{z+2\pi i} = e^z e^{2\pi i} = e^z$ for all z, it follows that $f(z + 2\pi i) = f(z)$. Because of this complex periodicity, all possible functional values of $f(z) = e^z$ are assumed in any infinite horizontal strip of width 2π. Thus, if we divide the complex plane into horizontal strips defined by $(2n - 1)\pi < y \leq (2n + 1)\pi$, $n = 0, \pm 1, \pm 2, \ldots$, then, as shown in Figure 16.17, for any point z in the strip $-\pi < y \leq \pi$, the values $f(z)$, $f(z + 2\pi i)$, $f(z - 2\pi i)$, $f(z + 4\pi i)$, and so on are the same. The strip $-\pi < y \leq \pi$ is called the **fundamental region** for the exponential function $f(z) = e^z$. The corresponding flow over the fundamental region is shown in Figure 16.18.

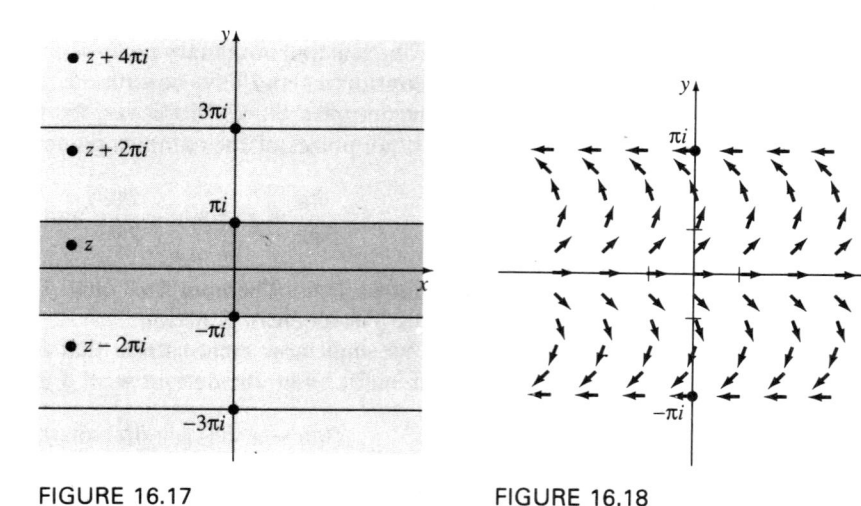

FIGURE 16.17

FIGURE 16.18

Polar Form of a Complex Number In Section 16.2 we saw that the complex number z could be written in polar form as $z = r(\cos \theta + i \sin \theta)$. Since $e^{i\theta} = \cos \theta + i \sin \theta$, we can now write the polar form of a complex number as

$$z = re^{i\theta}$$

For example, in polar form $z = 1 + i$ is $z = \sqrt{2}\, e^{\pi i/4}$.

▲ **Remark** In applying mathematics, mathematicians and engineers often approach the same problem in completely different ways. Consider, for example, the solution of Example 3 in Section 3.11. In this example we used strictly real analysis to find the steady-state current $i_p(t)$ in an *L-R-C* series circuit described by the differential equation

$$L\frac{d^2q}{dt^2} + R\frac{dq}{dt} + \frac{1}{C}q = E_0 \sin \gamma t$$

Electrical engineers often solve circuit problems such as this using complex analysis. To illustrate, let us first denote the imaginary unit $\sqrt{-1}$ by the symbol j to avoid confusion with the current i. Since current i is related to charge q by $i = dq/dt$, the differential equation is the same as

$$L\frac{di}{dt} + Ri + \frac{1}{C}q = E_0 \sin \gamma t$$

Moreover, the impressed voltage $E_0 \sin \gamma t$ can be replaced by $\text{Im}(E_0 e^{j\gamma t})$, where Im means the "imaginary part of." Because of this last form, the method of undetermined coefficients suggests that we assume a solution in the form of a constant multiple of complex exponential—that is, $i_p(t) = \text{Im}(Ae^{j\gamma t})$. We substitute this expression into the last differential equation, use the fact that q is an antiderivative of i, and equate coefficients of $e^{j\gamma t}$:

$$\left(jL\gamma + R + \frac{1}{jC\gamma}\right)A = E_0 \quad \text{gives} \quad A = \frac{E_0}{R + j\left(L\gamma - \dfrac{1}{C\gamma}\right)}$$

The quantity $Z = R + j(L\gamma - 1/C\gamma)$ is called the **complex impedance** of the circuit. Note that the modulus of the complex impedance, $|Z| = \sqrt{R^2 + (L\gamma - 1/C\gamma)^2}$, was denoted in Example 3 of Section 3.11 by the letter Z and called the impedance.

Now, in polar form the complex impedance is

$$Z = |Z|e^{j\theta} \quad \text{where} \quad \tan\theta = \frac{L\gamma - \dfrac{1}{C\gamma}}{R}$$

Hence, $A = E_0/Z = E_0/(|Z|e^{j\theta})$ and so the steady-state current can be written as

$$i_p(t) = \text{Im}\,\frac{E_0}{|Z|}e^{-j\theta}e^{j\gamma t}$$

The reader is encouraged to verify that this last expression is the same as (5) in Section 3.11.

16.6.2 Logarithmic Function

The logarithm of a complex number $z = x + iy$, $z \neq 0$, is defined as the inverse of the exponential function—that is,

$$w = \ln z \quad \text{if} \quad z = e^w \tag{5}$$

In (5) we note that $\ln z$ is not defined for $z = 0$, since there is no value of w for which $e^w = 0$. To find the real and imaginary parts of $\ln z$ we write $w = u + iv$ and use (3) and (5):

$$x + iy = e^{u+iv} = e^u(\cos v + i \sin v) = e^u \cos v + ie^u \sin v$$

The last equality implies $x = e^u \cos v$ and $y = e^u \sin v$. We can solve these two equations for u and v. First, by squaring and adding the equations, we find

$$e^{2u} = x^2 + y^2 = r^2 = |z|^2 \quad \text{and so} \quad u = \log_e |z|$$

where $\log_e |z|$ denotes the *real natural logarithm* of the modulus of z. Second, to solve for v we divide the two equations to obtain

$$\tan v = \frac{y}{x}$$

This last equation means that v is an argument of z, that is, $v = \theta = \arg z$. But since there is no unique argument of a given complex number $z = x + iy$, if θ is an argument of z, then so is $\theta + 2n\pi$, $n = 0, \pm 1, \pm 2, \ldots$.

DEFINITION 16.10 Logarithm of a Complex Number

For $z \neq 0$, and $\theta = \arg z$,

$$\ln z = \log_e |z| + i(\theta + 2n\pi), \quad n = 0, \pm 1, \pm 2, \ldots \tag{6}$$

As is clearly indicated in (6), there are infinitely many values of the logarithm of a complex number z. This should not be any great surprise since the exponential function is periodic.

In real calculus, logarithms of negative numbers are not defined. As the next example will show, this is not the case in complex calculus.

EXAMPLE 2 Find the values of (a) $\ln(-2)$, (b) $\ln i$, and (c) $\ln(-1-i)$.

Solution (a) With $\theta = \arg(-2) = \pi$ and $\log_e |-2| = 0.6932$, we have from (6)

$$\ln(-2) = 0.6932 + i(\pi + 2n\pi)$$

(b) With $\theta = \arg(i) = \pi/2$ and $\log_e|i| = \log_e 1 = 0$, we have from (6)

$$\ln i = i\left(\frac{\pi}{2} + 2n\pi\right)$$

In other words, $\ln i = \pi i/2, -3\pi i/2, 5\pi i/2, -7\pi i/2$, and so on.

(c) With $\theta = \arg(-1-i) = 5\pi/4$ and $\log_e|-1-i| = \log_e \sqrt{2} = 0.3466$, we have from (6)

$$\ln(-1-i) = 0.3466 + i\left(\frac{5\pi}{4} + 2n\pi\right) \qquad \square$$

EXAMPLE 3 Find all values of z such that $e^z = \sqrt{3} + i$.

Solution From (5), with the symbol w replaced by z, we have $z = \ln(\sqrt{3} + i)$. Now $|\sqrt{3} + i| = 2$ and $\tan\theta = 1/\sqrt{3}$ imply that $\arg(\sqrt{3} + i) = \pi/6$ and so (6) gives

$$z = \log_e 2 + i\left(\frac{\pi}{6} + 2n\pi\right) \quad \text{or} \quad z = 0.6931 + i\left(\frac{\pi}{6} + 2n\pi\right) \qquad \square$$

Principal Value It is interesting to note that as a consequence of (6) the logarithm of a positive real number has many values. For example, in real calculus $\log_e 5$ has only one value: $\log_e 5 = 1.6094$, whereas in complex calculus $\ln 5 = 1.6094 + 2n\pi i$. The value of $\ln 5$ corresponding to $n = 0$ is the same as the real logarithm $\log_e 5$ and is called the **principal value** of $\ln 5$. Recall that in Section 16.2 we stipulated that the principal argument of a complex number, written Arg z, lies in the interval $(-\pi, \pi]$. In general, we define the **principal value** of $\ln z$ as that complex logarithm corresponding to $n = 0$ and $\theta = \text{Arg } z$. To emphasize the principal value of the logarithm we shall adopt the notation Ln z. In other words,

$$\boxed{\text{Ln } z = \log_e|z| + i\,\text{Arg } z} \qquad (7)$$

Since Arg z is unique, there is only one value of Ln z for each $z \neq 0$.

EXAMPLE 4 The principal values of the logarithms in Example 2 are as follows:

(a) Since $\text{Arg}(-2) = \pi$, we need only set $n = 0$ in the result given in part (a) of Example 2:

$$\text{Ln}(-2) = 0.6932 + \pi i$$

(b) Similarly, since $\text{Arg}(i) = \pi/2$, we set $n = 0$ in the result in part (b) of Example 2 to obtain

$$\text{Ln } i = \frac{\pi}{2} i$$

(c) In part (c) of Example 2, $\arg(-1 - i) = 5\pi/4$ is not the principal argument of $z = -1 - i$. The argument of z that lies in the interval $(-\pi, \pi]$ is $\text{Arg}(-1 - i) = -3\pi/4$. Hence, it follows from (7) that

$$\text{Ln}(-1 - i) = 0.3466 - \frac{3\pi}{4} i \qquad \square$$

Up to this point we have avoided the use of the word *function* for the obvious reason that ln z defined in (6) is not a function in the strictest interpretation of that word. Nonetheless, it is customary to write $f(z) = \ln z$ and to refer to $f(z) = \ln z$ by the seemingly contradictory phrase *multiple-valued function*. Although we shall not pursue the details, (6) can be interpreted as an infinite collection of logarithmic functions (standard meaning of the word). Each function in the collection is called a **branch** of ln z. The function $f(z) = \text{Ln } z$ is then called the **principal branch** of ln z, or the **principal logarithmic function**. To minimize the confusion, we shall hereafter simply use the words *logarithmic function* when referring to either $f(z) = \ln z$ or $f(z) = \text{Ln } z$.

Some familiar properties of the logarithmic function hold in the complex case:

$$\boxed{\ln(z_1 z_2) = \ln z_1 + \ln z_2 \quad \text{and} \quad \ln\left(\frac{z_1}{z_2}\right) = \ln z_1 - \ln z_2} \qquad (8)$$

Equations (8) and (9) are to be interpreted in the sense that if values are assigned to two of the terms, then a correct value is assigned to the third term.

EXAMPLE 5 Suppose $z_1 = 1$ and $z_2 = -1$. Then if we take $\ln z_1 = 2\pi i$ and $\ln z_2 = \pi i$, we get

$$\ln(z_1 z_2) = \ln(-1) = \ln z_1 + \ln z_2 = 2\pi i + \pi i = 3\pi i$$

$$\ln\left(\frac{z_1}{z_2}\right) = \ln(-1) = \ln z_1 - \ln z_2 = 2\pi i - \pi i = \pi i \qquad \square$$

Just as (7) of Section 16.2 was not valid when arg z was replaced with Arg z, so too (8) is not true, in general, when ln z is replaced by Ln z. See Problems 45 and 46.

Analyticity The logarithmic function $f(z) = \text{Ln } z$ is not continuous at $z = 0$ since $f(0)$ is not defined. Moreover, $f(z) = \text{Ln } z$ is discontinuous at all

16.6 Exponential and Logarithmic Functions

points of the negative real axis. This is because the imaginary part of the function, $v = \text{Arg } z$, is discontinuous only at these points. To see this, suppose x_0 is a point on the negative real axis. As $z \to x_0$ from the upper half-plane, $\text{Arg } z \to \pi$, whereas if $z \to x_0$ from the lower half-plane, then $\text{Arg } z \to -\pi$. This means that $f(z) = \text{Ln } z$ is not analytic on the nonpositive real axis. However, $f(z) = \text{Ln } z$ is analytic throughout the domain D consisting of all the points in the complex plane except those on the nonpositive real axis. It is convenient to think of D as the complex plane from which the nonpositive real axis has been cut out. Since $f(z) = \text{Ln } z$ is the principal branch of $\ln z$, the nonpositive real axis is referred to as a **branch cut** for the function. See Figure 16.19. It is left as exercises to show that the Cauchy–Riemann equations are satisfied throughout this cut plane and that the derivative of $\text{Ln } z$ is given by

$$\frac{d}{dz} \text{Ln } z = \frac{1}{z} \qquad (9)$$

for all z in D.

Figure 16.20 shows $w = \text{Ln } z$ as a flow. Note that the vector field is not continuous along the branch cut.

Complex Powers Inspired by the identity $x^a = e^{a \ln x}$ in real variables, we can define complex powers of a complex number. If α is a complex number and $z = x + iy$, then z^α is defined by

$$z^\alpha = e^{\alpha \ln z}, \qquad z \neq 0 \qquad (10)$$

In general, z^α is multiple-valued since $\ln z$ is multiple-valued. However, in the special case when $\alpha = n$, $n = 0, \pm 1, \pm 2, \ldots$, (10) is single-valued since there is only one value for z^2, z^3, z^{-1}, and so on. To see that this is so, suppose $\alpha = 2$ and $z = re^{i\theta}$, where θ is any argument of z. Then

$$e^{2 \ln z} = e^{2(\ln r + i\theta)} = e^{2 \ln r + 2i\theta} = e^{2 \ln r} e^{2i\theta} = r^2 e^{i\theta} e^{i\theta} = (re^{i\theta})(re^{i\theta}) = z^2$$

If we use $\text{Ln } z$ in place of $\ln z$, then (10) gives the **principal value** of z^α.

EXAMPLE 6 Find the value of i^{2i}.

Solution With $z = i$, $\arg z = \pi/2$, and $\alpha = 2i$ it follows from (9) that

$$i^{2i} = e^{2i[\log_e 1 + i(\pi/2 + 2n\pi)]} = e^{-(1 + 4n)\pi}$$

where $n = 0, \pm 1, \pm 2, \ldots$. Inspection of the equation shows that i^{2i} is real for every value of n. Since $\pi/2$ is the principal argument of $z = i$, we obtain the principal value of i^{2i} for $n = 0$. To four rounded decimal places, this principal value is $i^{2i} = e^{-\pi} = 0.0043$. □

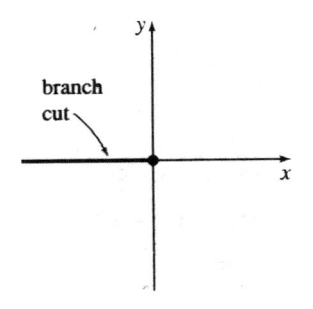

FIGURE 16.19

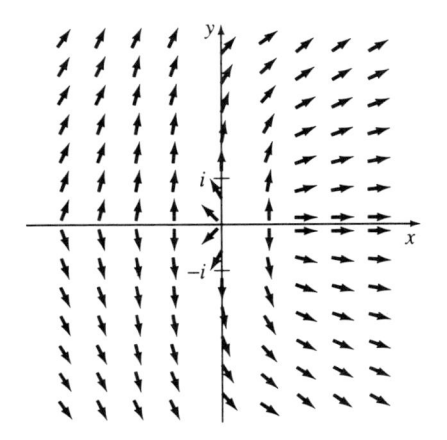

FIGURE 16.20

EXERCISES 16.6   Answers to odd-numbered problems begin on page A-87.

[16.6.1]

In Problems 1–10 express e^z in the form $a + ib$.

1. $z = \dfrac{\pi}{6}i$
2. $z = -\dfrac{\pi}{3}i$
3. $z = -1 + \dfrac{\pi}{4}i$
4. $z = 2 - \dfrac{\pi}{2}i$
5. $z = \pi + \pi i$
6. $z = -\pi + \dfrac{3\pi}{2}i$
7. $z = 1.5 + 2i$
8. $z = -0.3 + 0.5i$
9. $z = 5i$
10. $z = -0.23 - i$

In Problems 11 and 12 express the given number in the form $a + ib$.

11. $e^{1+5\pi i/4}e^{-1-\pi i/3}$
12. $\dfrac{e^{2+3\pi i}}{e^{-3+\pi i/2}}$

In Problems 13–16 use Definition 16.9 to express the given function in the form $f(z) = u + iv$.

13. $f(z) = e^{-iz}$
14. $f(z) = e^{2\bar{z}}$
15. $f(z) = e^{z^2}$
16. $f(z) = e^{1/z}$

In Problems 17–20 verify the given result.

17. $|e^z| = e^x$
18. $\dfrac{e^{z_1}}{e^{z_2}} = e^{z_1 - z_2}$
19. $e^{z+\pi i} = e^{z-\pi i}$
20. $(e^z)^n = e^{nz}$, n an integer

21. Show that $f(z) = e^{\bar{z}}$ is nowhere analytic.

22. (a) Use the result in Problem 15 to show that $f(z) = e^{z^2}$ is an entire function.
    (b) Verify that $u(x, y) = \text{Re}(e^{z^2})$ is a harmonic function.

[16.6.2]

In Problems 23–28 express $\ln z$ in the form $a + ib$.

23. $z = -5$
24. $z = -ei$
25. $z = -2 + 2i$
26. $z = 1 + i$
27. $z = \sqrt{2} + \sqrt{6}i$
28. $z = -\sqrt{3} + i$

In Problems 29–34 express $\text{Ln}\, z$ in the form $a + ib$.

29. $z = 6 - 6i$
30. $z = -e^3$
31. $z = -12 + 5i$
32. $z = 3 - 4i$
33. $z = (1 + \sqrt{3}i)^5$
34. $z = (1 + i)^4$

In Problems 35–38 find all values of z satisfying the given equation.

35. $e^z = 4i$
36. $e^{1/z} = -1$
37. $e^{z-1} = -ie^2$
38. $e^{2z} + e^z + 1 = 0$

In Problems 39–42 find all values of the given quantity.

39. $(-i)^{4i}$
40. $3^{i/\pi}$
41. $(1 + i)^{(1+i)}$
42. $(1 + \sqrt{3}i)^{3i}$

In Problems 43 and 44 find the principal value of the given quantity.

43. $(-1)^{(-2i/\pi)}$
44. $(1 - i)^{2i}$

45. If $z_1 = i$ and $z_2 = -1 + i$, verify that $\text{Ln}(z_1 z_2) \neq \text{Ln}\, z_1 + \text{Ln}\, z_2$.

46. Find two complex numbers z_1 and z_2 such that $\text{Ln}(z_1/z_2) \neq \text{Ln}\, z_1 - \text{Ln}\, z_2$.

47. Determine whether the given statement is true.
    (a) $\text{Ln}(-1 + i)^2 = 2\,\text{Ln}(-1 + i)$
    (b) $\text{Ln}\, i^3 = 3\,\text{Ln}\, i$
    (c) $\ln i^3 = 3 \ln i$

48. The laws of exponents hold for complex numbers α and β:
$$z^\alpha z^\beta = z^{\alpha+\beta} \qquad \dfrac{z^\alpha}{z^\beta} = z^{\alpha-\beta} \qquad (z^\alpha)^n = z^{n\alpha},\ n\ \text{an integer}$$
However, the last law is not valid if n is a complex number. Verify that (a) $(i^i)^2 = i^{2i}$, but (b) $(i^2)^i \neq i^{2i}$.

49. For complex numbers z satisfying $\text{Re}(z) > 0$, show that (7) can be written as
$$\text{Ln}\, z = \dfrac{1}{2}\log_e(x^2 + y^2) + i\tan^{-1}\dfrac{y}{x}$$

50. The function given in Problem 49 is analytic.
    (a) Verify that $u(x, y) = \log_e(x^2 + y^2)$ is a harmonic function.
    (b) Verify that $v(x, y) = \tan^{-1}(y/x)$ is a harmonic function.

16.7 TRIGONOMETRIC AND HYPERBOLIC FUNCTIONS

Trigonometric Functions If x is a real variable, then Euler's formula gives

$$e^{ix} = \cos x + i \sin x \quad \text{and} \quad e^{-ix} = \cos x - i \sin x$$

By subtracting and then adding these equations, we see that the real functions $\sin x$ and $\cos x$ can be expressed as a combination of exponential functions:

$$\sin x = \frac{e^{ix} - e^{-ix}}{2i} \qquad \cos x = \frac{e^{ix} + e^{-ix}}{2} \qquad (1)$$

Using (1) as a model, we now define the sine and cosine of a complex variable:

DEFINITION 16.11 Trigonometric Sine and Cosine

For any complex number $z = x + iy$,

$$\sin z = \frac{e^{iz} - e^{-iz}}{2i} \qquad \cos z = \frac{e^{iz} + e^{-iz}}{2} \qquad (2)$$

As in trigonometry, we define four additional trigonometric functions in terms of $\sin z$ and $\cos z$:

$$\tan z = \frac{\sin z}{\cos z} \qquad \cot z = \frac{1}{\tan z} \qquad \sec z = \frac{1}{\cos z} \qquad \csc z = \frac{1}{\sin z} \qquad (3)$$

When $y = 0$, each function in (2) and (3) reduces to its real counterpart.

Analyticity Since the exponential functions e^{iz} and e^{-iz} are entire functions, it follows that $\sin z$ and $\cos z$ are entire functions. Now, as we shall see shortly, $\sin z = 0$ only for the real numbers $z = n\pi$, n an integer, and $\cos z = 0$ only for the real numbers $z = (2n + 1)\pi/2$, n an integer. Thus, $\tan z$ and $\sec z$ are analytic except at the points $z = (2n + 1)\pi/2$, and $\cot z$ and $\csc z$ are analytic except at the points $z = n\pi$.

Derivatives Since $(d/dz)e^z = e^z$, it follows from the chain rule that $(d/dz)e^{iz} = ie^{iz}$ and $(d/dz)e^{-iz} = -ie^{-iz}$. Hence,

$$\frac{d}{dz}\sin z = \frac{d}{dz}\frac{e^{iz} - e^{-iz}}{2i} = \frac{e^{iz} + e^{-iz}}{2} = \cos z$$

In fact, it is readily shown that the forms of the derivatives of the complex trigonometric functions are the same as the real functions. We summarize the results:

$$\frac{d}{dz}\sin z = \cos z \qquad \frac{d}{dz}\cos z = -\sin z$$
$$\frac{d}{dz}\tan z = \sec^2 z \qquad \frac{d}{dz}\cot z = -\csc^2 z \qquad (4)$$
$$\frac{d}{dz}\sec z = \sec z \tan z \qquad \frac{d}{dz}\csc z = -\csc z \cot z$$

Identities The familiar trigonometric identities are also the same in the complex case.

$$\sin(-z) = -\sin z \qquad \cos(-z) = \cos z$$
$$\cos^2 z + \sin^2 z = 1$$
$$\sin(z_1 \pm z_2) = \sin z_1 \cos z_2 \pm \cos z_1 \sin z_2$$
$$\cos(z_1 \pm z_2) = \cos z_1 \cos z_2 \mp \sin z_1 \sin z_2$$
$$\sin 2z = 2 \sin z \cos z \qquad \cos 2z = \cos^2 z - \sin^2 z$$
$$\sin(z + 2\pi) = \sin z \qquad \cos(z + 2\pi) = \cos z \qquad \tan(z + \pi) = \tan z$$

The identities in the last line indicate that $\sin z$ and $\cos z$ are periodic with the real period 2π; $\tan z$ is periodic with period π.

Zeros To find the zeros of $\sin z$ and $\cos z$ we need to express both functions in the form $u + iv$. Before proceeding, recall from calculus that if y is real, then the hyperbolic sine and hyperbolic cosine are defined in terms of the real exponential functions e^y and e^{-y}:

$$\sinh y = \frac{e^y - e^{-y}}{2} \quad \text{and} \quad \cosh y = \frac{e^y + e^{-y}}{2} \qquad (5)$$

Now from Definition 16.11 and Euler's formula we find, after simplifying,

$$\sin z = \frac{e^{i(x+iy)} - e^{-i(x+iy)}}{2i} = \sin x \left(\frac{e^y + e^{-y}}{2}\right) + i \cos x \left(\frac{e^y - e^{-y}}{2}\right)$$

Thus, from (5) we have

$$\sin z = \sin x \cosh y + i \cos x \sinh y \qquad (6)$$

It is left as an exercise to show to that

$$\cos z = \cos x \cosh y - i \sin x \sinh y \qquad (7)$$

From (6), (7), and $\cosh^2 y = 1 + \sinh^2 y$, we find

$$|\sin z|^2 = \sin^2 x + \sinh^2 y \tag{8}$$

and
$$|\cos z|^2 = \cos^2 x + \sinh^2 y \tag{9}$$

Now a complex number z is zero if and only if $|z|^2 = 0$. Thus, if $\sin z = 0$, then from (8) we must have $\sin^2 x + \sinh^2 y = 0$. This implies that $\sin x = 0$ and $\sinh y = 0$, and so $x = n\pi$ and $y = 0$. Thus the only **zeros** of $\sin z$ are the real numbers $z = n\pi + 0i = n\pi$, $n = 0, \pm 1, \pm 2, \ldots$. Similarly it follows from (9) that $\cos z = 0$ only when $z = (2n + 1)\pi/2$, $n = 0, \pm 1, \pm 2, \ldots$.

EXAMPLE 1 From (6) we have, with the aid of a calculator,

$$\sin(2 + i) = \sin 2 \cosh 1 + i \cos 2 \sinh 1 = 1.4031 - 0.4891i \qquad \square$$

In ordinary trigonometry we are accustomed to the fact that $|\sin x| \leq 1$ and $|\cos x| \leq 1$. Inspection of (8) and (9) shows that these inequalities do not hold for the complex sine and cosine, since $\sinh y$ can range from $-\infty$ to ∞. In other words, it is perfectly feasible to have solutions for equations such as $\cos z = 10$.

EXAMPLE 2 Solve the equation $\cos z = 10$.

Solution From (2), $\cos z = 10$ is equivalent to $(e^{iz} + e^{-iz})/2 = 10$. Multiplying the last equation by e^{iz} then gives the quadratic equation in e^{iz}:

$$e^{2iz} - 20e^{iz} + 1 = 0$$

From the quadratic formula we find $e^{iz} = 10 \pm 3\sqrt{11}$. Thus, for $n = 0, \pm 1, \pm 2, \ldots$, we have $iz = \log_e(10 \pm 3\sqrt{11}) + 2n\pi i$. Dividing by i and utilizing $\log_e(10 - 3\sqrt{11}) = -\log_e(10 + 3\sqrt{11})$, we can express the solutions of the given equation as

$$z = 2n\pi \pm i \log_e(10 + 3\sqrt{11}) \qquad \square$$

Hyperbolic Functions We define the complex hyperbolic sine and cosine in a manner analogous to the real definitions given in (5):

DEFINITION 16.12 Hyperbolic Sine and Cosine

For any complex number $z = x + iy$,

$$\sinh z = \frac{e^z - e^{-z}}{2} \qquad \cosh z = \frac{e^z + e^{-z}}{2} \tag{10}$$

The hyperbolic tangent, cotangent, secant, and cosecant functions are defined in terms of sinh z and cosh z:

$$\tanh z = \frac{\sinh z}{\cosh z} \qquad \coth z = \frac{1}{\tanh z} \qquad \operatorname{sech} z = \frac{1}{\cosh z} \qquad \operatorname{csch} z = \frac{1}{\sinh z} \qquad (11)$$

The hyperbolic sine and cosine are entire functions, and the functions defined in (11) are analytic except at points where the denominators are zero. It is also easy to see from (10) that

$$\frac{d}{dz}\sinh z = \cosh z \quad \text{and} \quad \frac{d}{dz}\cosh z = \sinh z \qquad (12)$$

It is interesting to observe that, in contrast to real calculus, the trigonometric and hyperbolic functions are related in complex calculus. If we replace z by iz everywhere in (10) and compare the results with (2), we see that $\sinh(iz) = i \sin z$ and $\cosh(iz) = \cos z$. These equations enable us to express $\sin z$ and $\cos z$ in terms of $\sinh(iz)$ and $\cosh(iz)$, respectively. Similarly, by replacing z by iz in (2) we can express, in turn, $\sinh z$ and $\cosh z$ in terms of $\sin(iz)$ and $\cos(iz)$. We summarize the results:

$$\sin z = -i\sinh(iz) \qquad \cos z = \cosh(iz) \qquad (13)$$
$$\sinh z = -i\sin(iz) \qquad \cosh z = \cos(iz) \qquad (14)$$

Zeros The relationships given in (14) enable us to derive identities for the hyperbolic functions utilizing results for the trigonometric functions. For example, to express $\sinh z$ in the form $u + iv$ we write $\sinh z = -i\sin(iz)$ in the form $\sinh z = -i\sin(-y + ix)$ and use (6):

$$\sinh z = -i[\sin(-y)\cosh x + i\cos(-y)\sinh x]$$

Since $\sin(-y) = -\sin y$ and $\cos(-y) = \cos y$, the foregoing simplifies to

Similarly,
$$\sinh z = \sinh x \cos y + i \cosh x \sin y \qquad (15)$$
$$\cosh z = \cosh x \cos y + i \sinh x \sin y \qquad (16)$$

It also follows directly from (14) that the **zeros** of $\sinh z$ and $\cosh z$ are pure imaginary and are, respectively,

$$z = n\pi i \quad \text{and} \quad z = (2n+1)\frac{\pi i}{2}, \qquad n = 0, \pm 1, \pm 2, \ldots$$

Moreover, since $\sin z$ and $\cos z$ are periodic with real period 2π, the hyperbolic functions $\sinh z$ and $\cosh z$ are periodic with imaginary period $2\pi i$.

EXERCISES 16.7
Answers to odd-numbered problems begin on page A-88.

In Problems 1–12 express the given quantity in the form $a + ib$.

1. $\cos(3i)$
2. $\sin(-2i)$
3. $\sin\left(\dfrac{\pi}{4} + i\right)$
4. $\cos(2 - 4i)$
5. $\tan(i)$
6. $\cot\left(\dfrac{\pi}{2} + 3i\right)$
7. $\sec(\pi + i)$
8. $\csc(1 + i)$
9. $\cosh(\pi i)$
10. $\sinh\left(\dfrac{3\pi}{2}i\right)$
11. $\sinh\left(1 + \dfrac{\pi}{3}i\right)$
12. $\cosh(2 + 3i)$

In Problems 13 and 14 verify the given result.

13. $\sin\left(\dfrac{\pi}{2} + i \ln 2\right) = \dfrac{5}{4}$
14. $\cos\left(\dfrac{\pi}{2} + i \ln 2\right) = -\dfrac{3}{4}i$

In Problems 15–20 find all values of z satisfying the given equation.

15. $\sin z = 2$
16. $\cos z = -3i$
17. $\sinh z = -i$
18. $\sinh z = -1$
19. $\cos z = \sin z$
20. $\cos z = i \sin z$

In Problems 21 and 22 use the definition of equality of complex numbers to find all values of z satisfying the given equation.

21. $\cos z = \cosh 2$
22. $\sin z = i \sinh 2$
23. Prove that $\cos z = \cos x \cosh y - i \sin x \sinh y$.
24. Prove that $\sinh z = \sinh x \cos y + i \cosh x \sin y$.
25. Prove that $\cosh z = \cosh x \cos y + i \sinh x \sin y$.
26. Prove that $|\sinh z|^2 = \sin^2 y + \sinh^2 x$.
27. Prove that $|\cosh z|^2 = \cos^2 y + \sinh^2 x$.
28. Prove that $\cos^2 z + \sin^2 z = 1$.
29. Prove that $\cosh^2 z - \sinh^2 z = 1$.
30. Show that $\tan z = u + iv$, where

$$u = \frac{\sin 2x}{\cos 2x + \cosh 2y} \quad \text{and} \quad y = \frac{\sinh 2y}{\cos 2x + \cosh 2y}$$

31. Prove that $\tanh z$ is periodic with period πi.
32. Prove that (a) $\overline{\sin z} = \sin \bar{z}$ and (b) $\overline{\cos z} = \cos \bar{z}$.

16.8 INVERSE TRIGONOMETRIC AND HYPERBOLIC FUNCTIONS

As functions of a complex variable, both the trigonometric and hyperbolic functions are periodic. Consequently these functions do not possess inverses that are functions in the strictest interpretation of that word. The inverses of these analytic functions are multiple-valued functions. As we did in Section 16.6 in the discussion of the logarithmic function, we shall drop the adjective "multiple-valued."

Inverse Sine The **inverse sine** function, written as $\sin^{-1} z$ or arcsin z, is defined by

$$w = \sin^{-1} z \quad \text{if} \quad z = \sin w \tag{1}$$

The inverse sine can be expressed in terms of the logarithmic function. To see this we use (1) and the definition of the sine function:

$$\frac{e^{iw} - e^{-iw}}{2i} = z \quad \text{or} \quad e^{2iw} - 2ize^{iw} - 1 = 0$$

From the last equation and the quadratic formula we then obtain

$$e^{iw} = iz + (1 - z^2)^{1/2} \tag{2}$$

Note in (2) we did not use the customary symbolism $\pm\sqrt{1-z^2}$, since we know from Section 16.2 that $(1-z^2)^{1/2}$ is two-valued. Solving (2) for w then gives

$$\sin^{-1} z = -i \ln[iz + (1-z^2)^{1/2}] \tag{3}$$

Proceeding in a similar manner, we find the inverses of the cosine and tangent to be

$$\cos^{-1} z = -i \ln[z + i(1-z^2)^{1/2}] \tag{4}$$

$$\tan^{-1} z = \frac{i}{2} \ln \frac{i+z}{i-z} \tag{5}$$

EXAMPLE 1 Find all values of $\sin^{-1}\sqrt{5}$.

Solution From (3) we have

$$\sin^{-1}\sqrt{5} = -i \ln[\sqrt{5}\,i + (1 - (\sqrt{5})^2)^{1/2}]$$

With $(1 - (\sqrt{5})^2)^{1/2} = (-4)^{1/2} = \pm 2i$, the preceding expression becomes

$$\sin^{-1}\sqrt{5} = -i \ln[(\sqrt{5} \pm 2)i]$$

$$= -i\left[\log_e(\sqrt{5} \pm 2) + \left(\frac{\pi}{2} + 2n\pi\right)i\right], \quad n = 0, \pm 1, \pm 2, \ldots$$

The foregoing result can be simplified a little by noting that $\log_e(\sqrt{5} - 2) = \log_e(1/(\sqrt{5} + 2)) = -\log_e(\sqrt{5} + 2)$. Thus for $n = 0, \pm 1, \pm 2, \ldots$,

$$\sin^{-1}\sqrt{5} = \frac{\pi}{2} + 2n\pi \pm i \log_e(\sqrt{5} + 2) \tag{6}$$

□

To obtain particular values of, say, $\sin^{-1} z$, we must choose a specific root of $1 - z^2$ and a specific branch of the logarithm. For example, if we choose $(1 - (\sqrt{5})^2)^{1/2} = (-4)^{1/2} = 2i$ and the principal branch of the logarithm, then (6) gives the single value

$$\sin^{-1}\sqrt{5} = \frac{\pi}{2} - i \log_e(\sqrt{5} + 2)$$

16.8 Inverse Trigonometric and Hyperbolic Functions

Derivatives The derivatives of the three inverse trigonometric functions considered above can be found by implicit differentiation. To find the derivative of the inverse sine function $w = \sin^{-1}z$, we begin by differentiating $z = \sin w$:

$$\frac{d}{dz}z = \frac{d}{dz}\sin w \quad \text{gives} \quad \frac{dw}{dz} = \frac{1}{\cos w}$$

Using the trigonometric identity $\cos^2 w + \sin^2 w = 1$ (see Problem 28 in Exercises 16.7) in the form $\cos w = (1 - \sin^2 w)^{1/2} = (1 - z^2)^{1/2}$, we obtain

$$\boxed{\frac{d}{dz}\sin^{-1}z = \frac{1}{(1-z^2)^{1/2}}} \qquad (7)$$

Similarly, we find that

$$\boxed{\frac{d}{dz}\cos^{-1}z = \frac{-1}{(1-z^2)^{1/2}}} \qquad (8)$$

$$\boxed{\frac{d}{dz}\tan^{-1}z = \frac{1}{1+z^2}} \qquad (9)$$

It should be noted that the square roots used in (7) and (8) must be consistent with the square roots used in (3) and (4).

EXAMPLE 2 Find the derivative of $w = \sin^{-1}z$ at $z = \sqrt{5}$.

Solution In Example 1 if we use $(1 - (\sqrt{5})^2)^{1/2} = (-4)^{1/2} = 2i$, then that same root must be used in (7). The value of the derivative consistent with this choice is give by

$$\left.\frac{dw}{dz}\right|_{z=\sqrt{5}} = \frac{1}{(1-(\sqrt{5})^2)^{1/2}} = \frac{1}{(-4)^{1/2}} = \frac{1}{2i} = -\frac{1}{2}i \qquad \square$$

Inverse Hyperbolic Functions The inverse hyperbolic functions can also be expressed in terms of the logarithm. We summarize these results for the inverse hyperbolic sine, cosine, and tangent along with their derivatives:

$$\sinh^{-1}z = \ln[z + (z^2 + 1)^{1/2}] \qquad (10)$$

$$\cosh^{-1}z = \ln[z + (z^2 - 1)^{1/2}] \qquad (11)$$

$$\tanh^{-1}z = \frac{1}{2}\ln\frac{1+z}{1-z} \qquad (12)$$

$$\frac{d}{dz}\sinh^{-1}z = \frac{1}{(z^2+1)^{1/2}} \qquad (13)$$

$$\frac{d}{dz}\cosh^{-1}z = \frac{1}{(z^2-1)^{1/2}} \qquad (14)$$

$$\frac{d}{dz}\tanh^{-1}z = \frac{1}{1-z^2} \qquad (15)$$

EXAMPLE 3 Find all values of $\cosh^{-1}(-1)$.

Solution From (11) with $z = -1$, we get

$$\cosh^{-1}(-1) = \ln(-1) = \log_e 1 + (\pi + 2n\pi)i$$

Since $\log_e 1 = 0$, we have for $n = 0, \pm 1, \pm 2, \ldots,$

$$\cosh^{-1}(-1) = (2n+1)\pi i \qquad \square$$

EXERCISES 16.8   Answers to odd-numbered problems begin on page A-88.

In Problems 1–14 find all values of the given quantity.

1. $\sin^{-1}(-i)$
2. $\sin^{-1}\sqrt{2}$
3. $\sin^{-1}0$
4. $\sin^{-1}\frac{13}{5}$
5. $\cos^{-1}2$
6. $\cos^{-1}2i$
7. $\cos^{-1}\frac{1}{2}$
8. $\cos^{-1}\frac{5}{3}$
9. $\tan^{-1}1$
10. $\tan^{-1}3i$
11. $\sinh^{-1}\frac{4}{3}$
12. $\cosh^{-1}i$
13. $\tanh^{-1}(1+2i)$
14. $\tanh^{-1}(-\sqrt{3}i)$

SUMMARY

A **complex number** is a number of the form $z = x + iy$, where i is called the **imaginary unit** and satisfies $i^2 = -1$, x is the **real part** of z, and y is the **imaginary part** of z. Both $x = \text{Re}(z)$ and $y = \text{Im}(z)$ are real numbers. The number $\bar{z} = x - iy$ is called the **conjugate** of z. The real number $|z| = \sqrt{z\bar{z}}$ is the **modulus** of z. Geometrically we can picture a complex number either as a point in the complex or z-plane or as a vector whose initial point is the origin and whose terminal point is (x, y). A complex number can be written in **polar form** $z = r(\cos\theta + i\sin\theta)$, where $r = |z|$ and θ is an angle the vector z makes with the positive real axis called an **argument** of z. An argument of z satisfying $-\pi < \theta \leq \pi$ is called the **principal argument** and is denoted by Arg z.

A set of points in the complex plane is said to be an **open set** if every point in the set is an **interior point**—that is, if every point possesses some neighborhood that lies entirely within the set. An open set in which any pair of points can be joined by a polygonal line that lies entirely within the set is said to be **connected**. An open connected set is called a **domain**.

A **complex function** or a function of a complex variable z can be written as $w = f(z)$ or $f(z) = u(x, y) + iv(x, y)$, where u and v are real-valued functions. Although graphs of complex functions cannot be drawn, we can interpret $f(z)$ as a **transformation** or **mapping** between the z-plane and the w-plane. By considering the complex number $f(z)$ as a vector based at the point z, we can also interpret a complex function $w = f(z)$ as a **two-dimensional fluid flow**.

A complex function is said to be **analytic at a point** z if it is differentiable at z and at every point in some neighborhood of z. A function that is analytic at every point z is called an **entire function**. If a function $f(z) = u(x, y) + iv(x, y)$ is analytic in a domain D, then u and v necessarily satisfy the **Cauchy–Riemann equations**

$$\frac{\partial u}{\partial x} = \frac{\partial v}{\partial y} \quad \text{and} \quad \frac{\partial u}{\partial y} = -\frac{\partial v}{\partial x}$$

Furthermore, if u and v are continuous, have continuous first partial derivatives, and satisfy the Cauchy–Riemann equations at all points in a domain D, then $f(z) = u(x, y) + iv(x, y)$ is analytic in D.

CHAPTER 16 REVIEW EXERCISES

Answers to odd-numbered problems begin on page A-88.

Answer Problems 1–16 without referring back to the text. Fill in the blank or answer true/false.

1. $\text{Re}((1 + i)^{10}) = $ _____ and $\text{Im}((1 + i)^{10}) = $ _____.

2. If z is a point in the third quadrant, then $i\bar{z}$ is in the _____ quadrant.

3. If $z = 3 + 4i$, then $\text{Re}\left(\dfrac{z}{\bar{z}}\right) = $ _____.

4. $i^{127} - 5i^9 + 2i^{-1} = $ _____

5. If $z = \dfrac{4i}{-3 - 4i}$, then $|z| = $ _____.

6. Describe the region defined by $1 \le |z + 2| \le 3$.

7. $\text{Arg}(z + \bar{z}) = 0$ _____

8. If $z = \dfrac{5}{-\sqrt{3} + i}$, then $\text{Arg } z = $ _____.

9. If $e^z = 2i$, then $z = $ _____.

10. If $|e^z| = 1$, then z is a pure imaginary number. _____

11. The principal value of $(1 + i)^{(2 + i)}$ is _____.

12. If $f(z) = x^2 - 3xy - 5y^3 + i(4x^2y - 4x + 7y)$, then $f(-1 + 2i) = $ _____.

13. If the Cauchy–Riemann equations are satisfied at a point, then the function is necessarily analytic there. _____

14. $f(z) = e^z$ is periodic with period _____.

15. $\text{Ln}(-ie^3) = $ _____

16. $f(z) = \sin(x - iy)$ is nowhere analytic. _____

In Problems 17–20 write the given number in the form $a + ib$.

17. $i(2 - 3i)^2(4 + 2i)$

18. $\dfrac{3 - i}{2 + 3i} + \dfrac{2 - 2i}{1 + 5i}$

19. $\dfrac{(1 - i)^{10}}{(1 + i)^3}$

20. $4e^{\pi i/3} e^{-\pi i/4}$

In Problems 21–24 sketch the set of points in the complex plane satisfying the given inequality.

21. $\text{Im}(z^2) \le 2$

22. $\text{Im}(z + 5i) > 3$

23. $\dfrac{1}{|z|} \le 1$

24. $\text{Im}(z) < \text{Re}(z)$

25. Look up the definitions of conic sections in a calculus text. Now describe the set of points in the complex plane that satisfy the equation $|z - 2i| + |z + 2i| = 5$.

26. Let z and w be complex numbers such that $|z| = 1$ and $|w| \neq 1$. Prove that
$$\left|\frac{z - w}{1 - z\bar{w}}\right| = 1$$

In Problems 27 and 28 find all solutions of the given equation.

27. $z^4 = 1 - i$

28. $z^{3/2} = \dfrac{1}{2 - i}$

29. If $f(z) = z^{24} - 3z^{20} + 4z^{12} - 5z^6$, find $f\left(\dfrac{1 + i}{\sqrt{2}}\right)$.

30. Write $f(z) = \text{Im}(z - 3\bar{z}) + z\,\text{Re}(z^2) - 5z$ in the form $f(z) = u(x, y) + iv(x, y)$.

In Problems 31 and 32 find the image of the line $x = 1$ in the w-plane under the given mapping.

31. $f(z) = x^2 - y + i(y^2 - x)$

32. $f(z) = \dfrac{1}{z}$

In Problems 33–36 find all complex numbers for which the given statement is true.

33. $z = z^{-1}$

34. $\bar{z} = \dfrac{1}{z}$

35. $\bar{z} = -z$

36. $z^2 = (\bar{z})^2$

37. Show that the function $f(z) = -(2xy + 5x) + i(x^2 - 5y - y^2)$ is analytic for all z. Find $f'(z)$.

38. Determine whether the function $f(z) = x^3 + xy^2 - 4x + i(4y - y^3 - x^2 y)$ is differentiable. Is it analytic?

In Problems 39 and 40 verify the given equality.

39. $\text{Ln}(1 + i)(1 - i) = \text{Ln}(1 + i) + \text{Ln}(1 - i)$

40. $\text{Ln}\dfrac{1 + i}{1 - i} = \text{Ln}(1 + i) - \text{Ln}(1 - i)$

17

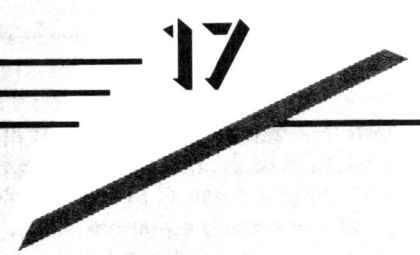

INTEGRATION IN THE COMPLEX PLANE

TOPICS TO REVIEW

line integrals (8.8)
path independence (8.9)
fundamental theorem of calculus

IMPORTANT CONCEPTS

contour
contour integral
ML-inequality
simply and multiply connected domains
Cauchy–Goursat theorem
deformation of contours
independence of path
antiderivative
fundamental theorem for contour integrals
Cauchy's integral formula
Cauchy's integral formula for derivatives
Liouville's theorem

17.1 Contour Integrals
17.2 Cauchy–Goursat Theorem
17.3 Independence of Path
17.4 Cauchy's Integral Formula
     Summary
     Chapter 17 Review Exercises

INTRODUCTION

To define an integral of a complex function f, we start with f defined along some curve or **contour** in the complex plane. We shall see in this section that the definition of a complex integral, its properties, and method of evaluation are quite similar to those of a real line integral in the plane. (See Section 8.8.)

17.1 CONTOUR INTEGRALS

Integration in the complex plane is defined in a manner similar to that of a line integral in the plane. In other words, we shall be dealing with an integral of a complex function $f(z)$ that is defined along a curve C in the complex plane. These curves are defined in terms of parametric equations $x = x(t)$, $y = y(t)$, $a \leq t \leq b$, where t is a real parameter. By using $x(t)$ and $y(t)$ as real and imaginary parts, we can also describe a curve C in the complex plane by means of a complex-valued function of a real variable t: $z(t) = x(t) + iy(t)$, $a \leq t \leq b$. For example, $x = \cos t$, $y = \sin t$, $0 \leq t \leq 2\pi$, describes a unit circle centered at the origin. This circle can also be described by $z(t) = \cos t + i \sin t$, or even more compactly by $z(t) = e^{it}$, $0 \leq t \leq 2\pi$. The same definitions of smooth curve, piecewise smooth curve, closed curve, and simple closed curve given in Section 8.8 carry over to this discussion. The student is encouraged to review these concepts. As before, we shall assume that the positive direction on C corresponds to increasing values of t. In complex variables, a piecewise smooth curve C is also called a **contour** or **path**. An integral of $f(z)$ on C is denoted by $\int_C f(z)\,dz$ or $\oint_C f(z)\,dz$ if the contour C is closed; it is referred to as a **contour integral** or a **complex line integral**.

$$f(z) = u(x, y) + iv(x, y)$$

1. Let f be defined at all points on a smooth curve C defined by $x = x(t)$, $y = y(t)$, $a \leq t \leq b$.
2. Divide C into n subarcs according to the partition $a = t_0 < t_1 < \cdots < t_n = b$ of $[a, b]$. The corresponding points on the curve C are $z_0 = x_0 + iy_0 = x(t_0) + iy(t_0)$, $z_1 = x_1 + iy_1 = x(t_1) + iy(t_1)$, $\ldots$, $z_n = x_n + iy_n = x(t_n) + iy(t_n)$. Let $\Delta z_k = z_k - z_{k-1}$, $k = 1, 2, \ldots, n$.
3. Let $\|P\|$ be the **norm** of the partition—that is, the maximum value of $|\Delta z_k|$.
4. Choose a point $z_k^* = x_k^* + iy_k^*$ on each subarc.
5. Form the sum $\sum_{k=1}^{n} f(z_k^*) \Delta z_k$.

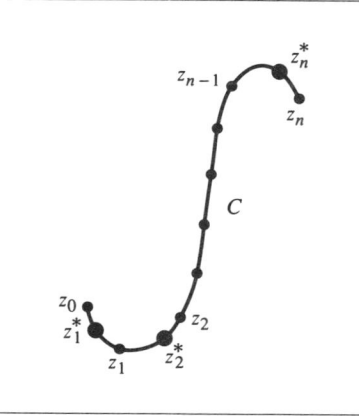

DEFINITION 17.1   Contour Integral

Let f be defined at points of a smooth curve C defined by $x = x(t)$, $y = y(t)$, $a \leq t \leq b$. The **contour integral** of f along C is

$$\int_C f(z)\,dz = \lim_{\|P\| \to 0} \sum_{k=1}^{n} f(z_k^*) \Delta z_k \qquad (1)$$

The limit in (1) exists if f is continuous at all points on C and C is either smooth or piecewise smooth. Consequently we shall, hereafter, assume these conditions as a matter of course.

17.1 Contour Integrals

A Method of Evaluation We shall turn now to the question of evaluating a contour integral. To facilitate the discussion let us suppress the subscripts and write (1) in the abbreviated form

$$\int_C f(z)\,dz = \lim \Sigma(u+iv)(\Delta x + i\,\Delta y) = \lim\{\Sigma(u\,\Delta x - v\,\Delta y) + i\,\Sigma(v\,\Delta x + u\,\Delta y)\}$$

This means

$$\int_C f(z)\,dz = \int_C u\,dx - v\,dy + i\int_C v\,dx + u\,dy \tag{2}$$

In other words, a contour integral $\int_C f(z)\,dz$ is a combination of two real-line integrals $\int_C u\,dx - v\,dy$ and $\int_C v\,dx + u\,dy$. Now, since $x = x(t)$ and $y = y(t)$, $a \le t \le b$, the right side of (2) is the same as

$$\int_a^b [u(x(t),y(t))x'(t) - v(x(t),y(t))y'(t)]\,dt + i\int_a^b [v(x(t),y(t))x'(t) + u(x(t),y(t))y'(t)]\,dt$$

But if we use $z(t) = x(t) + iy(t)$ to describe C, the last result is the same as $\int_a^b f(z(t))z'(t)\,dt$ when separated into two integrals. Thus we arrive at a practical means of evaluating a contour integral:

THEOREM 17.1 Evaluation of a Contour Integral

If f is continuous on a smooth curve C given by $z(t) = x(t) + iy(t)$, $a \le t \le b$, then

$$\int_C f(z)\,dz = \int_a^b f(z(t))z'(t)\,dt \tag{3}$$

If f is expressed in terms of the symbol z, then to evaluate $f(z(t))$ we simply replace the symbol z by $z(t)$. If f is not expressed in terms of z, then to evaluate $f(z(t))$ we replace x and y wherever they appear by $x(t)$ and $y(t)$, respectively.

EXAMPLE 1 Evaluate $\int_C \bar{z}\,dz$, where C is given by $x = 3t$, $y = t^2$, $-1 \le t \le 4$.

Solution We write $z(t) = 3t + it^2$ so that $z'(t) = 3 + 2it$ and $f(z(t)) = \overline{3t + it^2} = 3t - it^2$. Thus,

$$\int_C \bar{z}\,dz = \int_{-1}^4 (3t - it^2)(3 + 2it)\,dt$$

$$= \int_{-1}^4 (2t^3 + 9t)\,dt + i\int_{-1}^4 3t^2\,dt = 195 + 65i \qquad \square$$

EXAMPLE 2 Evaluate $\oint_C \frac{1}{z} dz$, where C is the circle $x = \cos t$, $y = \sin t$, $0 \le t \le 2\pi$.

Solution In this case $z(t) = \cos t + i \sin t = e^{it}$, $z'(t) = ie^{it}$, and $f(z) = 1/z = e^{-it}$. Hence,

$$\oint_C \frac{1}{z} dz = \int_0^{2\pi} (e^{-it}) ie^{it} dt = i \int_0^{2\pi} dt = 2\pi i$$

For some curves, the real variable x itself can be used as the parameter. For example, to evaluate $\int_C (8x^2 - iy) dz$ on $y = 5x$, $0 \le x \le 2$, we write $\int_C (8x^2 - iy) dz = \int_0^2 (8x^2 - 5ix)(1 + 5i) dx$ and integrate in the usual manner.

Properties The following properties of contour integrals are analogous to the properties of line integrals:

THEOREM 17.2

Suppose f and g are continuous in a domain D and C is a smooth curve lying entirely in D. Then:

(i) $\int_C k f(z) dz = k \int_C f(z) dz$, k a constant
(ii) $\int_C [f(z) + g(z)] dz = \int_C f(z) dz + \int_C g(z) dz$
(iii) $\int_C f(z) dz = \int_{C_1} f(z) dz + \int_{C_2} f(z) dz$, where C is the union of the smooth curves C_1 and C_2
(iv) $\int_{-C} f(z) dz = -\int_C f(z) dz$, where $-C$ denotes the curve having the opposite orientation of C.

EXAMPLE 3 Evaluate $\int_C (x^2 + iy^2) dz$, where C is the contour shown in Figure 17.1.

Solution In view of Theorem 17.2(iii) we write

$$\int_C (x^2 + iy^2) dz = \int_{C_1} (x^2 + iy^2) dz + \int_{C_2} (x^2 + iy^2) dz$$

Since the curve C_1 is defined by $y = x$, it makes sense to use x as a parameter. Therefore, $z(x) = x + ix$, $z'(x) = 1 + i$, $f(z(x)) = x^2 + ix^2$, and

$$\int_{C_1} (x^2 + iy^2) dz = \int_0^1 (x^2 + ix^2)(1 + i) dx$$

$$= (1 + i)^2 \int_0^1 x^2 dx = \frac{(1 + i)^2}{3} = \frac{2}{3} i$$

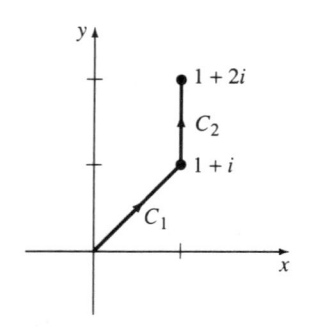

FIGURE 17.1

The curve C_2 is defined by $x = 1$, $1 \leq y \leq 2$. Using y as a parameter, we have $z(y) = 1 + iy$, $z'(y) = i$, and $f(z(y)) = 1 + iy^2$. Thus,

$$\int_{C_2} (x^2 + iy^2)\, dz = \int_1^2 (1 + iy^2) i\, dy$$

$$= -\int_1^2 y^2\, dy + i\int_1^2 dy = -\frac{7}{3} + i$$

Finally, we have $\int_C (x^2 + iy^2)\, dz = \frac{2}{3}i + (-\frac{7}{3} + i) = -\frac{7}{3} + \frac{5}{3}i$. □

There are times in the application of complex integration that it is useful to find an upper bound for the absolute value of a contour integral. In the next theorem we shall use the fact that the length of a plane curve is $s = \int_a^b \sqrt{[x'(t)]^2 + [y'(t)]^2}\, dt$. But if $z'(t) = x'(t) + iy'(t)$, then $|z'(t)| = \sqrt{[x'(t)]^2 + [y'(t)]^2}$ and consequently $s = \int_a^b |z'(t)|\, dt$.

THEOREM 17.3   A Bounding Theorem

If f is continuous on a smooth curve C and if $|f(z)| \leq M$ for all z on C, then $|\int_C f(z)\, dz| \leq ML$, where L is the length of C.

Proof From the triangle inequality (6) of Section 16.1 we can write

$$\left|\sum_{k=1}^n f(z_k^*) \Delta z_k\right| \leq \sum_{k=1}^n |f(z_k^*)||\Delta z_k| \leq M \sum_{k=1}^n |\Delta z_k| \quad (4)$$

Now, $|\Delta z_k|$ can be interpreted as the length of the chord joining the points z_k and z_{k-1}. Since the sum of the lengths of the chords cannot be greater than the length of C, (4) becomes $|\sum_{k=1}^n f(z_k^*) \Delta z_k| \leq ML$. Hence, as $\|P\| \to 0$, the last inequality yields $|\int_C f(z)\, dz| \leq ML$. ∎

Theorem 17.3 is used often in the theory of complex integration and is sometimes referred to as the **ML-inequality**.

EXAMPLE 4 Find an upper bound for the absolute value of $\oint_C \dfrac{e^z}{z+1}\, dz$, where C is the circle $|z| = 4$.

Solution First, the length s of the circle of radius 4 is 8π. Next, from the inequality (7) of Section 16.1, it follows that $|z + 1| \geq |z| - 1 = 4 - 1 = 3$, and so

$$\left|\frac{e^z}{z+1}\right| \leq \frac{|e^z|}{|z|-1} = \frac{|e^z|}{3} \quad (5)$$

In addition, $|e^z| = |e^x(\cos y + i \sin y)| = e^x$. For points on the circle $|z| = 4$ the maximum that x can be is 4, and so (5) becomes

$$\left| \frac{e^z}{z+1} \right| \leq \frac{e^4}{3}$$

Hence from Theorem 17.3 we have

$$\left| \oint_C \frac{e^z}{z+1} \, dz \right| \leq \frac{8\pi e^4}{3}$$

□

Circulation and Net Flux Let **T** and **N** denote the unit tangent vector and the unit normal vector to a positively oriented simple closed contour C. When we interpret the complex function $f(z) = u(x, y) + iv(x, y)$ as a vector, the line integrals

$$\oint_C f \cdot \mathbf{T} \, ds = \oint_C u \, dx + v \, dy \qquad (6)$$

and

$$\oint_C f \cdot \mathbf{N} \, ds = \oint_C u \, dy - v \, dx \qquad (7)$$

have special interpretations. The line integral in (6) is called the **circulation** around C and measures the tendency of the flow to rotate the curve C. See Section 8.8 for the derivation. The **net flux** across C is the difference between the rate at which fluid enters and the rate at which fluid leaves the region bounded by C. The net flux across C is given by the line integral in (7), and a nonzero value for $\oint_C f \cdot \mathbf{N} \, ds$ indicates the presence of sources or sinks for the fluid inside the curve C. Note that

$$\left(\oint_C f \cdot \mathbf{T} \, ds \right) + i \left(\oint_C f \cdot \mathbf{N} \, ds \right) = \oint_C (u - iv)(dx + i \, dy) = \oint_C \overline{f(z)} \, dz$$

and so

$$\text{circulation} = \text{Re}\left(\oint_C \overline{f(z)} \, dz \right) \qquad (8)$$

$$\text{net flux} = \text{Im}\left(\oint_C \overline{f(z)} \, dz \right) \qquad (9)$$

Thus, both of these key quantities may be found by computing a single complex integral.

EXAMPLE 5 Given the flow $f(z) = (1 + i)z$, compute the circulation around and the net flux across the circle $C: |z| = 1$.

Solution Since $\overline{f(z)} = (1 - i)\bar{z}$ and $z(t) = e^{it}$, $0 \leq t \leq 2\pi$, we have

$$\oint_C \overline{f(z)} \, dz = \int_0^{2\pi} (1 - i)e^{-it} i e^{it} \, dt = (1 + i) \int_0^{2\pi} dt = 2\pi(1 + i)$$

Thus, the circulation around C is 2π and the net flux across C is 2π. See Figure 17.2.

□

flow $f(z) = (1 + i)z$

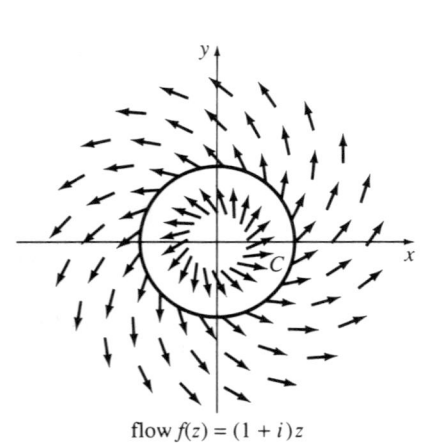

FIGURE 17.2

EXERCISES 17.1 Answers to odd-numbered problems begin on page A-88.

In Problems 1–16 evaluate the given integral along the indicated contour.

1. $\int_C (z+3)\, dz$, where C is $x=2t$, $y=4t-1$, $1 \le t \le 3$
2. $\int_C (2\bar{z}-z)\, dz$, where C is $x=-t$, $y=t^2+2$, $0 \le t \le 2$
3. $\int_C z^2\, dz$, where C is $z(t)=3t+2it$, $-2 \le t \le 2$
4. $\int_C (3z^2-2z)\, dz$, where C is $z(t)=t+it^2$, $0 \le t \le 1$
5. $\displaystyle\int_C \frac{1+z}{z}\, dz$, where C is the right half of the circle $|z|=1$ from $z=-i$ to $z=i$
6. $\int_C |z|^2\, dz$, where C is $x=t^2$, $y=1/t$, $1 \le t < 2$
7. $\oint_C \operatorname{Re}(z)\, dz$, where C is the circle $|z|=1$
8. $\oint_C \left(\dfrac{1}{(z+i)^3} - \dfrac{5}{z+i} + 8 \right) dz$, where C is the circle $|z+i|=1$, $0 \le t \le 2\pi$
9. $\int_C (x^2 + iy^3)\, dz$, where C is the straight line from $z=1$ to $z=i$
10. $\int_C (x^3 - iy^3)\, dz$, where C is the lower half of the circle $|z|=1$ from $z=-1$ to $z=1$
11. $\int_C e^z\, dz$, where C is the polygonal path consisting of the line segments from $z=0$ to $z=2$ and from $z=2$ to $z=1+\pi i$
12. $\int_C \sin z\, dz$, where C is the polygonal path consisting of the line segments from $z=0$ to $z=1$ and from $z=1$ to $z=1+i$
13. $\int_C \operatorname{Im}(z-i)\, dz$, where C is the polygonal path consisting of the circular arc along $|z|=1$ from $z=1$ to $z=i$ and the line segment from $z=i$ to $z=-1$
14. $\int_C dz$, where C is the left half of the ellipse $x^2/36 + y^2/4 = 1$ from $z=2i$ to $z=-2i$
15. $\oint_C ze^z\, dz$, where C is the square with vertices $z=0$, $z=1$, $z=1+i$, and $z=i$
16. $\int_C f(z)\, dz$, where $f(z)=\begin{cases} 2, & x<0 \\ 6x, & x>0 \end{cases}$ and C is the parabola $y=x^2$ from $z=-1+i$ to $z=1+i$

In Problems 17–20 evaluate the given integral along the contour C given in Figure 17.3.

17. $\oint_C x\, dz$
18. $\oint_C (2z-1)\, dz$
19. $\oint_C z^2\, dz$
20. $\oint_C \bar{z}^2\, dz$

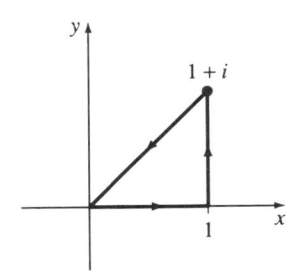

FIGURE 17.3

In Problems 21–24 evaluate $\int_C (z^2 - z + 2)$ from i to 1 along the indicated contours.

21.

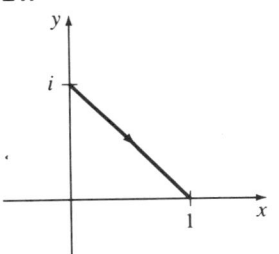

22.

FIGURE 17.4

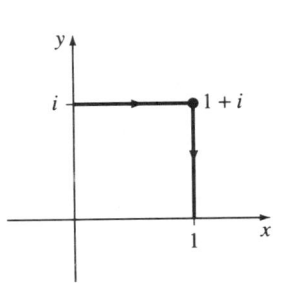

FIGURE 17.5

23.

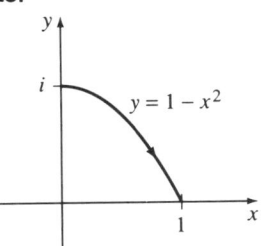

24.

FIGURE 17.6

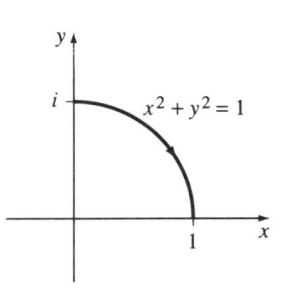

FIGURE 17.7

In Problems 25–28 find an upper bound for the absolute value of the given integral along the indicated contour.

25. $\oint_C \dfrac{e^z}{z^2+1}\, dz$, where C is the circle $|z|=5$

26. $\int_C \dfrac{1}{z^2-2i}\, dz$, where C is the right half of the circle $|z|=6$ from $z=-6i$ to $z=6i$

27. $\int_C (z^2 + 4)\, dz$, where C is the line segment from $z = 0$ to $z = 1 + i$

28. $\int_C \dfrac{1}{z^3}\, dz$, where C is one quarter of the circle $|z| = 4$ from $z = 4i$ to $z = 4$

29. (a) Use Definition 17.1 to show for any smooth curve C between z_0 and z_n that $\int_C dz = z_n - z_0$.

    (b) Use the result in part (a) to verify the answer to Problem 14.

30. Use Definition 17.1 to show for any smooth curve C between z_0 and z_n that $\int_C z\, dz = \tfrac{1}{2}(z_n^2 - z_0^2)$. [*Hint:* The integral exists, so choose $z_k^* = z_k$ and $z_k^* = z_{k-1}$.]

31. Use the results of Problems 29 and 30 to evaluate $\oint_C (6z + 4)\, dz$ where C is:

    (a) The straight line from $1 + i$ to $2 + 3i$
    (b) The closed contour $x^4 + y^4 = 4$

In Problems 32–35 compute the circulation and net flux for the given flow and the indicated closed contour.

32. $f(z) = 1/z$, where C is the circle $|z| = 2$

33. $f(z) = 2z$, where C is the circle $|z| = 1$

34. $f(z) = 1/(z - 1)$, where C is the circle $|z - 1| = 2$

35. $f(z) = \bar{z}$, where C is the square with vertices $z = 0$, $z = 1$, $z = 1 + i$, $z = i$

17.2 CAUCHY–GOURSAT THEOREM

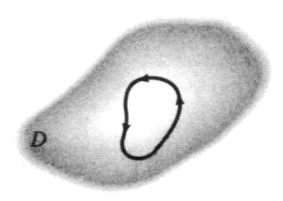

(a) simply connected domain

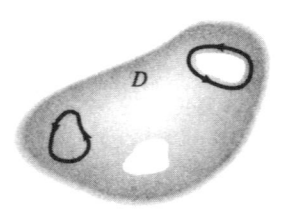

(b) multiply connected domain

FIGURE 17.8

Simply and Multiply Connected Domains In the discussion that follows we shall concentrate on contour integrals where the contour C is a **simple closed curve** with a positive (counterclockwise) orientation. Before doing this, we need to distinguish two kinds of domains. A domain D is said to be **simply connected** if every simple closed contour C lying entirely in D can be shrunk to a point without leaving D. In other words, in a simply connected domain every simple closed contour C lying entirely within it encloses only points of the domain D. Expressed yet another way, a simply connected domain has no "holes" in it. The entire complex plane is an example of a simply connected domain. A domain that is not simply connected is called a **multiply connected domain**; that is, a multiply connected domain has "holes" in it. See Figure 17.8. As in Section 8.9, we call a domain with one "hole" **doubly connected**, a domain with two "holes" **triply connected**, and so on.

Cauchy's Theorem In 1825 the French mathematician Louis-Augustin Cauchy proved one of the most important theorems in complex analysis. **Cauchy's theorem** says:

> *Suppose that a function f is analytic in a simply connected domain D and that f' is continuous in D. Then for every simple closed contour C in D, $\oint_C f(z)\, dz = 0$.*

The proof of this theorem is an immediate consequence of Green's theorem and the Cauchy–Riemann equations. Since f' is continuous throughout D, the real and imaginary parts of $f(z) = u + iv$ and their first partial derivatives are continuous throughout D. By (2) of Section 17.1 we write $\oint_C f(z)\, dz$ in terms of real-line integrals and use Green's theorem on each line integral:

$$\oint_C f(z)\, dz = \oint_C u(x,y)\, dx - v(x,y)\, dy + i \oint_C v(x,y)\, dx + u(x,y)\, dy$$

$$= \iint_D \left(-\frac{\partial v}{\partial x} - \frac{\partial u}{\partial y}\right) dA + i \iint_D \left(\frac{\partial u}{\partial x} - \frac{\partial v}{\partial y}\right) dA \qquad (1)$$

17.2 Cauchy–Goursat Theorem

Now since f is analytic, the Cauchy–Riemann equations, $\partial u/\partial x = \partial v/\partial y$ and $\partial u/\partial y = -\partial v/\partial x$, imply that the integrands in (1) are identically zero. Hence, we have $\oint_C f(z)\,dz = 0$.

In 1883 the French mathematician Edouard Goursat proved Cauchy's theorem without the assumption of continuity of f'. The resulting modified version of Cauchy's theorem is known as the **Cauchy–Goursat theorem**:

THEOREM 17.4 Cauchy–Goursat Theorem

Suppose a function f is analytic in a simply connected domain D. Then for every simple closed contour C in D, $\oint_C f(z)\,dz = 0$.

Since the interior of a simple closed contour is a simply connected domain, the Cauchy–Goursat theorem can be stated in the slightly more practical manner:

If f is analytic at all points within and on a simple closed contour C, then $\oint_C f(z)\,dz = 0$. (2)

EXAMPLE 1 Evaluate $\oint_C e^z\,dz$, where C is the curve shown in Figure 17.9.

Solution The function $f(z) = e^z$ is entire and C is a simple closed contour. It follows from the form of the Cauchy–Goursat theorem given in (2) that $\oint_C e^z\,dz = 0$. □

FIGURE 17.9

EXAMPLE 2 Evaluate $\oint_C \dfrac{dz}{z^2}$, where C is the ellipse $(x-2)^2 + \dfrac{(y-5)^2}{4} = 1$.

Solution The rational function $f(z) = 1/z^2$ is analytic everywhere except at $z = 0$. But $z = 0$ is not a point interior to or on the contour C. Thus, from (2) we have $\oint_C dz/z^2 = 0$. □

EXAMPLE 3 Given the flow $f(z) = \overline{\cos z}$, compute the circulation around and net flux across C, where C is the square with vertices $z = 1$, $z = i$, $z = -1$, and $z = -i$.

Solution We must compute $\oint_C \overline{f(z)}\,dz = \oint_C \cos z\,dz$ and then take the real and imaginary parts of the integral to find the circulation and net flux, respectively. The function $\cos z$ is analytic everywhere, and so $\oint_C \overline{f(z)}\,dz = 0$ from (2). The circulation and net flux are therefore both zero. Figure 17.10 shows the flow $f(z) = \overline{\cos z}$ and the contour C. □

flow $f(z) = \overline{\cos z}$

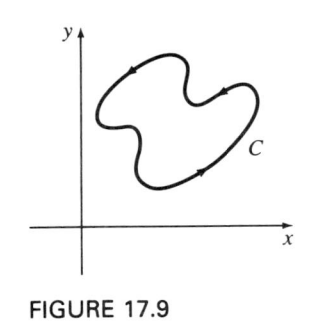

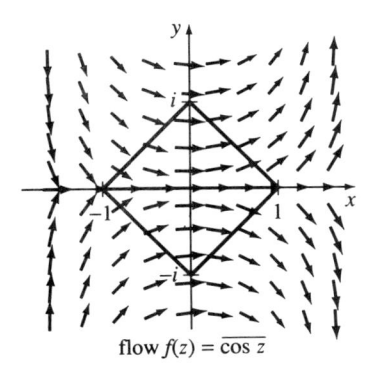

FIGURE 17.10

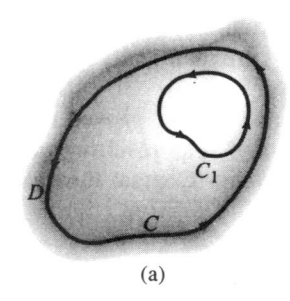

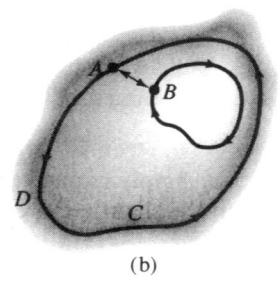

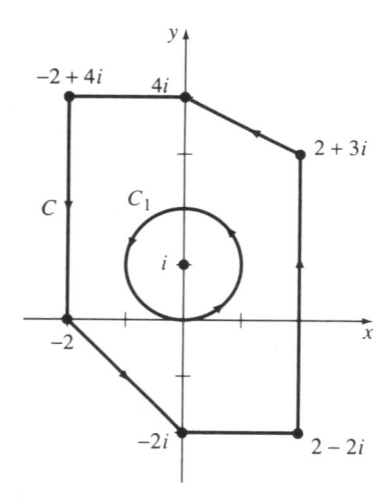

Cauchy–Goursat Theorem for Multiply Connected Domains If f is analytic in a multiply connected domain D, then we cannot conclude that $\oint_C f(z)\, dz = 0$ for every simple closed contour C in D. To begin, suppose D is a doubly connected domain and C and C_1 are simple closed contours such that C_1 surrounds the "hole" in the domain and is interior to C. See Figure 17.11(a). Suppose, also, that f is analytic on each contour and at each point interior to C but exterior to C_1. When we introduce the cut AB shown in Figure 17.11(b), the region bounded by the curves is simply connected. Now the integral from A to B has the opposite value of the integral from B to A, and so from (2) we have $\oint_C f(z)\, dz + \oint_{C_1} f(z)\, dz = 0$ or

$$\oint_C f(z)\, dz = \oint_{C_1} f(z)\, dz \qquad (3)$$

The last result is sometimes called the principle of **deformation of contours**, since we can think of the contour C_1 as a continuous deformation of the contour C. Under this deformation of contours the value of the integral does not change. Thus, on a practical level, (3) allows us to evaluate an integral over a complicated simple closed contour by replacing that contour with one that is more convenient.

EXAMPLE 4 Evaluate $\oint_C \dfrac{dz}{z-i}$, where C is the outer contour shown in Figure 17.12.

Solution In view of (3), we choose the more convenient circular contour C_1 in the figure. By taking the radius of the circle to be $r = 1$, we are guaranteed that C_1 lies within C. In other words, C_1 is the circle $|z - i| = 1$, which can be parameterized by $x = \cos t$, $y = 1 + \sin t$, $0 \leq t \leq 2\pi$, or equivalently by $z = i + e^{it}$, $0 \leq t \leq 2\pi$. From $z - i = e^{it}$ and $dz = ie^{it} dt$ we obtain

$$\oint_C \frac{dz}{z-i}\, dz = \oint_{C_1} \frac{dz}{z-i}\, dz = \int_0^{2\pi} \frac{ie^{it}}{e^{it}}\, dt = i\int_0^{2\pi} dt = 2\pi i \qquad \square$$

The result in Example 4 can be generalized. Using the principle of deformation of contours (2) and proceeding as in the example, we can show that if z_0 is any constant complex number interior to *any* simple closed contour C, then

$$\oint_C \frac{dz}{(z-z_0)^n} = \begin{cases} 2\pi i, & n = 1 \\ 0, & n \text{ an integer} \neq 1 \end{cases} \qquad (4)$$

The fact that the integral in (4) is zero when n is an integer $\neq 1$ follows only partially from the Cauchy–Goursat theorem. When n is zero or a negative integer, $1/(z-z_0)^n$ is a polynomial (for example, $n = -3$, $1/(z-z_0)^{-3} = (z-z_0)^3$) and therefore entire. Theorem 17.4 then implies $\oint_C dz/(z-z_0)^n = 0$. It is left as an exercise to show that the integral is still zero when n is a positive integer different from one. See Problem 22.

17.2 Cauchy–Goursat Theorem

EXAMPLE 5 Evaluate $\oint_C \dfrac{5z + 7}{z^2 + 2z - 3}\, dz$, where C is the circle $|z - 2| = 2$.

Solution Since the denominator factors as $z^2 + 2z - 3 = (z - 1)(z + 3)$, the integrand fails to be analytic at $z = 1$ and $z = -3$. Of these two points, only $z = 1$ lies within the contour C, which is a circle centered at $z = 2$ of radius $r = 2$. Now by partial fractions,

$$\frac{5z + 7}{z^2 + 2z - 3} = \frac{3}{z - 1} + \frac{2}{z + 3}$$

and so

$$\oint_C \frac{5z + 7}{z^2 + 2z - 3}\, dz = 3 \oint_C \frac{dz}{z - 1} + 2 \oint_C \frac{dz}{z + 3} \qquad (5)$$

In view of the result given in (4), the first integral in (5) has the value $2\pi i$. By the Cauchy–Goursat theorem the value of the second integral is zero. Hence, (5) becomes

$$\oint_C \frac{5z + 7}{z^2 + 2z - 3}\, dz = 3(2\pi i) + 2(0) = 6\pi i \qquad \square$$

If C, C_1, and C_2 are the simple closed contours shown in Figure 17.13 and if f is analytic on each of the three contours as well at each point interior to C but exterior to both C_1 and C_2, then by introducing cuts, we get from Theorem 17.4 that $\oint_C f(z)\, dz + \oint_{C_1} f(z)\, dz + \oint_{C_2}(z)\, dz = 0$. Hence,

$$\oint_C f(z)\, dz = \oint_{C_1} f(z)\, dz + \oint_{C_2} f(z)\, dz$$

The next theorem will summarize the general result for a multiply connected domain with n "holes":

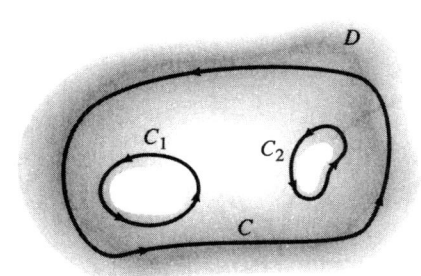

FIGURE 17.13

THEOREM 17.5 **Cauchy–Goursat Theorem for Multiply Connected Domains**

Suppose $C, C_1, \ldots, C_n$ are simple closed curves with a positive orientation such that $C_1, C_2, \ldots, C_n$ are interior to C but the regions interior to each C_k, $k = 1, 2, \ldots, n$, have no points in common. If f is analytic on each contour and at each point interior to C but exterior to all the C_k, $k = 1, 2, \ldots, n$, then

$$\oint_C f(z)\, dz = \sum_{k=1}^{n} \oint_{C_k} f(z)\, dz \qquad (6)$$

EXAMPLE 6 Evaluate $\oint_C \dfrac{dz}{z^2+1}$, where C is the circle $|z|=3$.

Solution In this case the denominator of the integrand factors as $z^2+1 = (z-i)(z+i)$. Consequently the integrand $1/(z^2+1)$ is not analytic at $z=i$ and $z=-i$. Both of these points lie within the contour C. Using partial fraction decomposition once more, we have

$$\frac{1}{z^2+1} = \frac{1/2i}{z-i} - \frac{1/2i}{z+i}$$

and
$$\oint_C \frac{dz}{z^2+1} = \frac{1}{2i} \oint_C \left[\frac{1}{z-i} - \frac{1}{z+i}\right] dz$$

We now surround the points $z=i$ and $z=-i$ by circular contours C_1 and C_2, respectively, that lie entirely within C. Specifically, the choice $|z-i| = \tfrac{1}{2}$ for C_1 and $|z+i| = \tfrac{1}{2}$ for C_2 will suffice. See Figure 17.14. From Theorem 17.5 we can then write

$$\oint_C \frac{dz}{z^2+1} = \frac{1}{2i}\oint_{C_1}\left[\frac{1}{z-i}-\frac{1}{z+i}\right]dz + \frac{1}{2i}\oint_{C_2}\left[\frac{1}{z-i}-\frac{1}{z+i}\right]dz$$
$$= \frac{1}{2i}\oint_{C_1}\frac{dz}{z-i} - \frac{1}{2i}\oint_{C_1}\frac{dz}{z+i} + \frac{1}{2i}\oint_{C_2}\frac{dz}{z-i} - \frac{1}{2i}\oint_{C_2}\frac{dz}{z+i} \quad (7)$$

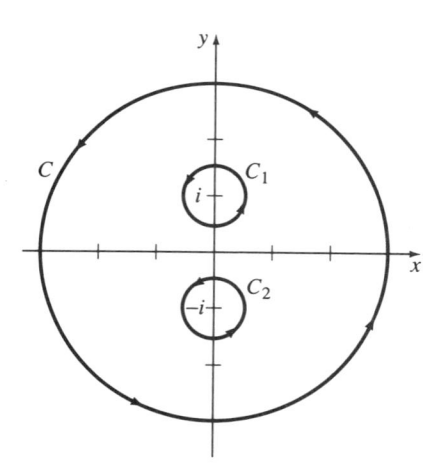

FIGURE 17.14

Because $1/(z+i)$ is analytic on C_1 and at each point in its interior and because $1/(z-i)$ is analytic on C_2 and at each point in its interior, it follows from (3) that the second and third integrals in (4) are zero. Moreover, it follows from (4), with $n=1$, that

$$\oint_{C_1} \frac{dz}{z-i} = 2\pi i \quad \text{and} \quad \oint_{C_2}\frac{dz}{z+i} = 2\pi i$$

17.2 Cauchy–Goursat Theorem

Thus (7) becomes $\oint_C \frac{dz}{z^2+1} = \pi - \pi = 0$

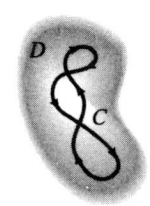

FIGURE 17.15

▲ **Remark** Throughout the foregoing discussion we assumed that C was a simple closed contour; in other words, C did not intersect itself. Although we shall not give the proof, it can be shown that the Cauchy–Goursat theorem is valid for any closed contour C in a simply connected domain D. As shown in Figure 17.15, the contour C is closed but not simple. Nevertheless, if f is analytic in D, then $\oint_C f(z)\, dz = 0$.

EXERCISES 17.2   Answers to odd-numbered problems begin on page A-88.

In Problems 1–8 prove that $\oint_C f(z)\, dz = 0$, where f is the given function and C is the unit circle $|z| = 1$.

1. $f(z) = z^3 - 1 + 3i$

2. $f(z) = z^2 + \frac{1}{z-4}$

3. $f(z) = \frac{z}{2z+3}$

4. $f(z) = \frac{z-3}{z^2+2z+2}$

5. $f(z) = \frac{\sin z}{(z^2-25)(z^2+9)}$

6. $f(z) = \frac{e^z}{2z^2+11z+15}$

7. $f(z) = \tan z$

8. $f(z) = \frac{z^2-9}{\cosh z}$

9. Evaluate $\oint_C \frac{1}{z}\, dz$, where C is the contour shown in Figure 17.16.

10. Evaluate $\oint_C \frac{5}{z+1+i}\, dz$, where C is the contour shown in Figure 17.17.

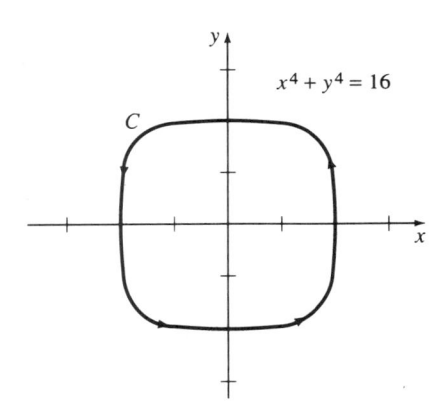

FIGURE 17.17

In Problems 11–20 use any of the results in this section to evaluate the given integral along the indicated closed contour(s).

11. $\oint_C \left(z + \frac{1}{z} \right) dz$; $|z| = 2$

12. $\oint_C \left(z + \frac{1}{z^2} \right) dz$; $|z| = 2$

13. $\oint_C \frac{z}{z^2 - \pi^2}\, dz$; $|z| = 3$

14. $\oint_C \frac{10}{(z+i)^4}\, dz$; $|z+i| = 1$

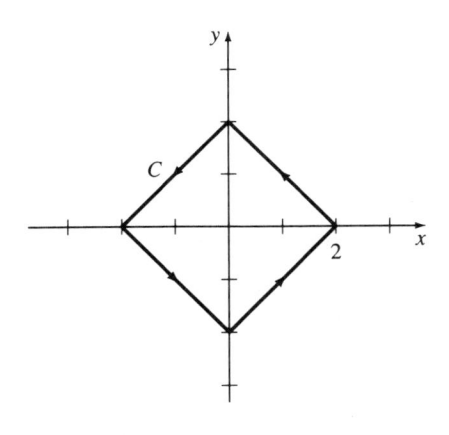

FIGURE 17.16

15. $\oint_C \dfrac{2z+1}{z^2+z} \, dz$; **(a)** $|z| = \tfrac{1}{2}$, **(b)** $|z| = 2$, **(c)** $|z - 3i| = 1$

16. $\oint_C \dfrac{2z}{z^2+3} \, dz$; **(a)** $|z| = 1$, **(b)** $|z - 2i| = 1$, **(c)** $|z| = 4$

17. $\oint_C \dfrac{-3z+2}{z^2-8z+12} \, dz$; **(a)** $|z - 5| = 2$, **(b)** $|z| = 9$

18. $\oint_C \left(\dfrac{3}{z+2} - \dfrac{1}{z-2i} \right) dz$; **(a)** $|z| = 5$, **(b)** $|z - 2i| = \tfrac{1}{2}$

19. $\oint_C \dfrac{z-1}{z(z-i)(z-3i)} \, dz$; $|z - i| = \tfrac{1}{2}$

20. $\oint_C \dfrac{1}{z^3 + 2iz^2} \, dz$; $|z| = 1$

21. Evaluate $\oint_C \dfrac{8z-3}{z^2-z} \, dz$, where C is the closed contour shown in Figure 17.18. [*Hint:* Express C as the union of two closed curves C_1 and C_2.]

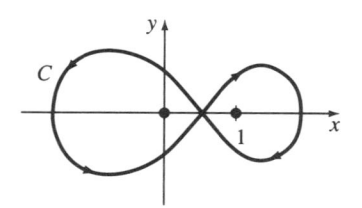

FIGURE 17.18

22. Suppose z_0 is any constant complex number interior to any simple closed contour C. Show that

$$\oint_C \dfrac{dz}{(z - z_0)^n} = \begin{cases} 2\pi i, & n = 1 \\ 0, & n \text{ a positive integer} \neq 1 \end{cases}$$

In Problems 23 and 24 evaluate the given integral by any means.

23. $\oint_C \left(\dfrac{e^z}{z+3} - 3\bar{z} \right) dz$, C is the unit circle $|z| = 1$

24. $\oint_C (z^3 + z^2 + \text{Re}(z)) \, dz$, C is the triangle with vertices $z = 0$, $z = 1 + 2i$, $z = 1$

17.3 INDEPENDENCE OF PATH

In real calculus, when a function f possesses an antiderivative F, a definite integral can be evaluated by the fundamental theorem of calculus:

$$\int_a^b f(x) \, dx = F(b) - F(a) \qquad (1)$$

Moreover, in Section 8.9 we saw that when a real line integral $\int_C P \, dx + Q \, dy$ is independent of the path—that is, depends on only the endpoints of the curve C—it can be evaluated by the fundamental theorem of line integrals (Theorem 8.8). It seems natural then to ask:

Can a contour integral $\int_C f(z) \, dz$ be independent of the path? Is there a complex version of the fundamental theorem (1)?

The answer to both of these questions is yes.

DEFINITION 17.2 Independence of the Path

Let z_0 and z_1 be points in a domain D. A contour integral $\int_C f(z) \, dz$ is said to be **independent of the path** if its value is the same for all contours C in D with an initial point z_0 and a terminal point z_1.

At the end of the preceding section we noted that the Cauchy–Goursat theorem also holds for closed contours, not just simple closed contours, in a

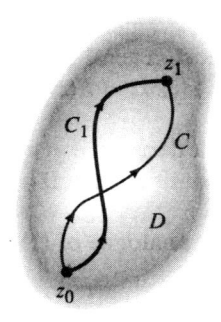

17.3 Independence of Path

simply connected domain D. Now suppose, as shown in Figure 17.19, that C and C_1 are two contours in a simply connected domain D, both with initial point z_0 and terminal point z_1. Note that C and $-C_1$ form a closed contour. Thus, if f is analytic in D, it follows from the Cauchy–Goursat theorem that

$$\int_C f(z)\,dz + \int_{-C_1} f(z)\,dz = 0 \tag{2}$$

But (2) is equivalent to

$$\int_C f(z)\,dz = \int_{C_1} f(z)\,dz \tag{3}$$

The result in (3) is also an example of the principle of deformation of contours introduced in (3) of Section 17.2. We shall summarize the last result as a theorem:

FIGURE 17.19

THEOREM 17.6

If f is an analytic function in a simply connected domain D, then $\int_C f(z)\,dz$ is independent of the path C.

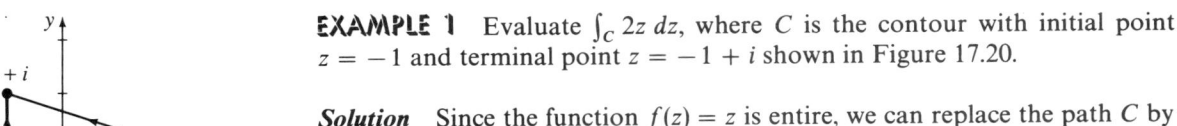

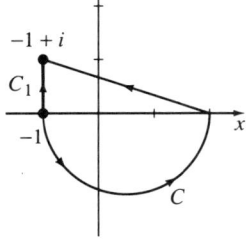

EXAMPLE 1 Evaluate $\int_C 2z\,dz$, where C is the contour with initial point $z = -1$ and terminal point $z = -1 + i$ shown in Figure 17.20.

Solution Since the function $f(z) = z$ is entire, we can replace the path C by any convenient contour C_1 joining $z = -1$ and $z = -1 + i$. In particular, by choosing C_1 to be the straight line segment $x = -1$, $0 \le y \le 1$, we have $z = -1 + iy$, $dz = i\,dy$. Therefore,

$$\int_C 2z\,dz = \int_{C_1} 2z\,dz = -2\int_0^1 y\,dy - 2i\int_0^1 dy = -1 - 2i \quad \square$$

FIGURE 17.20

A contour integral $\int_C f(z)\,dz$ that is independent of the path C is usually written $\int_{z_0}^{z_1} f(z)\,dz$, where z_0 and z_1 are the initial and terminal points of C. Hence in Example 1 we can write $\int_{-1}^{-1+i} 2z\,dz$.

There is an easier way to evaluate the contour integral in Example 1, but before proceeding we need another definition.

DEFINITION 17.3 Antiderivative

Suppose f is continuous in a domain D. If there exists a function F such that $F'(z) = f(z)$ for each z in D, then F is called an **antiderivative** of f.

For example, the function $F(z) = -\cos z$ is an antiderivative of $f(z) = \sin z$, since $F'(z) = \sin z$. As in real calculus, the most general antiderivative, or **indefinite integral**, of a function $f(z)$ is written $\int f(z)\,dz = F(z) + C$, where $F'(z) = f(z)$ and C is some complex constant.

Since an antiderivative F of a function f has a derivative at each point in a domain D, it is necessarily analytic and hence continuous in D (recall that differentiability implies continuity).

We are now in a position to prove the complex analogue of (1).

THEOREM 17.7 **Fundamental Theorem for Contour Integrals**

Suppose f is continuous in a domain D and F is an antiderivative of f in D. Then for any contour C in D with initial point z_0 and terminal point z_1,

$$\int_C f(z)\,dz = F(z_1) - F(z_0) \tag{4}$$

Proof We will prove (4) in the case when C is a smooth curve defined by $z = z(t)$, $a \le t \le b$. Using (3) of Section 17.1 and the fact that $F'(z) = f(z)$ for each z in D, we have

$$\int_C f(z)\,dz = \int_a^b f(z(t))z'(t)\,dt$$

$$= \int_a^b F'(z(t))z'(t)\,dt$$

$$= \int_a^b \frac{d}{dt} F(z(t))\,dt \quad \boxed{\text{chain rule}}$$

$$= F(z(t))\Big|_a^b$$

$$= F(z(b)) - F(z(a)) = F(z_1) - F(z_0) \quad \blacksquare$$

EXAMPLE 2 In Example 1 we saw that the integral $\int_C 2z\,dz$, where C is shown in Figure 17.20, is independent of the path. Now since $f(z) = 2z$ is an entire function, it is continuous. Moreover, $F(z) = z^2$ is an antiderivative of f, since $F'(z) = 2z$. Hence by (4) we have

$$\int_{-1}^{-1+i} 2z\,dz = z^2 \Big|_{-1}^{-1+i} = (-1+i)^2 - (-1)^2 = -1 - 2i \quad \square$$

17.3 Independence of Path

EXAMPLE 3 Evaluate $\int_C \cos z \, dz$, where C is any contour with initial point $z = 0$ and terminal point $z = 2 + i$.

Solution $F(z) = \sin z$ is an antiderivative of $f(z) = \cos z$, since $F'(z) = \cos z$. Therefore from (4) we have

$$\int_C \cos z \, dz = \int_0^{2+i} \cos z \, dz = \sin z \Big|_0^{2+i} = \sin(2+i) - \sin 0 = \sin(2+i)$$

If we desired a complex number of the form $a + ib$ for an answer, we can use $\sin(2 + i) = 1.4031 - 0.4891i$ (see Example 1 in Section 16.7). Hence,

$$\int_C \cos z \cdot dz = 1.4031 - 0.4891i \qquad \square$$

We can draw several immediate conclusions from Theorem 17.7. First, observe that if the contour C is closed, then $z_0 = z_1$ and consequently

$$\oint_C f(z) \, dz = 0 \tag{5}$$

Next, since the value of $\int_C f(z) \, dz$ depends on only the points z_0 and z_1, this value is the same for any contour C in D connecting these points. In other words:

If a continuous function f has an antiderivative F in D, then $\int_C f(z) \, dz$ is independent of the path. (6)

In addition we have the following sufficient condition for the existence of an antiderivative:

If f is continuous and $\int_C f(z) \, dz$ is independent of the path in a domain D, then f has an antiderivative everywhere in D. (7)

The last statement is important and deserves a proof. Assume that f is continuous, $\int_C f(z) \, dz$ is independent of the path in a domain D, and F is a function defined by

$$F(z) = \int_{z_0}^{z} f(s) \, ds$$

where s denotes a complex variable, z_0 is a fixed point in D, and z represents any point in D. We wish to show that $F'(z) = f(z)$; that is, F is an antiderivative of f in D. Now,

$$F(z + \Delta z) - F(z) = \int_{z_0}^{z + \Delta z} f(s) \, ds - \int_{z_0}^{z} f(s) \, ds = \int_{z}^{z + \Delta z} f(s) \, ds \tag{8}$$

Because D is a domain we can choose Δz so that $z + \Delta z$ is in D. Moreover, z and $z + \Delta z$ can be joined by a straight segment lying in D, as shown in Figure 17.21. This is the contour we use in the last integral in (8). With z fixed,

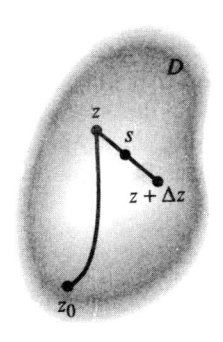

FIGURE 17.21

we can write*

$$f(z)\,\Delta z = f(z)\int_z^{z+\Delta z} ds = \int_z^{z+\Delta z} f(z)\,ds \quad \text{and} \quad f(z) = \frac{1}{\Delta z}\int_z^{z+\Delta z} f(z)\,ds \quad (9)$$

From (8) and (9) it follows that

$$\frac{F(z+\Delta z)-F(z)}{\Delta z} - f(z) = \frac{1}{\Delta z}\int_z^{z+\Delta z}[f(s)-f(z)]\,ds$$

Now f is continuous at the point z. This means that for any $\varepsilon > 0$ there exists a $\delta > 0$ so that $|f(s)-f(z)| < \varepsilon$ whenever $|s-z| < \delta$. Consequently if we choose Δz so that $|\Delta z| < \delta$, we have

$$\left|\frac{F(z+\Delta z)-F(z)}{\Delta z} - f(z)\right| = \left|\frac{1}{\Delta z}\int_z^{z+\Delta z}[f(s)-f(z)]\,ds\right|$$

$$= \left|\frac{1}{\Delta z}\right|\left|\int_z^{z+\Delta z}[f(s)-f(z)]\,ds\right| \le \frac{1}{|\Delta z|}\varepsilon|\Delta z| = \varepsilon$$

Hence, we have shown that

$$\lim_{\Delta z \to 0}\frac{F(z+\Delta z)-F(z)}{\Delta z} = f(z) \quad \text{or} \quad F'(z) = f(z)$$

If f is an analytic function in a simply connected domain D, it is necessarily continuous throughout D. This fact, when put together with the results in Theorem 17.6 and (7), leads to a theorem which states that an analytic function possesses an analytic antiderivative.

THEOREM 17.8 Existence of an Antiderivative

If f is analytic in a simply connected domain D, then f has an antiderivative in D; that is, there exists a function F such that $F'(z) = f(z)$ for all z in D.

In (9) of Section 16.6 we saw that $1/z$ is the derivative of $\operatorname{Ln} z$. This means that under some circumstances $\operatorname{Ln} z$ is an antiderivative of $1/z$. Care must be exercised in using this result. For example, suppose D is the entire complex plane without the origin. The function $1/z$ is analytic in this *multiply connected* domain. If C is any simple closed contour containing the origin, it does *not* follow from (5) that $\oint_C dz/z = 0$. In fact, from (4) of Section 17.2 with the identification $z_0 = 0$, we see that

$$\oint_C \frac{1}{z}\,dz = 2\pi i$$

---

*See Problem 29 in Exercises 17.1.

17.3 Independence of Path

In this case, Ln z is not an antiderivative of $1/z$ in D, since Ln z is not analytic in D. Recall that Ln z fails to be analytic on the nonpositive real axis (the branch cut of the principal branch of the logarithm).

EXAMPLE 4 Evaluate $\int_C \frac{1}{z} dz$, where C is the contour shown in Figure 17.22.

Solution Suppose that D is the simply connected domain defined by $x = \text{Re}(z) > 0$, $y = \text{Im}(z) > 0$. In this case, Ln z is an antiderivative of $1/z$, since both these functions are analytic in D. Hence by (4),

$$\int_3^{2i} \frac{1}{z} dz = \text{Ln } z \Big|_3^{2i} = \text{Ln } 2i - \text{Ln } 3$$

From (7) of Section 16.6, we have

$$\text{Ln } 2i = \log_e 2 + \frac{\pi}{2} i \quad \text{and} \quad \text{Ln } 3 = \log_e 3$$

and so

$$\int_3^{2i} \frac{1}{z} dz = \log_e \frac{2}{3} + \frac{\pi}{2} i = -0.4055 + 1.5708i \qquad \square$$

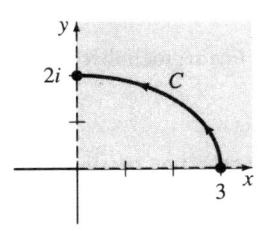

FIGURE 17.22

▲ **Remark** Suppose f and g are analytic in a simply connected domain D that contains the contour C. If z_0 and z_1 are the initial and terminal points of C, then the **integration by parts** formula is valid in D:

$$\int_{z_0}^{z_1} f(z) g'(z) \, dz = f(z) g(z) \Big|_{z_0}^{z_1} - \int_{z_0}^{z_1} f'(z) g(z) \, dz$$

This can be proved in a straightforward manner using Theorem 17.7 on the function $(d/dz)(fg)$. See Problems 21–24.

EXERCISES 17.3   Answers to odd-numbered problems begin on page A-88.

In Problems 1 and 2 evaluate the given integral, where C is the contour given in the figure, by **(a)** finding an alternative path of integration and **(b)** using Theorem 17.7.

1. $\int_C (4z - 1) \, dz$

2. $\int_C e^z \, dz$

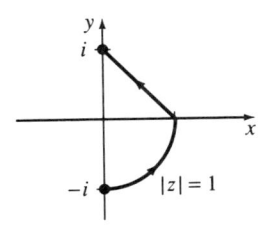

FIGURE 17.23

FIGURE 17.24

In Problems 3 and 4 evaluate the given integral along the indicated contour C.

3. $\int_C 2z\, dz$, where C is $z(t) = 2t^3 + i(t^4 - 4t^3 + 2)$, $-1 \le t \le 1$

4. $\int_C 6z^2\, dz$, where C is $z(t) = 2\cos^3 \pi t - i \sin^2 \frac{\pi}{4} t$, $0 \le t \le 2$

In Problems 5–24 use Theorem 17.7 to evaluate the given integral. Write each answer in the form $a + ib$.

5. $\int_0^{3+i} z^2\, dz$

6. $\int_{-2i}^{1} (3z^2 - 4z + 5i)\, dz$

7. $\int_{1-i}^{1+i} z^3\, dz$

8. $\int_{-3i}^{2i} (z^3 - z)\, dz$

9. $\int_{-i/2}^{1-i} (2z + 1)^2\, dz$

10. $\int_1^i (iz + 1)^3\, dz$

11. $\int_{i/2}^{i} e^{\pi z}\, dz$

12. $\int_{1-i}^{1+2i} ze^{z^2}\, dz$

13. $\int_{\pi}^{\pi + 2i} \sin \frac{z}{2}\, dz$

14. $\int_{1-2i}^{\pi i} \cos z\, dz$

15. $\int_{\pi i}^{2\pi i} \cosh z\, dz$

16. $\int_i^{1+(\pi/2)i} \sinh 3z\, dz$

17. $\int_C \frac{1}{z}\, dz$, C is the arc of the circle $z = 4e^{it}$, $-\pi/2 \le t \le \pi/2$

18. $\int_C \frac{1}{z}\, dz$, C is the straight line segment between $z = 1 + i$ and $z = 4 + 4i$

19. $\int_{-4i}^{4i} \frac{1}{z^2}\, dz$, C is any contour not passing through the origin.

20. $\int_{1-i}^{1+\sqrt{3}i} \left(\frac{1}{z} + \frac{1}{z^2} \right) dz$, C is any contour in the right half-plane $\text{Re}(z) > 0$

21. $\int_{\pi}^{i} e^z \cos z\, dz$

22. $\int_0^i z \sin z\, dz$

23. $\int_i^{1+i} ze^z\, dz$

24. $\int_0^{\pi i} z^2 e^z\, dz$

17.4 CAUCHY'S INTEGRAL FORMULA

In the discussion that follows we shall consider another important consequence of the Cauchy–Goursat theorem:

The value of an analytic function f at any point z_0 in a simply connected domain can be represented by a contour integral.

As a direct consequence of this last result, we shall then show:

An analytic function in a simply connected domain possesses derivatives of all orders.

We begin with the Cauchy integral formula. The idea in the next theorem is this: If f is analytic in a simply connected domain and z_0 is any point D, then the quotient $f(z)/(z - z_0)$ is *not* analytic in D. As a consequence, the integral of $f(z)/(z - z_0)$ around a simple closed contour C that contains z_0 is not necessarily zero but has, as we shall now see, the value $2\pi i f(z_0)$. This remarkable result indicates that the values of an analytic function f at points *inside* a simple closed contour C are determined by the values of f on the contour C.

> **THEOREM 17.9** **Cauchy's Integral Formula**
>
> Let f be analytic in a simply connected domain D, and let C be a simple closed contour lying entirely within D. If z_0 is any point within C, then
>
> $$f(z_0) = \frac{1}{2\pi i} \oint_C \frac{f(z)}{z - z_0}\, dz \qquad (1)$$

Proof Let D be a simply connected domain, C a simple closed contour in D, and z_0 an interior point of C. In addition, let C_1 be a circle centered at z_0 with radius small enough that it is interior to C. By the principle of deformation of contours, we can write

$$\oint_C \frac{f(z)}{z - z_0}\, dz = \oint_{C_1} \frac{f(z)}{z - z_0}\, dz \qquad (2)$$

We wish to show that the value of the integral on the right is $2\pi i f(z_0)$. To this end we add and subtract the constant $f(z_0)$ in the numerator:

$$\oint_{C_1} \frac{f(z)}{z - z_0}\, dz = \oint_{C_1} \frac{f(z_0) - f(z_0) + f(z)}{z - z_0}\, dz$$

$$= f(z_0) \oint_{C_1} \frac{dz}{z - z_0} + \oint_{C_1} \frac{f(z) - f(z_0)}{z - z_0}\, dz \qquad (3)$$

Now from (4) of Section 17.2 we know that

$$\oint_{C_1} \frac{dz}{z - z_0} = 2\pi i$$

Thus, (3) becomes

$$\oint_{C_1} \frac{f(z)}{z - z_0}\, dz = 2\pi i f(z_0) + \oint_{C_1} \frac{f(z) - f(z_0)}{z - z_0}\, dz \qquad (4)$$

Since f is continuous at z_0 for any arbitrarily small $\varepsilon > 0$, there exists a $\delta > 0$ such that $|f(z) - f(z_0)| < \varepsilon$ whenever $|z - z_0| < \delta$. In particular, if we choose the circle C_1 to be $|z - z_0| = \delta/2 < \delta$, then by the *ML*-inequality (Theorem 17.3) the absolute value of the integral on the right side of (4) satisfies

$$\left| \oint_{C_1} \frac{f(z) - f(z_0)}{z - z_0}\, dz \right| \le \frac{\varepsilon}{\delta/2} 2\pi \left(\frac{\delta}{2}\right) = 2\pi \varepsilon$$

In other words, the absolute value of the integral can be made arbitrarily small by taking the radius of the circle C_1 to be sufficiently small. This can happen only if the integral is zero. The Cauchy integral formula (1) follows from (4) by dividing both sides by $2\pi i$. ∎

The Cauchy integral formula (1) can be used to evaluate contour integrals. Since we often work problems without a simply connected domain explicitly defined, a more practical restatement of Theorem 17.9 is:

If f is analytic at all points within and on a simple closed contour C, and z_0 is any point interior to C, then $f(z_0) = \dfrac{1}{2\pi i} \oint_C \dfrac{f(z)}{z - z_0}\, dz.$ (5)

EXAMPLE 1 Evaluate $\oint_C \dfrac{z^2 - 4z + 4}{z + i}\, dz$, where C is the circle $|z| = 2$.

Solution First, we identify $f(z) = z^2 - 4z + 4$ and $z_0 = -i$ as a point within the circle C. Next, we observe that f is analytic at all points within and on the contour C. Thus by the Cauchy integral formula we obtain

$$\oint_C \frac{z^2 - 4z + 4}{z + i}\, dz = 2\pi i f(-i) = 2\pi i(3 + 4i) = 2\pi(-4 + 3i) \qquad \square$$

EXAMPLE 2 Evaluate $\oint_C \dfrac{z}{z^2 + 9}\, dz$, where C is the circle $|z - 2i| = 4$.

Solution By factoring the denominator as $z^2 + 9 = (z - 3i)(z + 3i)$, we see that $3i$ is the only point within the closed contour at which the integrand fails to be analytic. See Figure 17.25. Now by writing

$$\frac{z}{z^2 + 9} = \frac{\dfrac{z}{z + 3i}}{z - 3i}$$

we can identify $f(z) = z/(z + 3i)$. This function is analytic at all points within and on the contour C. From the Cauchy integral formula we then have

$$\oint_C \frac{z}{z^2 + 9}\, dz = \oint_C \frac{\dfrac{z}{z + 3i}}{z - 3i}\, dz = 2\pi i f(3i) = 2\pi i \frac{3i}{6i} = \pi i \qquad \square$$

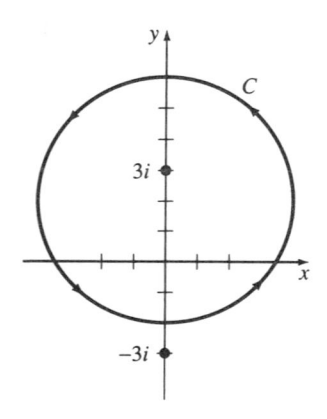

FIGURE 17.25

EXAMPLE 3 The complex function $f(z) = k/(\bar{z} - \bar{z}_1)$, where $k = a + ib$ and z_1 are complex numbers, gives rise to a flow in the domain $z \neq z_1$. If C is a simple closed contour containing $z = z_1$ in its interior, then from the Cauchy integral formula we have

$$\oint_C \overline{f(z)}\, dz = \oint_C \frac{a - ib}{z - z_1}\, dz = 2\pi i(a - ib)$$

Thus, the circulation around C is $2\pi b$ and the net flux across C is $2\pi a$. If z_1 were in the exterior of C, both the circulation and net flux would be zero by Cauchy's theorem.

17.4 Cauchy's Integral Formula

Note that when k is real, the circulation around C is zero but the net flux across C is $2\pi k$. The complex number z_1 is called a **source** for the flow when $k > 0$ and a **sink** when $k < 0$. Vector fields corresponding to these two cases are shown in Figure 17.26(a) and (b).

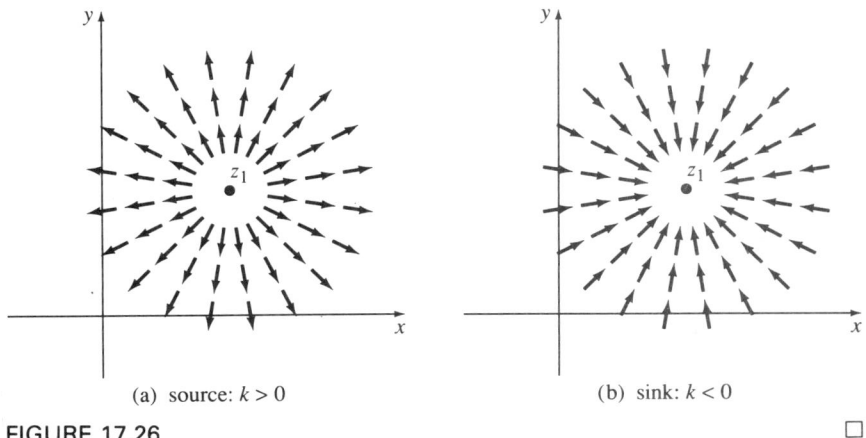

(a) source: $k > 0$ (b) sink: $k < 0$

FIGURE 17.26

Derivatives of Analytic Functions   We can now use Theorem 17.9 to prove that an analytic function possesses derivatives of all orders; that is, if f is analytic at a point z_0, then f', f'', f''', and so on are also analytic at z_0. Moreover, the values of the derivatives $f^{(n)}(z_0)$, $n = 1, 2, 3, \ldots$, are given by a formula similar to (1).

THEOREM 17.10   **Cauchy's Integral Formula for Derivatives**

Let f be analytic in a simply connected domain D, and let C be a simple closed contour lying entirely within D. If z_0 is any point interior to C, then

$$f^{(n)}(z_0) = \frac{n!}{2\pi i} \oint_C \frac{f(z)}{(z - z_0)^{n+1}} \, dz \tag{6}$$

Partial Proof   We will proof (6) only for the case $n = 1$. The remainder of the proof can be completed using the principle of mathematical induction.

We begin with the definition of the derivative and (1):

$$f'(z_0) = \lim_{\Delta z \to 0} \frac{f(z_0 + \Delta z) - f(z_0)}{\Delta z}$$

$$= \lim_{\Delta z \to 0} \frac{1}{2\pi i \, \Delta z} \left[\oint_C \frac{f(z)}{z - (z_0 + \Delta z)} \, dz - \oint_C \frac{f(z)}{z - z_0} \, dz \right]$$

$$= \lim_{\Delta z \to 0} \frac{1}{2\pi i} \oint_C \frac{f(z)}{(z - z_0 - \Delta z)(z - z_0)} \, dz$$

Before proceeding, let us set up some preliminaries. Since f is continuous on C, it is bounded; that is, there exists a real number M such that $|f(z)| \leq M$ for all points z on C. In addition, let L be the length of C and let δ denote the shortest distance between points on C and the point z_0. Thus for all points z on C, we have

$$|z - z_0| \geq \delta \quad \text{or} \quad \frac{1}{|z - z_0|^2} \leq \frac{1}{\delta^2}$$

Furthermore, if we choose $|\Delta z| \leq \delta/2$, then

$$|z - z_0 - \Delta z| \geq ||z - z_0| - |\Delta z|| \geq \delta - |\Delta z| \geq \frac{\delta}{2} \quad \text{and so} \quad \frac{1}{|z - z_0 - \Delta z|} \leq \frac{2}{\delta}$$

Now,

$$\left| \oint_C \frac{f(z)}{(z - z_0)^2} \, dz - \oint_C \frac{f(z)}{(z - z_0 - \Delta z)(z - z_0)} \, dz \right| = \left| \oint_C \frac{-\Delta z \, f(z)}{(z - z_0)^2 (z - z_0 - \Delta z)} \, dz \right| \leq \frac{2ML|\Delta z|}{\delta^3}$$

Because the last expression approaches zero as $\Delta z \to 0$, we have shown that

$$f'(z_0) = \lim_{\Delta z \to 0} \frac{f(z_0 + \Delta z) - f(z_0)}{\Delta z} = \frac{1}{2\pi i} \oint_C \frac{f(z)}{(z - z_0)^2} \, dz \qquad \blacksquare$$

If $f(z) = u(x, y) + iv(x, y)$ is analytic at a point, then its derivatives of all orders exist at that point and are continuous. Consequently, from

$$f'(z) = \frac{\partial u}{\partial x} + i \frac{\partial v}{\partial x} = \frac{\partial v}{\partial y} - i \frac{\partial u}{\partial y}$$

$$f''(z) = \frac{\partial^2 u}{\partial x^2} + i \frac{\partial^2 v}{\partial x^2} = \frac{\partial^2 v}{\partial y \, \partial x} - i \frac{\partial^2 u}{\partial y \, \partial x}$$

$$\vdots$$

we can conclude that the real functions u and v have continuous partial derivatives of all orders at a point of analyticity.

Like (1), (6) can sometimes be used to evaluate integrals.

EXAMPLE 4 Evaluate $\oint_C \dfrac{z + 1}{z^4 + 4z^3} \, dz$, where C is the circle $|z| = 1$.

Solution Inspection of the integrand shows that it is not analytic at $z = 0$ and $z = -4$, but only $z = 0$ lies within the closed contour. By writing the integrand as

$$\frac{z + 1}{z^4 + 4z^3} = \frac{\dfrac{z + 1}{z + 4}}{z^3}$$

17.4 Cauchy's Integral Formula

we can identify $z_0 = 0$, $n = 2$, and $f(z) = (z + 1)/(z + 4)$. By the quotient rule, $f''(z) = -6/(z + 4)^3$ and so by (6) we have

$$\oint_C \frac{z+1}{z^4 + 4z^3} \, dz = \frac{2\pi i}{2!} f''(0) = -\frac{3\pi}{32} i \qquad \square$$

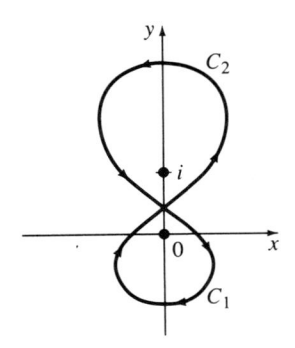

EXAMPLE 5 Evaluate $\oint_C \frac{z^2 + 3}{z(z-i)^2} \, dz$, where C is the contour shown in Figure 17.27.

Solution Although C is not a simple closed contour, we can think of it as the union of two simple closed contours C_1 and C_2 as indicated in Figure 17.27. By writing

$$\oint_C \frac{z^3 + 3}{z(z-i)^2} \, dz = \oint_{C_1} \frac{z^3 + 3}{z(z-i)^2} \, dz + \oint_{C_2} \frac{z^3 + 3}{z(z-i)^2} \, dz$$

$$= -\oint_{C_1} \frac{\frac{z^3+3}{(z-i)^2}}{z} \, dz + \oint_{C_2} \frac{\frac{z^3+3}{z}}{(z-i)^2} \, dz = -I_1 + I_2$$

we are in a position to use both (1) and (6).

To evaluate I_1, we identify $z_0 = 0$ and $f(z) = (z^3 + 3)/(z - i)^2$. By (1) it follows that

$$I_1 = \oint_{C_1} \frac{\frac{z^3+3}{(z-i)^2}}{z} \, dz = 2\pi i f(0) = -6\pi i$$

To evaluate I_2 we identify $z_0 = i$, $n = 1$, $f(z) = (z^3 + 3)/z$, and $f'(z) = (2z^3 - 3)/z^2$. From (6) we obtain

$$I_2 = \oint_{C_2} \frac{\frac{z^3+3}{z}}{(z-i)^2} \, dz = \frac{2\pi i}{1!} f'(i) = 2\pi i(3 + 2i) = 2\pi(-2 + 3i)$$

Finally we get

$$\oint_C \frac{z^3 + 3}{z(z-i)^2} \, dz = -I_1 + I_2 = 6\pi i + 2\pi(-2 + 3i) = 4\pi(-1 + 3i) \qquad \square$$

Liouville's Theorem If we take the contour C to be the circle $|z - z_0| = r$, it follows from (6) and the ML-inequality that

$$|f^{(n)}(z_0)| = \frac{n!}{2\pi} \left| \oint_C \frac{f(z)}{(z-z_0)^{n+1}} \, dz \right| \leq \frac{n!}{2\pi} M \frac{1}{r^{n+1}} 2\pi r = \frac{n!M}{r^n} \qquad (7)$$

where M is a real number such that $|f(z)| \leq M$ for all points z on C. The result in (7), called **Cauchy's inequality**, is used to prove the next result.

> **THEOREM 17.11 Liouville's Theorem**
>
> The only bounded entire functions are constants.

Proof Suppose f is an entire function and is bounded; that is, $|f(z)| \le M$ for all z. Then for any point z_0, (7) gives $|f'(z_0)| \le M/r$. By taking r arbitrarily large, we can make $|f'(z_0)|$ as small as we wish. This means $f'(z_0) = 0$ for all points z_0 in the complex plane. Hence, f must be a constant. ∎

Liouville's theorem enables us to prove, in turn, a result that is learned in elementary algebra: If $P(z)$ is a nonconstant polynomial, then the equation $P(z) = 0$ has at least one root. This result is known as the **fundamental theorem of algebra**. To prove it, let us suppose that $P(z) \ne 0$ for all z. This implies that the reciprocal of P,

$$f(z) = \frac{1}{P(z)}$$

is an entire function. Now since $|f(z)| \to 0$ as $|z| \to \infty$, the function f must be bounded for all finite z. It follows from Liouville's theorem that f is a constant and therefore P is a constant. But this is a contradiction to our underlying assumption that P was *not* a constant polynomial. We conclude that there must exist at least one number z for which $P(z) = 0$.

EXERCISES 17.4 *Answers to odd-numbered problems begin on page A-88.*

In Problems 1–24 use Theorems 17.9 and 17.10, when appropriate, to evaluate the given integral along the indicated closed contour(s).

1. $\oint_C \dfrac{4}{z - 3i}\, dz;\ |z| = 5$

2. $\oint_C \dfrac{z^2}{(z - 3i)^2}\, dz;\ |z| = 5$

3. $\oint_C \dfrac{e^z}{z - \pi i}\, dz;\ |z| = 4$

4. $\oint_C \dfrac{1 + 2e^z}{z}\, dz;\ |z| = 1$

5. $\oint_C \dfrac{z^2 - 3z + 4i}{z + 2i}\, dz;\ |z| = 3$

6. $\oint_C \dfrac{\cos z}{3z - \pi}\, dz;\ |z| = 1.1$

7. $\oint_C \dfrac{z^2}{z^2 + 4}\, dz;$ **(a)** $|z - i| = 2$, **(b)** $|z + 2i| = 1$

8. $\oint_C \dfrac{z^2 + 3z + 2i}{z^2 + 3z - 4}\, dz;$ **(a)** $|z| = 2$, **(b)** $|z + 5| = \tfrac{3}{2}$

9. $\oint_C \dfrac{z^2 + 4}{z^2 - 5iz - 4}\, dz;\ |z - 3i| = 1.3$

10. $\oint_C \dfrac{\sin z}{z^2 + \pi^2}\, dz;\ |z - 2i| = 2$

11. $\oint_C \dfrac{e^{z^2}}{(z - i)^3}\, dz;\ |z - i| = 1$

12. $\oint_C \dfrac{z}{(z + i)^4}\, dz;\ |z| = 2$

13. $\oint_C \dfrac{\cos 2z}{z^5}\, dz;\ |z| = 1$

14. $\oint_C \dfrac{e^{-z} \sin z}{z^3}\, dz;\ |z - 1| = 3$

15. $\oint_C \dfrac{2z+5}{z^2-2z}\,dz$; **(a)** $|z|=\tfrac{1}{2}$, **(b)** $|z+1|=2$, **(c)** $|z-3|=2$, **(d)** $|z+2i|=1$

16. $\oint_C \dfrac{z}{(z-1)(z-2)}\,dz$; **(a)** $|z|=\tfrac{1}{2}$, **(b)** $|z+1|=1$, **(c)** $|z-1|=\tfrac{1}{2}$, **(d)** $|z|=4$

17. $\oint_C \dfrac{z+2}{z^2(z-1-i)}\,dz$; **(a)** $|z|=1$, **(b)** $|z-1-i|=1$

18. $\oint_C \dfrac{1}{z^3(z-4)}\,dz$; **(a)** $|z|=1$, **(b)** $|z-2|=1$

19. $\oint_C \left(\dfrac{e^{2iz}}{z^4} - \dfrac{z^4}{(z-i)^3}\right) dz$; $|z|=6$

20. $\oint_C \left(\dfrac{\cosh z}{(z-\pi)^3} - \dfrac{\sin^2 z}{(2z-\pi)^3}\right) dz$; $|z|=3$

21. $\oint_C \dfrac{1}{z^3(z-1)^2}\,dz$; $|z-2|=5$

22. $\oint_C \dfrac{1}{z^2(z^2+1)}\,dz$; $|z-i|=\tfrac{3}{2}$

23. $\oint_C \dfrac{3z+1}{z(z-2)^2}\,dz$; C is given in Figure 17.28

24. $\oint_C \dfrac{e^{iz}}{(z^2+1)^2}\,dz$; C is given in Figure 17.29

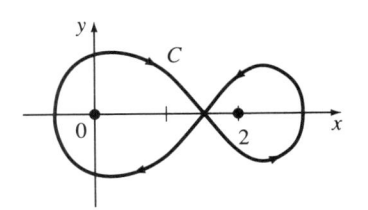

FIGURE 17.28

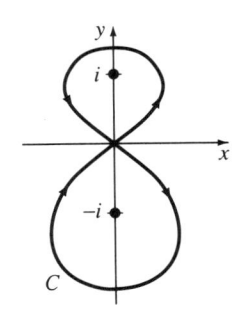

FIGURE 17.29

SUMMARY

An integral of a complex function f is called a **contour integral** and can be written as two real-line integrals. When we interpret the complex function $f(z) = u(x, y) + iv(x, y)$ as a flow, the **circulation** around a positively oriented simple closed contour C and the **net flux** across C are given by $\text{Re}(\oint_C \overline{f(z)}\,dz)$ and $\text{Im}(\oint_C \overline{f(z)}\,dz)$, respectively.

When f is analytic in a simply connected domain D, the **Cauchy–Goursat theorem** states that $\oint_C f(z)\,dz = 0$ for every closed contour C lying entirely within D. Moreover, in D a contour integral $\int_C f(z)\,dz$ is independent of the path C and can be evaluated by the **fundamental theorem for contour integrals**:

$$\int_C f(z)\,dz = F(z_1) - F(z_0)$$

where z_0 and z_1 are the initial point and terminal point of C, respectively, and F is an antiderivative of f in D. A function f that is analytic in a simply connected domain D has an analytic antiderivative F in D. If z_0 is any point within a simple closed contour C in D, then the value of the analytic function f at z_0 is determined by the values of f on C and is given by **Cauchy's**

integral formula:

$$f(z_0) = \frac{1}{2\pi i} \oint_C \frac{f(z)}{z - z_0} \, dz$$

The function f also possesses derivatives of all orders at the same point z_0. The values of these derivatives are given by the formula

$$f^{(n)}(z_0) = \frac{n!}{2\pi i} \oint_C \frac{f(z)}{(z - z_0)^{n+1}} \, dz$$

CHAPTER 17 REVIEW EXERCISES

Answers to odd-numbered problems begin on page A-88.

Answer Problems 1–12 without referring back to the text. Fill in the blank or answer true/false.

1. The sector defined by $-\pi/6 < \arg z < \pi/6$ is a simply connected domain. _____

2. If $\oint_C f(z) \, dz = 0$ for every simple closed contour C, then f is analytic within and on C. _____

3. The value of $\int_C \frac{z-2}{z} \, dz$ is the same for any path C in the right half-plane $\text{Re}(z) > 0$ between $z = 1 + i$ and $z = 10 + 8i$. _____

4. If g is entire, then $\oint_C \frac{g(z)}{z-i} \, dz = \oint_{C_1} \frac{g(z)}{z-i} \, dz$, where C is the circle $|z| = 3$ and C_1 is the ellipse $x^2 + y^2/9 = 1$. _____

5. If f is a polynomial and C is a simple closed curve, then
$$\oint_C f(z) \, dz = \underline{\quad\quad}.$$

6. If $f(z) = \oint_C \frac{\xi^2 + 6\xi - 2}{\xi - z} \, d\xi$, where C is $|z| = 3$, then $f(1+i) = \underline{\quad\quad}$.

7. If $f(z) = z^3 + e^z$ and C is the contour $z = 8e^{it}$, $0 \le t \le 2\pi$, then $\oint_C \frac{f(z)}{(z + \pi i)^3} \, dz = \underline{\quad\quad}$.

8. If f is entire and $|f(z)| \le 10$ for all z, then $f(z) = \underline{\quad\quad}$.

9. $\oint_C \frac{1}{(z - z_0)(z - z_1)} \, dz = 0$ for every simple closed contour C that encloses the points z_0 and z_1. _____

10. If f is analytic within and on the simple closed contour C and z_0 is a point within C, then $\oint_C \frac{f'(z)}{z - z_0} \, dz = \oint_C \frac{f(z)}{(z - z_0)^2} \, dz$. _____

11. $\oint_C z^n \, dz = \begin{cases} 0, & \text{if } n \underline{\quad} \\ 2\pi i, & \text{if } n \underline{\quad} \end{cases}$, where n is an integer and C is $|z| = 1$.

12. If $|f(z)| \le 2$ on $|z| = 3$, then $\left| \oint_C f(z) \, dz \right| \le \underline{\quad\quad}$.

In Problems 13–28 evaluate the given integral using the techniques considered in this chapter.

13. $\int_C (x + iy) \, dz$; C is the contour shown in Figure 17.30

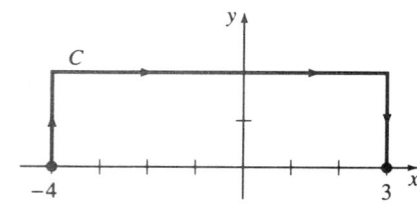

FIGURE 17.30

14. $\int_C (x - iy) \, dz$; C is the contour shown in Figure 17.30

15. $\int_C |z^2| \, dz$; C is $z(t) = t + it^2$, $0 \le t \le 2$

16. $\int_C e^{\pi z} \, dz$; C is the line segment from $z = i$ to $z = 1 + i$

17. $\oint_C e^{\pi z} \, dz$; C is the ellipse $x^2/100 + y^2/64 = 1$

18. $\displaystyle\int_{3i}^{1-i} (4z - 6) \, dz$

19. $\displaystyle\int_C \sin z \, dz$; C is $z(t) = t^4 + i(1 + t^3)^2$, $-1 \le t < 1$

20. $\displaystyle\int_C (4z^3 + 3z^2 + 2z + 1) \, dz$; C is the line segment from 0 to $2i$

21. $\oint_C (z^{-2} + z^{-1} + z + z^2) \, dz$; C is the circle $|z| = 1$

22. $\oint_C \dfrac{3z + 4}{z^2 - 1} \, dz$; C is the circle $|z| = 2$

23. $\oint_C \dfrac{e^{-2z}}{z^4} \, dz$; C is the circle $|z - 1| = 3$

24. $\oint_C \dfrac{\cos z}{z^3 - z^2} \, dz$; C is the circle $|z| = \tfrac{1}{2}$

25. $\oint_C \dfrac{1}{2z^2 + 7z + 3} \, dz$; C is the ellipse $x^2/4 + y^2 = 1$

26. $\oint_C z \csc z \, dz$; C is the rectangle with vertices $1 + i$, $1 - i$, $2 + i$, $2 - i$

27. $\oint_C \dfrac{z}{z + i} \, dz$; C is the contour shown in Figure 17.31

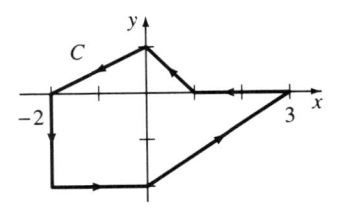

FIGURE 17.31

28. $\oint_C \dfrac{e^{i\pi z}}{2z^2 - 5z + 2} \, dz$; C is **(a)** $|z| = 1$, **(b)** $|z - 3| = 2$, **(c)** $|z + 3| = 2$

29. Let $f(z) = z^n g(z)$, where n is a positive integer, $g(z)$ is entire, and $g(z) \ne 0$ for all z. Let C be a circle with center at the origin. Evaluate $\oint_C \dfrac{f'(z)}{f(z)} \, dz$.

30. Let C be the straight line segment from i to $2 + i$. Show that
$$\left| \int_C \operatorname{Ln}(z + 1) \, dz \right| \le \log_e 10 + \dfrac{\pi}{2}$$

18

SERIES AND RESIDUES

TOPICS TO REVIEW

infinite series
tests for convergence
sum of a geometric series
binomial series

IMPORTANT CONCEPTS

sequence of complex numbers
infinite series of complex numbers
*n*th term test for divergence
absolute convergence
ratio test
root test
power series
center of a power series
radius of convergence
circle of convergence
Taylor series
singular point
isolated singular point
Laurent series
removable singularity
pole
essential singularity
residue
Cauchy's residue theorem

18.1 Sequences and Series
18.2 Taylor Series
18.3 Laurent Series
18.4 Zeros and Poles
18.5 Residues and Residue Theorem
18.6 Evaluation of Real Integrals
      Summary
      Chapter 18 Review Exercises

INTRODUCTION

Cauchy's integral formula for derivatives indicates that if a function f is analytic at a point z_0, then it possesses derivatives of all orders at that point. As a consequence of this result, we shall see that f can always be expanded in a power series centered at that point. On the other hand, if f fails to be analytic at a point z_0, we may still be able to expand it in a different kind of series known as a **Laurent series**. The notion of a Laurent series leads to the concept of a **residue** and this, in turn, leads to yet another way of evaluating complex integrals.

18.1 SEQUENCES AND SERIES

Much of the theory of complex sequences and series is analogous to that encountered in real calculus.

Sequences A **sequence** $\{z_n\}$ is a function whose domain is the set of positive integers; in other words, to each integer $n = 1, 2, 3, \ldots$ we assign a complex number z_n. For example, the sequence $\{1 + i^n\}$ is

$$\begin{array}{ccccc} 1+i, & 0, & 1-i, & 2, & 1+i, \ldots \\ \uparrow & \uparrow & \uparrow & \uparrow & \uparrow \\ n=1, & n=2, & n=3, & n=4, & n=5, \ldots \end{array} \quad (1)$$

If
$$\lim_{n \to \infty} z_n = L$$

we say the sequence $\{z_n\}$ is **convergent**. In other words, $\{z_n\}$ **converges** to the number L if, for each positive number ε, an N can be found such that $|z_n - L| < \varepsilon$ whenever $n > N$. As shown in Figure 18.1, when a sequence $\{z_n\}$ converges to L, all but a finite number of the terms of the sequence are within every ε-neighborhood of L. The sequence $\{1 + i^n\}$ illustrated in (1) is divergent, since the general term $z_n = 1 + i^n$ does not approach a fixed complex number as $n \to \infty$. Indeed, the first four terms of this sequence repeat endlessly as n increases.

FIGURE 18.1

EXAMPLE 1 The sequence $\left\{\dfrac{i^{n+1}}{n}\right\}$ converges, since

$$\lim_{n \to \infty} \frac{i^{n+1}}{n} = 0$$

As we see from
$$-1, -\frac{i}{2}, \frac{1}{3}, \frac{i}{4}, -\frac{1}{5}, \ldots$$

and Figure 18.2, the terms of the sequence spiral in toward the point $z = 0$. □

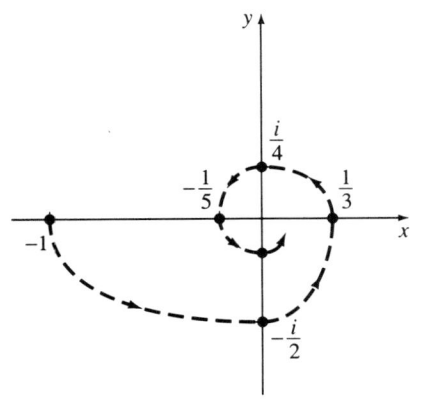

FIGURE 18.2

The following theorem should make intuitive sense:

THEOREM 18.1

A sequence $\{z_n\}$ converges to a complex number L if and only if $\mathrm{Re}(z_n)$ converges to $\mathrm{Re}(L)$ and $\mathrm{Im}(z_n)$ converges to $\mathrm{Im}(L)$.

EXAMPLE 2 The sequence $\left\{\dfrac{ni}{n+2i}\right\}$ converges to i. Note that $\operatorname{Re}(i) = 0$ and $\operatorname{Im}(i) = 1$. Then from

$$z_n = \frac{ni}{n+2i} = \frac{2n}{n^2+4} + i\frac{n^2}{n^2+4}$$

we see that $\operatorname{Re}(z_n) = 2n/(n^2+4) \to 0$ and $\operatorname{Im}(z_n) = n^2/(n^2+4) \to 1$ as $n \to \infty$. □

Series An **infinite series** of complex numbers

$$\sum_{k=1}^{\infty} z_k = z_1 + z_2 + z_3 + \cdots + z_n + \cdots$$

is **convergent** if the sequence of partial sums $\{S_n\}$, where

$$S_n = z_1 + z_2 + z_3 + \cdots + z_n$$

converges. If $S_n \to L$ as $n \to \infty$, we say that the **sum** of the series is L.

Geometric Series For the geometric series

$$\sum_{k=1}^{\infty} az^{k-1} = a + az + az^2 + \cdots + az^{n-1} + \cdots \tag{2}$$

the nth term of the sequence of partial sums is

$$S_n = a + az + az^2 + \cdots + az^{n-1} \tag{3}$$

By multiplying S_n by z and subtracting this result from S_n, we obtain $S_n - zS_n = a - az^n$. Solving for S_n gives

$$S_n = \frac{a(1-z^n)}{1-z} \tag{4}$$

Since $z^n \to 0$ as $n \to \infty$, whenever $|z| < 1$ we conclude from (4) that (2) converges to

$$\boxed{\dfrac{a}{1-z}}$$

when $|z| < 1$; the series diverges when $|z| \geq 1$. The special geometric series

$$\frac{1}{1-z} = 1 + z + z^2 + z^3 + \cdots \tag{5}$$

and

$$\frac{1}{1+z} = 1 - z + z^2 - z^3 + \cdots \tag{6}$$

18.1 Sequences and Series

valid for $|z| < 1$ will be of particular usefulness in the next two sections. In addition, we shall use

$$\frac{1-z^n}{1-z} = 1 + z + z^2 + z^3 + \cdots + z^{n-1} \tag{7}$$

in the alternative form

$$\frac{1}{1-z} = 1 + z + z^2 + z^3 + \cdots + z^{n-1} + \frac{z^n}{1-z} \tag{8}$$

in the proofs of the two principal theorems of this chapter.

EXAMPLE 3 The series

$$\sum_{k=1}^{\infty} \frac{(1+2i)^k}{5^k} = \frac{1+2i}{5} + \frac{(1+2i)^2}{5^2} + \frac{(1+2i)^3}{5^3} + \cdots$$

is a geometric series with $a = (1 + 2i)/5$ and $z = (1 + 2i)/5$. Since $|z| = \sqrt{5}/5 < 1$, the series converges and we can write

$$\sum_{k=1}^{\infty} \frac{(1+2i)^k}{5^k} = \frac{\frac{1+2i}{5}}{1 - \frac{1+2i}{5}} = \frac{i}{2} \qquad \square$$

THEOREM 18.2

If $\sum_{k=1}^{\infty} z_k$ converges, then $\lim_{n \to \infty} z_n = 0$.

An equivalent form of Theorem 18.2 is the familiar nth term test for divergence of an infinite series.

THEOREM 18.3 The nth Term Test for Divergence

If $\lim_{n \to \infty} z_n \neq 0$, then the series $\sum_{k=1}^{\infty} z_k$ diverges.

For example, the series $\sum_{k=1}^{\infty} (k + 5i)/k$ diverges since $a_n = (n + 5i)/n \to 1$ as $n \to \infty$. The geometric series (2) diverges when $|z| \geq 1$ since, in this case, $\lim_{n \to \infty} |z^n|$ does not exist.

DEFINITION 18.1 Absolute Convergence

An infinite series $\sum_{k=1}^{\infty} z_k$ is said be **absolute convergent** if $\sum_{k=1}^{\infty} |z_k|$ converges.

EXAMPLE 4 The series $\sum_{k=1}^{\infty} (i^k/k^2)$ is absolutely convergent since $|i^k/k^2| = 1/k^2$ and the real series $\sum_{k=1}^{\infty} (1/k^2)$ converges. Recall from calculus that a real series of the form $\sum_{k=1}^{\infty} (1/k^p)$ is called a *p*-series and converges for $p > 1$ and diverges for $p \leq 1$. □

As in real calculus:

Absolute convergence implies convergence.

Thus in Example 4, the series

$$\sum_{k=1}^{\infty} \frac{i^k}{k^2} = i - \frac{1}{2^2} - \frac{i}{3^2} + \cdots$$

converges.

The following two tests are the complex versions of the ratio and root tests that are encountered in calculus:

THEOREM 18.4 Ratio Test

Suppose $\sum_{k=1}^{\infty} z_k$ is a series of nonzero complex terms such that

$$\lim_{n \to \infty} \left| \frac{z_{n+1}}{z_n} \right| = L \qquad (9)$$

(*i*) If $L < 1$, then the series converges absolutely.
(*ii*) If $L > 1$ or $L = \infty$, then the series diverges.
(*iii*) If $L = 1$, the test is inconclusive.

THEOREM 18.5 Root Test

Suppose $\sum_{k=1}^{\infty} z_k$ is a series of complex terms such that

$$\lim_{n \to \infty} \sqrt[n]{|z_n|} = L \qquad (10)$$

(*i*) If $L < 1$, then the series converges absolutely.
(*ii*) If $L > 1$ or $L = \infty$, then the series diverges.
(*iii*) If $L = 1$, the test is inconclusive.

We are interested primarily in applying these tests to power series.

Power Series   The notion of a power series is important in the study of analytic functions. An infinite series of the form

$$\sum_{k=0}^{\infty} a_k(z - z_0)^k = a_0 + a_1(z - z_0) + a_2(z - z_0)^2 + \cdots \qquad (11)$$

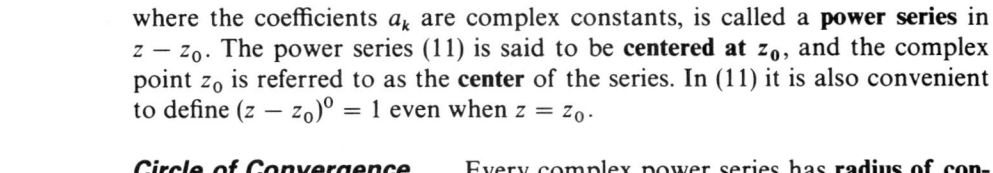

where the coefficients a_k are complex constants, is called a **power series** in $z - z_0$. The power series (11) is said to be **centered at** z_0, and the complex point z_0 is referred to as the **center** of the series. In (11) it is also convenient to define $(z - z_0)^0 = 1$ even when $z = z_0$.

Circle of Convergence

Every complex power series has **radius of convergence** R. Analogous to the concept of an *interval* of convergence in real calculus, when $0 < R < \infty$, a complex power series (11) has a **circle of convergence** defined by $|z - z_0| = R$. The power series converges absolutely for all z satisfying $|z - z_0| < R$ and diverges for $|z - z_0| > R$. See Figure 18.3. The radius R of convergence can be:

(*i*) zero (in which case (11) converges at only $z = z_0$),

(*ii*) a finite number (in which case (11) converges at all interior points of the circle $|z - z_0| = R$), or

(*iii*) ∞ (in which case (11) converges for all z).

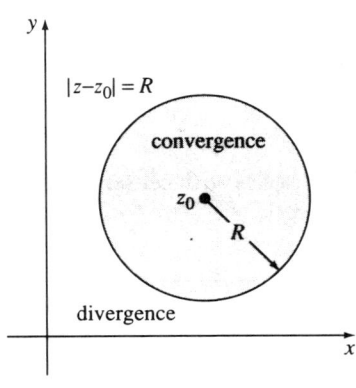

FIGURE 18.3

A power series may converge at some, all, or none of the points *on* the actual circle of convergence.

EXAMPLE 5 Consider the power series $\sum_{k=1}^{\infty} (z^{k+1}/k)$. By the ratio test (9),

$$\lim_{n \to \infty} \left| \frac{\frac{z^{n+2}}{n+1}}{\frac{z^{n+1}}{n}} \right| = \lim_{n \to \infty} \frac{n}{n+1} |z| = |z|$$

Thus the series converges absolutely for $|z| < 1$. The circle of convergence is $|z| = 1$ and the radius of convergence is $R = 1$. Note that on the circle of convergence, the series does not converge absolutely, since the series of absolute values is the well-known divergent harmonic series $\sum_{k=1}^{\infty} (1/k)$. Bear in mind this does not say, however, that the series diverges on the circle of convergence. In fact, at $z = -1$, $\sum_{k=1}^{\infty} ((-1)^{k+1}/k)$ is the convergent alternating harmonic series. Indeed, it can be shown that the series converges at all points on the circle $|z| = 1$ except at $z = 1$. □

It should be clear from Theorem 18.4 and Example 5 that for a power series $\sum_{k=0}^{\infty} a_k (z - z_0)^k$, the limit (9) depends on only the coefficients a_k. Thus, if

(*i*) $\lim_{n \to \infty} \left| \frac{a_{n+1}}{a_n} \right| = L \neq 0$, the radius of convergence is $R = 1/L$

(*ii*) $\lim_{n \to \infty} \left| \frac{a_{n+1}}{a_n} \right| = 0$, the radius of convergence is ∞

(*iii*) $\lim_{n \to \infty} \left| \frac{a_{n+1}}{a_n} \right| = \infty$, the radius of convergence is $R = 0$

Similar remarks can be made for the root test (10) by utilizing $\lim_{n \to \infty} \sqrt[n]{|a_n|}$.

EXAMPLE 6 Consider the power series $\sum_{k=1}^{\infty} \frac{(-1)^{k+1}(z-1-i)^k}{k!}$. Identifying $a_n = (-1)^{n+1}/n!$, we have

$$\lim_{n \to \infty} \left| \frac{\frac{(-1)^{n+2}}{(n+1)!}}{\frac{(-1)^{n+1}}{n!}} \right| = \lim_{n \to \infty} \frac{1}{n+1} = 0$$

Thus the radius of convergence is ∞; the power series with center $1 + i$ converges absolutely for all z. □

EXAMPLE 7 Consider the power series $\sum_{k=1}^{\infty} \left(\frac{6k+1}{2k+5}\right)^k (z-2i)^k$. With $a_n = \left(\frac{6n+1}{2n+5}\right)^n$, the root test in the form

$$\lim_{n \to \infty} \sqrt[n]{|a_n|} = \lim_{n \to \infty} \frac{6n+1}{2n+5} = 3$$

shows that the radius of convergence of the series is $R = \frac{1}{3}$. The circle of convergence is $|z - 2i| = \frac{1}{3}$; the series converges absolutely for $|z - 2i| < \frac{1}{3}$. □

EXERCISES 18.1   Answers to odd-numbered problems begin on page A-88.

In Problems 1–4 write out the first five terms of the given sequence.

1. $\{5i^n\}$
2. $\{2 + (-i)^n\}$
3. $\{1 + e^{n\pi i}\}$
4. $\{(1 + i)^n\}$ [*Hint:* Write in polar form.]

In Problems 5–10 determine whether the given sequence converges or diverges.

5. $\left\{\frac{3ni+2}{n+ni}\right\}$
6. $\left\{\frac{ni+2^n}{3ni+5^n}\right\}$
7. $\left\{\frac{(ni+2)^2}{n^2 i}\right\}$
8. $\left\{\frac{n(1+i^n)}{n+1}\right\}$
9. $\left\{\frac{n+i^n}{\sqrt{n}}\right\}$
10. $\{e^{1/n} + 2(\tan^{-1}n)i\}$

In Problems 11 and 12 show that the given sequence $\{z_n\}$ converges to a complex number L by computing $\lim_{n \to \infty} \text{Re}(z_n)$ and $\lim_{n \to \infty} \text{Im}(z_n)$.

11. $\left\{\frac{4n+3ni}{2n+i}\right\}$
12. $\left\{\left(\frac{1+i}{4}\right)^n\right\}$

In Problems 13 and 14 use the sequence of partial sums to show that the given series is convergent.

13. $\sum_{k=1}^{\infty} \left[\frac{1}{k+2i} - \frac{1}{k+1+2i}\right]$
14. $\sum_{k=2}^{\infty} \frac{i}{k(k+1)}$

In Problems 15–20 determine whether the given geometric series is convergent or divergent. If convergent, find its sum.

15. $\sum_{k=0}^{\infty} (1-i)^k$
16. $\sum_{k=1}^{\infty} 4i\left(\frac{1}{3}\right)^{k-1}$
17. $\sum_{k=1}^{\infty} \left(\frac{i}{2}\right)^k$
18. $\sum_{k=0}^{\infty} \frac{1}{2} i^k$
19. $\sum_{k=0}^{\infty} 3\left(\frac{2}{1+2i}\right)^k$
20. $\sum_{k=2}^{\infty} \frac{i^k}{(1+i)^{k-1}}$

In Problems 21–28 find the circle and radius of convergence of the given power series.

21. $\sum_{k=0}^{\infty} \frac{1}{(1-2i)^{k+1}} (z-2i)^k$

22. $\sum_{k=1}^{\infty} \frac{1}{k} \left(\frac{i}{1+i} \right)^k z^k$

23. $\sum_{k=1}^{\infty} \frac{(-1)^k}{k 2^k} (z - 1 - i)^k$

24. $\sum_{k=1}^{\infty} \frac{1}{k^2 (3+4i)^k} (z+3i)^k$

25. $\sum_{k=0}^{\infty} (1+3i)^k (z-i)^k$

26. $\sum_{k=1}^{\infty} \frac{z^k}{k^k}$

27. $\sum_{k=0}^{\infty} \frac{(z-4-3i)^k}{5^{2k}}$

28. $\sum_{k=0}^{\infty} (-1)^k \left(\frac{1+2i}{2} \right)^k (z+2i)^k$

29. Show that the power series $\sum_{k=1}^{\infty} \frac{(z-i)^k}{k 2^k}$ is not absolutely convergent on its circle of convergence. Determine at least one point on the circle of convergence at which the power series converges.

30. (a) Show that the power series $\sum_{k=1}^{\infty} \frac{z^k}{k^2}$ converges at every point on its circle of convergence.

    (b) Show that the power series $\sum_{k=1}^{\infty} k z^k$ diverges at every point on its circle of convergence.

18.2 TAYLOR SERIES

We saw in the last section that every power series has a radius of convergence R. The next three theorems will give some important facts about the nature of a power series within its circle of convergence $|z - z_0| = R$, $R \neq 0$.

THEOREM 18.6 Continuity

A power series $\sum_{k=0}^{\infty} a_k (z - z_0)^k$ represents a continuous function f within its circle of convergence $|z - z_0| = R$, $R \neq 0$.

THEOREM 18.7 Term-by-Term Integration

A power series $\sum_{k=0}^{\infty} a_k (z - z_0)^k$ can be integrated term by term within its circle of convergence $|z - z_0| = R$, $R \neq 0$, for every contour C lying entirely with the circle of convergence.

THEOREM 18.8 Term-by-Term Differentiation

A power series $\sum_{k=0}^{\infty} a_k (z - z_0)^k$ can be differentiated term by term within its circle of convergence $|z - z_0| = R$, $R \neq 0$.

Taylor Series Suppose a power series represents a function f for $|z - z_0| < R$, $R \neq 0$; that is,

$$f(z) = \sum_{k=0}^{\infty} a_k(z - z_0)^k = a_0 + a_1(z - z_0) + a_2(z - z_0)^2 + a_3(z - z_0)^3 + \cdots \quad (1)$$

It follows from Theorem 18.8 that the derivatives of f are

$$f'(z) = \sum_{k=1}^{\infty} k a_k(z - z_0)^{k-1} = a_1 + 2a_2(z - z_0) + 3a_3(z - z_0)^2 + \cdots \quad (2)$$

$$f''(z) = \sum_{k=2}^{\infty} k(k-1) a_k(z - z_0)^{k-2} = 2 \cdot 1 a_2 + 3 \cdot 2 a_3(z - z_0) + \cdots \quad (3)$$

$$f'''(z) = \sum_{k=3}^{\infty} k(k-1)(k-2) a_k(z - z_0)^{k-3} = 3 \cdot 2 \cdot 1 a_3 + \cdots \quad (4)$$

and so on. Each of the differentiated series has the same radius of convergence as the original series. Moreover, since the original power series represents a differentiable function f within its circle of convergence, we conclude that when $R \neq 0$,

a power series represents an analytic function within its circle of convergence.

There is a relationship between the coefficients a_k and the derivatives of f. Evaluating (1), (2), (3), and (4) at $z = z_0$ gives

$$f(z_0) = a_0, \quad f'(z_0) = 1! a_1, \quad f''(z_0) = 2! a_2, \quad \text{and} \quad f'''(z_0) = 3! a_3$$

respectively. In general, $f^{(n)}(z_0) = n! a_n$ or

$$a_n = \frac{f^{(n)}(z_0)}{n!}, \quad n \geq 0 \quad (5)$$

When $n = 0$ we interpret the zeroth derivative as $f(z_0)$ and $0! = 1$. Substituting (5) into (1) yields

$$\boxed{f(z) = \sum_{k=0}^{\infty} \frac{f^{(k)}(z_0)}{k!}(z - z_0)^k} \quad (6)$$

This series is called the **Taylor series** for f centered at z_0. A Taylor series with center $z_0 = 0$,

$$\boxed{f(z) = \sum_{k=0}^{\infty} \frac{f^{(k)}(0)}{k!} z^k} \quad (7)$$

is referred to as a **Maclaurin series**.

18.2 Taylor Series

We have just seen that a power series with a nonzero radius of convergence represents an analytic function. On the other hand, if we are given a function f that is analytic in some domain D, can we represent it by a power series of the form (6) and (7)? Since a power series converges in a circular domain, and a domain D is generally not circular, the question becomes: Can we expand f in *one or more* power series that are valid in circular domains that are all contained in D? The question will be answered in the affirmative by the next theorem.

THEOREM 18.9 Taylor's Theorem

Let f be analytic within a domain D and let z_0 be a point in D. Then f has the series representation

$$f(z) = \sum_{k=0}^{\infty} \frac{f^{(k)}(z_0)}{k!}(z - z_0)^k \qquad (8)$$

valid for the largest circle C with center at z_0 and radius R that lies entirely within D.

Proof Let z be a fixed point within the circle C and let s denote the variable of integration. The circle C is then described by $|s - z_0| = R$. See Figure 18.4. To begin, we use the Cauchy integral formula to obtain the value of f at z:

$$\begin{aligned} f(z) &= \frac{1}{2\pi i} \oint_C \frac{f(s)}{s - z}\, ds \\ &= \frac{1}{2\pi i} \oint_C \frac{f(s)}{(s - z_0) - (z - z_0)}\, ds \\ &= \frac{1}{2\pi i} \oint_C \frac{f(s)}{s - z_0} \left\{ \frac{1}{1 - \dfrac{z - z_0}{s - z_0}} \right\} ds \end{aligned} \qquad (9)$$

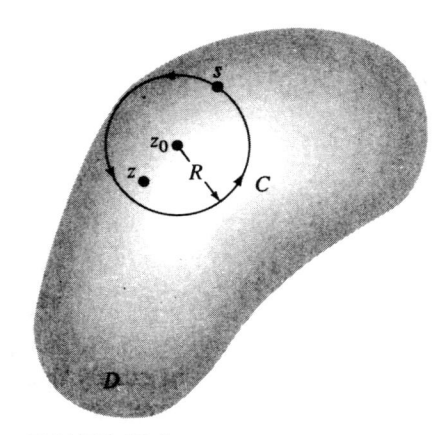

FIGURE 18.4

By replacing z by $(z - z_0)/(s - z_0)$ in (8) of Section 18.1, we have

$$\frac{1}{1 - \dfrac{z - z_0}{s - z_0}} = 1 + \frac{z - z_0}{s - z_0} + \left(\frac{z - z_0}{s - z_0}\right)^2 + \cdots + \left(\frac{z - z_0}{s - z_0}\right)^{n-1} + \frac{(z - z_0)^n}{(s - z)(s - z_0)^{n-1}}$$

and so (9) becomes

$$\begin{aligned} f(z) =\ & \frac{1}{2\pi i} \oint_C \frac{f(s)}{s - z_0}\, ds + \frac{z - z_0}{2\pi i} \oint_C \frac{f(s)}{(s - z_0)^2}\, ds + \frac{(z - z_0)^2}{2\pi i} \oint_C \frac{f(s)}{(s - z_0)^3}\, ds + \cdots \\ & + \frac{(z - z_0)^{n-1}}{2\pi i} \oint_C \frac{f(s)}{(s - z_0)^n}\, ds + \frac{(z - z_0)^n}{2\pi i} \oint_C \frac{f(s)}{(s - z)(s - z_0)^n}\, ds \end{aligned} \qquad (10)$$

Utilizing Cauchy's integral formula for derivatives, we can write (10) as

$$f(z) = f(z_0) + \frac{f'(z_0)}{1!}(z - z_0) + \frac{f''(z_0)}{2!}(z - z_0)^2 + \cdots + \frac{f^{(n-1)}(z_0)}{(n-1)!}(z - z_0)^{n-1} + R_n(z) \quad (11)$$

where
$$R_n(z) = \frac{(z - z_0)^n}{2\pi i} \oint_C \frac{f(s)}{(s - z)(s - z_0)^n} ds$$

Equation (11) is called Taylor's formula with remainder R_n. We now wish to show that $R_n(z) \to 0$ as $n \to \infty$. Since f is analytic in D, $|f(z)|$ has a maximum value M on the contour C. In addition, since z is inside C, we have $|z - z_0| < R$ and consequently,

$$|s - z| = |s - z_0 - (z - z_0)| \geq |s - z_0| - |z - z_0| = R - d$$

where $d = |z - z_0|$ is the distance from z and z_0. The ML-inequality then gives

$$|R_n(z)| = \left| \frac{(z - z_0)^n}{2\pi i} \oint_C \frac{f(s)}{(s - z)(s - z_0)^n} ds \right|$$

$$\leq \frac{d^n}{2\pi} \cdot \frac{M}{(R - d)R^n} \cdot 2\pi R = \frac{MR}{R - d} \left(\frac{d}{R} \right)^n$$

Because $d < R$, $(d/R)^n \to 0$ as $n \to \infty$, we conclude that $|R_n(z)| \to 0$ as $n \to \infty$. It follows that the infinite series

$$f(z_0) + \frac{f'(z_0)}{1!}(z - z_0) + \frac{f''(z_0)}{2!}(z - z_0)^2 + \cdots$$

converges to $f(z)$. In other words,

$$f(z) = \sum_{k=0}^{\infty} \frac{f^{(k)}(z_0)}{k!}(z - z_0)^k$$

is valid for any point z interior to C. ∎

We can find the radius of convergence of a Taylor series in exactly the same manner illustrated in Examples 5–7 of Section 18.1. However, we can simplify matters even further by noting that the radius of convergence is the *distance* from the center z_0 of the series to the nearest *isolated singularity* of f. We shall elaborate more on this concept in the next section, but an isolated singularity is a point at which f fails to be analytic but is, nonetheless, analytic at all other points throughout *some* neighborhood of the point. For example, $z = 5i$ is an isolated singularity of $f(z) = 1/(z - 5i)$. If the function f is entire, then the radius of convergence of a Taylor series centered at any point z_0 is necessarily infinite. Using (8) and the last fact, we can say that the Maclaurin series representations

$$e^z = 1 + \frac{z}{1!} + \frac{z^2}{2!} + \cdots = \sum_{k=0}^{\infty} \frac{z^k}{k!} \quad (12)$$

18.2 Taylor Series

$$\sin z = z - \frac{z^3}{3!} + \frac{z^5}{5!} - \cdots = \sum_{k=0}^{\infty} (-1)^k \frac{z^{2k+1}}{(2k+1)!} \qquad (13)$$

$$\cos z = 1 - \frac{z^2}{2!} + \frac{z^4}{4!} - \cdots = \sum_{k=0}^{\infty} (-1)^k \frac{z^{2k}}{(2k)!} \qquad (14)$$

are valid for all z.

If two power series with center z_0:

$$\sum_{k=0}^{\infty} a_k(z-z_0)^k \quad \text{and} \quad \sum_{k=0}^{\infty} b_k(z-z_0)^k$$

represent the same function and have the same nonzero radius of convergence, then $a_k = b_k$, $k = 0, 1, 2, \ldots$. Stated in another way, the power series expansion of a function with center z_0 is *unique*. On a practical level this means that a power series expansion of an analytic function f centered at z_0, irrespective of the method used to obtain it, is the Taylor series expansion of the function. For example, we can obtain (14) by simply differentiating (13) term by term. The Maclaurin series for e^{z^2} can be obtained by replacing the symbol z in (12) by z^2.

EXAMPLE 1 Find the Maclaurin expansion of $f(z) = 1/(1-z)^2$.

Solution We could, of course, begin by computing the coefficients using (8). However, we know from (5) of Section 18.1 that for $|z| < 1$,

$$\frac{1}{1-z} = 1 + z + z^2 + z^3 + \cdots \qquad (15)$$

Differentiating both sides of the last result with respect to z then yields

$$\frac{1}{(1-z)^2} = 1 + 2z + 3z^2 + \cdots = \sum_{k=1}^{\infty} kz^{k-1}$$

Since we are using Theorem 18.8, the radius of convergence of this last series is the same as the original series, $R = 1$. □

EXAMPLE 2 Expand $f(z) = 1/(1-z)$ in a Taylor series with center $z_0 = 2i$.

Solution We shall solve this problem in two ways. We begin by using (8). From the first several derivatives,

$$f'(z) = \frac{1}{(1-z)^2}, \quad f''(z) = \frac{2 \cdot 1}{(1-z)^3}, \quad f'''(z) = \frac{3 \cdot 2}{(1-z)^4}$$

we conclude that $f^{(n)}(z) = n!/(1-z)^{n+1}$ and so $f^{(n)}(2i) = n!/(1-2i)^{n+1}$. Thus from (8) we obtain the Taylor series

$$\frac{1}{1-z} = \sum_{k=0}^{\infty} \frac{1}{(1-2i)^{k+1}} (z-2i)^k \qquad (16)$$

Since the distance from the center $z_0 = 2i$ to the nearest singularity $z = 1$ is $\sqrt{5}$, we conclude that the circle of convergence for the power series in (16) is $|z - 2i| = \sqrt{5}$. This can be verified by the ratio test of the preceding section.

Alternative Solution In this solution we again use the geometric series (15). By adding and subtracting $2i$ in the denominator of $1/(1 - z)$, we can write

$$\begin{aligned}\frac{1}{1-z} &= \frac{1}{1 - z + 2i - 2i}\\ &= \frac{1}{1 - 2i - (z - 2i)}\\ &= \frac{1}{1-2i}\,\frac{1}{1 - \dfrac{z-2i}{1-2i}}\end{aligned}$$

We now write $\dfrac{1}{1 - \dfrac{z-2i}{1-2i}}$ as a power series by using (15) with the symbol z replaced by $(z - 2i)/(1 - 2i)$:

$$\begin{aligned}\frac{1}{1-z} &= \frac{1}{1-2i}\left[1 + \frac{z-2i}{1-2i} + \left(\frac{z-2i}{1-2i}\right)^2 + \left(\frac{z-2i}{1-2i}\right)^3 + \cdots\right]\\ &= \frac{1}{1-2i} + \frac{1}{(1-2i)^2}(z-2i) + \frac{1}{(1-2i)^3}(z-2i)^2 + \frac{1}{(1-2i)^4}(z-2i)^3 + \cdots\end{aligned}$$

The reader should verify that this last series is exactly the same as that in (16).

□

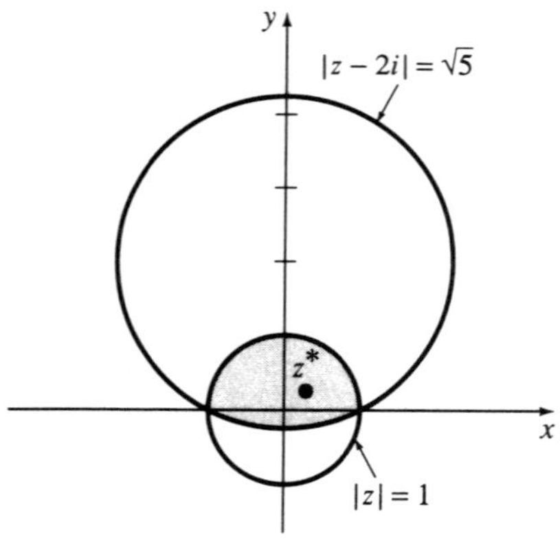

FIGURE 18.5

In (15) and (16) we represented the same function $1/(1 - z)$ by two different power series. The first series

$$\frac{1}{1-z} = 1 + z + z^2 + z^3 + \cdots$$

has center zero and radius of convergence one. The second series

$$\frac{1}{1-z} = \frac{1}{1-2i} + \frac{1}{(1-2i)^2}(z-2i) + \frac{1}{(1-2i)^3}(z-2i)^2 + \frac{1}{(1-2i)^4}(z-2i)^3 + \cdots$$

has center $2i$ and radius of convergence $\sqrt{5}$. The two different circles of convergence are illustrated in Figure 18.5. The interior of the intersection of the two circles (shaded) is the region where *both* series converge; in other words, at a specified point z^* in this region both series converge to same value $f(z^*) = 1/(1 - z^*)$. Outside the shaded region at least one of the two series must diverge.

EXERCISES 18.2 Answers to odd-numbered problems begin on page A-88.

In Problem 1–12 expand the given function in a Maclaurin series. Give the radius of convergence of each series.

1. $f(z) = \dfrac{z}{1+z}$

2. $f(z) = \dfrac{1}{4-2z}$

3. $f(z) = \dfrac{1}{(1+2z)^2}$

4. $f(z) = \dfrac{z}{(1-z)^3}$

5. $f(z) = e^{-2z}$

6. $f(z) = ze^{-z^2}$

7. $f(z) = \sinh z$

8. $f(z) = \cosh z$

9. $f(z) = \cos \dfrac{z}{2}$

10. $f(z) = \sin 3z$

11. $f(z) = \sin z^2$

12. $f(z) = \cos^2 z$ [*Hint:* Use a trigonometric identity.]

In Problems 13–22 expand the given function in a Taylor series centered at the indicated point. Give the radius of convergence of each series.

13. $f(z) = 1/z,\ z_0 = 1$

14. $f(z) = 1/z,\ z_0 = 1+i$

15. $f(z) = \dfrac{1}{3-z},\ z_0 = 2i$

16. $f(z) = \dfrac{1}{1+z},\ z_0 = -i$

17. $f(z) = \dfrac{z-1}{3-z},\ z_0 = 1$

18. $f(z) = \dfrac{1+z}{1-z},\ z_0 = i$

19. $f(z) = \cos z,\ z_0 = \pi/4$

20. $f(z) = \sin z,\ z_0 = \pi/2$

21. $f(z) = e^z,\ z = 3i$

22. $f(z) = (z-1)e^{-2z},\ z = 1$

In Problems 23 and 24 use (7) to find the first three nonzero terms of the Maclaurin series of the given function.

23. $f(z) = \tan z$

24. $f(z) = e^{1/(1+z)}$

In Problems 25 and 26 use partial fractions as an aid in obtaining the Maclaurin series for the given function. Give the radius of convergence of the series.

25. $f(z) = \dfrac{i}{(z-i)(z-2i)}$

26. $f(z) = \dfrac{z-7}{z^2 - 2z - 3}$

In Problems 27 and 28, without actually expanding, determine the radius of convergence of the Taylor series of the given function centered at the indicated point.

27. $f(z) = \dfrac{4+5z}{1+z^2},\ z_0 = 2+5i$

28. $f(z) = \cot z,\ z_0 = \pi i$

In Problems 29 and 30 expand the given function in the Taylor series centered at the indicated points. Give the radius of convergence of each series. Sketch the region within which both series converge.

29. $f(z) = \dfrac{1}{2+z},\ z_0 = -1,\ z_0 = i$

30. $f(z) = \dfrac{1}{z},\ z_0 = 1+i,\ z_0 = 3$

31. **(a)** Suppose the principal branch of the logarithm $f(z) = \text{Ln } z = \log_e|z| + i \text{ Arg } z$ is expanded in a Taylor series with center $z_0 = -1 + i$. Explain why $R = 1$ is the radius of the largest circle centered at $z_0 = -1 + i$ within which f is analytic.

(b) Show that within the circle $|z-(-1+i)| = 1$ the Taylor series for f is

$$\text{Ln } z = \dfrac{1}{2}\log_e 2 + \dfrac{3\pi}{4}i - \sum_{k=1}^{\infty} \dfrac{1}{k}\left(\dfrac{1+i}{2}\right)^k (z+1-i)^k$$

(c) Show that the radius of convergence for the power series in part **(b)** is $R = \sqrt{2}$. Explain why this does not contradict the result in part **(a)**.

32. **(a)** Consider the function $f(z) = \text{Ln}(1+z)$. What is the radius of the largest circle centered at the origin within which f is analytic?

(b) Expand f in a Maclaurin series. What is the radius of convergence of this series?

(c) Use the result in part **(b)** to find a Maclaurin series for $\text{Ln}(1-z)$.

(d) Find a Maclaurin series for $\text{Ln}\left(\dfrac{1+z}{1-z}\right)$.

In Problems 33 and 34 approximate the value of the given expression using the indicated number of terms of a Maclaurin series.

33. $e^{(1+i)/10}$, three terms

34. $\sin\left(\dfrac{1+i}{10}\right)$, two terms

35. In Section 14.1 we defined the error function as $\text{erf}(z) = (2/\sqrt{\pi}) \int_0^z e^{-t^2}\, dt$. Find a Maclaurin series for $\text{erf}(z)$.

36. Use the Maclaurin series for e^{iz} to prove Euler's formula for complex z:

$$e^{iz} = \cos z + i \sin z$$

18.3 LAURENT SERIES

Singularities If a complex function f fails to be analytic at $z = z_0$, then this point is said to be a **singularity** or a **singular point** of the function. For example, the complex numbers $z = 2i$ and $z = -2i$ are singularities of the function $f(z) = z/(z^2 + 4)$ since f is not defined at either of these points. Recall from Section 16.6 that the principal branch of the logarithm, $f(z) = \text{Ln}\, z$, is analytic at all points except points on the nonpositive x-axis. In other words, $z = 0$ as well as all negative real numbers are singular points of $f(z) = \text{Ln}\, z$. Recall from Section 18.2 that a point $z = z_0$ is said to be an **isolated singularity** of a function f if f is analytic throughout some neighborhood of z_0 but not at z_0 itself. For example, $z = 2i$ (as well as $z = -2i$) is an isolated singularity of of $f(z) = z/(z^2 + 4)$ since f is analytic at every point in the neighborhood defined by $|z - 2i| < 1$ except at $z = 2i$. In other words, if f is analytic in the deleted neighborhood $0 < |z - 2i| < 1$. The point $z = 0$ is not an isolated singularity of $f(z) = \text{Ln}\, z$ since every neighborhood of this point must contain points on the negative x-axis. We say that a singular point $z = z_0$ of a function f is **nonisolated** if every neighborhood of z_0 contains at least one singularity of f other than z_0.

If $z = z_0$ is a singularity of a function f, then f cannot be expanded in a power series with z_0 as its center. (Why?) However, about an isolated singularity $z = z_0$ it is possible to represent f by a new kind of series involving negative and nonnegative integer powers of $z - z_0$; that is,

$$f(z) = \cdots + \frac{a_{-2}}{(z - z_0)^2} + \frac{a_{-1}}{z - z_0} + a_0 + a_1(z - z_0) + a_2(z - z_0)^2 + \cdots$$

Using summation notation, we can write the last expression as the sum of two series

$$f(z) = \sum_{k=1}^{\infty} \frac{a_{-k}}{(z - z_0)^k} + \sum_{k=0}^{\infty} a_k(z - z_0)^k \quad (1)$$

The first series in (1) will converge for $|1/(z - z_0)| < r^*$ or equivalently for $|z - z_0| > 1/r^* = r$. The second series in (1) converges for $|z - z_0| < R$. Hence, the sum of the two series converges when z satisfies both these conditions—that is, when z is a point in an annular domain defined by $r < |z - z_0| < R$, $r < R$.

By summing over negative and nonnegative integers, we can write (1) compactly as

$$f(z) = \sum_{k=-\infty}^{\infty} a_k(z - z_0)^k$$

18.3 Laurent Series

EXAMPLE 1 The function $f(z) = (\sin z)/z^3$ is not analytic at $z = 0$ and hence cannot be expanded in a Maclaurin series. However, $\sin z$ is an entire function and from (13) of Section 18.2 we know that its Maclaurin series,

$$\sin z = z - \frac{z^3}{3!} + \frac{z^5}{5!} - \frac{z^7}{7!} + \cdots$$

converges for all z. Dividing this power series by z^3 gives the following series with negative and nonnegative integer powers of z:

$$f(z) = \frac{\sin z}{z^3} = \frac{1}{z^2} - \frac{1}{3!} + \frac{z^2}{5!} - \frac{z^4}{7!} + \cdots \quad (2)$$

This series converges for all z except $z = 0$, that is, for $0 < |z|$. □

A series representation of a function f that has the form given in (1), and (2) is such an example, is called a **Laurent series** or a **Laurent expansion** of f.

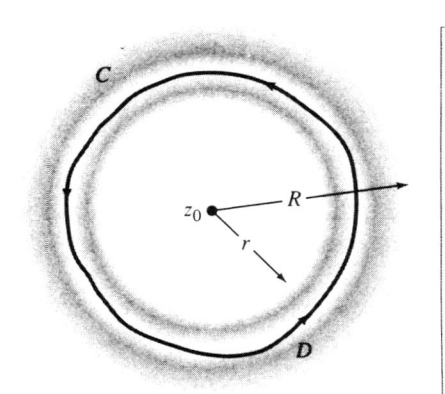

FIGURE 18.6

THEOREM 18.10 Laurent's Theorem

Let f be analytic within the annular domain D defined by $r < |z - z_0| < R$. Then f has the series representation

$$f(z) = \sum_{k=-\infty}^{\infty} a_k(z - z_0)^k \quad (3)$$

valid for $r < |z - z_0| < R$. The coefficients a_k are given by

$$a_k = \frac{1}{2\pi i} \oint_C \frac{f(s)}{(s - z_0)^{k+1}} ds, \quad k = 0, \pm 1, \pm 2, \ldots \quad (4)$$

where C is a simple closed curve that lies entirely within D and has z_0 in its interior (see Figure 18.6).

Proof Let C_1 and C_2 be concentric circles with center z_0 and radii r_1 and R_2, where $r < r_1 < R_2 < R$. Let z be a fixed point in D that also satisfies $r_1 < |z - z_0| < R_2$. See Figure 18.7. By introducing a cross cut between C_2 and C_1, we find from Cauchy's integral formula that

$$f(z) = \frac{1}{2\pi i} \oint_{C_2} \frac{f(s)}{s - z} ds - \frac{1}{2\pi i} \oint_{C_1} \frac{f(s)}{s - z} ds \quad (5)$$

Proceeding as in the proof of Theorem 18.9, we can write

$$\frac{1}{2\pi i} \oint_{C_2} \frac{f(s)}{s - z} ds = \sum_{k=0}^{\infty} a_k(z - z_0)^k \quad (6)$$

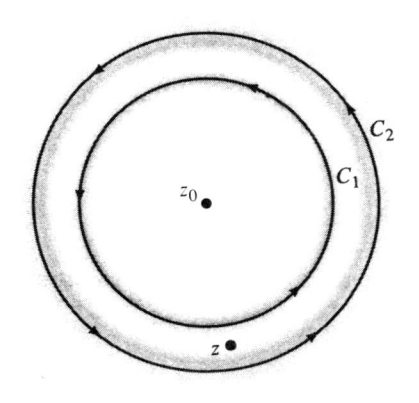

FIGURE 18.7

where
$$a_k = \frac{1}{2\pi i} \oint_{C_2} \frac{f(s)}{(s-z_0)^{k+1}} ds, \quad k = 0, 1, 2, \ldots \qquad (7)$$

Now using (5) and (8) of Section 18.1, we have

$$-\frac{1}{2\pi i} \oint_{C_1} \frac{f(s)}{s-z} ds = \frac{1}{2\pi i} \oint_{C_1} \frac{f(s)}{(z-z_0) - (s-z_0)} ds$$

$$= \frac{1}{2\pi i} \oint_{C_1} \frac{f(s)}{z-z_0} \left\{ \frac{1}{1 - \frac{s-z_0}{z-z_0}} \right\} ds$$

$$= \frac{1}{2\pi i} \oint_{C_1} \frac{f(s)}{z-z_0} \left\{ 1 + \frac{s-z_0}{z-z_0} + \left(\frac{s-z_0}{z-z_0}\right)^2 + \cdots \right.$$

$$\left. + \left(\frac{s-z_0}{z-z_0}\right)^{n-1} + \frac{(s-z_0)^n}{(z-s)(z-z_0)^{n-1}} \right\} ds \qquad (8)$$

$$= \sum_{k=1}^{n} \frac{a_{-k}}{(z-z_0)^k} + R_n(z)$$

where
$$a_{-k} = \frac{1}{2\pi i} \oint_{C_1} \frac{f(s)}{(s-z_0)^{-k+1}} ds, \quad k = 1, 2, 3, \ldots \qquad (9)$$

and
$$R_n(z) = \frac{1}{2\pi i (z-z_0)^n} \oint_{C_1} \frac{f(s)(s-z_0)^n}{z-s} ds$$

Now let d denote the distance from z to z_0, that is, $|z - z_0| = d$, and let M denote the maximum value of $|f(z)|$ on the contour C_1. Since $|s - z_0| = r_1$,

$$|z - s| = |z - z_0 - (s - z_0)| \geq |z - z_0| - |s - z_0| = d - r_1$$

The ML-inequality then gives

$$|R_n(z)| = \left| \frac{1}{2\pi i (z-z_0)^n} \oint_{C_1} \frac{f(s)(s-z_0)^n}{z-s} ds \right|$$

$$\leq \frac{1}{2\pi d^n} \cdot \frac{M r_1^n}{d - r_1} \cdot 2\pi r_1 = \frac{M r_1}{d - r_1} \left(\frac{r_1}{d}\right)^n$$

Because $r_1 < d$, $(r_1/d)^n \to 0$ as $n \to \infty$ and so $|R_n(z)| \to 0$ as $n \to \infty$. Thus we have shown that

$$-\frac{1}{2\pi i} \oint_{C_1} \frac{f(s)}{s-z} ds = \sum_{k=1}^{\infty} \frac{a_{-k}}{(z-z_0)^k} \qquad (10)$$

where the coefficients a_{-k} are given in (9). Combining (6) and (10), we see that (5) yields

$$f(z) = \sum_{k=0}^{\infty} a_k (z-z_0)^k + \sum_{k=1}^{\infty} \frac{a_{-k}}{(z-z_0)^k} \qquad (11)$$

Finally, by summing over nonnegative and negative integers, we can write (11) as

$$f(z) = \sum_{k=-\infty}^{\infty} a_k(z - z_0)^k$$

However, (7) and (9) can be written as a single integral:

$$a_k = \frac{1}{2\pi i} \oint_C \frac{f(z)}{(z - z_0)^{k+1}}, \quad k = 0, \pm 1, \pm 2, \ldots$$

where, in view of (3) of Section 17.2, we have replaced the contours C_1 and C_2 by any simple closed contour C in D with z_0 in its interior. ∎

In the case when $a_{-k} = 0$ for $k = 1, 2, 3, \ldots$, the Laurent series (3) is a Taylor series. Because of this, a Laurent expansion is a generalization of a Taylor series.

The annular domain in Theorem 18.10 defined by $r < |z - z_0| < R$ need not have the "ring" shape illustrated in Figure 18.7. Some other possible annular domains are: (i) $r = 0$, R finite; (ii) $r \neq 0$, $R \to \infty$; and (iii) $r = 0$, $R \to \infty$. In the first case, the series converges in the annular domain defined by $0 < |z - z_0| < R$. This is the interior of the circle $|z - z_0| = R$ except the point z_0. In the second case, the annular domain is defined by $r < |z - z_0|$; in other words, the domain consists of all points exterior to the circle $|z - z_0| = r$. In the third case, the domain is defined by $0 < |z - z_0|$. This represents the entire complex plane except the point z_0. The series we obtained in (2) is valid on this last type of domain.

In actual practice the formula in (4) for the coefficients of a Laurent series is seldom used. As a consequence, finding the Laurent series of a function in a specified annular domain is generally not an easy task. We often use the geometric series (5) and (6) of Section 18.1 or, as we did in Example 1, other known series. But regardless of how a Laurent expansion of a function f is obtained in a specified annular domain, it is *the* Laurent series; that is, the series we obtain is unique.

EXAMPLE 2 Expand $f(z) = 1/z(z - 1)$ in a Laurent series valid for (a) $0 < |z| < 1$, (b) $1 < |z|$, (c) $0 < |z - 1| < 1$, and (d) $1 < |z - 1|$.

Solution The four specified annular domains are shown in Figure 18.8. In parts (a) and (b) we want to represent f in a series involving only negative and nonnegative integer powers of z, whereas in parts (c) and (d) we want to represent f in a series involving negative and nonnegative integer powers of $z - 1$.

(a) By writing

$$f(z) = -\frac{1}{z}\frac{1}{1 - z}$$

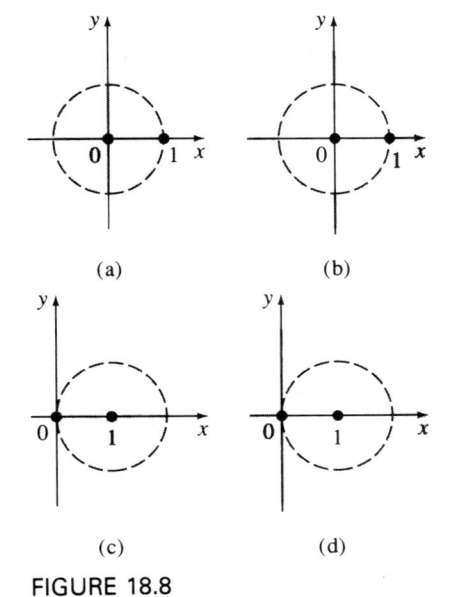

FIGURE 18.8

we can use (5) of Section 18.1:

$$f(z) = -\frac{1}{z}[1 + z + z^2 + z^3 + \cdots]$$

The series in the brackets converges for $|z| < 1$, but after this expression is multiplied by $1/z$, the resulting series

$$f(z) = -\frac{1}{z} - 1 - z - z^2 - \cdots$$

converges for $0 < |z| < 1$.

(b) To obtain a series that converges for $1 < |z|$ we start by constructing a series that converges for $|1/z| < 1$. To this end we write the given function f as

$$f(z) = \frac{1}{z^2} \frac{1}{1 - \frac{1}{z}}$$

and again use (5) of Section 18.1 with z replaced by $1/z$:

$$f(z) = \frac{1}{z^2}\left[1 + \frac{1}{z} + \frac{1}{z^2} + \frac{1}{z^3} + \cdots\right]$$

The series in the brackets converges for $|1/z| < 1$ or equivalently for $1 < |z|$. Thus, the required Laurent series is

$$f(z) = \frac{1}{z^2} + \frac{1}{z^3} + \frac{1}{z^4} + \frac{1}{z^5} + \cdots$$

(c) This is basically the same problem as part (a) except that we want all powers of $z - 1$. To that end we add and subtract 1 in the denominator and use (6) of Section 18.1 with z replaced by $z - 1$:

$$f(z) = \frac{1}{(1 - 1 + z)(z - 1)}$$

$$= \frac{1}{z - 1} \frac{1}{1 + (z - 1)}$$

$$= \frac{1}{z - 1}[1 - (z - 1) + (z - 1)^2 - (z - 1)^3 + \cdots]$$

$$= \frac{1}{z - 1} - 1 + (z - 1) - (z - 1)^2 + \cdots$$

The series in brackets converges for $|z - 1| < 1$ and so the last series converges for $0 < |z - 1| < 1$.

(d) Proceeding as in part (b), we write

$$f(z) = \frac{1}{z - 1} \frac{1}{1 + (z - 1)} = \frac{1}{(z - 1)^2} \frac{1}{1 + \frac{1}{z - 1}}$$

18.3 Laurent Series

$$= \frac{1}{(z-1)^2}\left[1 - \frac{1}{z-1} + \frac{1}{(z-1)^2} - \frac{1}{(z-1)^3} + \cdots\right]$$

$$= \frac{1}{(z-1)^2} - \frac{1}{(z-1)^3} + \frac{1}{(z-1)^4} - \frac{1}{(z-1)^5} + \cdots$$

Because the series within the brackets converges for $|1/(z-1)| < 1$, the final series converges for $1 < |z-1|$. □

EXAMPLE 3 Expand $f(z) = 1/(z-1)^2(z-3)$ in a Laurent series valid for (a) $0 < |z-1| < 2$ and (b) $0 < |z-3| < 2$.

Solution (a) As in parts (c) and (d) of Example 2, we want only powers of $z-1$, and so we need to express $z-3$ in terms of $z-1$. This can be done by writing

$$f(z) = \frac{1}{(z-1)^2(z-3)}$$

$$= \frac{1}{(z-1)^2}\frac{1}{-2+(z-1)}$$

$$= \frac{-1}{2(z-1)^2}\frac{1}{1-\frac{(z-1)}{2}}$$

and then using (5) of Section 18.1 with z replaced by $(z-1)/2$:

$$f(z) = \frac{-1}{2(z-1)^2}\left[1 + \frac{z-1}{2} + \frac{(z-1)^2}{2^2} + \frac{(z-1)^3}{2^3} + \cdots\right]$$

$$= -\frac{1}{2(z-1)^2} - \frac{1}{4(z-1)} - \frac{1}{8} - \frac{1}{16}(z-1) - \cdots \quad (12)$$

(b) To obtain powers of $z-3$ we write $z-1 = 2 + (z-3)$ and

$$f(z) = \frac{1}{(z-1)^2(z-3)}$$

$$= \frac{1}{z-3}[2+(z-3)]^{-2}$$

$$= \frac{1}{4(z-3)}\left[1 + \frac{z-3}{2}\right]^{-2}.$$

At this point we can expand $\left[1 + \frac{z-3}{2}\right]^{-2}$ in a power series using the general binomial theorem:

$$f(z) = \frac{1}{4(z-3)}\left[1 + \frac{(-2)}{1!}\left(\frac{z-3}{2}\right) + \frac{(-2)(-3)}{2!}\left(\frac{z-3}{2}\right)^2 + \frac{(-2)(-3)(-4)}{3!}\left(\frac{z-3}{2}\right)^3 + \cdots\right]$$

The binomial series is valid for $|(z-3)/2| < 1$ or $|z-3| < 2$. Multiplying this series by $1/4(z-3)$ gives a Laurent series that is valid for

$0 < |z - 3| < 2$:

$$f(z) = \frac{1}{4(z-3)} - \frac{1}{4} + \frac{3}{16}(z-3) - \frac{1}{8}(z-3)^2 + \cdots \qquad \square$$

EXAMPLE 4 Expand $f(z) = \dfrac{8z+1}{z(1-z)}$ in a Laurent series valid for $0 < |z| < 1$.

Solution By (5) of Section 18.1 we can write

$$f(z) = \frac{8z+1}{z(1-z)} = \frac{8z+1}{z}\frac{1}{1-z} = \left(8 + \frac{1}{z}\right)(1 + z + z^2 + z^3 + \cdots)$$

We then multiply the series by $8 + 1/z$ and collect like terms:

$$f(z) = \frac{1}{z} + 9 + 9z + 9z^2 + \cdots$$

The geometric series converges for $|z| < 1$. After multiplying by $8 + 1/z$, the resulting Laurent series is valid for $0 < |z| < 1$. $\qquad \square$

In the preceding examples the point at the center of the annular domain of validity for each Laurent series was an isolated singularity of the function f. A reexamination of Theorem 18.10 shows that this need not be the case.

EXAMPLE 5 Expand $f(z) = 1/z(z-1)$ in a Laurent series valid for $1 < |z - 2| < 2$.

Solution The specified annular domain is shown in Figure 18.9. The center of this domain, $z = 2$ is a point of analyticity of the function f. Our goal now is to find two series involving integer powers of $z - 2$: one converging for $1 < |z - 2|$ and the other converging for $|z - 2| < 2$. To accomplish this we start with the decomposition of f into partial fractions:

$$f(z) = -\frac{1}{z} + \frac{1}{z-1} = f_1(z) + f_2(z) \qquad (13)$$

Now, $\quad f_1(z) = -\dfrac{1}{z} = -\dfrac{1}{2 + z - 2}$

$$= -\frac{1}{2} \frac{1}{1 + \dfrac{z-2}{2}}$$

$$= -\frac{1}{2}\left[1 - \frac{z-2}{2} + \frac{(z-2)^2}{2^2} - \frac{(z-2)^3}{2^3} + \cdots\right]$$

$$= -\frac{1}{2} + \frac{z-2}{2^2} - \frac{(z-2)^2}{2^3} + \frac{(z-2)^3}{2^4} - \cdots$$

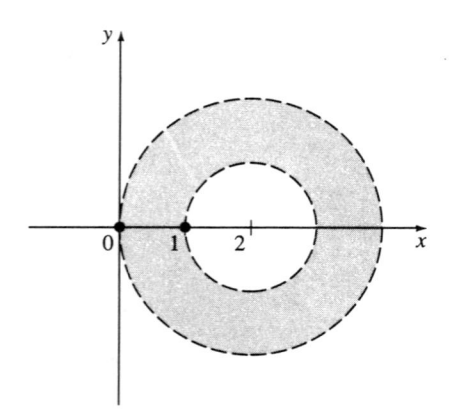

FIGURE 18.9

18.3 Laurent Series

This series converges for $|(z-2)/2| < 1$ or $|z-2| < 2$. Furthermore,

$$f_2(z) = \frac{1}{z-1} = \frac{1}{1+z-2}$$

$$= \frac{1}{z-2} \cdot \frac{1}{1 + \frac{1}{z-2}}$$

$$= \frac{1}{z-2}\left[1 - \frac{1}{z-2} + \frac{1}{(z-2)^2} - \frac{1}{(z-2)^3} + \cdots\right]$$

$$= \frac{1}{z-2} - \frac{1}{(z-2)^2} + \frac{1}{(z-2)^3} - \frac{1}{(z-2)^4} + \cdots$$

converges for $|1/(z-2)| < 1$ or $1 < |z-2|$. Substituting these two results in (13) then gives

$$f(z) = \cdots - \frac{1}{(z-2)^4} + \frac{1}{(z-2)^3} - \frac{1}{(z-2)^2} + \frac{1}{z-2} - \frac{1}{2}$$
$$+ \frac{z-2}{2^2} - \frac{(z-2)^2}{2^3} + \frac{(z-2)^3}{2^4} - \cdots$$

This representation is valid for $1 < |z-2| < 2$. □

EXAMPLE 6 Expand $f(z) = e^{3/z}$ in a Laurent series valid for $0 < |z|$.

Solution From (12) of Section 18.2 we know that for all finite z,

$$e^z = 1 + z + \frac{z^2}{2!} + \frac{z^3}{3!} + \cdots \tag{14}$$

By replacing z in (14) by $3/z$, $z \neq 0$, we obtain the Laurent series

$$e^{3/z} = 1 + \frac{3}{z} + \frac{3^2}{2!z^2} + \frac{3^3}{3!z^3} + \cdots$$

This series is valid for $0 < |z|$. □

In conclusion, we point out a result that will be of special interest to us in Sections 18.5 and 18.6. Replacing the complex variable s with the usual symbol z, we see that when $k = -1$, (4) for the Laurent series coefficients yields $a_{-1} = \frac{1}{2\pi i}\oint_C f(z)\,dz$ or, more important,

$$\oint_C f(z)\,dz = 2\pi i a_{-1} \tag{15}$$

EXERCISES 18.3   *Answers to odd-numbered problems begin on page A-89.*

In Problems 1–6 expand the given function in a Laurent series valid for the indicated annular domain.

1. $f(z) = \dfrac{\cos z}{z}, \; 0 < |z|$

2. $f(z) = \dfrac{z - \sin z}{z^5}, \; 0 < |z|$

3. $f(z) = e^{-1/z^2}, \; 0 < |z|$

4. $f(z) = \dfrac{1 - e^z}{z^2}, \; 0 < |z|$

5. $f(z) = \dfrac{e^z}{z - 1}, \; 0 < |z - 1|$

6. $f(z) = z \cos \dfrac{1}{z}, \; 0 < |z|$

In Problems 7–12 expand $f(z) = \dfrac{1}{z(z-3)}$ in a Laurent series valid for the indicated annular domain.

7. $0 < |z| < 3$
8. $|z| > 3$
9. $0 < |z - 3| < 3$
10. $|z - 3| > 3$
11. $1 < |z - 4| < 4$
12. $1 < |z + 1| < 4$

In Problems 13–16 expand $f(z) = \dfrac{1}{(z-1)(z-2)}$ in a Laurent series valid for the indicated annular domain.

13. $1 < |z| < 2$
14. $|z| > 2$
15. $0 < |z - 1| < 1$
16. $0 < |z - 2| < 1$

In Problems 17–20 expand $f(z) = \dfrac{z}{(z+1)(z-2)}$ in a Laurent series valid for the indicated annular domain.

17. $0 < |z + 1| < 3$
18. $|z + 1| > 3$
19. $1 < |z| < 2$
20. $0 < |z - 2| < 3$

In Problems 21 and 22 expand $f(z) = \dfrac{1}{z(1 - z)^2}$ in a Laurent series valid for the indicated annular domain.

21. $0 < |z| < 1$
22. $|z| > 1$

In Problems 23 and 24 expand $f(z) = \dfrac{1}{(z - 2)(z - 1)^3}$ in a Laurent series valid for the indicated annular domain.

23. $0 < |z - 2| < 1$
24. $0 < |z - 1| < 1$

In Problems 25 and 26 expand $f(z) = \dfrac{7z - 3}{z(z - 1)}$ in a Laurent series valid for the indicated annular domain.

25. $0 < |z| < 1$
26. $0 < |z - 1| < 1$

In Problems 27 and 28 expand $f(z) = \dfrac{z^2 - 2z + 2}{z - 2}$ in a Laurent series valid for the indicated annular domain.

27. $1 < |z - 1|$
28. $0 < |z - 2|$

18.4 ZEROS AND POLES

Principal Part   Suppose that $z = z_0$ is an isolated singularity of a function f and that

$$f(z) = \sum_{k=-\infty}^{\infty} a_k (z - z_0)^k = \sum_{k=1}^{\infty} \dfrac{a_{-k}}{(z - z_0)^k} + \sum_{k=0}^{\infty} a_k (z - z_0)^k \qquad (1)$$

is the Laurent series representation of f valid for $0 < |z - z_0| < R$. The part of the series involving negative powers of $z - z_0$, that is, $\sum_{k=1}^{\infty} a_{-k}/(z - z_0)^k$, is called the **principal part** of f at $z = z_0$. Isolated singular points are further classified according to whether the principal part contains zero, a finite number, or an infinite number of terms.

18.4 Zeros and Poles

(*i*) If the principal part is zero—that is, all the coefficients a_{-k} are zero—then $z = z_0$ is called a **removable singularity**.

(*ii*) If the principal part contains a finite number of nonzero terms, then $z = z_0$ is called a **pole**. If in this case a_{-n} is the last nonzero coefficient in the principal part, then $z = z_0$ is called a **pole of order n**. A pole of order 1 is called a **simple pole**.

(*iii*) If the principal part contains an infinite number of nonzero terms, then $z = z_0$ is called an **essential singularity**.

The following table summarizes the form of the Laurent series for a function f when $z = z_0$ is one of the above singularities.

$z = z_0$	**Laurent Series**
Removable singularity	$a_0 + a_1(z - z_0) + a_2(z - z_0)^2 + \cdots$
Pole of order n	$\dfrac{a_{-n}}{(z-z_0)^n} + \dfrac{a_{-(n-1)}}{(z-z_0)^{n-1}} + \cdots + \dfrac{a_{-1}}{z-z_0} + a_0 + a_1(z - z_0) + \cdots$
Simple pole	$\dfrac{a_{-1}}{z - z_0} + a_0 + a_1(z - z_0) + a_2(z - z_0)^2 + \cdots$
Essential singularity	$\cdots + \dfrac{a_{-2}}{(z-z_0)^2} + \dfrac{a_{-1}}{z - z_0} + a_0 + a_1(z - z_0) + a_2(z - z_0)^2 + \cdots$

EXAMPLE 1 Proceeding as we did in (2) of Section 18.3, we see from

$$\frac{\sin z}{z} = 1 - \frac{z^2}{3!} + \frac{z^4}{5!} - \cdots \qquad (2)$$

that $z = 0$ is a removable singularity of the function $f(z) = (\sin z)/z$. □

If a function f has a removable singularity at the point $z = z_0$, then we can always supply an appropriate definition for the value of $f(z_0)$ so that f becomes analytic at the point. For instance, since the right side of (2) is one at $z = 0$, it makes sense to define $f(0) = 1$. With this definition, the function $f(z) = (\sin z)/z$ in Example 1 is now analytic at $z = 0$.

EXAMPLE 2 (*a*) From

$$\underbrace{\frac{1}{z}}_{\text{principal part}} - \frac{z}{3!} + \frac{z^3}{5!} - \cdots$$

$$\frac{\sin z}{z^2} = \frac{1}{z} - \frac{z}{3!} + \frac{z^3}{5!} - \cdots$$

$0 < |z|$, we see that $a_{-1} \neq 0$ and so $z = 0$ is a simple pole of the function $f(z) = (\sin z)/z^2$. The function $f(z) = (\sin z)/z^3$ represented by the series in (2) of Section 18.3 has a pole of order 2 at $z = 0$.

(b) In Example 3 of Section 18.3 we showed that the Laurent expansion of $f(z) = 1/(z-1)^2(z-3)$ valid for $0 < |z-1| < 2$ was

$$f(z) = \overbrace{-\frac{1}{2(z-1)^2} - \frac{1}{4(z-1)}}^{\text{principal part}} - \frac{1}{8} - \frac{z-1}{16} - \cdots$$

Since $a_{-2} \neq 0$, we conclude that $z = 1$ is pole of order 2.

(c) From Example 6 of Section 18.3 we see from the Laurent series that the principal part of the function $f(z) = e^{3/z}$ contains an infinite number of terms. Thus, $z = 0$ is an essential singularity. □

In part (b) of Example 2 in Section 18.3 we showed that the Laurent series representation of $f(z) = 1/z(z-1)$ valid for $1 < |z|$ is

$$f(z) = \frac{1}{z^2} + \frac{1}{z^3} + \frac{1}{z^4} + \frac{1}{z^5} + \cdots$$

The point $z = 0$ is an isolated singularity of f and the Laurent series contains an infinite number of terms involving negative integer powers of z. Does this mean that $z = 0$ is an essential singularity of f? The answer is no. For this particular function, a reexamination of (1) shows that the Laurent series we are interested in is the one with the annular domain $0 < |z| < 1$. From part (a) of that same example we saw that

$$f(z) = -\frac{1}{z} - 1 - z - z^2 - \cdots$$

was valid for $0 < |z| < 1$. Thus we see that $z = 0$ is a simple pole.

Zeros Recall that z_0 is a **zero** of a function f if $f(z_0) = 0$. An analytic function f has a **zero of order n** at $z = z_0$ if

$$f(z_0) = 0, f'(z_0) = 0, f''(z_0) = 0, \ldots, f^{(n-1)}(z_0) = 0, \text{ but } f^{(n)}(z_0) \neq 0 \quad (3)$$

For example, for $f(z) = (z-5)^3$ we see that $f(5) = 0$, $f'(5) = 0$, $f''(5) = 0$, but $f'''(5) = 6$. Thus $z = 5$ is a zero of order 3. If an analytic function f has a zero of order n at $z = z_0$, it follows from (3) that the Taylor series expansion of f centered at z_0 must have the form

$$f(z) = a_n(z-z_0)^n + a_{n+1}(z-z_0)^{n+1} + a_{n+2}(z-z_0)^{n+2} + \cdots$$
$$= (z-z_0)^n[a_n + a_{n+1}(z-z_0) + a_{n+2}(z-z_0)^2 + \cdots]$$

where $a_n \neq 0$.

18.4 Zeros and Poles

EXAMPLE 3 The analytic function $f(z) = z \sin z^2$ has a zero at $z = 0$. By replacing z by z^2 in (13) of Section 18.2 we obtain

$$\sin z^2 = z^2 - \frac{z^6}{3!} + \frac{z^{10}}{5!} - \cdots$$

and so

$$f(z) = z \sin z^2 = z^3 \left[1 - \frac{z^4}{3!} + \frac{z^8}{5!} - \cdots \right]$$

Hence, $z = 0$ is a zero of order 3. □

A zero z_0 of a nontrivial analytic function f is *isolated* in the sense that there exists some neighborhood of z_0 for which $f(z) \neq 0$ at every point z in that neighborhood except at $z = z_0$. As a consequence, if z_0 is a zero of a nontrivial analytic function f, then the function $1/f(z)$ has an isolated singularity at the point $z = z_0$. The following result enables us, in some circumstances, to determine the poles of a function by inspection.

THEOREM 18.11

If the functions f and g are analytic at $z = z_0$ and f has a zero of order n at $z = z_0$ and $g(z_0) \neq 0$, then the function $F(z) = g(z)/f(z)$ has a pole of order n at $z = z_0$.

EXAMPLE 4 Inspection of the rational function

$$F(z) = \frac{2z + 5}{(z - 1)(z + 5)(z - 2)^4}$$

shows that the denominator has zeros of order 1 at $z = 1$ and $z = -5$, and a zero of order 4 at $z = 2$. Since the numerator is not zero at these points, it follows from Theorem 18.11 that f has simple poles at $z = 1$ and $z = -5$, and a pole of order 4 at $z = 2$. □

EXAMPLE 5 In Example 3 we saw that $z = 0$ is a zero of order 3 of $f(z) = z \sin z^2$. From Theorem 18.11 we conclude that the function $F(z) = 1/(z \sin z^2)$ has a pole of order 3 at $z = 0$. □

From the preceding discussion, it should be intuitively clear that if a function has a pole at $z = z_0$, then $|f(z)| \to \infty$ as $z \to z_0$ from any direction.

EXERCISES 18.4

In Problems 1 and 2 show that $z = 0$ is a removable singularity of the given function. Supply a definition of $f(0)$ so that f is analytic at $z = 0$.

1. $f(z) = \dfrac{e^{2z} - 1}{z}$

2. $f(z) = \dfrac{\sin 4z - 4z}{z^2}$

In Problems 3–8 determine the zeros and their orders for the given function.

3. $f(z) = (z + 2 - i)^2$

4. $f(z) = z^4 - 16$

5. $f(z) = z^4 + z^2$

6. $f(z) = z + \dfrac{9}{z}$

7. $f(z) = e^{2z} - e^z$

8. $f(z) = \sin^2 z$

In Problems 9–12 the indicated number is a zero of the given function. Use a Maclaurin or Taylor series to determine the order of the zero.

9. $f(z) = z(1 - \cos z^2);\ z = 0$

10. $f(z) = z - \sin z;\ z = 0$

11. $f(z) = 1 - e^{z-1};\ z = 1$

12. $f(z) = 1 - \pi i + z + e^z;\ z = \pi i$

In Problems 13–22 determine the order of the poles for the given function.

13. $f(z) = \dfrac{3z - 1}{z^2 + 2z + 5}$

14. $f(z) = 5 - \dfrac{6}{z^2}$

15. $f(z) = \dfrac{1 + 4i}{(z + 2)(z + i)^4}$

16. $f(z) = \dfrac{z - 1}{(z + 1)(z^3 + 1)}$

17. $f(z) = \tan z$

18. $f(z) = \dfrac{\cot \pi z}{z^2}$

19. $f(z) = \dfrac{1 - \cosh z}{z^4}$

20. $f(z) = \dfrac{e^z}{z^2}$

21. $f(z) = \dfrac{1}{1 - e^z}$

22. $f(z) = \dfrac{\sin z}{z^2 - z}$

23. Determine whether $z = 0$ is an isolated or nonisolated singularity of $f(z) = \tan(1/z)$.

24. Show that $z = 0$ is an essential singularity of $f(z) = z^3 \sin(1/z)$.

18.5 RESIDUES AND RESIDUE THEOREM

Residue We saw in the last section that if the complex function f has an isolated singularity at the point z_0, then f has a Laurent series representation

$$f(z) = \sum_{k=-\infty}^{\infty} a_k(z - z_0)^k = \cdots + \dfrac{a_{-2}}{(z - z_0)^2} + \dfrac{a_{-1}}{z - z_0} + a_0 + a_1(z - z_0) + \cdots$$

which is valid in some deleted neighborhood $0 < |z - z_0| < R$. The coefficient a_{-1} of $1/(z - z_0)$ is of special interest to us; a_{-1} is called the **residue** of f at z_0 and is denoted by $\text{Res}(f(z), z_0)$.

EXAMPLE 1 (a) In Example 2 of Section 18.4 we saw that $z = 1$ is a pole of order 2 of the function $f(z) = 1/(z - 1)^2(z - 3)$. From the Laurent series given in that example, we see that the coefficient of $1/(z - 1)$ is $a_{-1} = \text{Res}(f(z), 1) = -\tfrac{1}{4}$.

(b) Example 6 of Section 18.3 showed that $z = 0$ is an essential singularity of $f(z) = e^{3/z}$. From the Laurent series given in that example we see that the coefficient of $1/z$ is $a_{-1} = \text{Res}(f(z), 0) = 3$. □

18.5 Residues and Residue Theorem

Later on in this section we will see why the coefficient a_{-1} is so important. In the meantime we are going to examine ways of obtaining this complex number when z_0 is *pole* of a function f without the necessity of expanding f in a Laurent series at z_0. We begin with the residue at a simple pole.

THEOREM 18.12 Residue at a Simple Pole

If f has a simple pole at $z = z_0$, then

$$\text{Res}(f(z), z_0) = \lim_{z \to z_0} (z - z_0) f(z) \tag{1}$$

Proof Since $z = z_0$ is a simple pole, the Laurent expansion of f about that point has the form

$$f(z) = \frac{a_{-1}}{z - z_0} + a_0 + a_1(z - z_0) + a_2(z - z_0)^2 + \cdots$$

By multiplying both sides by $z - z_0$ and then taking the limit as $z \to z_0$, we obtain

$$\lim_{z \to z_0} (z - z_0) f(z) = \lim_{z \to z_0} [a_{-1} + a_0(z - z_0) + a_1(z - z_0)^2 + \cdots] = a_{-1} = \text{Res}(f(z), z_0) \quad \blacksquare$$

THEOREM 18.13 Residue at a Pole of Order n

If f has a pole of order n at $z = z_0$, then

$$\text{Res}(f(z), z_0) = \frac{1}{(n-1)!} \lim_{z \to z_0} \frac{d^{n-1}}{dz^{n-1}} (z - z_0)^n f(z) \tag{2}$$

Proof Since f is assumed to have a pole of order n, its Laurent expansion for $0 < |z - z_0| < R$ must have the form

$$f(z) = \frac{a_{-n}}{(z - z_0)^n} + \cdots + \frac{a_{-2}}{(z - z_0)^2} + \frac{a_{-1}}{z - z_0} + a_0 + a_1(z - z_0) + \cdots$$

We multiply the last expression by $(z - z_0)^n$:

$$(z - z_0)^n f(z) = a_{-n} + \cdots + a_{-2}(z - z_0)^{n-2} + a_{-1}(z - z_0)^{n-1} + a_0(z - z_0)^n + a_1(z - z_0)^{n+1} + \cdots$$

and then differentiate $n - 1$ times:

$$\frac{d^{n-1}}{dz^{n-1}}(z - z_0)^n f(z) = (n - 1)!a_{-1} + n!a_0(z - z_0) + \cdots \quad (3)$$

Since the all terms on the right side after the first involve positive integer powers of $z - z_0$, the limit of (3) as $z \to z_0$ is

$$\lim_{z \to z_0} \frac{d^{n-1}}{dz^{n-1}}(z - z_0)^n f(z) = (n - 1)!a_{-1}$$

Solving the last equation for a_{-1} gives (2). ∎

Note that (2) reduces to (1) when $n = 1$.

EXAMPLE 2 The function $f(z) = 1/(z - 1)^2(z - 3)$ has a simple pole at $z = 3$ and a pole of order 2 at $z = 1$. Use Theorems 18.12 and 18.13 to find the residues.

Solution Since $z = 3$ is a simple pole, we use (1):

$$\text{Res}(f(z), 3) = \lim_{z \to 3}(z - 3)f(z)$$

$$= \lim_{z \to 3} \frac{1}{(z - 1)^2} = \frac{1}{4}$$

Now at the pole of order 2 it follows from (2) that

$$\text{Res}(f(z), 1) = \frac{1}{1!} \lim_{z \to 1} \frac{d}{dz}(z - 1)^2 f(z)$$

$$= \lim_{z \to 1} \frac{d}{dz} \frac{1}{z - 3}$$

$$= \lim_{z \to 1} \frac{-1}{(z - 3)^2} = -\frac{1}{4} \quad \square$$

When f is not a rational function, calculating residues by means of (1) can sometimes be tedious. It is possible to devise alternative residue formulas. In particular, suppose a function f can be written as a quotient $f(z) = g(z)/h(z)$, where g and h are analytic at $z = z_0$. If $g(z_0) \neq 0$ and if the function h has a zero of order 1 at z_0, then f has a simple pole at $z = z_0$ and

$$\boxed{\text{Res}(f(z), z_0) = \frac{g(z_0)}{h'(z_0)}} \quad (4)$$

To see this last result, we use (1) and the facts that $h(z_0) = 0$ and that $\lim_{z \to z_0}(h(z) - h(z_0))/(z - z_0)$ is a definition of the derivative $h'(z_0)$:

18.5 Residues and Residue Theorem

$$\text{Res}(f(z), z_0) = \lim_{z \to z_0} (z - z_0) \frac{g(z)}{h(z)}$$

$$= \lim_{z \to z_0} \frac{g(z)}{\dfrac{h(z) - h(z_0)}{z - z_0}} = \frac{g(z_0)}{h'(z_0)}$$

Analogous formulas for residues at poles of order greater than 1 are complicated and will not be given.

EXAMPLE 3 The polynomial $z^4 + 1$ can be factored as $(z - z_1)(z - z_2) \times (z - z_3)(z - z_4)$, where $z_1, z_2, z_3,$ and z_4 are the four distinct roots of the equation $z^4 + 1 = 0$. It follows from Theorem 18.11 that the function

$$f(z) = \frac{1}{z^4 + 1}$$

has four simple poles. Now from (10) of Section 16.2 we have $z_1 = e^{\pi i/4}$, $z_2 = e^{3\pi i/4}$, $z_3 = e^{5\pi i/4}$, $z_4 = e^{7\pi i/4}$. To compute the residues we use (4) and Euler's formula:

$$\text{Res}(f(z), z_1) = \frac{1}{4z_1^3} = \frac{1}{4} e^{-3\pi i/4} = -\frac{1}{4\sqrt{2}} - \frac{1}{4\sqrt{2}} i$$

$$\text{Res}(f(z), z_2) = \frac{1}{4z_2^3} = \frac{1}{4} e^{-9\pi i/4} = \frac{1}{4\sqrt{2}} - \frac{1}{4\sqrt{2}} i$$

$$\text{Res}(f(z), z_3) = \frac{1}{4z_3^3} = \frac{1}{4} e^{-15\pi i/4} = \frac{1}{4\sqrt{2}} + \frac{1}{4\sqrt{2}} i$$

$$\text{Res}(f(z), z_4) = \frac{1}{4z_4^3} = \frac{1}{4} e^{-21\pi i/4} = -\frac{1}{4\sqrt{2}} + \frac{1}{4\sqrt{2}} i \qquad \square$$

Residue Theorem We come now to the reason for the importance of the residue concept. The next theorem states that under some circumstances, we can evaluate complex integrals $\oint_C f(z)\,dz$ by summing the residues at the isolated singularities of f within the closed contour C.

THEOREM 18.14 Cauchy's Residue Theorem

Let D be a simply connected domain and C a simple closed contour lying entirely within D. If a function f is analytic on and within C, except at a finite number of singular points $z_1, z_2, \ldots, z_n$ within C, then

$$\oint_C f(z)\,dz = 2\pi i \sum_{k=1}^{n} \text{Res}(f(z), z_k) \tag{5}$$

Proof Suppose $C_1, C_2, \ldots, C_n$ are circles centered at $z_1, z_2, \ldots, z_n$, respectively. Suppose further that each circle C_k has a radius r_k small enough so that $C_1, C_2, \ldots, C_n$ are mutually disjoint and are interior to the simple closed curve C. See Figure 18.10. Recall that (15) of Section 18.3 implies $\oint_{C_k} f(z)\, dz = 2\pi i \operatorname{Res}(f(z), z_k)$, and so Theorem 17.5 gives

$$\oint_C f(z)\, dz = \sum_{k=1}^n \oint_{C_k} f(z)\, dz = 2\pi i \sum_{k=1}^n \operatorname{Res}(f(z), z_k) \qquad \blacksquare$$

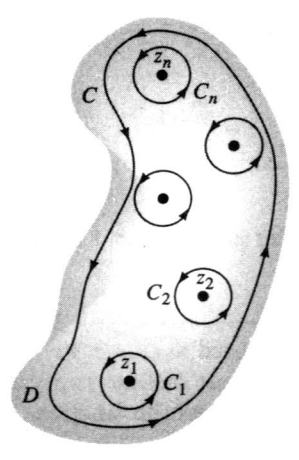

FIGURE 18.10

EXAMPLE 4 Evaluate $\oint_C \dfrac{1}{(z-1)^2(z-3)}\, dz$, where

(a) the contour C is the rectangle defined by $x = 0, x = 4, y = -1, y = 1$, and

(b) the contour C is the circle $|z| = 2$.

Solution (a) Since both poles $z = 1$ and $z = 3$ lie within the square, we have from (5) that

$$\oint_C \frac{1}{(z-1)^2(z-3)}\, dz = 2\pi i[\operatorname{Res}(f(z), 1) + \operatorname{Res}(f(z), 3)]$$

We found these residues in Examples 2 and 3, and so

$$\oint_C \frac{1}{(z-1)^2(z-3)}\, dz = 2\pi i\left[-\frac{1}{4} + \frac{1}{4}\right] = 0$$

(b) Since only the pole $z = 1$ lies within the circle $|z| = 2$, we have from (5) that

$$\oint_C \frac{1}{(z-1)^2(z-3)}\, dz = 2\pi i \operatorname{Res}(f(z), 1) = 2\pi i\left(-\frac{1}{4}\right) = -\frac{\pi}{2} i \qquad \square$$

EXAMPLE 5 Evaluate $\oint_C \dfrac{2z+6}{z^2+4}\, dz$, where the contour C is the circle $|z - i| = 2$.

Solution By writing $z^2 + 4 = (z - 2i)(z + 2i)$, we see that the integrand has simple poles at $-2i$ and $2i$. Now since only $2i$ lies within the contour C, it follows from (5) that

$$\oint_C \frac{2z+6}{z^2+4}\, dz = 2\pi i \operatorname{Res}(f(z), 2i)$$

But

$$\operatorname{Res}(f(z), 2i) = \lim_{z \to 2i} (z - 2i)\frac{2z+6}{(z-2i)(z+2i)}$$

$$= \frac{6 + 4i}{4i} = \frac{3 + 2i}{2i}$$

Hence, $\quad \oint_C \dfrac{2z+6}{z^2+4}\, dz = 2\pi i\left(\dfrac{3+2i}{2i}\right) = \pi(3 + 2i) \qquad \square$

18.5 Residues and Residue Theorem

EXAMPLE 6 Evaluate $\oint_C \dfrac{e^z}{z^4 + 5z^3}\, dz$, where the contour C is the circle $|z| = 2$.

Solution Since $z^4 + 5z^3 = z^3(z + 5)$ we see that the integrand has a pole of order 3 at $z = 0$ and a simple pole at $z = -5$. Since only $z = 0$ lies within the given contour, we have from (5) and (2),

$$\oint_C \frac{e^z}{z^4 + 5z^3}\, dz = 2\pi i\, \text{Res}(f(z), 0)$$

$$= 2\pi i \frac{1}{2!} \lim_{z \to 0} \frac{d^2}{dz^2} z^3 \frac{e^z}{z^3(z+5)}$$

$$= \pi i \lim_{z \to 0} \frac{(z^2 + 8z + 17)e^z}{(z+5)^3}$$

$$= \frac{17\pi}{125} i \qquad \square$$

EXAMPLE 7 Evaluate $\oint_C \tan z\, dz$, where the contour C is the circle $|z| = 2$.

Solution The integrand $\tan z = \sin z / \cos z$ has simple poles at the points where $\cos z = 0$. We saw in Section 16.7 that the only zeros for $\cos z$ are the real numbers $z = (2n+1)\pi/2$, $n = 0, \pm 1, \pm 2, \ldots$. Since only $-\pi/2$ and $\pi/2$ are within the circle $|z| = 2$, we have

$$\oint_C \tan z\, dz = 2\pi i \left[\text{Res}\left(f(z), -\frac{\pi}{2}\right) + \text{Res}\left(f(z), \frac{\pi}{2}\right) \right]$$

Now from (4) with $g(z) = \sin z$, $h(z) = \cos z$, and $h'(z) = -\sin z$, we see that

$$\text{Res}\left(f(z), -\frac{\pi}{2}\right) = \frac{\sin(-\pi/2)}{-\sin(-\pi/2)} = -1 \quad \text{and} \quad \text{Res}\left(f(z), \frac{\pi}{2}\right) = \frac{\sin(\pi/2)}{-\sin(\pi/2)} = -1$$

Therefore, $\oint_C \tan z\, dz = 2\pi i[-1 - 1] = -4\pi i$ $\qquad \square$

EXAMPLE 8 Evaluate $\oint_C e^{3/z}\, dz$, where the contour C is the circle $|z| = 1$.

Solution As we have seen, $z = 0$ is an essential singularity of the integrand $f(z) = e^{3/z}$ and so neither formula (1) nor (2) is applicable to find the residue of f at that point. Nevertheless, we saw in Example 1 that the Laurent series of f at $z = 0$ gives $\text{Res}(f(z), 0) = 3$. Hence from (5) we have

$$\oint_C e^{3/z}\, dz = 2\pi i\, \text{Res}(f(z), 0) = 6\pi i \qquad \square$$

Remark In the application of the limit formulas (1) and (2) for computing residues, the indeterminate form 0/0 may result. Although we are not going to prove it, it should be pointed out that **L'Hôpital's rule** is valid in complex analysis. If $f(z) = g(z)/h(z)$, where g and h are analytic at $z = z_0$, $g(z_0) = 0$, $h(z_0) = 0$, and $h'(z_0) \neq 0$, then

$$\lim_{z \to z_0} \frac{g(z)}{h(z)} = \frac{g'(z_0)}{h'(z_0)}$$

EXERCISES 18.5   Answers to odd-numbered problems begin on page A-89.

In Problems 1–6 use a Laurent series to find the indicated residue.

1. $f(z) = \dfrac{2}{(z-1)(z+4)}$; Res($f(z)$, 1)

2. $f(z) = \dfrac{1}{z^3(1-z)^3}$; Res($f(z)$, 0)

3. $f(z) = \dfrac{4z-6}{z(2-z)}$; Res($f(z)$, 0)

4. $f(z) = (z+3)^2 \sin\dfrac{2}{z+3}$; Res($f(z)$, -3)

5. $f(z) = e^{-2/z^2}$; Res($f(z)$, 0)

6. $f(z) = \dfrac{e^{-z}}{(z-2)^2}$; Res($f(z)$, 2)

In Problems 7–16 use (1), (2), or (4) to find the residue at each pole of the given function.

7. $f(z) = \dfrac{z}{z^2+16}$

8. $f(z) = \dfrac{4z+8}{2z-1}$

9. $f(z) = \dfrac{1}{z^4+z^3-2z^2}$

10. $f(z) = \dfrac{1}{(z^2-2z+2)^2}$

11. $f(z) = \dfrac{5z^2-4z+3}{(z+1)(z+2)(z+3)}$

12. $f(z) = \dfrac{2z-1}{(z-1)^4(z+3)}$

13. $f(z) = \dfrac{\cos z}{z^2(z-\pi)^3}$

14. $f(z) = \dfrac{e^z}{e^z-1}$

15. $f(z) = \sec z$

16. $f(z) = \dfrac{1}{z \sin z}$

In Problems 17–20 use Cauchy's residue theorem, where appropriate, to evaluate the given integral along the indicated contours.

17. $\oint_C \dfrac{1}{(z-1)(z+2)^2} dz$
    (a) $|z| = \tfrac{1}{2}$ (b) $|z| = \tfrac{3}{2}$
    (c) $|z| = 3$

18. $\oint_C \dfrac{z+1}{z^2(z-2i)} dz$
    (a) $|z| = 1$ (b) $|z-2i| = 1$
    (c) $|z-2i| = 4$

19. $\oint_C z^3 e^{-1/z^2} dz$
    (a) $|z| = 5$ (b) $|z+i| = 2$
    (c) $|z-3| = 1$

20. $\oint_C \dfrac{1}{z \sin z} dz$
    (a) $|z-2i| = 1$ (b) $|z-2i| = 3$
    (c) $|z| = 5$

In Problems 21–32 use Cauchy's residue theorem to evaluate the given integral along the indicated contour.

21. $\oint_C \dfrac{1}{z^2+4z+13} dz$, $C: |z-3i| = 3$

22. $\oint_C \dfrac{1}{z^3(z-1)^4} dz$, $C: |z-2| = \tfrac{3}{2}$

23. $\oint_C \dfrac{z}{z^4-1} dz$, $C: |z| = 2$

24. $\oint_C \dfrac{z}{(z+1)(z^2+1)}\, dz$, C is the ellipse $16x^2 + y^2 = 4$

25. $\oint_C \dfrac{ze^z}{z^2-1}\, dz$, $C: |z| = 2$

26. $\oint_C \dfrac{e^z}{z^3+2z^2}\, dz$, $C: |z| = 3$

27. $\oint_C \dfrac{\tan z}{z}\, dz$, $C: |z-1| = 2$

28. $\oint_C \dfrac{\cot \pi z}{z^2}\, dz$, $C: |z| = \tfrac{1}{2}$

29. $\oint_C \cot \pi z\, dz$, C is the rectangle defined by $x = \tfrac{1}{2}$, $x = \pi$, $y = -1$, $y = 1$

30. $\oint_C \dfrac{2z-1}{z^2(z^3+1)}\, dz$, C is the rectangle defined by $x = -2$, $x = 1$, $y = -\tfrac{1}{2}$, $y = 1$

31. $\oint_C \dfrac{e^{iz} + \sin z}{(z-\pi)^4}\, dz$, $C: |z-3| = 1$

32. $\oint_C \dfrac{\cos z}{(z-1)^2(z^2+9)}\, dz$, $C: |z-1| = 1$

18.6 EVALUATION OF REAL INTEGRALS

Residue theory is useful in evaluating real integrals of the forms

$$\int_0^{2\pi} F(\cos\theta, \sin\theta)\, d\theta \tag{1}$$

$$\int_{-\infty}^{\infty} f(x)\, dx \tag{2}$$

$$\int_{-\infty}^{\infty} f(x) \cos x\, dx \quad \text{and} \quad \int_{-\infty}^{\infty} f(x) \sin x\, dx \tag{3}$$

where F in (1) is a rational function of $\cos\theta$ and $\sin\theta$, and in (2) and (3) $f(x) = P(x)/Q(x)$ is a rational function such that the polynomials P and Q have no common factors.

Integrals of the Form $\int_0^{2\pi} F(\cos\theta, \sin\theta)\, d\theta$ The basic idea here is to convert an integral of form (1) into a complex integral where the contour C is the unit circle centered at the origin. As we saw on page 980, this contour can be parameterized by $z = \cos\theta + i\sin\theta = e^{i\theta}$, $0 \le \theta \le 2\pi$. Using

$$dz = ie^{i\theta}\, d\theta \qquad \cos\theta = \dfrac{e^{i\theta} + e^{-i\theta}}{2} \qquad \sin\theta = \dfrac{e^{i\theta} - e^{-i\theta}}{2i}$$

we replace, in turn, $d\theta$, $\cos\theta$, and $\sin\theta$ by

$$d\theta = \dfrac{dz}{iz} \qquad \cos\theta = \dfrac{1}{2}(z + z^{-1}) \qquad \sin\theta = \dfrac{1}{2i}(z - z^{-1}) \tag{4}$$

The integral in (1) then becomes

$$\oint_C F\!\left(\dfrac{1}{2}(z + z^{-1}), \dfrac{1}{2i}(z - z^{-1})\right) \dfrac{dz}{iz}$$

where C is $|z| = 1$.

EXAMPLE 1 Evaluate $\int_0^{2\pi} \frac{1}{(2+\cos\theta)^2} d\theta$.

Solution Using the substitutions in (4) and simplifying yield the contour integral

$$\frac{4}{i}\oint_C \frac{z}{(z^2+4z+1)^2} dz$$

With the aid of the quadratic formula we can write

$$f(z) = \frac{z}{(z^2+4z+1)^2} = \frac{z}{(z-z_0)^2(z-z_1)^2}$$

where $z_0 = -2-\sqrt{3}$ and $z_1 = -2+\sqrt{3}$. Since only z_1 is inside the unit circle C, we have

$$\oint_C \frac{z}{(z^2+4z+1)^2} dz = 2\pi i \, \text{Res}(f(z), z_1)$$

Now z_1 is a pole of order 2 and so from (2) of Section 18.5,

$$\text{Res}(f(z), z_1) = \lim_{z \to z_1} \frac{d}{dz}(z-z_1)^2 f(z)$$

$$= \lim_{z \to z_1} \frac{d}{dz} \frac{z}{(z-z_0)^2}$$

$$= \lim_{z \to z_1} -\frac{z+z_0}{(z-z_0)^3} = \frac{1}{6\sqrt{3}}$$

Hence,

$$\frac{4}{i}\oint_C \frac{z}{(z^2+4z+1)^2} dz = \frac{4}{i} \cdot 2\pi i \, \text{Res}(f(z), z_1)$$

$$= \frac{4}{i} \cdot 2\pi i \cdot \frac{1}{6\sqrt{3}}$$

and finally

$$\int_0^{2\pi} \frac{1}{(2+\cos\theta)^2} d\theta = \frac{4\pi}{3\sqrt{3}} \qquad \square$$

Integrals of the Form $\int_{-\infty}^{\infty} f(x)\,dx$ When f is continuous on $(-\infty, \infty)$, recall from calculus that the improper integral $\int_{-\infty}^{\infty} f(x)\,dx$ is defined in terms of two distinct limits:

$$\int_{-\infty}^{\infty} f(x)\,dx = \lim_{r \to \infty} \int_{-r}^{0} f(x)\,dx + \lim_{R \to \infty} \int_0^{R} f(x)\,dx \qquad (5)$$

18.6 Evaluation of Real Integrals

If both limits in (5) exist, the integral is said to be **convergent**; if one or both of the limits fail to exist, the integral is **divergent**. In the event that we know (a priori) that an integral $\int_{-\infty}^{\infty} f(x)\, dx$ converges, we can evaluate it by means of a single limiting process:

$$\int_{-\infty}^{\infty} f(x)\, dx = \lim_{R \to \infty} \int_{-R}^{R} f(x)\, dx \qquad (6)$$

It is important to note that the symmetric limit in (6) may exist even though the improper integral is divergent. For example, the integral $\int_{-\infty}^{\infty} x\, dx$ is divergent since $\lim_{R \to \infty} \int_{0}^{R} x\, dx = \lim_{R \to \infty} R^2/2 = \infty$. However, using (6), we obtain

$$\lim_{R \to \infty} \int_{-R}^{R} x\, dx = \lim_{R \to \infty} \left[\frac{R^2}{2} - \frac{(-R)^2}{2} \right] = 0 \qquad (7)$$

The limit in (6) is called the **Cauchy principal value** of the integral and is written

$$\text{P.V.} \int_{-\infty}^{\infty} f(x)\, dx = \lim_{R \to \infty} \int_{-R}^{R} f(x)\, dx$$

In (7) we have shown that P.V. $\int_{-\infty}^{\infty} x\, dx = 0$. To summarize, when an integral of form (2) converges, its Cauchy principal value is the same as the value of the integral. If the integral diverges, it may still possess a Cauchy principal value.

To evaluate an integral $\int_{-\infty}^{\infty} f(x)\, dx$, where $f(x) = P(x)/Q(x)$ is continuous on $(-\infty, \infty)$, by residue theory we replace x by the complex variable z and integrate the complex function f over a closed contour C that consists of the interval $[-R, R]$ on the real axis and a semicircle C_R of radius large enough to enclose all the poles of $f(z) = P(z)/Q(z)$ in the upper-half plane $\text{Re}(z) > 0$. See Figure 18.11. By Theorem 18.14 we have

$$\oint_C f(z)\, dz = \int_{C_R} f(z)\, dz + \int_{-R}^{R} f(x)\, dx = 2\pi i \sum_{k=1}^{n} \text{Res}(f(z), z_k)$$

where z_k, $k = 1, 2, \ldots, n$, denotes poles in the upper half-plane. If we can show that the integral $\int_{C_R} f(z)\, dz \to 0$ as $R \to \infty$, then we have

$$\text{P.V.} \int_{-\infty}^{\infty} f(x)\, dx = \lim_{R \to \infty} \int_{-R}^{R} f(x)\, dx = 2\pi i \sum_{k=1}^{n} \text{Res}(f(z), z_k) \qquad (8)$$

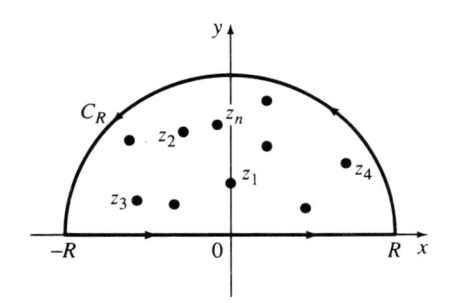

FIGURE 18.11

EXAMPLE 2 Evaluate the Cauchy principal value of

$$\int_{-\infty}^{\infty} \frac{1}{(x^2 + 1)(x^2 + 9)}\, dx$$

Solution Let $f(z) = 1/(z^2 + 1)(z^2 + 9)$. Since

$$(z^2 + 1)(z^2 + 9) = (z - i)(z + i)(z - 3i)(z + 3i)$$

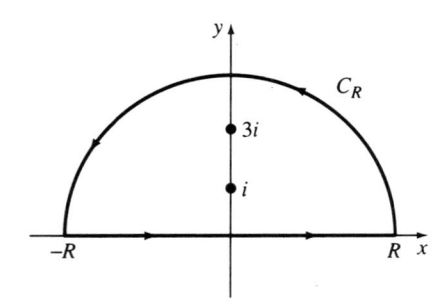

we let C be the closed contour consisting of the interval $[-R, R]$ on the x-axis and the semicircle C_R of radius $R > 3$. As seen from Figure 18.12,

$$\oint_C \frac{1}{(z^2+1)(z^2+9)} dz = \int_{-R}^{R} \frac{1}{(x^2+1)(x^2+9)} dx + \int_{C_R} \frac{1}{(z^2+1)(z^2+9)} dz$$

$$= I_1 + I_2$$

and $\qquad I_1 + I_2 = 2\pi i [\text{Res}(f(z), i) + \text{Res}(f(z), 3i)]$

At the simple poles $z = i$ and $z = 3i$ we find, respectively,

$$\text{Res}(f(z), i) = \frac{1}{16i} \quad \text{and} \quad \text{Res}(f(z), 3i) = -\frac{1}{48i}$$

so that $\qquad I_1 + I_2 = 2\pi i \left[\frac{1}{16i} - \frac{1}{48i} \right] = \frac{\pi}{12}$ $\qquad$ (9)

We now want to let $R \to \infty$ in (9). Before doing this, we note that on C_R,

$$|(z^2+1)(z^2+9)| = |z^2+1||z^2+9| \geq ||z|^2 - 1|||z|^2 - 9| = (R^2 - 1)(R^2 - 9)$$

and so from the ML-inequality of Section 17.1 we can write

$$|I_2| = \left| \int_{C_R} \frac{1}{(z^2+1)(z^2+9)} dz \right| \leq \frac{\pi R}{(R^2-1)(R^2-9)}$$

This last result shows that $|I_2| \to 0$ as $R \to \infty$, and so we conclude that $\lim_{R \to \infty} I_2 = 0$. It follows from (9) that $\lim_{R \to \infty} I_1 = \pi/12$; in other words,

$$\lim_{R \to \infty} \int_{-R}^{R} \frac{1}{(x^2+1)(x^2+9)} dx = \frac{\pi}{12} \quad \text{or} \quad \text{P.V.} \int_{-\infty}^{\infty} \frac{1}{(x^2+1)(x^2+9)} dx = \frac{\pi}{12} \quad \square$$

It is often tedious to have to show that the contour integral along C_R approaches zero as $R \to \infty$. Sufficient conditions under which this is always true are as follows:

THEOREM 18.15

Suppose $f(z) = P(z)/Q(z)$, where the degree of $P(z)$ is n and the degree of $Q(z)$ is $m \geq n + 2$. If C_R is a semicircular contour $z = Re^{i\theta}$, $0 \leq \theta \leq \pi$, then $\int_{C_R} f(z) dz \to 0$ as $R \to \infty$.

In other words, the integral along C_R approaches zero as $R \to \infty$ when the denominator of f is of a power at least 2 more than its numerator. The proof of this fact follows in the same manner as in Example 2. Notice in that example that the conditions stipulated in Theorem 18.15 are satisfied, since the degree of $P(z) = 1$ is 0 and the degree of $Q(z) = (z^2+1)(z^2+9)$ is 4.

EXAMPLE 3 Evaluate the Cauchy principal value of $\int_{-\infty}^{\infty} \frac{1}{x^4 + 1} dx$.

Solution By inspection of the integrand, we see that the conditions given in Theorem 18.15 are satisfied. Moreover, we know from Example 3 of Section 18.5 that f has simple poles in the upper half-plane at $z_1 = e^{\pi i/4}$ and $z_2 = e^{3\pi i/4}$. We also saw in that example that the residues at these poles are

$$\text{Res}(f(z), z_1) = -\frac{1}{4\sqrt{2}} - \frac{1}{4\sqrt{2}}i \quad \text{and} \quad \text{Res}(f(z), z_2) = \frac{1}{4\sqrt{2}} - \frac{1}{4\sqrt{2}}i$$

Thus, by (8),

$$\text{P.V.} \int_{-\infty}^{\infty} \frac{1}{x^4 + 1} dx = 2\pi i[\text{Res}(f(z), z_1) + \text{Res}(f(z), z_2)] = \frac{\pi}{\sqrt{2}} \quad \square$$

Integrals of the Forms $\int_{-\infty}^{\infty} f(x) \cos \alpha x\, dx$ or $\int_{-\infty}^{\infty} f(x) \sin \alpha x\, dx$
We encountered integrals of this type in Section 14.4 in the study of Fourier transforms. Accordingly $\int_{-\infty}^{\infty} f(x) \cos \alpha x\, dx$ and $\int_{-\infty}^{\infty} f(x) \sin \alpha x\, dx$, $\alpha > 0$, are referred to as **Fourier integrals**. Fourier integrals appear as the real and imaginary parts in the improper integral $\int_{-\infty}^{\infty} f(x)e^{i\alpha x}\, dx$. Using Euler's formula $e^{i\alpha x} = \cos \alpha x + i \sin \alpha x$, we get

$$\int_{-\infty}^{\infty} f(x)e^{i\alpha x}\, dx = \int_{-\infty}^{\infty} f(x) \cos \alpha x\, dx + i \int_{-\infty}^{\infty} f(x) \sin \alpha x\, dx \quad (10)$$

whenever both integrals on the right side converge. When $f(x) = P(x)/Q(x)$ is continuous on $(-\infty, \infty)$ we can evaluate both Fourier integrals at the same time by considering the integral $\int_C f(z)e^{i\alpha z}\, dz$, where $\alpha > 0$ and the contour C again consists of the interval $[-R, R]$ on the real axis and a semicircular contour C_R with radius large enough to enclose the poles of $f(z)$ in the upper half-plane.

Before proceeding we give, without proof, sufficient conditions under which the contour integral along C_R approaches zero as $R \to \infty$:

THEOREM 18.16

Suppose $f(z) = P(z)/Q(z)$, where the degree of $P(z)$ is n and the degree of $Q(z)$ is $m \geq n + 1$. If C_R is a semicircular contour $z = Re^{i\theta}$, $0 \leq \theta \leq \pi$, and $\alpha > 0$, then $\int_{C_R} (P(z)/Q(z))e^{i\alpha z}\, dz \to 0$ as $R \to \infty$.

EXAMPLE 4 Evaluate the Cauchy principal value of $\int_0^{\infty} \frac{x \sin x}{x^2 + 9} dx$.

Solution First note that the limits of integration are not from $-\infty$ to ∞ as required by the method. This can be rectified by observing that since the integrand is an even function of x, we can write

$$\int_0^\infty \frac{x \sin x}{x^2 + 9} \, dx = \frac{1}{2} \int_{-\infty}^\infty \frac{x \sin x}{x^2 + 9} \, dx \tag{11}$$

With $\alpha = 1$ we now form the contour integral

$$\oint_C \frac{z}{z^2 + 9} e^{iz} \, dz$$

where C is the same contour shown in Figure 18.12. By Theorem 18.14,

$$\int_{C_R} \frac{z}{z^2 + 9} e^{iz} \, dz + \int_{-R}^R \frac{x}{x^2 + 9} e^{ix} \, dx = 2\pi i \, \text{Res}(f(z)e^{iz}, 3i)$$

where $f(z) = z/(z^2 + 9)$. From (4) of Section 18.5,

$$\text{Res}(f(z)e^{iz}, 3i) = \left.\frac{z e^{iz}}{2z}\right|_{z=3i} = \frac{e^{-3}}{2}$$

Hence, in view of Theorem 18.16 we conclude $\int_{C_R} f(z)e^{iz} \, dz \to 0$ as $R \to \infty$ and so

$$\text{P.V.} \int_{-\infty}^\infty \frac{x}{x^2 + 9} e^{ix} \, dx = 2\pi i \left(\frac{e^{-3}}{2}\right) = \frac{\pi}{e^3} i$$

But by (10),

$$\int_{-\infty}^\infty \frac{x}{x^2 + 9} e^{ix} \, dx = \int_{-\infty}^\infty \frac{x \cos x}{x^2 + 9} \, dx + i \int_{-\infty}^\infty \frac{x \sin x}{x^2 + 9} \, dx = \frac{\pi}{e^3} i$$

Equating real and imaginary parts in the last line gives the bonus result

$$\text{P.V.} \int_{-\infty}^\infty \frac{x \cos x}{x^2 + 9} \, dx = 0 \quad \text{along with} \quad \text{P.V.} \int_{-\infty}^\infty \frac{x \sin x}{x^2 + 9} \, dx = \frac{\pi}{e^3}$$

Finally, in view of (11) we obtain the value of the prescribed integral:

$$\int_0^\infty \frac{x \sin x}{x^2 + 9} \, dx = \frac{1}{2} \text{P.V.} \int_{-\infty}^\infty \frac{x \sin x}{x^2 + 9} \, dx = \frac{\pi}{2e^3} \quad \square$$

Indented Contours The improper integrals of form (2) and (3) that we have considered up to this point were continuous on the interval $[-\infty, \infty]$. In other words, the complex function $f(z) = P(z)/Q(z)$ did not have poles on the real axis. In the event f has poles on the real axis, we must modify the procedure used in Examples 2–4. For example, to evaluate $\int_{-\infty}^\infty f(x) \, dx$ by residues when $f(z)$ has a pole at $z = c$, where c is a real number, we use an **indented contour** as illustrated in Figure 18.13. The symbol C_r denotes a semicircular contour centered at $z = c$ oriented in the *positive* direction. The next theorem is important to this discussion:

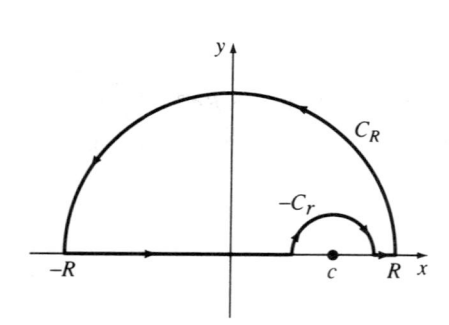

FIGURE 18.13

THEOREM 18.17

Suppose f has a simple pole $z = c$ on the real axis. If C_r is the contour defined by $z = c + re^{i\theta}$, $0 \le \theta \le \pi$, then

$$\lim_{r \to 0} \int_{C_r} f(z)\, dz = \pi i\, \text{Res}(f(z), c)$$

Proof Since f has a simple pole at $z = c$, its Laurent series is

$$f(z) = \frac{a_{-1}}{z - c} + g(z)$$

where $a_{-1} = \text{Res}(f(z), c)$ and g is analytic at c. Using the Laurent series and the parameterization of C_r, we have

$$\int_{C_r} f(z)\, dz = a_{-1} \int_0^\pi \frac{ire^{i\theta}}{re^{i\theta}}\, d\theta + ir \int_0^\pi g(c + re^{i\theta})e^{i\theta}\, d\theta = I_1 + I_2 \quad (12)$$

First, we see that

$$I_1 = a_{-1} \int_0^\pi \frac{ire^{i\theta}}{re^{i\theta}}\, d\theta = a_{-1} \int_0^\pi i\, d\theta = \pi i a_{-1} = \pi i\, \text{Res}(f(z), c)$$

Next, g is analytic at c and so it is continuous at this point and bounded in a neighborhood of the point; that is, there exists an $M > 0$ for which $|g(c + re^{i\theta})| \le M$. Hence,

$$|I_2| = \left| ir \int_0^\pi g(c + re^{i\theta})e^{i\theta}\, d\theta \right| \le r \int_0^\pi M\, d\theta = \pi r M$$

It follows from this last inequality that $\lim_{r \to 0} |I_2| = 0$ and consequently $\lim_{r \to 0} I_2 = 0$. By taking the limit of (12) as $r \to 0$, we have proved the theorem. ∎

EXAMPLE 5 Evaluate the Cauchy principal value of $\displaystyle\int_{-\infty}^{\infty} \frac{\sin x}{x(x^2 - 2x + 2)}\, dx$.

Solution Since the integral is of form (3), we consider the contour integral $\oint_C e^{iz}\, dz / z(z^2 - 2z + 2)$. The function $f(z) = 1/z(z^2 - 2z + 2)$ has simple poles at $z = 0$ and at $z = 1 + i$ in the upper half-plane. The contour C shown in Figure 18.14 is indented at the origin. Adopting an obvious notation, we have

$$\oint_C = \int_{C_R} + \int_{-R}^{-r} + \int_{-C_r} + \int_r^R = 2\pi i\, \text{Res}(f(z)e^{iz}, 1 + i) \quad (13)$$

where $\int_{-C_r} = -\int_{C_r}$. Taking the limits of (13) as $R \to \infty$ and as $r \to 0$, we find

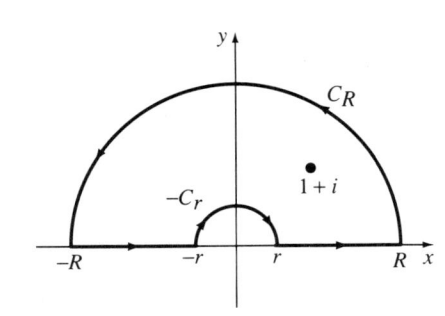

FIGURE 18.14

from Theorems 18.16 and 18.17 that

$$\text{P.V.} \int_{-\infty}^{\infty} \frac{e^{ix}}{x(x^2 - 2x + 2)} \, dx - \pi i \, \text{Res}(f(z)e^{iz}, 0) = 2\pi i \, \text{Res}(f(z)e^{iz}, 1 + i)$$

Now,

$$\text{Res}(f(z)e^{iz}, 0) = \frac{1}{2} \quad \text{and} \quad \text{Res}(f(z)e^{iz}, 1 + i) = -\frac{e^{-1+i}}{4}(1 + i)$$

Therefore,

$$\text{P.V.} \int_{-\infty}^{\infty} \frac{e^{ix}}{x(x^2 - 2x + 2)} \, dx = \pi i \left(\frac{1}{2}\right) + 2\pi i \left(-\frac{e^{-1+i}}{4}(1 + i)\right)$$

Using $e^{-1+i} = e^{-1}(\cos 1 + i \sin 1)$, simplifying, and then equating real and imaginary parts, we get from the last equality

$$\text{P.V.} \int_{-\infty}^{\infty} \frac{\cos x}{x(x^2 - 2x + 2)} \, dx = \frac{\pi}{2} e^{-1}(\sin 1 + \cos 1)$$

and

$$\text{P.V.} \int_{-\infty}^{\infty} \frac{\sin x}{x(x^2 - 2x + 2)} \, dx = \frac{\pi}{2}[1 + e^{-1}(\sin 1 - \cos 1)] \quad \square$$

EXERCISES 18.6 Answers to odd-numbered problems begin on page A-90.

In Problems 1–10 evaluate the given trigonometric integral.

1. $\int_0^{2\pi} \frac{1}{1 + 0.5 \sin \theta} \, d\theta$

2. $\int_0^{2\pi} \frac{1}{10 - 6 \cos \theta} \, d\theta$

3. $\int_0^{2\pi} \frac{\cos \theta}{3 + \sin \theta} \, d\theta$

4. $\int_0^{2\pi} \frac{1}{1 + 3 \cos^2 \theta} \, d\theta$

5. $\int_0^{\pi} \frac{1}{2 - \cos \theta} \, d\theta$ [Hint: Let $t = 2\pi - \theta$.]

6. $\int_0^{\pi} \frac{1}{1 + \sin^2 \theta} \, d\theta$

7. $\int_0^{2\pi} \frac{\sin^2 \theta}{5 + 4 \cos \theta} \, d\theta$

8. $\int_0^{2\pi} \frac{\cos^2 \theta}{3 - \sin \theta} \, d\theta$

9. $\int_0^{2\pi} \frac{\cos 2\theta}{5 - 4 \cos \theta} \, d\theta$

10. $\int_0^{2\pi} \frac{1}{\cos \theta + 2 \sin \theta + 3} \, d\theta$

In Problems 11–30 evaluate the Cauchy principal value of the given improper integral.

11. $\int_{-\infty}^{\infty} \frac{1}{x^2 - 2x + 2} \, dx$

12. $\int_{-\infty}^{\infty} \frac{1}{x^2 - 6x + 25} \, dx$

13. $\int_{-\infty}^{\infty} \frac{1}{(x^2 + 4)^2} \, dx$

14. $\int_{-\infty}^{\infty} \frac{x^2}{(x^2 + 1)^2} \, dx$

15. $\int_{-\infty}^{\infty} \frac{1}{(x^2 + 1)^3} \, dx$

16. $\int_{-\infty}^{\infty} \frac{x}{(x^2 + 4)^3} \, dx$

17. $\int_{-\infty}^{\infty} \frac{2x^2 - 1}{x^4 + 5x^2 + 4} \, dx$

18. $\int_{-\infty}^{\infty} \frac{dx}{(x^2 + 1)^2(x^2 + 9)}$

19. $\int_0^{\infty} \frac{x^2 + 1}{x^4 + 1} \, dx$

20. $\int_0^{\infty} \frac{1}{x^6 + 1} \, dx$

21. $\int_{-\infty}^{\infty} \frac{\cos x}{x^2 + 1} \, dx$

22. $\int_{-\infty}^{\infty} \frac{\cos 2x}{x^2 + 1} \, dx$

23. $\int_{-\infty}^{\infty} \frac{x \sin x}{x^2 + 1} \, dx$

24. $\int_0^{\infty} \frac{\cos x}{(x^2 + 4)^2} \, dx$

25. $\int_0^{\infty} \frac{\cos 3x}{(x^2 + 1)^2} \, dx$

26. $\int_{-\infty}^{\infty} \frac{\sin x}{x^2 + 4x + 5} \, dx$

27. $\int_0^{\infty} \frac{\cos 2x}{x^4 + 1} \, dx$

28. $\int_0^{\infty} \frac{x \sin x}{x^4 + 1} \, dx$

29. $\int_{-\infty}^{\infty} \frac{\cos x}{(x^2 + 1)(x^2 + 9)} \, dx$

30. $\int_0^\infty \dfrac{x \sin x}{(x^2+1)(x^2+4)}\,dx$

In Problems 31 and 32 use an indented contour and residues to establish the given result.

31. P.V. $\int_{-\infty}^\infty \dfrac{\sin x}{x}\,dx = \pi$

32. P.V. $\int_{-\infty}^\infty \dfrac{\sin x}{x(x^2+1)}\,dx = \pi(1 - e^{-1})$

33. Establish the general result

$$\int_0^\pi \dfrac{d\theta}{(a+\cos\theta)^2} = \dfrac{a\pi}{(\sqrt{a^2-1})^3}, \quad a > 1$$

and use this formula to verify the answer in Example 1.

34. Establish the general result

$$\int_0^{2\pi} \dfrac{\sin^2\theta}{a + b\cos\theta}\,d\theta = \dfrac{2\pi}{b^2}(a - \sqrt{a^2-b^2}), \quad a > b > 0$$

and use this formula to verify the answer to Problem 7.

35. Use the contour shown in Figure 18.15 to show that

$$\text{P.V.} \int_{-\infty}^\infty \dfrac{e^{ax}}{1+e^x}\,dx = \dfrac{\pi}{\sin a\pi}, \quad 0 < a < 1$$

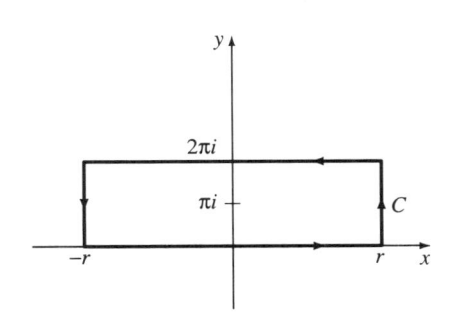

FIGURE 18.15

36. The steady-state temperature $u(x, y)$ in a semi-infinite plate is determined from

$$\dfrac{\partial^2 u}{\partial x^2} + \dfrac{\partial^2 u}{\partial y^2} = 0, \quad 0 < x < \pi, \quad y > 0$$

$$u(0, y) = 0, \quad u(\pi, y) = \dfrac{2y}{y^4 + 4}, \quad y > 0$$

$$u(x, 0) = 0, \quad 0 < x < \pi$$

Use a Fourier transform and the residue method to show that

$$u(x, y) = \int_0^\infty \dfrac{e^{-\alpha}\sin\alpha\sinh\alpha x}{\sinh\alpha\pi}\sin\alpha y\,d\alpha$$

SUMMARY

In Chapter 16 we saw that a function f is analytic at a point z_0 if it is differentiable at z_0 and at all points in some neighborhood of z_0. In this chapter we considered a new characterization of an analytic function; that is, a function f is analytic at z_0 if and only if it can be expanded in a power series centered at z_0 with a radius of convergence $R \neq 0$.

We also encountered a new type of series expansion for a complex function. In particular, if $z = z_0$ is an **isolated singularity** of a function f, then f can be expanded in a **Laurent series**

$$f(z) = \sum_{k=1}^\infty \dfrac{a_{-k}}{(z-z_0)^k} + \sum_{k=0}^\infty a_k(z-z_0)^k$$

which is valid in some annular domain defined by $r < |z - z_0| < R$. If the Laurent series for f is valid for $0 < |z - z_0| < R$, then $\Sigma_{k=1}^\infty a_{-k}/(z-z_0)^k$ is called the **principal part** of f. If all the coefficients in the principal part are zero, then z_0 is a **removable singularity**; if the principal part contains only a finite number of nonzero terms, then z_0 is a **pole**; and finally, if the principal part contains an infinite number of nonzero terms, then z_0 is an **essential singularity**. In the case of a pole or an essential singularity, the coefficient a_{-1} is called the **residue** of f at z_0. The fact that the residue of f at z_0 is defined

CHAPTER 18 REVIEW EXERCISES
Answers to odd-numbered problems begin on page A-90.

Answer Problems 1–12 without referring back to the text. Fill in the blank or answer true/false.

1. A function f is analytic at a point z_0 if f can be expanded in a convergent power series centered at z_0. _____

2. A power series represents a continuous function at every point within and on its circle of convergence. _____

3. For $f(z) = 1/(z - 3)$, the Laurent series valid for $|z| > 3$ is $z^{-1} + 3z^{-2} + 9z^{-3} + \cdots$. Since there are an infinite number of negative powers of $z = z - 0$, $z = 0$ is an essential singularity. _____

4. The only possible singularities of a rational function are poles. _____

5. The function $f(z) = e^{1/(z-1)}$ has an essential singularity at $z = 1$. _____

6. The function $f(z) = z/(e^z - 1)$ has a removable singularity at $z = 0$. _____

7. The function $f(z) = z(e^z - 1)$ possesses a zero of order 2 at $z = 0$. _____

8. The function $f(z) = (z + 5)/(z^3 \sin^2 z)$ has a pole of order _____ at $z = 0$.

9. If $f(z) = \cot \pi z$, then $\text{Res}(f(z), 0) =$ _____.

10. The Laurent series of f valid for $0 < |z - 1|$ is given by
$$(z-1)^{-3} - \frac{(z-1)^{-1}}{3!} + \frac{(z-1)}{5!} - \frac{(z-1)^3}{7!} + \cdots$$
From this series we see that f has a pole of order _____ at $z = 1$ and $\text{Res}(f(z), 1) =$ _____.

11. The circle of convergence of the power series
$$\sum_{k=1}^{\infty} \frac{(z-i)^k}{(2+i)^{k+1}} \text{ is } \underline{\hspace{2cm}}.$$

12. The power series $\sum_{k=1}^{\infty} \frac{z^k}{2^{k+1}}$ converges at $z = 2i$. _____

13. Find a Maclaurin expansion of $f(z) = e^z \cos z$. [*Hint:* Use the identity $\cos z = (e^{iz} + e^{-iz})/2$.]

14. Show that the function $f(z) = 1/\sin(\pi/z)$ has an infinite number of singular points. Are any of these isolated singular points?

In Problems 15–18 use known results as an aid in expanding the given function in a Laurent series valid for the indicated annular region.

15. $f(z) = \dfrac{1 - e^{iz}}{z^4}, \; 0 < |z|$

16. $f(z) = e^{z/(z-2)}, \; 0 < |z - 2|$

17. $f(z) = (z - i)^2 \sin \dfrac{1}{z - i}, \; 0 < |z - i|$

18. $f(z) = \dfrac{1 - \cos z^2}{z^5}, \; 0 < |z|$

19. Expand $f(z) = \dfrac{2}{z^2 - 4z + 3}$ in an appropriate series valid for
 (a) $|z| < 1$ (b) $1 < |z| < 3$ (c) $|z| > 3$
 (d) $0 < |z - 1| < 2$

20. Expand $f(z) = \dfrac{1}{(z-5)^2}$ in an appropriate series valid for
 (a) $|z| < 5$ (b) $|z| > 5$ (c) $0 < |z - 5|$

In Problems 21–30 use Cauchy's residue theorem to evaluate the given integral along the indicated contour.

21. $\oint_C \dfrac{2z + 5}{z(z+2)(z-1)^4} \, dz, \; C: |z + 2| = \tfrac{5}{2}$

22. $\oint_C \dfrac{z^2}{(z-1)^3(z^2+4)} \, dz, \; C$ is the ellipse $x^2/4 + y^2 = 1$

23. $\oint_C \dfrac{1}{2 \sin z - 1} \, dz, \; C: |z - \tfrac{1}{2}| = \tfrac{1}{3}$

24. $\oint_C \dfrac{z+1}{\sinh z} \, dz, \; C$ is the rectangle defined by $x = -1$, $x = 1, y = 4, y = -1$

25. $\oint_C \dfrac{e^{2z}}{z^4 + 2z^3 + 2z^2} \, dz, \; C: |z| = 4$

26. $\oint_C \dfrac{1}{z^4 - 2z^2 + 4}\, dz$, C is the square defined by $x = -2$, $x = 2$, $y = 0$, $y = 1$

27. $\oint_C \dfrac{1}{z(e^z - 1)}\, dz$, $C\colon |z| = 1$ [*Hint:* Use the Maclaurin series for $z(e^z - 1)$.]

28. $\oint_C \dfrac{z}{(z + 1)(z - 1)^{10}}\, dz$, $C\colon |z - 1| = 3$

29. $\oint_C \left[ze^{3/z} + \dfrac{\sin z}{z^2(z - \pi)^3}\right] dz$, $C\colon |z| = 6$

30. $\oint_C \csc \pi z\, dz$, C is the rectangle defined by $x = -\tfrac{1}{2}$, $x = \tfrac{5}{2}$, $y = -1$, $y = 1$

In Problems 31 and 32 evaluate the Cauchy principal value of the given improper integral.

31. $\displaystyle\int_{-\infty}^{\infty} \dfrac{x^2}{(x^2 + 2x + 2)(x^2 + 1)^2}\, dx$

32. $\displaystyle\int_{-\infty}^{\infty} \dfrac{a \cos x + x \sin x}{x^2 + a^2}\, dx$, $a > 0$ [*Hint:* Consider $e^{iz}/(z - ai)$.]

In Problems 33 and 34 evaluate the given trigonometric integral.

33. $\displaystyle\int_0^{2\pi} \dfrac{\cos^2 \theta}{2 + \sin \theta}\, d\theta$

34. $\displaystyle\int_0^{2\pi} \dfrac{\cos 3\theta}{5 - 4 \cos \theta}\, d\theta$

35. Use an indented contour to show that
$$\text{P.V.} \int_0^{\infty} \dfrac{1 - \cos x}{x^2}\, dx = \dfrac{\pi}{2}.$$

36. Show that $\int_0^{\infty} e^{-a^2 x^2} \cos bx\, dx = e^{-b^2/4a^2} \sqrt{\pi}/2a$ by considering the complex integral $\oint_C e^{-a^2 z^2} e^{ibz}\, dz$ along the contour C shown in Figure 18.16. Use the known result $\int_{-\infty}^{\infty} e^{-a^2 x^2}\, dx = \sqrt{\pi}/a$.

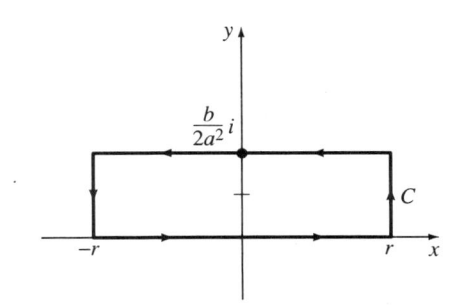

FIGURE 18.16

37. The Laurent expansion of $f(z) = e^{(u/2)(z - 1/z)}$ valid for $0 < |z|$ can be shown to be $f(z) = \Sigma_{k=-\infty}^{\infty} J_k(u) z^k$, where $J_k(u)$ is the Bessel function of the first kind of order k. Use (4) of Section 18.3 and the contour $C\colon |z| = 1$ to show that the coefficients $J_k(u)$ are given by

$$J_k(u) = \dfrac{1}{2\pi} \int_0^{2\pi} \cos(kt - u \sin t)\, dt$$

19

CONFORMAL MAPPINGS AND APPLICATIONS

TOPICS TO REVIEW

functions as mappings (16.4)
boundary-value problems and Laplace's equation (12.5)

IMPORTANT CONCEPTS

mapping
planar transformation
translation and rotation
conformal mapping
Dirichlet problem
linear fractional transformation
cross-ratio
Riemann mapping theorem
Schwarz–Christoffel transformation
Poisson integral formulas
complex potential
equipotential lines
stream function
streamlines

19.1 Complex Functions As Mappings
19.2 Conformal Mapping and the Dirichlet Problem
19.3 Linear Fractional Transformations
19.4 Schwarz–Christoffel Transformations
[O] 19.5 Poisson Integral Formulas
19.6 Applications
Summary
Chapter 19 Review Exercises

INTRODUCTION

A **conformal mapping** is an analytic function that maps a given region in the z-plane to another region in the w-plane so that the angles between curves in the z-plane are preserved. In this chapter we shall study the mapping properties of the elementary functions introduced in Chapter 16 and develop two new classes of special mappings: **linear fractional transformations** and **Schwarz–Christoffel transformations**.

In earlier chapters we used Fourier series and integral transforms to solve boundary-value problems involving Laplace's equation. Conformal mapping methods can be used to transfer known solutions to Laplace's equation from one region to another. In addition, fluid flows around obstacles and through channels can be determined using conformal mappings.

19.1 COMPLEX FUNCTIONS AS MAPPINGS

Mappings Between Planes In Chapter 16 we emphasized the algebraic definitions and properties of complex functions. In order to give a geometric interpretation of a complex function $w = f(z)$, we place a z-plane and a w-plane side by side and imagine that a point $z = x + iy$ in the domain of the definition of f has mapped (or transformed) to the point $w = f(z)$ in the second plane. Thus, the complex function $w = f(z) = u(x, y) + iv(x, y)$ may be considered as the **planar transformation**

$$u = u(x, y) \qquad v = v(x, y)$$

and $w = f(z)$ is called the **image** of z under f.

Figure 19.1 indicates the images of a finite number of complex numbers in the region R. More useful information is obtained by finding the image of the region R together with the images of a family of curves lying inside R. Common choices for the curves are families of lines, families of circles, and the system of level curves for the real and imaginary parts of f.

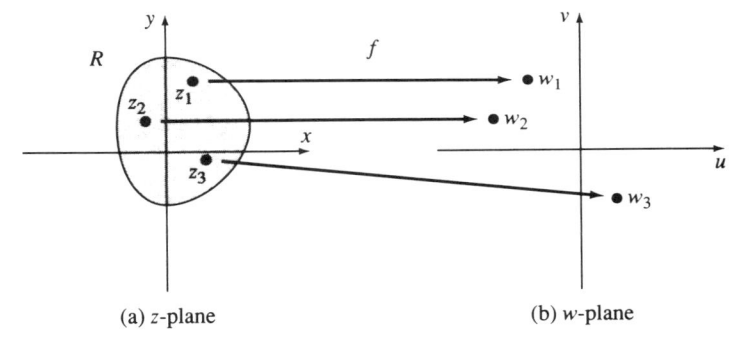

(a) z-plane   (b) w-plane

FIGURE 19.1

Note that if $z(t) = x(t) + iy(t)$, $a \leq t \leq b$, describes a curve C in the region, then $w = f(z(t))$, $a \leq t \leq b$, is a parametric representation of the corresponding curve C' in the w-plane. In addition, a point z on the level curve $u(x, y) = a$ will be mapped to a point w that lies on the vertical line $u = a$, and a point z on the level curve $v(x, y) = b$ will be mapped to a point w that lies on the horizontal line $v = b$.

EXAMPLE 1 The horizontal strip $0 \leq y \leq \pi$ lies in the fundamental region of the exponential function $f(z) = e^z$. A vertical line segment $x = a$ in this region can be described by $z(t) = a + it$, $0 \leq t \leq \pi$, and so $w = f(z(t)) = e^a e^{it}$. Thus, the image is a semicircle with center at $w = 0$ and with radius $r = e^a$. Similarly, a horizontal line $y = b$ can be parameterized by $z(t) = t + ib$, $-\infty < t < \infty$, and so $w = f(z(t)) = e^t e^{ib}$. Since Arg $w = b$ and $|w| = e^t$, the image is a ray emanating from the origin, and since $0 \leq$ Arg $w \leq \pi$, the image

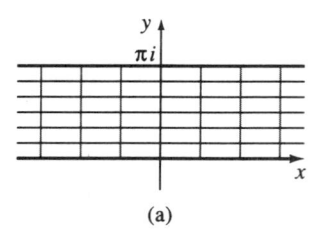

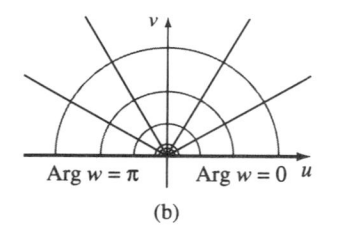

FIGURE 19.2

of the entire horizontal strip is the upper half-plane $v \geq 0$. Note that the horizontal lines $y = 0$ and $y = \pi$ are mapped onto the positive and negative u-axis, respectively. See Figure 19.2 for the mapping by $f(z) = e^z$.

From $w = e^x e^{iy}$, we can conclude that $|w| = e^x$ and $y = \text{Arg } w$. Hence, $z = x + iy = \log_e|w| + i \text{ Arg } w = \text{Ln } w$. The inverse function $f^{-1}(w) = \text{Ln } w$ therefore maps the upper half-plane $v \geq 0$ back to the horizontal strip $0 \leq y \leq \pi$. □

EXAMPLE 2 The complex function $f(z) = 1/z$ has domain $z \neq 0$ and real and imaginary parts $u(x, y) = x/(x^2 + y^2)$ and $v(x, y) = -y/(x^2 + y^2)$, respectively. When $a \neq 0$, a level curve $u(x, y) = a$ can be written as

$$x^2 - \frac{1}{a}x + y^2 = 0 \quad \text{or} \quad \left(x - \frac{1}{2a}\right)^2 + y^2 = \left(\frac{1}{2a}\right)^2$$

The level curve is therefore a circle with its center on the x-axis and passing through the origin. A point z on this circle other than zero is mapped to a point w on the line $u = a$. Likewise, the level curve $v(x, y) = b$, $b \neq 0$, can be written as

$$x^2 + \left(y + \frac{1}{2b}\right)^2 = \left(\frac{1}{2b}\right)^2$$

and a point z on this circle is mapped to a point w on the line $v = b$. Figure 19.3 shows the mapping by $f(z) = 1/z$. Part (a) shows the two collections

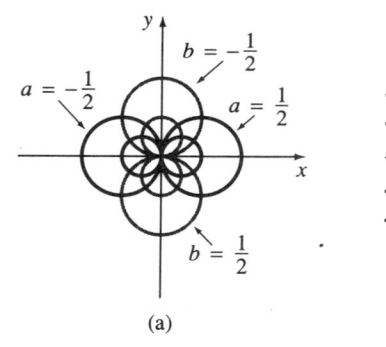

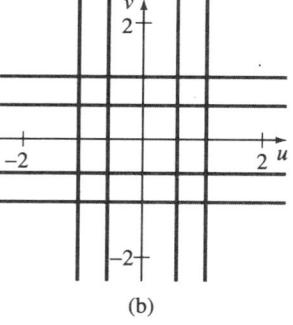

FIGURE 19.3

19.1 Complex Functions as Mappings

of circular level curves, and part (b) shows their corresponding images in the w-plane.

Since $w = 1/z$, we have $z = 1/w$. Thus $f^{-1}(w) = 1/w$, and so $f = f^{-1}$. We can therefore conclude that f maps the horizontal line $y = b$ to the circle $u^2 + (v + 1/2b)^2 = (1/2b)^2$, and f maps the vertical line $x = a$ to the circle $(u - 1/2a)^2 + v^2 = (1/2a)^2$. □

Translation and Rotation The elementary function $f(z) = z + z_0$ may be interpreted as a **translation** in the z-plane. To see this, we let $z = x + iy$ and $z_0 = h + ik$. Since $w = f(z) = (x + h) + i(y + k)$, the point (x, y) has been translated h units in the horizontal direction and k units in the vertical direction to the new position at $(x + h, y + k)$. In particular, the origin O has been mapped to $z_0 = h + ik$.

The elementary function $g(z) = e^{i\theta_0}z$ may be interpreted as a **rotation** through θ_0 degrees, for if $z = re^{i\theta}$, then $w = g(z) = re^{i(\theta + \theta_0)}$. Note that if the complex mapping $h(z) = e^{i\theta_0}z + z_0$ is applied to a region R that is centered at the origin, then image region R' may be obtained by first rotating R through θ_0 degrees and then translating the center to the new position z_0. See Figure 19.4 for the mapping by $h(z) = e^{i\theta_0}z + z_0$.

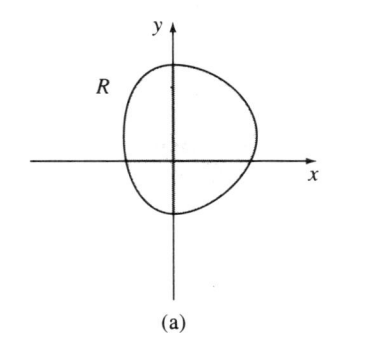

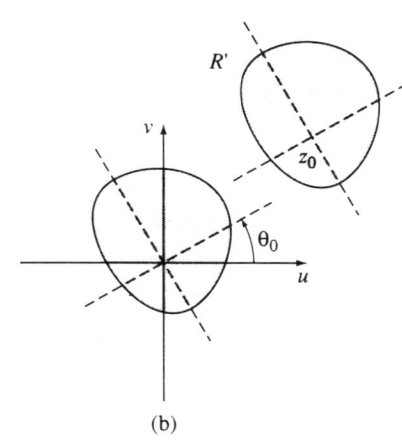

FIGURE 19.4

EXAMPLE 3 Find a complex function that maps the horizontal strip $-1 \leq y \leq 1$ onto the vertical strip $2 \leq x \leq 4$.

Solution If the horizontal strip $-1 \leq y \leq 1$ is rotated through 90°, the vertical strip $-1 \leq x \leq 1$ results, and the vertical strip $2 \leq x \leq 4$ can be obtained by shifting this vertical strip 3 units to the right. See Figure 19.5. Since $e^{i\pi/2} = i$, we obtain $h(z) = iz + 3$ as the desired complex mapping. □

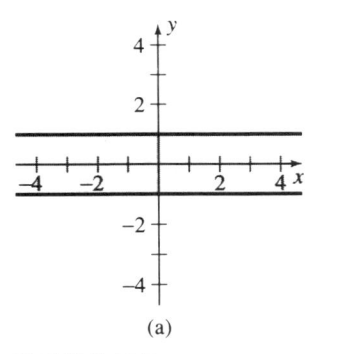

FIGURE 19.5

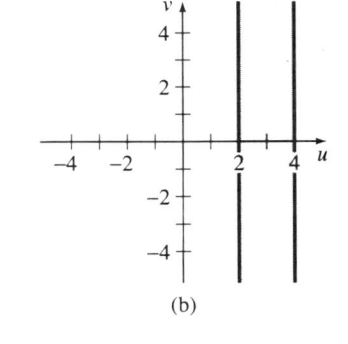

Magnification A **magnification** is a complex function of the form $f(z) = \alpha z$, where α is a fixed positive real number. Note that $|w| = |\alpha z| = \alpha|z|$, and so f changes the length (but not the direction) of the complex number z by a fixed factor α. If $g(z) = az + b$ and $a = r_0 e^{i\theta_0}$, then the vector z is rotated through θ_0 degrees, magnified by a factor of r_0, and then translated using b.

EXAMPLE 4 Find a complex function that maps the disk $|z| \leq 1$ onto the disk $|w - (1 + i)| \leq \frac{1}{2}$.

Solution We must first contract the radius of the disk by a factor of $\frac{1}{2}$ and then translate its center to the point $1 + i$. Therefore, $w = f(z) = z/2 + (1 + i)$ maps $|z| \leq 1$ to the disk $|w - (1 + i)| \leq \frac{1}{2}$. □

Power Functions A complex function of the form $f(z) = z^\alpha$, where α is a fixed positive real number, is called a **real power function**. Figure 19.6 shows the effect of the complex function $f(z) = z^\alpha$ on the angular wedge $0 \leq \text{Arg } z \leq \theta_0$. If $z = re^{i\theta}$, then $w = f(z) = r^\alpha e^{i\alpha\theta}$. Hence, $0 \leq \text{Arg } w \leq \alpha\theta_0$ and the opening of the wedge is changed by a factor of α. It is not hard to show that a circular arc with center at the origin is mapped to a similar circular arc, and rays emanating from the origin are mapped to similar rays.

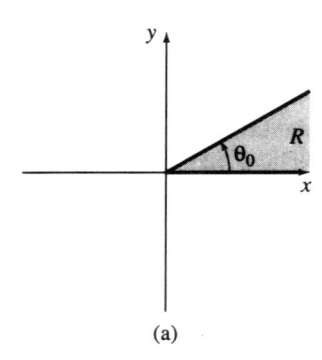

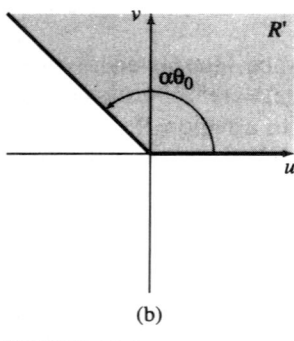

FIGURE 19.6

EXAMPLE 5 Find a complex function that maps the upper half-plane $y \geq 0$ onto the wedge $0 \leq \text{Arg } w \leq \pi/4$.

Solution The upper half-plane $y \geq 0$ can also be described by the inequality $0 \leq \text{Arg } z \leq \pi$. We must therefore find a complex mapping which reduces the angle $\theta_0 = \pi$ by a factor of $\alpha = \frac{1}{4}$. Hence, $f(z) = z^{1/4}$. □

Successive Mappings To find a complex mapping between two regions R and R', it is often convenient to first map R onto a third region R'' and then find a complex mapping from R'' onto R'. More precisely, if $\zeta = f(z)$ maps R onto R'', and $w = g(\zeta)$ maps R'' onto R', then the composite function $w = g(f(z))$ maps R onto R'. See Figure 19.7 for a diagram of successive mappings.

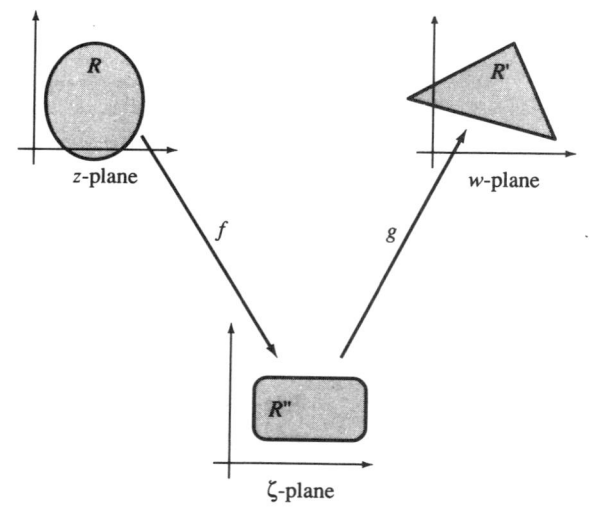

FIGURE 19.7

EXAMPLE 6 Find a complex function that maps the horizontal strip $0 \le y \le \pi$ onto the wedge $0 \le \text{Arg } w \le \pi/4$.

Solution We saw in Example 1 that the complex function $f(z) = e^z$ mapped the horizontal strip $0 \le y \le \pi$ onto the upper half-plane $0 \le \text{Arg } \zeta \le \pi$. From Example 5, the upper half-plane $0 \le \text{Arg } \zeta \le \pi$ is mapped onto the wedge $0 \le \text{Arg } w \le \pi/4$ by $g(\zeta) = \zeta^{1/4}$. It therefore follows that the composite function $w = g(f(z)) = g(e^z) = e^{z/4}$ maps the horizontal strip $0 \le y \le \pi$ onto the wedge $0 \le \text{Arg } w \le \pi/4$. □

EXAMPLE 7 Find a complex function that maps the wedge $\pi/4 \le \text{Arg } z \le 3\pi/4$ onto the upper half-plane $v \ge 0$.

Solution We first rotate the wedge $\pi/4 \le \text{Arg } z \le 3\pi/4$ so that it is in the standard position shown in Figure 19.6. If $\zeta = f(z) = e^{-i\pi/4}z$, then the image of this wedge is the wedge R'' defined by $0 \le \text{Arg } \zeta \le \pi/2$. The real power function $w = g(\zeta) = \zeta^2$ expands the opening of R'' by a factor of 2 to give the upper half-plane $0 \le \text{Arg } w \le \pi$ as its image. Therefore, $w = g(f(z)) = (e^{-i\pi/4}z)^2 = -iz^2$ is the desired mapping. □

In Sections 19.2–19.4 we will expand our knowledge of complex mappings and show how they can be used to solve Laplace's equation in the plane.

EXERCISES 19.1
Answers to odd-numbered problems begin on page A-90.

In Problems 1–10 a curve in the z-plane and a complex mapping $w = f(z)$ are given. In each case, find the image curve in the w-plane.

1. $y = x$ under $w = 1/z$
2. $y = 1$ under $w = 1/z$
3. Hyperbola $xy = 1$ under $w = z^2$
4. Hyperbola $x^2 - y^2 = 4$ under $w = z^2$
5. Semicircle $|z| = 1$, $y > 0$, under $w = \text{Ln } z$
6. Ray $\theta = \pi/4$ under $w = \text{Ln } z$
7. Ray $\theta = \theta_0$ under $w = z^{1/2}$
8. Circular arc $r = 2$, $0 \le \theta \le \pi/2$, under $w = z^{1/2}$
9. Curve $e^x \cos y = 1$ under $w = e^z$
10. Circle $|z| = 1$ under $w = z + 1/z$

In Problems 11–20 a region in the z-plane and a complex mapping $w = f(z)$ are given. In each case, find the image region in the w-plane.

11. First quadrant under $w = 1/z$
12. Strip $0 \le y \le 1$ under $w = 1/z$
13. Strip $\pi/4 \le y \le \pi/2$ under $w = e^z$
14. Rectangle $0 \le x \le 1$, $0 \le y \le \pi$ under $w = e^z$
15. Circle $|z| = 1$ under $w = z + 4i$
16. Circle $|z| = 1$ under $w = 2z - 1$
17. Strip $0 \le y \le 1$ under $w = iz$
18. First quadrant under $w = (1 + i)z$
19. Wedge $0 \le \text{Arg } z \le \pi/4$ under $w = z^3$
20. Wedge $0 \le \text{Arg } z \le \pi/4$ under $w = z^{1/2}$

In Problems 21–30 find a complex mapping from the given region R in the z-plane to the image region R' in the w-plane.

21. Strip $1 \le y \le 4$ to the strip $0 \le u \le 3$
22. Strip $1 \le y \le 4$ to the strip $0 \le v \le 3$
23. Disk $|z - 1| \le 1$ to the disk $|w| \le 2$
24. Strip $-1 \le x \le 1$ to the strip $-1 \le v \le 1$

25. Wedge $\pi/4 \leq \text{Arg } z \leq \pi/2$ to the upper half-plane $v \geq 0$

26. Strip $0 \leq y \leq 4$ to the upper half-plane $v \geq 0$

27. Strip $0 \leq y \leq \pi$ to the wedge $0 \leq \text{Arg } w \leq 3\pi/2$

28. Wedge $0 \leq \text{Arg } z \leq 3\pi/2$ to the half-plane $u \geq 2$

29.

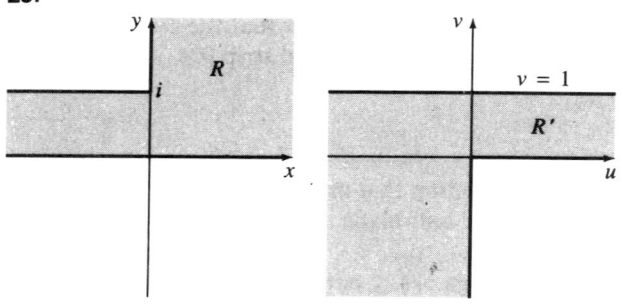

FIGURE 19.8

30.

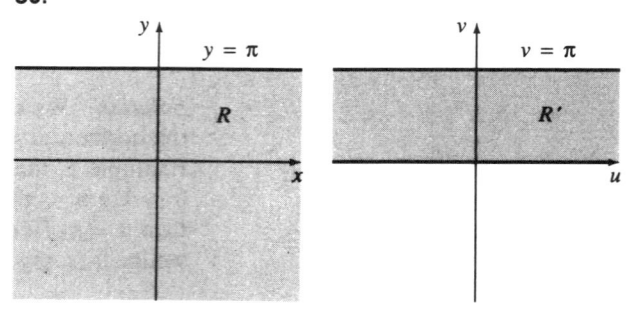

FIGURE 19.9

19.2 CONFORMAL MAPPING AND THE DIRICHLET PROBLEM

Angle-Preserving Mappings A complex mapping $w = f(z)$ defined on a domain D is called **conformal at $z = z_0$** in D when f preserves the angles between any two curves in D that intersect at z_0. More precisely, if C_1 and C_2 intersect in D at z_0, and C'_1 and C'_2 are the corresponding images in the w-plane, we require that the angle θ between C_1 and C_2 equal the angle ϕ between C'_1 and C'_2. See Figure 19.10.

These angles can be computed in terms of tangent vectors to the curves. If z'_1 and z'_2 denote tangent vectors to curves C_1 and C_2, respectively, then, applying the law of cosines to the triangle determined by z'_1 and z'_2, we have

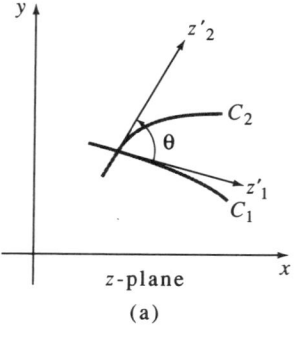

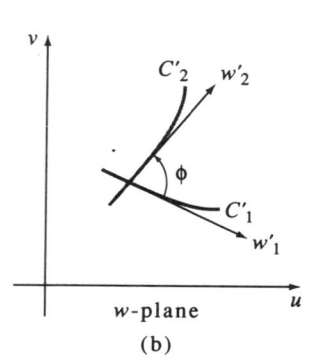

FIGURE 19.10

19.2 Conformal Mapping and the Dirichlet Problem

$$|z'_1 - z'_2|^2 = |z'_1|^2 + |z'_2|^2 - 2|z'_1||z'_2|\cos\theta \text{ or}$$

$$\theta = \cos^{-1}\left(\frac{|z'_1|^2 + |z'_2|^2 - |z'_1 - z'_2|^2}{2|z'_1||z'_2|}\right) \tag{1}$$

Likewise, if w'_1 and w'_2 denote tangent vectors to curves C'_1 and C'_2, respectively, then

$$\phi = \cos^{-1}\left(\frac{|w'_1|^2 + |w'_2|^2 - |w'_1 - w'_2|^2}{2|w'_1||w'_2|}\right) \tag{2}$$

The next theorem will give a simple condition which guarantees that $\theta = \phi$.

THEOREM 19.1*   **Conformal Mapping**

If $f(z)$ is analytic in the domain D and $f'(z_0) \neq 0$, then f is conformal at $z = z_0$.

Proof If a curve C in D is parameterized by $z = z(t)$, then $w = f(z(t))$ describes the image curve in the w-plane. Applying the chain rule to $w = f(z(t))$ gives $w' = f'(z(t))z'(t)$. If curves C_1 and C_2 intersect in D at z_0, then $w'_1 = f'(z_0)z'_1$ and $w'_2 = f'(z_0)z'_2$. Since $f'(z_0) \neq 0$, we can use (2) to obtain

$$\phi = \cos^{-1}\left(\frac{|f'(z_0)z'_1|^2 + |f'(z_0)z'_2|^2 - |f'(z_0)z'_1 - f'(z_0)z'_2|^2}{2|f'(z_0)z'_1||f'(z_0)z'_2|}\right)$$

We can apply the laws of absolute value to factor out $|f'(z_0)|^2$ in the numerator and denominator and obtain

$$\phi = \cos^{-1}\left(\frac{|z'_1|^2 + |z'_2|^2 - |z'_1 - z'_2|^2}{2|z'_1||z'_2|}\right)$$

Therefore, from (1), $\phi = \theta$. ∎

EXAMPLE 1 The analytic function $f(z) = e^z$ is conformal at all points in the z-plane, since $f'(z) = e^z$ is never zero. □

EXAMPLE 2 The analytic function $g(z) = z^2$ is conformal at all points except $z = 0$ since $g'(z) = 2z \neq 0$ for $z \neq 0$. From Figure 19.6 we see that $g(z)$ doubles the angle formed by the two rays at the origin. □

---

*It is also possible to prove that f preserves the *sense of direction* between the tangent vectors.

If $f'(z_0) = 0$ but $f''(z_0) \neq 0$, it is possible to show that f *doubles* the angle between any two curves in D that intersect at $z = z_0$. The next two examples will introduce two important complex mappings that are conformal at all but a finite number of points in their domains.

EXAMPLE 3 The vertical strip $-\pi/2 \leq x \leq \pi/2$ is called the fundamental region of the trigonometric function $w = \sin z$. A vertical line $x = a$ in the interior of this region can be described by $z(t) = a + it$, $-\infty < t < \infty$. From (6) in Section 16.7, we have

$$\sin z = \sin x \cosh y + i \cos x \sinh y$$

and so
$$u + iv = \sin(a + it) = \sin a \cosh t + i \cos a \sinh t$$

From the identity $\cosh^2 t - \sinh^2 t = 1$, it follows that

$$\frac{u^2}{\sin^2 a} - \frac{v^2}{\cos^2 a} = 1$$

The image of the vertical line $x = a$ is therefore a hyperbola with $\pm \sin a$ as u-intercepts, and since $-\pi/2 < a < \pi/2$, the hyperbola crosses the u-axis between $u = -1$ and $u = 1$. Note that if $a = -\pi/2$, $w = -\cosh t$, and so the line $x = -\pi/2$ is mapped onto the interval $(-\infty, -1]$ on the negative u-axis. Likewise, the line $x = \pi/2$ is mapped onto the interval $[1, \infty)$ on the positive u-axis.

A similar argument establishes that the horizontal line segment described by $z(t) = t + ib$, $-\pi/2 < t < \pi/2$, is mapped onto either the upper portion or the lower portion of the ellipse

$$\frac{u^2}{\cosh^2 b} + \frac{v^2}{\sinh^2 b} = 1$$

according to whether $b > 0$ or $b < 0$. These results are summarized in Figure 19.11, which shows the mapping by $f(z) = \sin z$. Note that we have carefully used capital letters to indicate where portions of the boundary are mapped. Thus, for example, boundary segment AB is transformed to $A'B'$.

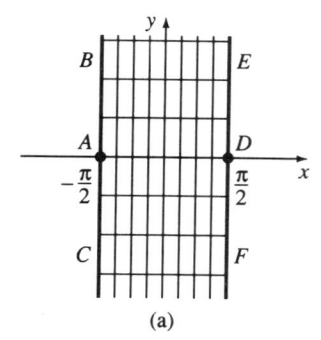

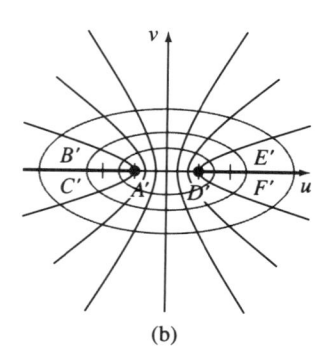

FIGURE 19.11

Since $f'(z) = \cos z$, f is conformal at all points in the region except $z = \pm\pi/2$. The hyperbolas and ellipses are therefore orthogonal since they are images of the orthogonal families of horizontal segments and vertical lines. Note that the 180° angle at $z = -\pi/2$ formed by segments AB and AC is doubled to form a single line segment at $w = -1$. □

EXAMPLE 4 The complex mapping $f(z) = z + 1/z$ is conformal at all values of z except $z = \pm 1$ and $z = 0$. In particular the function is conformal for all values of z in the upper half-plane that satisfy $|z| > 1$. If $z = re^{i\theta}$, then $w = re^{i\theta} + (1/r)e^{-i\theta}$, and so

$$u = \left(r + \frac{1}{r}\right)\cos\theta$$
$$v = \left(r - \frac{1}{r}\right)\sin\theta \tag{3}$$

Note that if $r = 1$, then $v = 0$ and $u = 2\cos\theta$. Therefore, the semicircle $z = e^{it}$, $0 \le t \le \pi$, is mapped to the segment $[-2, 2]$ on the u-axis. It follows from (3) that if $r > 1$, the semicircle $z = re^{it}$, $0 \le t \le \pi$, is mapped onto the upper half of the ellipse $u^2/a^2 + v^2/b^2 = 1$, where $a = r + 1/r$ and $b = r - 1/r$. See Figure 19.12 for the mapping by $f(z) = z + 1/z$.

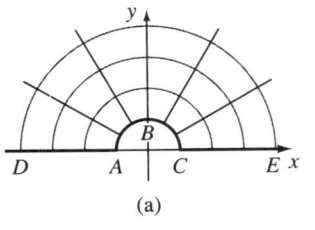

FIGURE 19.12

For a fixed value of θ, the ray $z = te^{i\theta}$, for $t \ge 1$, is mapped to the portion of the hyperbola $u^2/\cos^2\theta - v^2/\sin^2\theta = 4$ in the upper half-plane $v \ge 0$. This follows from (3), since

$$\frac{u^2}{\cos^2\theta} - \frac{v^2}{\sin^2\theta} = \left(t + \frac{1}{t}\right)^2 - \left(t - \frac{1}{t}\right)^2 = 4$$

Since f is conformal for $|z| > 1$ and a ray $\theta = \theta_0$ intersects a circle $|z| = r$ at a right angle, the hyperbolas and ellipses in the w-plane are orthogonal. □

Conformal Mappings Using Tables Conformal mappings are given in Appendix III. The mappings have been categorized as elementary mappings (E-1 to E-9), mappings to half-planes (H-1 to H-6), mappings to circular regions (C-1 to C-5), and miscellaneous mappings (M-1 to M-10). Some of these complex mappings will be derived in Sections 19.3 and 19.4.

The entries indicate not only the images of the region R but also the images of various portions of the boundary of R. This will be especially useful when we attempt to solve boundary-value problems using conformal maps. You should use the appendix much like you use integral tables to find antiderivatives. In some cases a single entry can be used to find a conformal mapping between two given regions R and R'. In other cases successive transformations may be required to map R to R'.

EXAMPLE 5 Use the conformal mappings in Appendix III to find a conformal mapping between the strip $0 \leq y \leq 2$ and the upper half-plane $v \geq 0$. What is the image of the negative x-axis?

Solution An appropriate mapping may be obtained directly from entry H-2. Letting $a = 2$ then $f(z) = e^{\pi z/2}$ and noting the positions of E, D, E', and D' in the figure, we can map the negative x-axis onto the interval $(0, 1)$ on the u-axis. □

EXAMPLE 6 Use the conformal mappings in Appendix III to find a conformal mapping between the strip $0 \leq y \leq 2$ and the disk $|w| \leq 1$. What is the image of the negative x-axis?

Solution Appendix III does not have an entry that maps the strip $0 \leq y \leq 2$ directly onto the disk. In Example 5 the strip was mapped by $f(z) = e^{\pi z/2}$ onto the upper half-plane and, from entry C-4, the complex mapping $w = \dfrac{i - \zeta}{i + \zeta}$ maps the upper half-plane to the disk $|w| \leq 1$. Therefore, $w = g(f(z)) = \dfrac{i - e^{\pi z/2}}{i + e^{\pi z/2}}$ maps the strip $0 \leq y \leq 2$ onto the disk $|w| \leq 1$.

The negative x-axis is first mapped to the interval $(0, 1)$ in the ζ-plane, and from the position of points C and C' in C-4, the interval $(0, 1)$ is mapped to the circular arc $w = e^{i\theta}$, $0 < \theta < \pi/2$, in the w-plane. □

Harmonic Functions and the Dirichlet Problem A bounded harmonic function $u = u(x, y)$ that takes on prescribed values on the entire boundary of a region R is called a solution to a **Dirichlet problem** on R. In Chapters 12–14 we introduced a number of techniques for solving Laplace's equation in the plane, and we interpreted the solution to a Dirichlet problem as the steady-state temperature distribution in the interior of R that results from the fixed temperatures on the boundary.

There are at least two disadvantages to the Fourier series and integral transform methods presented in Chapters 12–14. The methods work for only simple regions in the plane and the solutions typically take the form of either infinite series or improper integrals. As such, they are difficult to evaluate. In Section 16.5 we saw that the real and imaginary parts of an analytic function are both harmonic. Since we have a large stockpile of analytic functions, we can find **closed-form solutions** to many Dirichlet problems and use these solutions to sketch the isotherms and lines of flow of the temperature distribution.

We will next show how conformal mappings can be used to solve a Dirichlet problem in a region R once the solution to the corresponding Dirichlet problem in the image region R' is known. The method depends on the following theorem:

THEOREM 19.2   Transformation Theorem for Harmonic Functions

Let f be an analytic function that maps a domain D onto a domain D'. If U is harmonic in D', then the real-valued function $u(x, y) = U(f(z))$ is harmonic in D.

Proof   We will give a proof for the special case in which D' is simply connected. If U has a harmonic conjugate V in D', then $H = U + iV$ is analytic in D', and so the composite function $H(f(z)) = U(f(z)) + iV(f(z))$ is analytic in D. By Theorem 16.4, it follows that the real part $U(f(z))$ is harmonic in D, and the proof is complete.

To establish that U has a harmonic conjugate, let $g(z) = \partial U/\partial x - i\, \partial U/\partial y$. The first Cauchy–Riemann equation $(\partial/\partial x)(\partial U/\partial x) = (\partial/\partial y)(-\partial U/\partial y)$ is equivalent to Laplace's equation $\partial^2 U/\partial x^2 + \partial^2 U/\partial y^2 = 0$, which is satisfied because U is harmonic in D'. The second Cauchy–Riemann equation $(\partial/\partial y)(\partial U/\partial x) = -(\partial/\partial x)(-\partial U/\partial y)$ is equivalent to the equality of the second-order mixed partial derivatives. Therefore $g(z)$ is analytic in the simply connected domain D' and so, by Theorem 17.8, has an antiderivative $G(z)$. If $G(z) = U_1 + iV_1$, then $g(z) = G'(z) = \partial U_1/\partial x - i\, \partial U_1/\partial y$. Since $g(z) = \partial U/\partial x - i\, \partial U/\partial y$, it follows that U and U_1 have equal first partial derivatives. Therefore, $H = U + iV_1$ is analytic in D', and so U has a harmonic conjugate in D'.   ∎

Theorem 19.2 can be used to solve a Dirichlet problem in a region R by transforming the problem to a region R' in which the solution U either is apparent or has been found by prior methods (including the Fourier series and integral transform methods of Chapters 12–14). The key steps are summarized next.

Solving Dirichlet Problems Using Conformal Mapping

(*1*)   Find a conformal mapping $w = f(z)$ that transforms the original region R onto the image region R'. The region R' may be a region for which many explicit solutions to Dirichlet problems are known.

(*2*)   Transfer the boundary conditions from the boundary of R to the boundary of R'. The value of u at a boundary point ξ of R is assigned as the value of U at the corresponding boundary point $f(\xi)$. See Figure 19.13 for an illustration of transferring boundary conditions.

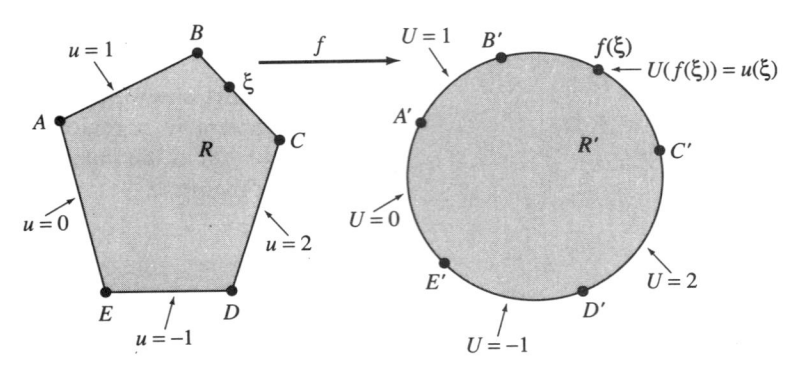

1062   CHAPTER 19   Conformal Mappings and Applications

FIGURE 19.13

(3) Solve the corresponding Dirichlet problem in R'. The solution U may be apparent from the simplicity of the problem in R' or may be found using Fourier or integral transform methods. (Additional methods will be presented in Sections 19.3 and 19.5.).

(4) The solution to the original Dirichlet problem is $u(x, y) = U(f(z))$.

EXAMPLE 7   The function $U(u, v) = (1/\pi)$ Arg w is harmonic in the upper half-plane $v > 0$ since it is the imaginary part of the analytic function $g(w) = (1/\pi)$ Ln w. Use this function to solve the Dirichlet problem in Figure 19.14(a).

Solution   The analytic function $f(z) = \sin z$ maps the original region to the upper half-plane $v \geq 0$ and maps the boundary segments to the segments shown in Figure 19.14(b). The harmonic function $U(u, v) = (1/\pi)$ Arg w satisfies the transferred boundary conditions $U(u, 0) = 0$ for $u > 0$ and $U(u, 0) = 1$ for $u < 0$. Therefore, $u(x, y) = U(\sin z) = (1/\pi)$ Arg$(\sin z)$ is the solution to the original problem. If $\tan^{-1}(v/u)$ is chosen to lie between 0 and π, the solution can also be written as

$$u(x, y) = \frac{1}{\pi} \tan^{-1}\left(\frac{\cos x \sinh y}{\sin x \cosh y}\right)$$

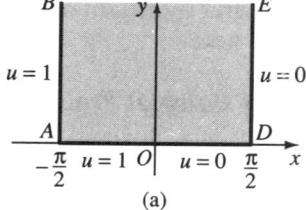

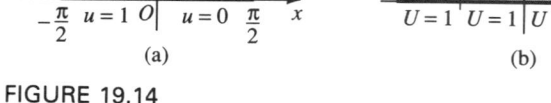

FIGURE 19.14

EXAMPLE 8   From C-1 in Appendix III of conformal mappings, the analytic function $f(z) = (z - a)/(az - 1)$, where $a = (7 + 2\sqrt{6})/5$, maps the region outside the two open disks $|z| < 1$ and $|z - \frac{5}{2}| < \frac{1}{2}$ onto the circular region $r_0 \leq |w| \leq 1$, where $r_0 = 5 - 2\sqrt{6}$. Figure 19.15(a) shows the original Dirichlet problem, and Figure 19.15(b) shows the transferred boundary conditions.

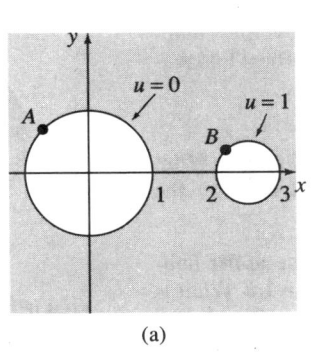

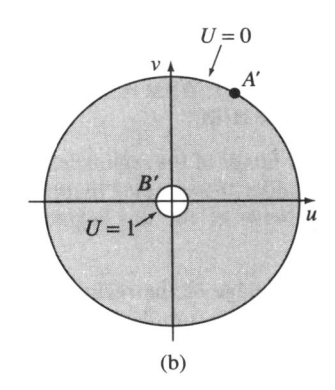

In Problem 10 in Exercises 13.1 we discovered that $U(r, \theta) = (\log_e r)/(\log_e r_0)$ is the solution to the new Dirichlet problem. From Theorem 19.2 we can conclude that the solution to the original boundary-value problem is

$$u(x, y) = \frac{1}{\log_e(5 - 2\sqrt{6})} \log_e \left| \frac{z - (7 + 2\sqrt{6})/5}{(7 + 2\sqrt{6})z/5 - 1} \right|$$

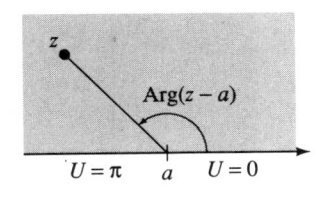

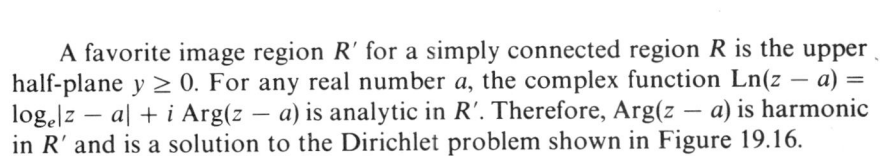

A favorite image region R' for a simply connected region R is the upper half-plane $y \geq 0$. For any real number a, the complex function $\text{Ln}(z - a) = \log_e|z - a| + i\,\text{Arg}(z - a)$ is analytic in R'. Therefore, $\text{Arg}(z - a)$ is harmonic in R' and is a solution to the Dirichlet problem shown in Figure 19.16.

It follows that the solution in R' to the Dirichlet problem with

$$U(x, 0) = \begin{cases} c_0, & a < x < b \\ 0, & \text{otherwise} \end{cases}$$

is the harmonic function $U(x, y) = (c_0/\pi)(\text{Arg}(z - b) - \text{Arg}(z - a))$. A large number of Dirichlet problems in the upper half-plane $y \geq 0$ can be solved by adding together harmonic functions of this form.

FIGURE 19.16

EXERCISES 19.2
Answers to odd-numbered problems begin on page A-90.

In Problems 1–6 determine where the given complex mapping is conformal.

1. $f(z) = z^3 - 3z + 1$
2. $f(z) = \cos z$
3. $f(z) = z + e^z + 1$
4. $f(z) = z + \text{Ln } z + 1$
5. $f(z) = (z^2 - 1)^{1/2}$
6. $f(z) = \pi i - \frac{1}{2}[\text{Ln}(z + 1) + \text{Ln}(z - 1)]$

In Problems 7–10 use the results in Examples 3 and 4.

7. Use the identity $\cos z = \sin(\pi/2 - z)$ to find the image of the strip $0 \leq x \leq \pi$ under the complex mapping $w = \cos z$. What is the image of a horizontal line in the strip?

8. Use the identity $\sinh z = -i \sin(iz)$ to find the image of the strip $-\pi/2 \leq y \leq \pi/2$, $x \geq 0$, under the complex mapping $w = \sinh z$. What is the image of a vertical line segment in the strip?

9. Find the image of the region defined by $-\pi/2 \leq x \leq \pi/2$, $y \geq 0$, under the complex mapping $w = (\sin z)^{1/4}$. What is the image of the line segment $[-\pi/2, \pi/2]$ on the x-axis?

10. Find the image of the region $|z| \leq 1$ in the upper half-plane under the complex mapping $w = z + 1/z$. What is the image of the line segment $[-1, 1]$ on the x-axis?

In Problems 11–18 use the conformal mappings in Appendix III to find a conformal mapping from the given region R in the z-plane onto the target region R' in the w-plane, and find the image of the given boundary curve.

11.

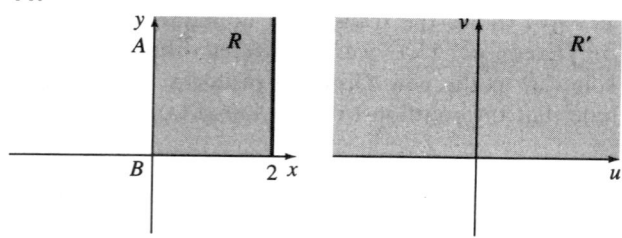

FIGURE 19.17

12.

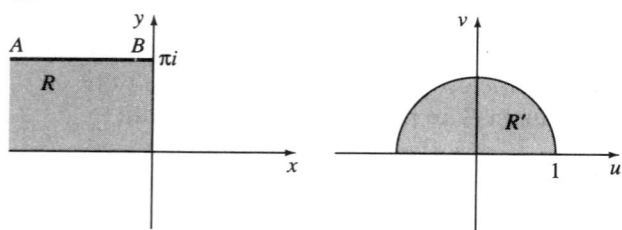

FIGURE 19.18

13.

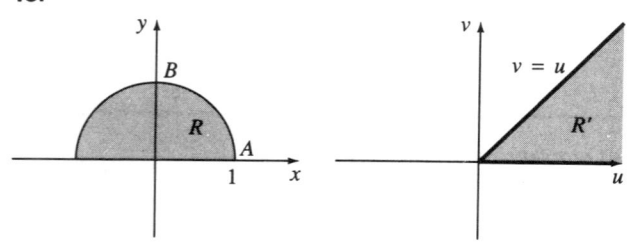

FIGURE 19.19

14.

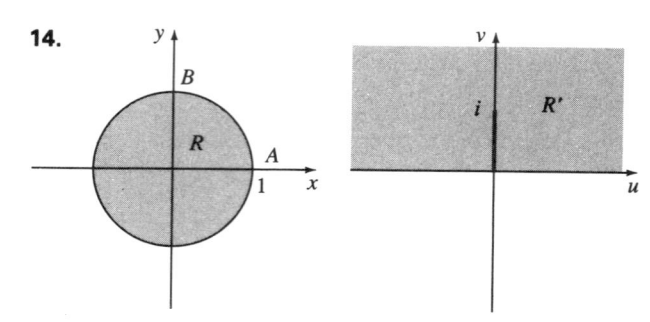

FIGURE 19.20

15.

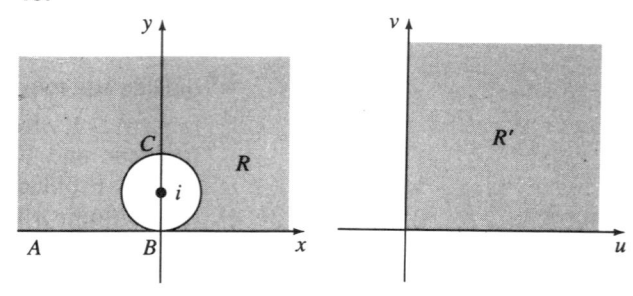

FIGURE 19.21

16.

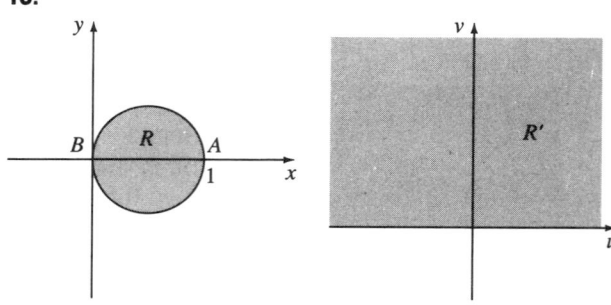

FIGURE 19.22

17.

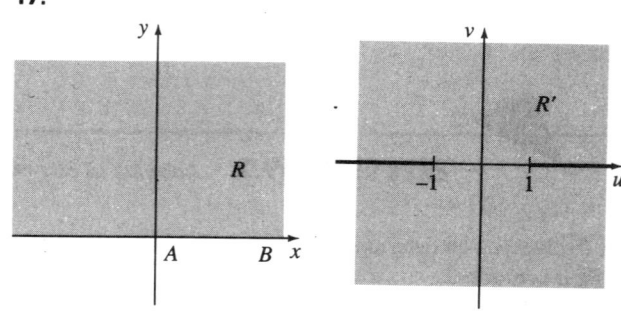

FIGURE 19.23

18.

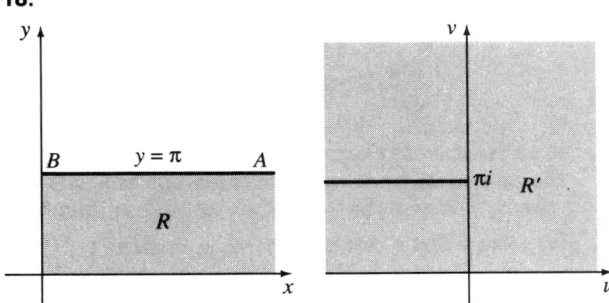

FIGURE 19.24

In Problems 19–22 use an appropriate conformal mapping and the harmonic function $U = (1/\pi) \operatorname{Arg} w$ to solve the given Dirichlet problem.

19.

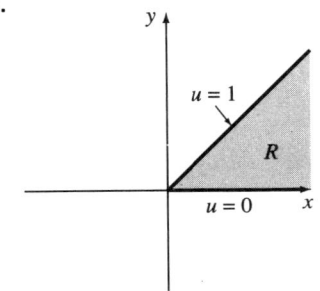

FIGURE 19.25

20.

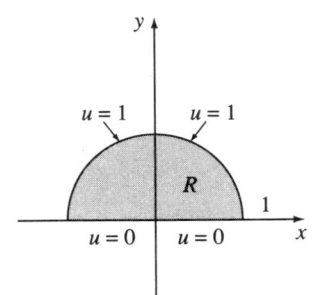

FIGURE 19.26

21.

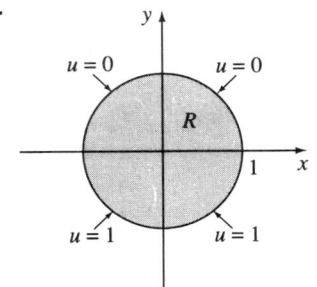

FIGURE 19.27

22.

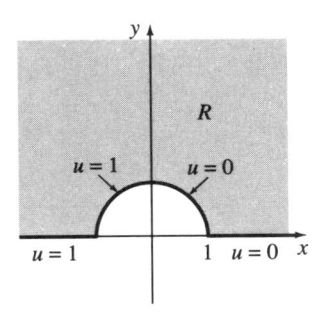

FIGURE 19.28

In Problems 23–26 use an appropriate conformal mapping and the harmonic function $U = (c_0/\pi)[\operatorname{Arg}(w - 1) - \operatorname{Arg}(w + 1)]$ to solve the given Dirichlet problem.

23.

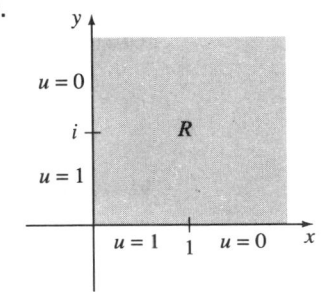

FIGURE 19.29

24.

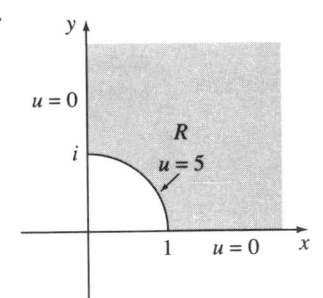

FIGURE 19.30

25.

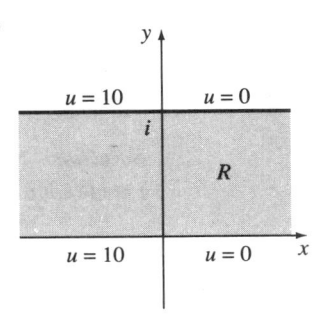

FIGURE 19.31

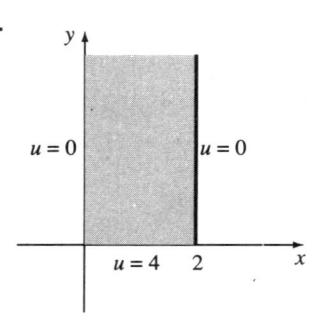

26.

FIGURE 19.32 (region with $u=0$ on left and right sides, $u=4$ on bottom at $x=2$)

27. A real-valued function $\phi(x, y)$ is called **biharmonic** in a domain D when the fourth-order differential equation

$$\frac{\partial^4 \phi}{\partial x^4} + 2 \frac{\partial^4 \phi}{\partial x^2 \partial y^2} + \frac{\partial^4 \phi}{\partial x^4} = 0$$

at all points in D. Examples of biharmonic functions are the Airy stress function in the mechanics of solids and velocity potentials in the analysis of viscous fluid flow.

(a) Show that if ϕ is biharmonic in D, then $u = \partial^2\phi/\partial x^2 + \partial^2\phi/\partial y^2$ is harmonic in D.

(b) If $g(z)$ is analytic in D and $\phi(x, y) = \text{Re}(\bar{z}g(z))$, show that ϕ is biharmonic in D.

19.3 LINEAR FRACTIONAL TRANSFORMATIONS

A **linear fractional transformation** is a complex function of the form

$$T(z) = \frac{az + b}{cz + d}$$

where a, b, c, and d are complex constants. Since

$$T'(z) = \frac{ad - bc}{(cz + d)^2}$$

T is conformal at z provided $\Delta = ad - bc \neq 0$ and $z \neq -d/c$. (If $\Delta = 0$, then $T'(z) = 0$ and $T(z)$ would be a constant function.) Linear fractional transformations are circle preserving in a sense that we will make precise in this section, and, as we saw in Example 8 of Section 19.2, they can be useful in solving Dirichlet problems in regions bounded by circles.

Note that when $c \neq 0$, $T(z)$ has a simple pole at $z_0 = -d/c$ and so

$$\lim_{z \to z_0} |T(z)| = \infty$$

We will write $T(z_0) = \infty$ as shorthand for this limit. In addition, if $c \neq 0$, then

$$\lim_{|z| \to \infty} T(z) = \lim_{|z| \to \infty} \frac{a + b/z}{c + d/z} = \frac{a}{c}$$

and we write $T(\infty) = a/c$.

EXAMPLE 1 If $T(z) = (2z + 1)/(z - i)$, compute $T(0)$, $T(\infty)$, and $T(i)$.

Solution Note that $T(0) = 1/(-i) = i$ and $T(\infty) = \lim_{|z| \to \infty} T(z) = 2$. Since $z = i$ is a simple pole for $T(z)$, we have $\lim_{z \to i} |T(z)| = \infty$ and we write $T(i) = \infty$. □

Circle-Preserving Property If $c = 0$, the linear fractional transformation reduces to a linear function $T(z) = Az + B$. We saw in Section 19.1 that such a complex mapping can be considered as the composite of a rotation, magnification, and translation. As such, a linear function will map a circle in the z-plane to a circle in the w-plane. When $c \neq 0$, we can divide $cz + d$ into $az + d$ to obtain

$$w = \frac{az + b}{cz + d} = \frac{bc - ad}{c} \frac{1}{cz + d} + \frac{a}{c} \qquad (1)$$

If we let $A = (bc - ad)/c$ and $B = a/c$, $T(z)$ can be written as the composite of transformations:

$$z_1 = cz + d \qquad z_2 = \frac{1}{z_1} \qquad w = Az_2 + B \qquad (2)$$

A general linear fractional transformation can therefore be written as the composite of two linear functions and the inversion $w = 1/z$. Note that if $|z - z_1| = r$ and $w = 1/z$, then

$$\left|\frac{1}{w} - \frac{1}{w_1}\right| = \frac{|w - w_1|}{|w||w_1|} = r$$

or
$$|w - w_1| = (r|w_1|)|w - 0| \qquad (3)$$

It is not hard to show that the set of all points w that satisfy

$$|w - w_1| = \lambda |w - w_2| \qquad (4)$$

is a line when $\lambda = 1$ and is a circle when $\lambda > 0$ and $\lambda \neq 1$. It follows from (3) that the image of the circle $|z - z_1| = r$ under the inversion $w = 1/z$ is a circle except when $r = 1/|w_1| = |z_1|$. In the latter case, the original circle passes through the origin and the image is a line. See Figure 19.3. From (2), we can deduce the following theorem:

THEOREM 19.3 Circle-Preserving Property

A linear fractional transformation maps a circle in the z-plane to either a line or a circle in the w-plane. The image is a line if and only if the original circle passes through a pole of the linear fractional transformation.

Proof We have shown that a linear function maps a circle to a circle, whereas an inversion maps a circle to a circle or a line. It follows from (2) that a circle in the z-plane will be mapped to either a circle or a line in the w-plane. If the original circle passes through a pole z_0, then $T(z_0) = \infty$ and so the image is unbounded. Therefore, the image of such a circle must be a line. If the original circle does not pass through z_0, then the image is bounded and must be a circle. ∎

EXAMPLE 2 Find the images of the circles $|z| = 1$ and $|z| = 2$ under $T(z) = (z + 2)/(z - 1)$. What are the images of the interiors of these circles?

Solution The circle $|z| = 1$ passes through the pole $z_0 = 1$ of the linear fractional transformation and so the image is a line. Since $T(-1) = -\frac{1}{2}$ and $T(i) = -\frac{1}{2} - \frac{3}{2}i$, we can conclude that the image is the line $u = -\frac{1}{2}$. The image of the interior $|z| < 1$ is either the half-plane $u < -\frac{1}{2}$ or the half-plane $u > -\frac{1}{2}$. Using $z = 0$ as a *test point*, $T(0) = -2$, and so the image is the half-plane $u < -\frac{1}{2}$.

The circle $|z| = 2$ does not pass through the pole and so the image is a circle. For $|z| = 2$,

$$|\bar{z}| = 2 \quad \text{and} \quad \overline{T(z)} = \overline{\left(\frac{z+2}{z-1}\right)} = \frac{\bar{z}+2}{\bar{z}-1} = T(\bar{z})$$

Therefore, $\overline{T(z)}$ is a point on the image circle and so the image circle is symmetric with respect to the u-axis. Since $T(-2) = 0$ and $T(2) = 4$, the center of the circle is $w = 2$ and the image is the circle $|w - 2| = 2$ (see Figure 19.33). The image of the interior $|z| < 2$ is either the interior or the exterior of the image circle $|w - 2| = 2$. Since $T(0) = -2$, we can conclude that the image is $|w - 2| > 2$.

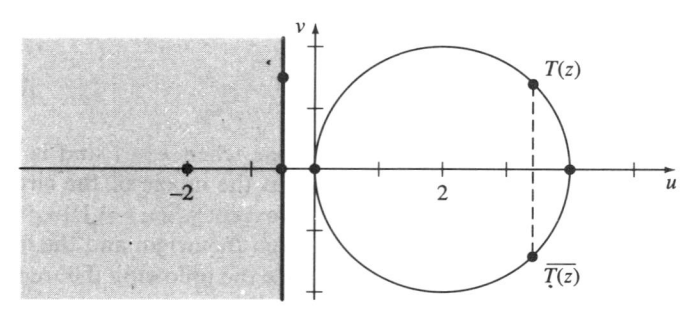

FIGURE 19.33

Constructing Special Mappings In order to use linear fractional transformations to solve Dirichlet problems, we must construct special functions that map a given circular region R to a target region R' in which the corresponding Dirichlet problem is solvable. Since a circular boundary is determined by three of its points, we must find a linear fractional transformation $w = T(z)$ that maps three given points z_1, z_2, and z_3 on the boundary of R to three points w_1, w_2, and w_3 on the boundary of R'. In addition, the interior of R' must be the image of the interior of R. See Figure 19.34.

Matrix Methods Matrix methods can be used to simplify many of the computations. We can associate the matrix

$$\mathbf{A} = \begin{bmatrix} a & b \\ c & d \end{bmatrix}$$

19.3 Linear Fractional Transformations

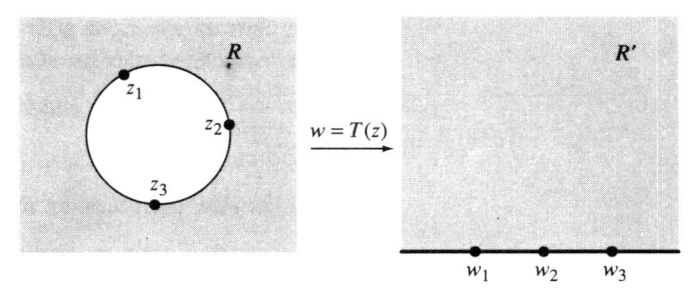

FIGURE 19.34

with $T(z) = (az + b)/(cz + d)$.* If $T_1(z) = (a_1 z + b_1)/(c_1 z + d_1)$ and $T_2(z) = (a_2 z + b_2)/(c_2 z + d_2)$, then the composite function $T(z) = T_2(T_1(z))$ is given by $T(z) = (az + b)/(cz + d)$, where

$$\begin{bmatrix} a & b \\ c & d \end{bmatrix} = \begin{bmatrix} a_2 & b_2 \\ c_2 & d_2 \end{bmatrix} \begin{bmatrix} a_1 & b_1 \\ c_1 & d_1 \end{bmatrix} \tag{5}$$

If $w = T(z) = (az + b)/(cz + d)$, we can solve for z to obtain $z = (dw - b)/(-cw + a)$. Therefore the inverse of the linear fractional transformation T is $T^{-1}(w) = (dw - b)/(-cw + a)$ and we associate the matrix

$$\text{adj } \mathbf{A} = \begin{bmatrix} d & -b \\ -c & a \end{bmatrix} \tag{6}$$

with T^{-1}. The matrix adj $\mathbf{A}$ is the adjoint matrix of $\mathbf{A}$ (see Section 7.5), the matrix for T.

EXAMPLE 3 If $T(z) = \dfrac{2z - 1}{z + 2}$ and $S(z) = \dfrac{z - i}{iz - 1}$, find $S^{-1}(T(z))$.

Solution From (5) and (6), we have $S^{-1}(T(z)) = (az + b)/(cz + d)$, where

$$\begin{bmatrix} a & b \\ c & d \end{bmatrix} = \text{adj}\left(\begin{bmatrix} 1 & -i \\ i & -1 \end{bmatrix} \right) \begin{bmatrix} 2 & -1 \\ 1 & 2 \end{bmatrix}$$

$$= \begin{bmatrix} -1 & i \\ -i & 1 \end{bmatrix} \begin{bmatrix} 2 & -1 \\ 1 & 2 \end{bmatrix} = \begin{bmatrix} -2 + i & 1 + 2i \\ 1 - 2i & 2 + i \end{bmatrix}$$

Therefore, $S^{-1}(T(z)) = \dfrac{(-2 + i)z + 1 + 2i}{(1 - 2i)z + 2 + i}$. □

Triples to Triples The linear fractional transformation

$$T(z) = \frac{z - z_1}{z - z_3} \frac{z_2 - z_3}{z_2 - z_1}$$

---

*The matrix $\mathbf{A}$ is not unique since the numerator and denominator in $T(z)$ can be multiplied by a nonzero constant.

has a zero at $z = z_1$, a pole at $z = z_3$, and $T(z_2) = 1$. Therefore $T(z)$ maps three *distinct* complex numbers $z_1, z_2,$ and z_3 to 0, 1, and ∞, respectively. The term $\dfrac{z - z_1}{z - z_3} \dfrac{z_2 - z_3}{z_2 - z_1}$ is called the **cross-ratio** of the complex numbers $z, z_1, z_2,$ and z_3.

Likewise, the complex mapping $S(w) = \dfrac{w - w_1}{w - w_3} \dfrac{w_2 - w_3}{w_2 - w_1}$ sends $w_1, w_2,$ and w_3 to 0, 1, and ∞, and so S^{-1} maps 0, 1, and ∞ to $w_1, w_2,$ and w_3. It follows that the linear fractional transformation $w = S^{-1}(T(z))$ maps the triple $z_1, z_2,$ and z_3 to $w_1, w_2,$ and w_3. From $w = S^{-1}(T(z))$, we have $S(w) = T(z)$ and we can conclude that

$$\frac{w - w_1}{w - w_3} \frac{w_2 - w_3}{w_2 - w_1} = \frac{z - z_1}{z - z_3} \frac{z_2 - z_3}{z_2 - z_1} \qquad (7)$$

In constructing a linear fractional transformation that maps the triple $z_1, z_2,$ and z_3 to $w_1, w_2,$ and w_3, we can use matrix methods to compute $w = S^{-1}(T(z))$. Alternatively, we can substitute into (7) and solve the resulting equation for w.

EXAMPLE 4 Construct a linear fractional transformation that maps the points $1, i,$ and -1 on the circle $|z| = 1$ to the points $-1, 0,$ and 1 on the real axis.

Solution Substituting into (7), we have

$$\frac{w + 1}{w - 1} \frac{0 - 1}{0 - (-1)} = \frac{z - 1}{z + 1} \frac{i + 1}{i - 1}$$

or

$$-\frac{w + 1}{w - 1} = -i\frac{z - 1}{z + 1}$$

Solving for w, we obtain $w = -i(z - i)/(z + i)$. Alternatively, we could use the matrix method to compute $w = S^{-1}(T(z))$. $\square$

When $z_k = \infty$ plays the role of one of the points in a triple, the definition of the cross-ratio is changed by replacing each factor that contains z_k by one. For example, if $z_2 = \infty$, both $z_2 - z_3$ and $z_2 - z_1$ are replaced by one, giving $(z - z_1)/(z - z_3)$ as the cross-ratio.

EXAMPLE 5 Construct a linear fractional transformation that maps the points $\infty, 0,$ and 1 on the real axis to the points $1, i,$ and -1 on the circle $|w| = 1$.

Solution Since $z_1 = \infty$, the terms $z - z_1$ and $z_2 - z_1$ in the cross product are replaced by 1. It follows that

$$\frac{w - 1}{w + 1} \frac{i + 1}{i - 1} = \frac{1}{z - 1} \frac{0 - 1}{1}$$

19.3 Linear Fractional Transformations

or
$$S(w) = -i\frac{w-1}{w+1} = \frac{-1}{z-1} = T(z)$$

If we use the matrix method to find $w = S^{-1}(T(z))$, then

$$\begin{bmatrix} a & b \\ c & d \end{bmatrix} = \text{adj}\left(\begin{bmatrix} -i & i \\ 1 & 1 \end{bmatrix}\right)\begin{bmatrix} 0 & -1 \\ 1 & -1 \end{bmatrix} = \begin{bmatrix} -i & -1+i \\ -i & 1+i \end{bmatrix}$$

and so $w = \dfrac{-iz - 1 + i}{-iz + 1 + i} = \dfrac{z - 1 - i}{z - 1 + i}$. □

EXAMPLE 6 Solve the Dirichlet problem in Figure 19.35(a) using conformal mapping by constructing a linear fractional transformation that maps the given region into the upper half-plane.

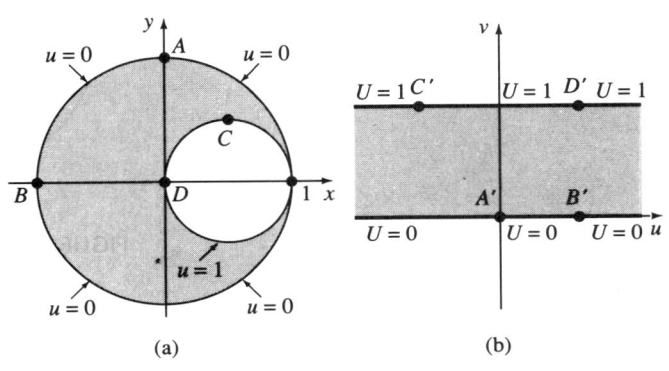

FIGURE 19.35

Solution The boundary circles $|z| = 1$ and $|z - \frac{1}{2}| = \frac{1}{2}$ each pass through $z = 1$. We can therefore map each boundary circle to a line by selecting a linear fractional transformation that has $z = 1$ as a pole. If we further require that $T(i) = 0$ and $T(-1) = 1$, then

$$T(z) = \frac{z - i}{z - 1} \frac{-1 - 1}{-1 - i} = (1 - i)\frac{z - i}{z - 1}$$

Since $T(0) = 1 + i$ and $T(\frac{1}{2} + \frac{1}{2}i) = -1 + i$, T maps the interior of the circle $|z| = 1$ onto the upper half-plane and maps the circle $|z - \frac{1}{2}| = \frac{1}{2}$ onto the line $v = 1$. Figure 19.35(b) shows the transferred boundary conditions.

The harmonic function $U(u, v) = v$ is the solution to the simplified Dirichlet problem in the w-plane, and so, by Theorem 19.2, $u(x, y) = U(T(z))$ is the solution to the original Dirichlet problem in the z-plane.

Since the imaginary part of $T(z) = (1 - i)\dfrac{z - i}{z - 1}$ is $\dfrac{1 - x^2 - y^2}{(x - 1)^2 + y^2}$, the solution is given by

$$u(x, y) = \frac{1 - x^2 - y^2}{(x - 1)^2 + y^2}$$

The level curves $u(x, y) = c$ can be written as

$$\left(x - \frac{c}{1+c}\right)^2 + y^2 = \left(\frac{1}{1+c}\right)^2$$

and are therefore circles which pass through $z = 1$. See Figure 19.36. These level curves may be interpreted as the isotherms of the steady-state temperature distribution induced by the boundary temperatures.

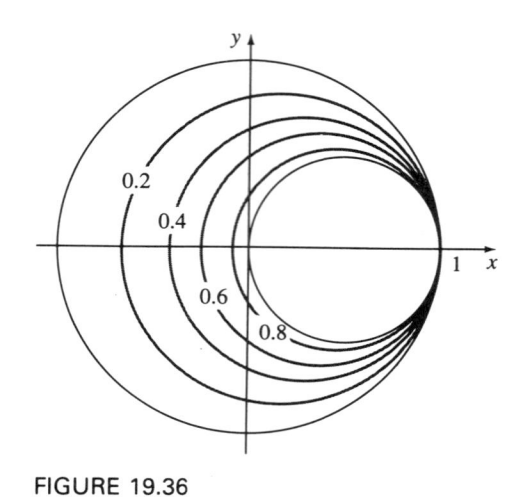

FIGURE 19.36

EXERCISES 19.3   Answers to odd-numbered problems begin on page A-91.

In Problems 1–4 a linear fractional transformation is given.
(a) Compute $T(0)$, $T(1)$, and $T(\infty)$.
(b) Find the images of the circles $|z| = 1$ and $|z - 1| = 1$.
(c) Find the image of the disk $|z| \leq 1$.

1. $T(z) = \dfrac{i}{z}$

2. $T(z) = \dfrac{1}{z - 1}$

3. $T(z) = \dfrac{z + 1}{z - 1}$

4. $T(z) = \dfrac{z - i}{z}$

In Problems 5–8 use the matrix method to compute $S^{-1}(w)$ and $S^{-1}(T(z))$ for each pair of linear fractional transformations.

5. $T(z) = \dfrac{z}{iz - 1}$ and $S(z) = \dfrac{iz + 1}{z - 1}$

6. $T(z) = \dfrac{iz}{z - 2i}$ and $S(z) = \dfrac{2z + 1}{z + 1}$

7. $T(z) = \dfrac{2z - 3}{z - 3}$ and $S(z) = \dfrac{z - 2}{z - 1}$

8. $T(z) = \dfrac{z - 1 + i}{iz - 2}$ and $S(z) = \dfrac{(2 - i)z}{z - 1 - i}$

In Problems 9–16 construct a linear fractional transformation that maps the given triple z_1, z_2, and z_3 to the triple w_1, w_2, and w_3.

9. $-1, 0, 2$ to $0, 1, \infty$

10. $i, 0, -i$ to $0, 1, \infty$

11. $0, 1, \infty$ to $0, i, 2$

12. $0, 1, \infty$ to $1 + i, 0, 1 - i$

13. $-1, 0, 1$ to $i, \infty, 0$

14. $-1, 0, 1$ to $\infty, -i, 1$

15. $1, i, -i$ to $-1, 0, 3$

16. $1, i, -i$ to $i, -i, 1$

17. Use the results in Example 2 and the harmonic function $U = (\log_e r)/(\log_e r_0)$ to solve the Dirichlet problem in Figure 19.37. Explain why the level curves must be circles.

18. Use the linear fractional transformation that maps $-1, 1, 0$ to $0, 1, \infty$ to solve the Dirichlet problem in Figure 19.38. Explain why, with one exception, all level curves must be circles. Which level curve is a line?

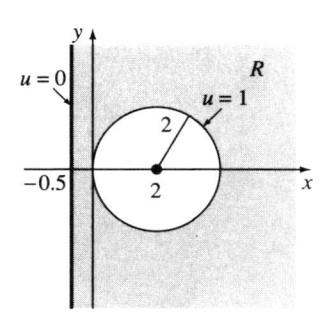

FIGURE 19.37

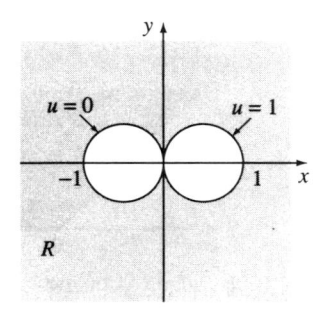

FIGURE 19.38

19. Derive the conformal mapping H-1 in the conformal mappings in Appendix III.

20. Derive the conformal mapping H-5 in conformal mappings in Appendix III by first mapping $1, i, -1$ to $\infty, i, 0$.

21. Show that the composite of two linear fractional transformations is a linear fractional transformation and verify (5).

22. If $w_1 \neq w_2$ and $\lambda > 0$, show that the set of all points w that satisfy $|w - w_1| = \lambda |w - w_2|$ is a line when $\lambda = 1$ and is a circle when $\lambda \neq 1$. [*Hint:* Write as $|w - w_1|^2 = \lambda^2 |w - w_2|^2$ and expand.]

19.4 SCHWARZ–CHRISTOFFEL TRANSFORMATIONS

Mapping to Polygonal Regions If D' is a simply connected domain with at least one boundary point, then the famous **Riemann mapping theorem** asserts the existence of an analytic function g that conformally maps the unit open disk $|z| < 1$ onto D'. The Riemann mapping theorem is a pure existence theorem which does not specify a formula for the conformal mapping. Since the upper half-plane $y > 0$ can be conformally mapped onto this disk using a linear fractional transformation, it follows that there exists a conformal mapping f between the upper half-plane and D'. In particular, there are analytic functions that map the upper half-plane onto polygonal regions of the types shown in Figure 19.39. Unlike the Riemann mapping theorem, the Schwarz–Christoffel formula specifies a form for the derivative $f'(z)$ of a conformal mapping from the upper half-plane to a bounded or unbounded polygonal region.

Special Cases To motivate the general Schwarz–Christoffel formula, we first examine the effect of the mapping $f(z) = (z - x_1)^{\alpha/\pi}$, $0 < \alpha < 2\pi$, on the upper half-plane $y \geq 0$. This mapping is the composite of the translation $\zeta = z - x_1$ and the real power function $w = \zeta^{\alpha/\pi}$. Since $w = \zeta^{\alpha/\pi}$ changes the angle in a wedge by a factor of α/π, the interior angle in the image region is $(\alpha/\pi)\pi = \alpha$. See Figure 19.40.

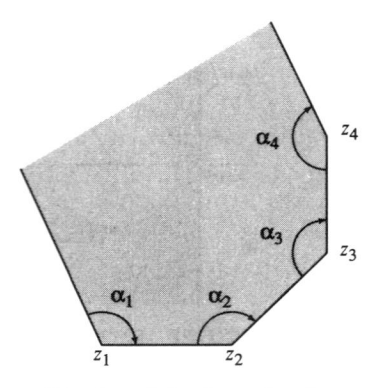

1074 CHAPTER 19 Conformal Mappings and Applications

(a) bounded polygonal region
(b) unbounded polygonal region
FIGURE 19.39

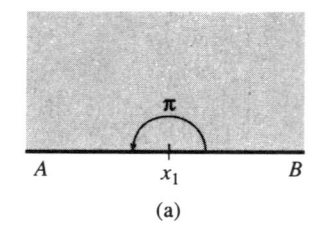

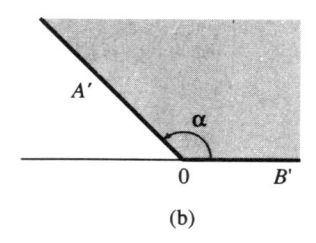

(a)
(b)
FIGURE 19.40

Note that $f'(z) = A(z - x_1)^{(\alpha/\pi) - 1}$ for $A = \alpha/\pi$. Next assume that $f(z)$ is a function which is analytic in the upper half-plane and which has the derivative

$$f'(z) = A(z - x_1)^{(\alpha_1/\pi) - 1}(z - x_2)^{(\alpha_2/\pi) - 1} \qquad (1)$$

where $x_1 < x_2$. In determining the images of line segments on the x-axis, we will use the fact that a curve $w = w(t)$ in the w-plane is a line segment when the argument of its tangent vector $w'(t)$ is constant. From (1), an argument of $f'(t)$ is given by

$$\arg f'(t) = \operatorname{Arg} A + \left(\frac{\alpha_1}{\pi} - 1\right) \operatorname{Arg}(t - x_1) + \left(\frac{\alpha_2}{\pi} - 1\right) \operatorname{Arg}(t - x_2) \qquad (2)$$

Since $\operatorname{Arg}(t - x) = \pi$ for $t < x$, we can find the variation of $\arg f'(t)$ along the x-axis. The results are shown in the following table:

Interval	arg $f'(t)$	Change in argument
$(-\infty, x_1)$	$\operatorname{Arg} A + (\alpha_1 - \pi) + (\alpha_2 - \pi)$	0
(x_1, x_2)	$\operatorname{Arg} A + (\alpha_2 - \pi)$	$\pi - \alpha_1$
(x_2, ∞)	$\operatorname{Arg} A$	$\pi - \alpha_2$

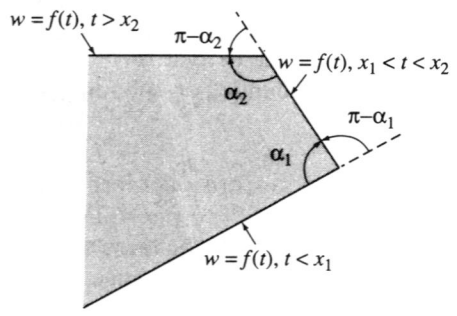

FIGURE 19.41

Since $\arg f'(t)$ is constant on the intervals in the table, the images are line segments, and Figure 19.41 shows the image of the upper half-plane. Note that the interior angles of the polygonal image region are α_1 and α_2. This discussion generalizes to produce the Schwarz–Christoffel formula.

General Formula

> **THEOREM 19.4 Schwarz–Christoffel Formula**
>
> Let $f(z)$ be a function which is analytic in the upper half-plane $y > 0$ and which has the derivative
>
> $$f'(z) = A(z - x_1)^{(\alpha_1/\pi) - 1}(z - x_2)^{(\alpha_2/\pi) - 1} \cdots (z - x_n)^{(\alpha_n/\pi) - 1} \quad (3)$$
>
> where $x_1 < x_2 < \cdots < x_n$ and each α_i satisfies $0 < \alpha_i < 2\pi$. Then $f(z)$ maps the upper half-plane $y \geq 0$ to a polygonal region with interior angles $\alpha_1, \alpha_2, \ldots, \alpha_n$.

In applying this formula to a particular polygonal target region, the reader should carefully note the following comments:

(1) One can select the location of three of the points x_k on the x-axis. A judicious choice can simplify the computation of $f(z)$. The selection of the remaining points depends on the shape of the target polygon.

(2) A general formula for $f(z)$ is

$$f(z) = A\left(\int (z - x_1)^{(\alpha_1/\pi) - 1}(z - x_2)^{(\alpha_2/\pi) - 1} \cdots (z - x_n)^{(\alpha_n/\pi) - 1} \, dz\right) + B$$

and therefore $f(z)$ may be considered as the composite of the conformal mapping

$$g(z) = \int (z - x_1)^{(\alpha_1/\pi) - 1}(z - x_2)^{(\alpha_2/\pi) - 1} \cdots (z - x_n)^{(\alpha_n/\pi) - 1} \, dz$$

and the linear function $w = Az + B$. The linear function $w = Az + B$ allows us to magnify, rotate, and translate the image polygon produced by $g(z)$. (See Section 19.1.)

(3) If the polygonal region is bounded, only $n - 1$ of the n interior angles should be included in the Schwarz–Christoffel formula. As an illustration, the interior angles $\alpha_1, \alpha_2, \alpha_3$, and α_4 are sufficient to determine the Schwarz–Christoffel formula for the pentagon shown in Figure 19.39(a).

EXAMPLE 1 Use the Schwarz–Christoffel formula to construct a conformal mapping from the upper half-plane to the strip $|v| \leq 1$, $u \geq 0$.

Solution We may select $x_1 = -1$ and $x_2 = 1$ on the x-axis, and we will construct a conformal mapping f with $f(-1) = -i$ and $f(1) = i$. See Figure 19.42. Since $\alpha_1 = \alpha_2 = \pi/2$, the Schwarz–Christoffel formula (3) gives

$$f'(z) = A(z + 1)^{-1/2}(z - 1)^{-1/2} = A\frac{1}{(z^2 - 1)^{1/2}} = \frac{A}{i}\frac{1}{(1 - z^2)^{1/2}}$$

CHAPTER 19 Conformal Mappings and Applications

FIGURE 19.42

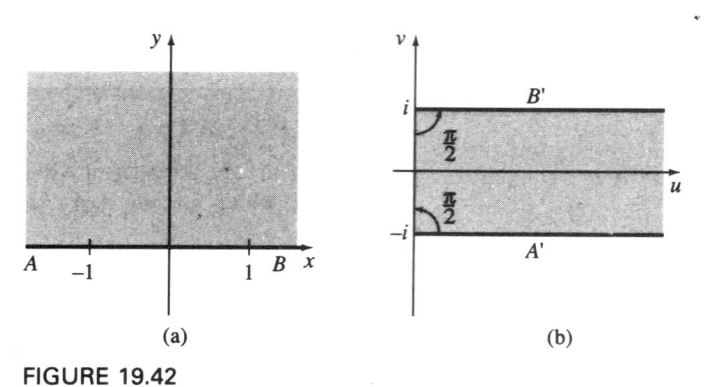

Therefore, $f(z) = -Ai \sin^{-1} z + B$. Since $f(-1) = -i$ and $f(1) = i$, we obtain, respectively,

$$-i = Ai\frac{\pi}{2} + B$$

$$i = -Ai\frac{\pi}{2} + B$$

and conclude that $B = 0$ and $A = -2/\pi$. Thus, $f(z) = (2/\pi)i \sin^{-1} z$. □

EXAMPLE 2 Use the Schwarz–Christoffel formula to construct a conformal mapping from the upper half-plane to the region shown in Figure 19.43(b).

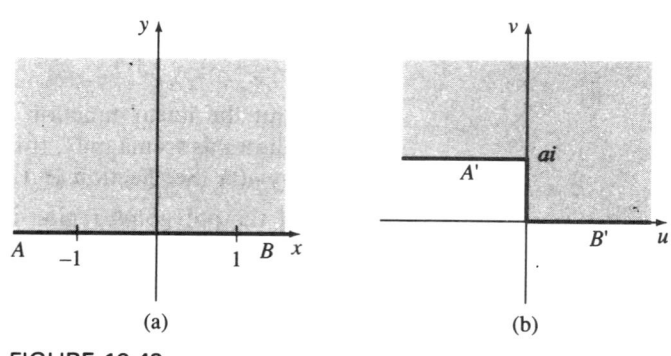

FIGURE 19.43

Solution We again select $x_1 = -1$ and $x_2 = 1$, and we will require that $f(-1) = ai$ and $f(1) = 0$. Since $\alpha_1 = 3\pi/2$ and $\alpha_2 = \pi/2$, the Schwarz–Christoffel formula (3) gives

$$f'(z) = A(z + 1)^{1/2}(z - 1)^{-1/2}$$

If we write $f'(z)$ as $A(z/(z^2 - 1)^{1/2} + 1/(z^2 - 1)^{1/2})$, it follows that

$$f(z) = A[(z^2 - 1)^{1/2} + \cosh^{-1} z] + B$$

19.4 Schwarz–Christoffel Transformations

Note that $\cosh^{-1}(-1) = \pi i$ and $\cosh^{-1} 1 = 0$, and so $ai = f(-1) = A(\pi i) + B$ and $0 = f(1) = B$. Therefore, $A = a/\pi$ and $f(z) = (a/\pi)[(z^2 - 1)^{1/2} + \cosh^{-1} z]$. □

The next example will show that it may not always be possible to find $f(z)$ in terms of elementary functions.

EXAMPLE 3 Use the Schwarz–Christoffel formula to construct a conformal mapping from the upper half-plane to the interior of the equilateral triangle shown in Figure 19.44(b).

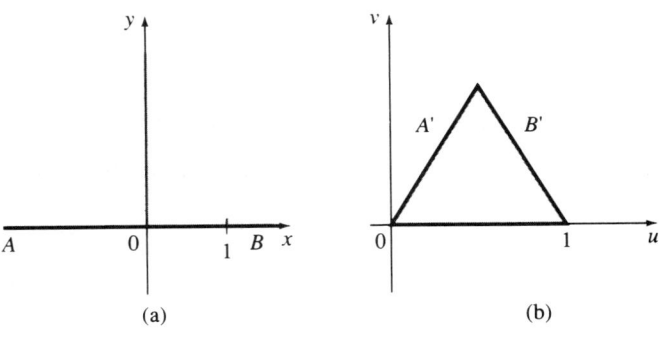

FIGURE 19.44

Solution Since the polygonal region is bounded, only two of the three 60° interior angles should be included in the Schwarz–Christoffel formula. If $x_1 = 0$ and $x_2 = 1$, we obtain $f'(z) = Az^{-2/3}(z-1)^{-2/3}$. It is not possible to evaluate $f(z)$ in terms of elementary functions; however, we can use Theorem 17.8 to construct the antiderivative

$$f(z) = A \int_0^z \frac{1}{s^{2/3}(s-1)^{2/3}} \, ds + B$$

If we require that $f(0) = 0$ and $f(1) = 1$, it follows that $B = 0$ and

$$1 = A \int_0^1 \frac{1}{x^{2/3}(x-1)^{2/3}} \, dx$$

It can be shown that this last integral is $\Gamma(\frac{1}{3})$, where Γ denotes the gamma function. Therefore, the required conformal mapping is

$$f(z) = \frac{1}{\Gamma(\frac{1}{3})} \int_0^z \frac{1}{s^{2/3}(s-1)^{2/3}} \, ds \qquad \square$$

The Schwarz–Christoffel formula can sometimes be used to suggest a possible conformal mapping from the upper half-plane onto a nonpolygonal region R'. A key first step is to approximate R' by polygonal regions. This will be illustrated in the final example.

EXAMPLE 4 Use the Schwarz–Christoffel formula to construct a conformal mapping from the upper half-plane to the upper half-plane with the horizontal line $v = \pi$, $u \leq 0$, deleted.

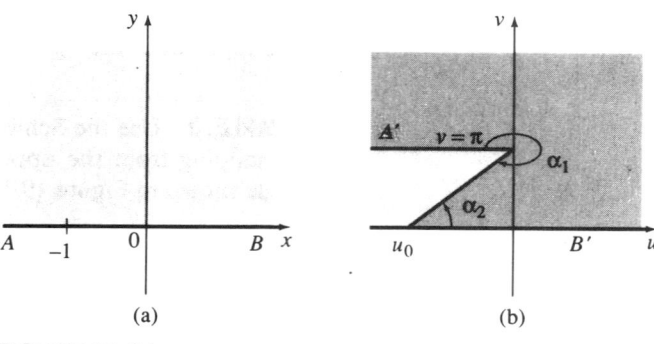

FIGURE 19.45

Solution The nonpolygonal target region can be approximated by a polygonal region by adjoining a line segment from $w = \pi i$ to a point u_0 on the negative u-axis. See Figure 19.45(b). If we require that $f(-1) = \pi i$ and $f(0) = u_0$, the Schwarz–Christoffel transformation satisfies

$$f'(z) = A(z+1)^{(\alpha_1/\pi)-1} z^{(\alpha_2/\pi)-1}$$

Note that as u_0 approaches $-\infty$, the interior angles α_1 and α_2 approach 2π and 0, respectively. This suggests we examine conformal mappings that satisfy $w' = A(z+1)^1 z^{-1} = A(1 + 1/z)$ or $w = A(z + \text{Ln } z) + B$.

We will first determine the image of the upper half-plane under $g(z) = z + \text{Ln } z$ and then translate the image region if needed. For t real,

$$g(t) = t + \log_e|t| + i \text{ Arg } t$$

If $t < 0$, $\text{Arg } t = \pi$ and $u(t) = t + \log_e|t|$ varies from $-\infty$ to -1. It follows that $w = g(t)$ moves along the line $v = \pi$ from $-\infty$ to -1. When $t > 0$, $\text{Arg } t = 0$ and $u(t)$ varies from $-\infty$ to ∞. Therefore g maps the positive x-axis onto the u-axis. We can conclude that $g(z) = z + \text{Ln } z$ maps the upper half-plane onto the upper half-plane with the horizontal line $v = \pi$, $u \leq -1$, deleted. Therefore, $w = z + \text{Ln } z + 1$ maps the upper half-plane onto the original target region. □

Many of the conformal mappings in Appendix III can be derived using the Schwarz–Christoffel formula, and we will show in Section 19.6 that these mappings are especially useful in analyzing two-dimensional fluid flows.

EXERCISES 19.4 Answers to odd-numbered problems begin on page A-91.

In Problems 1–4 use (2) to describe the image of the upper half-plane $y \geq 0$ under the conformal mapping $w = f(z)$ that satisfies the given conditions. Do *not* attempt to find $f(z)$.

1. $f'(z) = (z-1)^{-1/2}$, $f(1) = 0$

2. $f'(z) = (z+1)^{-1/3}$, $f(-1) = 0$

3. $f'(z) = (z+1)^{-1/2}(z-1)^{1/2}$, $f(-1) = 0$

4. $f'(z) = (z+1)^{-1/2}(z-1)^{-3/4}$, $f(-1) = 0$

In Problems 5–8 find $f'(z)$ for the given polygonal region using $x_1 = -1$, $x_2 = 0$, $x_3 = 1$, $x_4 = 2$, and so on. Do *not* attempt to find $f(z)$.

5. $f(-1) = 0$, $f(0) = 1$

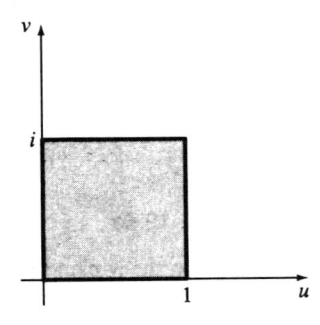

FIGURE 19.46

6. $f(-1) = -1$, $f(0) = 0$

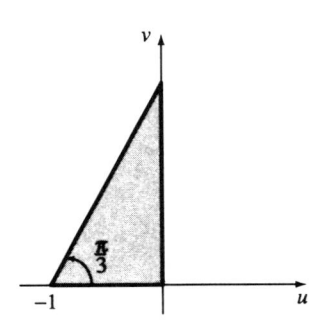

FIGURE 19.47

7. $f(-1) = -1$, $f(0) = 1$

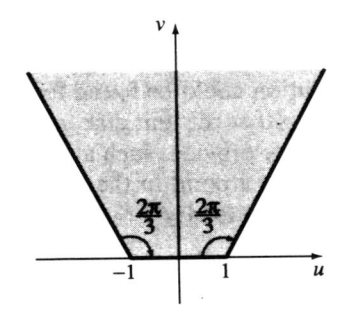

FIGURE 19.48

8. $f(-1) = i$, $f(0) = 0$

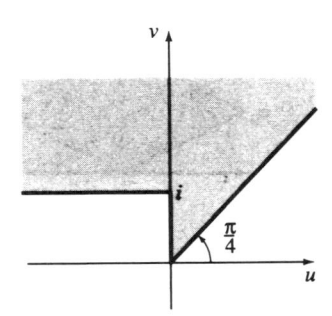

FIGURE 19.49

9. Use the Schwarz–Christoffel formula to construct a conformal mapping from the upper half-plane $y \geq 0$ to the region in Figure 19.50. Require that $f(-1) = \pi i$ and $f(1) = 0$.

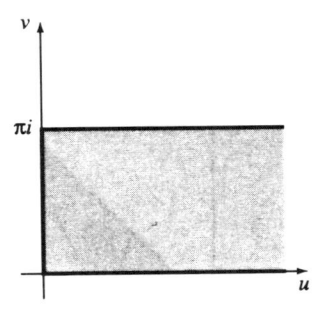

FIGURE 19.50

10. Use the Schwarz–Christoffel formula to construct a conformal mapping from the upper half-plane $y \geq 0$ to the region in Figure 19.51. Require that $f(-1) = -ai$ and $f(1) = ai$.

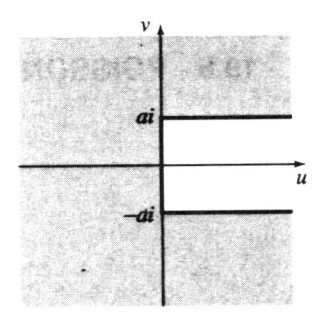

FIGURE 19.51

11. Use the Schwarz–Christoffel formula to construct a conformal mapping from the upper half-plane $y \geq 0$ to the horizontal strip $0 \leq v \leq \pi$ by first approximating the

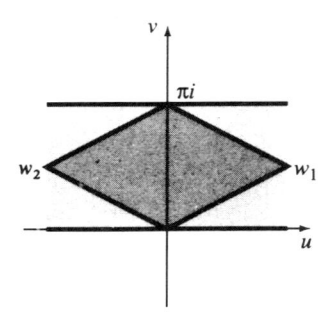

FIGURE 19.52

strip by the polygonal region shown in Figure 19.52. Require that $f(-1) = \pi i$, $f(0) = w_2 = -\overline{w_1}$, and $f(1) = 0$, and let $w_1 \to \infty$ in the horizontal direction.

12. Use the Schwarz–Christoffel formula to construct a conformal mapping from the upper half-plane $y \geq 0$ to the wedge $0 \leq \text{Arg } w \leq \pi/4$ by first approximating the wedge by the region shown in Figure 19.53. Require that $f(0) = 0$ and $f(1) = 1$ and let $\theta \to 0$.

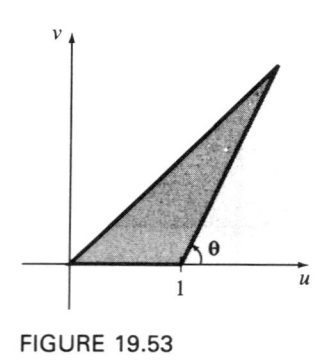

FIGURE 19.53

13. Verify M-4 in Appendix III by first approximating the region R' by the polygonal region shown in Figure 19.54. Require that $f(-1) = -u_1$, $f(0) = ai$, and $f(1) = u_1$ and let $u_1 \to 0$ along the u-axis.

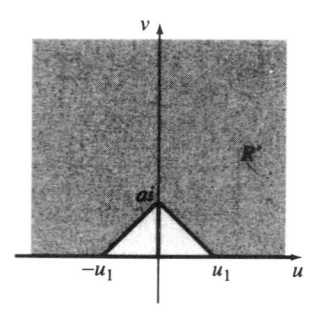

FIGURE 19.54

14. Show that if a curve in the w-plane is parameterized by $w = w(t)$, $a \leq t \leq b$, and $\arg w'(t)$ is constant, then the curve is a line segment. [*Hint:* If $w(t) = u(t) + iv(t)$, then $\tan(\arg w'(t)) = dv/du$.]

[O] 19.5 POISSON INTEGRAL FORMULAS

The success of the conformal mapping method depends on the recognition of the solution to the new Dirichlet problem in the image region R'. It would therefore be helpful if a general solution could be found for Dirichlet problems in either the upper half-plane $y \geq 0$ or the unit disk $|z| \leq 1$. The **Poisson integral formula** for the upper half-plane provides such a solution by expressing the value of a harmonic function at a point in the interior of the upper half-plane in terms of its values on the boundary $y = 0$.

Formulas for the Upper Half-Plane To develop the formula, we first assume that the boundary function is given by $u(x, 0) = f(x)$, where $f(x)$ is the step function indicated in Figure 19.55. The solution of the corresponding

19.5 Poisson Integral Formulas

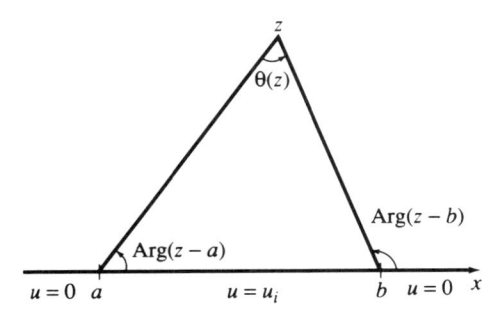

FIGURE 19.55

Dirichlet problem in the upper half-plane is

$$u(x, y) = \frac{u_i}{\pi}[\text{Arg}(z - b) - \text{Arg}(z - a)] \tag{1}$$

Since $\text{Arg}(z - b)$ is an exterior angle in the triangle formed by z, a, and b, $\text{Arg}(z - b) = \theta(z) + \text{Arg}(z - a)$, where $0 < \theta(z) < \pi$, and we can write

$$u(x, y) = \frac{u_i}{\pi}\theta(z) = \frac{u_i}{\pi}\text{Arg}\left(\frac{z - b}{z - a}\right) \tag{2}$$

The superposition principle can be used to solve the more general Dirichlet problem in Figure 19.56. If $u(x, 0) = u_i$ for $x_{i-1} \leq x \leq x_i$ and $u(x, 0) = 0$ outside of the interval $[a, b]$, then from (1),

$$u(x, y) = \sum_{i=1}^{n} \frac{u_i}{\pi}[\text{Arg}(z - x_i) - \text{Arg}(z - x_{i-1})] = \frac{1}{\pi}\sum_{i=1}^{n} u_i\theta_i(z) \tag{3}$$

Note that $\text{Arg}(z - t) = \tan^{-1}(y/(x - t))$, where $\tan^{-1}$ is selected between 0 and π, and therefore $(d/dt)\text{Arg}(z - t) = y/((x - t)^2 + y^2)$. From (3),

$$u(x, y) = \frac{1}{\pi}\sum_{i=1}^{n}\int_{x_{i-1}}^{x_i} u_i \frac{d}{dt}\text{Arg}(z - t)\, dt = \frac{1}{\pi}\sum_{i=1}^{n}\int_{x_{i-1}}^{x_i}\frac{u_i y}{(x - t)^2 + y^2}\, dt$$

Since $u(x, 0) = 0$ outside of the interval $[a, b]$, we have

$$u(x, y) = \frac{y}{\pi}\int_{-\infty}^{\infty}\frac{u(t, 0)}{(x - t)^2 + y^2}\, dt \tag{4}$$

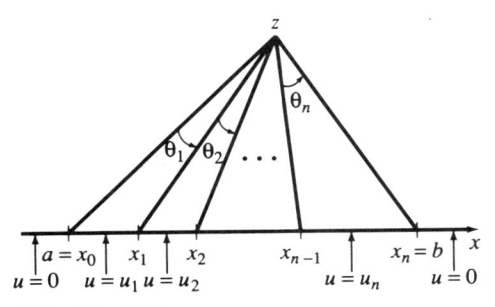

FIGURE 19.56

A bounded piecewise continuous function can be approximated by step functions, and therefore our discussion suggests that (4) is the general solution to the Dirichlet problem in the upper half-plane. This is the content of Theorem 19.5:

THEOREM 19.5 Poisson Integral Formula for the Upper Half-Plane

Let $u(x, 0)$ be a piecewise continuous function on every finite interval and bounded on $-\infty < x < \infty$. Then the function defined by

$$u(x, y) = \frac{y}{\pi} \int_{-\infty}^{\infty} \frac{u(t, 0)}{(x - t)^2 + y^2} \, dt$$

is the solution of the corresponding Dirichlet problem on the upper half-plane $y \geq 0$.

There are a few functions for which it is possible to evaluate the integral in (4), but in general numerical methods are required to evaluate the integral.

EXAMPLE 1 Find the solution of the Dirichlet problem in the upper half-plane that satisfies the boundary condition $u(x, 0) = x$ when $|x| < 1$, and $u(x, 0) = 0$ otherwise.

Solution By the Poisson integral formula,

$$u(x, y) = \frac{y}{\pi} \int_{-1}^{1} \frac{t}{(x - t)^2 + y^2} \, dt$$

Using the substitution $s = x - t$, we can show that

$$u(x, y) = \frac{1}{\pi} \left[\frac{y}{2} \log_e((x - t)^2 + y^2) - x \tan^{-1}\left(\frac{x - t}{y}\right) \right]\bigg|_{t=-1}^{t=1}$$

which can be simplified to

$$u(x, y) = \frac{y}{2\pi} \log_e\left[\frac{(x - 1)^2 + y^2}{(x + 1)^2 + y^2}\right] + \frac{x}{\pi}\left[\tan^{-1}\left(\frac{x + 1}{y}\right) - \tan^{-1}\left(\frac{x - 1}{y}\right)\right] \quad \square$$

In most of the examples and exercises $u(x, 0)$ is a step function and we will use the integrated solution (3) rather than (4). If the first interval is $(-\infty, x_1)$, then the term $\text{Arg}(z - x_1) - \text{Arg}(z - a)$ in the sum should be replaced by $\text{Arg}(z - x_1)$. Likewise, if the last interval is (x_{n-1}, ∞), then $\text{Arg}(z - b) - \text{Arg}(z - x_{n-1})$ should be replaced by $\pi - \text{Arg}(z - x_{n-1})$.

EXAMPLE 2 The conformal mapping $f(z) = z + 1/z$ maps the region in the upper half-plane and outside the circle $|z| = 1$ onto the upper half-plane $v \geq 0$. Use this mapping and the Poisson integral formula to solve the Dirichlet problem shown in Figure 19.57(a).

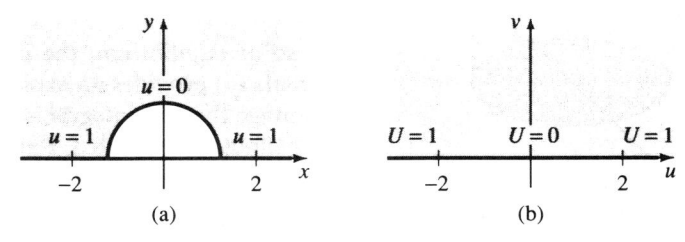

FIGURE 19.57

Solution Using the results of Example 4 in Section 19.2, we can transfer the boundary conditions to the w-plane. See Figure 19.57(b). Since $U(u, 0)$ is a step function, we will use the integrated solution (3) rather than the Poisson integral. The solution to the new Dirichlet problem is

$$U(u, v) = \frac{1}{\pi} \operatorname{Arg}(w + 2) + \frac{1}{\pi}[\pi - \operatorname{Arg}(w - 2)] = 1 + \frac{1}{\pi} \operatorname{Arg}\left(\frac{w + 2}{w - 2}\right)$$

and therefore

$$u(x, y) = U\left(z + \frac{1}{z}\right) = 1 + \frac{1}{\pi} \operatorname{Arg}\left(\frac{z + 1/z + 2}{z + 1/z - 2}\right)$$

which can be simplified to $u(x, y) = 1 + \dfrac{1}{\pi} \operatorname{Arg}\left(\dfrac{z + 1}{z - 1}\right)^2$. □

Formula for the Unit Disk A Poisson integral formula can also be developed to solve the general Dirichlet problem for the unit disk:

THEOREM 19.6 Poisson Integral Formula for the Unit Disk

Let $u(e^{i\theta})$ be bounded and piecewise continuous for $-\pi \leq \theta \leq \pi$. Then the solution to the corresponding Dirichlet problem on the open unit disk $|z| < 1$ is given by

$$u(x, y) = \frac{1}{2\pi} \int_{-\pi}^{\pi} u(e^{it}) \frac{1 - |z|^2}{|e^{it} - z|^2} \, dt \qquad (5)$$

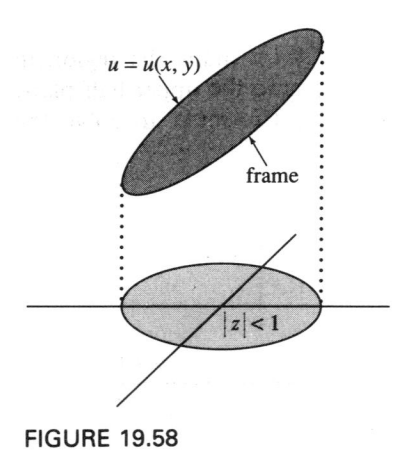

Geometric Interpretation Figure 19.58 shows a thin membrane (such as a soap film) that has been stretched across a frame defined by $u = u(e^{i\theta})$. The displacement u in the direction perpendicular to the z-plane satisfies the **two-dimensional wave equation**

$$a^2\left(\frac{\partial^2 u}{\partial x^2} + \frac{\partial^2 u}{\partial y^2}\right) = \frac{\partial^2 u}{\partial t^2}$$

and so at equilibrium, the displacement function $u = u(x, y)$ is harmonic. Formula (5) provides an explicit solution for the displacement u and has the advantage that the integral is over the finite interval $[-\pi, \pi]$. When the integral cannot be evaluated, standard numerical integration procedures can be used to estimate $u(x, y)$ at a fixed point $z = x + iy$ with $|z| < 1$.

EXAMPLE 3 A frame for a membrane is defined by $u(e^{i\theta}) = |\theta|$ for $-\pi \leq \theta \leq \pi$. Estimate the equilibrium displacement of the membrane at $(-0.5, 0)$, $(0, 0)$, and $(0.5, 0)$.

Solution Using (5), we get $u(x, y) = \frac{1}{2\pi}\int_{-\pi}^{\pi} |t|\frac{1-|z|^2}{|e^{it}-z|^2}\,dt$. When $(x, y) = (0, 0)$, we get

$$u(x, y) = \frac{1}{2\pi}\int_{-\pi}^{\pi} |t|\,dt = \frac{\pi}{2}$$

For the other two values of (x, y), the integral is not elementary and must be estimated using a numerical integration procedure. Using Simpson's rule, we obtain (to four decimal places) $u(-0.5, 0) = 2.2269$ and $u(0.5, 0) = 0.9147$. □

Fourier Series Form The Poisson integral formula for the unit disk is actually a compact way of writing the Fourier series solution to Laplace's equation that we developed in Chapter 13. To see this, first note that $u_n(r, \theta) = r^n \cos n\theta$ and $v_n(r, \theta) = r^n \sin n\theta$ are each harmonic, since these functions are the real and imaginary parts of z^n. If a_0, a_n, and b_n are chosen to be the Fourier coefficients of $u(e^{i\theta})$ for $-\pi < \theta < \pi$, then, by the superposition principle,

$$u(r, \theta) = \frac{a_0}{2} + \sum_{n=1}^{\infty}(a_n r^n \cos n\theta + b_n r^n \sin n\theta) \qquad (6)$$

is harmonic and $u(1, \theta) = (a_0/2) + \Sigma_{n=1}^{\infty}(a_n \cos n\theta + b_n \sin n\theta) = u(e^{i\theta})$. Since the solution of the Dirichlet problem is also given by (5), we have

$$u(r, \theta) = \frac{1}{2\pi}\int_{-\pi}^{\pi} u(e^{it})\frac{1-r^2}{|e^{it}-re^{i\theta}|^2}\,dt = \frac{a_0}{2} + \sum_{n=1}^{\infty}(a_n r^n \cos n\theta + b_n r^n \sin n\theta)$$

EXAMPLE 4 Find the solution of the Dirichlet problem in the unit disk satisfying the boundary condition $u(e^{i\theta}) = \sin 4\theta$. Sketch the level curve $u = 0$.

19.5 Poisson Integral Formulas

Solution Rather than working with the Poisson integral (5), we will use the Fourier series solution (6), which reduces to $u(r, \theta) = r^4 \sin 4\theta$. Therefore, $u = 0$ if and only if $\sin 4\theta = 0$. This implies $u = 0$ on the lines $x = 0$, $y = 0$, and $y = \pm x$.

If we switch to rectangular coordinates, $u(x, y) = 4xy(x^2 - y^2)$. The surface $u(x, y) = 4xy(x^2 - y^2)$, the frame $u(e^{i\theta}) = \sin 4\theta$, and the system of level curves were sketched using graphics software and are shown in Figure 19.59.

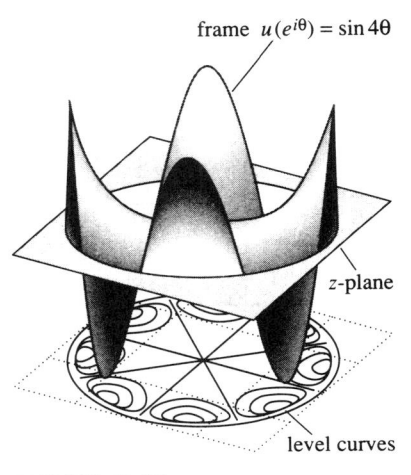

FIGURE 19.59

EXERCISES 19.5 *Answers to odd-numbered problems begin on page A-91.*

In Problems 1–4 use the integrated solution (3) to the Poisson integral formula to solve the given Dirichlet problem in the upper half-plane.

1.

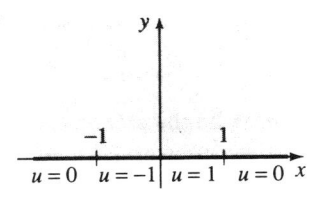

$u = 0 \quad u = -1 \mid u = 1 \quad u = 0$ at $-1, 1$ on x-axis

FIGURE 19.60

2.

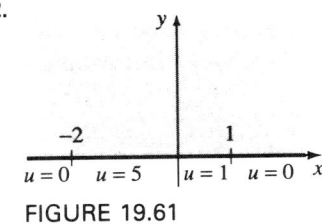

$u = 0 \quad u = 5 \mid u = 1 \quad u = 0$ at $-2, 1$ on x-axis

FIGURE 19.61

3.

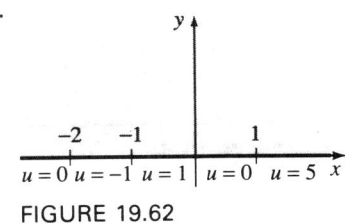

$u = 0 \quad u = -1 \quad u = 1 \mid u = 0 \quad u = 5$ at $-2, -1, 1$ on x-axis

FIGURE 19.62

4.

$u = 1 \quad u = -1 \quad u = 1 \mid u = 1 \quad u = 0$ at $-2, -1, 1$ on x-axis

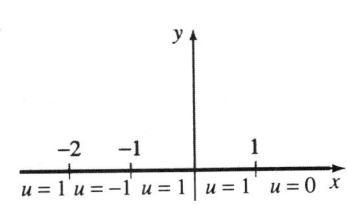

FIGURE 19.63

5. Find the solution of the Dirichlet problem in the upper half-plane that satisfies the boundary condition $u(x, 0) = x^2$ when $0 < x < 1$, and $u(x, 0) = 0$ otherwise.

6. Find the solution of the Dirichlet problem in the upper half-plane that satisfies the boundary condition $u(x, 0) = \cos x$. [*Hint:* Let $s = t - x$ and use the Section 18.6 formulas

$$\int_{-\infty}^{\infty} \frac{\cos s}{s^2 + a^2}\, ds = \frac{\pi e^{-a}}{a} \qquad \int_{-\infty}^{\infty} \frac{\sin s}{s^2 + a^2}\, ds = 0$$

for $a > 0$.]

In Problems 7–10 solve the given Dirichlet problem by finding a conformal mapping from the given region onto the upper half-plane $v \geq 0$.

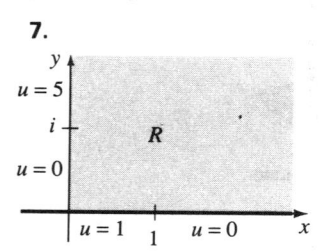

7.

FIGURE 19.64

8.

FIGURE 19.65

9.

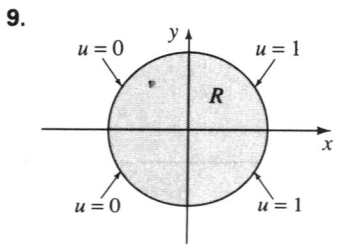

FIGURE 19.66

10.

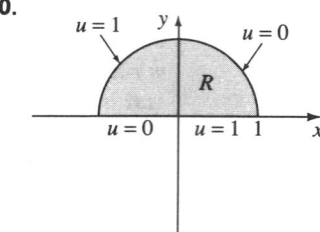

FIGURE 19.67

11. A frame for a membrane is defined by $u(e^{i\theta}) = \theta^2/\pi^2$ for $-\pi \leq \theta \leq \pi$. Use the Poisson integral formula for the unit disk to estimate the equilibrium displacement of the membrane at $(-0.5, 0)$, $(0, 0)$, and $(0.5, 0)$.

12. A frame for a membrane is defined by $u(e^{i\theta}) = e^{-|\theta|}$ for $-\pi \leq \theta \leq \pi$. Use the Poisson integral formula for the unit disk to estimate the equilibrium displacement of the membrane at $(-0.5, 0)$, $(0, 0)$, and $(0.5, 0)$.

13. Use the Poisson integral formula for the unit disk to show that $u(0, 0)$ is the average value of the function $u = u(e^{i\theta})$ on the boundary $|z| = 1$.

In Problems 14 and 15 solve the given Dirichlet problem for the unit disk using the Fourier series form of the Poisson integral formula, and sketch the system of level curves.

14. $u(e^{i\theta}) = \cos 2\theta$ 15. $u(e^{i\theta}) = \sin\theta + \cos\theta$

19.6 APPLICATIONS

In Sections 19.2, 19.3, and 19.5 we demonstrated how Laplace's equation can be solved with conformal mapping methods and we interpreted a solution $u = u(x, y)$ of the Dirichlet problem as either the steady-state temperature at the point (x, y) or the equilibrium displacement of a membrane at the point (x, y). Laplace's equation is a fundamental partial differential equation that arises in a variety of contexts. In this section we will establish a general relationship between vector fields and analytic functions and use our conformal mapping techniques to solve problems involving electrostatic force fields and two-dimensional fluid flows.

Vector Fields A vector field $\mathbf{F}(x, y) = P(x, y)\mathbf{i} + Q(x, y)\mathbf{j}$ in a domain D can also be expressed in the complex form

$$\mathbf{F}(x, y) = P(x, y) + iQ(x, y)$$

19.6 Applications

and thought of as a complex function. Recall from Chapter 8 that div $\mathbf{F} = \partial P/\partial x + \partial Q/\partial y$ and curl $\mathbf{F} = (\partial Q/\partial x - \partial P/\partial y)\mathbf{k}$. If we require that both div $\mathbf{F} = 0$ and curl $\mathbf{F} = \mathbf{0}$, then

$$\frac{\partial P}{\partial x} = -\frac{\partial Q}{\partial y} \quad \text{and} \quad \frac{\partial P}{\partial y} = \frac{\partial Q}{\partial x} \tag{1}$$

This set of equations is reminiscent of the Cauchy–Riemann criterion for analyticity presented in Theorem 16.3 and suggests that we examine the complex function $g(z) = P(x, y) - iQ(x, y)$.

THEOREM 19.7   Vector Fields and Analyticity

(i) Suppose that $\mathbf{F}(x, y) = P(x, y) + iQ(x, y)$ is a vector field in a domain D and $P(x, y)$ and $Q(x, y)$ are continuous and have continuous first partial derivatives in D. If div $\mathbf{F} = 0$ and curl $\mathbf{F} = \mathbf{0}$, then the complex function

$$g(z) = P(x, y) - iQ(x, y)$$

is analytic in D.

(ii) Conversely, if $g(z)$ is analytic in D, then $\mathbf{F}(x, y) = \overline{g(z)}$ defines a vector field in D for which div $\mathbf{F} = 0$ and curl $\mathbf{F} = \mathbf{0}$.

Proof If $u(x, y)$ and $v(x, y)$ denote the real and imaginary parts of $g(z)$, then $u = P$ and $v = -Q$. Therefore the equations in (1) are equivalent to the equations

$$\frac{\partial u}{\partial x} = -\frac{\partial(-v)}{\partial y} \quad \text{and} \quad \frac{\partial u}{\partial y} = \frac{\partial(-v)}{\partial x}$$

that is,
$$\frac{\partial u}{\partial x} = \frac{\partial v}{\partial y} \quad \text{and} \quad \frac{\partial u}{\partial y} = -\frac{\partial v}{\partial x} \tag{2}$$

The equations in (2) are the Cauchy–Riemann equations for analyticity. ∎

EXAMPLE 1 The vector field defined by $\mathbf{F}(x, y) = (-kq/|z - z_0|^2)(z - z_0)$ may be interpreted as the electric field produced by a wire that is perpendicular to the z-plane at $z = z_0$ and carries a charge of q coulombs per unit length. The corresponding complex function is

$$g(z) = \frac{-kq}{|z - z_0|^2} \overline{(z - z_0)} = \frac{-kq}{\overline{z - z_0}}$$

Since $g(z)$ is analytic for $z \neq z_0$, div $\mathbf{F} = 0$ and curl $\mathbf{F} = \mathbf{0}$. □

EXAMPLE 2 The complex function $g(z) = Az$, $A > 0$, is analytic in the first quadrant and therefore gives rise to the vector field $\mathbf{V}(x, y) = \overline{g(z)} = Ax - iAy$, which satisfies div $\mathbf{V} = 0$ and curl $\mathbf{V} = \mathbf{0}$. We will show toward the end of this section that $\mathbf{V}(x, y)$ may be interpreted as the velocity of a fluid that moves around the corner produced by the boundary of the first quadrant. □

The physical interpretation of the conditions div $\mathbf{F} = 0$ and curl $\mathbf{F} = \mathbf{0}$ depends on the setting. If $\mathbf{F}(x, y)$ represents the force in an electric field which acts on a unit test charge placed at (x, y), then, by Theorem 8.9, curl $\mathbf{F} = \mathbf{0}$ if and only if the field is **conservative**. The work done in transporting a test charge between two points in D must be independent of the path.

If C is a simple closed contour that lies in D, *Gauss's law* asserts that the line integral $\oint_C (\mathbf{F} \cdot \mathbf{n})\, ds$ is proportional to the total charge enclosed by the curve C. If D is simply connected and all the electric charge is distributed on the boundary of D, then $\oint_C (\mathbf{F} \cdot \mathbf{n})\, ds = 0$ for any simple closed contour in D. By the divergence theorem in the form (1) of Section 8.16,

$$\oint_C (\mathbf{F} \cdot \mathbf{n})\, ds = \iint_R \operatorname{div} \mathbf{F}\, dA \tag{3}$$

where R is the region enclosed by C, and we can conclude that div $\mathbf{F} = 0$ in D. Conversely, if div $\mathbf{F} = 0$ in D, the double integral is zero and therefore the domain D contains no charge.

Potential Functions Suppose that $\mathbf{F}(x, y)$ is a vector field in a *simply connected* domain D with both div $\mathbf{F} = 0$ and curl $\mathbf{F} = \mathbf{0}$. By Theorem 17.8, the analytic function $g(z) = P(x, y) - iQ(x, y)$ has an antiderivative

$$G(z) = \phi(x, y) + i\psi(x, y) \tag{4}$$

in D, which is called a **complex potential** for the vector field $\mathbf{F}$. Note that $g(z) = G'(z) = \dfrac{\partial \phi}{\partial x}(x, y) + i\dfrac{\partial \psi}{\partial x}(x, y) = \dfrac{\partial \phi}{\partial x}(x, y) - i\dfrac{\partial \phi}{\partial y}(x, y)$ and so

$$\frac{\partial \phi}{\partial x} = P \quad \text{and} \quad \frac{\partial \phi}{\partial y} = Q \tag{5}$$

Therefore, $\mathbf{F} = \nabla \phi$ and, as in Section 8.9, the harmonic function ϕ is called a (real) **potential function** for $\mathbf{F}$.* When the potential ϕ is specified on the boundary of a region R, we can use conformal mapping techniques to solve the resulting Dirichlet problem. The equipotential lines $\phi(x, y) = c$ can be sketched and the vector field $\mathbf{F}$ can be determined using (5).

---

*If $\mathbf{F}$ is an electric field, the electric potential function Φ is defined to be $-\phi$ and $\mathbf{F} = -\nabla \Phi$.

19.6 Applications

EXAMPLE 3 The potential ϕ in the half-plane $x \geq 0$ satisfies the boundary conditions $\phi(0, y) = 0$ and $\phi(x, 0) = 1$ for $x \geq 1$ (see Figure 19.68(a)). Determine a complex potential, the equipotential lines, and the force field **F**.

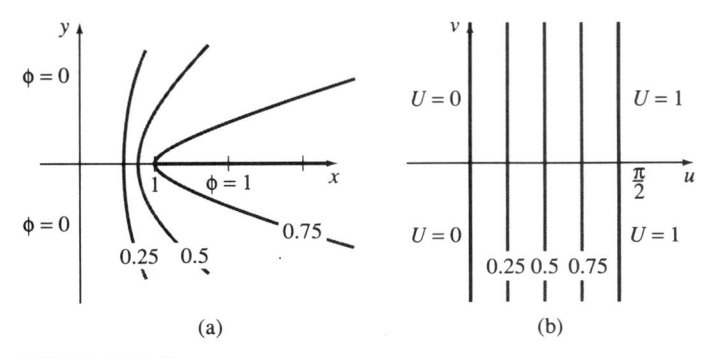

FIGURE 19.68

Solution We saw in Example 3 of Section 19.2 that the analytic function $z = \sin w$ maps the strip $0 \leq u \leq \pi/2$ in the w-plane to the region R in question. Therefore $f(z) = \sin^{-1} z$ maps R onto the strip, and Figure 19.68(b) shows the transferred boundary conditions. The simplified Dirichlet problem has the solution $U(u, v) = \frac{2}{\pi} u$, and so $\phi(x, y) = U(\sin^{-1} z) = \text{Re}(\frac{2}{\pi} \sin^{-1} z)$ is the potential function on D, and $G(z) = \frac{2}{\pi} \sin^{-1} z$ is a complex potential for the force field **F**.

Note that the equipotential lines $\phi = c$ are the images of the equipotential lines $U = c$ in the w-plane under the inverse mapping $z = \sin w$. In Example 3 of Section 19.2 we showed that the vertical line $u = a$ is mapped onto a branch of the hyperbola

$$\frac{x^2}{\sin^2 a} - \frac{y^2}{\cos^2 a} = 1$$

Since the equipotential line $U = c$, $0 < c < 1$, is the vertical line $u = \frac{\pi}{2} c$, it follows that the equipotential line $\phi = c$ is the right branch of the hyperbola

$$\frac{x^2}{\sin^2(\pi c/2)} - \frac{y^2}{\cos^2(\pi c/2)} = 1$$

Since $\mathbf{F} = \overline{G'(z)}$ and $(d/dz) \sin^{-1} z = 1/(1 - z^2)^{1/2}$, the force field is given by

$$\mathbf{F} = \frac{2}{\pi} \overline{\frac{1}{(1 - z^2)^{1/2}}} = \frac{2}{\pi} \frac{1}{(1 - \bar{z}^2)^{1/2}} \qquad \square$$

Steady-State Fluid Flow The vector $\mathbf{V}(x, y) = P(x, y) + iQ(x, y)$ may also be interpreted as the velocity vector of a two-dimensional steady-state fluid flow at a point (x, y) in a domain D. The velocity at all points in the domain is therefore independent of time and all movement takes place in planes that are parallel to a z-plane.

The physical interpretation of the conditions div **V** = 0 and curl **V** = **0** was discussed in Section 8.7. Recall that if curl **V** = **0** in *D*, the flow is called **irrotational**. If a small circular paddle wheel is placed in the fluid, the net angular velocity on the boundary of the wheel is zero and so the wheel will not rotate. If div **V** = 0 in *D*, the flow is called **incompressible**. In a simply connected domain *D*, an incompressible flow has the special property that the amount of fluid in the interior of any simple closed contour *C* is independent of time. The rate at which fluid enters the interior of *C* matches the rate at which it leaves, and consequently there can be no fluid sources or sinks at points in *D*.

If div **V** = 0 and curl **V** = **0**, **V** has a **complex velocity potential**

$$G(z) = \phi(x, y) + i\psi(x, y)$$

that satisfies $\overline{G'(z)} = \mathbf{V}$. In this setting, special importance is placed on the level curves $\psi(x, y) = c$. If $z(t) = x(t) + iy(t)$ is the path of a particle (such as a small cork) which has been placed in the fluid, then

$$\begin{aligned} \frac{dx}{dt} &= P(x, y) \\ \frac{dy}{dt} &= Q(x, y) \end{aligned} \tag{6}$$

Hence, $dy/dx = Q(x, y)/P(x, y)$ or $-Q(x, y)\, dx + P(x, y)\, dy = 0$. This differential equation is exact, since div **V** = 0 implies $\partial(-Q)/\partial y = \partial P/\partial x$. By the Cauchy–Riemann equations, $\partial \psi/\partial x = -\partial \phi/\partial y = -Q$ and $\partial \psi/\partial y = \partial \phi/\partial x = P$, and therefore all solutions of (6) satisfy $\psi(x, y) = c$. The function $\psi(x, y)$ is therefore called a **stream function** and the level curves $\psi(x, y) = c$ are **streamlines** for the flow.

EXAMPLE 4 The *uniform flow* in the upper half-plane is defined by $\mathbf{V}(x, y) = A(1, 0)$, where *A* is a fixed positive constant. Note that $|\mathbf{V}| = A$ and so a particle in the fluid moves at a constant speed. A complex potential for the vector field is $G(z) = Az = Ax + iAy$, and so the streamlines are the horizontal lines $Ay = c$. See Figure 19.69(a). Note that the boundary $y = 0$ of the region is itself a streamline. □

EXAMPLE 5 The analytic function $G(z) = z^2$ gives rise to the vector field $\mathbf{V}(x, y) = \overline{G'(z)} = (2x, -2y)$ in the first quadrant. Since $z^2 = x^2 - y^2 + i(2xy)$, the stream function is $\psi(x, y) = 2xy$ and the streamlines are the hyperbolas $2xy = c$. This flow, called *flow around a corner*, is depicted in Figure 19.69(b). As in Example 4, the boundary lines $x = 0$ and $y = 0$ in the first quadrant are themselves streamlines. □

Constructing Special Flows The process of constructing an irrotational and incompressible flow that remains inside a given region *R* is called **streamlining**. Since the streamlines are described by $\psi(x, y) = c$, two distinct streamlines do not intersect. Therefore if the boundary is itself a streamline,

19.6 Applications

FIGURE 19.69

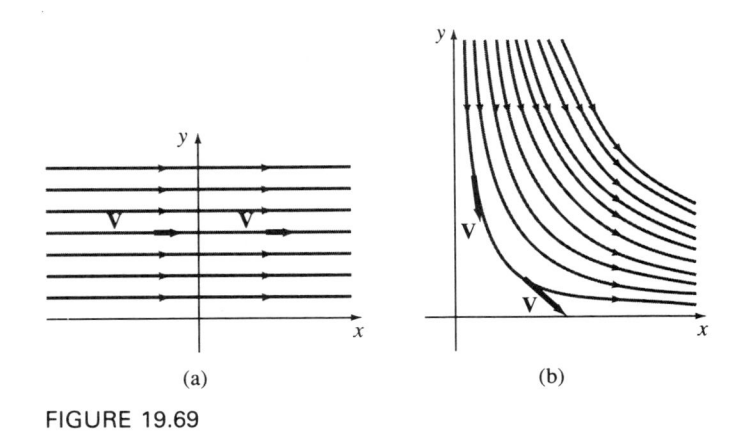

a particle which starts inside R cannot leave R. This is the content of the following theorem:

THEOREM 19.8 Streamlining

Suppose that $G(z) = \phi(x, y) + i\psi(x, y)$ is analytic in a region R and $\psi(x, y)$ is constant on the boundary of R. Then $\mathbf{V}(x, y) = \overline{G'(z)}$ defines an irrotational and incompressible fluid flow in R. Moreover, if a particle is placed inside R, its path $z = z(t)$ remains in R.

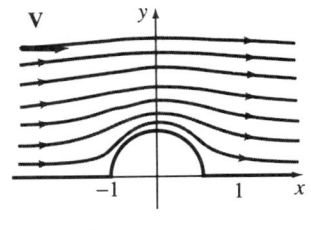

FIGURE 19.70

EXAMPLE 6 The analytic function $G(z) = z + 1/z$ maps the region R in the upper half-plane and outside the circle $|z| = 1$ onto the upper half-plane $v \geq 0$. The boundary of R is mapped onto the u-axis, and so $v = \psi(x, y) = y - y/(x^2 + y^2)$ is zero on the boundary of R. Figure 19.70 shows the streamlines of the resulting flow. The velocity field is given by $\overline{G'(z)} = 1 - 1/\bar{z}^2$, and so

$$\overline{G'(re^{i\theta})} = 1 - \frac{1}{r^2} e^{2i\theta}$$

It follows that $\mathbf{V} \approx (1, 0)$ for large values of r, and so the flow is approximately uniform at large distances from the circle $|z| = 1$. The resulting flow in the region R is called *flow around a cylinder*. The mirror image of the flow can be adjoined to give a flow around a complete cylinder. □

If R is a polygonal region, we can use the Schwarz–Christoffel formula to find a conformal mapping $z = f(w)$ from the upper half-plane R' onto R. The inverse function $G(z) = f^{-1}(z)$ maps the boundary of R onto the u-axis. Therefore, if $G(z) = \phi(x, y) + i\psi(x, y)$, then $\psi(x, y) = 0$ on the boundary of R. Note that the streamlines $\psi(x, y) = c$ in the z-plane are the images of the horizontal lines $v = c$ in the w-plane under $z = f(w)$.

EXAMPLE 7 The analytic function $f(w) = w + \text{Ln } w + 1$ maps the upper half-plane $v \geq 0$ to the upper half-plane $y \geq 0$ with the horizontal line $y = \pi$, $x \leq 0$, deleted. See Example 4 in Section 19.4. If $G(z) = f^{-1}(z) = \phi(x, y) + i\psi(x, y)$, then $G(z)$ maps R onto the upper half-plane and maps the boundary of R onto the u-axis. Therefore, $\psi(x, y) = 0$ on the boundary of R.

It is not possible to find an explicit formula for the stream function $\psi(x, y)$. The streamlines, however, are the images of the horizontal lines $v = c$ under $z = f(w)$. If we write $w = t + ic, c > 0$, then the streamlines can be represented in the parametric form

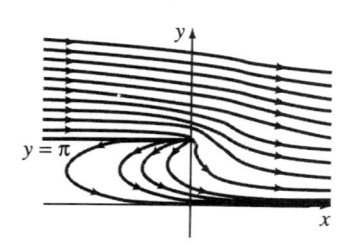

$$z = f(t + ic) = t + ic + \text{Ln}(t + ic) + 1$$

that is,
$$x = t + 1 + \frac{1}{2}\log_e(t^2 + c^2)$$
$$y = c + \text{Arg}(t + ic)$$

FIGURE 19.71

Graphing software was used to generate the streamlines in Figure 19.71. □

A stream function $\psi(x, y)$ is harmonic but, unlike a solution to a Dirichlet problem, we do *not* require $\psi(x, y)$ to be bounded (see Examples 4–6) or to assume a fixed set of constants on the boundary. Therefore there may be many different stream functions for a given region that satisfy Theorem 19.8. This will be illustrated in the final example.

EXAMPLE 8 The analytic function $f(w) = w + e^w + 1$ maps the horizontal strip $0 \leq v \leq \pi$ onto the region R shown in Figure 19.71. Therefore, $G(z) = f^{-1}(z) = \phi(x, y) + i\psi(x, y)$ maps R back to the strip and, from M-1 in the conformal mappings in Appendix III, maps the boundary line $y = 0$ onto the u-axis and maps the boundary line $y = \pi$, $x \leq 0$, onto the horizontal line $v = \pi$. Therefore, $\psi(x, y)$ is constant on the boundary of R.

The streamlines are the images of the horizontal lines $v = c, 0 < c < \pi$, under $z = f(w)$. As in Example 7, a parametric representation of the streamlines is

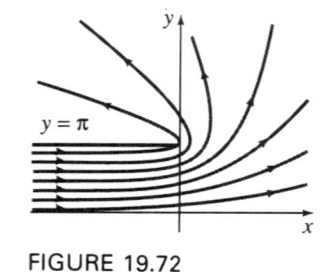

$$z = f(t + ic) = t + ic + e^{t+ic} + 1$$

or
$$x = t + 1 + e^t \cos c$$
$$y = c + e^t \sin c$$

FIGURE 19.72

The streamlines are shown in Figure 19.72. Unlike the flow in Example 7, the fluid appears to emerge from the strip $0 \leq y \leq \pi, x \leq 0$. □

EXERCISES 19.6 Answers to odd-numbered problems begin on page A-91.

In Problems 1–4 verify that div $\mathbf{F} = 0$ and curl $\mathbf{F} = \mathbf{0}$ for the given vector field $\mathbf{F}(x, y)$ by examining the corresponding complex function $g(z) = P(x, y) - iQ(x, y)$. Find a complex potential for the vector field and sketch the equipotential lines.

1. $\mathbf{F}(x, y) = (\cos \theta_0)\mathbf{i} + (\sin \theta_0)\mathbf{j}$

2. $\mathbf{F}(x, y) = -y\mathbf{i} - x\mathbf{j}$

3. $\mathbf{F}(x, y) = \dfrac{x}{x^2 + y^2}\mathbf{i} + \dfrac{y}{x^2 + y^2}\mathbf{j}$

4. $\mathbf{F}(x, y) = \dfrac{x^2 - y^2}{(x^2 + y^2)^2}\mathbf{i} + \dfrac{2xy}{(x^2 + y^2)^2}\mathbf{j}$

5. The potential ϕ on the wedge $0 \leq \operatorname{Arg} z \leq \pi/4$ satisfies the boundary conditions $\phi(x, 0) = 0$ and $\phi(x, x) = 1$ for $x > 0$. Determine a complex potential, the equipotential lines, and the corresponding force field $\mathbf{F}$.

6. Use the conformal mapping $f(z) = 1/z$ to determine a complex potential, the equipotential lines, and the corresponding force field $\mathbf{F}$ for the potential ϕ that satisfies the boundary conditions shown in Figure 19.73.

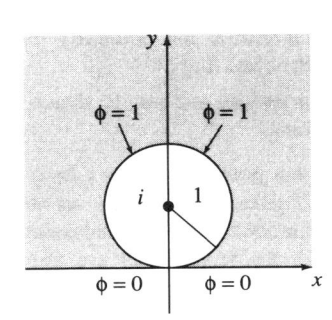

FIGURE 19.73

7. The potential ϕ on the semicircle $|z| \leq 1$, $y \geq 0$, satisfies the boundary conditions $\phi(x, 0) = 0$, $-1 < x < 1$, and $\phi(e^{i\theta}) = 1$, $0 < \theta < \pi$. Show that

$$\phi(x, y) = \frac{1}{\pi} \operatorname{Arg}\left(\frac{z-1}{z+1}\right)^2$$

and use the mapping properties of linear fractional transformations to explain why the equipotential lines are arcs of circles.

8. Use the conformal mapping C-1 in Appendix III to find the potential ϕ in the region outside the two circles $|z| = 1$ and $|z - 3| = 1$ if the potential is kept at zero on $|z| = 1$ and one on $|z - 3| = 1$. Use the mapping properties of linear fractional transformations to explain why the equipotential lines are, with one exception, circles.

In Problems 9–14 a complex velocity potential $G(z)$ is defined on a region R.
 (a) Find the stream function and verify that the boundary of R is a streamline.
 (b) Find the corresponding velocity vector field $\mathbf{V}(x, y)$.
 [O] (c) With the aid of a graphics calculator or graphics software, sketch the streamlines of the flow.

9. $G(z) = z^4$

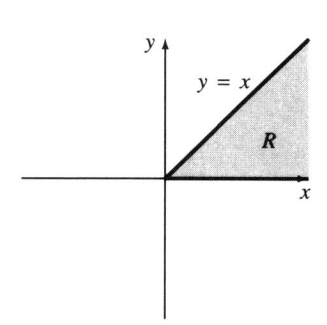

FIGURE 19.74

10. $G(z) = z^{2/3}$

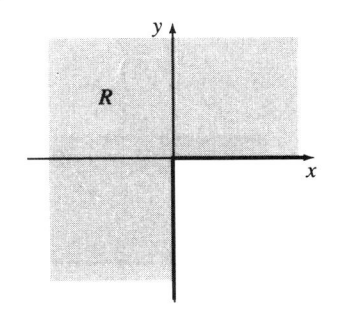

FIGURE 19.75

11. $G(z) = \sin z$

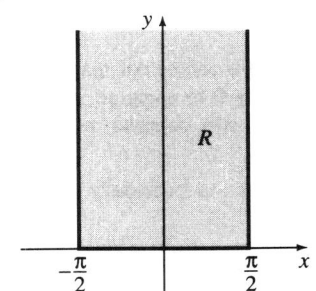

FIGURE 19.76

12. $G(z) = i \sin^{-1} z$

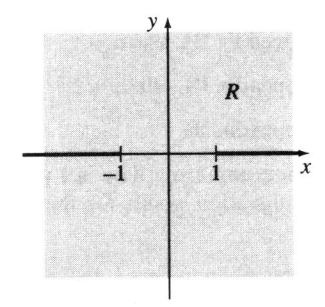

FIGURE 19.77

13. $G(z) = z^2 + 1/z^2$

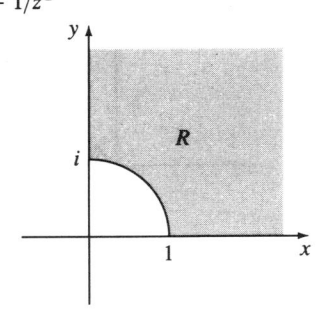

FIGURE 19.78

14. $G(z) = e^z$

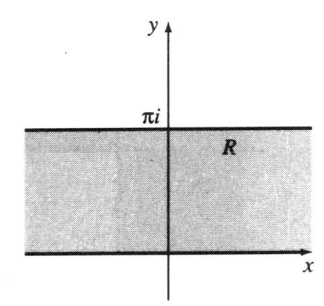

FIGURE 19.79

In Problems 15–18 a conformal mapping $z = f(w)$ from the upper half-plane $v \geq 0$ to a region R in the z-plane is given and the flow in R with complex potential $G(z) = f^{-1}(z)$ is constructed.

(a) Verify that the boundary of R is a streamline for the flow.

(b) Find a parametric representation for the streamlines of the flow.

[O] (c) With the aid of a graphics calculator or graphics software, sketch the streamlines of the flow.

15. M-9 in Appendix III

16. M-4 in Appendix III. Use $a = 1$

17. M-2 in Appendix III. Use $a = 1$

18. M-5 in Appendix III

19. A stagnation point in a flow is a point at which $\mathbf{V} = \mathbf{0}$. Find all stagnation points for the flows in Examples 5 and 6.

20. For any two real numbers k and x_1, the function $G(z) = k \operatorname{Ln}(z - x_1)$ is analytic in the upper half-plane and therefore is a complex potential for a flow. The real number x_1 is called a *sink* when $k < 0$ and a *source* for the flow when $k > 0$.

(a) Show that the streamlines are rays emanating from x_1.

(b) Show that $\mathbf{V} = (k/|z - x_1|^2)(z - x_1)$ and conclude that the flow is directed toward x_1 precisely when $k < 0$.

21. If $f(z)$ is a conformal mapping from a domain D onto the upper half-plane, a flow with a source at a point ζ_0 on the boundary of D is defined by the complex potential $G(z) = k \operatorname{Ln}(f(z) - f(\zeta_0))$, where $k > 0$. Determine the streamlines for a flow in the first quadrant with a source at $\zeta_0 = 1$ and $k = 1$.

22. (a) Construct a flow on the horizontal strip $0 < y < \pi$ with a sink at the boundary point $\zeta_0 = 0$. [*Hint:* See Problem 21.]

[O] (b) Use graphing software to sketch the streamlines of the flow.

23. The complex potential $G(z) = k \operatorname{Ln}(z - 1) - k \operatorname{Ln}(z + 1)$ with $k > 0$ gives rise to a flow on the upper half-plane with a single source at $z = 1$ and a single sink at $z = -1$. Show that the streamlines are the family of circles $x^2 + (y - c)^2 = 1 + c^2$. See Figure 19.80.

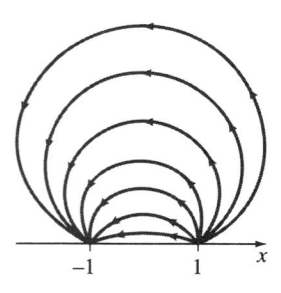

FIGURE 19.80

24. The flow with velocity vector $\mathbf{V} = (a + ib)/\bar{z}$ is called a *vortex* at $z = 0$, and the geometric nature of the streamlines depends on the choice of a and b.

(a) Show that if $z = x(t) + iy(t)$ is the path of a particle, then

$$\frac{dx}{dt} = \frac{ax - by}{x^2 + y^2}$$

$$\frac{dy}{dt} = \frac{bx + ay}{x^2 + y^2}$$

(b) Change to polar coordinates to establish that $dr/dt = a/r$ and $d\theta/dt = b/r^2$, and conclude that $r = ce^{a\theta/b}$ for $b \neq 0$. [*Hint:* See (2) of Section 10.1.]

(c) Conclude that the logarithmic spirals in part (b) spiral inward if and only if $a < 0$, and the curves are traversed clockwise if and only if $b < 0$. See Figure 19.81.

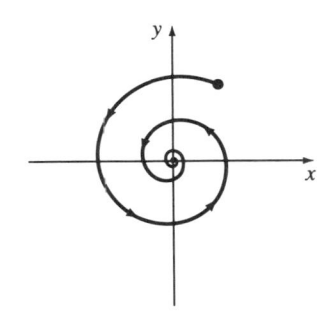

FIGURE 19.81

SUMMARY

A complex function $w = f(z)$ can be considered as a **mapping** between a region R in the z-plane and a region R' in the w-plane. When $f(z)$ is analytic inside R and $f'(z_0) \neq 0$, f is **conformal at $z = z_0$**; that is, f preserves the angles between any two curves in R which intersect at z_0. The mapping properties of analytic functions are conveniently summarized in **tables of conformal mappings**, and such a table can often be used to construct a conformal mapping between a given region R and a target region R'.

A bounded harmonic function $u = u(x, y)$ that takes on prescribed values on the boundary of a region R is called a solution to a **Dirichlet problem** on R. Conformal mapping methods can be used to transfer known solutions from one region to another. If f is an analytic function that maps R onto R', then the value of u at a boundary point ξ of R is assigned as the value of U at the corresponding boundary point $f(\xi)$ of R'. If U is the known solution to the new Dirichlet problem in R', then $u(x, y) = U(f(z))$ is harmonic in R and is the solution to the original Dirichlet problem.

A **linear fractional transformation** is a complex function of the form $T(z) = (az + b)/(cz + d)$, where a, b, c, and d are complex constants and $ad - bc \neq 0$. A linear fractional transformation T maps a circle in the z-plane to either a circle or a line in the w-plane. The image is a line when the pole of T lies on the circle. A linear fractional transformation that maps three given points z_1, z_2, and z_3 in the z-plane to three given points w_1, w_2, and w_3 in the w-plane may be obtained by solving the **cross-ratio** equation

$$\frac{w - w_1}{w - w_3} \frac{w_2 - w_3}{w_2 - w_1} = \frac{z - z_1}{z - z_3} \frac{z_2 - z_3}{z_2 - z_1}$$

for w.

A **Schwarz–Christoffel formula** $w = f(z)$ maps the upper half-plane $y \geq 0$ onto either a bounded or an unbounded polygonal region. If the polygonal region has interior angles $\alpha_1, \alpha_2, \ldots, \alpha_n$, where $0 < \alpha_i < 2\pi$, then the derivative of the Schwarz–Christoffel transformation satisfies the equation

$$f'(z) = A(z - x_1)^{(\alpha_1/\pi) - 1}(z - x_2)^{(\alpha_2/\pi) - 1} \cdots (z - x_n)^{(\alpha_n/\pi) - 1}$$

Three of the points x_k on the x-axis may be chosen arbitrarily, and, if the polygonal region is bounded, only $n - 1$ of the n interior angles should be included in the formula. The Schwarz–Christoffel formula can sometimes be used to suggest a possible conformal mapping from the upper half-plane onto a nonpolygonal region R'.

The **Poisson integral formulas** provide general solutions to Dirichlet problems in the upper half-plane $y \geq 0$ or the unit disk $|z| \leq 1$:

$$\text{half-plane:} \quad u(x, y) = \frac{y}{\pi} \int_{-\infty}^{\infty} \frac{u(t, 0)}{(x - t)^2 + y^2} \, dt$$

$$\text{unit disk:} \quad u(x, y) = \frac{1}{2\pi} \int_{-\pi}^{\pi} u(e^{it}) \frac{1 - |z|^2}{|e^{it} - z|^2} \, dt$$

When $u(t, 0)$ is a step function, it is easier to use the integrated solution

$$u(x, y) = \sum_{i=1}^{n} \frac{u_i}{\pi} [\text{Arg}(z - x_i) - \text{Arg}(z - x_{i-1})]$$

A vector field $\mathbf{F}(x, y) = P(x, y) + iQ(x, y)$ in a domain D which satisfies div $\mathbf{F} = 0$ and curl $\mathbf{F} = \mathbf{0}$ is of the form $\mathbf{F}(x, y) = \overline{g(z)}$, where $g(z)$ is analytic in D. An antiderivative $G(z) = \phi(x, y) + i\psi(x, y)$ for $g(z)$ is called a **complex potential** for the vector field $\mathbf{F}$, and the level curves $\phi(x, y) = c$ of the harmonic function ϕ are called **equipotential lines**. If $P(x, y) + iQ(x, y)$ is interpreted as the velocity vector of a two-dimensional steady-state fluid flow, then the level curves $\psi(x, y) = c$ are **streamlines** for the flow.

An irrotational and incompressible flow that remains inside a given region R can be constructed by finding an analytic function $G(z) = \phi(x, y) + i\psi(x, y)$ on R for which $\psi(x, y)$ is constant on the boundary of R. This process is called **streamlining**.

CHAPTER 19 REVIEW EXERCISES
Answers to odd-numbered problems begin on page A-93.

Answer Problems 1–10 without referring back to the text. Fill in the blank or answer true/false.

1. Under the complex mapping $f(z) = z^2$, the curve $xy = 2$ is mapped onto the line _____.

2. The complex mapping $f(z) = -iz$ is a rotation through _____ degrees.

3. The image of the upper half-plane $y \geq 0$ under the complex mapping $f(z) = z^{2/3}$ is _____.

4. The analytic function $f(z) = \cosh z$ is conformal except at $z =$ _____.

5. If $w = f(z)$ is an analytic function that maps a domain D onto the upper half-plane $v > 0$, then the function $u = \text{Arg}(f(z))$ is harmonic in D. _____

6. Is the image of the circle $|z - 1| = 1$ under the complex mapping $T(z) = (z - 1)/(z - 2)$ a circle or a line? _____

7. The linear fractional transformation $T(z) = \dfrac{z - z_1}{z - z_3} \dfrac{z_2 - z_3}{z_2 - z_1}$ maps the triple z_1, z_2, and z_3 to _____.

8. If $f'(z) = z^{-1/2}(z + 1)^{-1/2}(z - 1)^{-1/2}$, then $f(z)$ maps the upper half-plane $y > 0$ onto the interior of a rectangle. _____

9. If $\mathbf{F}(x, y) = P(x, y)\mathbf{i} + Q(x, y)\mathbf{j}$ is a vector field in a domain D with div $\mathbf{F} = 0$ and curl $\mathbf{F} = \mathbf{0}$, then the complex function $g(z) = P(x, y) + iQ(x, y)$ is analytic in D. _____

10. If $G(z) = \phi(x, y) + i\psi(x, y)$ is analytic in a region R and $\mathbf{V}(x, y) = \overline{iG'(z)}$, then the streamlines of the corresponding flow are described by $\phi(x, y) = c$. _____

11. Find the image of the first quadrant under the complex mapping $w = \text{Ln } z = \log_e |z| + i \text{ Arg } z$. What are images of the rays $\theta = \theta_0$ that lie in the first quadrant?

In Problems 12 and 13 use the conformal mappings in Appendix III to find a conformal mapping from the given region R in the z-plane onto the target region R' in the w-plane, and find the image of the given boundary curve.

12.

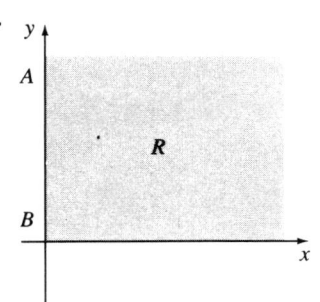

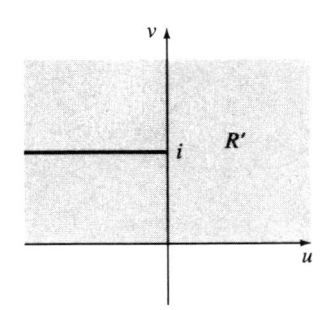

FIGURE 19.82

13.

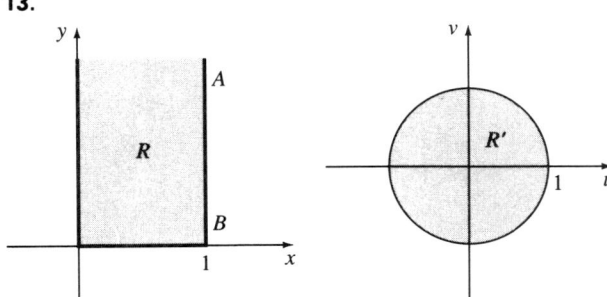

FIGURE 19.83

In Problems 14 and 15 use an appropriate conformal mapping to solve the given Dirichlet problem.

14.

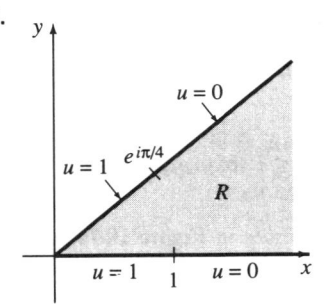

FIGURE 19.84

15.

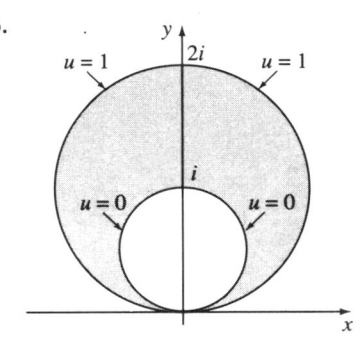

FIGURE 19.85

16. Derive conformal mapping C-4 in Appendix III by constructing the linear fractional transformation that maps $1, -1, \infty$ to $i, -i, -1$.

17. (a) Approximate the region R' in M-9 by the polygonal region shown in Figure 19.86. Require that $f(-1) = u_1$, $f(0) = \pi i/2$, and $f(1) = u_1 + \pi i$.

(b) Show that when $u_1 \to \infty$,
$$f'(z) = Az(z + 1)^{-1}(z - 1)^{-1} = \frac{1}{2} A \left[\frac{1}{z+1} + \frac{1}{z-1} \right].$$

(c) If we require that $\text{Im}(f(t)) = 0$ for $t < -1$, $\text{Im}(f(t)) = \pi$ for $t > 1$, and $f(0) = \pi i/2$, conclude that $f(z) = \pi i - \frac{1}{2}[\text{Ln}(z + 1) + \text{Ln}(z - 1)]$.

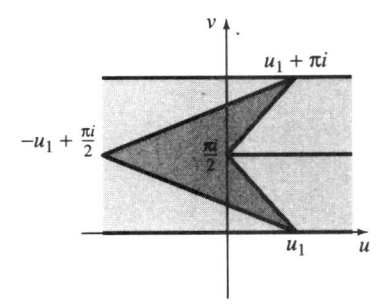

FIGURE 19.86

18. (a) Find the solution $u(x, y)$ of the Dirichlet problem in the upper half-plane $y \geq 0$ that satisfies the boundary condition $u(x, 0) = \sin x$. [*Hint:* See Problem 6 in Exercises 19.5.]

(b) Find the solution $u(x, y)$ of the Dirichlet problem in the unit disk $|z| \leq 1$ that satisfies the boundary condition $u(e^{i\theta}) = \sin \theta$.

19. Explain why the streamlines in Figure 19.72 may also be interpreted as the equipotential lines of the potential ϕ that satisfies $\phi(x, 0) = 0$ for $-\infty < x < \infty$ and $\phi(x, \pi) = 1$ for $x < 0$.

20. Verify that the boundary of the region R defined by $y^2 \geq 4(1 - x)$ is a streamline for the fluid flow with complex potential $G(z) = i(z^{1/2} - 1)$. Sketch the streamlines of the flow.

APPENDICES

I GAMMA FUNCTION

II TABLE OF LAPLACE TRANSFORMS

III CONFORMAL MAPPINGS

IV BASIC PROGRAMS FOR NUMERICAL METHODS IN CHAPTER 15

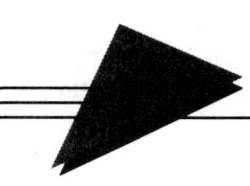

APPENDIX I

Gamma Function

GAMMA FUNCTION

Euler's integral definition of the **gamma function**[*] is

$$\Gamma(x) = \int_0^\infty t^{x-1} e^{-t} \, dt \tag{1}$$

Convergence of the integral requires that $x - 1 > -1$, or $x > 0$. The recurrence relation

$$\boxed{\Gamma(x + 1) = x\Gamma(x)} \tag{2}$$

which we saw in Section 5.4, can be obtained from (1) by employing integration by parts. Now when $x = 1$,

$$\Gamma(1) = \int_0^\infty e^{-t} \, dt = 1$$

and thus (2) gives
$$\Gamma(2) = 1\Gamma(1) = 1$$
$$\Gamma(3) = 2\Gamma(2) = 2 \cdot 1$$
$$\Gamma(4) = 3\Gamma(3) = 3 \cdot 2 \cdot 1$$

and so on. In this manner it is seen that when n is a positive integer,

$$\boxed{\Gamma(n + 1) = n!}$$

For this reason the gamma function is often called the **generalized factorial function**.

---

[*]This function was first defined by Leonhard Euler in his text *Institutiones calculi integralis* published in 1768.

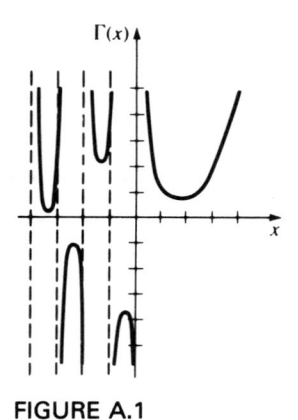

Although the integral form (1) does not converge for $x < 0$, it can be shown by means of alternative definitions that the gamma function is defined for all real and complex numbers except $x = -n$, $n = 0, 1, 2, \ldots$. As a consequence, (2) is actually valid for $x \neq -n$. Considered as a function of a real variable x, the graph of $\Gamma(x)$ is as given in Figure A.1. Observe that the nonpositive integers correspond to the vertical asymptotes of the graph.

In Problems 27–33 of Exercises 5.4 we utilized the fact that $\Gamma(\tfrac{1}{2}) = \sqrt{\pi}$. This result can be derived from (1) by setting $x = \tfrac{1}{2}$:

$$\Gamma(\tfrac{1}{2}) = \int_0^\infty t^{-1/2} e^{-t}\, dt \tag{3}$$

FIGURE A.1

By letting $t = u^2$, we can write (3) as

$$\Gamma(\tfrac{1}{2}) = 2\int_0^\infty e^{-u^2}\, du$$

But

$$\int_0^\infty e^{-u^2}\, du = \int_0^\infty e^{-v^2}\, dv$$

and so

$$[\Gamma(\tfrac{1}{2})]^2 = \left(2\int_0^\infty e^{-u^2}\, du\right)\left(2\int_0^\infty e^{-v^2}\, dv\right)$$

$$= 4\int_0^\infty \int_0^\infty e^{-(u^2+v^2)}\, du\, dv$$

Switching to polar coordinates $u = r\cos\theta$, $v = r\sin\theta$ enables us to evaluate the double integral:

$$4\int_0^\infty \int_0^\infty e^{-(u^2+v^2)}\, du\, dv = 4\int_0^{\pi/2} \int_0^\infty e^{-r^2} r\, dr\, d\theta = \pi$$

Hence,

$$[\Gamma(\tfrac{1}{2})]^2 = \pi \quad \text{or} \quad \Gamma(\tfrac{1}{2}) = \sqrt{\pi}$$

EXAMPLE 1 Evaluate $\Gamma(-\tfrac{1}{2})$.

Solution In view of (2), it follows that with $x = -\tfrac{1}{2}$,

$$\Gamma\!\left(\frac{1}{2}\right) = -\frac{1}{2}\Gamma\!\left(-\frac{1}{2}\right)$$

Therefore,

$$\Gamma\!\left(-\frac{1}{2}\right) = -2\Gamma\!\left(\frac{1}{2}\right) = -2\sqrt{\pi} \qquad \square$$

APPENDIX I EXERCISES *Answers to odd-numbered problems begin on page A-93.*

1. Evaluate
   - (a) $\Gamma(5)$
   - (b) $\Gamma(7)$
   - (c) $\Gamma(-\tfrac{3}{2})$
   - (d) $\Gamma(-\tfrac{5}{2})$

2. Use (1) and the fact that $\Gamma(\tfrac{6}{5}) = 0.92$ to evaluate
$$\int_0^\infty x^5 e^{-x^5}\, dx.$$
[*Hint:* Let $t = x^5$.]

3. Use (1) and the fact that $\Gamma(\frac{2}{3}) = 0.89$ to evaluate $\int_0^\infty x^4 e^{-x^3}\, dx$.

4. Evaluate $\int_0^1 x^3 \left(\ln \frac{1}{x}\right)^3 dx$. [*Hint:* Let $t = -\ln x$.]

5. Use the fact that $\Gamma(x) > \int_0^1 t^{x-1} e^{-t}\, dt$ to show that $\Gamma(x)$ is unbounded as $x \to 0^+$.

6. Use (1) to derive (2) for $x > 0$.

APPENDIX II

Table of Laplace Transforms

$f(t)$	$\mathscr{L}\{f(t)\} = F(s)$
1. 1	$\dfrac{1}{s}$
2. t	$\dfrac{1}{s^2}$
3. t^n	$\dfrac{n!}{s^{n+1}}$, n a positive integer
4. $t^{-1/2}$	$\sqrt{\dfrac{\pi}{s}}$
5. $t^{1/2}$	$\dfrac{\sqrt{\pi}}{2s^{3/2}}$
6. t^α	$\dfrac{\Gamma(\alpha+1)}{s^{\alpha+1}}$, $\alpha > -1$
7. $\sin kt$	$\dfrac{k}{s^2 + k^2}$
8. $\cos kt$	$\dfrac{s}{s^2 + k^2}$
9. $\sin^2 kt$	$\dfrac{2k^2}{s(s^2 + 4k^2)}$
10. $\cos^2 kt$	$\dfrac{s^2 + 2k^2}{s(s^2 + 4k^2)}$
11. e^{at}	$\dfrac{1}{s - a}$
12. $\sinh kt$	$\dfrac{k}{s^2 - k^2}$
13. $\cosh kt$	$\dfrac{s}{s^2 - k^2}$

(continues)

APPENDIX II Table of Laplace Transforms

$f(t)$	$\mathscr{L}\{f(t)\} = F(s)$
14. $\sinh^2 kt$	$\dfrac{2k^2}{s(s^2 - 4k^2)}$
15. $\cosh^2 kt$	$\dfrac{s^2 - 2k^2}{s(s^2 - 4k^2)}$
16. $e^{at}t$	$\dfrac{1}{(s-a)^2}$
17. $e^{at}t^n$	$\dfrac{n!}{(s-a)^{n+1}}$, n a positive integer
18. $e^{at}\sin kt$	$\dfrac{k}{(s-a)^2 + k^2}$
19. $e^{at}\cos kt$	$\dfrac{s-a}{(s-a)^2 + k^2}$
20. $e^{at}\sinh kt$	$\dfrac{k}{(s-a)^2 - k^2}$
21. $e^{at}\cosh kt$	$\dfrac{s-a}{(s-a)^2 - k^2}$
22. $t\sin kt$	$\dfrac{2ks}{(s^2+k^2)^2}$
23. $t\cos kt$	$\dfrac{s^2 - k^2}{(s^2+k^2)^2}$
24. $\sin kt + kt\cos kt$	$\dfrac{2ks^2}{(s^2+k^2)^2}$
25. $\sin kt - kt\cos kt$	$\dfrac{2k^3}{(s^2+k^2)^2}$
26. $t\sinh kt$	$\dfrac{2ks}{(s^2-k^2)^2}$
27. $t\cosh kt$	$\dfrac{s^2+k^2}{(s^2-k^2)^2}$
28. $\dfrac{e^{at} - e^{bt}}{a-b}$	$\dfrac{1}{(s-a)(s-b)}$
29. $\dfrac{ae^{at} - be^{bt}}{a-b}$	$\dfrac{s}{(s-a)(s-b)}$
30. $1 - \cos kt$	$\dfrac{k^2}{s(s^2+k^2)}$
31. $kt - \sin kt$	$\dfrac{k^3}{s^2(s^2+k^2)}$
32. $\cos at - \cos bt$	$\dfrac{s(b^2 - a^2)}{(s^2+a^2)(s^2+b^2)}$
33. $\sin kt \sinh kt$	$\dfrac{2k^2 s}{s^4 + 4k^4}$

(continues)

	$f(t)$	$\mathscr{L}\{f(t)\} = F(s)$
34.	$\sin kt \cosh kt$	$\dfrac{k(s^2 + 2k^2)}{s^4 + 4k^4}$
35.	$\cos kt \sinh kt$	$\dfrac{k(s^2 - 2k^2)}{s^4 + 4k^4}$
36.	$\cos kt \cosh kt$	$\dfrac{s^3}{s^4 + 4k^4}$
37.	$\delta(t)$	1
38.	$\delta(t - a)$	e^{-as}
39.	$\mathscr{U}(t - a)$	$\dfrac{e^{-as}}{s}$
40.	$J_0(kt)$	$\dfrac{1}{\sqrt{s^2 + k^2}}$
41.	$\dfrac{e^{bt} - e^{at}}{t}$	$\ln \dfrac{s - a}{s - b}$
42.	$\dfrac{2(1 - \cos at)}{t}$	$\ln \dfrac{s^2 + a^2}{s^2}$
43.	$\dfrac{2(1 - \cosh at)}{t}$	$\ln \dfrac{s^2 - a^2}{s^2}$
44.	$\dfrac{\sin at}{t}$	$\arctan\left(\dfrac{a}{s}\right)$
45.	$\dfrac{\sin at \cos bt}{t}$	$\dfrac{1}{2}\arctan\dfrac{a+b}{s} + \dfrac{1}{2}\arctan\dfrac{a-b}{s}$
46.	$\dfrac{1}{\sqrt{\pi t}} e^{-a^2/4t}$	$\dfrac{e^{-a\sqrt{s}}}{\sqrt{s}}$
47.	$\dfrac{a}{2\sqrt{\pi t^3}} e^{-a^2/4t}$	$e^{-a\sqrt{s}}$
48.	$\operatorname{erfc}\left(\dfrac{a}{2\sqrt{t}}\right)$	$\dfrac{e^{-a\sqrt{s}}}{s}$
49.	$2\sqrt{\dfrac{t}{\pi}} e^{-a^2/4t} - a\operatorname{erfc}\left(\dfrac{a}{2\sqrt{t}}\right)$	$\dfrac{e^{-a\sqrt{s}}}{s\sqrt{s}}$
50.	$e^{ab}e^{b^2t} \operatorname{erfc}\left(b\sqrt{t} + \dfrac{a}{2\sqrt{t}}\right)$	$\dfrac{e^{-a\sqrt{s}}}{\sqrt{s}(\sqrt{s} + b)}$
51.	$-e^{ab}e^{b^2t} \operatorname{erfc}\left(b\sqrt{t} + \dfrac{a}{2\sqrt{t}}\right) + \operatorname{erfc}\left(\dfrac{a}{2\sqrt{t}}\right)$	$\dfrac{be^{-a\sqrt{s}}}{s(\sqrt{s} + b)}$
52.	$e^{at}f(t)$	$F(s - a)$
53.	$f(t - a)\mathscr{U}(t - a)$	$e^{-as}F(s)$
54.	$f^{(n)}(t)$	$s^n F(s) - s^{(n-1)}f(0) - \cdots - f^{(n-1)}(0)$
55.	$t^n f(t)$	$(-1)^n \dfrac{d^n}{ds^n} F(s)$
56.	$\displaystyle\int_0^t f(\tau)g(t - \tau)\,d\tau$	$F(s)G(s)$

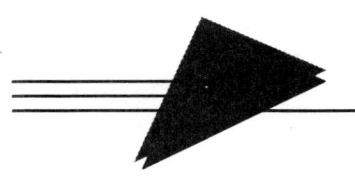

APPENDIX III

Conformal Mappings

Elementary Mappings

E-1

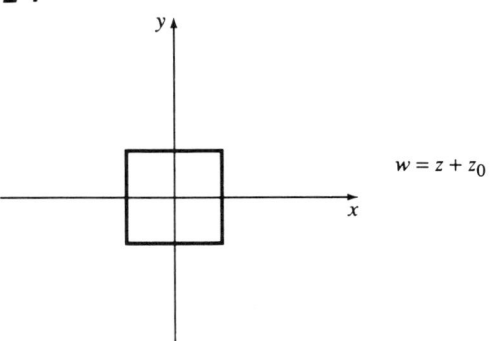

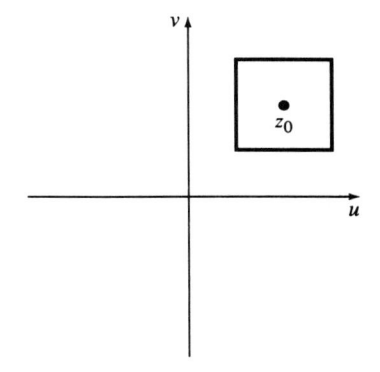

$$w = z + z_0$$

E-2

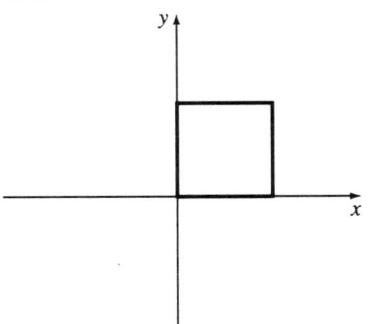

$$w = e^{i\theta}z$$

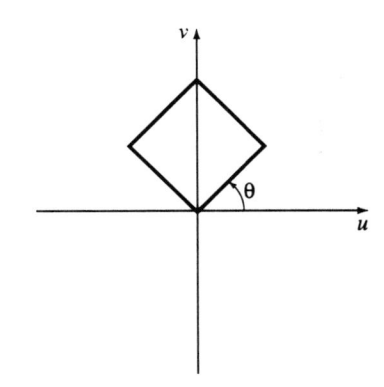

APPENDIX III Conformal Mappings

E-3

$w = \alpha z, \ \alpha > 0$

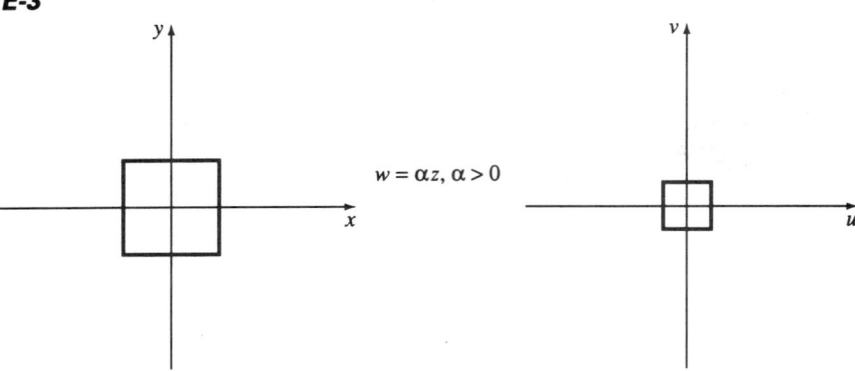

E-4

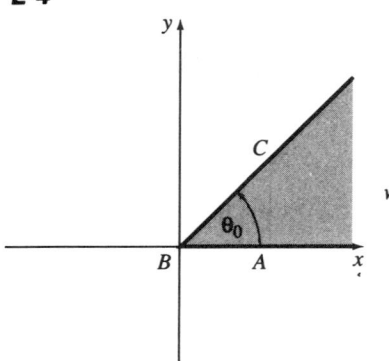

$w = z^\alpha, \ \alpha > 0$

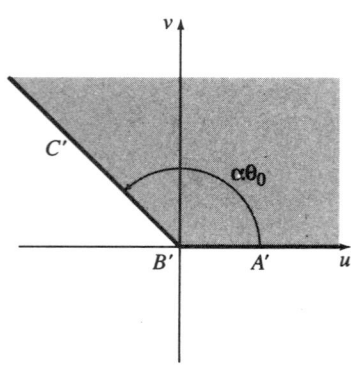

E-5

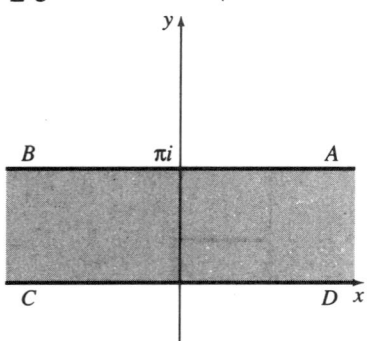

$w = e^z$
$z = \operatorname{Ln} w$

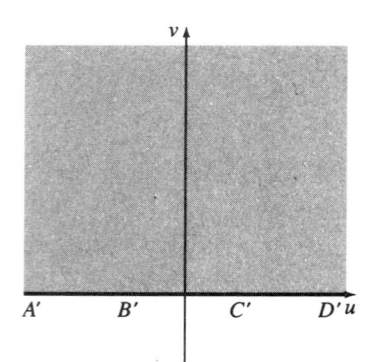

APPENDIX III Conformal Mappings

E-6

$w = \sin z$
$z = \sin^{-1} w$

E-7

$w = \dfrac{1}{z}$

E-8

$w = \log_e |z| + i \operatorname{Arg} z$
$a > 1$

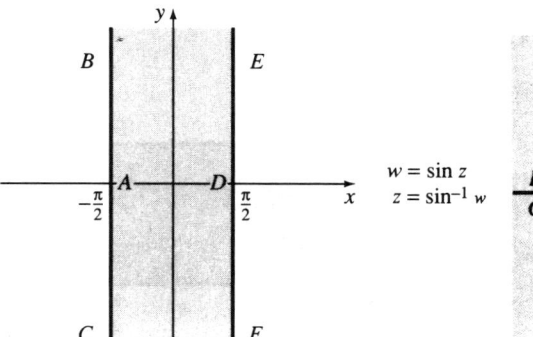

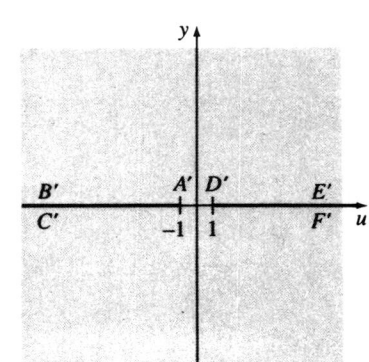

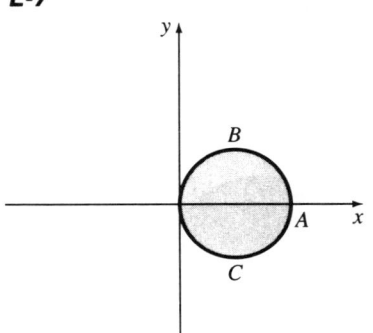

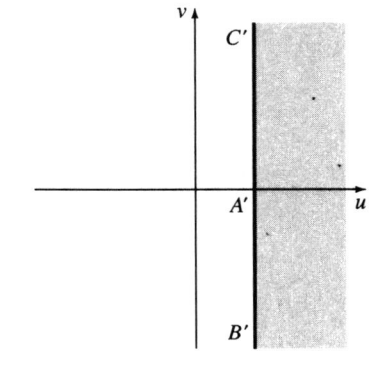

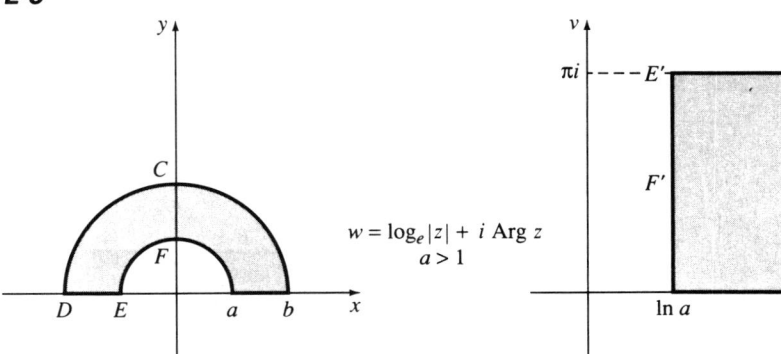

APPENDIX III Conformal Mappings

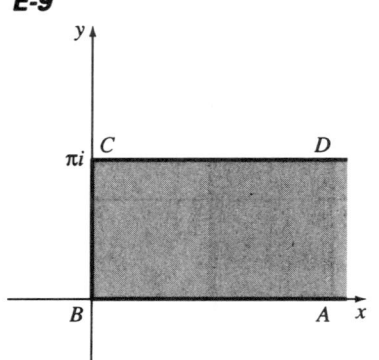

E-9

$w = \cosh z$

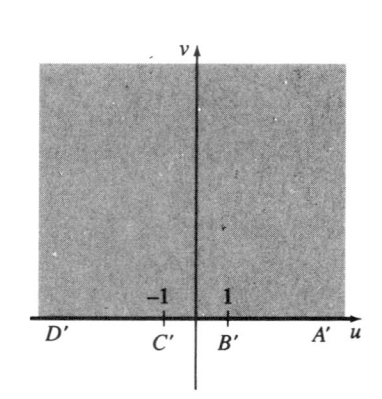

Mappings to Half-Planes

H-1

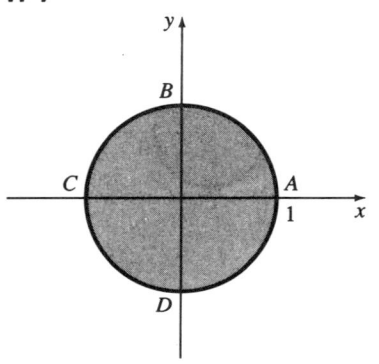

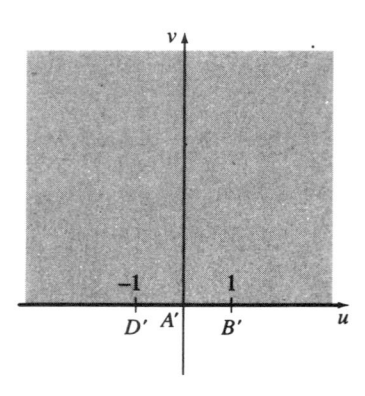

$w = i\dfrac{1-z}{1+z}$

H-2

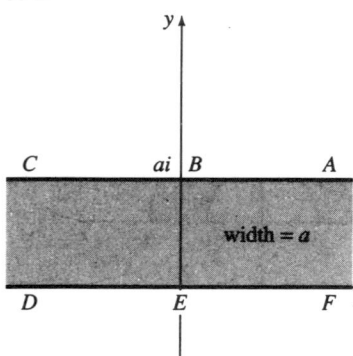

width = a

$w = e^{\pi z/a}$

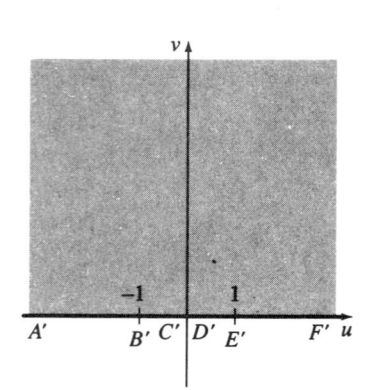

APPENDIX III Conformal Mappings

H-3

$$w = \frac{a}{2}\left(z + \frac{1}{z}\right)$$

H-4

width = a

$$w = \cos\left(\frac{\pi z}{a}\right)$$

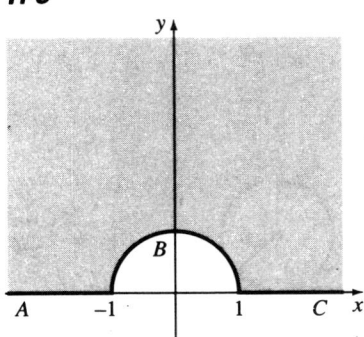

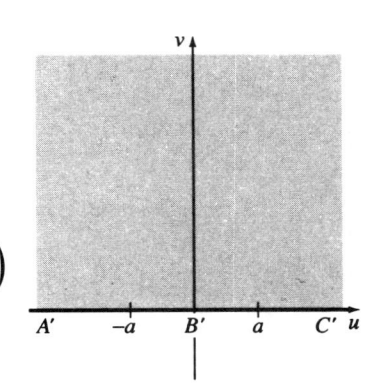

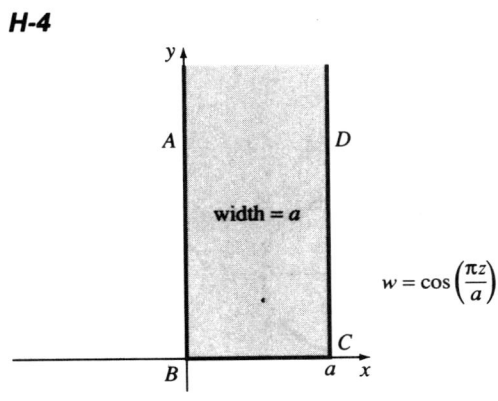

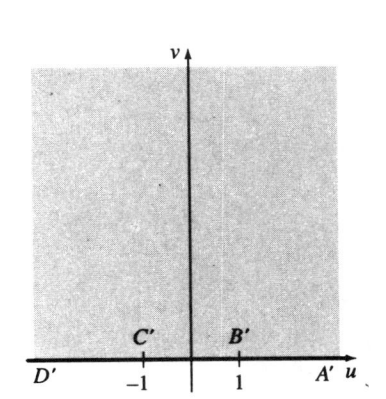

H-5

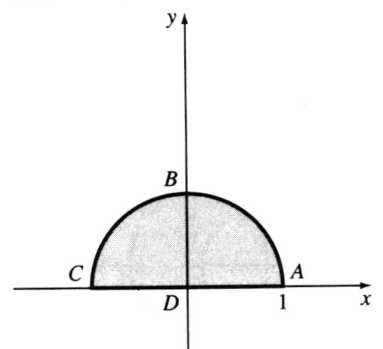

$$w = \left(\frac{1+z}{1-z}\right)^2$$

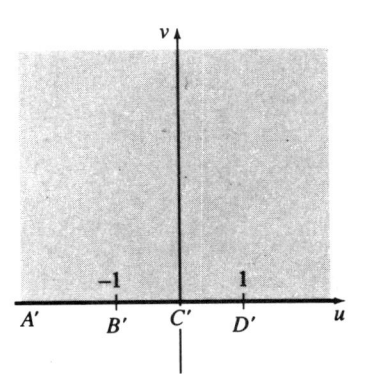

H-6

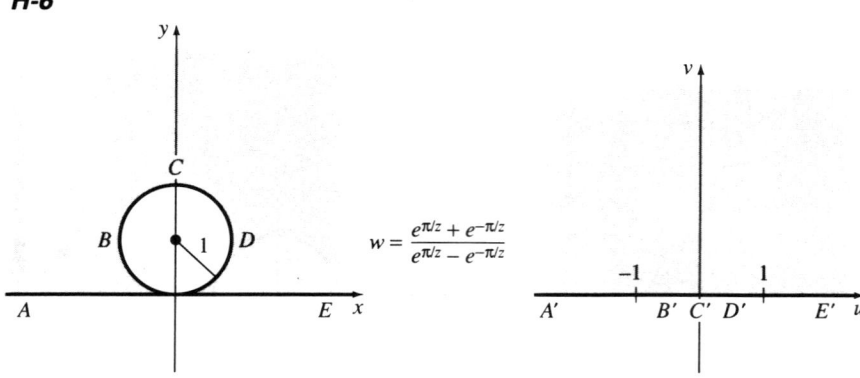

$$w = \frac{e^{\pi/z} + e^{-\pi/z}}{e^{\pi/z} - e^{-\pi/z}}$$

Mappings to Circular Regions

C-1

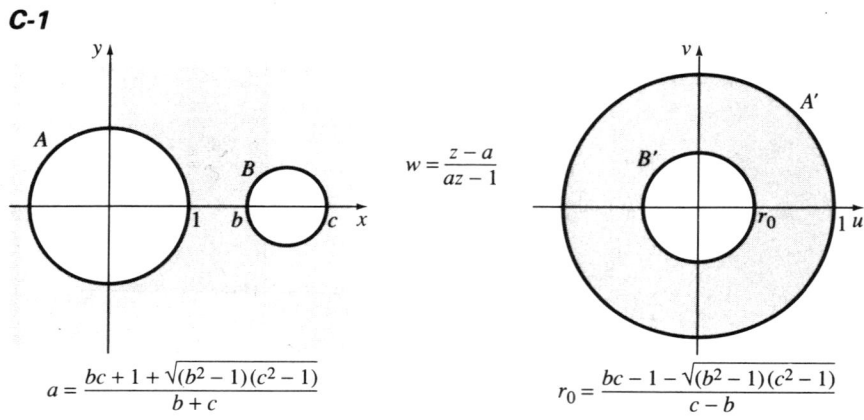

$$w = \frac{z-a}{az-1}$$

$$a = \frac{bc + 1 + \sqrt{(b^2-1)(c^2-1)}}{b+c}$$

$$r_0 = \frac{bc - 1 - \sqrt{(b^2-1)(c^2-1)}}{c-b}$$

C-2

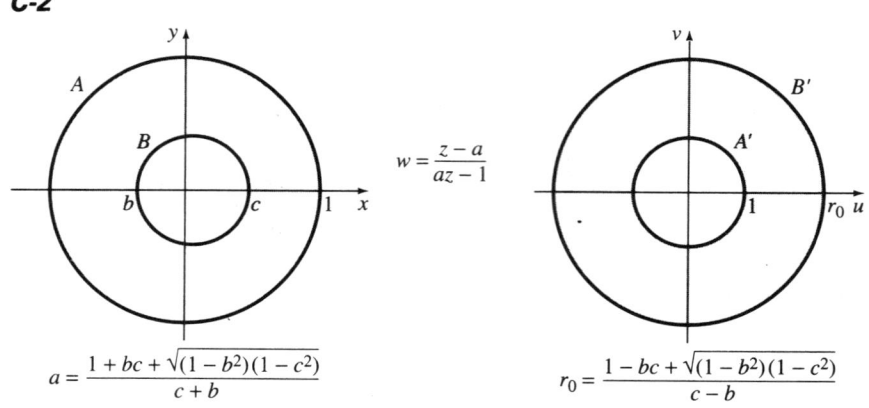

$$w = \frac{z-a}{az-1}$$

$$a = \frac{1 + bc + \sqrt{(1-b^2)(1-c^2)}}{c+b}$$

$$r_0 = \frac{1 - bc + \sqrt{(1-b^2)(1-c^2)}}{c-b}$$

C-3

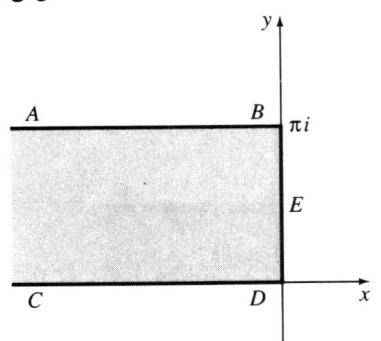

$w = e^z$

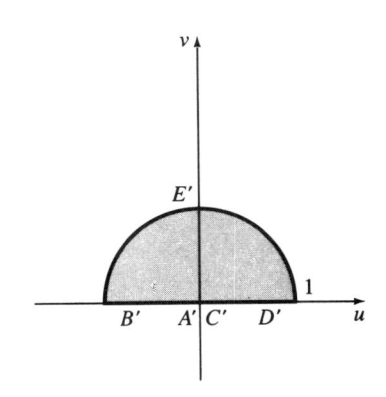

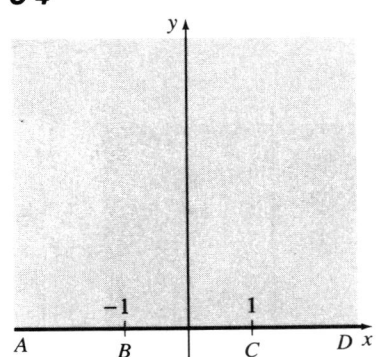

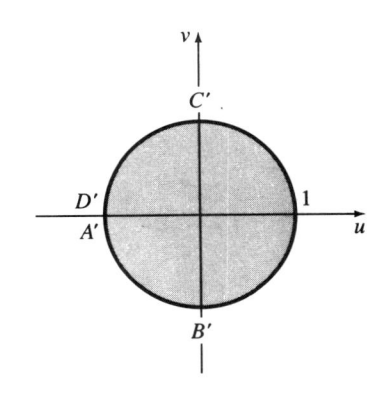

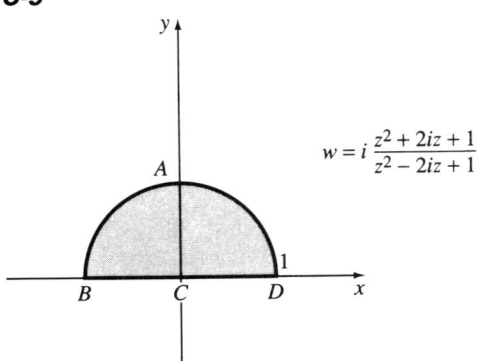

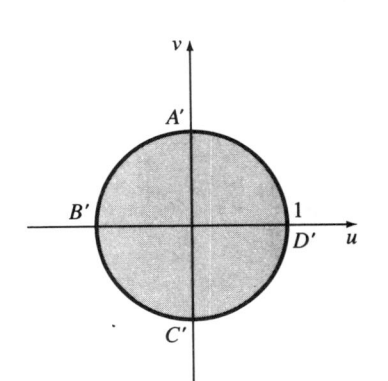

C-4

$w = \dfrac{i-z}{i+z}$

C-5

$w = i\dfrac{z^2 + 2iz + 1}{z^2 - 2iz + 1}$

APPENDIX III Conformal Mappings

Miscellaneous Mappings

M-1

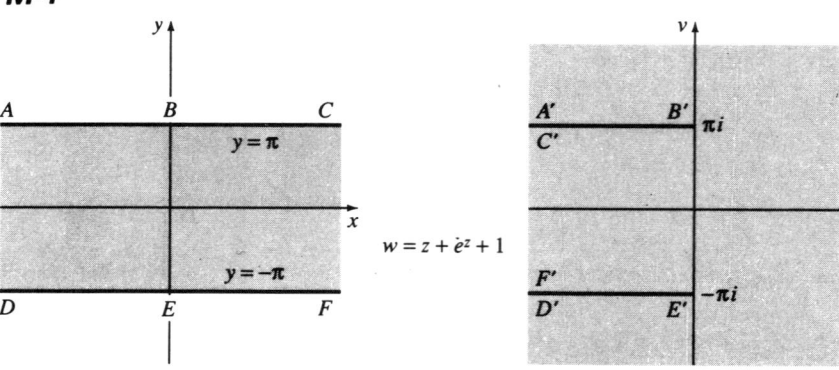

$$w = z + e^z + 1$$

M-2

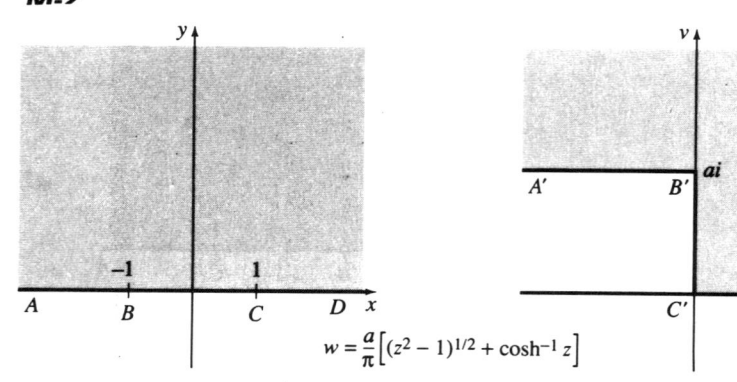

$$w = \frac{a}{\pi}\left[(z^2 - 1)^{1/2} + \cosh^{-1} z\right]$$

M-3

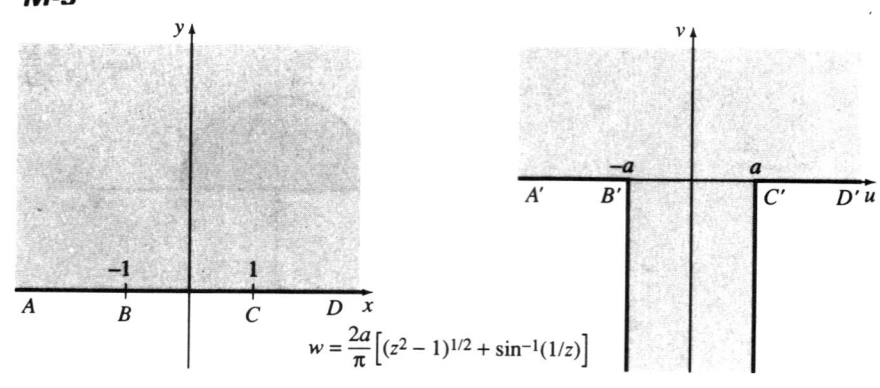

$$w = \frac{2a}{\pi}\left[(z^2 - 1)^{1/2} + \sin^{-1}(1/z)\right]$$

APPENDIX III Conformal Mappings

M-4

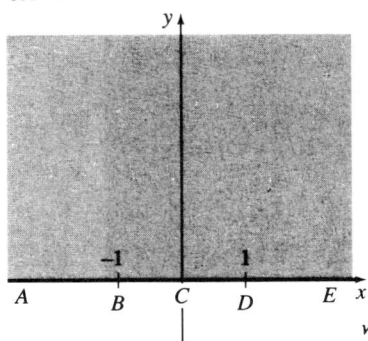

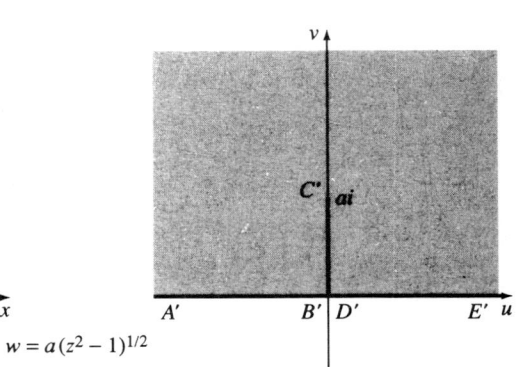

$w = a(z^2 - 1)^{1/2}$

M-5

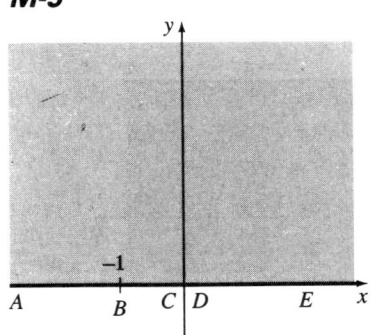

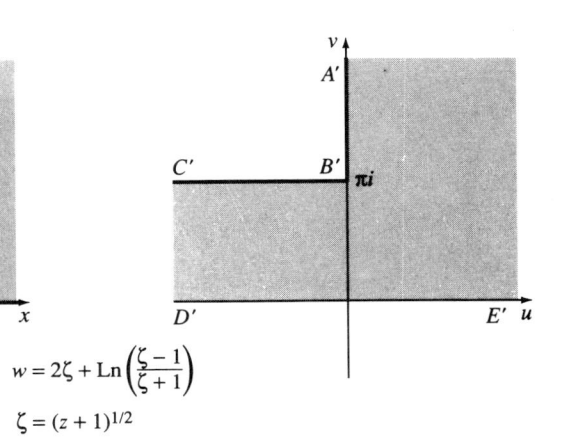

$w = 2\zeta + \text{Ln}\left(\dfrac{\zeta - 1}{\zeta + 1}\right)$

$\zeta = (z + 1)^{1/2}$

M-6

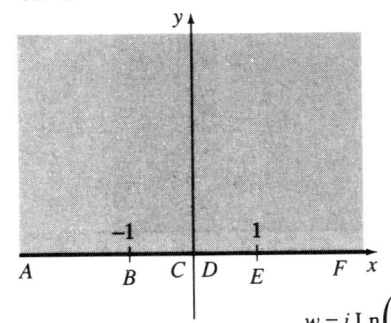

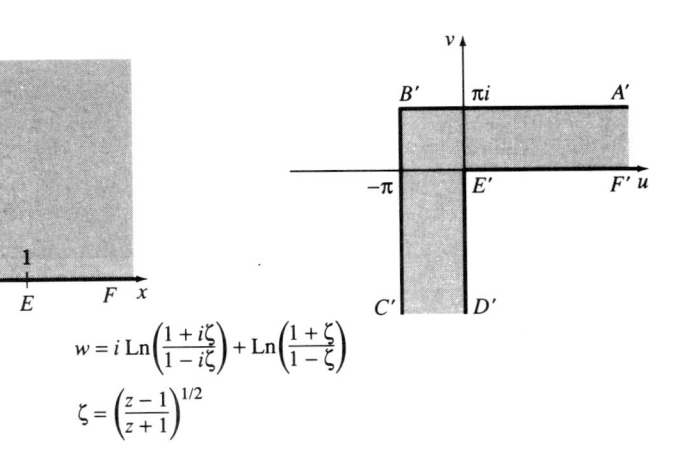

$w = i\,\text{Ln}\left(\dfrac{1 + i\zeta}{1 - i\zeta}\right) + \text{Ln}\left(\dfrac{1 + \zeta}{1 - \zeta}\right)$

$\zeta = \left(\dfrac{z - 1}{z + 1}\right)^{1/2}$

M-7

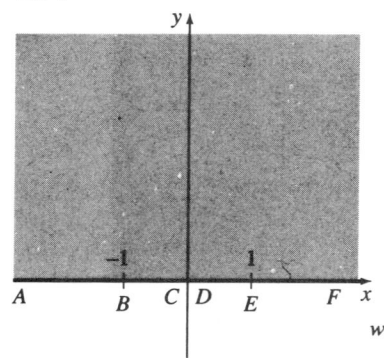

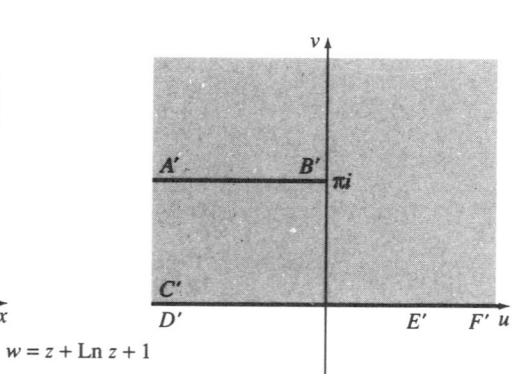

$w = z + \operatorname{Ln} z + 1$

M-8

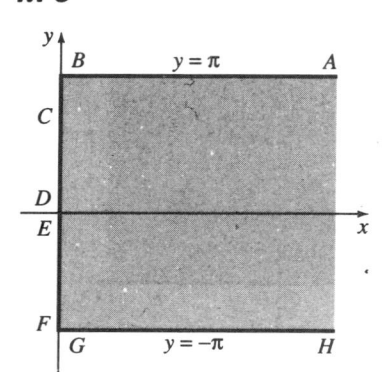

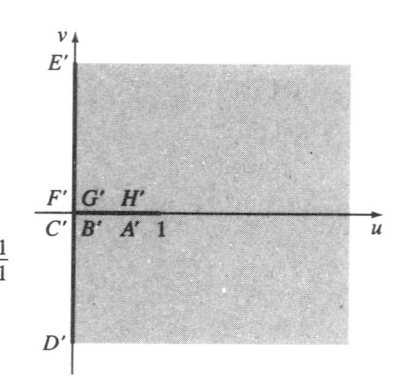

$w = \dfrac{e^z + 1}{e^z - 1}$

M-9

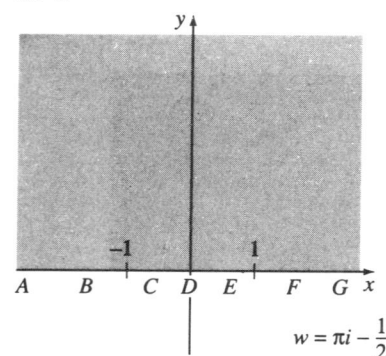

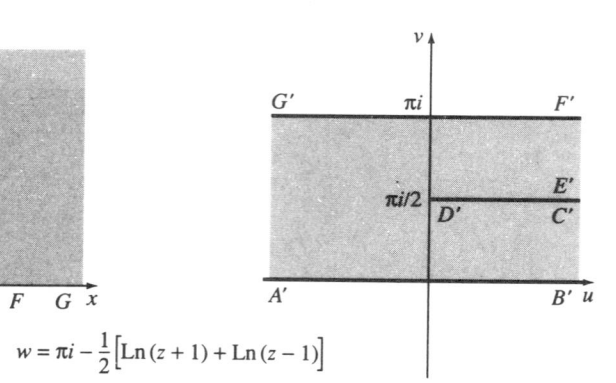

$w = \pi i - \dfrac{1}{2}\left[\operatorname{Ln}(z+1) + \operatorname{Ln}(z-1)\right]$

M-10

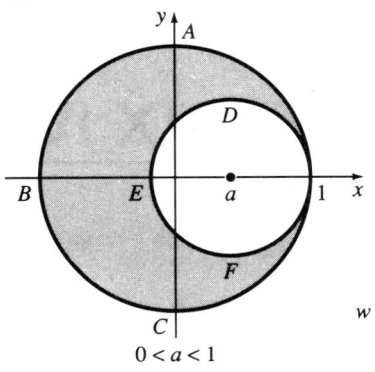

$$w = (1-i)\frac{z-i}{z-1}$$

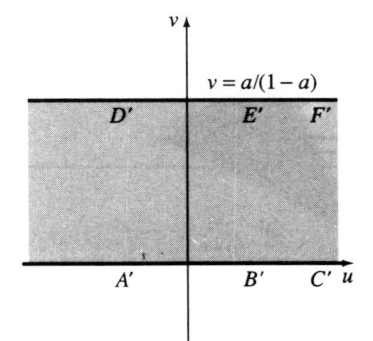

$0 < a < 1$

$v = a/(1-a)$

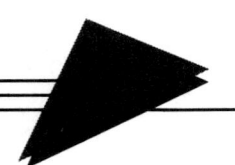

APPENDIX IV

BASIC Programs for Numerical Methods in Chapter 15

1. Newton's Method

```
10  REM NEWTON'S METHOD FOR FINDING ROOTS
20  INPUT "THE INITIAL VALUE IS ";X0
30  DEF FNY(X) = ...
40  DEF FND(X) = ...
50  X1 = X0 - FNY(X0)/FND(X0)
60  PRINT X1
70  LET X0 = X1
80  GO TO 50
90  END
```

Those familiar with BASIC should provide an exit from the loop in the foregoing program.

2. Trapezoidal Rule

```
10   REM EVALUATION OF DEFINITE INTEGRALS
     VIA THE TRAPEZOIDAL RULE
20   DEF FNY(X) = ...
30   INPUT "WHAT IS THE INTERVAL? ";A,B
40   INPUT "HOW MANY SUBDIVISIONS? ";N
50   LET H = (B - A)/N
60   LET T = FNY(A) + FNY(B)
70   FOR X = A + H TO B - H/2 STEP H
80   LET T = T + 2 * FNY(X)
90   NEXT X
100  LET T = T * H/2
110  PRINT "USING ";N; "SUBDIVISIONS, THE
     TRAPEZOID RULE YIELDS ";T
120  END
```

3. Simpson's Rule

```
10   REM EVALUATION OF DEFINITE INTEGRALS
     USING SIMPSON'S RULE
20   DEF FNY(X) = ...
30   INPUT "WHAT IS THE INTERVAL? ";A,B
40   INPUT "HOW MANY SUBDIVISIONS? ";N
50   LET H = (B − A)/N
60   LET S = FNY(A) + FNY(B)
70   FOR X = A + H TO B − H/2 STEP 2 * H
80   LET S = S + 4 * FNY(X)
90   NEXT X
100  FOR X = A + 2 * H TO B − 3 * H/2 STEP 2 * H
110  LET S = S + 2 * FNY(X)
120  NEXT X
130  LET S = H * S/3
140  PRINT "USING ";N; "SUBDIVISIONS,
     SIMPSON'S RULE YIELDS ";S
150  END
```

4. Euler's Method

```
100  REM EULER'S METHOD TO SOLVE Y' = FNF(X, Y)
110  REM DEFINE FNF(X, Y) HERE
120  REM GET INPUTS
130  PRINT
140  INPUT "STEP SIZE =",H
150  INPUT "NUMBER OF STEPS =",N
160  INPUT "XO =",X
170  INPUT "YO =",Y
180  PRINT
190  REM SET UP TABLE
200  PRINT "X", "Y"
210  PRINT
220  PRINT X, Y
230  REM COMPUTE X AND Y VALUES
240  FOR I = 1 TO N
250  Y = Y + H * FNF(X, Y)
260  X = X + H
270  PRINT X, Y
280  NEXT I
290  END
```

5. Improved Euler's Method

```
100  REM IMPROVED EULER'S METHOD TO SOLVE Y' = FNF(X, Y)
110  REM DEFINE FNF(X, Y) HERE
120  REM GET INPUTS
130  PRINT
140  INPUT "STEP SIZE =",H
```

```
150  INPUT "NUMBER OF STEPS =",N
160  INPUT "XO =",X
170  INPUT "YO =",Y
180  PRINT
190  REM SET UP TABLE
200  PRINT "X", "Y"
210  PRINT
220  PRINT X, Y
230  REM COMPUTE X AND Y VALUES
240  FOR I = 1 TO N
250  FVAL = FNF(X, Y)
260  Y = Y + H * (FVAL + FNF(X + H, Y + H * FVAL))/2
270  X = X + H
280  PRINT X, Y
290  NEXT I
300  END
```

6. Three-Term Taylor Method

```
100  REM THREE-TERM TAYLOR'S METHOD TO SOLVE Y' = FNF(X, Y)
110  REM THE DERIVATIVE OF FNF(X, Y) IS DENOTED BY FNDF(X, Y)
120  REM DEFINE FNF(X, Y) HERE
130  REM DEFINE FNDF(X, Y) HERE
140  REM GET INPUTS
150  PRINT
160  INPUT "STEP SIZE =",H
170  INPUT "NUMBER OF STEPS =",N
180  INPUT "XO =",X
190  INPUT "YO =",Y
200  PRINT
210  REM SET UP TABLE
220  PRINT "X", "Y"
230  PRINT
240  PRINT X, Y
250  REM COMPUTE X AND Y VALUES
260  FOR I = 1 TO N
270  FVAL = FNF(X, Y)
280  Y = Y + H * FNF(X, Y) + H * H * FNDF(X, Y)/2
290  X = X + H
300  PRINT X, Y
310  NEXT I
320  END
```

7. Runge–Kutta Method

```
100  REM RUNGE-KUTTA METHOD TO SOLVE Y' = FNF(X, Y)
110  REM DEFINE FNF(X, Y) HERE
120  REM GET INPUTS
130  PRINT
140  INPUT "STEP SIZE =",H
150  INPUT "NUMBER OF STEPS =",N
160  INPUT "XO =",X
170  INPUT "YO =",Y
```

```
180  PRINT
190  REM SET UP TABLE
200  PRINT "X", "Y"
210  PRINT
220  PRINT X, Y
230  REM COMPUTE X AND Y VALUES
240  FOR I = 1 TO N
250  K1 = H * FNF(X, Y)
260  K2 = H * FNF(X + H/2, Y + K1/2)
270  K3 = H * FNF(X + H/2, Y + K2/2)
280  K4 = H * FNF(X + H, Y + K3)
290  Y = Y + (K1 + 2 * K2 + 2 * K3 + K4)/6
300  X = X + H
310  PRINT X, Y
320  NEXT I
330  END
```

8. Gauss–Seidel Iteration

```
10   REM GAUSS-SEIDEL ITERATION TO SOLVE THE SYSTEM X = AX + B
20   REM WHERE THE MATRIX A HAS ZEROS ON THE MAIN DIAGONAL
30   INPUT "THE NUMBER OF UNKNOWNS IS ",N
40   INPUT "THE TOLERANCE IS ",TOL
50   DIM A(N,N)
60   DIM B(N)
70   DIM X(N)
80   DIM Y(N)
90   PRINT "INPUT THE COEFFICIENTS OF THE MATRIX A"
100  FOR I = 1 TO N
110  FOR J = 1 TO N
120  PRINT "A("I", "J") =";
130  INPUT "", A(I, J)
140  NEXT J
150  NEXT I
160  REM PRINT "INPUT THE CONSTANT COEFFICIENTS"
170  FOR I = 1 TO N
180  PRINT "B("I") =";
190  INPUT "", B(I)
200  NEXT I
210  PRINT "INPUT THE INITIAL GUESS"
220  FOR I = 1 TO N
230  PRINT "X("I") =";
240  INPUT "", X(I)
250  NEXT I
260  REM COMPUTE THE NEXT ITERATE
270  FOR I = 1 TO N
280  Y(I) = X(I)
290  X(I) = 0
300  FOR J = 1 TO N
310  X(I) = X(I) + A(I, J)*X(J)
320  NEXT J
330  X(I) = X(I) + B(I)
```

```
340  NEXT I
350  REM CHECK THE TOLERANCE
360  FOR I = 1 TO N
370  IF ABS(X(I)-Y(I)) > TOL GOTO 270
380  NEXT I
390  REM DISPLAY THE SOLUTION
400  PRINT "THE SOLUTION WITH TOLERANCE " TOL " IS"
410  FOR I = 1 TO N
420  PRINT " X("I") = "X(I)
430  NEXT I
440  END
```

9. Numerical Solution of the One-Dimensional Heat Equation

```
10   REM NUMERICAL SOLUTION OF THE ONE-DIMENSIONAL HEAT EQUATION
20   REM DEFINE THE INITIAL TEMPERATURE FUNCTION FNF(X) BELOW
30   DEF FNF(X) =
40   REM GET INPUTS
50   INPUT "THE COEFFICIENT IN THE PDE IS C =",C
60   INPUT "THE LENGTH OF THE ROD IS A =",A
70   INPUT "THE ENDING TIME IS T =",T
80   INPUT "THE CONSTANT LEFT-HAND TEMPERATURE IS T1 =",T1
90   INPUT "THE CONSTANT RIGHT-HAND TEMPERATURE IS T2 =",T2
100  INPUT "THE NUMBER OF X SUBINTERVALS IS N =",N
110  INPUT "THE NUMBER OF T SUBINTERVALS IS M =",M
120  REM COMPUTE H, K, AND L
130  H = A/N: K = T/M: L = C*K/H^2
140  REM FORMAT THE TABLE HEADER AND PRINT THE FIRST LINE
150  PRINT "TIME";
160  FOR I = 1 TO N-1
170  PRINT TAB(2+9*I);"X=";
180  PRINT USING "#.##";I*H;
190  NEXT I
200  PRINT: PRINT STRING$(10*N-1,45)
210  PRINT USING "##.##"; 0;
220  FOR I = 1 TO N-1
230  PRINT TAB(9*I);
240  PRINT USING "###.####";FNF(I*H);
250  NEXT I
260  PRINT
270  DIM U(N+1,M+1)
280  REM INITIALIZE THE U(I,J) VALUES
290  FOR I = 1 TO N-1: U(I,0) = FNF(I*H): NEXT I
300  FOR J = 0 TO M: U(0,J) = T1: U(N,J) = T2: NEXT J
310  REM COMPUTE THE U(I,J) VALUES
320  FOR J = 0 TO M-1
330  PRINT USING "##.##";(J+1)*K;
340  FOR I = 1 TO N-1
350  U(I,J+1) = L*U(I+1,J) + (1-2*L)*U(I,J) + L*U(I-1,J)
360  PRINT TAB(9*I);
```

```
370   PRINT USING "###.####";U(I,J+1);
380   NEXT I
390   PRINT
400   NEXT J
410   END
```

10. Crank–Nicholson Method

```
10    REM  CRANK–NICHOLSON METHOD FOR THE
      ONE-DIMENSIONAL HEAT EQUATION
20    REM DEFINE THE INITIAL TEMPERATURE
      FUNCTION FNF(X) BELOW
30    DEF FNF(X) =
40    REM GET INPUTS
50    INPUT "THE COEFFICIENT IN THE PDE IS C =",C
60    INPUT "THE LENGTH OF THE ROD IS A =",A
70    INPUT "THE ENDING TIME IS T =",T
80    INPUT "THE CONSTANT LEFT-HAND TEMPERATURE IS T1 =",T1
90    INPUT "THE CONSTANT RIGHT-HAND TEMPERATURE IS T2 =",T2
100   INPUT "THE NUMBER OF X SUBINTERVALS IS N =",N
110   INPUT "THE NUMBER OF T SUBINTERVALS IS M =",M
120   REM COMPUTE H, K, AND L
130   H = A/N: K = T/M: L = C*K/H^2
140   REM FORMAT THE TABLE HEADER AND PRINT THE FIRST LINE
150   PRINT "TIME";
160   FOR I = 1 TO N−1
170   PRINT TAB(2+9*I);"X =";
180   PRINT USING "#.##";I*H;
190   NEXT I
200   PRINT: PRINT STRING$(10*N−1,45)
210   PRINT USING "##.##"; 0;
220   FOR I = 1 TO N−1
230   PRINT TAB(9*I)
240   PRINT USING "###.####"; FNF(I*H);
250   NEXT I
260   PRINT
270   DIM U(N+1,M+1): DIM B(N−1): DIM R(N−1): DIM S(N−1)
280   REM INITIALIZE THE U(I,J) VALUES
290   FOR I = 1 TO N−1: U(I,0) = FNF(I*H): NEXT I
300   FOR J = 0 TO M: U(0,J) = T1: U(N,J) = T2: NEXT J
310   REM COMPUTE THE U(I,J) VALUES
320   REM THE SYSTEM OF EQUATIONS IS SOLVED USING
      GAUSS ELIMINATION
330   REM ADAPTED TO THE APPROPRIATE TRIDIAGONAL MATRIX
340   ALPHA = 2*(1+1/L): BETA = 2*(1−1/L)
350   R(1) = ALPHA
360   FOR I = 2 TO N−1: R(I) = ALPHA − 1/R(I−1): NEXT I
370   FOR J = 0 TO M−1
380   B(1) = U(2,J) − BETA*U(1,J) + U(0,J) + U(0,J+1)
390   FOR I = 2 TO N−2: B(I) = U(I+1,J) − BETA*U(I,J) + U(I−1,J): NEXT I
400   B(N−1) = U(N,J) − BETA*U(N−1,J) + U(N−2,J) + U(N,J+1)
410   S(1) = B(1)
420   FOR I = 2 TO N−1: S(I) = B(I) + (1/R(I−1))*S(I−1): NEXT I
```

```
430  U(N-1,J+1) = S(N-1)/R(N-1)
440  FOR I = N-2 TO 1 STEP -1: U(I,J+1) = (1/R(I))*(S(I) + U(I+1,J+1)): NEXT I
450  PRINT USING "##.##"; (J+1)*K;
460  FOR I = 1 TO N-1
470  PRINT TAB(9*I);
480  PRINT USING "###.####"; U(I,J+1);
490  NEXT I
500  PRINT
510  NEXT J
520  END
```

11. Numerical Solution of the One-Dimensional Wave Equation

```
10   REM NUMERICAL SOLUTION OF THE ONE-DIMENSIONAL
     WAVE EQUATION
20   REM WITH 0 < X < A AND U(0,T) = U(A,T) = 0 FOR T > 0
30   REM DEFINE THE INITIAL POSITION FUNCTION FNF(X) BELOW
40   DEF FNF(X) =
50   REM DEFINE THE INITIAL VELOCITY FUNCTION FNG(X) BELOW
60   DEF FNG(X) =
70   REM GET INPUTS
80   INPUT "THE COEFFICIENT IN THE PDE IS THE SQUARE OF C =",C
90   INPUT "THE LENGTH OF THE X-INTERVAL IS A =",A
100  INPUT "THE ENDING TIME IS T =",T
110  INPUT "THE NUMBER OF X SUBINTERVALS IS N =",N
120  INPUT "THE NUMBER OF T SUBINTERVALS IS M =",M
130  REM COMPUTE H, K, AND L
140  H = A/N: K = T/M: L = C*K/H
150  REM FORMAT THE TABLE HEADER AND PRINT THE FIRST LINE
160  PRINT " TIME";
170  FOR I = 1 TO N-1
180  PRINT TAB(2+9*I);"X=";
190  PRINT USING "#.##";I*H;
200  NEXT I
210  PRINT: PRINT STRING$(10*N-1,45)
220  PRINT USING "##.##"; 0;
230  FOR I = 1 TO N-1
240  PRINT TAB(9*I)
250  PRINT USING "###.####"; FNF(I*H);
260  NEXT I
270  PRINT
280  DIM U(N+1,M+1)
290  REM INITIALIZE THE U(0,J) AND U(A,J) VALUES
300  FOR J = 0 TO M: U(0,J) = 0: U(A,J) = 0: NEXT J
310  REM INITIALIZE THE U(I,0) VALUES
320  FOR I = 1 TO N-1: U(I,0) = FNF(I*H): NEXT I
330  REM COMPUTE THE U(I,1) VALUES
340  PRINT USING "##.##"; K;
350  FOR I = 1 TO N-1
360  U(I,1) = (L^2/2)*(U(I+1,0) + U(I-1,0)) + (1 - L^2)*U(I,0) + K*FNG(I*H)
370  PRINT TAB(9*I);
```

```
380  PRINT USING "###.####"; U(I,1)
390  NEXT I
400  PRINT
410  REM COMPUTE THE U(I,J) KVALUES
420  FOR J = 1 TO M−1
430  PRINT USING "##.##"; (J+1)*K;
440  FOR I = 1 TO N−1
450  U(I,J+1) = L^2*U(I+1,J) + 2*(1−L^2)*U(I,J) + L^2*U(I−1,J) − U(I,J−1)
460  PRINT TAB(9*I);
470  PRINT USING "###.####"; U(I,J+1);
480  NEXT I
490  PRINT
500  NEXT J
510  END
```

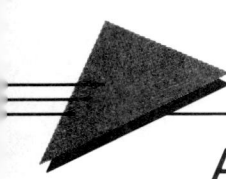

ANSWERS TO ODD-NUMBERED PROBLEMS

Exercises 1.1, Page 11
1. Linear, second-order
3. Nonlinear, first-order
5. Linear, fourth-order
7. Nonlinear, second-order
9. Linear, third-order
43. $k = -\frac{1}{4}$ 45. $y = -1$
47. $m = 2$ and $m = 3$ 49. $m = \dfrac{1 \pm \sqrt{5}}{2}$
51. Yes; yes
53. $y = 0$; no real solution; $y = \pm 1$

Exercises 1.2, Page 17
1. $xy' = y - 2$ 3. $y' + y = 0$
5. $2(x + 1)y' = y$
7. $(x^2 - y)y' = xy$ 9. $y'' - y' = 0$
11. $y'' + \omega^2 y = 0$ 13. $y'' - k^2 y = 0$
15. $y'' - 8y' + 16y = 0$ 17. $xy'' + y' = 0$
19. $y''' - 6y'' + 11y' - 6y = 0$
21. $(1 + \cos\theta)\dfrac{dr}{d\theta} + r\sin\theta = 0$ 23. $xy' - y = 0$
25. $2xyy' = y^2 - x^2$ 27. $2xy' = y$
29. $yy'' + (y')^2 = 0$

Exercises 1.3, Page 29
1. $\dfrac{dv}{dt} + \dfrac{k}{m}v = g$
3. $k = gR^2$; $\dfrac{d^2r}{dt^2} - \dfrac{gR^2}{r^2} = 0$; $v\dfrac{dv}{dr} - \dfrac{gR^2}{r^2} = 0$
5. $L\dfrac{di}{dt} + Ri = E(t)$ 7. $\dfrac{dh}{dt} = -\dfrac{\pi}{750}\sqrt{h}$
9. $\dfrac{dh}{dt} = -\dfrac{1}{30\sqrt{h}(10-h)}$ 11. $\dfrac{dx}{dt} + kx = r, k > 0$
13. $mx'' = -k\cos\theta$ $my'' = -mg - k\sin\theta$
$\quad\;\; = -k\cdot\dfrac{1}{v}\dfrac{dx}{dt}$ $\quad\;\; = -mg - k\cdot\dfrac{1}{v}\dfrac{dy}{dt}$
$\quad\;\; = -|c|\dfrac{dx}{dt}$ $\quad\;\; = -mg - |c|\dfrac{dy}{dt}$
15. Using $\tan\phi = \dfrac{x}{y}$, $\tan\left(\dfrac{\pi}{2} - \theta\right) = \dfrac{dy}{dx}$,

$\tan\theta = \dfrac{dx}{dy}$, and $\tan\phi = \tan 2\theta = \dfrac{2\tan\theta}{1 - \tan^2\theta}$,
we obtain $x\left(\dfrac{dx}{dy}\right)^2 + 2y\dfrac{dx}{dy} = x$.
17. By combining Newton's second law of motion with his law of gravitation, we obtain

$$m\dfrac{d^2y}{dt^2} = -k_1\dfrac{mM}{y^2}$$

where M is the mass of the earth and k_1 is a constant of proportionality. Dividing by m gives

$$\dfrac{d^2y}{dt^2} = -\dfrac{k}{y^2}$$

where $k = k_1 M$. The constant k is gR^2, where R is the radius of the earth. This follows from the fact that on the surface of the earth $y = R$, so that

$$k_1\dfrac{mM}{R^2} = mg$$

$$k_1 M = gR^2 \quad \text{or} \quad k = gR^2$$

If $t = 0$ is the time at which burnout occurs, then

$$y(0) = R + y_B$$

where y_B is the distance from the earth's surface to the rocket at the time of burnout, and

$$y'(0) = V_B$$

is the corresponding velocity at that time.

19. $\dfrac{dy}{dx} = -\dfrac{y}{\sqrt{s^2 - y^2}}$

Chapter 1 Review Exercises, Page 32
1. Ordinary, first-order, nonlinear
3. Partial, second-order   9. $y = x^2$
11. $y = \dfrac{x^2}{2}$ 13. $y = 0; y = e^x$
15. $x < 0$ or $x > 1$
19. $(x - 2)y' = y - 1$

A-29

21. $\dfrac{dh}{dt} = -\dfrac{25\sqrt{2g}}{16\pi} h^{-3/2}$

Exercises 2.1, Page 38

1. Half-planes defined by either $y > 0$ or $y < 0$
3. Half-planes defined by either $x > 0$ or $x < 0$
5. The regions defined by either $y > 2$, $y < -2$, or $-2 < y < 2$
7. Any region not containing (0, 0)
9. The entire xy-plane   11. $y = 0, y = x^3$
13. There is some interval around $x = 0$ on which the unique solution is $y = 0$.
15. $y = 0, y = x$. No, the given function is nondifferentiable at $x = 0$.
17. Yes   19. No

Exercises 2.2, Page 44

1. $y = -\tfrac{1}{5} \cos 5x + c$ 3. $y = \tfrac{1}{3} e^{-3x} + c$
5. $y = x + 5 \ln|x+1| + c$ 7. $y = cx^4$
9. $y^{-2} = 2x^{-1} + c$
11. $-3 + 3x \ln|x| = xy^3 + cx$
13. $-3e^{-2y} = 2e^{3x} + c$ 15. $2 + y^2 = c(4 + x^2)$
17. $y^2 = x - \ln|x + 1| + c$
19. $\dfrac{x^3}{3} \ln x - \dfrac{1}{9} x^3 = \dfrac{y^2}{2} + 2y + \ln|y| + c$
21. $S = ce^{kr}$ 23. $\dfrac{P}{1-P} = ce^t$ or $P = \dfrac{ce^t}{1+ce^t}$
25. $4 \cos y = 2x + \sin 2x + c$
27. $-2 \cos x + e^y + ye^{-y} + e^{-y} = c$
29. $(e^x + 1)^{-2} + 2(e^y + 1)^{-1} = c$
31. $(y+1)^{-1} + \ln|y+1| = \dfrac{1}{2} \ln\left|\dfrac{x+1}{x-1}\right| + c$
33. $y - 5 \ln|y + 3| = x - 5 \ln|x + 4| + c$
or $\left(\dfrac{y+3}{x+4}\right)^5 = c_1 e^{y-x}$
35. $-\cot y = \cos x + c$ 37. $y = \sin\left(\dfrac{x^2}{2} + c\right)$
39. $-y^{-1} = \tan^{-1}(e^x) + c$
41. $(1 + \cos x)(1 + e^y) = 4$
43. $\sqrt{y^2 + 1} = 2x^2 + \sqrt{2}$
45. $x = \tan(4y - 3\pi/4)$ 47. $xy = e^{-(1+1/x)}$
49. $y = 3 \dfrac{1-e^{6x}}{1+e^{6x}}$; $y = 3$; $y = 3 \dfrac{2 - e^{6x-2}}{2 + e^{6x-2}}$
51. $y = 1$ 53. $y = 1$
55. $y = 1 + \dfrac{1}{10} \tan \dfrac{x}{10}$
57. $y = -x - 1 + \tan(x + c)$
59. $2y - 2x + \sin 2(x + y) = c$
61. $4(y - 2x + 3) = (x + c)^2$

Exercises 2.3, Page 51

1. Homogeneous of degree 3
3. Homogeneous of degree 2
5. Not homogeneous
7. Homogeneous of degree 0
9. Homogeneous of degree -2
11. $x \ln|x| + y = cx$
13. $(x - y) \ln|x - y| = y + c(x - y)$
15. $x + y \ln|x| = cy$
17. $\ln(x^2 + y^2) + 2 \tan^{-1}(y/x) = c$
19. $4x = y(\ln|y| - c)^2$ 21. $y^9 = c(x^3 + y^3)^2$
23. $(y/x)^2 = 2 \ln|x| + c$ 25. $e^{2x/y} = 8 \ln|y| + c$
27. $x \cos(y/x) = c$ 29. $y + x = cx^2 e^{y/x}$
31. $y^3 + 3x^3 \ln|x| = 8x^3$ 33. $y^2 = 4x(x + y)^2$
35. $\ln|x| = e^{y/x} - 1$
37. $4x \ln\left|\dfrac{y}{x}\right| + x \ln x + y - x = 0$
39. $3x^{3/2} \ln x + 3x^{1/2} y + 2y^{3/2} = 5x^{3/2}$
41. $(x + y) \ln|y| + x = 0$
43. $\ln|y| = -2(1 - x/y)^{1/2} + \sqrt{2}$
45. $(y + 1)^2 + 2(y + 1)(x - 2) - (x - 2)^2 = c$
47. By homogeneity the equation can be written as

$$M\left(\dfrac{x}{y}, 1\right) dx + N\left(\dfrac{x}{y}, 1\right) dy = 0$$

With $v = x/y$, it follows that

$$M(v, 1)(v\, dy + y\, dv) + N(v, 1)\, dy = 0$$
$$[vM(v, 1) + N(v, 1)]\, dy + yM(v, 1)\, dv = 0$$

or
$$\dfrac{dy}{y} + \dfrac{M(v, 1)\, dv}{vM(v, 1) + N(v, 1)} = 0$$

49. $\dfrac{dy}{dx} = -\dfrac{M(x, y)}{N(x, y)} = -\dfrac{y^n M(x/y, 1)}{y^n N(x/y, 1)}$
$= -\dfrac{M(x/y, 1)}{N(x/y, 1)} = G\left(\dfrac{x}{y}\right)$

Exercises 2.4, Page 57

1. $x^2 - x + \tfrac{3}{2} y^2 + 7y = c$
3. $\tfrac{5}{2} x^2 + 4xy - 2y^4 = c$
5. $x^2 y^2 - 3x + 4y = c$
7. Not exact, but homogeneous
9. $xy^3 + y^2 \cos x - \tfrac{1}{2} x^2 = c$ 11. Not exact
13. $xy - 2xe^x + 2e^x - 2x^3 = c$
15. $x + y + xy - 3 \ln|xy| = c$
17. $x^3 y^3 - \tan^{-1} 3x = c$
19. $-\ln|\cos x| + \cos x \sin y = c$
21. $y - 2x^2 y - y^2 - x^4 = c$
23. $x^4 y - 5x^3 - xy + y^3 = c$
25. $\tfrac{1}{3} x^3 + x^2 y + xy^2 - y = \tfrac{4}{3}$
27. $4xy + x^2 - 5x + 3y^2 - y = 8$
29. $y^2 \sin x - x^3 y - x^2 + y \ln y - y = 0$
31. $k = 10$ 33. $k = 1$
35. $M(x, y) = ye^{xy} + y^2 - (y/x^2) + h(x)$
37. $M(x, y) = 6xy^3$
$N(x, y) = 4y^3 + 9x^2 y^2$
$\partial M/\partial y = 18xy^2 = \partial N/\partial x$
Solution is $3x^2 y^3 + y^4 = c$.

A-30

39. $M(x, y) = -x^2y^2 \sin x + 2xy^2 \cos x$
$N(x, y) = 2x^2y \cos x$
$\partial M/\partial y = -2x^2y \sin x + 4xy \cos x = \partial N/\partial x$
Solution is $x^2y^2 \cos x = c$.

41. $M(x, y) = 2xy^2 + 3x^2$
$N(x, y) = 2x^2y$
$\partial M/\partial y = 4xy = \partial N/\partial x$
Solution is $x^2y^2 + x^3 = c$.

43. A separable first-order differential equation can be written $h(y)\, dy - g(x)\, dx = 0$. When $M(x, y) = -g(x)$ and $N(x, y) = h(y)$, it follows that $\partial M/\partial y = 0 = \partial N/\partial x$.

Exercises 2.5, Page 65
1. $y = ce^{5x}, -\infty < x < \infty$
3. $y = \frac{1}{3} + ce^{-4x}, -\infty < x < \infty$
5. $y = \frac{1}{4}e^{3x} + ce^{-x}, -\infty < x < \infty$
7. $y = \frac{1}{3} + ce^{-x^3}, -\infty < x < \infty$
9. $y = x^{-1} \ln x + cx^{-1}, 0 < x < \infty$
11. $x = -\frac{4}{5}y^2 + cy^{-1/2}, 0 < y < \infty$
13. $y = -\cos x + \frac{\sin x}{x} + \frac{c}{x}, 0 < x < \infty$
15. $y = \frac{c}{e^x + 1}, -\infty < x < \infty$
17. $y = \sin x + c \cos x, -\pi/2 < x < \pi/2$
19. $y = \frac{1}{7}x^3 - \frac{1}{5}x + cx^{-4}, 0 < x < \infty$
21. $y = \frac{1}{2x^2}e^x + \frac{c}{x^2}e^{-x}, 0 < x < \infty$
23. $y = \sec x + c \csc x, 0 < x < \pi/2$
25. $x = \frac{1}{2}e^y - \frac{1}{2y}e^y + \frac{1}{4y^2}e^y + \frac{c}{y^2}e^{-y}, 0 < y < \infty$
27. $y = e^{-3x} + \frac{c}{x}e^{-3x}, 0 < x < \infty$
29. $x = 2y^6 + cy^4, 0 < y < \infty$
31. $y = e^{-x}\ln(e^x + e^{-x}) + ce^{-x}, -\infty < x < \infty$
33. $x = \frac{1}{y} + \frac{c}{y}e^{-y^2}, 0 < y < \infty$
35. $(\sec\theta + \tan\theta)r = \theta - \cos\theta + c, -\pi/2 < \theta < \pi/2$
37. $y = \frac{5}{3}(x+2)^{-1} + c(x+2)^{-4}, -2 < x < \infty$
39. $y = 10 + ce^{-\sinh x}, -\infty < x < \infty$
41. $y = 4 - 2e^{-5x}, -\infty < x < \infty$
43. $i(t) = E/R + (i_0 - E/R)e^{-Rt/L}, -\infty < t < \infty$
45. $y = \sin x \cos x - \cos x, -\pi/2 < x < \pi/2$
47. $T(t) = 50 + 150e^{kt}, -\infty < t < \infty$
49. $(x+1)y = x\ln x - x + 21, 0 < x < \infty$
51. $y = \frac{2x}{x-2}, 2 < x < \infty$
53. $x = \frac{1}{2}y + 8/y, 0 < y < \infty$
55. $y = \begin{cases} \frac{1}{2}(1 - e^{-2x}), & 0 \le x \le 3 \\ \frac{1}{2}(e^6 - 1)e^{-2x}, & x > 3 \end{cases}$
57. $y = \begin{cases} \frac{1}{2} + \frac{3}{2}e^{-x^2}, & 0 \le x < 1 \\ (\frac{1}{2}e + \frac{3}{2})e^{-x^2}, & x \ge 1 \end{cases}$

Exercises 2.6, Page 69
1. $y^3 = 1 + cx^{-3}$ **3.** $y^{-3} = x + \frac{1}{3} + ce^{3x}$
5. $e^{x/y} = cx$ **7.** $y^{-3} = -\frac{9}{5}x^{-1} + \frac{49}{5}x^{-6}$
9. $x^{-1} = 2 - y^2 - e^{-y^2/2}$, the equation is Bernoulli in the variable x.
11. $y = 2 + \dfrac{1}{ce^{-3x} - 1/3}$ **13.** $y = \dfrac{2}{x} + \dfrac{1}{cx^{-3} - x/4}$
15. $y = -e^x + \dfrac{1}{ce^{-x} - 1}$ **17.** $y = -2 + \dfrac{1}{ce^{-x} - 1}$
19. $y = cx + 1 - \ln c; y = 2 + \ln x$
21. $y = cx - c^3; 27y^2 = 4x^3$
23. $y = cx - e^c; y = x \ln x - x$

Exercises 2.7, Page 73
1. $x^2e^{2y} = 2x \ln x - 2x + c$
3. $e^{-x} = y \ln|y| + cy$ **5.** $-e^{-y/x^4} = x^2 + c$
7. $x^2 + y^2 = x - 1 + ce^{-x}$
9. $\ln(\tan y) = x + cx^{-1}$
11. $x^3y^3 = 2x^3 - 9\ln|x| + c$
13. $e^y = -e^{-x}\cos x + ce^{-x}$
15. $y^2 \ln x = ye^y - e^y + c$
17. $y = \ln|\cos(c_1 - x)| + c_2$
19. $y = -\dfrac{1}{c_1}(1 - c_1^2 x^2)^{1/2} + c_2$
21. The given equation is a Clairaut equation in $u = y'$. The solution is $y = c_1 x^2/2 + x + c_1^3 x + c_2$.
23. $y = c_1 + c_2 x^2$ **25.** $\frac{1}{3}y^3 - c_1 y = x + c_2$
27. $y = -\sqrt{1 - x^2}$

Exercises 2.8, Page 76
1. $y_1(x) = 1 - x$
$y_2(x) = 1 - \dfrac{x}{1!} + \dfrac{x^2}{2!}$
$y_3(x) = 1 - \dfrac{x}{1!} + \dfrac{x^2}{2!} - \dfrac{x^3}{3!}$
$y_4(x) = 1 - \dfrac{x}{1!} + \dfrac{x^2}{2!} - \dfrac{x^3}{3!} + \dfrac{x^4}{4!}$
$y_n(x) \to e^{-x}$ as $n \to \infty$
3. $y_1(x) = 1 + x^2$
$y_2(x) = 1 + \dfrac{x^2}{1!} + \dfrac{x^4}{2!}$
$y_3(x) = 1 + \dfrac{x^2}{1!} + \dfrac{x^4}{2!} + \dfrac{x^6}{3!}$
$y_4(x) = 1 + \dfrac{x^2}{1!} + \dfrac{x^4}{2!} + \dfrac{x^6}{3!} + \dfrac{x^8}{4!}$
$y_n(x) \to e^{x^2}$ as $n \to \infty$
5. $y_1(x) = y_2(x) = y_3(x) = y_4(x) = 0;$
$y_n(x) \to 0$ as $n \to \infty$
7. $y_1(x) = x$
$y_2(x) = x + \frac{1}{3}x^3$
$y_3(x) = x + \frac{1}{3}x^3 + \frac{2}{15}x^5 + \frac{1}{63}x^7;$
$y = \tan x;$

The Maclaurin series expansion of tan x is
$x + \frac{1}{3}x^3 + \frac{2}{15}x^5 + \frac{17}{315}x^7 + \cdots, |x| < \pi/2$.

Exercises 2.9, Page 80

1. $x^2 + y^2 = c_2^2$ 3. $2y^2 + x^2 = c_2$
5. $2\ln|y| = x^2 + y^2 + c_2$ 7. $y^2 = 2x + c_2$
9. $2x^2 + 3y^2 = c_2$ 11. $x^3 + y^3 = c_2$
13. $y^2 \ln|y| + x^2 = c_2 y^2$ 15. $y^2 - x^2 = c_2 x$
17. $2y^2 = 2\ln|x| + x^2 + c_2$
19. $y = \frac{1}{4} - \frac{1}{6}x^2 + c_2 x^{-4}$ 21. $2y^3 = 3x^2 + c_2$
23. $2\ln(\cosh y) + x^2 = c_2$ 25. $y^{5/3} = x^{5/3} + c_2$
27. $y = 2 - x + 3e^{-x}$ 29. $r = c_2 \sin\theta$
31. $r^2 = c_2 \cos 2\theta$ 33. $r = c_2 \csc\theta$
35. Let β be the angle of inclination, measured from the positive x-axis, of the tangent line to a member of the given family, and ϕ the angle of inclination of the tangent to a trajectory. At the point where the curves intersect, the angle between the tangents is α. From the accompanying figures we conclude that there exist two possible cases and that $\phi = \beta \pm \alpha$. Thus, the slope of the tangent line to a

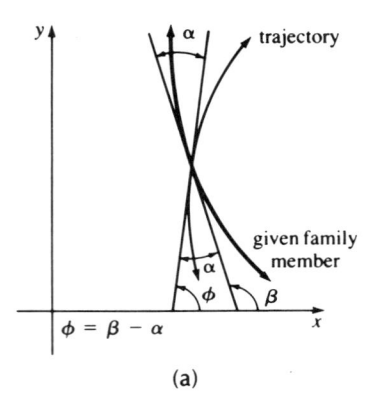

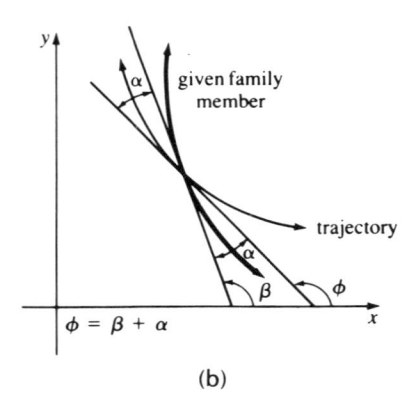

trajectory is

$$\frac{dy}{dx} = \tan\phi = \tan(\beta \pm \alpha)$$

$$= \frac{\tan\beta \pm \tan\alpha}{1 \mp \tan\beta \tan\alpha}$$

$$= \frac{f(x, y) \pm \tan\alpha}{1 \mp f(x, y) \tan\alpha}$$

37. $\mp \frac{2}{\sqrt{3}} \tan^{-1}\left(\frac{y}{x}\right) + \ln c_2(x^2 + y^2) = 0$

39. Since the given equation is quadratic in c_1, it follows from the quadratic formula that

$$c_1 = -x \pm \sqrt{x^2 + y^2}$$

Differentiating this last expression and solving for dy/dx give

$$\frac{dy}{dx} = \frac{-x + \sqrt{x^2 + y^2}}{y}$$

and

$$\frac{dy}{dx} = \frac{-x - \sqrt{x^2 + y^2}}{y}$$

These two equations correspond to choosing $c_1 > 0$ and $c_1 < 0$ in the given family, respectively. Forming the product of these derivatives yields

$$\left(\frac{dy}{dx}\right)_{(1)} \cdot \left(\frac{dy}{dx}\right)_{(2)} = \frac{x^2 - x^2 - y^2}{y^2} = -1$$

This shows that the family is self-orthogonal.

41. The differential equation of the orthogonal family is $(x - y)\,dx + (x + y)\,dy = 0$. The verification follows by substituting $x = c_2 e^{-t}\cos t$ and $y = c_2 e^{-t}\sin t$ into the equation.

Exercises 2.10, Page 89

1. 7.9 years; 10 years   3. 760   5. 11 h
7. 136.5 h
9. $I(15) = 0.00098 I_0$ or $I(15)$ is approximately 0.1% of I_0.
11. 15,600 years
13. $T(1) = 36.67$ degrees; approximately 3.06 minutes
15. $i(t) = \frac{3}{5} - \frac{3}{5}e^{-500t}$; $i \to \frac{3}{5}$ as $t \to \infty$
17. $q(t) = \frac{1}{100} - \frac{1}{100}e^{-50t}$; $i(t) = \frac{1}{2}e^{-50t}$
19. $i(t) = \begin{cases} 60 - 60e^{-t/10}, & 0 \le t \le 20 \\ 60(e^2 - 1)e^{-t/10}, & t > 20 \end{cases}$
21. $A(t) = 200 - 170e^{-t/50}$
23. $A(t) = 1000 - 1000e^{-t/100}$ 25. 64.38 lb
27. $v(t) = \frac{mg}{k} + \left(v_0 - \frac{mg}{k}\right)e^{-kt/m}$;

$v \to \frac{mg}{k}$ as $t \to \infty$;

$s(t) = \frac{mg}{k}t - \frac{m}{k}\left(v_0 - \frac{mg}{k}\right)e^{-kt/m}$

$+ \frac{m}{k}\left(v_0 - \frac{mg}{k}\right) + s_0$

29. $E(t) = E_0 e^{-(t-t_1)/RC}$
31. $P(t) = P_0 e^{(k_1 - k_2)t}$;
$k_1 > k_2$, births surpass deaths so population

increases. $k_1 = k_2$, a constant population since number of births equals number of deaths. $k_1 < k_2$, deaths surpass births so population decreases.

33. From $r^2\, d\theta = \dfrac{L}{M}\, dt$ we get

$$A = \frac{1}{2}\int_{\theta_1}^{\theta_2} r^2\, d\theta = \frac{1}{2}\frac{L}{M}\int_a^b dt$$

$$= \frac{1}{2}\frac{L}{M}(b - a)$$

Exercises 2.11, Page 98

1. 1834; 2000 **3.** 1,000,000; 52.9 months
5. (a) Separating variables gives

$$\frac{dP}{P(a - b\ln P)} = dt$$

so that

$$-(1/b)\ln|a - b\ln P| = t + c_1$$

$$a - b\ln P = c_2 e^{-bt} \qquad (e^{-bc_1} = c_2)$$

$$\ln P = (a/b) - ce^{-bt} \qquad (c_2/b = c)$$

$$P(t) = e^{a/b}\cdot e^{-ce^{-bt}}$$

(b) If $P(0) = P_0$, then

$$P_0 = e^{a/b}e^{-c} = e^{a/b - c}$$

and so $\quad \ln P_0 = (a/b) - c$

$$c = (a/b) - \ln P_0$$

7. 29.3 grams; $X \to 60$ as $t \to \infty$; 0 grams of A and 30 grams of B.
9. For $\alpha \neq \beta$ the differential equation separates as

$$\frac{1}{\alpha - \beta}\left[-\frac{1}{\alpha - X} + \frac{1}{\beta - X}\right]dx = k\, dt$$

It follows immediately that

$$\frac{1}{\alpha - \beta}[\ln|\alpha - X| - \ln|\beta - X|] = kt + c$$

or $\quad \dfrac{1}{\alpha - \beta}\ln\left|\dfrac{\alpha - X}{\beta - X}\right| = kt + c$

For $\alpha = \beta$ the equation can be written as

$$(\alpha - X)^{-2}\, dX = k\, dt$$

It follows that $(\alpha - X)^{-1} = kt + c$ or

$$X = \alpha - \frac{1}{kt + c}$$

11. $v^2 = (2gR^2/y) + v_0^2 - 2gR$.
We note that as y increases, v decreases. In particular, if $v_0^2 - 2gR < 0$, then there must be some value of y for $v = 0$; the rocket stops and returns to earth under the influence of gravity. However, if $v_0^2 - 2gR \geq 0$, then $v > 0$ for all values of y. Hence we should have $v_0 \geq \sqrt{2gR}$. Using the values $R = 4000$ mi, $g = 32$ ft/s^2, 1 ft = 1/5280 mi, 1 s = 1/3600 h, we get $v_0 \geq 25{,}067$ mi/h.

13. Using the condition $y'(1) = 0$, we find

$$\frac{dy}{dx} = \frac{1}{2}[x^{v_1/v_2} - x^{-v_1/v_2}]$$

Now, if $v_1 = v_2$, then $y = \tfrac{1}{4}x^2 - \tfrac{1}{2}\ln x - \tfrac{1}{4}$; if $v_1 \neq v_2$, then

$$y = \frac{1}{2}\left[\frac{x^{1+(v_1/v_2)}}{1 + \dfrac{v_1}{v_2}} - \frac{x^{1-(v_1/v_2)}}{1 - \dfrac{v_1}{v_2}}\right] + \frac{v_1 v_2}{v_2^2 - v_1^2}$$

15. $2h^{1/2} = -\tfrac{1}{25}t + 2\sqrt{20};\; t = 50\sqrt{20}$ s
17. To evaluate the indefinite integral of the left side of

$$\frac{\sqrt{100 - y^2}}{y}\, dy = -dx$$

we use the substitution $y = 10\cos\theta$. It follows that

$$x = 10\ln\left(\frac{10 + \sqrt{100 - y^2}}{y}\right) - \sqrt{100 - y^2}$$

19. Under the substitution $w = x^2$, the differential equation becomes

$$w = y\frac{dw}{dy} + \frac{1}{4}\left(\frac{dw}{dy}\right)^2$$

which is Clairaut's equation. The solution is

$$x^2 = cy + \frac{c^2}{4}$$

If $2c_1 = c$, then we recognize

$$x^2 = 2c_1 y + c_1^2$$

as describing a family of parabolas.
21. $-\gamma \ln y + \delta y = \alpha \ln x - \beta x + c$
23. (a) The equation $2\dfrac{d^2\theta}{dt^2}\dfrac{d\theta}{dt} + 2\dfrac{g}{l}\sin\theta\,\dfrac{d\theta}{dt} = 0$ is the same as

$$\frac{d}{dt}\left(\frac{d\theta}{dt}\right)^2 + 2\frac{g}{l}\sin\theta\,\frac{d\theta}{dt} = 0$$

Integrating this last equation with respect to t and using the initial conditions give the result.
(b) From (a),

$$dt = \sqrt{\frac{l}{2g}}\,\frac{d\theta}{\sqrt{\cos\theta - \cos\theta_0}}$$

A-33

Integrating this last equation gives the time for the pendulum to move from $\theta = \theta_0$ to $\theta = 0$:

$$t = \sqrt{\frac{l}{2g}} \int_0^{\theta_0} \frac{d\theta}{\sqrt{\cos\theta - \cos\theta_0}}$$

The period is the total time T to go from $\theta = \theta_0$ to $\theta = -\theta_0$ and back again to $\theta = \theta_0$. This is

$$T = 4\sqrt{\frac{l}{2g}} \int_0^{\theta_0} \frac{d\theta}{\sqrt{\cos\theta - \cos\theta_0}}$$

$$= 2\sqrt{\frac{2l}{g}} \int_0^{\theta_0} \frac{d\theta}{\sqrt{\cos\theta - \cos\theta_0}}$$

Chapter 2 Review Exercises, Page 102

1. $x^2 + y^2 > 25$ and $x^2 + y^2 < 25$ 3. False
5. (a) Linear in x
   (b) Homogeneous, exact, linear in y
   (c) Clairaut
   (d) Bernoulli in x
   (e) Separable
   (f) Separable, Ricatti
   (g) Linear in x
   (h) Homogeneous
   (i) Bernoulli
   (j) Homogeneous, exact, Bernoulli
   (k) Separable, homogeneous, exact, linear in x and in y
   (l) Exact, linear in y
   (m) Homogeneous
   (n) Separable
   (o) Clairaut
   (p) Ricatti
7. $2y^2 \ln y - y^2 = 4xe^x - 4e^x - 1$
9. $2y^2 + x^2 = 9x^6$ 11. $e^{xy} - 4y^3 = 5$
13. $y = \frac{1}{4} - 320(x^2 + 4)^{-4}$
15. $y = \dfrac{1}{x^4 - x^4 \ln|x|}$ 17. $x^2 - \sin\dfrac{1}{y^2} = c$
19. $y_1(x) = 1 + x + \frac{1}{3}x^3$
    $y_2(x) = 1 + x + x^2 + \frac{2}{3}x^3 + \frac{1}{6}x^4 + \frac{2}{15}x^5 + \frac{1}{63}x^7$
21. $y^3 + 3/x = c_2$ 23. $2(y-2)^2 + (x-1)^2 = c_2^2$
25. $P(45) = 8.99$ billion
27. $x(t) = \dfrac{\alpha c_1 e^{\alpha k_1 t}}{1 + c_1 e^{\alpha k_1 t}}$, $y(t) = c_2(1 + c_1 e^{\alpha k_1 t})^{k_2/k_1}$
29. Approximately 1565 s
31. Approximately 14.87 min; 287.2 lb

Exercises 3.1, Page 124

1. $y = \frac{1}{2}e^x - \frac{1}{2}e^{-x}$ 3. $y = \frac{3}{5}e^{4x} + \frac{2}{5}e^{-x}$
5. $y = 3x - 4x \ln x$ 7. $y = 0, y = x^2$
9. (a) $y = e^x \cos x - e^x \sin x$
   (b) No solution
   (c) $y = e^x \cos x + e^{-\pi/2} e^x \sin x$
   (d) $y = c_2 e^x \sin x$, where c_2 is arbitrary

11. $(-\infty, 2)$ 13. $\lambda = n, n = 1, 2, 3, \ldots$
15. Dependent   17. Dependent
19. Dependent   21. Independent
23. $W(x^{1/2}, x^2) = \frac{3}{2}x^{3/2} \neq 0$ on $(0, \infty)$
25. $W(\sin x, \csc x) = -2 \cot x$;
    $W = 0$ only at $x = \pi/2$ in the interval.
27. $W(e^x, e^{-x}, e^{4x}) = -30e^{4x} \neq 0$ on $(-\infty, \infty)$
29. No
31. (a) $y'' - 2y^3 = \dfrac{2}{x^3} - 2\left(\dfrac{1}{x}\right)^3 = 0$

    (b) $y'' - 2y^3 = \dfrac{2c}{x^3} - 2\dfrac{c^3}{x^3} = \dfrac{2}{x^3}c(1-c^2) \neq 0$
    for $c \neq 0, \pm 1$
33. The functions satisfy the differential equation and are linearly independent on the interval since $W(e^{-3x}, e^{4x}) = 7e^x \neq 0$; $y = c_1 e^{-3x} + c_2 e^{4x}$.
35. The functions satisfy the differential equation and are linearly independent on the interval since $W(e^x \cos 2x, e^x \sin 2x) = 2e^{2x} \neq 0$; $y = c_1 e^x \cos 2x + c_2 e^x \sin 2x$.
37. The functions satisfy the differential equation and are linearly independent on the interval since $W(x^3, x^4) = x^6 \neq 0$; $y = c_1 x^3 + c_2 x^4$.
39. The functions satisfy the differential equation and are linearly independent on the interval since $W(x, x^{-2}, x^{-2} \ln x) = 9x^{-6} \neq 0$; $y = c_1 x + c_2 x^{-2} + c_3 x^{-2} \ln x$.
41. e^{2x} and e^{5x} form a fundamental set of solutions of the homogeneous equation; $6e^x$ is a particular solution of the nonhomogeneous equation.
43. e^{2x} and xe^{2x} form a fundamental set of solutions of the homogeneous equation; $x^2 e^{2x} + x - 2$ is a particular solution of the nonhomogeneous equation.
45. (a) The accompanying graphs show that y_1 and y_2 are not multiples of each other. Also,

$$x^2 y_1'' - 4x y_1' + 6y_1 = x^2(6x) - 4x(3x^2) + 6x^3$$
$$= 12x^3 - 12x^3 = 0$$

For $x \geq 0$ the demonstration that y_2 is a solution of the equation is exactly as given above for y_1. For $x < 0$, $y_2 = -x^3$ and so

$$x^2 y_2'' - 4x y_2' + 6y_2 = x^2(-6x)$$
$$-4x(-3x^2) + 6(-x^3)$$
$$= -12x^3 + 12x^3 = 0$$

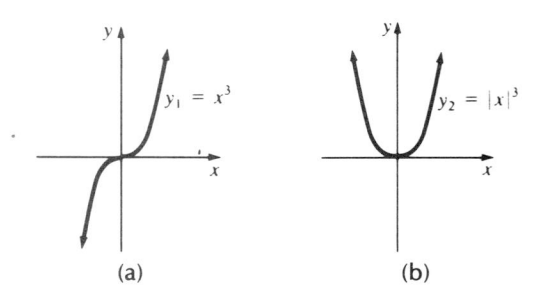

(a) $y_1 = x^3$ (b) $y_2 = |x|^3$

A-34

(b) For $x \geq 0$,
$$W(y_1, y_2) = \begin{vmatrix} x^3 & x^3 \\ 3x^2 & 3x^2 \end{vmatrix} = 3x^5 - 3x^5 = 0$$

For $x < 0$,
$$W(y_1, y_2) = \begin{vmatrix} x^3 & -x^3 \\ 3x^2 & -3x^2 \end{vmatrix} = -3x^5 + 3x^5 = 0$$

Thus $W(y_1, y_2) = 0$ for every real value of x.
(c) No, $a_2(x) = x^2$ is zero at $x = 0$.
(d) Since $Y_1 = y_1$, we need only show

$$x^2 Y_2'' - 4xY_2' + 6Y_2 = x^2(2) - 4x(2x) + 6x^2$$
$$= 8x^2 - 8x^2 = 0$$

and $W(x^3, x^2) = -x^4$. Thus Y_1 and Y_2 are linearly independent solutions on the interval.
(e) $Y_1 = x^3$, $Y_2 = x^2$, or $y_2 = |x|^3$
(f) Neither; we form a general solution on an interval for which $a_2(x) \neq 0$ for every x in the interval. The linear combination

$$y = c_1 Y_1 + c_2 Y_2$$

would be a general solution of the equation on, say, the interval $(0, \infty)$.

Exercises 3.2, Page 130

1. $y_2 = e^{-5x}$
3. $y_2 = xe^{2x}$
5. $y_2 = \sin 4x$
7. $y_2 = \sinh x$
9. $y_2 = xe^{2x/3}$
11. $y_2 = x^4 \ln|x|$
13. $y_2 = 1$
15. $y_2 = x^2 + x + 2$
17. $y_2 = x \cos(\ln x)$
19. $y_2 = x$
21. $y_2 = x \ln x$
23. $y_2 = x^3$
25. $y_2 = x^2$
27. $y_2 = 3x + 2$
29. $y_2 = \frac{1}{2}[\tan x \sec x + \ln|\sec x + \tan x|]$
31. $y_2 = e^{2x}$, $y_p = -\frac{1}{2}$
33. $y_2 = e^{2x}$, $y_p = \frac{5}{2}e^{3x}$

Exercises 3.3, Page 137

1. $y = c_1 + c_2 e^{-x/4}$
3. $y = c_1 e^{-6x} + c_2 e^{6x}$
5. $y = c_1 \cos 3x + c_2 \sin 3x$
7. $y = c_1 e^{3x} + c_2 e^{-2x}$
9. $y = c_1 e^{-4x} + c_2 x e^{-4x}$
11. $y = c_1 e^{(-3+\sqrt{29})x/2} + c_2 e^{(-3-\sqrt{29})x/2}$
13. $y = c_1 e^{2x/3} + c_2 e^{-x/4}$
15. $y = e^{2x}(c_1 \cos x + c_2 \sin x)$
17. $y = e^{-x/3}\left(c_1 \cos \frac{\sqrt{2}}{3} x + c_2 \sin \frac{\sqrt{2}}{3} x\right)$
19. $y = c_1 + c_2 e^{-x} + c_3 e^{5x}$
21. $y = c_1 e^x + e^{-x/2}\left(c_2 \cos \frac{\sqrt{3}}{2} x + c_3 \sin \frac{\sqrt{3}}{2} x\right)$
23. $y = c_1 e^{-x} + c_2 e^{3x} + c_3 x e^{3x}$
25. $y = c_1 e^x + e^{-x}(c_2 \cos x + c_3 \sin x)$
27. $y = c_1 e^{-x} + c_2 x e^{-x} + c_3 x^2 e^{-x}$

29. $y = c_1 + c_2 x + e^{-x/2}\left(c_3 \cos \frac{\sqrt{3}}{2} x + c_4 \sin \frac{\sqrt{3}}{2} x\right)$
31. $y = c_1 \cos \frac{\sqrt{3}}{2} x + c_2 \sin \frac{\sqrt{3}}{2} x$
$\qquad + c_3 x \cos \frac{\sqrt{3}}{2} x + c_4 x \sin \frac{\sqrt{3}}{2} x$
33. $y = c_1 + c_2 e^{-2x} + c_3 e^{2x} + c_4 \cos 2x + c_5 \sin 2x$
35. $y = c_1 e^x + c_2 x e^x + c_3 e^{-x} + c_4 x e^{-x} + c_5 e^{-5x}$
37. $y = 2 \cos 4x - \frac{1}{2} \sin 4x$
39. $y = -\frac{3}{4} e^{-5x} + \frac{3}{4} e^{-x}$
41. $y = -e^{x/2} \cos(x/2) + e^{x/2} \sin(x/2)$
43. $y = 0$
45. $y = e^{2(x-1)} - e^{x-1}$
47. $y = \frac{5}{36} - \frac{5}{36} e^{-6x} + \frac{1}{6} x e^{-6x}$
49. $y = -\frac{1}{6} e^{2x} + \frac{1}{6} e^{-x} \cos \sqrt{3} x - \frac{\sqrt{3}}{6} e^{-x} \sin \sqrt{3} x$
51. $y = 2 - 2e^x + 2xe^x - \frac{1}{2} x^2 e^x$
53. $y = e^{5x} - xe^{5x}$
55. $y = -2 \cos x$
57. $\dfrac{d^3 y}{dx^3} + 6 \dfrac{d^2 y}{dx^2} - 15 \dfrac{dy}{dx} - 100 y = 0$
59. $y = c_1 e^x + e^{4x}(c_2 \cos x + c_3 \sin x)$
61. $y'' - 3y' - 18y = 0$
63. $y''' - 7y'' = 0$
65. $y = e^{-\sqrt{2}x/2}\left(c_1 \cos \frac{\sqrt{2}}{2} x + c_2 \sin \frac{\sqrt{2}}{2} x\right)$
$\qquad + e^{\sqrt{2}x/2}\left(c_3 \cos \frac{\sqrt{2}}{2} x + c_4 \sin \frac{\sqrt{2}}{2} x\right)$

Exercises 3.4, Page 147

1. $y = c_1 e^{-x} + c_2 e^{-2x} + 3$
3. $y = c_1 e^{5x} + c_2 x e^{5x} + \frac{6}{5} x + \frac{3}{5}$
5. $y = c_1 e^{-2x} + c_2 x e^{-2x} + x^2 - 4x + \frac{7}{2}$
7. $y = c_1 \cos \sqrt{3} x + c_2 \sin \sqrt{3} x + (-4x^2 + 4x - \frac{4}{3}) e^{3x}$
9. $y = c_1 + c_2 e^x + 3x$
11. $y = c_1 e^{x/2} + c_2 x e^{x/2} + 12 + \frac{1}{2} x^2 e^{x/2}$
13. $y = c_1 \cos 2x + c_2 \sin 2x - \frac{3}{4} x \cos 2x$
15. $y = c_1 \cos x + c_2 \sin x - \frac{1}{2} x^2 \cos x + \frac{1}{2} x \sin x$
17. $y = c_1 e^x \cos 2x + c_2 e^x \sin 2x + \frac{1}{4} x e^x \sin 2x$
19. $y = c_1 e^{-x} + c_2 x e^{-x} - \frac{1}{2} \cos x + \frac{12}{25} \sin 2x$
$\qquad - \frac{9}{25} \cos 2x$
21. $y = c_1 + c_2 x + c_3 e^{6x} - \frac{1}{4} x^2 - \frac{6}{37} \cos x + \frac{1}{37} \sin x$
23. $y = c_1 e^x + c_2 x e^x + c_3 x^2 e^x - x - 3 - \frac{2}{3} x^3 e^x$
25. $y = c_1 \cos x + c_2 \sin x + c_3 x \cos x + c_4 x \sin x$
$\qquad + x^2 - 2x - 3$
27. $y_p = 4 + \frac{4}{3} \cos 2x$
29. $y = \sqrt{2} \sin 2x - \frac{1}{2}$
31. $y = -200 + 200 e^{-x/5} - 3x^2 + 30x$
33. $y = -10 e^{-2x} \cos x + 9 e^{-2x} \sin x + 7 e^{-4x}$
35. $x = \dfrac{F_0}{2\omega^2} \sin \omega t - \dfrac{F_0}{2\omega} t \cos \omega t$
37. $y = -\dfrac{1}{6} \cos x - \dfrac{\pi}{4} \sin x + \dfrac{1}{2} x \sin x + \dfrac{1}{3} \sin 2x$
39. $y = 11 - 11 e^x + 9 x e^x + 2x - 12 x^2 e^x + \frac{1}{2} e^{5x}$

A-35

Exercises 3.5, Page 156

1. $(D+5)y = 9\sin x$
3. $(3D^2 - 5D + 1)y = e^x$
5. $(D^3 - 4D^2 + 5D)y = 4x$
7. $(3D-2)(3D+2)$
9. $(D-6)(D+2)$
11. $D(D+5)^2$
13. $(D-1)(D-2)(D+5)$
15. $D(D+2)(D^2 - 2D + 4)$
17. D^4
19. $D(D-2)$
21. $D^2 + 4$
23. $D^3(D^2 + 16)$
25. $(D+1)(D-1)^3$
27. $D(D^2 - 2D + 5)$
29. $1, x, x^2, x^3, x^4$
31. $e^{6x}, e^{-3x/2}$
33. $\cos\sqrt{5}x, \sin\sqrt{5}x$
35. $1, e^{5x}, xe^{5x}$
37. $y = c_1 e^{-3x} + c_2 e^{3x} - 6$
39. $y = c_1 + c_2 e^{-x} + 3x$
41. $y = c_1 e^{-2x} + c_2 x e^{-2x} + \frac{1}{2}x + 1$
43. $y = c_1 + c_2 x + c_3 e^{-x} + \frac{2}{3}x^4 - \frac{8}{3}x^3 + 8x^2$
45. $y = c_1 e^{-3x} + c_2 e^{4x} + \frac{1}{7}x e^{4x}$
47. $y = c_1 e^{-x} + c_2 e^{3x} - e^x + 3$
49. $y = c_1 \cos 5x + c_2 \sin 5x + \frac{1}{4}\sin x$
51. $y = c_1 e^{-3x} + c_2 x e^{-3x} - \frac{1}{49}x e^{4x} + \frac{2}{343}e^{4x}$
53. $y = c_1 e^{-x} + c_2 e^x + \frac{1}{6}x^3 e^x - \frac{1}{4}x^2 e^x + \frac{1}{4}x e^x - 5$
55. $y = e^x(c_1 \cos 2x + c_2 \sin 2x) + \frac{1}{3}e^x \sin x$
57. $y = c_1 \cos 5x + c_2 \sin 5x - 2x \cos 5x$
59. $y = e^{-x/2}\left(c_1 \cos \frac{\sqrt{3}}{2}x + c_2 \sin \frac{\sqrt{3}}{2}x\right) + \sin x + 2\cos x - x\cos x$
61. $y = c_1 + c_2 x + c_3 e^{-8x} + \frac{11}{256}x^2 + \frac{7}{32}x^3 - \frac{1}{16}x^4$
63. $y = c_1 e^x + c_2 x e^x + c_3 x^2 e^x + \frac{1}{6}x^3 e^x + x - 13$
65. $y = c_1 + c_2 x + c_3 e^x + c_4 x e^x + \frac{1}{2}x^2 e^x + \frac{1}{2}x^2$
67. $y = c_1 e^{x/2} + c_2 e^{-x/2} + c_3 \cos\frac{x}{2} + c_4 \sin\frac{x}{2} + \frac{1}{8}x e^{x/2}$
69. $y = \frac{5}{8}e^{-8x} + \frac{5}{8}e^{8x} - \frac{1}{4}$
71. $y = -\frac{41}{125} + \frac{41}{125}e^{5x} - \frac{1}{10}x^2 + \frac{9}{25}x$
73. $y = -\pi \cos x - \frac{11}{3}\sin x - \frac{8}{3}\cos 2x + 2x\cos x$
75. $y = 2e^{2x}\cos 2x - \frac{3}{64}e^{2x}\sin 2x + \frac{1}{8}x^3 + \frac{3}{16}x^2 + \frac{3}{32}x$
77. $y_p = Ae^x + Be^x \cos 2x + Ce^x \sin 2x + Exe^x \cos 2x + Fxe^x \sin 2x$
79. The operators do not commute.

Exercises 3.6, Page 164

1. $y = c_1 \cos x + c_2 \sin x + x \sin x + \cos x \ln|\cos x|; (-\pi/2, \pi/2)$
3. $y = c_1 \cos x + c_2 \sin x + \frac{1}{2}\sin x - \frac{1}{2}x\cos x$
$= c_1 \cos x + c_3 \sin x - \frac{1}{2}x\cos x; (-\infty, \infty)$
5. $y = c_1 \cos x + c_2 \sin x + \frac{1}{2} - \frac{1}{6}\cos 2x; (-\infty, \infty)$
7. $y = c_1 e^x + c_2 e^{-x} + \frac{1}{4}xe^x - \frac{1}{4}xe^{-x}$
$= c_1 e^x + c_2 e^{-x} + \frac{1}{2}x \sinh x; (-\infty, \infty)$
9. $y = c_1 e^{2x} + c_2 e^{-2x}$
$+ \frac{1}{4}\left(e^{2x}\ln|x| - e^{-2x}\int_{x_0}^{x}\frac{e^{4t}}{t}dt\right),$
$x_0 > 0; (0, \infty)$
11. $y = c_1 e^{-x} + c_2 e^{-2x} + (e^{-x} + e^{-2x}) \times \ln(1+e^x); (-\infty, \infty)$
13. $y = c_1 e^{-2x} + c_2 e^{-x} - e^{-2x} \sin e^x; (-\infty, \infty)$
15. $y = c_1 e^x + c_2 x e^x - \frac{1}{2}e^x \ln(1+x^2) + xe^x \tan^{-1}x; (-\infty, \infty)$
17. $y = c_1 e^{-x} + c_2 x e^{-x} + \frac{1}{2}x^2 e^{-x} \ln x - \frac{3}{4}x^2 e^{-x}; (0, \infty)$
19. $y = c_1 e^x \cos 3x + c_2 e^x \sin x - \frac{1}{27}e^x \cos 3x \ln|\sec 3x + \tan 3x|; (-\pi/6, \pi/6)$
21. $y = c_1 + c_2 \cos x + c_3 \sin x - \ln|\cos x| - \sin x \ln|\sec x + \tan x|; (-\pi/2, \pi/2)$
23. $y = c_1 e^x + c_2 e^{2x} + c_3 e^{-x} + \frac{1}{8}e^{3x}; (-\infty, \infty)$
25. $y = \frac{1}{4}e^{-x/2} + \frac{3}{4}e^{x/2} + \frac{1}{8}x^2 e^{x/2} - \frac{1}{4}xe^{x/2}$
27. $y = \frac{4}{9}e^{-4x} + \frac{25}{36}e^{2x} - \frac{1}{4}e^{-2x} + \frac{1}{9}e^{-x}$
29. $y = c_1 x + c_2 x \ln x + \frac{2}{3}x(\ln x)^3$
31. $y = c_1 x^{-1/2} \cos x + c_2 x^{-1/2} \sin x + x^{-1/2}$
33. $y = c_1 + c_2 e^x + c_3 e^{-x} + \frac{1}{4}x^2 e^x - \frac{3}{4}xe^x$

Exercises 3.7, Page 171

1. $x = c_1 e^t + c_2 t e^t$
$y = (c_1 - c_2)e^t + c_2 t e^t$
3. $x = c_1 \cos t + c_2 \sin t + t + 1$
$y = c_1 \sin t - c_2 \cos t + t - 1$
5. $x = \frac{1}{2}c_1 \sin t + \frac{1}{2}c_2 \cos t - 2c_3 \sin \sqrt{6}t - 2c_4 \cos \sqrt{6}t$
$y = c_1 \sin t + c_2 \cos t + c_3 \sin \sqrt{6}t + c_4 \cos \sqrt{6}t$
7. $x = c_1 e^{2t} + c_2 e^{-2t} + c_3 \sin 2t + c_4 \cos 2t + \frac{1}{5}e^t$
$y = c_1 e^{2t} + c_2 e^{-2t} - c_3 \sin 2t - c_4 \cos 2t - \frac{1}{5}e^t$
9. $x = c_1 - c_2 \cos t + c_3 \sin t + \frac{17}{15}e^{3t}$
$y = c_1 + c_2 \sin t + c_3 \cos t - \frac{4}{15}e^{3t}$
11. $x = c_1 e^t + c_2 e^{-t/2} \cos \frac{\sqrt{3}}{2}t + c_3 e^{-t/2} \sin \frac{\sqrt{3}}{2}t$
$y = \left(-\frac{3}{2}c_2 - \frac{\sqrt{3}}{2}c_3\right)e^{-t/2}\cos\frac{\sqrt{3}}{2}t + \left(\frac{\sqrt{3}}{2}c_2 - \frac{3}{2}c_3\right)e^{-t/2}\sin\frac{\sqrt{3}}{2}t$
13. $x = c_1 e^{4t} + \frac{4}{3}e^t$
$y = -\frac{3}{4}c_1 e^{4t} + c_2 + 5e^t$
15. $x = c_1 + c_2 t + c_3 e^t + c_4 e^{-t} - \frac{1}{2}t^2$
$y = (c_1 - c_2 + 2) + (c_2 + 1)t + c_4 e^{-t} - \frac{1}{2}t^2$
17. $x = c_1 e^t + c_2 e^{-t/2}\sin\frac{\sqrt{3}}{2}t + c_3 e^{-t/2}\cos\frac{\sqrt{3}}{2}t$
$y = c_1 e^t + \left(-\frac{1}{2}c_2 - \frac{\sqrt{3}}{2}c_3\right)e^{-t/2}\sin\frac{\sqrt{3}}{2}t + \left(\frac{\sqrt{3}}{2}c_2 - \frac{1}{2}c_3\right)e^{-t/2}\cos\frac{\sqrt{3}}{2}t$
$z = c_1 e^t + \left(-\frac{1}{2}c_2 + \frac{\sqrt{3}}{2}c_3\right)e^{-t/2}\sin\frac{\sqrt{3}}{2}t + \left(-\frac{\sqrt{3}}{2}c_2 - \frac{1}{2}c_3\right)e^{-t/2}\cos\frac{\sqrt{3}}{2}t$
19. $x = -6c_1 e^{-t} - 3c_2 e^{-2t} + 2c_3 e^{3t}$
$y = c_1 e^{-t} + c_2 e^{-2t} + c_3 e^{3t}$
$z = 5c_1 e^{-t} + c_2 e^{-2t} + c_3 e^{3t}$

21. $x = -c_1 e^{-t} + c_2 + \frac{1}{3}t^3 - 2t^2 + 5t$
    $y = c_1 e^{-t} + 2t^2 - 5t + 5$
23. $x = e^{-3t+3} - te^{-3t+3}$
    $y = -e^{-3t+3} + 2te^{-3t+3}$
25. $Dx - Dy = 0$
    $(D - 1)x - y = 0$

Exercises 3.8, Page 178

1. A weight of 4 lb ($\frac{1}{8}$ slug) attached to a spring is released from a point 3 units above the equilibrium position with an initial upward velocity of 2 ft/s. The spring constant is 3 lb/ft.
3. $x(t) = 2\sqrt{2} \sin\left(5t - \frac{\pi}{4}\right)$
5. $x(t) = \sqrt{5} \sin(\sqrt{2}t + 3.6052)$
7. $x(t) = \frac{\sqrt{101}}{10} \sin(10t + 1.4711)$ 9. 8 lb
11. $\sqrt{2}\pi/8$ 13. $x(t) = -\frac{1}{4} \cos 4\sqrt{6}t$
15. (a) $x(\pi/12) = -1/4; x(\pi/8) = -1/2;$
    $x(\pi/6) = -1/4; x(\pi/4) = 1/2;$
    $x(9\pi/32) = \sqrt{2}/4$
    (b) 4 ft/s; downward
    (c) $t = (2n + 1)\pi/16, n = 0, 1, 2, \ldots$
17. (a) The 20-kg mass
    (b) The 20-kg mass; the 50-kg mass
    (c) $t = n\pi, n = 0, 1, 2, \ldots$; at the equilibrium position; the 50-kg mass is moving upward whereas the 20-kg mass is moving upward when n is even and downward when n is odd.
19. $x(t) = \frac{1}{2} \cos 2t + \frac{3}{4} \sin 2t$
    $= \frac{\sqrt{13}}{4} \sin(2t + 0.5880)$
21. (a) $x(t) = -\frac{2}{3} \cos 10t + \frac{1}{2} \sin 10t$
    $= \frac{5}{6} \sin(10t - 0.927)$
    (b) 5/6 ft; $\pi/5$
    (c) 15 cycles
    (d) 0.721 s
    (e) $(2n + 1)\pi/20 + 0.0927, n = 0, 1, 2, \ldots$
    (f) $x(3) = -0.597$ ft
    (g) $x'(3) = -5.814$ ft/s
    (h) $x''(3) = 59.702$ ft/s^2
    (i) $\pm 8\frac{1}{3}$ ft/s
    (j) $0.1451 + n\pi/5; 0.3545 + n\pi/5, n = 0, 1, 2, \ldots$
    (k) $0.3545 + n\pi/5, n = 0, 1, 2, \ldots$
23. 120 lb/ft; $x(t) = \frac{\sqrt{3}}{12} \sin 8\sqrt{3}t$
25. Using $x(t) = c_1 \cos \omega t + c_2 \sin \omega t, x(0) = x_0$ and $x'(0) = v_0$, we find $c_1 = x_0$ and $c_2 = v_0/\omega$. The result follows from $A = \sqrt{c_1^2 + c_2^2}$.
27. $x(t) = 2\sqrt{2} \cos\left(5t + \frac{5\pi}{4}\right)$
29. When $\omega t + \phi = (2m + 1)\pi/2, |x''| = A\omega^2$. But $T = 2\pi/\omega$ implies $\omega = 2\pi/T$ and $\omega^2 = 4\pi^2/T^2$.

Therefore, the magnitude of the acceleration is $|x''| = 4\pi^2 A/T^2$.

Exercises 3.9, Page 188

1. A 2-lb weight is attached to a spring whose constant is 1 lb/ft. The system is damped with a resisting force numerically equal to 2 times the instantaneous velocity. The weight starts from the equilibrium position with an upward velocity of 1.5 ft/s.
3. Mass above equilibrium position; heading upward
5. Mass below equilibrium position; heading upward
7. $\frac{1}{4}$ s; $\frac{1}{2}$ s, $x(\frac{1}{2}) = e^{-2}$; that is, the weight is approximately 0.14 ft below the equilibrium position.
9. (a) $x(t) = \frac{4}{3}e^{-2t} - \frac{1}{3}e^{-8t}$
   (b) $x(t) = -\frac{2}{3}e^{-2t} + \frac{5}{3}e^{-8t}$
11. (a) $x(t) = e^{-2t}[-\cos 4t - \frac{1}{2} \sin 4t]$
    (b) $x(t) = \frac{\sqrt{5}}{2} e^{-2t} \sin(4t + 4.249)$
    (c) $t = 1.294$ s
13. (a) $\beta > \frac{5}{2}$
    (b) $\beta = \frac{5}{2}$
    (c) $0 < \beta < \frac{5}{2}$
15. $x(t) = \frac{2}{7}e^{-7t} \sin 7t$ 17. $v_0 > 2$ ft/s
19. Suppose $\gamma = \sqrt{\omega^2 - \lambda^2}$. Then the derivative of $x(t) = Ae^{-\lambda t} \sin(\gamma t + \phi)$ is
    $$x'(t) = Ae^{-\lambda t}[\gamma \cos(\gamma t + \phi) - \lambda \sin(\gamma t + \phi)]$$
    So $x'(t) = 0$ implies $\tan(\gamma t + \phi) = \gamma/\lambda$, from which it follows that
    $$t = \frac{1}{\gamma}\left[\tan^{-1}\frac{\gamma}{\lambda} + k\pi - \phi\right]$$
    The difference between the t values between two successive maxima (or minima) is then
    $$t_{k+2} - t_k = (k + 2)\left(\frac{\pi}{\gamma}\right) - k\left(\frac{\pi}{\gamma}\right) = \frac{2\pi}{\gamma}$$
21. $t_{k+1}^* - t_k^* = \frac{(2k + 3)\pi/2 - \phi}{\sqrt{\omega^2 - \lambda^2}} - \frac{(2k + 1)\pi/2 - \phi}{\sqrt{\omega^2 - \lambda^2}}$
    $= \frac{\pi}{\sqrt{\omega^2 - \lambda^2}}$
23. Let the quasi period $2\pi/\sqrt{\omega^2 - \lambda^2}$ be denoted by T_q. From (15) we find
    $$\frac{x_n}{x_{n+2}} = \frac{x(t)}{x(t + T_q)}$$
    $$= \frac{e^{-\lambda t} \sin(\sqrt{\omega^2 - \lambda^2}t + \phi)}{e^{-\lambda(t + T_q)} \sin(\sqrt{\omega^2 - \lambda^2}(t + T_q) + \phi)}$$
    $$= e^{\lambda T_q}$$

A-37

since
$$\sin(\sqrt{\omega^2 - \lambda^2}\,t + \phi) = \sin(\sqrt{\omega^2 - \lambda^2}(t + T_q) + \phi)$$
Therefore,
$$\ln\left(\frac{x_n}{x_{n+2}}\right) = \lambda T_q = \frac{2\pi\lambda}{\sqrt{\omega^2 - \lambda^2}}$$

Exercises 3.10, Page 196

1. $x(t) = e^{-t/2}\left(-\frac{4}{3}\cos\frac{\sqrt{47}}{2}t - \frac{64}{3\sqrt{47}}\sin\frac{\sqrt{47}}{2}t\right)$
   $\quad + \frac{10}{3}(\cos 3t + \sin 3t)$
3. $x(t) = \frac{1}{4}e^{-4t} + te^{-4t} - \frac{1}{4}\cos 4t$
5. $x(t) = -\frac{1}{2}\cos 4t + \frac{9}{4}\sin 4t + \frac{1}{2}e^{-2t}\cos 4t$
   $\quad - 2e^{-2t}\sin 4t$
7. $m\dfrac{d^2x}{dt^2} = -k(x-h) - \beta\dfrac{dx}{dt}$ or
   $\dfrac{d^2x}{dt^2} + 2\lambda\dfrac{dx}{dt} + \omega^2 x = \omega^2 h(t)$, where
   $$2\lambda = \frac{\beta}{m} \quad\text{and}\quad \omega^2 = \frac{k}{m}$$
9. (a) $x(t) = \frac{2}{3}\sin 4t - \frac{1}{3}\sin 8t$
   (b) $t = n\pi/4,\ n = 0, 1, 2, \ldots$
   (c) $t = \pi/6 + n\pi/2,\ n = 0, 1, 2, \ldots$
   and $t = \pi/3 + n\pi/2,\ n = 0, 1, 2, \ldots$
   (d) $\sqrt{3}/2$ cm, $-\sqrt{3}/2$ cm
   (e)

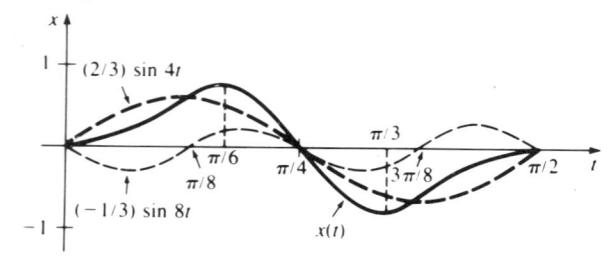

   [Graph showing $x(t)$, with dashed curves labeled $(2/3)\sin 4t$ and $(-1/3)\sin 8t$, and markings at $\pi/8$, $\pi/6$, $\pi/4$, $\pi/3$, $3\pi/8$, $\pi/2$]

11. (a) $g'(\gamma) = 0$ implies $\gamma(\gamma^2 - \omega^2 + 2\lambda^2) = 0$ so that either $\gamma = 0$ or $\gamma = \sqrt{\omega^2 - 2\lambda^2}$. The first derivative test can be used to verify that $g(\gamma)$ is a maximum at the latter value.
    (b) $g(\sqrt{\omega^2 - 2\lambda^2}) = F_0/2\lambda\sqrt{\omega^2 - \lambda^2}$
13. $x_p = -5\cos 2t + 5\sin 2t$
    $\quad = 5\sqrt{2}\sin\left(2t - \dfrac{\pi}{4}\right)$
15. (a) $x(t) = x_c + x_p$
    $\quad = c_1\cos\omega t + c_2\sin\omega t + \dfrac{F_0}{\omega^2 - \gamma^2}\cos\gamma t$
    where the initial conditions imply that
    $c_1 = -F_0/(\omega^2 - \gamma^2)$ and $c_2 = 0$

(b) By L'Hôpital's rule, the given limit is the same as
$$\lim_{\gamma\to\omega}\frac{F_0(-t\sin\gamma t)}{-2\gamma} = \frac{F_0}{2\omega}t\sin\omega t$$

17. $x(t) = -\cos 2t - \frac{1}{8}\sin 2t + \frac{3}{4}t\sin 2t + \frac{5}{4}t\cos 2t$
19. (a) Recall that
    $$\cos(u - v) = \cos u\cos v + \sin u\sin v$$
    $$\cos(u + v) = \cos u\cos v - \sin u\sin v$$
    Subtracting gives
    $$\sin u\sin v = \tfrac{1}{2}[\cos(u - v) - \cos(u + v)]$$
    Setting $u = \tfrac{1}{2}(\gamma - \omega)t$ and $v = \tfrac{1}{2}(\gamma + \omega)t$ then gives
    $$\sin\tfrac{1}{2}(\gamma - \omega)t\,\sin\tfrac{1}{2}(\gamma + \omega)t = \tfrac{1}{2}[\cos\omega t - \cos\gamma t]$$
    from which the result follows.
    (b) For small ε, $\gamma \approx \omega$ so $\gamma + \omega \approx 2\gamma$ and therefore
    $$\frac{-2F_0}{(\omega + \gamma)(\omega - \gamma)}\sin\tfrac{1}{2}(\gamma - \omega)t\,\sin\tfrac{1}{2}(\gamma + \omega)t$$
    $$\approx \frac{F_0}{2\gamma\varepsilon}\sin\varepsilon t\,\sin\tfrac{1}{2}(2\gamma)t$$
    (c) By L'Hôpital's rule the given limit is the same as
    $$\lim_{\varepsilon\to 0}\frac{F_0 t\cos\varepsilon t\,\sin\gamma t}{2\gamma} = \frac{F_0}{2\gamma}t\sin\gamma t = \frac{F_0}{2\omega}t\sin\omega t$$

Exercises 3.11, Page 203

1. $q(t) = -\frac{15}{4}\cos 4t + \frac{15}{4}$; $i(t) = 15\sin 4t$
3. Underdamped   5. 4.568 coulombs; 0.0509 s
7. $q(t) = 10 - 10e^{-3t}(\cos 3t + \sin 3t)$;
   $i(t) = 60e^{-3t}\sin 3t$; 10.432 coulombs
9. $i_p = \frac{100}{13}\cos t - \frac{150}{13}\sin t$
13. $q(t) = -\frac{1}{2}e^{-10t}(\cos 10t + \sin 10t) + \frac{3}{2}$; $\frac{3}{2}$ coulombs
15. Show that $dZ/dC = 0$ when $C = 1/L\gamma^2$. At this value, Z is a minimum and, correspondingly, the amplitude E_0/Z is a maximum.
17. $q(t) = \left(q_0 - \dfrac{E_0 C}{1 - \gamma^2 LC}\right)\cos\dfrac{t}{\sqrt{LC}}$
    $\quad + \sqrt{LC}\,i_0\sin\dfrac{t}{\sqrt{LC}} + \dfrac{E_0 C}{1 - \gamma^2 LC}\cos\gamma t$
    $i(t) = i_0\cos\dfrac{t}{\sqrt{LC}}$
    $\quad - \dfrac{1}{\sqrt{LC}}\left(q_0 - \dfrac{E_0 C}{1 - \gamma^2 LC}\right)\sin\dfrac{t}{\sqrt{LC}}$
    $\quad - \dfrac{E_0 C\gamma}{1 - \gamma^2 LC}\sin\gamma t$

A-38

19. $\theta(t) = \frac{1}{2}\cos 4t + \frac{\sqrt{3}}{2}\sin 4t$; 1; $\pi/2$; $2/\pi$

Chapter 3 Review Exercises, Page 206

1. $y = 0$
3. False, the functions $f_1(x) = 0$ and $f_2(x) = e^x$ are linearly dependent on $(-\infty, \infty)$, but f_2 is not a constant multiple of f_1.
5. $(-\infty, 0)$; $(0, \infty)$ **7.** False
9. $y_p = A + Bxe^x$ **11.** 8 ft **13.** $\frac{5}{4}$ m
15. False; there could be an impressed force driving the system.
17. Overdamped **19.** $y_2 = \sin 2x$
21. $y = c_1 e^{(1+\sqrt{3})x} + c_2 e^{(1-\sqrt{3})x}$
23. $y = c_1 + c_2 e^{-5x} + c_3 x e^{-5x}$
25. $y = c_1 e^{-x/3} + e^{-3x/2}\left(c_2 \cos\frac{\sqrt{7}}{2}x + c_3 \sin\frac{\sqrt{7}}{2}x\right)$
27. $y = e^{3x/2}\left(c_1 \cos\frac{\sqrt{11}}{2}x + c_2 \sin\frac{\sqrt{11}}{2}x\right)$
$\quad + \frac{4}{5}x^3 + \frac{36}{25}x^2 + \frac{46}{125}x - \frac{222}{625}$
29. $y = c_1 + c_2 e^{2x} + c_3 e^{3x} + \frac{1}{5}\sin x - \frac{1}{5}\cos x + \frac{4}{3}x$
31. $y = e^x(c_1 \cos x + c_2 \sin x)$
$\quad - e^x \cos x \ln|\sec x + \tan x|$
33. $y = \frac{1}{2}\cos x + \frac{1}{2}\sin x + \frac{1}{2}\sec x$
35. $x(t) = -\frac{2}{3}e^{-2t} + \frac{1}{3}e^{-4t}$ **37.** $0 < m \leq 2$
39. $x(t) = e^{-4t}$
$\quad \times \left(\frac{26}{17}\cos 2\sqrt{2}t + \frac{28\sqrt{2}}{17}\sin 2\sqrt{2}t\right) + \frac{8}{17}e^{-t}$
41. (a) $q(t) = -\frac{1}{150}\sin 100t + \frac{1}{75}\sin 50t$
   (b) $i(t) = -\frac{2}{3}\cos 100t + \frac{2}{3}\cos 50t$
   (c) $t = n\pi/50$, $n = 0, 1, 2, \ldots$

Exercises 4.1, Page 216

1. $\frac{2}{s}e^{-s} - \frac{1}{s}$ **3.** $\frac{1}{s^2} - \frac{1}{s^2}e^{-s}$ **5.** $\frac{1 + e^{-s\pi}}{s^2 + 1}$
7. $\frac{e^{-s}}{s} + \frac{e^{-s}}{s^2}$ **9.** $\frac{1}{s} - \frac{1}{s^2} + \frac{e^{-s}}{s^2}$
11. $\frac{e^7}{s-1}$ **13.** $\frac{1}{(s-4)^2}$ **15.** $\frac{1}{s^2 + 2s + 2}$
17. $\frac{s^2 - 1}{(s^2 + 1)^2}$ **19.** $\frac{48}{s^5}$ **21.** $\frac{4}{s^2} - \frac{10}{s}$
23. $\frac{2}{s^3} + \frac{6}{s^2} - \frac{3}{s}$ **25.** $\frac{6}{s^4} + \frac{6}{s^3} + \frac{3}{s^2} + \frac{1}{s}$
27. $\frac{1}{s} + \frac{1}{s-4}$ **29.** $\frac{1}{s} + \frac{2}{s-2} + \frac{1}{s-4}$
31. $\frac{8}{s^3} - \frac{15}{s^2+9}$

33. Use $\sinh kt = \dfrac{e^{kt} - e^{-kt}}{2}$ to show that
$$\mathscr{L}\{\sinh kt\} = \frac{k}{s^2 - k^2}$$
35. $\dfrac{1}{2(s-2)} - \dfrac{1}{2s}$ **37.** $\dfrac{2}{s^2 + 16}$
39. $\dfrac{1}{2}\left(\dfrac{s}{s^2+9} + \dfrac{s}{s^2+1}\right)$ **41.** $\dfrac{1}{2}\left(\dfrac{3}{s^2+9} - \dfrac{1}{s^2+1}\right)$
43. The result follows by letting $u = st$ in
$$\mathscr{L}\{t^\alpha\} = \int_0^\infty t^\alpha e^{-st}\,dt.$$
45. $\dfrac{\frac{1}{2}\Gamma(\frac{1}{2})}{s^{3/2}} = \dfrac{\sqrt{\pi}}{2s^{3/2}}$
47. On $0 \leq t \leq 1$, $e^{-st} \geq e^{-s}(s > 0)$. Therefore,
$$\int_0^1 e^{-st}\frac{1}{t^2}\,dt \geq e^{-s}\int_0^1 \frac{1}{t^2}\,dt$$
The latter integral diverges.

Exercises 4.2, Page 223

1. $\frac{1}{2}t^2$ **3.** $t - 2t^4$ **5.** $1 + 3t + \frac{3}{2}t^2 + \frac{1}{6}t^3$
7. $t - 1 + e^{2t}$ **9.** $\frac{1}{4}e^{-t/4}$ **11.** $\frac{5}{7}\sin 7t$
13. $\cos\dfrac{t}{2}$ **15.** $\frac{1}{4}\sinh 4t$
17. $2\cos 3t - 2\sin 3t$ **19.** $\frac{1}{3} - \frac{1}{3}e^{-3t}$
21. $\frac{3}{4}e^{-3t} + \frac{1}{4}e^t$ **23.** $0.3e^{0.1t} + 0.6e^{-0.2t}$
25. $\frac{1}{2}e^{2t} - e^{3t} + \frac{1}{2}e^{6t}$ **27.** $-\frac{1}{3}e^{-t} + \frac{8}{15}e^{2t} - \frac{1}{5}e^{-3t}$
29. $\frac{1}{4}t - \frac{1}{8}\sin 2t$ **31.** $-\frac{1}{4}e^{-2t} + \frac{1}{4}\cos 2t + \frac{1}{4}\sin 2t$
33. $\frac{1}{3}\sin t - \frac{1}{6}\sin 2t$ **35.** $1/s$

Exercises 4.3, Page 239

1. $\dfrac{1}{(s-10)^2}$ **3.** $\dfrac{6}{(s+2)^4}$
5. $\dfrac{3}{(s-1)^2+9}$ **7.** $\dfrac{3}{(s-5)^2-9}$
9. $\dfrac{1}{(s-2)^2} + \dfrac{2}{(s-3)^2} + \dfrac{1}{(s-4)^2}$
11. $\dfrac{1}{2}\left[\dfrac{1}{s+1} - \dfrac{s+1}{(s+1)^2+4}\right]$ **13.** $\frac{1}{2}t^2 e^{-2t}$
15. $e^{3t}\sin t$ **17.** $e^{-2t}\cos t - 2e^{-2t}\sin t$
19. $e^{-t} - te^{-t}$ **21.** $5 - t - 5e^{-t} - 4te^{-t} - \frac{3}{2}t^2 e^{-t}$
23. $\dfrac{e^{-s}}{s^2}$ **25.** $\dfrac{e^{-2s}}{s^2} + 2\dfrac{e^{-2s}}{s}$ **27.** $\dfrac{s}{s^2+4}e^{-\pi s}$
29. $\dfrac{6e^{-s}}{(s-1)^4}$ **31.** $\frac{1}{2}(t-2)^2 \mathcal{U}(t-2)$
33. $-\sin t\,\mathcal{U}(t - \pi)$
35. $\mathcal{U}(t-1) - e^{-(t-1)}\mathcal{U}(t-1)$
37. $\dfrac{s^2 - 4}{(s^2+4)^2}$ **39.** $\dfrac{6s^2 + 2}{(s^2-1)^3}$

A-39

41. $\dfrac{12s - 24}{[(s-2)^2 + 36]^2}$ **43.** $\tfrac{1}{2} t \sin t$

45. (c)   **47.** (f)   **49.** (a)

51. $f(t) = 2 - 4\mathscr{U}(t - 3)$; $\mathscr{L}\{f(t)\} = \dfrac{2}{s} - \dfrac{4}{s} e^{-3s}$

53. $f(t) = t^2 \mathscr{U}(t - 1)$
$= (t - 1)^2 \mathscr{U}(t - 1)$
$+ 2(t - 1)\mathscr{U}(t - 1) + \mathscr{U}(t - 1)$;

$\mathscr{L}\{f(t)\} = 2\dfrac{e^{-s}}{s^3} + 2\dfrac{e^{-s}}{s^2} + \dfrac{e^{-s}}{s}$

55. $f(t) = t - t\mathscr{U}(t - 2)$
$= t - (t - 2)\mathscr{U}(t - 2)$
$- 2\mathscr{U}(t - 2)$;

$\mathscr{L}\{f(t)\} = \dfrac{1}{s^2} - \dfrac{e^{-2s}}{s^2} - 2\dfrac{e^{-2s}}{s}$

57. $f(t) = \mathscr{U}(t - a) - \mathscr{U}(t - b)$; $\mathscr{L}\{f(t)\} = \dfrac{e^{-as}}{s} - \dfrac{e^{-bs}}{s}$

59.

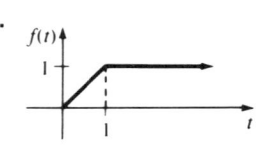

61. $\dfrac{e^{-t} - e^{3t}}{t}$ **63.** $\dfrac{\sin 2t}{t}$

65. Since $f'(t) = e^t$, $f(0) = 1$, it follows from Theorem 4.8 that $\mathscr{L}\{e^t\} = s\mathscr{L}\{e^t\} - 1$. Solving gives $\mathscr{L}\{e^t\} = 1/(s - 1)$.

67. $\dfrac{s + 1}{s[(s + 1)^2 + 1]}$ **69.** $\dfrac{1}{s^2(s - 1)}$

71. $\dfrac{3s^2 + 1}{s^2(s^2 + 1)^2}$ **73.** $\dfrac{6}{s^5}$ **75.** $\dfrac{48}{s^8}$

77. $\dfrac{s - 1}{(s + 1)[(s - 1)^2 + 1]}$

79. $\displaystyle\int_0^t f(\tau) e^{-5(t-\tau)} d\tau$ **81.** $1 - e^{-t}$

83. $-\tfrac{1}{3} e^{-t} + \tfrac{1}{3} e^{2t}$ **85.** $\tfrac{1}{4} t \sin 2t$

87. $\dfrac{(1 - e^{-as})^2}{s(1 - e^{-2as})} = \dfrac{1 - e^{-as}}{s(1 + e^{-as})}$

89. $\dfrac{a}{s}\left(\dfrac{1}{bs} - \dfrac{1}{e^{bs} - 1}\right)$

91. $\dfrac{\coth(\pi s/2)}{s^2 + 1}$

93. $\dfrac{1}{s^2 + 1}$

95. The result follows from letting $u = t - \tau$ in the first integral.

97. The result follows from $\cosh at = (e^{at} + e^{-at})/2$ and the first translation theorem.

99. The result follows from letting $u = at$, $a > 0$.

101. $\ln\dfrac{s + 1}{s - 1}$

Exercises 4.4, Page 252

1. $y = -1 + e^t$ **3.** $y = te^{-4t} + 2e^{-4t}$

5. $y = \tfrac{4}{3} e^{-t} - \tfrac{1}{3} e^{-4t}$

7. $y = \tfrac{1}{9} t + \tfrac{2}{27} - \tfrac{2}{27} e^{3t} + \tfrac{10}{9} te^{3t}$

9. $y = \tfrac{1}{20} t^5 e^{2t}$

11. $y = \cos t - \tfrac{1}{2} \sin t - \tfrac{1}{2} t \cos t$

13. $y = \tfrac{1}{2} - \tfrac{1}{2} e^t \cos t + \tfrac{1}{2} e^t \sin t$

15. $y = -\tfrac{8}{9} e^{-t/2} + \tfrac{1}{9} e^{-2t} + \tfrac{5}{18} e^t + \tfrac{1}{2} e^{-t}$

17. $y = \cos t$

19. $y = [5 - 5e^{-(t-1)}]\mathscr{U}(t - 1)$

21. $y = -\tfrac{1}{4} + \tfrac{1}{2} t + \tfrac{1}{4} e^{-2t} - \tfrac{1}{4}\mathscr{U}(t - 1)$
$- \tfrac{1}{2}(t - 1)\mathscr{U}(t - 1)$
$+ \tfrac{1}{4} e^{-2(t-1)}\mathscr{U}(t - 1)$

23. $y = \cos 2t - \tfrac{1}{6} \sin 2(t - 2\pi)\mathscr{U}(t - 2\pi)$
$+ \tfrac{1}{3} \sin(t - 2\pi)\mathscr{U}(t - 2\pi)$

25. $y = \sin t + [1 - \cos(t - \pi)]\mathscr{U}(t - \pi)$
$- [1 - \cos(t - 2\pi)]\mathscr{U}(t - 2\pi)$

27. $y = (e + 1)te^{-t} + (e - 1)e^{-t}$

29. $f(t) = \sin t$

31. $f(t) = -\tfrac{1}{8} e^{-t} + \tfrac{1}{8} e^t + \tfrac{3}{4} te^t + \tfrac{1}{4} t^2 e^t$

33. $f(t) = e^{-t}$

35. $f(t) = \tfrac{3}{8} e^{2t} + \tfrac{1}{8} e^{-2t} + \tfrac{1}{2} \cos 2t + \tfrac{1}{4} \sin 2t$

37. $y = \sin t - \tfrac{1}{2} t \sin t$

39. $i(t) = 20{,}000[te^{-100t} - (t - 1)e^{-100(t-1)}\mathscr{U}(t - 1)]$

41. $q(t) = \dfrac{E_0 C}{1 - kRC}(e^{-kt} - e^{-t/RC})$;

$q(t) = \dfrac{E_0}{R} te^{-t/RC}$

43. $q(t) = \tfrac{2}{5}\mathscr{U}(t - 3) - \tfrac{2}{5} e^{-5(t-3)}\mathscr{U}(t - 3)$

45. $i(t) = \dfrac{1}{101} e^{-10t} - \dfrac{1}{101} \cos t + \dfrac{10}{101} \sin t$
$- \dfrac{10}{101} e^{-10(t - 3\pi/2)}\mathscr{U}\left(t - \dfrac{3\pi}{2}\right)$
$+ \dfrac{10}{101} \cos\left(t - \dfrac{3\pi}{2}\right)\mathscr{U}\left(t - \dfrac{3\pi}{2}\right)$
$+ \dfrac{1}{101} \sin\left(t - \dfrac{3\pi}{2}\right)\mathscr{U}\left(t - \dfrac{3\pi}{2}\right)$

47. $i(t) = \dfrac{t}{R} + \dfrac{L}{R^2}(e^{-Rt/L} - 1)$
$+ \dfrac{1}{R} \displaystyle\sum_{n=1}^{\infty}(e^{-R(t-n)/L} - 1)\mathscr{U}(t - n)$;

for $0 \leq t < 2$,

$i(t) = \begin{cases} \dfrac{t}{R} + \dfrac{L}{R^2}(e^{-Rt/L} - 1), & 0 \leq t < 1 \\[6pt] \dfrac{t}{R} + \dfrac{L}{R^2}(e^{-Rt/L} - 1) \\[6pt] \quad + \dfrac{1}{R}(e^{-R(t-1)/L} - 1), & 1 \leq t < 2 \end{cases}$

49. $q(t) = \tfrac{3}{5} e^{-10t} + 6te^{-10t} - \tfrac{3}{5} \cos 10t$;
$i(t) = -60te^{-10t} + 6 \sin 10t$;
steady-state current is $6 \sin 10t$

A-40

51. $q(t) = \dfrac{E_0}{L\left(k^2 + \dfrac{1}{LC}\right)}\left[e^{-kt} - \cos\left(\dfrac{t}{\sqrt{LC}}\right)\right]$
$+ \dfrac{kE_0\sqrt{C/L}}{k^2 + \dfrac{1}{LC}}\sin\left(\dfrac{t}{\sqrt{LC}}\right)$

53. $x(t) = -\dfrac{3}{2}e^{-7t/2}\cos\dfrac{\sqrt{15}}{2}t - \dfrac{7\sqrt{15}}{10}e^{-7t/2}\sin\dfrac{\sqrt{15}}{2}t$

55. $y(x) = \dfrac{w_0}{EI}\left(\dfrac{L^2}{4}x^2 - \dfrac{L}{6}x^3 + \dfrac{1}{24}x^4\right);$
$\dfrac{17w_0 L^4}{384EI};\; \dfrac{w_0 L^4}{8EI}$

57. $y(x) = \dfrac{w_0 L^2}{16EI}x^2 - \dfrac{w_0 L}{12EI}x^3 + \dfrac{w_0}{24EI}x^4$
$- \dfrac{w_0}{24EI}\left(x - \dfrac{L}{2}\right)^4 \mathcal{U}\left(x - \dfrac{L}{2}\right)$

59. $y = \tfrac{1}{3}t^3 + \tfrac{1}{2}ct^2$

Exercises 4.5, Page 259

1. $y = e^{3(t-2)}\mathcal{U}(t-2)$
3. $y = \sin t + \sin t\,\mathcal{U}(t - 2\pi)$
5. $y = -\cos t\,\mathcal{U}\left(t - \dfrac{\pi}{2}\right) + \cos t\,\mathcal{U}\left(t - \dfrac{3\pi}{2}\right)$
7. $y = \tfrac{1}{2} - \tfrac{1}{2}e^{-2t} + [\tfrac{1}{2} - \tfrac{1}{2}e^{-2(t-1)}]\mathcal{U}(t-1)$
9. $y = e^{-2(t-2\pi)}\sin t\,\mathcal{U}(t - 2\pi)$
11. $y = e^{-2t}\cos 3t + \tfrac{2}{3}e^{-2t}\sin 3t$
$+ \tfrac{1}{3}e^{-2(t-\pi)}\sin 3(t-\pi)\mathcal{U}(t-\pi)$
$+ \tfrac{1}{3}e^{-2(t-3\pi)}\sin 3(t - 3\pi)\mathcal{U}(t - 3\pi)$
13. $y(x) = \begin{cases} \dfrac{P_0}{EI}\left(\dfrac{L}{4}x^2 - \dfrac{1}{6}x^3\right), & 0 \le x < \dfrac{L}{2} \\ \dfrac{P_0 L^2}{4EI}\left(\dfrac{1}{2}x - \dfrac{L}{12}\right), & \dfrac{L}{2} \le x \le L \end{cases}$
17. $y = e^{-t}\cos t + e^{-(t-3\pi)}\sin t\,\mathcal{U}(t - 3\pi)$
19. $i(t) = \dfrac{1}{L}e^{-Rt/L}$; no

Exercises 4.6, Page 265

1. $x = -\tfrac{1}{3}e^{-2t} + \tfrac{1}{3}e^t$
$y = \tfrac{1}{3}e^{-2t} + \tfrac{2}{3}e^t$
3. $x = -\cos 3t - \tfrac{5}{3}\sin 3t$
$y = 2\cos 3t - \tfrac{7}{3}\sin 3t$
5. $x = -2e^{3t} + \tfrac{5}{2}e^{2t} - \tfrac{1}{2}$
$y = \tfrac{8}{3}e^{3t} - \tfrac{5}{2}e^{2t} - \tfrac{1}{6}$
7. $x = -\tfrac{1}{2}t - \tfrac{3}{4}\sqrt{2}\sin\sqrt{2}t$
$y = -\tfrac{1}{2}t + \tfrac{3}{4}\sqrt{2}\sin\sqrt{2}t$
9. $x = 8 + \dfrac{2}{3!}t^3 + \dfrac{1}{4!}t^4$
$y = -\dfrac{2}{3!}t^3 + \dfrac{1}{4!}t^4$
11. $x = \tfrac{1}{2}t^2 + t + 1 - e^{-t}$
$y = -\tfrac{1}{3} + \tfrac{1}{3}e^{-t} + \tfrac{1}{3}te^{-t}$

13. $x_1 = \tfrac{1}{5}\sin t + \dfrac{2\sqrt{6}}{15}\sin\sqrt{6}t + \tfrac{2}{5}\cos t - \tfrac{2}{5}\cos\sqrt{6}t$
$x_2 = \tfrac{2}{5}\sin t - \dfrac{\sqrt{6}}{15}\sin\sqrt{6}t + \tfrac{4}{5}\cos t + \tfrac{1}{5}\cos\sqrt{6}t$

15. (b) $i_2 = \tfrac{100}{9} - \tfrac{100}{9}e^{-900t}$
$i_3 = \tfrac{80}{9} - \tfrac{80}{9}e^{-900t}$
(c) $i_1 = 20 - 20e^{-900t}$

17. $i_2 = -\tfrac{20}{13}e^{-2t} + \tfrac{375}{1469}e^{-15t} + \tfrac{145}{113}\cos t + \tfrac{85}{113}\sin t$
$i_3 = \tfrac{30}{13}e^{-2t} + \tfrac{250}{1469}e^{-15t} - \tfrac{280}{113}\cos t + \tfrac{810}{113}\sin t$

19. $i_1 = \tfrac{6}{5} - \tfrac{6}{5}e^{-100t}\cos 100t$
$i_2 = \tfrac{6}{5} - \tfrac{6}{5}e^{-100t}\cos 100t - \tfrac{6}{5}e^{-100t}\sin 100t$

21. (b) $q(t) = 50e^{-t}\sin(t-1)\mathcal{U}(t-1)$

Chapter 4 Review Exercises, Page 268

1. $\dfrac{1}{s^2} - \dfrac{2}{s^2}e^{-s}$ **3.** False **5.** True
7. $\dfrac{1}{s+7}$ **9.** $\dfrac{2}{s^2 + 4}$ **11.** $\dfrac{4s}{(s^2+4)^2}$
13. $\tfrac{1}{6}t^5$ **15.** $\tfrac{1}{2}t^2 e^{5t}$
17. $e^{5t}\cos 2t + \tfrac{5}{2}e^{5t}\sin 2t$
19. $\cos\pi(t-1)\mathcal{U}(t-1) + \sin\pi(t-1)\mathcal{U}(t-1)$
21. -5 **23.** $e^{-ks}F(s-a)$
25. (a) $f(t) = t - (t-1)\mathcal{U}(t-1) - \mathcal{U}(t-4)$
(b) $\mathscr{L}\{f(t)\} = \dfrac{1}{s^2} - \dfrac{1}{s^2}e^{-s} - \dfrac{1}{s}e^{-4s}$
(c) $\mathscr{L}\{e^t f(t)\} = \dfrac{1}{(s-1)^2} - \dfrac{1}{(s-1)^2}e^{-(s-1)}$
$- \dfrac{1}{s-1}e^{-4(s-1)}$

27. (a) $f(t) = 2 + (t-2)\mathcal{U}(t-2)$
(b) $\mathscr{L}\{f(t)\} = \dfrac{2}{s} + \dfrac{1}{s^2}e^{-2s}$
(c) $\mathscr{L}\{e^t f(t)\} = \dfrac{2}{s-1} + \dfrac{1}{(s-1)^2}e^{-2(s-1)}$

29. $y = 5te^t + \tfrac{1}{2}t^2 e^t$
31. $y = 5\mathcal{U}(t-\pi) - 5e^{2(t-\pi)}\cos\sqrt{2}(t-\pi)\mathcal{U}(t-\pi)$
$+ 5\sqrt{2}e^{2(t-\pi)}\sin\sqrt{2}(t-\pi)\mathcal{U}(t-\pi)$
33. $y = -\tfrac{2}{125} - \tfrac{2}{25}t - \tfrac{1}{5}t^2 + \tfrac{127}{125}e^{5t}$
$- [-\tfrac{37}{125} - \tfrac{12}{25}(t-1) - \tfrac{1}{5}(t-1)^2$
$+ \tfrac{37}{125}e^{5(t-1)}]\mathcal{U}(t-1)$
35. $y = 1 + t + \tfrac{1}{2}t^2$
37. $x = -\tfrac{1}{4} + \tfrac{9}{8}e^{-2t} + \tfrac{1}{8}e^{2t}$
$y = t + \tfrac{9}{4}e^{-2t} - \tfrac{1}{4}e^{2t}$
39. $i(t) = -9 + 2t + 9e^{-t/5}$
41. $y(x) = \dfrac{w_0}{12EIL}\left[-\dfrac{1}{5}x^5 + \dfrac{L}{2}x^4 - \dfrac{L^2}{2}x^3 + \dfrac{L^3}{4}x^2\right.$
$\left. + \dfrac{1}{5}\left(x - \dfrac{L}{2}\right)^5 \mathcal{U}\left(x - \dfrac{L}{2}\right)\right]$

Exercises 5.1, Page 277

1. $y = c_1 x^{-1} + c_2 x^2$
3. $y = c_1 + c_2 \ln x$
5. $y = c_1 \cos(2 \ln x) + c_2 \sin(2 \ln x)$
7. $y = c_1 x^{(2-\sqrt{6})} + c_2 x^{(2+\sqrt{6})}$
9. $y_1 = c_1 \cos(\frac{1}{5} \ln x) + c_2 \sin(\frac{1}{5} \ln x)$
11. $y = c_1 x^{-2} + c_2 x^{-2} \ln x$
13. $y = x[c_1 \cos(\ln x) + c_2 \sin(\ln x)]$
15. $y = x^{-1/2}\left[c_1 \cos\left(\frac{\sqrt{3}}{6} \ln x\right) + c_2 \sin\left(\frac{\sqrt{3}}{6} \ln x\right)\right]$
17. $y = c_1 x^3 + c_2 \cos(\sqrt{2} \ln x) + c_3 \sin(\sqrt{2} \ln x)$
19. $y = c_1 x^{-1} + c_2 x^2 + c_3 x^4$
21. $y = c_1 + c_2 x + c_3 x^2 + c_4 x^{-3}$
23. $y = 2 - 2x^{-2}$
25. $y = \cos(\ln x) + 2 \sin(\ln x)$
27. $y = 2(-x)^{1/2} - 5(-x)^{1/2} \ln(-x)$
29. $y = c_1 + c_2 \ln x + x^2/4$
31. $y = c_1 x^{-1/2} + c_2 x^{-1} + \frac{1}{15}x^2 - \frac{1}{6}x$
33. $y = c_1 x + c_2 x \ln x + x(\ln x)^2$
35. $y = c_1 x^{-1} + c_2 x^{-8} + \frac{1}{30}x^2$
37. $y = x^2[c_1 \cos(3 \ln x) + c_2 \sin(3 \ln x)] + \frac{4}{13} + \frac{3}{10}x$
39. $y = c_1 x^2 + c_2 x^{-10} - \frac{1}{7}x^{-3}$
41. $u(r) = \left(\frac{u_0 - u_1}{b - a}\right)\frac{ab}{r} + \frac{u_1 b - u_0 a}{b - a}$
43. $y = c_1(x-1)^{-1} + c_2(x-1)^4$
45. $y = c_1 \cos(\ln(x+2)) + c_2 \sin(\ln(x+2))$

Exercises 5.2, Page 289

1. $y = ce^{-x}$; $y = c_0 \sum_{n=0}^{\infty} \frac{(-1)^n}{n!} x^n$
3. $y = ce^{x^3/3}$; $y = c_0 \sum_{n=0}^{\infty} \frac{1}{n!}\left(\frac{x^3}{3}\right)^n$
5. $y = c/(1-x)$; $y = c_0 \sum_{n=0}^{\infty} x^n$
7. $y = C_1 \cos x + C_2 \sin x$;
$y = c_0 \sum_{n=0}^{\infty} \frac{(-1)^n}{(2n)!} x^{2n} + c_1 \sum_{n=0}^{\infty} \frac{(-1)^n}{(2n+1)!} x^{2n+1}$
9. $y = C_1 + C_2 e^x$;
$y = c_0 + c_1 \sum_{n=1}^{\infty} \frac{x^n}{n!} = c_0 - c_1 + c_1 \sum_{n=0}^{\infty} \frac{x^n}{n!}$
$= c_0 - c_1 + c_1 e^x$
11. $y_1(x) = c_0\left[1 + \frac{1}{3 \cdot 2}x^3 + \frac{1}{6 \cdot 5 \cdot 3 \cdot 2}x^6\right.$
$\left. + \frac{1}{9 \cdot 8 \cdot 6 \cdot 5 \cdot 3 \cdot 2}x^9 + \cdots\right]$
$y_2(x) = c_1\left[x + \frac{1}{4 \cdot 3}x^4 + \frac{1}{7 \cdot 6 \cdot 4 \cdot 3}x^7\right.$
$\left. + \frac{1}{10 \cdot 9 \cdot 7 \cdot 6 \cdot 4 \cdot 3}x^{10} + \cdots\right]$
13. $y_1(x) = c_0\left[1 - \frac{1}{2}x^2 - \frac{3}{4!}x^4 - \frac{21}{6!}x^6 - \cdots\right]$

$y_2(x) = c_1\left[x + \frac{1}{3!}x^3 + \frac{5}{5!}x^5 + \frac{45}{7!}x^7 + \cdots\right]$

15. $y_1(x) = c_0\left[1 - \frac{1}{3!}x^3 + \frac{4^2}{6!}x^6 - \frac{7^2 \cdot 4^2}{9!}x^9 + \cdots\right]$
$y_2(x) = c_1\left[x - \frac{2^2}{4!}x^4 + \frac{5^2 \cdot 2^2}{7!}x^7\right.$
$\left. - \frac{8^2 \cdot 5^2 \cdot 2^2}{10!}x^{10} + \cdots\right]$

17. $y_1(x) = c_0$; $y_2(x) = c_1 \sum_{n=1}^{\infty} \frac{1}{n}x^n$
19. $y_1(x) = c_0 \sum_{n=0}^{\infty} x^{2n}$; $y_2(x) = c_1 \sum_{n=0}^{\infty} x^{2n+1}$
21. $y_1(x) = c_0\left[1 + \frac{1}{4}x^2 - \frac{7}{4 \cdot 4!}x^4 + \frac{23 \cdot 7}{8 \cdot 6!}x^6 - \cdots\right]$
$y_2(x) = c_1\left[x - \frac{1}{6}x^3 + \frac{14}{2 \cdot 5!}x^5 - \frac{34 \cdot 14}{4 \cdot 7!}x^7 - \cdots\right]$
23. $y_1(x) = c_0[1 + \frac{1}{2}x^2 + \frac{1}{6}x^3 + \frac{1}{6}x^4 + \cdots]$
$y_2(x) = c_1[x + \frac{1}{2}x^2 + \frac{1}{2}x^3 + \frac{1}{4}x^4 + \cdots]$
25. $y(x) = -2[1 + \frac{1}{2!}x^2 + \frac{1}{3!}x^3 + \frac{1}{4!}x^4 + \cdots] + 6x$
$= 8x - 2e^x$
27. $y(x) = 3 - 12x^2 + 4x^4$
29. $y_1(x) = c_0[1 - \frac{1}{6}x^3 + \frac{1}{120}x^5 + \cdots]$
$y_2(x) = c_1[x - \frac{1}{12}x^4 + \frac{1}{180}x^6 + \cdots]$
31. $y_1(x) = c_0[1 - \frac{1}{2}x^2 + \frac{1}{6}x^3 - \frac{1}{40}x^5 + \cdots]$
$y_2(x) = c_1[x - \frac{1}{6}x^3 + \frac{1}{12}x^4 - \frac{1}{60}x^5 + \cdots]$
33. $y_1(x) = c_0\left[1 + \frac{1}{3!}x^3 + \frac{4}{6!}x^6 + \frac{7 \cdot 4}{9!}x^9 + \cdots\right]$
$+ c_1\left[x + \frac{2}{4!}x^4 + \frac{5 \cdot 2}{7!}x^7 + \frac{8 \cdot 5 \cdot 2}{10!}x^{10} + \cdots\right]$
$+ \frac{1}{2!}x^2 + \frac{3}{5!}x^5 + \frac{6 \cdot 3}{8!}x^8 + \frac{9 \cdot 6 \cdot 3}{11!}x^{11} + \cdots$
35. For $n = 1$: $y = x$;
for $n = 2$: $y = 1 - 2x^2$

Exercises 5.3, Page 305

1. $x = 0$, irregular singular point
3. $x = -3$, regular singular point; $x = 3$, irregular singular point
5. $x = 0, 2i, -2i$, regular singular points
7. $x = -3, 2$, regular singular points
9. $x = 0$, irregular singular point; $x = -5, 5, 2$, regular points
11. $r_1 = \frac{3}{2}, r_2 = 0$;
$y(x) = C_1 x^{3/2}\left[1 - \frac{2}{5}x + \frac{2^2}{7 \cdot 5 \cdot 2}x^2\right.$
$\left. - \frac{2^3}{9 \cdot 7 \cdot 5 \cdot 3!}x^3 + \cdots\right]$
$+ C_2\left[1 + 2x - 2x^2 + \frac{2^3}{3 \cdot 3!}x^3 - \cdots\right]$

A-42

13. $r_1 = \frac{7}{8}, r_2 = 0$;
$$y(x) = C_1 x^{7/8}\left[1 - \frac{2}{15}x + \frac{2^2}{23 \cdot 15 \cdot 2}x^2 - \frac{2^3}{31 \cdot 23 \cdot 15 \cdot 3!}x^3 + \cdots\right]$$
$$+ C_2\left[1 - 2x + \frac{2^2}{9 \cdot 2}x^2 - \frac{2^3}{17 \cdot 9 \cdot 3!}x^3 + \cdots\right]$$

15. $r_1 = \frac{1}{3}, r_2 = 0$;
$$y(x) = C_1 x^{1/3}\left[1 + \frac{1}{3}x + \frac{1}{3^2 \cdot 2}x^2 + \frac{1}{3^3 \cdot 3!}x^3 + \cdots\right]$$
$$+ C_2\left[1 + \frac{1}{2}x + \frac{1}{5 \cdot 2}x^2 + \frac{1}{8 \cdot 5 \cdot 2}x^3 + \cdots\right]$$

17. $r_1 = \frac{5}{2}, r_2 = 0$;
$$y(x) = C_1 x^{5/2}\left[1 + \frac{2 \cdot 2}{7}x + \frac{2^2 \cdot 3}{9 \cdot 7}x^2 + \frac{2^3 \cdot 4}{11 \cdot 9 \cdot 7}x^3 + \cdots\right]$$
$$+ C_2\left[1 + \frac{1}{3}x - \frac{1}{6}x^2 - \frac{1}{6}x^3 - \cdots\right]$$

19. $r_1 = \frac{2}{3}, r_2 = \frac{1}{3}$;
$$y(x) = C_1 x^{2/3}[1 - \tfrac{1}{2}x + \tfrac{5}{28}x^2 - \tfrac{1}{21}x^3 + \cdots]$$
$$+ C_2 x^{1/3}[1 - \tfrac{1}{2}x + \tfrac{1}{5}x^2 - \tfrac{7}{120}x^3 + \cdots]$$

21. $r_1 = 1, r_2 = -\frac{1}{2}$;
$$y(x) = C_1 x\left[1 + \frac{1}{5}x + \frac{1}{5 \cdot 7}x^2 + \frac{1}{5 \cdot 7 \cdot 9}x^3 + \cdots\right]$$
$$+ C_2 x^{-1/2}\left[1 + \frac{1}{2}x + \frac{1}{2 \cdot 4}x^2 + \frac{1}{2 \cdot 4 \cdot 6}x^3 + \cdots\right]$$

23. $r_1 = 0, r_2 = -1$;
$$y(x) = C_1 x^{-1}\sum_{n=0}^{\infty}\frac{1}{(2n)!}x^{2n} + C_2 x^{-1}\sum_{n=0}^{\infty}\frac{1}{(2n+1)!}x^{2n+1}$$
$$= \frac{1}{x}[C_1 \cosh x + C_2 \sinh x]$$

25. $r_1 = 4, r_2 = 0$;
$$y(x) = C_1\left[1 + \frac{2}{3}x + \frac{1}{3}x^2\right] + C_2 \sum_{n=0}^{\infty}(n+1)x^{n+4}$$

27. $r_1 = r_2 = 0$;
$$y(x) = C_1 y_1(x) + C_2\left[y_1(x)\ln x + y_1(x) \times \left(-x + \frac{1}{4}x^2 - \frac{1}{3 \cdot 3!}x^3 + \frac{1}{4 \cdot 4!}x^4 - \cdots\right)\right]$$
where $y_1(x) = \sum_{n=0}^{\infty}\frac{1}{n!}x^n = e^x$

29. $r_1 = r_2 = 0$;
$$y(x) = C_1 y_1(x) + C_2[y_1(x)\ln x + y_1(x) \times (2x + \tfrac{5}{4}x^2 + \tfrac{23}{27}x^3 + \cdots)]$$
where $y_1(x) = \sum_{n=0}^{\infty}\frac{(-1)^n}{(n!)^{2n}}x^n$

31. $r_1 = r_2 = 1$;
$$y(x) = C_1 xe^{-x} + C_2 xe^{-x} \times \left[\ln x + x + \frac{1}{4}x^2 + \frac{1}{3 \cdot 3!}x^3 + \cdots\right]$$

33. $r_1 = 2, r_2 = 0$;
$$y(x) = C_1 x^2 + C_2\left[\frac{1}{2}x^2 \ln x - \frac{1}{2} + x - \frac{1}{3!}x^3 + \cdots\right]$$

35. The method of Frobenius yields only the trivial solution $y(x) = 0$.

37. $y(x) = 1 + \sum_{n=1}^{\infty}(-1)^n \frac{p(p-1)\cdots(p-n+1)}{(n!)^2}x^n$

39. There is a regular singular point at ∞.

41. There is a regular singular point at ∞.

Exercises 5.4, Page 316

1. $y = c_1 J_{1/3}(x) + c_2 J_{-1/3}(x)$

3. $y = c_1 J_{5/2}(x) + c_2 J_{-5/2}(x)$

5. $y = c_1 J_0(x) + c_2 Y_0(x)$

7. $y = c_1 J_2(3x) + c_2 Y_2(3x)$

9. After the change of variables is used, the differential equation becomes
$$x^2 v'' + xv' + (\lambda^2 x^2 - \tfrac{1}{4})v = 0$$
Since the solution of the last equation is
$$v = c_1 J_{1/2}(\lambda x) + c_2 J_{-1/2}(\lambda x)$$
we find
$$y = c_1 x^{-1/2}J_{1/2}(\lambda x) + c_2 x^{-1/2}J_{-1/2}(\lambda x)$$

11. After substituting into the differential equation, we find
$$xy'' + (1 + 2n)y' + xy = x^{-n-1}[x^2 J_n'' + xJ_n' + (x^2 - n^2)J_n]$$
$$= x^{-n-1} \cdot 0 = 0$$

13. From Problem 10 with $n = \frac{1}{2}$, we find $y = x^{1/2}J_{1/2}(x)$; from Problem 11 with $n = -\frac{1}{2}$, we find $y = x^{1/2}J_{-1/2}(x)$.

15. From Problem 10 with $n = -1$, we find $y = x^{-1}J_{-1}(x)$; from Problem 11 with $n = 1$, we find $y = x^{-1}J_1(x)$, but since $J_{-1}(x) = -J_1(x)$, no new solution results.

17. From Problem 12 with $\lambda = 1$ and $v = \pm\frac{3}{2}$, we find $y = \sqrt{x}J_{3/2}(x)$ and $y = \sqrt{x}J_{-3/2}(x)$.

A-43

19. Using the hint, we can write

$$xJ'_\nu(x) = -\nu \sum_{n=0}^\infty \frac{(-1)^n}{n!\Gamma(1+\nu+n)}\left(\frac{x}{2}\right)^{2n+\nu}$$
$$+ 2\sum_{n=0}^\infty \frac{(-1)^n(n+\nu)}{n!(n+\nu)\Gamma(n+\nu)}\left(\frac{x}{2}\right)^{2n+\nu}$$
$$= -\nu \sum_{n=0}^\infty \frac{(-1)^n}{n!\Gamma(1+\nu+n)}\left(\frac{x}{2}\right)^{2n+\nu}$$
$$+ x\sum_{n=0}^\infty \frac{(-1)^n}{n!\Gamma(n+\nu)}\left(\frac{x}{2}\right)^{2n+\nu-1}$$
$$= -\nu J_\nu(x) + xJ_{\nu-1}(x)$$

21. Subtracting the equations:

$$xJ'_\nu(x) = \nu J_\nu(x) - xJ_{\nu+1}(x)$$
$$xJ'_\nu(x) = -\nu J_\nu(x) + xJ_{\nu-1}(x)$$

gives $\quad 2\nu J_\nu(x) = xJ_{\nu+1}(x) + xJ_{\nu-1}(x)$

23. From (15), $(d/dr)[rJ_1(r)] = rJ_0(r)$. Therefore,

$$\int_0^x rJ_0(r)\,dr = \int_0^x \frac{d}{dr}[rJ_1(r)]\,dr = rJ_1(r)\Big|_0^x = xJ_1(x).$$

25. $J'_0(x) = J_{-1}(x)$ follows from (15) with $\nu = 0$; $J_{-1}(x) = -J_1(x)$ follows from property (i) with $m=1$.

27. $J_{-1/2}(x) = \sqrt{\dfrac{2}{\pi x}}\cos x$

29. $J_{-3/2}(x) = \sqrt{\dfrac{2}{\pi x}}\left[-\sin x - \dfrac{\cos x}{x}\right]$

31. $J_{-5/2}(x) = \sqrt{\dfrac{2}{\pi x}}\left[\dfrac{3}{x}\sin x + \left(\dfrac{3}{x^2}-1\right)\cos x\right]$

33. $J_{-7/2}(x) = \sqrt{\dfrac{2}{\pi x}}\left[\left(1-\dfrac{15}{x^2}\right)\sin x + \left(\dfrac{6}{x}-\dfrac{15}{x^3}\right)\cos x\right]$

35. $y = c_1 I_\nu(x) + c_2 I_{-\nu}(x)$, $\nu \neq$ integer

37. Since $1/\Gamma(1-m+n) = 0$ when $n \leq m-1$, m a positive integer,

$$J_{-m}(x) = \sum_{n=0}^\infty \frac{(-1)^n}{n!\Gamma(1-m+n)}\left(\frac{x}{2}\right)^{2n-m}$$
$$= \sum_{n=m}^\infty \frac{(-1)^n}{n!\Gamma(1-m+n)}\left(\frac{x}{2}\right)^{2n-m}$$
$$J_{-m}(x) = \sum_{k=0}^\infty \frac{(-1)^{k+m}}{(k+m)!\Gamma(1+k)}\left(\frac{x}{2}\right)^{2k+m} \quad (n=k+m)$$
$$= (-1)^m \sum_{k=0}^\infty \frac{(-1)^k}{\Gamma(1+k+m)k!}\left(\frac{x}{2}\right)^{2k+m}$$
$$= (-1)^m J_m(x)$$

39. (a) $P_6(x) = \frac{1}{16}(231x^6 - 315x^4 + 105x^2 - 5)$
$P_7(x) = \frac{1}{16}[429x^7 - 693x^5 + 315x^3 - 35x]$

(b) $y = P_6(x)$ satisfies $(1-x^2)y'' - 2xy' + 42y = 0$.
$y = P_7(x)$ satisfies $(1-x^2)y'' - 2xy' + 56y = 0$.

41. If $x = \cos\theta$, then $\dfrac{dy}{d\theta} = \dfrac{dy}{dx}\dfrac{dx}{d\theta} = -\sin\theta\,\dfrac{dy}{dx}$ and

$\dfrac{d^2y}{d\theta^2} = \sin^2\theta\,\dfrac{d^2y}{dx^2} - \cos\theta\,\dfrac{dy}{dx}$. Now, the original equation can be written as

$$\frac{d^2y}{d\theta^2} + \frac{\cos\theta}{\sin\theta}\frac{dy}{d\theta} + n(n+1)y = 0$$

and so

$$\sin^2\theta\,\frac{d^2y}{dx^2} - 2\cos\theta\,\frac{dy}{dx} + n(n+1)y = 0$$

Since $x = \cos\theta$ and $\sin^2\theta = 1 - \cos^2\theta = 1 - x^2$, we obtain

$$(1-x^2)\frac{d^2y}{dx^2} - 2x\frac{dy}{dx} + n(n+1)y = 0$$

43. By the binomial theorem we have formally

$$(1 - 2xt + t^2)^{-1/2} = 1 + \frac{1}{2}(2xt - t^2) + \frac{1\cdot 3}{2^2 2!}$$
$$\times (2xt - t^2)^2 + \cdots$$

Grouping by powers of t, we then find

$$(1 - 2xt + t^2)^{-1/2} = 1\cdot t^0 + x\cdot t + \tfrac{1}{2}(3x^2 - 1)t^2 + \cdots$$
$$= P_0(x)t^0 + P_1(x)t + P_2(x)t^2 + \cdots$$

45. For $k=1$, $P_2(x) = \tfrac{1}{2}[3xP_1(x) - P_0(x)]$
$= \tfrac{1}{2}(3x^2 - 1)$
For $k=2$, $P_3(x) = \tfrac{1}{3}[5xP_2(x) - 2P_1(x)]$
$= \tfrac{1}{3}[5x\cdot\tfrac{1}{2}(3x^2-1) - 2x]$
$= \tfrac{1}{2}(5x^3 - 3x)$
For $k=3$, $P_4(x) = \tfrac{1}{4}[7xP_3(x) - 3P_2(x)]$
$= \tfrac{1}{4}[7x\cdot\tfrac{1}{2}(5x^3 - 3x) - \tfrac{3}{2}(3x^2 - 1)]$
$= \tfrac{1}{8}(35x^4 - 30x^2 + 3)$

47. For $n = 0, 1, 2, 3$, the values of the integral are $2, \tfrac{2}{3}$, $\tfrac{2}{5}$, and $\tfrac{2}{7}$, respectively. In general,

$$\int_{-1}^1 P_n^2(x)\,dx = \frac{2}{2n+1}, \quad n = 0, 1, 2, \ldots$$

49. y_2 is obtained from (4) of Section 4.2.

Chapter 5 Review Exercises, Page 319

1. $y = c_1 x^{-1/3} + c_2 x^{1/2}$
3. $y(x) = c_1 x^2 + c_2 x^3 + x^4 - x^2 \ln x$
5. The singular points are $x = 0$, $x = -1 + \sqrt{3}i$, $x = -1 - \sqrt{3}i$. All other finite values of x, real or complex, are ordinary points.
7. RSP $x = 0$; ISP $x = 5$

9. RSP $x = -3, x = 3$; ISP $x = 0$ **11.** $|x| < \infty$

13. $y_1(x) = c_0\left[1 - \frac{1}{3 \cdot 2}x^3 + \frac{1}{6 \cdot 5 \cdot 3 \cdot 2}x^6 - \frac{1}{9 \cdot 8 \cdot 6 \cdot 5 \cdot 3 \cdot 2}x^9 + \cdots\right]$

$y_2(x) = c_1\left[x - \frac{1}{4 \cdot 3}x^4 + \frac{1}{7 \cdot 6 \cdot 4 \cdot 3}x^7 - \frac{1}{10 \cdot 9 \cdot 7 \cdot 6 \cdot 4 \cdot 3}x^{10} + \cdots\right]$

15. $y_1(x) = c_0[1 + \frac{3}{2}x^2 + \frac{1}{2}x^3 + \frac{5}{8}x^4 + \cdots]$
$y_2(x) = c_1[x + \frac{1}{2}x^3 + \frac{1}{4}x^4 + \cdots]$

17. $r_1 = 1, r_2 = -\frac{1}{2}$;

$y(x) = C_1 x\left[1 + \frac{1}{5}x + \frac{1}{7 \cdot 5 \cdot 2}x^2 + \frac{1}{9 \cdot 7 \cdot 5 \cdot 3 \cdot 2}x^3 + \cdots\right]$

$+ C_2 x^{-1/2}\left[1 - x - \frac{1}{2}x^2 - \frac{1}{3^2 \cdot 2}x^3 - \cdots\right]$

19. $r_1 = 3, r_2 = 0$;

$y_1(x) = C_3\left[x^3 + \frac{5}{4}x^4 + \frac{11}{8}x^5 + \cdots\right]$

$y(x) = C_1 y_1(x) + C_2\left[-\frac{1}{36}y_1(x)\ln x + y_1(x) \times \left(-\frac{1}{3}\frac{1}{x^3} + \frac{1}{4}\frac{1}{x^2} + \frac{1}{16}\frac{1}{x} + \cdots\right)\right]$

21. $r_1 = r_2 = 0; y(x) = C_1 e^x + C_2 e^x \ln x$

23. $y(x) = c_0\left[1 - \frac{1}{2^2}x^2 + \frac{1}{2^4(1 \cdot 2)^2}x^4 - \frac{1}{2^6(1 \cdot 2 \cdot 3)^2}x^6 + \cdots\right]$

Exercises 6.1, Page 330

1. $6\mathbf{i} + 12\mathbf{j}; \mathbf{i} + 8\mathbf{j}; 3\mathbf{i}; \sqrt{65}; 3$
3. $\langle 12, 0\rangle; \langle 4, -5\rangle; \langle 4, 5\rangle; \sqrt{41}; \sqrt{41}$
5. $-9\mathbf{i} + 6\mathbf{j}; -3\mathbf{i} + 9\mathbf{j}; -3\mathbf{i} - 5\mathbf{j}; 3\sqrt{10}; \sqrt{34}$
7. $-6\mathbf{i} + 27\mathbf{j}; 0; -4\mathbf{i} + 18\mathbf{j}; 0; 2\sqrt{85}$
9. $\langle 6, -14\rangle; \langle 2, 4\rangle$ **11.** $10\mathbf{i} - 12\mathbf{j}; 12\mathbf{i} - 17\mathbf{j}$
13. $\langle 20, 52\rangle; \langle -2, 0\rangle$
15. $2\mathbf{i} + 5\mathbf{j}$ **17.** $2\mathbf{i} + 2\mathbf{j}$

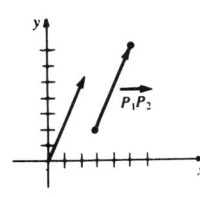

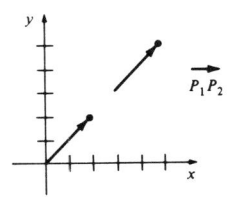

19. $(1, 18)$ **21.** (a), (b), (c), (e), (f)
23. $\langle 6, 15\rangle$ **25.** $\left\langle\frac{1}{\sqrt{2}}, \frac{1}{\sqrt{2}}\right\rangle; \left\langle-\frac{1}{\sqrt{2}}, -\frac{1}{\sqrt{2}}\right\rangle$

27. $\langle 0, -1\rangle; \langle 0, 1\rangle$ **29.** $\left\langle\frac{5}{13}, \frac{12}{13}\right\rangle$
31. $\frac{6}{\sqrt{58}}\mathbf{i} + \frac{14}{\sqrt{58}}\mathbf{j}$ **33.** $\left\langle -3, -\frac{15}{2}\right\rangle$
35.   **37.** $-(\mathbf{a} + \mathbf{b})$

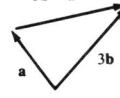

41. $\mathbf{a} = \frac{5}{2}\mathbf{b} - \frac{1}{2}\mathbf{c}$ **43.** $\pm\frac{1}{\sqrt{2}}(\mathbf{i} + \mathbf{j})$
45. (b) approximately 31°
47. $\mathbf{F} = \frac{qQ}{4\pi\varepsilon_0}\frac{1}{L\sqrt{L^2 + a^2}}\mathbf{i}$

Exercises 6.2, Page 337

1.–5.

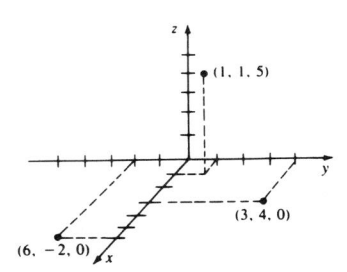

7. The set $\{(x, y, 5) \mid x, y \text{ real numbers}\}$ is a plane perpendicular to the z-axis, 5 units above the xy-plane.
9. The set $\{(2, 3, z) \mid z \text{ a real number}\}$ is a line perpendicular to the xy-plane at $(2, 3, 0)$.
11. $(0, 0, 0), (2, 0, 0), (2, 5, 0), (0, 5, 0), (0, 0, 8), (2, 0, 8), (2, 5, 8), (0, 5, 8)$
13. $(-2, 5, 0), (-2, 0, 4), (0, 5, 4); (-2, 5, -2); (3, 5, 4)$
15. The union of the coordinate planes
17. The point $(-1, 2, -3)$
19. The union of the planes $z = -5$ and $z = 5$
21. $\sqrt{70}$ **23.** 7; 5   **25.** Right triangle
27. Isosceles   **29.** $d(P_1, P_2) + d(P_1, P_3) = d(P_2, P_3)$
31. 6 or -2 **33.** $(4, \frac{1}{2}, \frac{3}{2})$
35. $P_1(-4, -11, 10)$ **37.** $\langle -3, -6, 1\rangle$
39. $\langle 2, 1, 1\rangle$ **41.** $\langle 2, 4, 12\rangle$
43. $\langle -11, -41, -49\rangle$ **45.** $\sqrt{139}$ **47.** 6
49. $\langle -\frac{2}{3}, \frac{1}{3}, -\frac{2}{3}\rangle$ **51.** $4\mathbf{i} - 4\mathbf{j} + 4\mathbf{k}$
53.

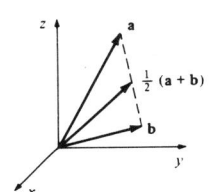

A-45

Exercises 6.3, Page 346

1. $25\sqrt{2}$ 3. 12 5. -16 7. 48
9. 29 11. 25 13. $\langle -\frac{2}{5}, \frac{4}{5}, 2 \rangle$
15. (a) and (f), (c) and (d), (b) and (e)
17. $\langle \frac{4}{9}, -\frac{1}{3}, 1 \rangle$ 21. 1.11 radians or 63.43°
23. 1.89 radians or 108.43°
25. $\cos \alpha = 1/\sqrt{14}$, $\cos \beta = 2/\sqrt{14}$, $\cos \gamma = 3/\sqrt{14}$; $\alpha = 74.5°$, $\beta = 57.69°$, $\gamma = 36.7°$
27. $\cos \alpha = \frac{1}{2}$, $\cos \beta = 0$, $\cos \gamma = -\sqrt{3}/2$; $\alpha = 60°$, $\beta = 90°$, $\gamma = 150°$
29. 0.955 radian or 54.74°; 0.616 radian or 35.26°
31. $\alpha = 58.19°$, $\beta = 42.45°$, $\gamma = 65.06°$ 33. $\frac{5}{7}$
35. $-6\sqrt{11}/11$ 37. $72\sqrt{109}/109$
39. $\langle -\frac{21}{5}, \frac{28}{5} \rangle$; $\langle -\frac{4}{5}, -\frac{3}{5} \rangle$
41. $\langle -\frac{12}{7}, \frac{6}{7}, \frac{4}{7} \rangle$; $\langle \frac{5}{7}, -\frac{20}{7}, \frac{45}{7} \rangle$ 43. $\langle \frac{72}{25}, \frac{96}{25} \rangle$
45. 1000 ft-lb 47. 0; 150 N-m
49. Approximately 1.80 angstroms

Exercises 6.4, Page 354

1. $-5\mathbf{i} - 5\mathbf{j} + 3\mathbf{k}$ 3. $\langle -12, -2, 6 \rangle$
5. $-5\mathbf{i} + 5\mathbf{k}$ 7. $\langle -3, 2, 3 \rangle$ 9. 0
11. $6\mathbf{i} + 14\mathbf{j} + 4\mathbf{k}$ 13. $-3\mathbf{i} - 2\mathbf{j} - 5\mathbf{k}$
17. $-\mathbf{i} + \mathbf{j} + \mathbf{k}$ 19. $2\mathbf{k}$ 21. $\mathbf{i} + 2\mathbf{j}$
23. $-24\mathbf{k}$ 25. $5\mathbf{i} - 5\mathbf{j} - \mathbf{k}$ 27. 0
29. $\sqrt{41}$ 31. $-\mathbf{j}$ 33. 0 35. 6
37. $12\mathbf{i} - 9\mathbf{j} + 18\mathbf{k}$ 39. $-4\mathbf{i} + 3\mathbf{j} - 6\mathbf{k}$
41. $-21\mathbf{i} + 16\mathbf{j} + 22\mathbf{k}$ 43. -10
45. 14 square units 47. $\frac{1}{2}$ square unit
49. $\frac{7}{2}$ square units 51. 10 cubic units
53. Coplanar
55. 32; in the xy-plane, 30° from the positive x-axis in the direction of the negative y-axis; $16\sqrt{3}\mathbf{i} - 16\mathbf{j}$
57. $\mathbf{A} = \mathbf{i} - \mathbf{k}$, $\mathbf{B} = \mathbf{j} - \mathbf{k}$, $\mathbf{C} = 2\mathbf{k}$

Exercises 6.5, Page 363

1. $\langle x, y, z \rangle = \langle 1, 2, 1 \rangle + t\langle 2, 3, -3 \rangle$
3. $\langle x, y, z \rangle = \langle \frac{1}{2}, -\frac{1}{2}, 1 \rangle + t\langle -2, 3, -\frac{3}{2} \rangle$
5. $\langle x, y, z \rangle = \langle 1, 1, -1 \rangle + t\langle 5, 0, 0 \rangle$
7. $x = 2 + 4t$, $y = 3 - 4t$, $z = 5 + 3t$
9. $x = 1 + 2t$, $y = -2t$, $z = -7t$
11. $x = 4 + 10t$, $y = \frac{1}{2} + \frac{3}{4}t$, $z = \frac{1}{3} + \frac{1}{6}t$
13. $\dfrac{x-1}{9} = \dfrac{y-4}{10} = \dfrac{z+9}{7}$
15. $\dfrac{x+7}{11} = \dfrac{z-5}{-4}$, $y = 2$
17. $x = 5$, $\dfrac{y-10}{9} = \dfrac{z+2}{12}$
19. $x = 4 + 3t$, $y = 6 + \frac{1}{2}t$, $z = -7 - \frac{3}{2}t$; $\dfrac{x-4}{3} = 2y - 12 = \dfrac{2z + 14}{-3}$
21. $x = 5t$, $y = 9t$, $z = 4t$; $\dfrac{x}{5} = \dfrac{y}{9} = \dfrac{z}{4}$
23. $x = 6 + 2t$, $y = 4 - 3t$, $z = -2 + 6t$

25. $x = 2 + t$, $y = -2$, $z = 15$
27. Both lines pass through the origin and have parallel direction vectors.
29. $(0, 5, 15)$, $(5, 0, \frac{15}{2})$, $(10, -5, 0)$
31. $(2, 3, -5)$ 33. Lines do not intersect.
35. 40.37° 37. $x = 4 - 6t$, $y = 1 + 3t$, $z = 6 + 3t$
39. $2x - 3y + 4z = 19$ 41. $5x - 3z = 51$
43. $6x + 8y - 4z = 11$ 45. $5x - 3y + z = 2$
47. $3x - 4y + z = 0$ 49. The points are collinear.
51. $x + y - 4z = 25$ 53. $z = 12$
55. $-3x + y + 10z = 18$ 57. $9x - 7y + 5z = 17$
59. $6x - 2y + z = 12$
61. Orthogonal; (a) and (d), (b) and (c), (d) and (f), (b) and (e); parallel: (a) and (f), (c) and (e)
63. (c) and (d) 65. $x = 2 + t$, $y = \frac{1}{2} - t$, $z = t$
67. $x = \frac{1}{2} - \frac{1}{2}t$, $y = \frac{1}{2} - \frac{3}{2}t$, $z = t$
69. $(-5, 5, 9)$ 71. $(1, 2, -5)$
73. $x = 5 + t$, $y = 6 + 3t$, $z = -12 + t$
75. $3x - y - 2z = 10$
77. 79.
81.

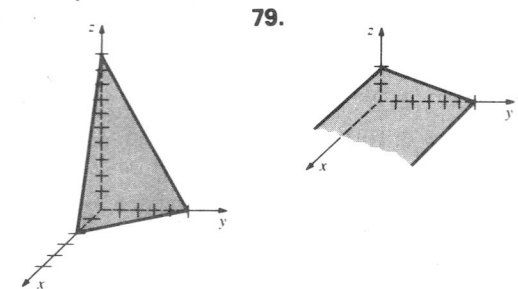

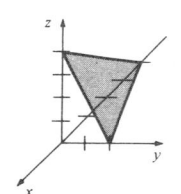

Exercises 6.6, Page 372

1. Not a vector space, axiom (vi) is not satisfied
3. Not a vector space, axiom (x) is not satisfied
5. Vector space
7. Not a vector space, axiom (ii) is not satisfied
9. Vector space 11. A subspace
13. Not a subspace 15. A subspace
17. A subspace 19. Not a subspace
23. (b) $\mathbf{a} = 7\mathbf{u}_1 - 12\mathbf{u}_2 + 8\mathbf{u}_3$
25. Linearly dependent
27. Linearly independent
29. f is discontinuous at $x = -1$ and at $x = -3$.
31. $2\sqrt{\frac{2}{3}}\pi^{2/3}$, $\sqrt{\pi}$

Chapter 6 Review Exercises, Page 374

1. True 3. False 5. True 7. True
9. True 11. $9\mathbf{i} + 2\mathbf{j} + 2\mathbf{k}$ 13. $5\mathbf{i}$

A-46

15. 14   **17.** $-6\mathbf{i} + \mathbf{j} - 7\mathbf{k}$ **19.** $(4, 7, 5)$
21. $(5, 6, 3)$ **23.** $-36\sqrt{2}$
25. $12, -8,$ and 6 **27.** $3\sqrt{10}/2$ **29.** 2 units
31. $(\mathbf{i} - \mathbf{j} - 3\mathbf{k})/\sqrt{11}$ **33.** 2
35. $\frac{26}{9}\mathbf{i} + \frac{7}{9}\mathbf{j} + \frac{20}{9}\mathbf{k}$ **37.** Sphere; plane
39. $\dfrac{x-7}{4} = \dfrac{y-3}{-2} = \dfrac{z+5}{6}$
41. The direction vectors are orthogonal and the point of intersection is $(3, -3, 0)$.
43. $14x - 5y - 3z = 0$ **45.** $30\sqrt{2}$ N-m
47. Approximately 153 lb
49. Not a vector space   **51.** A subspace; $1, x$

Exercises 7.1, Page 385

1. 2×4 **3.** 3×3 **5.** 3×4
7. Not equal   **9.** Not equal
11. $x = 2, y = 4$ **13.** $c_{23} = 9, c_{12} = 12$
15. $\begin{bmatrix} 2 & 11 \\ 2 & -1 \end{bmatrix}; \begin{bmatrix} -6 & 1 \\ 14 & -19 \end{bmatrix}; \begin{bmatrix} 2 & 28 \\ 12 & -12 \end{bmatrix}$
17. $\begin{bmatrix} -11 & 6 \\ 17 & -22 \end{bmatrix}; \begin{bmatrix} -32 & 27 \\ -4 & -1 \end{bmatrix};$
$\begin{bmatrix} 19 & -18 \\ -30 & 31 \end{bmatrix}; \begin{bmatrix} 19 & 6 \\ 3 & 22 \end{bmatrix}$
19. $\begin{bmatrix} 9 & 24 \\ 3 & 8 \end{bmatrix}; \begin{bmatrix} 3 & 8 \\ -6 & -16 \end{bmatrix}; \begin{bmatrix} 0 & 0 \\ 0 & 0 \end{bmatrix}; \begin{bmatrix} -4 & -5 \\ 8 & 10 \end{bmatrix}$
21. $180; \begin{bmatrix} 4 & 8 & 10 \\ 8 & 16 & 20 \\ 10 & 20 & 25 \end{bmatrix}; \begin{bmatrix} 6 \\ 12 \\ -5 \end{bmatrix}$
23. $\begin{bmatrix} 7 & 38 \\ 10 & 75 \end{bmatrix}; \begin{bmatrix} 7 & 38 \\ 10 & 75 \end{bmatrix}$ **25.** $\begin{bmatrix} -14 \\ 1 \end{bmatrix}$
27. $\begin{bmatrix} -38 \\ -2 \end{bmatrix}$ **29.** 4×5
37. $\mathbf{A} = \begin{bmatrix} 1 & 2 \\ 0 & 0 \end{bmatrix}, \mathbf{B} = \begin{bmatrix} 2 & -4 \\ -1 & 2 \end{bmatrix}$
39. $\mathbf{AB}$ is not necessarily the same as $\mathbf{BA}$.
41. $a_{11}x_1 + a_{12}x_2 = b_1$
$a_{21}x_1 + a_{22}x_2 = b_2$
45. (b) $M_R = \begin{bmatrix} \cos\beta & 0 & -\sin\beta \\ 0 & 1 & 0 \\ \sin\beta & 0 & \cos\beta \end{bmatrix},$
$M_P = \begin{bmatrix} 1 & 0 & 0 \\ 0 & \cos\alpha & \sin\alpha \\ 0 & -\sin\alpha & \cos\alpha \end{bmatrix}$
(c) $x_S = \dfrac{3\sqrt{2} + 2\sqrt{3} - \sqrt{6} + 6}{8} \approx 1.4072$
$y_S = \dfrac{3\sqrt{2} + 2\sqrt{3} - 3\sqrt{6} + 2}{8} \approx 0.2948$
$z_S = \dfrac{\sqrt{2} + \sqrt{6}}{8} \approx 0.9659$

Exercises 7.2, Page 398

1. $x_1 = 4, x_2 = -7$ **3.** $x_1 = -\frac{2}{3}, x_2 = \frac{1}{3}$
5. $x_1 = 0, x_2 = 4, x_3 = -1$
7. $x_1 = -t, x_2 = t, x_3 = 0$ **9.** Inconsistent
11. $x_1 = 0, x_2 = 0, x_3 = 0$
13. $x_1 = -2, x_2 = -2, x_3 = 4$
15. $x_1 = 1, x_2 = 2 - t, x_3 = t$
17. $x_1 = 0, x_2 = 1, x_3 = 1, x_4 = 0$
19. Inconsistent
21. $x_1 = 0.3, x_2 = -0.12, x_3 = 4.1$
23. $2\text{Na} + 2\text{H}_2\text{O} \rightarrow 2\text{NaOH} + \text{H}_2$
25. $\text{Fe}_3\text{O}_4 + 4\text{C} \rightarrow 3\text{Fe} + 4\text{CO}$
27. $3\text{Cu} + 8\text{HNO}_3 \rightarrow 3\text{Cu(NO}_3)_2 + 4\text{H}_2\text{O} + 2\text{NO}$
29. $i_1 = \frac{35}{9}, i_2 = \frac{38}{9}, i_3 = \frac{1}{3}$
31. Interchange row 1 and row 2 in $\mathbf{I}_3$.
33. Multiply the second row of $\mathbf{I}_3$ by c and add to the third row.
35. $\mathbf{EA} = \begin{bmatrix} a_{21} & a_{22} & a_{23} \\ a_{11} & a_{12} & a_{13} \\ a_{31} & a_{32} & a_{33} \end{bmatrix}$
37. $\mathbf{EA} = \begin{bmatrix} a_{11} & a_{12} & a_{13} \\ a_{21} & a_{22} & a_{23} \\ ca_{21} + a_{31} & ca_{22} + a_{32} & ca_{23} + a_{33} \end{bmatrix}$

Exercises 7.3, Page 405

1. 9   **3.** 1   **5.** 2   **7.** 10   **9.** -7
11. 17   **13.** $\lambda^2 - 3\lambda - 4$ **15.** -48
17. 62   **19.** 0   **21.** -85
23. $-x + 2y - z$ **25.** -104 **27.** 48
29. $\lambda_1 = -5, \lambda_2 = 7$

Exercises 7.4, Page 412

1. Theorem 7.8   **3.** Theorem 7.11
5. Theorem 7.9   **7.** Theorem 7.7
9. Theorem 7.5   **11.** -5 **13.** -5
15. 80   **17.** -105 **23.** 0   **25.** -15
27. -9 **29.** 0   **31.** 16

Exercises 7.5, Page 424

3. $\begin{bmatrix} \frac{1}{9} & \frac{1}{9} \\ -\frac{4}{9} & \frac{5}{9} \end{bmatrix}$ **5.** $\begin{bmatrix} \frac{1}{6} & 0 \\ \frac{1}{4} & \frac{1}{2} \end{bmatrix}$
7. $\begin{bmatrix} -\frac{1}{2} & \frac{1}{2} & \frac{1}{2} \\ -\frac{1}{8} & \frac{1}{4} & -\frac{3}{8} \\ \frac{3}{8} & -\frac{1}{4} & \frac{1}{8} \end{bmatrix}$ **9.** $\begin{bmatrix} \frac{7}{15} & -\frac{13}{30} & -\frac{8}{15} \\ \frac{1}{15} & -\frac{2}{15} & \frac{1}{15} \\ \frac{2}{15} & \frac{7}{30} & \frac{2}{15} \end{bmatrix}$
11. $\begin{bmatrix} \frac{1}{3} & 0 & 0 \\ 0 & \frac{1}{6} & 0 \\ 0 & 0 & -\frac{1}{2} \end{bmatrix}$ **13.** $\begin{bmatrix} \frac{2}{9} & \frac{7}{9} & -\frac{4}{3} & -\frac{4}{3} \\ -\frac{1}{27} & \frac{1}{27} & \frac{2}{9} & -\frac{1}{9} \\ \frac{10}{27} & \frac{17}{27} & -\frac{2}{9} & -\frac{17}{9} \\ \frac{4}{27} & -\frac{4}{27} & \frac{1}{9} & \frac{4}{9} \end{bmatrix}$
15. $\begin{bmatrix} \frac{1}{6} & \frac{1}{12} \\ 0 & \frac{1}{4} \end{bmatrix}$ **17.** $\begin{bmatrix} -\frac{1}{4} & \frac{1}{4} \\ \frac{5}{12} & -\frac{1}{12} \end{bmatrix}$

19. Singular matrix

21. $\begin{bmatrix} 0 & \frac{2}{3} & \frac{1}{3} \\ 0 & -\frac{1}{3} & -\frac{2}{3} \\ \frac{1}{3} & -\frac{2}{3} & 0 \end{bmatrix}$

23. $\begin{bmatrix} 5 & 6 & -3 \\ 2 & 2 & -1 \\ -1 & -1 & 1 \end{bmatrix}$

25. $\begin{bmatrix} -\frac{1}{2} & -\frac{2}{3} & -\frac{1}{6} & \frac{7}{6} \\ 1 & \frac{1}{3} & \frac{1}{3} & -\frac{4}{3} \\ 0 & -\frac{1}{3} & -\frac{1}{3} & \frac{1}{3} \\ -\frac{1}{2} & 1 & \frac{1}{2} & \frac{1}{2} \end{bmatrix}$

27. $\begin{bmatrix} -\frac{1}{3} & \frac{1}{3} \\ -1 & \frac{10}{3} \end{bmatrix}$

29. $\begin{bmatrix} -2 & 3 \\ 3 & -4 \end{bmatrix}$

31. $x = 5$

35. By Theorem 7.10, det $\mathbf{AB}$ = det $\mathbf{A}$ det $\mathbf{B}$. Since det $\mathbf{A} \neq 0$ and det $\mathbf{B} \neq 0$, it follows that det $\mathbf{AB} \neq 0$. By Theorem 7.16, $\mathbf{AB}$ is nonsingular.

37. The result follows from det $(\mathbf{AA}^{-1})$ = det $\mathbf{I}$.

39. Multiply $\mathbf{AB} = \mathbf{0}$ by $\mathbf{A}^{-1}$.   41. No

43. $x_1 = 6, x_2 = -2$ 45. $x_1 = \frac{3}{4}, x_2 = -\frac{1}{2}$

47. $x_1 = 2, x_2 = 4, x_3 = -6$

49. $x_1 = 21, x_2 = 1, x_3 = -11$

51. $x_1 = \frac{9}{10}, x_2 = \frac{13}{20}$;
    $x_1 = 6, x_2 = 16$;
    $x_1 = -2, x_2 = -7$

53. System has only the trivial solution.

55. System has nontrivial solutions.

57. $i_1 = \dfrac{-R_3 E_2 + R_3 E_1 + R_2 E_1 - R_2 E_3}{R_3 R_1 + R_3 R_2 + R_1 R_2}$,

$i_2 = \dfrac{R_3 E_2 - R_3 E_1 - R_1 E_3 + R_1 E_2}{R_3 R_1 + R_3 R_2 + R_1 R_2}$,

$i_3 = \dfrac{-R_2 E_1 + R_1 E_3 + R_2 E_3 - R_1 E_2}{R_3 R_1 + R_3 R_2 + R_1 R_2}$

Exercises 7.6, Page 429

1. $x_1 = -\frac{3}{5}, x_2 = \frac{6}{5}$ 3. $x_1 = 0.1, x_2 = -0.3$
5. $x = 4, y = -7$ 7. $x_1 = -4, x_2 = 4, x_3 = -5$
9. $u = 4, v = \frac{3}{2}, w = 1$ 11. $k = \frac{6}{5}$
13. $T_1 \approx 450.8$ lb, $T_2 \approx 423$ lb

Exercises 7.7, Page 436

1. $\mathbf{K}_3, \lambda = -1$ 3. $\mathbf{K}_3, \lambda = 0$
5. $\mathbf{K}_2, \lambda = 3; \mathbf{K}_3, \lambda = 1$
7. $\lambda_1 = 6, \lambda_2 = 1, \mathbf{K}_1 = \begin{bmatrix} 2 \\ 7 \end{bmatrix}, \mathbf{K}_2 = \begin{bmatrix} 1 \\ 1 \end{bmatrix}$
9. $\lambda_1 = \lambda_2 = -4, \mathbf{K}_1 = \begin{bmatrix} 1 \\ -4 \end{bmatrix}$
11. $\lambda_1 = 3i, \lambda_2 = -3i$,
    $\mathbf{K}_1 = \begin{bmatrix} 1 - 3i \\ 5 \end{bmatrix}, \mathbf{K}_2 = \begin{bmatrix} 1 + 3i \\ 5 \end{bmatrix}$
13. $\lambda_1 = 4, \lambda_2 = -5$,
    $\mathbf{K}_1 = \begin{bmatrix} 1 \\ 0 \end{bmatrix}, \mathbf{K}_2 = \begin{bmatrix} -8 \\ 9 \end{bmatrix}$

15. $\lambda_1 = 0, \lambda_2 = 4, \lambda_3 = -4$,
    $\mathbf{K}_1 = \begin{bmatrix} 9 \\ 45 \\ 25 \end{bmatrix}, \mathbf{K}_2 = \begin{bmatrix} 1 \\ 1 \\ 1 \end{bmatrix}, \mathbf{K}_3 = \begin{bmatrix} 1 \\ 9 \\ 1 \end{bmatrix}$

17. $\lambda_1 = \lambda_2 = \lambda_3 = -2$,
    $\mathbf{K}_1 = \begin{bmatrix} 2 \\ -1 \\ 0 \end{bmatrix}, \mathbf{K}_2 = \begin{bmatrix} 0 \\ 0 \\ 1 \end{bmatrix}$

19. $\lambda_1 = -1, \lambda_2 = i, \lambda_3 = -i$,
    $\mathbf{K}_1 = \begin{bmatrix} 1 \\ -1 \\ 1 \end{bmatrix}, \mathbf{K}_2 = \begin{bmatrix} -1 \\ i \\ i \end{bmatrix}, \mathbf{K}_3 = \begin{bmatrix} -1 \\ -i \\ -i \end{bmatrix}$

21. $\lambda_1 = 1, \lambda_2 = 5, \lambda_3 = -7$,
    $\mathbf{K}_1 = \begin{bmatrix} 1 \\ 0 \\ 0 \end{bmatrix}, \mathbf{K}_2 = \begin{bmatrix} 1 \\ 2 \\ 0 \end{bmatrix}, \mathbf{K}_3 = \begin{bmatrix} -1 \\ -2 \\ 4 \end{bmatrix}$

23. For $\mathbf{A}$, $\lambda_1 = 4, \lambda_2 = 6$,
    $\mathbf{K}_1 = \begin{bmatrix} 1 \\ -1 \end{bmatrix}, \mathbf{K}_2 = \begin{bmatrix} 1 \\ 1 \end{bmatrix}$

25. $\lambda_1 = 0, \lambda_2 = 6$

Exercises 7.8, Page 443

1. (b) $\lambda_1 = -4, \lambda_2 = -1, \lambda_3 = 16$
3. (b) $\lambda_1 = 18, \lambda_2 = \lambda_3 = -8$ 5. Orthogonal
7. Orthogonal   9. Not orthogonal

11. $\begin{bmatrix} \frac{1}{\sqrt{2}} & \frac{1}{\sqrt{2}} \\ -\frac{1}{\sqrt{2}} & \frac{1}{\sqrt{2}} \end{bmatrix}$

13. $\begin{bmatrix} \frac{3}{\sqrt{10}} & \frac{1}{\sqrt{10}} \\ -\frac{1}{\sqrt{10}} & \frac{3}{\sqrt{10}} \end{bmatrix}$

15. $\begin{bmatrix} -\frac{1}{\sqrt{2}} & \frac{1}{\sqrt{2}} & 0 \\ 0 & 0 & 1 \\ \frac{1}{\sqrt{2}} & \frac{1}{\sqrt{2}} & 0 \end{bmatrix}$

17. $\begin{bmatrix} -\frac{3}{\sqrt{11}} & \frac{1}{\sqrt{66}} & \frac{1}{\sqrt{6}} \\ \frac{1}{\sqrt{11}} & -\frac{4}{\sqrt{66}} & \frac{2}{\sqrt{6}} \\ \frac{1}{\sqrt{11}} & \frac{7}{\sqrt{66}} & \frac{1}{\sqrt{6}} \end{bmatrix}$

19. $a = -\frac{4}{5}, b = \frac{3}{5}$

21. (b) $\begin{bmatrix} 4 \\ 1 \\ -1 \end{bmatrix}, \begin{bmatrix} 0 \\ 1 \\ 1 \end{bmatrix}, \begin{bmatrix} -1 \\ 2 \\ -2 \end{bmatrix}$

23. Use $(\mathbf{AB})^T = \mathbf{B}^T \mathbf{A}^T$.

Exercises 7.9, Page 453

1. $\mathbf{P} = \begin{bmatrix} -3 & 1 \\ 1 & 1 \end{bmatrix}, \mathbf{D} = \begin{bmatrix} 1 & 0 \\ 0 & 5 \end{bmatrix}$
3. Not diagonalizable
5. $\mathbf{P} = \begin{bmatrix} 13 & 1 \\ 2 & 1 \end{bmatrix}, \mathbf{D} = \begin{bmatrix} -7 & 0 \\ 0 & 4 \end{bmatrix}$
7. $\mathbf{P} = \begin{bmatrix} 1 & 1 \\ -1 & 1 \end{bmatrix}, \mathbf{D} = \begin{bmatrix} \frac{1}{3} & 0 \\ 0 & \frac{2}{3} \end{bmatrix}$
9. $\mathbf{P} = \begin{bmatrix} 1 & 1 \\ -i & i \end{bmatrix}, \mathbf{D} = \begin{bmatrix} -i & 0 \\ 0 & i \end{bmatrix}$
11. $\mathbf{P} = \begin{bmatrix} 1 & 0 & 1 \\ 0 & 1 & 1 \\ 0 & 0 & 1 \end{bmatrix}, \mathbf{D} = \begin{bmatrix} 1 & 0 & 0 \\ 0 & -1 & 0 \\ 0 & 0 & 2 \end{bmatrix}$
13. $\mathbf{P} = \begin{bmatrix} 1 & 1 & 1 \\ 0 & 1 & 0 \\ -1 & 1 & 1 \end{bmatrix}, \mathbf{D} = \begin{bmatrix} 0 & 0 & 0 \\ 0 & 1 & 0 \\ 0 & 0 & 2 \end{bmatrix}$
15. Not diagonalizable
17. $\mathbf{P} = \begin{bmatrix} 0 & 1+\sqrt{5} & 1-\sqrt{5} \\ 0 & 2 & 2 \\ 1 & 0 & 0 \end{bmatrix}$,
$\mathbf{D} = \begin{bmatrix} 1 & 0 & 0 \\ 0 & \sqrt{5} & 0 \\ 0 & 0 & -\sqrt{5} \end{bmatrix}$
19. $\mathbf{P} = \begin{bmatrix} -3 & -1 & -1 & 1 \\ 0 & 1 & 0 & 0 \\ -3 & 0 & 0 & 1 \\ 1 & 0 & 1 & 0 \end{bmatrix}$,
$\mathbf{D} = \begin{bmatrix} 2 & 0 & 0 & 0 \\ 0 & 2 & 0 & 0 \\ 0 & 0 & 1 & 0 \\ 0 & 0 & 0 & -1 \end{bmatrix}$
21. $\mathbf{P} = \begin{bmatrix} \frac{1}{\sqrt{2}} & \frac{1}{\sqrt{2}} \\ -\frac{1}{\sqrt{2}} & \frac{1}{\sqrt{2}} \end{bmatrix}, \mathbf{D} = \begin{bmatrix} 0 & 0 \\ 0 & 2 \end{bmatrix}$
23. $\mathbf{P} = \begin{bmatrix} -\frac{\sqrt{10}}{\sqrt{14}} & \frac{\sqrt{10}}{\sqrt{35}} \\ \frac{2}{\sqrt{14}} & \frac{5}{\sqrt{35}} \end{bmatrix}, \mathbf{D} = \begin{bmatrix} 3 & 0 \\ 0 & 10 \end{bmatrix}$
25. $\mathbf{P} = \begin{bmatrix} -\frac{1}{\sqrt{2}} & \frac{1}{\sqrt{2}} & 0 \\ \frac{1}{\sqrt{2}} & \frac{1}{\sqrt{2}} & 0 \\ 0 & 0 & 1 \end{bmatrix}, \mathbf{D} = \begin{bmatrix} -1 & 0 & 0 \\ 0 & 1 & 0 \\ 0 & 0 & 1 \end{bmatrix}$
27. $\mathbf{P} = \begin{bmatrix} \frac{2}{3} & \frac{2}{3} & \frac{1}{3} \\ \frac{2}{3} & -\frac{1}{3} & -\frac{2}{3} \\ \frac{1}{3} & -\frac{2}{3} & \frac{2}{3} \end{bmatrix}, \mathbf{D} = \begin{bmatrix} 3 & 0 & 0 \\ 0 & 6 & 0 \\ 0 & 0 & 9 \end{bmatrix}$
29. $\mathbf{P} = \begin{bmatrix} 0 & \frac{1}{\sqrt{2}} & \frac{1}{\sqrt{2}} \\ 1 & 0 & 0 \\ 0 & -\frac{1}{\sqrt{2}} & \frac{1}{\sqrt{2}} \end{bmatrix}, \mathbf{D} = \begin{bmatrix} 1 & 0 & 0 \\ 0 & -6 & 0 \\ 0 & 0 & 8 \end{bmatrix}$

31. Ellipse; using
$\mathbf{X} = \begin{bmatrix} \frac{1}{\sqrt{2}} & \frac{1}{\sqrt{2}} \\ -\frac{1}{\sqrt{2}} & \frac{1}{\sqrt{2}} \end{bmatrix} \mathbf{X}'$
we get $X^2/4 + Y^2/6 = 1$.

33. Hyperbola; using
$\mathbf{X} = \begin{bmatrix} \frac{1}{\sqrt{5}} & -\frac{2}{\sqrt{5}} \\ \frac{2}{\sqrt{5}} & \frac{1}{\sqrt{5}} \end{bmatrix} \mathbf{X}'$
we get $X^2/4 - Y^2/4 = 1$.

35. $\mathbf{A} = \begin{bmatrix} 4 & -1 \\ 2 & 1 \end{bmatrix}$ 39. $\mathbf{A}^5 = \begin{bmatrix} 21 & 11 \\ 22 & 10 \end{bmatrix}$

Exercises 7.10, Page 457

1. (a) $\begin{bmatrix} 35 & 15 & 38 & 36 & 0 \\ 27 & 10 & 26 & 20 & 0 \end{bmatrix}$
3. (a) $\begin{bmatrix} 48 & 64 & 120 & 107 & 40 \\ 32 & 40 & 75 & 67 & 25 \end{bmatrix}$
5. (a) $\begin{bmatrix} 31 & 44 & 15 & 61 & 50 & 49 & 41 \\ 24 & 29 & 15 & 47 & 35 & 31 & 21 \\ 1 & -15 & 15 & 0 & -15 & -5 & -19 \end{bmatrix}$
7. STUDY_HARD
9. MATH_IS_IMPORTANT_
11. DAD_I_NEED_MONEY_TODAY
13. (a) $\mathbf{B}' = \begin{bmatrix} 15 & 22 & 20 & 8 & 23 & 6 & 21 & 22 \\ 10 & 22 & 18 & 23 & 25 & 2 & 23 & 25 \\ 3 & 26 & 26 & 14 & 23 & 16 & 26 & 12 \end{bmatrix}$

Exercises 7.11, Page 464

1. [0 1 1 0]   3. [0 0 0 1 1]
5. [1 0 1 0 1 0 0 1]   7. [1 0 0]
9. Parity error   11. [1 0 0 1 1]
13. [0 0 1 0 1 1 0]
15. [0 1 0 0 1 0 1]
17. [1 1 0 0 1 1 0]
19. Code word; [0 0 0 0]   21. [0 0 0 1]

A-49

23. Code word; [1 1 1 1]   **25.** [1 0 0 1]
27. [1 0 1 0]
29. $2^7 = 128$; $2^4 = 16$;
[0 0 0 0 0 0 0], [0 1 0 0 1 0 1],
[0 1 1 0 0 1 1], [0 1 0 1 0 1 0],
[0 1 1 1 1 0 0], [0 0 1 0 1 1 0],
[0 0 1 1 0 0 1], [0 0 0 1 1 1 1],
[1 0 0 0 0 1 1], [1 1 0 0 1 1 0],
[1 0 1 0 1 0 1], [1 0 0 1 1 0 0],
[1 1 1 0 0 0 0], [1 1 0 1 0 0 1],
[1 0 1 1 0 1 0], [1 1 1 1 1 1 1]

Exercises 7.12, Page 468
1. $y = 0.4x + 0.6$ **3.** $y = 1.1x - 0.3$
5. $y = 1.3571x + 1.9286$
7. $v = -0.84T + 234$, 116.4, 99.6

Chapter 7 Review Exercises, Page 469

1. $\begin{bmatrix} 2 & 3 & 4 \\ 3 & 4 & 5 \\ 4 & 5 & 6 \\ 5 & 6 & 7 \end{bmatrix}$ **3.** $\begin{bmatrix} 3 & 4 \\ 6 & 8 \end{bmatrix}$, $[11]$ **5.** False

7. $\frac{5}{8}, -5$ **9.** 0   **11.** False   **13.** True
15. False   **17.** True   **19.** False
23. (a) $\begin{bmatrix} 1 & 1 \\ -1 & -1 \end{bmatrix}$ **25.** $x_1 = -\frac{1}{2}, x_2 = 7, x_3 = \frac{1}{2}$
29. 240   **31.** Trivial solution only
33. $I_2 + 10HNO_3 \rightarrow 2HIO_3 + 10NO_2 + 4H_2O$
35. $x_1 = -\frac{1}{2}, x_2 = \frac{1}{4}, x_3 = \frac{2}{3}$
37. $x = X \cos\theta - Y \sin\theta$,
$y = X \sin\theta + Y \cos\theta$
39. $x_1 = 7, x_2 = 5, x_3 = 23$
41. $\lambda_1 = 5, \lambda_2 = -1$, $\mathbf{K}_1 = \begin{bmatrix} 1 \\ 2 \end{bmatrix}$, $\mathbf{K}_2 = \begin{bmatrix} 1 \\ -1 \end{bmatrix}$
43. $\lambda_1 = \lambda_2 = -1, \lambda_3 = 8$,
$\mathbf{K}_1 = \begin{bmatrix} 1 \\ -2 \\ 0 \end{bmatrix}$, $\mathbf{K}_2 = \begin{bmatrix} 1 \\ 0 \\ -1 \end{bmatrix}$, $\mathbf{K}_3 = \begin{bmatrix} 2 \\ 1 \\ 2 \end{bmatrix}$
45. $\lambda_1 = \lambda_2 = -3, \lambda_3 = 5$,
$\mathbf{K}_1 = \begin{bmatrix} -2 \\ 1 \\ 0 \end{bmatrix}$, $\mathbf{K}_2 = \begin{bmatrix} 3 \\ 0 \\ 1 \end{bmatrix}$, $\mathbf{K}_3 = \begin{bmatrix} 1 \\ 2 \\ -1 \end{bmatrix}$
47. $\begin{bmatrix} \frac{1}{\sqrt{6}} \\ -\frac{2}{\sqrt{6}} \\ \frac{1}{\sqrt{6}} \end{bmatrix}$ **49.** Hyperbola

51. $\begin{bmatrix} 204 & 13 & 208 & 55 & 124 & 120 & 105 \\ 185 & 12 & 188 & 50 & 112 & 108 & 96 \\ & & 214 & 50 & 6 & 138 & 19 & 210 \\ & & 194 & 45 & 6 & 126 & 18 & 189 \end{bmatrix}$

53. HELP_IS_ON_THE_WAY
55. [1 1 0 0 1]; parity error

Exercises 8.1, Page 481
1.   **3.**

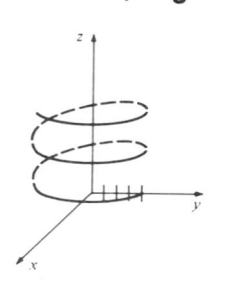

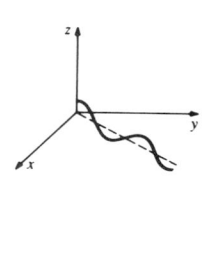

5. 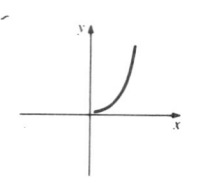   **7.**

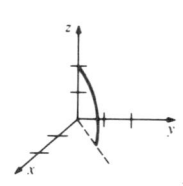

9. 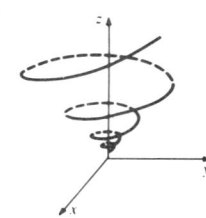   **11.** 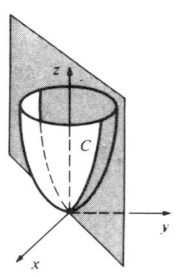
$\mathbf{r}(t) = t\mathbf{i} + t\mathbf{j} + 2t^2\mathbf{k}$

13.

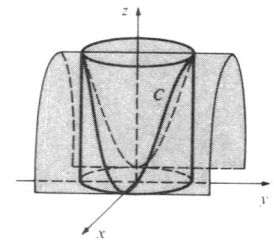

$\mathbf{r}(t) = 3\cos t\mathbf{i} + 3\sin t\mathbf{j} + 9\sin^2 t\mathbf{k}$
15. $2\mathbf{i} - 32\mathbf{j}$
17. $(1/t)\mathbf{i} - (1/t^2)\mathbf{j}$; $-(1/t^2)\mathbf{i} + (2/t^3)\mathbf{j}$
19. $\langle e^{2t}(2t+1), 3t^2, 8t-1 \rangle$; $\langle 4e^{2t}(t+1), 6t, 8 \rangle$
21.   **23.**

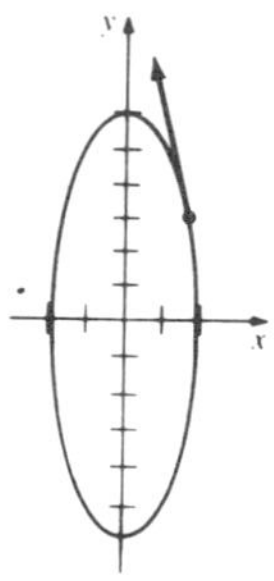

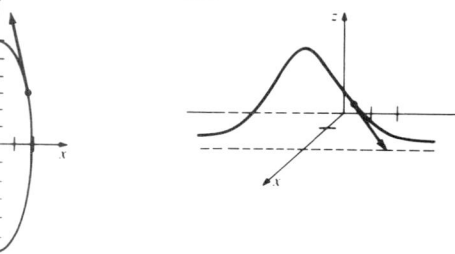

A-50

25. $x = 2 + t, y = 2 + 2t, z = \frac{8}{3} + 4t$
27. $\mathbf{r}'(t) \times \mathbf{r}''(t)$ **29.** $\mathbf{r}(t) \cdot [\mathbf{r}'(t) \times \mathbf{r}'''(t)]$
31. $2\mathbf{r}'_1(2t) - (1/t^2)\mathbf{r}'_2(1/t)$ **33.** $\frac{3}{2}\mathbf{i} + 9\mathbf{j} + 15\mathbf{k}$
35. $e^t(t-1)\mathbf{i} + \frac{1}{2}e^{-2t}\mathbf{j} + \frac{1}{2}e^{t^2}\mathbf{k} + \mathbf{c}$
37. $(6t+1)\mathbf{i} + (3t^2 - 2)\mathbf{j} + (t^3 + 1)\mathbf{k}$
39. $(2t^3 - 6t + 6)\mathbf{i} + (7t - 4t^{3/2} - 3)\mathbf{j} + (t^2 - 2t)\mathbf{k}$
41. $2\sqrt{a^2 + c^2}\pi$ **43.** $\sqrt{6}(e^{3\pi} - 1)$
45. $a\cos(s/a)\mathbf{i} + a\sin(s/a)\mathbf{j}$
47. Differentiate $\mathbf{r}(t) \cdot \mathbf{r}(t) = c^2$.

Exercises 8.2, Page 486

1. Speed is $\sqrt{5}$.

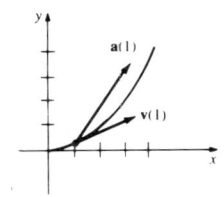

3. Speed is 2. 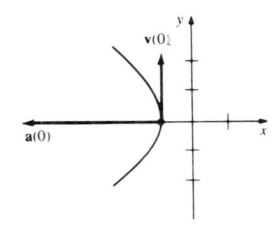   **5.** Speed is $\sqrt{5}$.

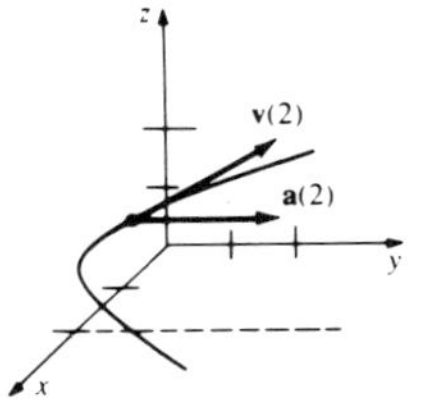

7. Speed is $\sqrt{14}$.

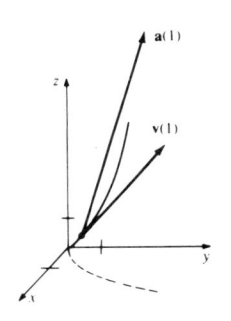

9. $(0, 0, 0)$ and $(25, 115, 0)$;
$\mathbf{v}(0) = -2\mathbf{j} - 5\mathbf{k}, \mathbf{a}(0) = 2\mathbf{i} + 2\mathbf{k}$,
$\mathbf{v}(5) = 10\mathbf{i} + 73\mathbf{j} + 5\mathbf{k}, \mathbf{a}(5) = 2\mathbf{i} + 30\mathbf{j} + 2\mathbf{k}$
11. $\mathbf{r}(t) = (-16t^2 + 240t)\mathbf{j} + 240\sqrt{3}\,t\mathbf{i}$ and
$x(t) = 240\sqrt{3}\,t, y(t) = -16t^2 + 240t$; 900 ft;
approximately 6235 ft; 480 ft/s
13. 72.11 ft/s
15. 97.98 ft/s
17. Assume that (x_0, y_0) are the coordinates of the center of the target at $t = 0$. Then $\mathbf{r}_p = \mathbf{r}_t$ when $t = x_0/(v_0 \cos\theta) = y_0/(v_0 \sin\theta)$. This implies $\tan\theta = y_0/x_0$. In other words, aim directly at the target at $t = 0$.
21. 191.33 lb

23. $y = -\dfrac{g}{2v_0^2 \cos^2\theta}x^2 + (\tan\theta)x + s_0$ is an equation of a parabola.

25. $\mathbf{r}(t) = k_1 e^{2t^3}\mathbf{i} + \dfrac{1}{2t^2 + k_2}\mathbf{j} + (k_3 e^{t^2} - 1)\mathbf{k}$

27. Since $\mathbf{F}$ is directed along $\mathbf{r}$, we must have $\mathbf{F} = c\mathbf{r}$ for some constant c. Hence $\boldsymbol{\tau} = \mathbf{r} \times (c\mathbf{r}) = c(\mathbf{r} \times \mathbf{r}) = \mathbf{0}$. If $\boldsymbol{\tau} = \mathbf{0}$, then $d\mathbf{L}/dt = \mathbf{0}$. This implies that $\mathbf{L}$ is a constant.

Exercises 8.3, Page 493

1. $\mathbf{T} = (\sqrt{5}/5)(-\sin t\mathbf{i} + \cos t\mathbf{j} + 2\mathbf{k})$
3. $\mathbf{T} = (a^2 + c^2)^{-1/2}(-a\sin t\mathbf{i} + a\cos t\mathbf{j} + c\mathbf{k})$,
$\mathbf{N} = -\cos t\mathbf{i} - \sin t\mathbf{j}$,
$\mathbf{B} = (a^2 + c^2)^{-1/2}(c\sin t\mathbf{i} - c\cos t\mathbf{j} + a\mathbf{k})$,
$\kappa = a/(a^2 + c^2)$
5. $3\sqrt{2}x - 3\sqrt{2}y + 4z = 3\pi$
7. $4t/\sqrt{1 + 4t^2}, 2/\sqrt{1 + 4t^2}$
9. $2\sqrt{6}, 0, t > 0$ **11.** $2t/\sqrt{1 + t^2}, 2/\sqrt{1 + t^2}$
13. $0, 5$ **15.** $-\sqrt{3}e^{-t}, 0$
17. $\kappa = \dfrac{\sqrt{b^2c^2\sin^2 t + a^2c^2\cos^2 t + a^2b^2}}{(a^2\sin^2 t + b^2\cos^2 t + c^2)^{3/2}}$
23. $\kappa = 2, \rho = \frac{1}{2}; \kappa = 2/\sqrt{125} \approx 0.18$,
$\rho = \sqrt{125}/2 \approx 5.59$; the curve is sharper at $(0, 0)$
25. κ is close to zero.

Exercises 8.4, Page 500

1.   **3.**

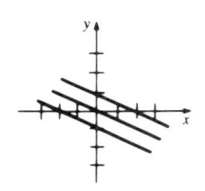

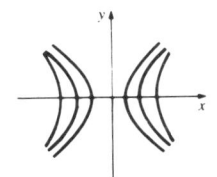

5.

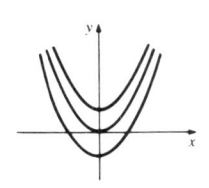

7. Elliptical cylinders   **9.** Ellipsoids
11.

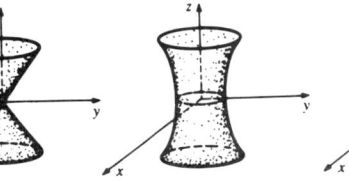

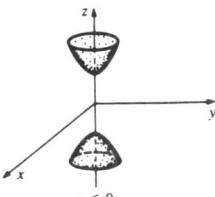

$c = 0$ $c > 0$ $c < 0$

13. $\partial z/\partial x = 2x - y^2, \partial z/\partial y = -2xy + 20y^4$

15. $\partial z/\partial x = 20x^3y^3 - 2xy^6 + 30x^4$,
    $\partial z/\partial y = 15x^4y^2 - 6x^2y^5 - 4$
17. $\partial z/\partial x = 2x^{-1/2}/(3y^2 + 1)$,
    $\partial z/\partial y = -24\sqrt{x}y/(3y^2 + 1)^2$
19. $\partial z/\partial x = -3x^2(x^3 - y^2)^{-2}$,
    $\partial z/\partial y = 2y(x^3 - y^2)^{-2}$
21. $\partial z/\partial x = -10 \cos 5x \sin 5x$,
    $\partial z/\partial y = 10 \sin 5y \cos 5y$
23. $f_x = e^{x^3y}(3x^3y + 1)$, $f_y = x^4e^{x^3y}$
25. $f_x = 7y/(x + 2y)^2$, $f_y = -7x/(x + 2y)^2$
27. $g_u = 8u/(4u^2 + 5v^3)$, $g_v = 15v^2/(4u^2 + 5v^3)$
29. $w_x = x^{-1/2}y$, $w_y = 2\sqrt{x} - (y/z)e^{y/z} - e^{y/z}$,
    $w_z = (y^2/z^2)e^{y/z}$
31. $F_u = 2uw^2 - v^3 - vwt^2 \sin(ut^2)$,
    $F_v = -3uv^2 + w \cos(ut^2)$, $F_x = 128x^7t^4$,
    $F_t = -2uvwt \sin(ut^2) + 64x^8t^3$
39. $\partial z/\partial x = 3x^2v^2e^{uv^2} + 2uve^{uv^2}$, $\partial z/\partial y = -4yuve^{uv^2}$
41. $\partial z/\partial u = 16u^3 - 40y(2u - v)$,
    $\partial z/\partial v = -96v^2 + 20y(2u - v)$
43. $\partial w/\partial t = -3u(u^2 + v^2)^{1/2}e^{-t} \sin \theta$
    $- 3v(u^2 + v^2)^{1/2}e^{-t} \cos \theta$,
    $\partial w/\partial \theta = 3u(u^2 + v^2)^{1/2}e^{-t} \cos \theta$
    $- 3v(u^2 + v^2)^{1/2}e^{-t} \sin \theta$
45. $\partial R/\partial u = s^2t^4e^{v^2} - 4rst^4uve^{-u^2} + 8rs^2t^3uv^2e^{u^2v^2}$,
    $\partial R/\partial v = 2s^2t^4uve^{v^2} + 2rst^4e^{-u^2} + 8rs^2t^3u^2ve^{u^2v^2}$
47. $\dfrac{\partial w}{\partial t} = \dfrac{xu}{(x^2 + y^2)^{1/2}(rs + tu)} + \dfrac{y \cosh rs}{u(x^2 + y^2)^{1/2}}$,
    $\dfrac{\partial w}{\partial r} = \dfrac{xs}{(x^2 + y^2)^{1/2}(rs + tu)} + \dfrac{sty \sinh rs}{u(x^2 + y^2)^{1/2}}$,
    $\dfrac{\partial w}{\partial u} = \dfrac{xt}{(x^2 + y^2)^{1/2}(rs + tu)} - \dfrac{ty \cosh rs}{u^2(x^2 + y^2)^{1/2}}$
49. $dz/dt = (4ut - 4vt^{-3})/(u^2 + v^2)$
51. $dw/dt|_{t=\pi} = -2$ 57. 5.31 cm²/s

Exercises 8.5, Page 509
1. $(2x - 3x^2y^2)\mathbf{i} + (-2x^3y + 4y^3)\mathbf{j}$
3. $(y^2/z^3)\mathbf{i} + (2xy/z^3)\mathbf{j} - (3xy^2/z^4)\mathbf{k}$
5. $4\mathbf{i} - 32\mathbf{j}$ 7. $2\sqrt{3}\mathbf{i} - 8\mathbf{j} - 4\sqrt{3}\mathbf{k}$
9. $\sqrt{3}x + y$ 11. $\frac{15}{2}(\sqrt{3} - 2)$
13. $-1/2\sqrt{10}$ 15. $98/\sqrt{5}$ 17. $-3\sqrt{2}$
19. -1 21. $-12/\sqrt{17}$
23. $\sqrt{2}\mathbf{i} + (\sqrt{2}/2)\mathbf{j}$, $\sqrt{5/2}$
25. $-2\mathbf{i} + 2\mathbf{j} - 4\mathbf{k}$, $2\sqrt{6}$
27. $-8\sqrt{\pi/6}\mathbf{i} - 8\sqrt{\pi/6}\mathbf{j}$, $-8\sqrt{\pi/3}$
29. $-\frac{3}{8}\mathbf{i} - 12\mathbf{j} - \frac{2}{3}\mathbf{k}$, $-\sqrt{83,281}/24$
31. $\pm 31/\sqrt{17}$
33. $\mathbf{u} = \frac{3}{5}\mathbf{i} - \frac{4}{5}\mathbf{j}$; $\mathbf{u} = \frac{4}{5}\mathbf{i} + \frac{3}{5}\mathbf{j}$; $\mathbf{u} = -\frac{4}{5}\mathbf{i} - \frac{3}{5}\mathbf{j}$
35. $D_\mathbf{u}f = (9x^2 + 3y^2 - 18xy^2 - 6x^2y)/\sqrt{10}$;
    $D_\mathbf{u}F = (-6x^2 - 54y^2 + 54x + 6y - 72xy)/10$
37. $(2, 5), (-2, 5)$ 39. $-16\mathbf{i} - 4\mathbf{j}$
41. $x = 3e^{-4t}$, $y = 4e^{-2t}$

43. One possible function is
$$f(x, y) = x^3 - \tfrac{2}{3}y^3 + xy^3 + e^{xy}$$

Exercises 8.6, Page 514
1. 3. 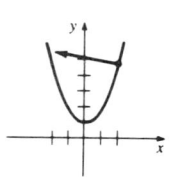
5. 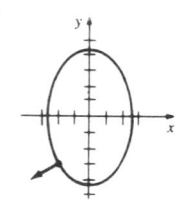 7. 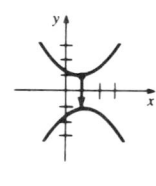
9. 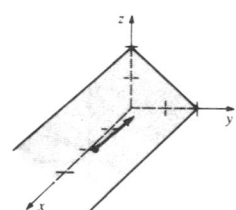 11. 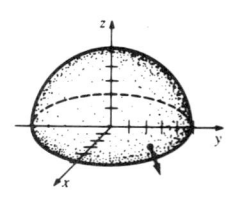
13. $(-4, -1, 17)$ 15. $-2x + 2y + z = 9$
17. $6x - 2y - 9z = 5$ 19. $6x - 8y + z = 50$
21. $2x + y - \sqrt{2}z = (4 + 5\pi)/4$
23. $\sqrt{2}x + \sqrt{2}y - z = 2$
25. $(1/\sqrt{2}, \sqrt{2}, 3/\sqrt{2}), (-1/\sqrt{2}, -\sqrt{2}, -3/\sqrt{2})$
27. $(-2, 0, 5), (-2, 0, -3)$
33. $x = 1 + 2t$, $y = -1 - 4t$, $z = 1 + 2t$
35. $(x - \tfrac{1}{2})/4 = (y - \tfrac{1}{3})/6 = -(z - 3)$

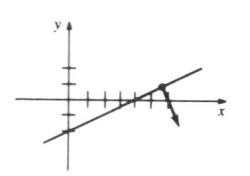

Exercises 8.7, Page 520
1.

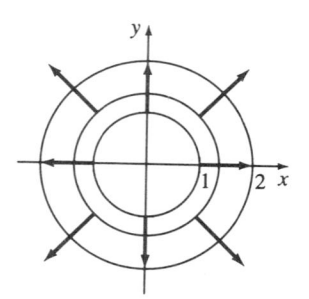

A-52

3.

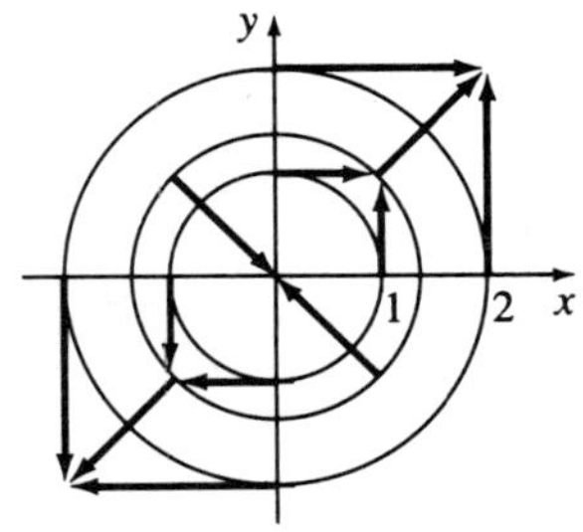

5.

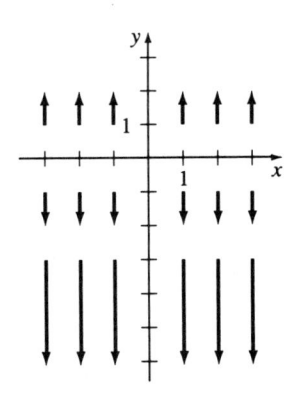

7. $(x - y)\mathbf{i} + (x - y)\mathbf{j}; 2z$ 9. $\mathbf{0}; 4y + 8z$
11. $(4y^3 - 6xz^2)\mathbf{i} + (2z^3 - 3x^2)\mathbf{k}; 6xy$
13. $(3e^{-z} - 8yz)\mathbf{i} - xe^{-z}\mathbf{j}; e^{-z} + 4z^2 - 3ye^{-z}$
15. $(xy^2e^y + 2xye^y + x^3yze^z + x^3ye^z)\mathbf{i} - y^2e^y\mathbf{j} + (-3x^2yze^z - xe^x)\mathbf{k}; xye^x + ye^x - x^3ze^z$
35. $2\mathbf{i} + (1 - 8y)\mathbf{j} + 8z\mathbf{k}$
45. div $\mathbf{F} = 1 \neq 0$. If there existed a vector field $\mathbf{G}$ such that $\mathbf{F} = $ curl $\mathbf{G}$, then necessarily div $\mathbf{F} = $ div curl $\mathbf{G} = 0$.

Exercises 8.8, Page 531

1. $-125/3\sqrt{2}; -250(\sqrt{2} - 4)/12; \frac{125}{2}$
3. $3; 6; 3\sqrt{5}$ 5. $-1; (\pi - 2)/2; \pi^2/8; \sqrt{2}\pi^2/8$
7. 21 9. 30 11. 1 13. 1
15. 460 17. $\frac{26}{9}$ 19. $-\frac{64}{3}$ 21. $-\frac{8}{3}$
23. 0 25. $\frac{123}{2}$ 27. 70 29. $-\frac{19}{8}$
31. e 33. -4 35. 0
37. On each curve the line integral has the value $\frac{208}{3}$.
41. $\bar{x} = \frac{3}{2}, \bar{y} = 2/\pi$

Exercises 8.9, Page 540

1. $\frac{16}{3}$ 3. 14 5. 3 7. 330
9. 1096 11. $\phi = x^4y^3 + 3x + y$
13. Not a gradient field   15. $\phi = \frac{1}{4}x^4 + xy + \frac{1}{4}y^4$
17. $3 + e^{-1}$ 19. 63 21. $8 + 2e^3$
23. 16 25. $\pi - 4$ 27. $\phi = (Gm_1m_2)/|\mathbf{r}|$

Exercises 8.10, Page 550

1. $24y - 20e^y$ 3. $x^2e^{3x^2} - x^2e^x$ 5. $\frac{x}{2}\ln 5$
7. $2 - \sin y$

9.

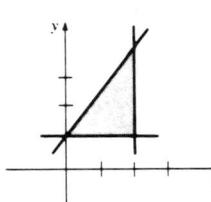

11.

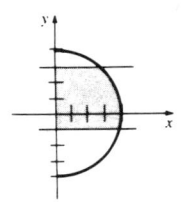

13. $\frac{1}{21}$ 15. $\frac{25}{84}$ 17. 96
19. $2\ln 2 - 1$ 21. $\frac{14}{3}$ 23. (c), 16π
25. 18 27. 2π 29. 4 31. $30\ln 6$
33. $15\pi/4$ 35. $(2^{3/2} - 1)/18$ 37. $\frac{2}{3}\sin 8$
39. $\pi/8$ 41. $\bar{x} = \frac{8}{3}, \bar{y} = 2$ 43. $\bar{x} = 3, \bar{y} = \frac{3}{2}$
45. $\bar{x} = \frac{17}{21}, \bar{y} = \frac{55}{147}$ 47. $\bar{x} = 0, \bar{y} = \frac{4}{7}$
49. $\bar{x} = (3e^4 + 1)/[4(e^4 - 1)],$
    $\bar{y} = 16(e^5 - 1)/[25(e^4 - 1)]$
51. $\frac{1}{105}$ 53. $4k/9$ 55. $\frac{256}{21}$ 57. $\frac{941}{10}$
59. $a\sqrt{10}/5$ 61. $ab^3\pi/4; a^3b\pi/4; b/2; a/2$
63. $ka^4/6$ 65. $16\sqrt{2}k/3$ 67. $a\sqrt{3}/3$

Exercises 8.11, Page 557

1. $27\pi/2$ 3. $(4\pi - 3\sqrt{3})/6$ 5. $25\pi/3$
7. $(2\pi/3)(15^{3/2} - 7^{3/2})$ 9. $\frac{5}{4}$
11. $\bar{x} = 13/3\pi, \bar{y} = 13/3\pi$ 13. $\bar{x} = \frac{12}{5}, \bar{y} = 3\sqrt{3}/2$
15. $\bar{x} = (4 + 3\pi)/6, \bar{y} = \frac{4}{3}$ 17. $\pi a^4 k/4$
19. $(ka/12)(15\sqrt{3} - 4\pi)$ 21. $\pi a^4 k/2$
23. $4k$ 25. 9π 27. $(\pi/4)(e - 1)$
29. $3\pi/8$ 31. 250
33. Approximately 1450 m^3 35. $\sqrt{\pi}/2$

Exercises 8.12, Page 564

1. 3 3. 0 5. 75π 7. 48π
9. $\frac{56}{3}$ 11. $\frac{2}{3}$ 13. $\frac{1}{8}$
15. $(\mathbf{b} \times \mathbf{a}) \times$ (area of region bounded by C)
19. $3a^2\pi/8$ 23. $45\pi/2$ 25. π
27. $27\pi/2$ 29. $3\pi/2$ 33. 3π

Exercises 8.13, Page 574

1. $3\sqrt{29}$ 3. $10\pi/3$ 5. $(\pi/6)(17^{3/2} - 1)$
7. $25\pi/6$ 9. $2a^2(\pi - 2)$ 11. $8a^2$
13. $2\pi a(c_2 - c_1)$ 15. $\frac{26}{3}$ 17. 0
19. 972π 21. $(3^{5/2} - 2^{7/2} + 1)/15$
23. $9(17^{3/2} - 1)$ 25. $12\sqrt{14}$ 27. $k\sqrt{3}/12$
29. 18 31. 28π 33. 8π 35. $5\pi/2$
37. $-8\pi a^3$ 39. $4\pi kq$ 41. $(1, \frac{2}{3}, 2)$

Exercises 8.14, Page 581

1. -40π 3. $\frac{45}{2}$ 5. $\frac{3}{2}$ 7. -3
9. $-3\pi/2$ 11. π 13. -152π
15. 112 17. Take the surface to be $z = 0$; $81\pi/4$

A-53

Exercises 8.15, Page 594

1. 48 3. 36 5. $\pi - 2$ 7. $\frac{1}{4}e^2 - \frac{1}{2}e$
9. 50

11. $\int_0^4 \int_0^{2-(x/2)} \int_{x+2y}^4 F(x, y, z)\, dz\, dy\, dx;$

$\int_0^2 \int_{2y}^4 \int_0^{z-2y} F(x, y, z)\, dx\, dz\, dy;$

$\int_0^4 \int_0^{z/2} \int_0^{z-2y} F(x, y, z)\, dx\, dy\, dz;$

$\int_0^4 \int_x^4 \int_0^{(z-x)/2} F(x, y, z)\, dy\, dz\, dx;$

$\int_0^4 \int_0^z \int_0^{(z-x)/2} F(x, y, z)\, dy\, dx\, dz$

13. $\int_0^2 \int_{x^3}^8 \int_0^4 dz\, dy\, dx;$

$\int_0^8 \int_0^4 \int_0^{\sqrt[3]{y}} dx\, dz\, dy;$

$\int_0^4 \int_0^2 \int_{x^3}^8 dy\, dx\, dz$

15.

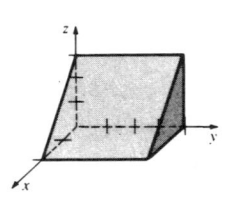

17. 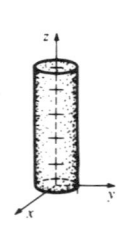   19.

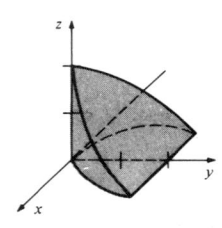

21. $16\sqrt{2}$ 23. 16π
25. $\bar{x} = \frac{4}{5}, \bar{y} = \frac{32}{7}, \bar{z} = \frac{8}{3}$
27. $\bar{x} = 0, \bar{y} = 2, \bar{z} = 0$

29. $\int_{-1}^1 \int_{-\sqrt{1-x^2}}^{\sqrt{1-x^2}} \int_{2y+2}^{8-y} (x + y + 4)\, dz\, dy\, dx$

31. $2560k/3; \sqrt{80/9}$ 33. $k/30$
35. $(-10/\sqrt{2}, 10/\sqrt{2}, 5)$ 37. $(\sqrt{3}/2, \frac{3}{2}, -4)$
39. $(\sqrt{2}, -\pi/4, -9)$ 41. $(2\sqrt{2}, 2\pi/3, 2)$
43. $r^2 + z^2 = 25$ 45. $r^2 - z^2 = 1$
47. $z = x^2 + y^2$ 49. $x = 5$
51. $(2\pi/3)(64 - 12^{3/2})$ 53. $625\pi/2$
55. $(0, 0, 3a/8)$ 57. $8\pi k/3$
59. $(\sqrt{3}/3, \frac{1}{3}, 0); (\frac{2}{3}, \pi/6, 0)$
61. $(-4, 4, 4\sqrt{2}); (4\sqrt{2}, 3\pi/4, 4\sqrt{2})$
63. $(5\sqrt{2}, \pi/2, 5\pi/4)$ 65. $(\sqrt{2}, \pi/4, \pi/6)$
67. $\rho = 8$ 69. $\phi = \pi/6, \phi = 5\pi/6$

71. $x^2 + y^2 + z^2 = 100$ 73. $z = 2$
75. $9\pi(2 - \sqrt{2})$ 77. $2\pi/9$ 79. $(0, 0, \frac{7}{6})$
81. πk

Exercises 8.16, Page 603

1. $\frac{3}{2}$ 3. $12a^5\pi/5$ 5. 256π 7. $62\pi/5$
9. $4\pi(b - a)$ 11. 128 13. $\pi/2$

Exercises 8.17, Page 611

1. $(0, 0), (-2, 8), (16, 20), (14, 28)$
3.

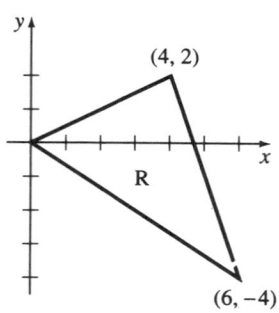

5. 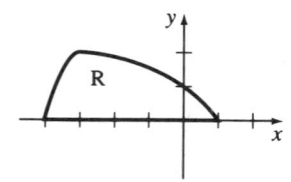   7. $-2v$

9. $-1/3u^2$
11.

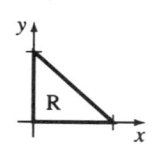

$(0, 0)$ is the image of every point on the boundary $u = 0$.

13. 16 15. $\frac{1}{2}$ 17. $\frac{1}{4}(b - a)(d - c)$
19. $\frac{1}{2}(1 - \ln 2)$ 21. $\frac{315}{4}$ 23. $\frac{1}{4}(e - e^{-1})$
25. 126 27. $\frac{5}{2}(b - a) \ln \frac{d}{c}$ 29. $15\pi/2$

Chapter 8 Review Exercises, Page 614

1. True 3. True 5. False 7. True
9. False 11. False 13. True
15. True 17. True

19. $\nabla \phi = -\dfrac{x}{(x^2 + y^2)^{3/2}} \mathbf{i} - \dfrac{y}{(x^2 + y^2)^{3/2}} \mathbf{j}$

21. $v(1) = 6\mathbf{i} + \mathbf{j} + 2\mathbf{k}, v(4) = 6\mathbf{i} + \mathbf{j} + 8\mathbf{k},$
$a(t) = 2\mathbf{k}$ for all t
23. $\mathbf{i} + 4\mathbf{j} + (3\pi/4)\mathbf{k}$

25.

27. $(6x^2 - 2y^2 - 8xy)/\sqrt{40}$ **29.** $2; -2/\sqrt{2}; 4$
31. $4\pi x + 3y - 12z = 4\pi - 6\sqrt{3}$
33. $\int_0^1 \int_x^{2x} \sqrt{1-x^2}\, dy\, dx;$
$\int_0^1 \int_{y/2}^{y} \sqrt{1-x^2}\, dx\, dy + \int_1^2 \int_{y/2}^{1} \sqrt{1-x^2}\, dx\, dy; \frac{1}{3}$
35. $41k/1512$ **37.** 8π **39.** $6xy$ **41.** 0
43. $56\sqrt{2}\pi^3/3$ **45.** 12 **47.** $2 + 2/3\pi$
49. $\pi^2/2$ **51.** $(\ln 3)(17^{3/2} - 5^{3/2})/12$
53. $-4\pi c$ **55.** 0 **57.** 125π **59.** 3π
61. $\frac{5}{3}$ **63.** 0

Exercises 9.1, Page 624

1. $x_1' = x_2$
$x_2' = -4x_1 + 3x_2 + \sin 3t$

3. $x_1' = x_2$
$x_2' = x_3$
$x_3' = 10x_1 - 6x_2 + 3x_3 + t^2 + 1$

5. $x_1' = x_2$
$x_2' = x_3$
$x_3' = x_4$
$x_4' = -x_1 - 4x_2 + 2x_3 + t$

7. $x_1' = x_2$
$x_2' = \dfrac{t}{t+1} x_1$

9. $x' = -2x + y + 5t$
$y' = 2x + y - 2t$

11. $Dx = t^2 + 5t - 2$
$Dy = -x + 5t - 2$

13. The system is degenerate.

15. $Dx = u$
$Dy = v$
$Du = w$
$Dv = 10t^2 - 4u + 3v$
$Dw = 4x + 4v - 3w$

17. $x_1 = \frac{25}{2} e^{-t/25} + \frac{25}{2} e^{-3t/25}$
$x_2 = \frac{25}{4} e^{-t/25} - \frac{25}{4} e^{-3t/25}$

19. $x_1' = \frac{1}{50} x_2 - \frac{3}{50} x_1$
$x_2' = \frac{3}{50} x_1 - \frac{7}{100} x_2 + \frac{1}{100} x_3$
$x_3' = \frac{1}{20} x_2 - \frac{1}{20} x_3$

Exercises 9.2, Page 628

1. $\begin{bmatrix} -5e^{-t} \\ -2e^{-t} \\ 7e^{-t} \end{bmatrix}$ **3.** $4\begin{bmatrix} 1 \\ -1 \end{bmatrix} e^{2t} - 12 \begin{bmatrix} 2 \\ 1 \end{bmatrix} e^{-3t}$

5. $\begin{bmatrix} -2\sin t \\ \dfrac{\sin^2 t}{\cos t} \end{bmatrix} e^t + \begin{bmatrix} 2\cos t - 2\sin t \\ -\sin t - \cos t \end{bmatrix} e^t \ln|\cos t|$

7. $\begin{bmatrix} 4e^{4t} & -\pi \sin \pi t \\ 2 & 6t \end{bmatrix}; \begin{bmatrix} \frac{1}{4}e^8 - \frac{1}{4} & 0 \\ 4 & 6 \end{bmatrix};$
$\begin{bmatrix} \frac{1}{4}e^{4t} - \frac{1}{4} & (1/\pi)\sin \pi t \\ t^2 & t^3 - t \end{bmatrix}$

9. $\mathbf{X}' = \begin{bmatrix} 3 & -5 \\ 4 & 8 \end{bmatrix} \mathbf{X}$, where $\mathbf{X} = \begin{bmatrix} x \\ y \end{bmatrix}$

11. $\mathbf{X}' = \begin{bmatrix} -3 & 4 & -9 \\ 6 & -1 & 0 \\ 10 & 4 & 3 \end{bmatrix} \mathbf{X}$, where $\mathbf{X} = \begin{bmatrix} x \\ y \\ z \end{bmatrix}$

13. $\mathbf{X}' = \begin{bmatrix} 1 & -1 & 1 \\ 2 & 1 & -1 \\ 1 & 1 & 1 \end{bmatrix} \mathbf{X} + \begin{bmatrix} 0 \\ -3t^2 \\ t^2 \end{bmatrix} + \begin{bmatrix} t \\ 0 \\ -t \end{bmatrix}$
$+ \begin{bmatrix} -1 \\ 0 \\ 2 \end{bmatrix}$, where $\mathbf{X} = \begin{bmatrix} x \\ y \\ z \end{bmatrix}$

15. $\dfrac{dx}{dt} = 4x + 2y + e^t$
$\dfrac{dy}{dt} = -x + 3y - e^t$

17. $\dfrac{dx}{dt} = x - y + 2z + e^{-t} - 3t$
$\dfrac{dy}{dt} = 3x - 4y + z + 2e^{-t} + t$
$\dfrac{dz}{dt} = -2x + 5y + 6z + 2e^{-t} - t$

19. $\det \mathbf{A}(t) = 2e^{3t} \ne 0$ for every value of t;
$\mathbf{A}^{-1}(t) = \dfrac{1}{2e^{3t}} \begin{bmatrix} 3e^{4t} & -e^{4t} \\ -4e^{-t} & 2e^{-t} \end{bmatrix}$

Exercises 9.3, Page 640

7. Yes; $W(\mathbf{X}_1, \mathbf{X}_2) = -2e^{-8t} \ne 0$ implies $\mathbf{X}_1$ and $\mathbf{X}_2$ are linearly independent on $(-\infty, \infty)$.

9. No; $W(\mathbf{X}_1, \mathbf{X}_2, \mathbf{X}_3) = \begin{vmatrix} 1+t & 1 & 3+2t \\ -2+2t & -2 & -6+4t \\ 4+2t & 4 & 12+4t \end{vmatrix} = 0$
for every t. The solution vectors are linearly dependent on $(-\infty, \infty)$. Note that $\mathbf{X}_3 = 2\mathbf{X}_1 + \mathbf{X}_2$.

17. $\Phi(t) = \begin{bmatrix} e^{2t} & e^{7t} \\ -2e^{2t} & 3e^{7t} \end{bmatrix}$,
$\Phi^{-1}(t) = \dfrac{1}{5e^{9t}} \begin{bmatrix} 3e^{7t} & -e^{7t} \\ 2e^{2t} & e^{2t} \end{bmatrix}$

19. $\Phi(t) = \begin{bmatrix} -e^t & -te^t \\ 3e^t & 3te^t - e^t \end{bmatrix}$,
$\Phi^{-1}(t) = \dfrac{1}{e^{2t}} \begin{bmatrix} 3te^t - e^t & te^t \\ -3e^t & -e^t \end{bmatrix}$

21. $\Psi(t) = \begin{bmatrix} \frac{3}{5}e^{2t} + \frac{2}{5}e^{7t} & -\frac{1}{5}e^{2t} + \frac{1}{5}e^{7t} \\ -\frac{6}{5}e^{2t} + \frac{6}{5}e^{7t} & \frac{2}{5}e^{2t} + \frac{3}{5}e^{7t} \end{bmatrix}$

23. $\Psi(t) = \begin{bmatrix} 3te^t + e^t & te^t \\ -9te^t & -3te^t + e^t \end{bmatrix}$

25. $\mathbf{X}(t_0) = \mathbf{\Phi}(t_0)\mathbf{C}$ implies $\mathbf{C} = \mathbf{\Phi}^{-1}(t_0)\mathbf{X}(t_0)$. Substituting in $\mathbf{X} = \mathbf{\Phi}(t)\mathbf{C}$ gives $\mathbf{X} = \mathbf{\Phi}(t)\mathbf{\Phi}^{-1}(t_0)\mathbf{X}_0$.

27. Comparing $\mathbf{X} = \mathbf{\Phi}(t)\mathbf{\Phi}^{-1}(t_0)\mathbf{X}_0$ and $\mathbf{X} = \mathbf{\Psi}(t)\mathbf{X}_0$ implies $[\mathbf{\Psi}(t) - \mathbf{\Phi}(t)\mathbf{\Phi}^{-1}(t_0)]\mathbf{X}_0 = \mathbf{0}$. Since this last equation is to hold for any $\mathbf{X}_0$, we conclude $\mathbf{\Psi}(t) = \mathbf{\Phi}(t)\mathbf{\Phi}^{-1}(t_0)$.

Exercises 9.4, Page 654

1. $\mathbf{X} = c_1 \begin{bmatrix} 1 \\ 2 \end{bmatrix} e^{5t} + c_2 \begin{bmatrix} 1 \\ -1 \end{bmatrix} e^{-t}$

3. $\mathbf{X} = c_1 \begin{bmatrix} 2 \\ 1 \end{bmatrix} e^{-3t} + c_2 \begin{bmatrix} 2 \\ 5 \end{bmatrix} e^t$

5. $\mathbf{X} = c_1 \begin{bmatrix} 5 \\ 2 \end{bmatrix} e^{8t} + c_2 \begin{bmatrix} 1 \\ 4 \end{bmatrix} e^{-10t}$

7. $\mathbf{X} = c_1 \begin{bmatrix} 1 \\ 0 \\ 0 \end{bmatrix} e^t + c_2 \begin{bmatrix} 2 \\ 3 \\ 1 \end{bmatrix} e^{2t} + c_3 \begin{bmatrix} 1 \\ 0 \\ 2 \end{bmatrix} e^{-t}$

9. $\mathbf{X} = c_1 \begin{bmatrix} -1 \\ 0 \\ 1 \end{bmatrix} e^{-t} + c_2 \begin{bmatrix} 1 \\ 4 \\ 3 \end{bmatrix} e^{3t} + c_3 \begin{bmatrix} 1 \\ -1 \\ 3 \end{bmatrix} e^{-2t}$

11. $\mathbf{X} = c_1 \begin{bmatrix} 4 \\ 0 \\ -1 \end{bmatrix} e^{-t} + c_2 \begin{bmatrix} -12 \\ 6 \\ 5 \end{bmatrix} e^{-t/2} + c_3 \begin{bmatrix} 4 \\ 2 \\ -1 \end{bmatrix} e^{-3t/2}$

13. $\mathbf{X} = 3 \begin{bmatrix} 1 \\ 1 \end{bmatrix} e^{t/2} + 2 \begin{bmatrix} 0 \\ 1 \end{bmatrix} e^{-t/2}$

15. $\mathbf{X} = c_1 \begin{bmatrix} \cos t \\ 2\cos t + \sin t \end{bmatrix} e^{4t} + c_2 \begin{bmatrix} \sin t \\ 2\sin t - \cos t \end{bmatrix} e^{4t}$

17. $\mathbf{X} = c_1 \begin{bmatrix} \cos t \\ -\cos t - \sin t \end{bmatrix} e^{4t} + c_2 \begin{bmatrix} \sin t \\ -\sin t + \cos t \end{bmatrix} e^{4t}$

19. $\mathbf{X} = c_1 \begin{bmatrix} 5\cos 3t \\ 4\cos 3t + 3\sin 3t \end{bmatrix}$
$+ c_2 \begin{bmatrix} 5\sin 3t \\ 4\sin 3t - 3\cos 3t \end{bmatrix}$

21. $\mathbf{X} = c_1 \begin{bmatrix} 1 \\ 0 \\ 0 \end{bmatrix} + c_2 \begin{bmatrix} -\cos t \\ \cos t \\ \sin t \end{bmatrix} + c_3 \begin{bmatrix} \sin t \\ -\sin t \\ \cos t \end{bmatrix}$

23. $\mathbf{X} = c_1 \begin{bmatrix} 0 \\ 2 \\ 1 \end{bmatrix} e^t + c_2 \begin{bmatrix} \sin t \\ \cos t \\ \cos t \end{bmatrix} e^t + c_3 \begin{bmatrix} \cos t \\ -\sin t \\ -\sin t \end{bmatrix} e^t$

25. $\mathbf{X} = \begin{bmatrix} 28 \\ -5 \\ 25 \end{bmatrix} e^{2t} + c_2 \begin{bmatrix} 5\cos 3t \\ -4\cos 3t - 3\sin 3t \\ 0 \end{bmatrix} e^{-2t}$
$+ c_3 \begin{bmatrix} 5\sin 3t \\ -4\sin 3t + 3\cos 3t \\ 0 \end{bmatrix} e^{-2t}$

27. $\mathbf{X} = -\begin{bmatrix} 25 \\ -7 \\ 6 \end{bmatrix} e^t - \begin{bmatrix} \cos 5t - 5\sin 5t \\ \cos 5t \\ \cos 5t \end{bmatrix}$
$+ 6 \begin{bmatrix} 5\cos 5t + \sin 5t \\ \sin 5t \\ \sin 5t \end{bmatrix}$

29. $\mathbf{X} = c_1 \begin{bmatrix} 1 \\ 3 \end{bmatrix} + c_2 \left\{ \begin{bmatrix} 1 \\ 3 \end{bmatrix} t + \begin{bmatrix} \frac{1}{4} \\ -\frac{1}{4} \end{bmatrix} \right\}$

31. $\mathbf{X} = c_1 \begin{bmatrix} 1 \\ 1 \end{bmatrix} e^{2t} + c_2 \left\{ \begin{bmatrix} 1 \\ 1 \end{bmatrix} te^{2t} + \begin{bmatrix} -\frac{1}{3} \\ 0 \end{bmatrix} e^{2t} \right\}$

33. $\mathbf{X} = c_1 \begin{bmatrix} 1 \\ 1 \\ 1 \end{bmatrix} e^t + c_2 \begin{bmatrix} 1 \\ 1 \\ 0 \end{bmatrix} e^{2t} + c_3 \begin{bmatrix} 1 \\ 0 \\ 1 \end{bmatrix} e^{2t}$

35. $\mathbf{X} = c_1 \begin{bmatrix} -4 \\ -5 \\ 2 \end{bmatrix} + c_2 \begin{bmatrix} 2 \\ 0 \\ -1 \end{bmatrix} e^{5t}$
$+ c_3 \left\{ \begin{bmatrix} 2 \\ 0 \\ -1 \end{bmatrix} te^{5t} + \begin{bmatrix} -\frac{1}{2} \\ -\frac{1}{2} \\ -1 \end{bmatrix} e^{5t} \right\}$

37. $\mathbf{X} = c_1 \begin{bmatrix} 0 \\ 1 \\ 1 \end{bmatrix} e^t + c_2 \left\{ \begin{bmatrix} 0 \\ 1 \\ 1 \end{bmatrix} te^t + \begin{bmatrix} 0 \\ 1 \\ 0 \end{bmatrix} e^t \right\}$
$+ c_3 \left\{ \begin{bmatrix} 0 \\ 1 \\ 1 \end{bmatrix} \frac{t^2}{2} e^t + \begin{bmatrix} 0 \\ 1 \\ 0 \end{bmatrix} te^t + \begin{bmatrix} \frac{1}{2} \\ 0 \\ 0 \end{bmatrix} e^t \right\}$

39. $\mathbf{X} = -7 \begin{bmatrix} 2 \\ 1 \end{bmatrix} e^{4t} + 13 \begin{bmatrix} 2t+1 \\ t+1 \end{bmatrix} e^{4t}$

41. $\mathbf{X} = \begin{bmatrix} \frac{6}{5}e^{5t} - \frac{1}{5}e^{-5t} \\ \frac{2}{5}e^{5t} + \frac{3}{5}e^{-5t} \end{bmatrix}$

43. $\mathbf{X} = c_1 t^2 \begin{bmatrix} 3 \\ 1 \end{bmatrix} + c_2 t^4 \begin{bmatrix} 1 \\ 1 \end{bmatrix}$

Exercises 9.5, Page 658

1. $\mathbf{X} = \begin{bmatrix} 3c_1 e^{7t} - 2c_2 e^{-4t} \\ c_1 e^{7t} + 3c_2 e^{-4t} \end{bmatrix}$

3. $\mathbf{X} = \begin{bmatrix} c_1 e^{t/2} + c_2 e^{3t/2} \\ -2c_1 e^{t/2} + 2c_2 e^{3t/2} \end{bmatrix}$

5. $\mathbf{X} = \begin{bmatrix} -c_1 e^{-4t} + c_2 e^{2t} \\ c_1 e^{-4t} + c_2 e^{2t} \\ c_2 e^{2t} + c_3 e^{6t} \end{bmatrix}$

7. $\mathbf{X} = \begin{bmatrix} c_1 e^{-t} - c_2 e^{2t} - c_3 e^{2t} \\ c_1 e^{-t} + c_2 e^{2t} \\ c_1 e^{-t} + c_3 e^{2t} \end{bmatrix}$

9. $\mathbf{X} = \begin{bmatrix} c_1 e^t + 2c_2 e^{2t} + 3c_3 e^{3t} \\ c_1 e^t + 2c_2 e^{2t} + 4c_3 e^{3t} \\ c_1 e^t + 2c_2 e^{2t} + 5c_3 e^{3t} \end{bmatrix}$

Exercises 9.6, Page 665

1. $\mathbf{X} = c_1 \begin{bmatrix} -1 \\ 1 \end{bmatrix} e^{-t} + c_2 \begin{bmatrix} -3 \\ 1 \end{bmatrix} e^{t} + \begin{bmatrix} -1 \\ 3 \end{bmatrix}$

3. $\mathbf{X} = c_1 \begin{bmatrix} 1 \\ -1 \end{bmatrix} e^{-2t} + c_2 \begin{bmatrix} 1 \\ 1 \end{bmatrix} e^{4t} + \begin{bmatrix} -\frac{1}{4} \\ \frac{3}{4} \end{bmatrix} t^2 + \begin{bmatrix} \frac{1}{4} \\ -\frac{1}{4} \end{bmatrix} t + \begin{bmatrix} -2 \\ \frac{3}{4} \end{bmatrix}$

5. $\mathbf{X} = c_1 \begin{bmatrix} 1 \\ -3 \end{bmatrix} e^{3t} + c_2 \begin{bmatrix} 1 \\ 9 \end{bmatrix} e^{7t} + \begin{bmatrix} \frac{55}{36} \\ -\frac{19}{4} \end{bmatrix} e^{t}$

7. $\mathbf{X} = c_1 \begin{bmatrix} 1 \\ 0 \\ 0 \end{bmatrix} e^{t} + c_2 \begin{bmatrix} 1 \\ 1 \\ 0 \end{bmatrix} e^{2t} + c_3 \begin{bmatrix} 1 \\ 2 \\ 2 \end{bmatrix} e^{5t} - \begin{bmatrix} \frac{3}{2} \\ \frac{7}{2} \\ 2 \end{bmatrix} e^{4t}$

9. $\mathbf{X} = 13 \begin{bmatrix} 1 \\ -1 \end{bmatrix} e^{t} + 2 \begin{bmatrix} -4 \\ 6 \end{bmatrix} e^{2t} + \begin{bmatrix} -9 \\ 6 \end{bmatrix}$

11. $\mathbf{X} = c_1 \begin{bmatrix} 1 \\ 1 \end{bmatrix} + c_2 \begin{bmatrix} 3 \\ 2 \end{bmatrix} e^{t} - \begin{bmatrix} 11 \\ 11 \end{bmatrix} t - \begin{bmatrix} 15 \\ 10 \end{bmatrix}$

13. $\mathbf{X} = c_1 \begin{bmatrix} 2 \\ 1 \end{bmatrix} e^{t/2} + c_2 \begin{bmatrix} 10 \\ 3 \end{bmatrix} e^{3t/2}$
$- \begin{bmatrix} \frac{13}{2} \\ \frac{13}{4} \end{bmatrix} t e^{t/2} - \begin{bmatrix} \frac{15}{2} \\ \frac{9}{4} \end{bmatrix} e^{t/2}$

15. $\mathbf{X} = c_1 \begin{bmatrix} 2 \\ 1 \end{bmatrix} e^{t} + c_2 \begin{bmatrix} 1 \\ 1 \end{bmatrix} e^{2t} + \begin{bmatrix} 3 \\ 3 \end{bmatrix} e^{t} + \begin{bmatrix} 4 \\ 2 \end{bmatrix} t e^{t}$

17. $\mathbf{X} = c_1 \begin{bmatrix} 4 \\ 1 \end{bmatrix} e^{3t} + c_2 \begin{bmatrix} -2 \\ 1 \end{bmatrix} e^{-3t} + \begin{bmatrix} -12 \\ 0 \end{bmatrix} t - \begin{bmatrix} \frac{4}{3} \\ \frac{4}{3} \end{bmatrix}$

19. $\mathbf{X} = c_1 \begin{bmatrix} 1 \\ -1 \end{bmatrix} e^{t} + c_2 \begin{bmatrix} -t \\ \frac{1}{2} - t \end{bmatrix} e^{t} + \begin{bmatrix} \frac{1}{2} \\ -2 \end{bmatrix} e^{-t}$

21. $\mathbf{X} = c_1 \begin{bmatrix} \cos t \\ \sin t \end{bmatrix} + c_2 \begin{bmatrix} \sin t \\ -\cos t \end{bmatrix}$
$+ \begin{bmatrix} \cos t \\ \sin t \end{bmatrix} t + \begin{bmatrix} -\sin t \\ \cos t \end{bmatrix} \ln|\cos t|$

23. $\mathbf{X} = c_1 \begin{bmatrix} \cos t \\ \sin t \end{bmatrix} e^{t} + c_2 \begin{bmatrix} \sin t \\ -\cos t \end{bmatrix} e^{t} + \begin{bmatrix} \cos t \\ \sin t \end{bmatrix} t e^{t}$

25. $\mathbf{X} = c_1 \begin{bmatrix} \cos t \\ -\sin t \end{bmatrix} + c_2 \begin{bmatrix} \sin t \\ \cos t \end{bmatrix} + \begin{bmatrix} \cos t \\ -\sin t \end{bmatrix} t$
$+ \begin{bmatrix} -\sin t \\ \sin t \tan t \end{bmatrix} - \begin{bmatrix} \sin t \\ \cos t \end{bmatrix} \ln|\cos t|$

27. $\mathbf{X} = c_1 \begin{bmatrix} 2 \sin t \\ \cos t \end{bmatrix} e^{t} + c_2 \begin{bmatrix} 2 \cos t \\ -\sin t \end{bmatrix} e^{t} + \begin{bmatrix} 3 \sin t \\ \frac{3}{2} \cos t \end{bmatrix} t e^{t}$
$+ \begin{bmatrix} \cos t \\ -\frac{1}{2} \sin t \end{bmatrix} e^{t} \ln|\sin t|$
$+ \begin{bmatrix} 2 \cos t \\ -\sin t \end{bmatrix} e^{t} \ln|\cos t|$

29. $\mathbf{X} = c_1 \begin{bmatrix} 1 \\ -1 \\ 0 \end{bmatrix} + c_2 \begin{bmatrix} 1 \\ 1 \\ 0 \end{bmatrix} e^{2t} + c_3 \begin{bmatrix} 0 \\ 0 \\ 1 \end{bmatrix} e^{3t}$
$+ \begin{bmatrix} -\frac{1}{4} e^{2t} + \frac{1}{2} t e^{2t} \\ -e^{t} + \frac{1}{4} e^{2t} + \frac{1}{2} t e^{2t} \\ \frac{1}{2} t^2 e^{3t} \end{bmatrix}$

31. $\mathbf{X} = \begin{bmatrix} 2 \\ 2 \end{bmatrix} t e^{2t} + \begin{bmatrix} -1 \\ 1 \end{bmatrix} e^{2t} + \begin{bmatrix} -2 \\ 2 \end{bmatrix} t e^{4t} + \begin{bmatrix} 2 \\ 0 \end{bmatrix} e^{4t}$

33. $\mathbf{X} = \begin{bmatrix} -2 \\ 4 \end{bmatrix} e^{2t} + \begin{bmatrix} 7 \\ -9 \end{bmatrix} e^{7t} + \begin{bmatrix} 20 \\ 60 \end{bmatrix} t e^{7t}$

35. $\begin{bmatrix} i_1 \\ i_2 \end{bmatrix} = 2 \begin{bmatrix} 1 \\ 3 \end{bmatrix} e^{-2t} + \frac{6}{29} \begin{bmatrix} 3 \\ -1 \end{bmatrix} e^{-12t}$
$+ \begin{bmatrix} \frac{332}{29} \\ \frac{276}{29} \end{bmatrix} \sin t - \begin{bmatrix} \frac{76}{29} \\ \frac{168}{29} \end{bmatrix} \cos t$

37. $\mathbf{X} = \begin{bmatrix} 20 + c_1 e^{-t} + 2c_2 e^{-2t} \\ 53 + 3c_1 e^{-t} + 7c_2 e^{-2t} \end{bmatrix}$

39. $\mathbf{X} = \begin{bmatrix} \frac{1}{2} t^2 - \frac{41}{10} t - \frac{41}{100} + c_1 + c_2 e^{10t} \\ -\frac{1}{2} t^2 + \frac{39}{10} t - \frac{41}{100} - c_1 + c_2 e^{10t} \end{bmatrix}$

Exercises 9.7, Page 669

1. $\begin{bmatrix} \cosh t & \sinh t \\ \sinh t & \cosh t \end{bmatrix}$

3. $\mathbf{X} = \begin{bmatrix} \cosh t & \sinh t \\ \sinh t & \cosh t \end{bmatrix} \begin{bmatrix} c_1 \\ c_2 \end{bmatrix}$
$= c_1 \begin{bmatrix} \cosh t \\ \sinh t \end{bmatrix} + c_2 \begin{bmatrix} \sinh t \\ \cosh t \end{bmatrix}$

5. $\mathbf{X} = c_1 \begin{bmatrix} \cosh t \\ \sinh t \end{bmatrix} + c_2 \begin{bmatrix} \sinh t \\ \cosh t \end{bmatrix} - \begin{bmatrix} 1 \\ 1 \end{bmatrix}$

7. $\mathbf{X} = c_1 \begin{bmatrix} 1 \\ 0 \end{bmatrix} e^{t} + c_2 \begin{bmatrix} 0 \\ 1 \end{bmatrix} e^{2t} + \begin{bmatrix} -t - 1 \\ \frac{1}{2} e^{4t} \end{bmatrix}$

11. $\mathbf{X} = \begin{bmatrix} \frac{3}{2} e^{3t} - \frac{1}{2} e^{5t} & -\frac{1}{2} e^{3t} + \frac{1}{2} e^{5t} \\ \frac{3}{2} e^{3t} - \frac{3}{2} e^{5t} & -\frac{1}{2} e^{3t} + \frac{3}{2} e^{5t} \end{bmatrix} \begin{bmatrix} c_1 \\ c_2 \end{bmatrix}$

Chapter 9 Review Exercises, Page 671

1. $\begin{bmatrix} t^3 + 3t^2 + 5t - 2 \\ -t^3 - t + 2 \\ 4t^3 + 12t^2 + 8t + 1 \end{bmatrix}; \begin{bmatrix} 3t^2 + 6t + 5 \\ -3t^2 - 1 \\ 12t^2 + 24t + 8 \end{bmatrix}$

3. $Dx = u$
$Dy = v$
$Du = -2u + v - 2x - \ln t + 10t - 4$
$Dv = -u - x + 5t - 2$

5. $\mathbf{X} = c_1 \begin{bmatrix} 1 \\ -1 \end{bmatrix} e^{t} + c_2 \left\{ \begin{bmatrix} 1 \\ -1 \end{bmatrix} t e^{t} + \begin{bmatrix} 0 \\ 1 \end{bmatrix} e^{t} \right\}$

7. $\mathbf{X} = c_1 \begin{bmatrix} \cos 2t \\ -\sin 2t \end{bmatrix} e^{t} + c_2 \begin{bmatrix} \sin 2t \\ \cos 2t \end{bmatrix} e^{t}$

9. $\mathbf{X} = c_1 \begin{bmatrix} -1 \\ 1 \\ 0 \end{bmatrix} + c_2 \begin{bmatrix} -1 \\ 0 \\ 1 \end{bmatrix} + c_3 \begin{bmatrix} 1 \\ 1 \\ 1 \end{bmatrix} e^{3t}$

11. $\mathbf{X} = c_1 \begin{bmatrix} 1 \\ 0 \end{bmatrix} e^{2t} + c_2 \begin{bmatrix} 4 \\ 1 \end{bmatrix} e^{4t} + \begin{bmatrix} 16 \\ -4 \end{bmatrix} t + \begin{bmatrix} 11 \\ -1 \end{bmatrix}$

13. $\mathbf{X} = c_1 \begin{bmatrix} \cos t \\ \cos t - \sin t \end{bmatrix} + c_2 \begin{bmatrix} \sin t \\ \sin t + \cos t \end{bmatrix} - \begin{bmatrix} 1 \\ 1 \end{bmatrix}$
$+ \begin{bmatrix} \sin t \\ \sin t + \cos t \end{bmatrix} \ln|\csc t - \cot t|$

Exercises 10.1, Page 679

1. $x' = y$
   $y' = -9\sin x$; critical points at $(\pm n\pi, 0)$

3. $x' = y$
   $y' = x^2 + y(x^3 - 1)$; critical point at $(0, 0)$

5. $x' = y$
   $y' = \varepsilon x^3 - x$; critical points at $(0, 0)$, $\left(\frac{1}{\sqrt{\varepsilon}}, 0\right), \left(-\frac{1}{\sqrt{\varepsilon}}, 0\right)$

7. $(0, 0)$ and $(-1, -1)$ 9. $(0, 0)$ and $(\frac{4}{3}, \frac{4}{3})$

11. $(0, 0), (10, 0), (0, 16)$, and $(4, 12)$

13. $(0, y)$, y arbitrary

15. $(0, 0), (0, 1), (0, -1), (1, 0), (-1, 0)$

17. (a) $x = c_1 e^{5t} - c_2 e^{-t}$ (b) $x = -2e^{-t}$
       $y = 2c_1 e^{5t} + c_2 e^{-t}$ $y = 2e^{-t}$
    (c) [graph showing point $(-2, 2)$ with trajectory toward origin]

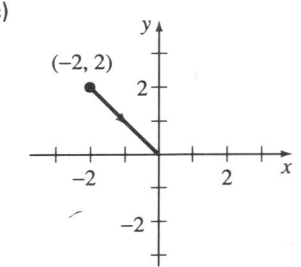

19. (a) $x = c_1(4\cos 3t - 3\sin 3t) + c_2(4\sin 3t + 3\cos 3t)$
       $y = c_1(5\cos 3t) + c_2(5\sin 3t)$
    (b) $x = 4\cos 3t - 3\sin 3t$
       $y = 5\cos 3t$
    (c) [elliptical trajectory through $(4, 5)$]

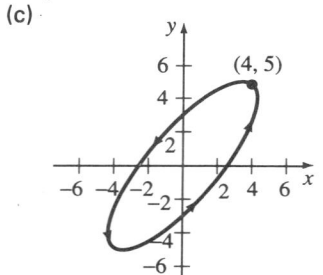

21. (a) $x = c_1(\sin t - \cos t)e^{4t} + c_2(-\sin t - \cos t)e^{4t}$
       $y = 2c_1(\cos t)e^{4t} + 2c_2(\sin t)e^{4t}$
    (b) $x = (\sin t - \cos t)e^{4t}$
       $y = 2(\cos t)e^{4t}$
    (c) [graph from $(-1, 2)$]

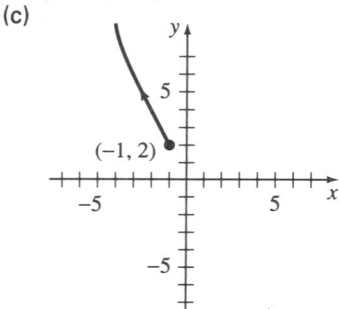

23. $r = \dfrac{1}{\sqrt[4]{4t + c_1}}, \theta = t + c_2; r = 4\dfrac{1}{\sqrt[4]{1024t + 1}}$,
    $\theta = t$; the solution spirals toward the origin as t increases.

25. $r = \dfrac{1}{\sqrt{1 + c_1 e^{-2t}}}, \theta = t + c_2; r = 1, \theta = t$
    (or $x = \cos t$ and $y = \sin t$) is the solution which
    satisfies $\mathbf{X}(0) = (1, 0); r = \dfrac{1}{\sqrt{1 - \frac{3}{4}e^{-2t}}}, \theta = t$ is
    the solution which satisfies $\mathbf{X}(0) = (2, 0)$. This
    solution spirals toward the circle $r = 1$ as t increases.

27. If $\mathbf{X}(t) = (x(t), y(t))$ is a solution,
    $\dfrac{d}{dt} f(x(t), y(t)) = \dfrac{\partial f}{\partial x} x'(t) + \dfrac{\partial f}{\partial y} y'(t) = QP - PQ = 0$.
    Therefore, $f(x(t), y(t)) = c$ for some constant c.

Exercises 10.2, Page 687

1. If $\mathbf{X}(0) = \mathbf{X}_0$ lies on the line $y = 2x$, then $\mathbf{X}(t)$
   approaches $(0, 0)$ along this line. For all other initial
   conditions, $\mathbf{X}(t)$ approaches $(0, 0)$ from the direction
   determined by the line $y = -x/2$.

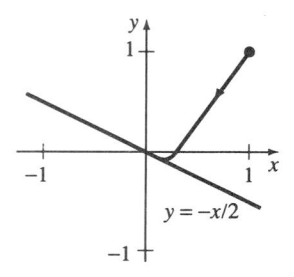

   [graph showing lines $y = 2x$ and $y = -x/2$]

3. All solutions are unstable spirals which become
   unbounded as t increases.

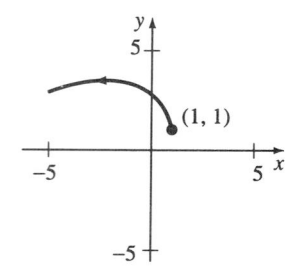

   [graph showing spiral from $(1, 1)$]

5. All solutions approach $(0, 0)$ from the direction
   specified by the line $y = x$.

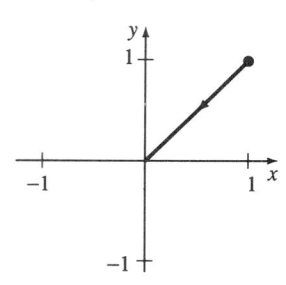

   [graph showing line $y = x$]

A-58

7. If $\mathbf{X}(0) = \mathbf{X}_0$ lies on the line $y = 3x$, then $\mathbf{X}(t)$ approaches $(0, 0)$ along this line. For all other initial conditions, $\mathbf{X}(t)$ becomes unbounded and $y = x$ serves as the asymptote.

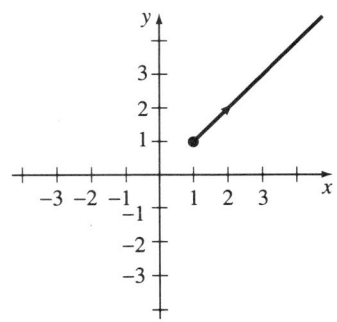

9. Saddle point 11. Saddle point
13. Degenerate stable node 15. Stable spiral
17. $|\mu| < 1$
19. $\mu < -1$ for a saddle point; $-1 < \mu < 3$ for an unstable spiral point
21. $\mathbf{X}(t) = \mathbf{X}_C(t) - \mathbf{A}^{-1}\mathbf{F}$ and $\mathbf{X}_C(t)$ approaches $(0, 0)$ as t increases.

Exercises 10.3, Page 696

1. $r = r_0 e^{\alpha t}$
3. $x = 0$ is unstable; $x = n + 1$ is asymptotically stable.
5. $T = T_0$ is unstable.
7. $x = \alpha$ is unstable; $x = \beta$ is asymptotically stable.
9. $P = a/b$ is unstable; $P = c$ is asymptotically stable
11. $(\tfrac{1}{2}, 1)$ stable spiral point
13. $(\sqrt{2}, 0)$ and $(-\sqrt{2}, 0)$ saddle points; $(\tfrac{1}{2}, -\tfrac{7}{4})$ stable spiral point
15. $(1, 1)$ stable node, $(1, -1)$ saddle, $(2, 2)$ saddle, $(2, -2)$ unstable spiral point
17. $(0, -1)$ saddle; $(0, 0)$ unclassified; $(0, 1)$ stable but unable to classify further
19. $(0, 0)$ unstable node, $(10, 0)$ saddle point, $(0, 16)$ saddle point, $(4, 12)$ stable node
21. $\theta = 0$ is a saddle point. It is not possible to classify either $\theta = \pi/3$ or $\theta = -\pi/3$.
23. It is not possible to classify $x = 0$.
25. It is not possible to classify $x = 0$, but $x = 1/\sqrt{\varepsilon}$ and $x = -1/\sqrt{\varepsilon}$ are saddle points.
27. $\tau = 0$ and $\Delta = \beta$. Therefore, $(0, 0)$ is a saddle point if $\beta < 0$.
29. $(0, 0)$ is a stable spiral point; for $\mathbf{X}_1$, $\Delta = -6.07298$

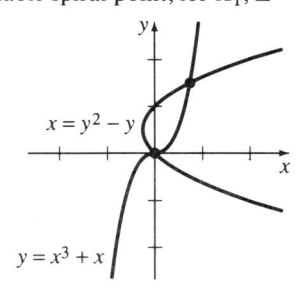

31. The solution satisfying $\mathbf{X}(0) = (x_0, 0)$ with $x_0 > 0$ is $y^2 = x_0^4 - x^4$, and if $-x_0 < x < x_0$, there are two corresponding values of y.
33. (a) At both $(1, 0)$ and $(-1, 0)$, $\tau = 0$ and $\Delta = 4$ and so no conclusions can be drawn.
    (b) The family of circles $(x - c)^2 + y^2 = c^2 - 1$ is shown in the figure.

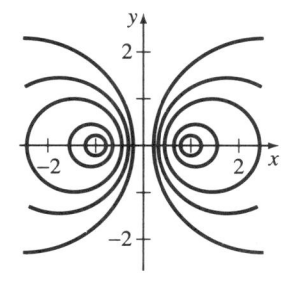

35. $|v_0| < \tfrac{1}{2}\sqrt{2}$
37. If $\beta > 0$, $(0, 0)$ is the only critical point and is stable. If $\beta < 0$, $(0, 0)$, $(\hat{x}, 0)$, and $(-\hat{x}, 0)$, where $\hat{x}^2 = -\alpha/\beta$, are critical points. $(0, 0)$ is stable, whereas $(\hat{x}, 0)$ and $(-\hat{x}, 0)$ are each saddle points.

Exercises 10.4, Page 705

1. $|\omega_0| < \sqrt{\dfrac{3g}{l}}$
5. (a) First show that $y^2 = v_0^2 + g \ln\left(\dfrac{1 + x^2}{1 + x_0^2}\right)$.
7. If $x_m < x_1 < x_M$, then $F(x_1) > F(x_m) = F(x_M)$. Letting $x = x_1$, we get
$$G(y) = \frac{c_0}{F(x_1)} = \frac{F(x_m)G(a/b)}{F(x_1)} < G\left(\frac{a}{b}\right)$$
and from Figure 10.28(b) there are two solutions y_1 and y_2 to $G(y) = c_0/F(x_1)$ that satisfy $y_1 < a/b < y_2$.
9. (a) The new critical point is $(d/c - \varepsilon_2/c, a/b + \varepsilon_1/b)$. Note that the new differential equation is also a Lotka–Volterra predator-prey differential equation, and so the new critical point is a center.
   (b) Yes
11. If we let $c = 1 - (K_1/K_2)\alpha_{21}$, show that $\tau = -r_1 + r_2 c$, $\Delta = -r_1 r_2 c$, and $\tau^2 - 4\Delta = (cr_2 + r_1)^2$.
13. Show that $\alpha_{12}\alpha_{21} > 1$ and conclude that $\Delta < 0$.
15. (a) $(0, 0)$ is the only critical point.
    (b) $\tau = 0$, $\Delta = k/m$, and so $(0, 0)$ is a center, a stable spiral point, or an unstable spiral point. Since $(0, 0)$ must be asymptotically stable (from physical considerations), $(0, 0)$ must be a stable spiral point.
17. If $(0, 0)$ were a stable spiral point, there would exist an x with more than two corresponding

values of y. This is impossible, since $y^2 + 2F(x) = c$ has at most two solutions for a given value of x.

19. The critical points are $(0, \beta)$ and $(\hat{x}, \hat{y})$, where $\hat{y} = 1/(\alpha - 1)$ and $\hat{x} = \alpha(\beta - \hat{y})$. The critical point $(\hat{x}, \hat{y})$ will lie in the first quadrant provided $\alpha > 1$ and $\beta - \hat{y} > 0$; that is, $\beta > 1/(\alpha - 1)$. Show that $\Delta = ((\alpha - 1)^2/\alpha^2)\hat{x}$, $\tau = -(\Delta + 1)$, and $\tau^2 - 4\Delta = (\Delta - 1)^2$ to conclude that $(\hat{x}, \hat{y})$ is a stable node.

Exercises 10.5, Page 717

1. The system has no critical points.

3. $\dfrac{\partial P}{\partial x} + \dfrac{\partial Q}{\partial y} = -2 < 0$

5. $\dfrac{\partial P}{\partial x} + \dfrac{\partial Q}{\partial y} = -\mu + 9y^2 > 0$ if $\mu < 0$.

7. The single critical point $(0, 0)$ is a saddle point.

9. $\delta(x, y) = e^{-y/2}$

11. $\dfrac{\partial P}{\partial x} + \dfrac{\partial Q}{\partial y} = 4(1 - x^2 - 3y^2) > 0$ for $x^2 + 3y^2 < 1$

13. Use $\delta(x, y) = 1/(xy)$ and show that
$$\dfrac{\partial(\delta P)}{\partial x} + \dfrac{\partial(\delta Q)}{\partial y} = -\dfrac{r}{Kx}.$$

15. If $\mathbf{n} = (-2x, -2y)$, show that $\mathbf{V} \cdot \mathbf{n} = 2(x - y)^2 + 2y^4$.

17. Yes; the sole critical point $(0, 0)$ lies outside the invariant region $\tfrac{1}{16} \leq x^2 + y^2 \leq 1$, and so Theorem 10.8(ii) applies.

19. $\mathbf{V} \cdot \mathbf{n} = 2y^2(1 - x^2) \geq 2y^2(1 - r^2)$ and $\partial P/\partial x + \partial Q/\partial y = x^2 - 1 < 0$. The sole critical point is $(0, 0)$ and this critical point is a stable spiral point. Therefore, Theorem 10.9(ii) applies.

21. (a) $\dfrac{\partial P}{\partial x} + \dfrac{\partial Q}{\partial y} = 2xy - 1 - x^2 \leq 2x - 1 - x^2$
$= -(x - 1)^2 \leq 0$
(b) $\lim_{t \to \infty} \mathbf{X}(t) = (\tfrac{3}{2}, \tfrac{2}{9})$, a stable spiral point

Chapter 10 Review Exercises, Page 720

1. True **3.** A center or a saddle point
5. False **7.** False **9.** True

11. $r = \dfrac{1}{\sqrt[3]{3t + 1}}$, $\theta = t$, the solution curve spirals toward the origin.

13. Center; degenerate stable node

15. Stable node for $\mu < -2$; stable spiral point for $-2 < \mu < 0$; unstable spiral point for $0 < \mu < 2$; unstable node for $\mu > 2$

17. Show that $y^2 = (1 + x_0^2 - x^2)^2 - 1$.

19. $\dfrac{\partial P}{\partial x} + \dfrac{\partial Q}{\partial y} = 1$

21. (a) *Hint:* Use the Bendixson negative criterion.
(d) In (b), $(0, 0)$ is a stable spiral point when $\beta < 2ml\sqrt{g/l - \omega^2}$. In (c), $(\hat{x}, 0)$ and $(-\hat{x}, 0)$ are stable spiral points when
$\beta < 2ml\sqrt{\omega^2 - g^2/(\omega^2l^2)}$.

Exercises 11.1, Page 731

1. $\displaystyle\int_{-2}^{2} x \cdot x^2 \, dx = \dfrac{x^4}{4}\bigg|_{-2}^{2} = 4 - 4 = 0$

3. $\displaystyle\int_{0}^{2} e^x(xe^{-x} - e^{-x}) \, dx = \int_{0}^{2} (x - 1) \, dx$
$= \dfrac{x^2}{2} - x \bigg|_{0}^{2}$
$= 0$

5. $\displaystyle\int_{-\pi/2}^{\pi/2} x \cos 2x \, dx = [\tfrac{1}{2}x \sin 2x + \tfrac{1}{4}\cos 2x]_{-\pi/2}^{\pi/2}$
$= -\tfrac{1}{4} + \tfrac{1}{4} = 0$

7. $\displaystyle\int_{0}^{\pi/2} \sin(2m + 1)x \sin(2n + 1)x \, dx$
$= \dfrac{1}{2}\int_{0}^{\pi/2} [\cos 2(m - n)x - \cos 2(m + n + 1)x] \, dx$
$= \dfrac{1}{4}\left[\dfrac{\sin 2(m - n)x}{m - n} - \dfrac{\sin 2(m + n + 1)x}{m + n + 1}\right]_{0}^{\pi/2}$
$= 0, m \neq n; \sqrt{\pi/2}$

9. $\displaystyle\int_{0}^{\pi} \sin mx \sin nx \, dx$
$= \dfrac{1}{2}[\cos(m - n)x - \cos(m + n)x] \, dx$
$= \dfrac{1}{2}\left[\dfrac{\sin(m - n)x}{m - n} - \dfrac{\sin(m + n)x}{m + n}\right]_{0}^{\pi}$
$= 0, m \neq n; \sqrt{\pi/2}$

11. $\displaystyle\int_{0}^{p} \cos \dfrac{n\pi}{p} x \, dx = \dfrac{p}{n\pi} \sin \dfrac{n\pi}{p} x \bigg|_{0}^{p} = 0, n \neq 0;$
$\displaystyle\int_{0}^{p} \cos \dfrac{m\pi}{p} x \cos \dfrac{n\pi}{p} x \, dx$
$= \dfrac{1}{2}\int_{0}^{p} \left[\cos \dfrac{(m - n)\pi}{p} x + \cos \dfrac{(m + n)\pi}{p} x\right] dx$
$= \dfrac{p}{2\pi}\left[\dfrac{\sin \dfrac{(m - n)}{p} x}{m - n} + \dfrac{\sin \dfrac{(m + n)\pi}{p} x}{m + n}\right]_{0}^{p}$
$= 0, m \neq n; \|1\| = \sqrt{p}, \left\|\cos \dfrac{n\pi}{p} x\right\| = \sqrt{p/2}$

13. For example,
$\displaystyle\int_{-\infty}^{\infty} e^{-x^2} H_0(x) H_1(x) \, dx = \int_{-\infty}^{\infty} e^{-x^2}(2x) \, dx$
$= -\int_{-\infty}^{0} e^{-x^2}(-2x \, dx) - \int_{0}^{\infty} e^{-x^2}(-2x \, dx)$
$= -e^{-x^2}\bigg|_{-\infty}^{0} - e^{-x^2}\bigg|_{0}^{\infty} = -1 - (-1) = 0$

The results $\displaystyle\int_{-\infty}^{\infty} e^{-x^2} H_0(x) H_2(x) \, dx = 0$

and $\int_{-\infty}^{\infty} e^{-x^2} H_1(x) H_2(x) \, dx = 0$

follow from integration by parts.

15. $\int_a^b \phi_n(x) \, dx = \int_a^b \phi_0(x) \phi_n(x) \, dx$
$= 0 \quad \text{for } n = 1, 2, 3, \ldots$

17. $\|\phi_m(x) + \phi_n(x)\|^2$
$= \int_a^b [\phi_m(x) + \phi_n(x)]^2 \, dx$
$= \int_a^b \phi_m^2(x) \, dx + 2 \underbrace{\int_a^b \phi_m(x) \phi_n(x) \, dx}_{\text{zero by orthogonality}}$
$\quad + \int_a^b \phi_n^2(x) \, dx$
$= \int_a^b \phi_m^2(x) \, dx + \int_a^b \phi_n^2(x) \, dx$
$= \|\phi_m(x)\|^2 + \|\phi_n(x)\|^2$

19. The nontrivial function $f(x) = x^2$ is orthogonal to each function in the orthogonal set.

Exercises 11.2, Page 736

1. $f(x) = \frac{1}{2} + \frac{1}{\pi} \sum_{n=1}^{\infty} \frac{1 - (-1)^n}{n} \sin nx$

3. $f(x) = \frac{3}{4} + \sum_{n=1}^{\infty} \left\{ \frac{(-1)^n - 1}{n^2 \pi^2} \cos n\pi x - \frac{1}{n\pi} \sin n\pi x \right\}$

5. $f(x) = \frac{\pi^2}{6} + \sum_{n=1}^{\infty} \left\{ \frac{2(-1)^n}{n^2} \cos nx \right.$
$\left. + \left(\frac{(-1)^{n+1} \pi}{n} + \frac{2}{\pi n^3} [(-1)^n - 1] \right) \sin nx \right\}$

7. $f(x) = \pi + 2 \sum_{n=1}^{\infty} \frac{(-1)^{n+1}}{n} \sin nx$

9. $f(x) = \frac{1}{\pi} + \frac{1}{2} \sin x + \frac{1}{\pi} \sum_{n=2}^{\infty} \frac{(-1)^n + 1}{1 - n^2} \cos nx$

11. $f(x) = -\frac{1}{4} + \frac{1}{\pi} \sum_{n=1}^{\infty} \left\{ -\frac{1}{n} \sin \frac{n\pi}{2} \cos \frac{n\pi}{2} x \right.$
$\left. + \frac{3}{n} \left(1 - \cos \frac{n\pi}{2} \right) \sin \frac{n\pi}{2} x \right\}$

13. $f(x) = \frac{9}{4} + 5 \sum_{n=1}^{\infty} \left\{ \frac{(-1)^n - 1}{n^2 \pi^2} \cos \frac{n\pi}{5} x \right.$
$\left. + \frac{(-1)^{n+1}}{n\pi} \sin \frac{n\pi}{5} x \right\}$

15. $f(x) = \frac{2 \sinh \pi}{\pi} \left[\frac{1}{2} + \sum_{n=1}^{\infty} \frac{(-1)^n}{1 + n^2} (\cos nx - n \sin nx) \right]$

17. At the endpoint $x = \pi$, the series will converge to
$\frac{f(\pi -) + f(-\pi +)}{2} = \frac{\pi^2}{2}$

Substituting $x = \pi$ into the series gives
$\frac{\pi^2}{6} + 2 \sum_{n=1}^{\infty} \frac{1}{n^2}$

Equating the two results then yields
$\frac{\pi^2}{6} = \sum_{n=1}^{\infty} \frac{1}{n^2} = 1 + \frac{1}{2^2} + \frac{1}{3^2} + \frac{1}{4^2} + \cdots$

Now at $x = 0$ the series converges to
$f(0) = 0 = \frac{\pi^2}{6} + \sum_{n=1}^{\infty} \frac{2(-1)^n}{n^2}$

This implies
$\frac{\pi^2}{12} = \sum_{n=1}^{\infty} \frac{(-1)^{n+1}}{n^2} = 1 - \frac{1}{2^2} + \frac{1}{3^2} - \frac{1}{4^2} + \cdots$

19. Set $x = \pi/2$.

Exercises 11.3, Page 744

1. Odd **3.** Neither even nor odd **5.** Even
7. Odd **9.** Neither even nor odd

11. $f(x) = \frac{2}{\pi} \sum_{n=1}^{\infty} \frac{1 - (-1)^n}{n} \sin nx$

13. $f(x) = \frac{\pi}{2} + \frac{2}{\pi} \sum_{n=1}^{\infty} \frac{(-1)^n - 1}{n^2} \cos nx$

15. $f(x) = \frac{1}{3} + \frac{4}{\pi^2} \sum_{n=1}^{\infty} \frac{(-1)^n}{n^2} \cos n\pi x$

17. $f(x) = \frac{2\pi^2}{3} + 4 \sum_{n=1}^{\infty} \frac{(-1)^{n+1}}{n^2} \cos nx$

19. $f(x) = \frac{2}{\pi} \sum_{n=1}^{\infty} \frac{1 - (-1)^n(1 + \pi)}{n} \sin nx$

21. $f(x) = \frac{3}{4} + \frac{4}{\pi^2} \sum_{n=1}^{\infty} \frac{\cos \frac{n\pi}{2} - 1}{n^2} \cos \frac{n\pi}{2} x$

23. $f(x) = \frac{2}{\pi} + \frac{2}{\pi} \sum_{n=2}^{\infty} \frac{1 + (-1)^n}{1 - n^2} \cos nx$

25. $f(x) = \frac{1}{2} + \frac{2}{\pi} \sum_{n=1}^{\infty} \frac{\sin \frac{n\pi}{2}}{n} \cos n\pi x$

$f(x) = \frac{2}{\pi} \sum_{n=1}^{\infty} \frac{1 - \cos \frac{n\pi}{2}}{n} \sin n\pi x$

27. $f(x) = \frac{2}{\pi} + \frac{4}{\pi} \sum_{n=1}^{\infty} \frac{(-1)^n}{1 - 4n^2} \cos 2nx$

$f(x) = \frac{8}{\pi} \sum_{n=1}^{\infty} \frac{n}{4n^2 - 1} \sin 2nx$

29. $f(x) = \frac{\pi}{4} + \frac{2}{\pi} \sum_{n=1}^{\infty} \frac{2 \cos \frac{n\pi}{2} - (-1)^n - 1}{n^2} \cos nx$

A-61

$$f(x) = \frac{4}{\pi} \sum_{n=1}^{\infty} \frac{\sin \frac{n\pi}{2}}{n^2} \sin nx$$

31. $f(x) = \frac{3}{4} + \frac{4}{\pi^2} \sum_{n=1}^{\infty} \frac{\cos \frac{n\pi}{2} - 1}{n^2} \cos \frac{n\pi}{2} x$

$f(x) = \sum_{n=1}^{\infty} \left\{ \frac{4}{n^2\pi^2} \sin \frac{n\pi}{2} - \frac{2}{n\pi} (-1)^n \right\} \sin \frac{n\pi}{2} x$

33. $f(x) = \frac{5}{6} + \frac{2}{\pi^2} \sum_{n=1}^{\infty} \frac{3(-1)^n - 1}{n^2} \cos n\pi x$

$f(x) = 4 \sum_{n=1}^{\infty} \left\{ \frac{(-1)^{n+1}}{n\pi} + \frac{(-1)^n - 1}{n^3\pi^3} \right\} \sin n\pi x$

35. $f(x) = \frac{4\pi^2}{3} + 4 \sum_{n=1}^{\infty} \left\{ \frac{1}{n^2} \cos nx - \frac{\pi}{n} \sin nx \right\}$

37. $f(x) = \frac{3}{2} - \frac{1}{\pi} \sum_{n=1}^{\infty} \frac{1}{n} \sin 2n\pi x$

39. $x_p(t) = \frac{10}{\pi} \sum_{n=1}^{\infty} \frac{1 - (-1)^n}{n(10 - n^2)} \sin nt$

41. $x_p(t) = \frac{\pi^2}{18} + 16 \sum_{n=1}^{\infty} \frac{1}{n^2(n^2 - 48)} \cos nt$

43. $y(x) = \frac{2w_0 L^4}{EI\pi^5} \sum_{n=1}^{\infty} \frac{(-1)^{n+1}}{n^5} \sin \frac{n\pi}{L} x$

45. Let f and g be even functions. Define
$F(x) = f(x)g(x)$. Then
$$F(-x) = f(-x)g(-x) = f(x)g(x) = F(x)$$

47. Let f and g be even functions. Define
$F(x) = f(x) + g(x)$. Then
$$F(-x) = f(-x) + g(-x)$$
$$= f(x) + g(x) = F(x)$$

49. Let f be an odd function. Then
$$\int_{-a}^{a} f(x) \, dx = \int_{-a}^{0} f(x) \, dx + \int_{0}^{a} f(x) \, dx = I_1 + I_2$$

In I_1 let $-x = t$ and $-dx = dt$ so that
$$I_1 = \int_{a}^{0} f(-t)(-dt) = \int_{0}^{a} f(-t) \, dt$$
$$= -\int_{0}^{a} f(t) \, dt = -I_2$$

Therefore $I_1 + I_2 = 0$.

51. Adding the results of Problems 13 and 14 and dividing by 2 give

$$\frac{\pi}{4} + \sum_{n=1}^{\infty} \left\{ \frac{(-1)^n - 1}{\pi n^2} \cos nx + \frac{(-1)^{n+1}}{n} \sin nx \right\}$$

53. $f(x, y) = \frac{1}{4} + \frac{1}{\pi^2} \sum_{m=1}^{\infty} \frac{(-1)^m - 1}{m^2} \cos m\pi x$

$+ \frac{1}{\pi^2} \sum_{n=1}^{\infty} \frac{(-1)^n - 1}{n^2} \cos n\pi y$

$+ \frac{4}{\pi^4} \sum_{m=1}^{\infty} \sum_{n=1}^{\infty} \frac{[(-1)^m - 1][(-1)^n - 1]}{m^2 n^2}$

$\times \cos m\pi x \cos n\pi y$

Exercises 11.4, Page 754

1. $\lambda = n^2, n = 1, 2, 3, \ldots; y = \sin nx$
3. $\lambda = (2n - 1)^2 \pi^2 / 4L^2, n = 1, 2, 3, \ldots;$
$y = \cos((2n - 1)\pi x / 2L)$
5. $\lambda = n^2, n = 0, 1, 2, \ldots; y = \cos nx$
7. $\lambda_n = x_n^2$, where the x_n, $n = 1, 2, 3, \ldots,$ are the consecutive positive roots of $\cot \sqrt{\lambda} = \sqrt{\lambda}$;
$y = \cos \sqrt{\lambda_n} x$
9. $\lambda = n^2\pi^2/25, n = 1, 2, 3, \ldots; y = e^{-x} \sin(n\pi x/5)$
11. $\lambda = n\pi/L, n = 1, 2, 3, \ldots; y = \sin(n\pi x/L)$
15. $\lambda = n^2, n = 1, 2, 3, \ldots; y = \sin(n \ln x);$

$\frac{d}{dx}[xy'] + \lambda \frac{1}{x} y = 0;$

$\int_{1}^{e^\pi} \frac{1}{x} \sin(m \ln x) \sin(n \ln x) \, dx = 0, m \neq n$

17. $\frac{1}{2}[1 + \sin^2 \sqrt{\lambda_n}]$

19. $\frac{d}{dx}[xe^{-x} y'] + ne^{-x} y = 0;$

$\int_{0}^{\infty} e^{-x} L_m(x) L_n(x) \, dx = 0, m \neq n$

21. $\lambda_1 = 4.1159, \lambda_2 = 24.1393,$
$\lambda_3 = 63.6592, \lambda_4 = 122.8883$

23. Since $r(a) = r(b)$ and

$y_n(a) = y_n(b) \quad y_m(a) = y_m(b)$
$y'_n(a) = y'_n(b) \quad y'_m(a) = y'_m(b)$

the right-hand member of (8) becomes

$r(a)[y_m(a)y'_n(a) - y_n(a)y'_m(a)$
$\qquad - y_m(a)y'_n(a) + y_n(a)y'_m(a)] = 0$

Exercises 11.5, Page 761

1. 1.3, 2.3, 3.4, 4.3

3. $f(x) = \sum_{i=1}^{\infty} \frac{1}{\lambda_i J_1(2\lambda_i)} J_0(\lambda_i x)$

5. $f(x) = 4 \sum_{i=1}^{\infty} \frac{\lambda_i J_1(2\lambda_i)}{(4\lambda_i^2 + 1) J_0^2(2\lambda_i)} J_0(\lambda_i x)$

7. $f(x) = 20 \sum_{i=1}^{\infty} \frac{\lambda_i J_2(4\lambda_i)}{(2\lambda_i^2 + 1) J_1^2(4\lambda_i)} J_1(\lambda_i x)$

A-62

9. $f(x) = \dfrac{9}{2} - 4 \sum_{i=1}^{\infty} \dfrac{J_2(3\lambda_i)}{\lambda_i^2 J_0^2(3\lambda_i)} J_0(\lambda_i x)$

11. $f(x) = \tfrac{1}{3} P_0(x) + \tfrac{2}{3} P_2(x)$

13. $f(x) = \tfrac{1}{4} P_0(x) + \tfrac{1}{2} P_1(x) + \tfrac{5}{16} P_2(x) - \tfrac{3}{32} P_4(x) + \cdots$

15. Use $\cos 2\theta = 2\cos^2\theta - 1$.

19. $f(x) = \tfrac{1}{2} P_0(x) + \tfrac{5}{8} P_2(x) - \tfrac{3}{16} P_4(x) + \cdots$; $f(x) = |x|$ on $(-1, 1)$

Chapter 11 Review Exercises, Page 763

1. True   **3.** Cosine   **5.** $\tfrac{3}{2}$ **7.** False

9. $\dfrac{1}{\sqrt{1-x^2}}$; $-1 \le x \le 1$

11. $\displaystyle\int_0^L \sin\dfrac{(2m+1)\pi}{2L} x \sin\dfrac{(2n+1)\pi}{2L} x \, dx$

$= \dfrac{1}{2} \int_0^L \left[\cos\dfrac{(m-n)\pi}{L} x - \cos\dfrac{(m+n+1)\pi}{L} x \right] dx$

$= \dfrac{L}{2\pi} \left[\dfrac{\sin\dfrac{(m-n)\pi}{L} x}{m-n} - \dfrac{\sin\dfrac{(m+n+1)\pi}{L} x}{m+n+1} \right]_0^L$

$= 0, m \ne n$

13. $f(x) = \dfrac{1}{2} + \dfrac{2}{\pi} \sum_{n=1}^{\infty} \left\{ \dfrac{1}{n^2\pi}[(-1)^n - 1]\cos n\pi x + \dfrac{2}{n}(-1)^n \sin n\pi x \right\}$

15. $f(x) = 1 - e^{-1} + 2 \sum_{n=1}^{\infty} \dfrac{1-(-1)^n e^{-1}}{1+n^2\pi^2} \cos n\pi x$

17. $\lambda = (2n-1)^2 \pi^2/36$, $n = 1, 2, 3, \ldots$; $y = \cos\left(\dfrac{2n-1}{2} \pi \ln x\right)$

19. $f(x) = \dfrac{1}{4} \sum_{i=1}^{\infty} \dfrac{J_1(2\lambda_i)}{\lambda_i J_1^2(4\lambda_i)} J_0(\lambda_i x)$

Exercises 12.1, Page 770

1. The possible cases can be summarized in one form $u = c_1 e^{c_2(x+y)}$, where c_1 and c_2 are constants.

3. $u = c_1 e^{y + c_2(x-y)}$ **5.** $u = c_1(xy)^{c_2}$

7. Not separable

9. $u = e^{-t}(A_1 e^{k\lambda^2 t} \cosh \lambda x + B_1 e^{k\lambda^2 t} \sinh \lambda x)$
$u = e^{-t}(A_2 e^{-k\lambda^2 t} \cos \lambda x + B_2 e^{-k\lambda^2 t} \sin \lambda x)$
$u = (c_7 x + c_8) c_9 e^{-t}$

11. $u = (c_1 \cosh \lambda x + c_2 \sinh \lambda x)(c_3 \cosh \lambda at + c_4 \sinh \lambda at)$
$u = (c_5 \cos \lambda x + c_6 \sin \lambda x)(c_7 \cos \lambda at + c_8 \sin \lambda at)$
$u = (c_9 x + c_{10})(c_{11} t + c_{12})$

13. $u = (c_1 \cosh \lambda x + c_2 \sinh \lambda x)(c_3 \cos \lambda y + c_4 \sin \lambda y)$
$u = (c_5 \cos \lambda x + c_6 \sin \lambda x)(c_7 \cosh \lambda y + c_8 \sinh \lambda y)$
$u = (c_9 x + c_{10})(c_{11} y + c_{12})$

15. For $\lambda^2 > 0$ there are three possibilities:

$u = (c_1 \cosh \lambda x + c_2 \sinh \lambda x)(c_3 \cosh \sqrt{1-\lambda^2}\, y + c_4 \sinh \sqrt{1-\lambda^2}\, y), \lambda^2 < 1$

$u = (c_1 \cosh \lambda x + c_2 \sinh \lambda x)(c_3 \cos \sqrt{\lambda^2-1}\, y + c_4 \sin \sqrt{\lambda^2-1}\, y), \lambda^2 > 1$

$u = (c_1 \cosh x + c_2 \sinh x) \times (c_3 y + c_4), \lambda^2 = 1$

The results for the case $-\lambda^2 < 0$ are similar. For $\lambda^2 = 0$ we have

$u = (c_1 x + c_2)(c_3 \cosh y + c_4 \sinh y)$

17. Elliptic   **19.** Parabolic   **21.** Hyperbolic

23. Parabolic   **25.** Hyperbolic

27. Using $-\lambda^2$ as a separation constant, we obtain

$T' + k\lambda^2 T = 0$
$rR'' + R' + \lambda^2 r R = 0$

This last equation can be written as

$r^2 R'' + rR' + \lambda^2 r^2 R = 0$

which we recognize as Bessel's equation with $v = 0$. The solution of the respective equations are as indicated in the problem.

29. $u = e^{n(-3x+y)}$, $u = e^{n(2x+y)}$

31. The equation $x^2 + 4y^2 = 4$ defines an ellipse. The partial differential equation is hyperbolic outside the ellipse, parabolic on the ellipse, and elliptic inside the ellipse.

Exercises 12.2, Page 777

1. $k \dfrac{\partial^2 u}{\partial x^2} = \dfrac{\partial u}{\partial t}, 0 < x < L, t > 0$

$u(0, t) = 0, \left.\dfrac{\partial u}{\partial x}\right|_{x=L} = 0, t > 0$

$u(x, 0) = f(x), 0 < x < L$

3. $k \dfrac{\partial^2 u}{\partial x^2} = \dfrac{\partial u}{\partial t}, 0 < x < L, t > 0$

$u(0, t) = 100, \left.\dfrac{\partial u}{\partial x}\right|_{x=L} = -hu(L, t), t > 0$

$u(x, 0) = f(x), 0 < x < L$

5. $a^2 \dfrac{\partial^2 u}{\partial x^2} = \dfrac{\partial^2 u}{\partial t^2}, 0 < x < L, t > 0$

$u(0, t) = 0, u(L, t) = 0, t > 0$

$u(x, 0) = x(L - x)$, $\dfrac{\partial u}{\partial t}\bigg|_{t=0} = 0$, $0 < x < L$

7. $a^2 \dfrac{\partial^2 u}{\partial x^2} - 2\beta \dfrac{\partial u}{\partial t} = \dfrac{\partial^2 u}{\partial t^2}$, $0 < x < L$, $t > 0$

$u(0, t) = 0$, $u(L, t) = \sin \pi t$, $t > 0$

$u(x, 0) = f(x)$, $\dfrac{\partial u}{\partial t}\bigg|_{t=0} = 0$, $0 < x < L$

9. $\dfrac{\partial^2 u}{\partial x^2} + \dfrac{\partial^2 u}{\partial y^2} = 0$, $0 < x < 4$, $0 < y < 2$

$\dfrac{\partial u}{\partial x}\bigg|_{x=0} = 0$, $u(4, y) = f(y)$, $0 < y < 2$

$\dfrac{\partial u}{\partial y}\bigg|_{y=0} = 0$, $u(x, 2) = 0$, $0 < x < 4$

Exercises 12.3, Page 779

1. $u(x, t) = \dfrac{2}{\pi} \sum_{n=1}^{\infty} \left(\dfrac{-\cos \dfrac{n\pi}{2} + 1}{n} \right) e^{-k(n^2\pi^2/L^2)t} \sin \dfrac{n\pi}{L} x$

3. $u(x, t) = \dfrac{1}{L} \int_0^L f(x)\, dx$

$+ \dfrac{2}{L} \sum_{n=1}^{\infty} \left(\int_0^L f(x) \cos \dfrac{n\pi}{L} x\, dx \right)$

$\times e^{-k(n^2\pi^2/L^2)t} \cos \dfrac{n\pi}{L} x$

5. $u(x, t) = e^{-ht} \left[\dfrac{1}{L} \int_0^L f(x)\, dx \right.$

$+ \dfrac{2}{L} \sum_{n=1}^{\infty} \left(\int_0^L f(x) \cos \dfrac{n\pi}{L} x\, dx \right)$

$\left. \times e^{-k(n^2\pi^2/L^2)t} \cos \dfrac{n\pi}{L} x \right]$

Exercises 12.4, Page 782

1. $u(x, t) = \dfrac{L^2}{\pi^3} \sum_{n=1}^{\infty} \dfrac{1 - (-1)^n}{n^3} \cos \dfrac{n\pi a}{L} t \sin \dfrac{n\pi}{L} x$

3. $u(x, t) = \dfrac{6\sqrt{3}}{\pi^2} \left(\cos \dfrac{\pi a}{L} t \sin \dfrac{\pi}{L} x \right.$

$- \dfrac{1}{5^2} \cos \dfrac{5\pi a}{L} t \sin \dfrac{5\pi}{L} x$

$\left. + \dfrac{1}{7^2} \cos \dfrac{7\pi a}{L} t \sin \dfrac{7\pi}{L} x - \cdots \right)$

5. $u(x, t) = \dfrac{1}{a} \sin at \sin x$

7. $u(x, t) = \dfrac{8h}{\pi^2} \sum_{n=1}^{\infty} \dfrac{\sin \dfrac{n\pi}{2}}{n^2} \cos \dfrac{n\pi a}{L} t \sin \dfrac{n\pi}{L} x$

9. $u(x, t) = e^{-\beta t} \sum_{n=1}^{\infty} A_n \{\cos q_n t + \dfrac{\beta}{q_n} \sin q_n t\} \sin nx$

where $A_n = \dfrac{2}{\pi} \int_0^\pi f(x) \sin nx\, dx$ and $q_n = \sqrt{n^2 - \beta^2}$

11. $u(x, t) = \sum_{n=1}^{\infty} \left(A_n \cos \dfrac{n^2\pi^2}{L^2} at + B_n \sin \dfrac{n^2\pi^2}{L^2} at \right)$

$\times \sin \dfrac{n\pi}{L} x$

where $A_n = \dfrac{2}{L} \int_0^L f(x) \sin \dfrac{n\pi}{L} x\, dx$

$B_n = \dfrac{2L}{n^2\pi^2 a} \int_0^L g(x) \sin \dfrac{n\pi}{L} x\, dx$

15. $u(x, t) = \sin x \cos 2at + t$

17. $u(x, t) = \dfrac{1}{2a} \sin 2x \sin 2at$

Exercises 12.5, Page 788

1. $u(x, y) = \dfrac{2}{a} \sum_{n=1}^{\infty} \left(\dfrac{1}{\sinh \dfrac{n\pi}{a} b} \int_0^a f(x) \sin \dfrac{n\pi}{a} x\, dx \right)$

$\times \sinh \dfrac{n\pi}{a} y \sin \dfrac{n\pi}{a} x$

3. $u(x, y) = \dfrac{2}{a} \sum_{n=1}^{\infty} \left(\dfrac{1}{\sinh \dfrac{n\pi}{a} b} \int_0^a f(x) \sin \dfrac{n\pi}{a} x\, dx \right)$

$\times \sinh \dfrac{n\pi}{a}(b - y) \sin \dfrac{n\pi}{a} x$

5. $u(x, y) = \dfrac{1}{2} x + \dfrac{2}{\pi^2} \sum_{n=1}^{\infty} \dfrac{1 - (-1)^n}{n^2 \sinh n\pi} \sinh n\pi x \cos n\pi y$

7. $u(x, y) = \dfrac{2}{\pi} \sum_{n=1}^{\infty} \dfrac{[1 - (-1)^n]}{n}$

$\times \dfrac{n \cosh nx + \sinh nx}{n \cosh n\pi + \sinh n\pi} \sin ny$

9. $u(x, y) = \dfrac{2}{\pi} \sum_{n=1}^{\infty} \left(\int_0^\pi f(x) \sin nx\, dx \right) e^{-ny} \sin nx$

11. $u(x, y) = \sum_{n=1}^{\infty} \left(A_n \cosh \dfrac{n\pi}{a} y + B_n \sinh \dfrac{n\pi}{a} y \right) \sin \dfrac{n\pi}{a} x$

where $A_n = \dfrac{2}{a} \int_0^a f(x) \sin \dfrac{n\pi}{a} x\, dx$ and

$B_n = \dfrac{1}{\sinh \dfrac{n\pi}{a} b} \left(\dfrac{2}{a} \int_0^a g(x) \sin \dfrac{n\pi}{a} x\, dx - A_n \cosh \dfrac{n\pi}{a} b \right)$

13. $u = u_1 + u_2$ where

$u_1(x, y) = \dfrac{2}{\pi} \sum_{n=1}^{\infty} \dfrac{1 - (-1)^n}{n \sinh n\pi} \sinh ny \sin nx$

$u_2(x, y) = \dfrac{2}{\pi} \sum_{n=1}^{\infty} \dfrac{[1 - (-1)^n]}{n}$

$\times \dfrac{\sinh nx + \sinh n(\pi - x)}{\sinh n\pi} \sin ny$

Exercises 12.6, Page 790

1. $u(x, t) = 100 + \dfrac{200}{\pi} \sum_{n=1}^{\infty} \dfrac{(-1)^n - 1}{n} e^{-kn^2\pi^2 t} \sin n\pi x$

3. $u(x, t) = u_0 - \dfrac{r}{2k} x(x - 1) + 2 \sum_{n=1}^{\infty} \left[\dfrac{u_0}{n\pi} + \dfrac{r}{kn^3\pi^3} \right]$
   $\times [(-1)^n - 1] e^{-kn^2\pi^2 t} \sin n\pi x$

5. $u(x, t) = \psi(x) + \sum_{n=1}^{\infty} A_n e^{-kn^2\pi^2 t} \sin n\pi x$, where

   $\psi(x) = \dfrac{A}{k\beta^2} [-e^{-\beta x} + (e^{-\beta} - 1)x + 1]$

   $A_n = 2 \int_0^1 [f(x) - \psi(x)] \sin n\pi x \, dx$

7. $\psi(x) = u_0 \left[1 - \dfrac{\sinh \sqrt{h/k}\, x}{\sinh \sqrt{h/k}} \right]$

9. $u(x, t) = \dfrac{A}{6a^2}(x - x^3)$
   $+ \dfrac{2A}{a^2\pi^3} \sum_{n=1}^{\infty} \dfrac{(-1)^n}{n^3} \cos n\pi a t \sin n\pi x$

11. $u(x, y) = (u_0 - u_1) y + u_1$
    $+ \dfrac{2}{\pi} \sum_{n=1}^{\infty} \dfrac{u_0(-1)^n - u_1}{n} e^{-n\pi x} \sin n\pi y$

Exercises 12.7, Page 795

1. $u(x, t) = 2h \sum_{n=1}^{\infty} \dfrac{\sin \lambda_n}{\lambda_n [h + \sin^2 \lambda_n]} e^{-k\lambda_n^2 t} \cos \lambda_n x$,
   where the λ_n are the consecutive positive roots of $\cot \lambda = \lambda/h$

3. $u(x, y) = \sum_{n=1}^{\infty} A_n \sinh \lambda_n y \sin \lambda_n x$,
   where $A_n = \dfrac{2h}{\sinh \lambda_n b [ah + \cos^2 \lambda_n a]} \int_0^a f(x) \sin \lambda_n x \, dx$
   and the λ_n are the consecutive positive roots of $\tan \lambda a = -\lambda/h$

5. $u(x, t) = \sum_{n=1}^{\infty} A_n e^{-k(2n-1)^2 \pi^2 t/4L^2} \sin\left(\dfrac{2n-1}{2L}\right) \pi x$,
   where $A_n = \dfrac{2}{L} \int_0^L f(x) \sin\left(\dfrac{2n-1}{2L}\right) \pi x \, dx$

7. $u(x, y) = \dfrac{4u_0}{\pi} \sum_{n=1}^{\infty} \dfrac{1}{(2n-1) \cosh\left(\dfrac{2n-1}{2}\right)\pi}$
   $\times \cosh\left(\dfrac{2n-1}{2}\right)\pi x \sin\left(\dfrac{2n-1}{2}\right)\pi y$

Exercises 12.8, Page 798

1. $u(x, y, t) = \sum_{m=1}^{\infty} \sum_{n=1}^{\infty} A_{mn} e^{-k(m^2 + n^2)t} \sin mx \sin ny$,
   where $A_{mn} = \dfrac{4u_0}{mn\pi^2} [1 - (-1)^m][1 - (-1)^n]$

3. $u(x, y, t)$
   $= \sum_{m=1}^{\infty} \sum_{n=1}^{\infty} A_{mn} \sin mx \sin ny \cos a\sqrt{m^2 + n^2}\, t$,
   where $A_{mn} = \dfrac{16}{m^3 n^3 \pi^2} [(-1)^m - 1][(-1)^n - 1]$

5. $u(x, y, z) = \sum_{m=1}^{\infty} \sum_{n=1}^{\infty} A_{mn} \sinh \omega_{mn} z \sin \dfrac{m\pi}{a} x \sin \dfrac{n\pi}{b} y$,
   where $\omega_{mn} = \sqrt{(m\pi/a)^2 + (n\pi/b)^2}$ and
   $A_{mn} = \dfrac{4}{ab \sinh(c\omega_{mn})} \int_0^b \int_0^a f(x, y)$
   $\times \sin \dfrac{m\pi}{a} x \sin \dfrac{n\pi}{b} y \, dx \, dy$

Chapter 12 Review Exercises, Page 799

1. $u = c_1 e^{(c_2 x + y/c_2)}$

3. $\psi(x) = u_0 + \dfrac{(u_1 - u_0)}{1 + \pi} x$

5. $u(x, t) = \dfrac{2h}{\pi^2 a} \sum_{n=1}^{\infty} \dfrac{\left(\cos \dfrac{n\pi}{4} - \cos \dfrac{3n\pi}{4} \right)}{n^2}$
   $\times \sin n\pi a t \sin n\pi x$

7. $u(x, y) = \dfrac{100}{\pi} \sum_{n=1}^{\infty} \dfrac{1 - (-1)^n}{n \sinh n\pi} \sinh nx \sin ny$

9. $u(x, y) = \dfrac{100}{\pi} \sum_{n=1}^{\infty} \dfrac{1 - (-1)^n}{n} e^{-nx} \sin ny$

11. $u(x, t) = e^{-t} \sin x$

Exercises 13.1, Page 805

1. $u(r, \theta) = \dfrac{u_0}{2} + \dfrac{u_0}{\pi} \sum_{n=1}^{\infty} \dfrac{1 - (-1)^n}{n} r^n \sin n\theta$

3. $u(r, \theta) = \dfrac{2\pi^2}{3} - 4 \sum_{n=1}^{\infty} \dfrac{r^n}{n^2} \cos n\theta$

5. $u(r, \theta) = A_0 + \sum_{n=1}^{\infty} r^{-n}(A_n \cos n\theta + B_n \sin n\theta)$, where
   $A_0 = \dfrac{1}{2\pi} \int_0^{2\pi} f(\theta) \, d\theta$
   $A_n = \dfrac{c^n}{\pi} \int_0^{2\pi} f(\theta) \cos n\theta \, d\theta$
   $B_n = \dfrac{c^n}{\pi} \int_0^{2\pi} f(\theta) \sin n\theta \, d\theta$

7. $u(r, \theta) = \dfrac{1}{2} + \dfrac{2}{\pi} \sum_{n=1}^{\infty} \dfrac{\sin \dfrac{n\pi}{2}}{n} \left(\dfrac{r}{c}\right)^{2n} \cos 2n\theta$

9. $u(r, \theta) = A_0 \ln\left(\dfrac{r}{b}\right) + \sum_{n=1}^{\infty} \left[\left(\dfrac{b}{r}\right)^n - \left(\dfrac{r}{b}\right)^n \right]$
   $\times [A_n \cos n\theta + B_n \sin n\theta]$, where
   $A_0 \ln\left(\dfrac{a}{b}\right) = \dfrac{1}{2\pi} \int_0^{2\pi} f(\theta) \, d\theta$
   $\left[\left(\dfrac{b}{a}\right)^n - \left(\dfrac{a}{b}\right)^n \right] A_n = \dfrac{1}{\pi} \int_0^{2\pi} f(\theta) \cos n\theta \, d\theta$

A-65

$$\left[\left(\frac{b}{a}\right)^n - \left(\frac{a}{b}\right)^n\right] B_n = \frac{1}{\pi} \int_0^{2\pi} f(\theta) \sin n\theta \, d\theta$$

11. $u(r, \theta) = \dfrac{u_0}{\pi} \theta + \dfrac{2u_0}{\pi} \sum_{n=1}^{\infty} \dfrac{r^n}{n} \sin n\theta$

Exercises 13.2, Page 811

1. $u(r, t) = \dfrac{2}{ac} \sum_{n=1}^{\infty} \dfrac{\sin \lambda_n at J_0(\lambda_n r)}{\lambda_n^2 J_1(\lambda_n c)}$

3. $u(r, z) = u_0 \sum_{n=1}^{\infty} \dfrac{\sinh \lambda_n (4-z) J_0(\lambda_n r)}{\lambda_n \sinh 4\lambda_n J_1(2\lambda_n)}$

5. $u(r, t) = \sum_{n=1}^{\infty} A_n J_0(\lambda_n r) e^{-k\lambda_n^2 t}$, where

$A_n = \dfrac{2}{c^2 J_1^2(\lambda_n c)} \int_0^c r J_0(\lambda_n r) f(r) \, dr$

7. $u(r, t) = \sum_{n=1}^{\infty} A_n J_0(\lambda_n r) e^{-k\lambda_n^2 t}$, where

$A_n = \dfrac{2\lambda_n^2}{(\lambda_n^2 + h^2) J_0^2(\lambda_n)} \int_0^1 r J_0(\lambda_n r) f(r) \, dr$

9. $u(r, t) = 100 + 50 \sum_{n=1}^{\infty} \dfrac{J_1(\lambda_n) J_0(\lambda_n r)}{\lambda_n J_1^2(2\lambda_n)} e^{-\lambda_n^2 t}$

11. $u(x, t) = \sum_{n=1}^{\infty} A_n \cos(\lambda_n \sqrt{g} \, t) J_0(2\lambda_n \sqrt{x})$, where

$A_n = \dfrac{2}{L J_1^2(2\lambda_n \sqrt{L})} \int_0^{\sqrt{L}} v J_0(2\lambda_n v) f(v^2) \, dv$

Exercises 13.3, Page 814

1. $u(r, \theta) = 50 \left[\dfrac{1}{2} P_0(\cos \theta) + \dfrac{3}{4}\left(\dfrac{r}{a}\right) P_1(\cos \theta) \right.$
$\left. - \dfrac{7}{16}\left(\dfrac{r}{a}\right)^3 P_3(\cos \theta) + \dfrac{11}{32}\left(\dfrac{r}{a}\right)^5 P_5(\cos \theta) + \cdots \right]$

3. $u(r, \theta) = \dfrac{r}{a} \cos \theta$

5. $u(r, \theta) = \sum_{n=0}^{\infty} A_n \dfrac{b^{2n+1} - r^{2n+1}}{b^{2n+1} r^{n+1}} P_n(\cos \theta)$, where

$\dfrac{b^{2n+1} - a^{2n+1}}{b^{2n+1} a^{n+1}} A_n = \dfrac{2n+1}{2} \int_0^{\pi} f(\theta) P_n(\cos \theta) \sin \theta \, d\theta$

7. $u(r, \theta) = \sum_{n=0}^{\infty} A_n r^{2n} P_{2n}(\cos \theta)$, where

$a^{2n} A_n = (4n+1) \int_0^{\pi/2} f(\theta) P_{2n}(\cos \theta) \sin \theta \, d\theta$

Chapter 13 Review Exercises, Page 816

1. $u(r, \theta) = \dfrac{2u_0}{\pi} \sum_{n=1}^{\infty} \dfrac{1 - (-1)^n}{n} \left(\dfrac{r}{c}\right)^n \sin n\theta$

3. $u(r, \theta) = \dfrac{4u_0}{\pi} \sum_{n=1}^{\infty} \dfrac{1 - (-1)^n}{n^3} \sin n\theta$

5. $u(r, \theta) = \dfrac{2u_0}{\pi} \sum_{n=1}^{\infty} \dfrac{r^{4n} + r^{-4n}}{2^{4n} + 2^{-4n}} \dfrac{1 - (-1)^n}{n} \sin 4n\theta$

7. $u(r, t) = 2e^{-ht} \sum_{n=1}^{\infty} \dfrac{J_0(\lambda_n r)}{\lambda_n J_1(\lambda_n)} e^{-\lambda_n^2 t}$

9. $u(r, z) = 50 \sum_{n=1}^{\infty} \dfrac{\cosh \lambda_n z J_0(\lambda_n r)}{\lambda_n \cosh 4\lambda_n J_1(2\lambda_n)}$

11. $u(r, \theta) = 100\left[\dfrac{3}{2} r P_1(\cos \theta) - \dfrac{7}{8} r^3 P_3(\cos \theta) + \dfrac{11}{16} r^5 P_5(\cos \theta) + \cdots\right]$

Exercises 14.1, Page 820

1. (a) Let $\tau = u^2$ in the integral $\text{erf}(\sqrt{t})$.

(b) $\mathscr{L}\{\text{erf}(\sqrt{t})\} = \dfrac{1}{\sqrt{\pi}} \mathscr{L}\{1\} \mathscr{L}\{t^{-1/2} e^{-t}\}$

$= \dfrac{1}{\sqrt{\pi}} \dfrac{1}{s} \left(\dfrac{\sqrt{\pi}}{s^{1/2}}\right)\bigg|_{s \to s+1}$

$= \dfrac{1}{s\sqrt{s+1}}$

3. $\mathscr{L}\{e^t \text{erf}(\sqrt{t})\} = \mathscr{L}\{\text{erf}(\sqrt{t})\}|_{s \to s-1}$

$= \dfrac{1}{s\sqrt{s+1}}\bigg|_{s \to s-1}$

$= \dfrac{1}{(s-1)\sqrt{s}}$

5. $\mathscr{L}\left\{e^{-Gt/C} \text{erf}\left(\dfrac{x}{2}\sqrt{\dfrac{RC}{t}}\right)\right\}$

$= \mathscr{L}\left\{e^{-Gt/C}\left[1 - \text{erfc}\left(\dfrac{x}{2}\sqrt{\dfrac{RC}{t}}\right)\right]\right\}$

$= \dfrac{1}{s + G/C} - \dfrac{e^{-x\sqrt{RC}\sqrt{s}}}{s}\bigg|_{s \to s + G/C}$

$= \dfrac{1}{s + G/C} - \dfrac{e^{-x\sqrt{RC}\sqrt{s+G/C}}}{s + G/C}$

$= \dfrac{C}{Cs + G}\{1 - e^{-x\sqrt{RCs + RG}}\}$

7. $y(t) = e^{\pi t} \text{erfc}(\sqrt{\pi t})$

9. Use the property $\int_0^b - \int_0^a = \int_0^b + \int_a^0$.

Exercises 14.2, Page 825

1. $u(x, t) = A \cos \dfrac{a\pi}{L} t \sin \dfrac{\pi}{L} x$

3. $u(x, t) = f\left(t - \dfrac{x}{a}\right) \mathscr{U}\left(t - \dfrac{x}{a}\right)$

5. $u(x, t) = \left[\dfrac{1}{2} g\left(t - \dfrac{x}{a}\right)^2 + A \sin \omega\left(t - \dfrac{x}{a}\right)\right]$
$\times \mathscr{U}\left(t - \dfrac{x}{a}\right) - \dfrac{1}{2} gt^2$

7. $u(x, t) = a\dfrac{F_0}{E} \sum_{n=0}^{\infty} (-1)^n \left\{\left(t - \dfrac{2nL + L - x}{a}\right)\right.$
$\times \mathscr{U}\left(t - \dfrac{2nL + L - x}{a}\right)$
$\left. - \left(t - \dfrac{2nL + L + x}{a}\right) \mathscr{U}\left(t - \dfrac{2nL + L + x}{a}\right)\right\}$

9. $u(x, t) = (t - x) \sinh(t - x)\mathcal{U}(t - x) + xe^{-x}\cosh t - e^{-x}t \sinh t$

11. $u(x, t) = u_0 \operatorname{erfc}\left(\dfrac{x}{2\sqrt{kt}}\right)$

13. $u(x, t) = u_1 + (u_0 - u_1) \operatorname{erfc}\left(\dfrac{x}{2\sqrt{t}}\right)$

15. $u(x, t) = u_0\left[1 - \left\{\operatorname{erfc}\left(\dfrac{x}{2\sqrt{t}}\right) - e^{x+t}\operatorname{erfc}\left(\sqrt{t} + \dfrac{x}{2\sqrt{t}}\right)\right\}\right]$

17. $u(x, t) = \dfrac{x}{2\sqrt{\pi}} \displaystyle\int_0^t \dfrac{f(t - \tau)}{\tau^{3/2}} e^{-x^2/4\tau} d\tau$

19. $u(x, t) = 60 + 40 \operatorname{erfc}\left(\dfrac{x}{2\sqrt{t - 2}}\right)\mathcal{U}(t - 2)$

21. $u(x, t) = 100\left[-e^{1-x+t}\operatorname{erfc}\left(\sqrt{t} + \dfrac{1-x}{2\sqrt{t}}\right) + \operatorname{erfc}\left(\dfrac{1-x}{2\sqrt{t}}\right)\right]$

23. $u(x, t) = u_0 + u_0 e^{-(\pi^2/L^2)t}\sin\left(\dfrac{\pi}{L}x\right)$

25. $u(x, t) = u_0 - u_0 \displaystyle\sum_{n=0}^{\infty}(-1)^n\left[\operatorname{erfc}\left(\dfrac{2n+1-x}{2\sqrt{kt}}\right) + \operatorname{erfc}\left(\dfrac{2n+1+x}{2\sqrt{kt}}\right)\right]$

27. $u(x, t) = u_0 e^{-Gt/C} \operatorname{erf}\left(\dfrac{x}{2}\sqrt{\dfrac{RC}{t}}\right)$

29. $u(x, t) = A\sqrt{\dfrac{k}{\pi t}}e^{-x^2/4kt}$; an impulse of heat, or flash burn, takes place at $x = 0$.

Exercises 14.3, Page 834

1. $f(x) = \dfrac{1}{\pi}\displaystyle\int_0^{\infty} \dfrac{\sin \alpha \cos \alpha x + 3(1 - \cos \alpha)\sin \alpha x}{\alpha}d\alpha$

3. $f(x) = \dfrac{1}{\pi}\displaystyle\int_0^{\infty}[A(\alpha)\cos \alpha x + B(\alpha)\sin \alpha x]\, d\alpha$,
where $A(\alpha) = (3\alpha \sin 3\alpha + \cos 3\alpha - 1)/\alpha^2$
$B(\alpha) = (\sin 3\alpha - 3\alpha \cos 3\alpha)/\alpha^2$

5. $f(x) = \dfrac{1}{\pi}\displaystyle\int_0^{\infty}\dfrac{\cos \alpha x + \alpha \sin \alpha x}{1 + \alpha^2}d\alpha$

7. $f(x) = \dfrac{10}{\pi}\displaystyle\int_0^{\infty}\dfrac{(1 - \cos \alpha)\sin \alpha x}{\alpha}d\alpha$

9. $f(x) = \dfrac{2}{\pi}\displaystyle\int_0^{\infty}\dfrac{(\pi\alpha \sin \pi\alpha + \cos \pi\alpha - 1)\cos \alpha x}{\alpha^2}d\alpha$

11. $f(x) = \dfrac{4}{\pi}\displaystyle\int_0^{\infty}\dfrac{\alpha \sin \alpha x}{4 + \alpha^4}d\alpha$

13. $f(x) = \dfrac{2k}{\pi}\displaystyle\int_0^{\infty}\dfrac{\cos \alpha x}{k^2 + \alpha^2}d\alpha$

$f(x) = \dfrac{2}{\pi}\displaystyle\int_0^{\infty}\dfrac{\alpha \sin \alpha x}{k^2 + \alpha^2}d\alpha$

15. $f(x) = \dfrac{2}{\pi}\displaystyle\int_0^{\infty}\dfrac{(4 - \alpha^2)\cos \alpha x}{(4 + \alpha^2)^2}d\alpha$

$f(x) = \dfrac{8}{\pi}\displaystyle\int_0^{\infty}\dfrac{\alpha \sin \alpha x}{(4 + \alpha^2)^2}d\alpha$

17. $f(x) = \dfrac{2}{\pi}\dfrac{1}{1 + x^2}$, $x > 0$

19. Let $x = 2$ in (7). Use a trigonometric identity and replace α by x. In part (b) make the change of variable $2x = kt$.

Exercises 14.4, Page 839

1. $u(x, t) = \dfrac{1}{\pi}\displaystyle\int_{-\infty}^{\infty}\dfrac{e^{-k\alpha^2 t}}{1 + \alpha^2}e^{-i\alpha x}d\alpha$

$= \dfrac{1}{\pi}\displaystyle\int_{-\infty}^{\infty}\dfrac{\cos \alpha x}{1 + \alpha^2}e^{-k\alpha^2 t}d\alpha$

3. $u(x, t) = \dfrac{1}{\sqrt{1 + 4kt}}e^{-x^2/(1+4kt)}$

5. $u(x, t) = \dfrac{2u_0}{\pi}\displaystyle\int_0^{\infty}\dfrac{1 - e^{-k\alpha^2 t}}{\alpha}\sin \alpha x\, d\alpha$

7. $u(x, t) = \dfrac{2}{\pi}\displaystyle\int_0^{\infty}\dfrac{1 - \cos \alpha}{\alpha}e^{-k\alpha^2 t}\sin \alpha x\, d\alpha$

9. $u(x, t) = \dfrac{2}{\pi}\displaystyle\int_0^{\infty}\dfrac{\sin \alpha}{\alpha}e^{-k\alpha^2 t}\cos \alpha x\, d\alpha$

11. $u(x, t) = \dfrac{1}{2\pi}\displaystyle\int_{-\infty}^{\infty}\left(F(\alpha)\cos \alpha at + G(\alpha)\dfrac{\sin \alpha at}{\alpha a}\right)e^{-i\alpha x}d\alpha$

13. $u(x, y) = \dfrac{2}{\pi}\displaystyle\int_0^{\infty}\dfrac{\sinh \alpha(\pi - x)}{(1 + \alpha^2)\sinh \alpha\pi}\cos \alpha y\, d\alpha$

15. $u(x, y) = \dfrac{100}{\pi}\displaystyle\int_0^{\infty}\dfrac{\sin \alpha}{\alpha}e^{-\alpha y}\cos \alpha x\, d\alpha$

17. $u(x, y) = \dfrac{2}{\pi}\displaystyle\int_0^{\infty}F(\alpha)\dfrac{\sinh \alpha(2 - y)}{\sinh 2\alpha}\sin \alpha x\, d\alpha$

19. $u(x, y) = \dfrac{2}{\pi}\displaystyle\int_0^{\infty}\dfrac{\alpha}{1 + \alpha^2}[e^{-\alpha x}\sin \alpha y + e^{-\alpha y}\sin \alpha x]\, d\alpha$

21. $u(x, y) = \dfrac{1}{2\sqrt{\pi}}\displaystyle\int_{-\infty}^{\infty}\dfrac{e^{-\alpha^2/4}\cosh \alpha y}{\cosh \alpha}e^{-i\alpha x}d\alpha$

$= \dfrac{1}{2\sqrt{\pi}}\displaystyle\int_{-\infty}^{\infty}\dfrac{e^{-\alpha^2/4}\cosh \alpha y}{\cosh \alpha}\cos \alpha x\, d\alpha$

Chapter 14 Review Exercises, Page 842

1. $u(x, y) = \dfrac{2}{\pi}\displaystyle\int_0^{\infty}\dfrac{\sinh \alpha y}{\alpha(1 + \alpha^2)\cosh \alpha\pi}\cos \alpha x\, d\alpha$

3. $u(x, t) = u_0 e^{-ht}\operatorname{erf}\left(\dfrac{x}{2\sqrt{t}}\right)$

5. $u(x, t) = \displaystyle\int_0^t \operatorname{erfc}\left(\dfrac{x}{2\sqrt{\tau}}\right)d\tau$

7. $u(x, t) = \dfrac{u_0}{2\pi} \displaystyle\int_{-\infty}^{\infty} \dfrac{\sin \alpha(\pi - x) + \sin \alpha x}{\alpha} e^{-k\alpha^2 t} \, d\alpha$

9. $u(x, y) = \dfrac{100}{\pi} \displaystyle\int_{0}^{\infty} \left(\dfrac{1 - \cos \alpha}{\alpha}\right)[e^{-\alpha x} \sin \alpha y + 2e^{-\alpha y} \sin \alpha x] \, d\alpha$

11. $u(x, y) = \dfrac{2}{\pi} \displaystyle\int_{0}^{\infty} \left(\dfrac{B \cosh \alpha y}{(1 + \alpha^2) \sinh \alpha \pi} + \dfrac{A}{\alpha}\right) \sin \alpha x \, d\alpha$

Exercises 15.1, Page 851

1. One real root **3.** No real roots
5. 3.1623 **7.** 1.5874 **9.** 0.6823
11. $-1.1414, 1.1414$ **13.** 0, 0.8767
15. 2.4981 **17.** 1.6560 **19.** 0.7297
21. 0.0915 **23.** 0.0337, 44.494; 44.497
25. 1.8955 **29.**

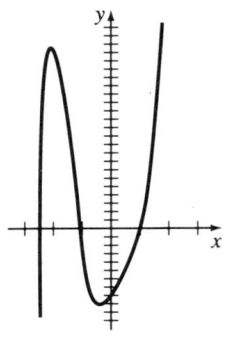

1.000, -1.2494, -2.6638

31.

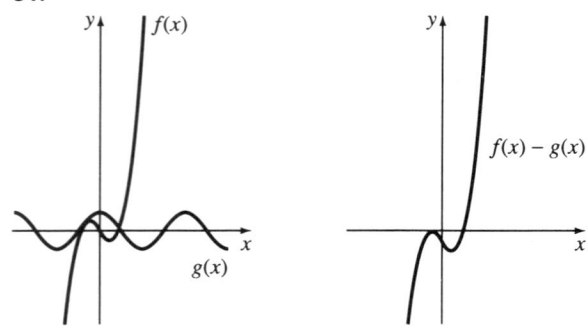

The graphs in parts (a) and (b) suggest that there is a negative and a positive root. A computer zoom-in of the graph of $f(x) - g(x)$ in the neighborhood of the suspected "negative root" shows that there is no root. Newton's method gives the approximation 1.4645 for the positive root.

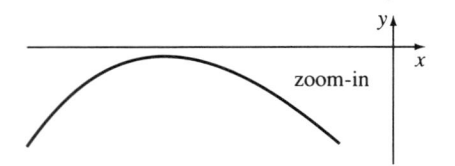

Exercises 15.2, Page 864

1. 78, $M_3 = 77.25$ **3.** 22, $T_3 = 22.5$
5. 1.7564, 1.8667 **7.** 1.1475, 1.1484
9. 0.4393, 0.4228 **11.** 0.4470, 0.4900
13. $\frac{26}{3}$, 8.6611 **15.** 1.6222 **17.** 0.7854
19. 0.4339 **21.** 11.1053 **23.** $n \geq 8$
25. 1.11
27. For Simpson's rule: $n \geq 26$; for trapezoidal rule: $n \geq 366$
29. Trapezoidal rule gives 1.10.
31. For $n = 2$ and $n = 4$ the midpoint rule gives 36, which is the exact value of the integral.
33. $\frac{2}{3}$; $M_8 = \frac{21}{32}$; $T_8 = \frac{11}{16}$; $E_8 = \frac{1}{96}$ for midpoint rule and $E_8 = \frac{1}{48}$ for trapezoidal rule. The error for the midpoint rule is one-half the error for the trapezoidal rule.
35. Approximately 7.1
37. Approximately 4976 gal
39.

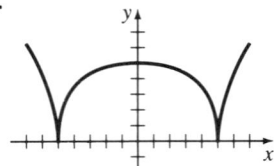

41. 4028

Exercises 15.3, Page 870

1. A family of vertical lines $x = c - 4$
3. A family of hyperbolas $x^2 - y^2 = c$
5. A family of circles $x^2 + (y + 1)^2 = c^2$ with center at $(0, -1)$
7. A family of hyperbolas $xy + y^2 = c$
9. A family of straight lines $y = c(x - 2) + 1$ passing through $(2, 1)$
11.

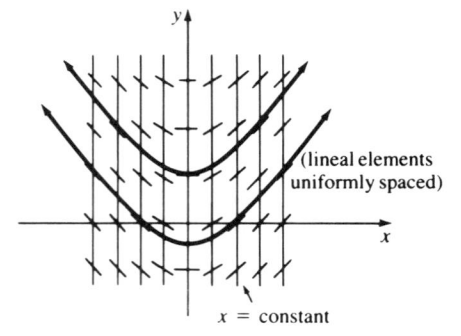

13.

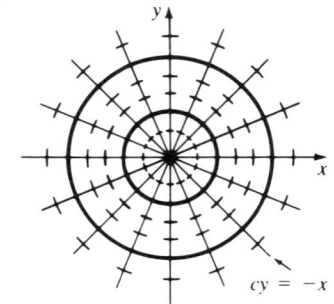

A-68

15.

(lineal elements not uniformly spaced)

$xy = c$

(lineal elements uniformly spaced)

(b)

x_n	y_n
1.00	5.0000
1.05	4.4000
1.10	3.8950
1.15	3.4707
1.20	3.1151
1.25	2.8179
1.30	2.5702
1.35	2.3647
1.40	2.1950
1.45	2.0557
1.50	1.9424

17.

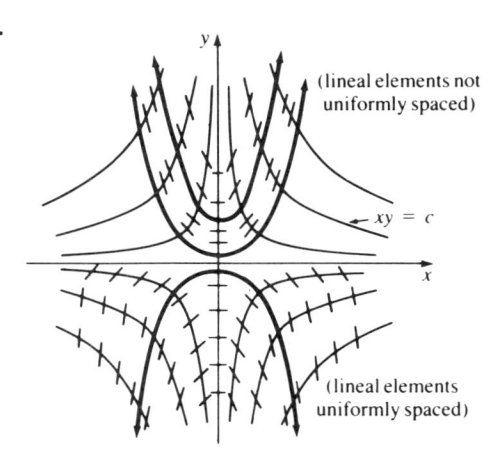

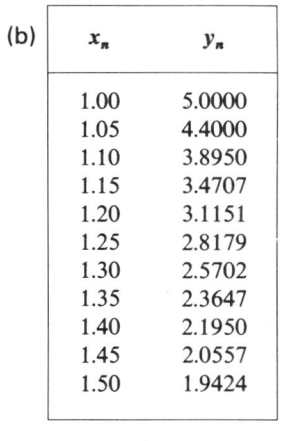

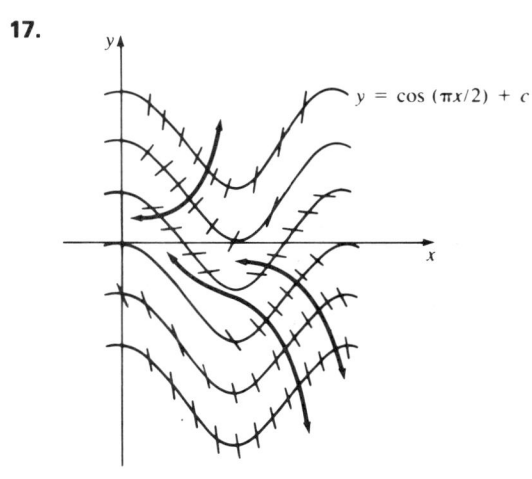

$y = \cos(\pi x/2) + c$

5. (a)

x_n	y_n
0.00	0.0000
0.10	0.1000
0.20	0.2010
0.30	0.3050
0.40	0.4143
0.50	0.5315

(b)

x_n	y_n
0.00	0.0000
0.05	0.0500
0.10	0.1001
0.15	0.1506
0.20	0.2018
0.25	0.2538
0.30	0.3070
0.35	0.3617
0.40	0.4183
0.45	0.4770
0.50	0.5384

19. $y = \dfrac{\alpha - c\gamma}{c\delta - \beta} x$ **21.** $3x + 2y = -\tfrac{3}{2}$

23. $y = \pm\sqrt{2}x$ **25.** $y = 4x;\ y = -x$

Exercises 15.4, Page 876

1. $y = 1 - x + \tan(x + \pi/4)$

3. (a)

x_n	y_n
1.00	5.0000
1.10	3.8000
1.20	2.9800
1.30	2.4260
1.40	2.0582
1.50	1.8207

7. (a)

x_n	y_n
0.00	0.0000
0.10	0.1000
0.20	0.1905
0.30	0.2731
0.40	0.3492
0.50	0.4198

A-69

(b)

x_n	y_n
0.00	0.0000
0.05	0.0500
0.10	0.0976
0.15	0.1429
0.20	0.1863
0.25	0.2278
0.30	0.2676
0.35	0.3058
0.40	0.3427
0.45	0.3782
0.50	0.4124

9. (a)

x_n	y_n
0.00	0.5000
0.10	0.5250
0.20	0.5431
0.30	0.5548
0.40	0.5613
0.50	0.5639

(b)

x_n	y_n
0.00	0.5000
0.05	0.5125
0.10	0.5232
0.15	0.5322
0.20	0.5395
0.25	0.5452
0.30	0.5496
0.35	0.5527
0.40	0.5547
0.45	0.5559
0.50	0.5565

11. (a)

x_n	y_n
1.00	1.0000
1.10	1.0000
1.20	1.0191
1.30	1.0588
1.40	1.1231
1.50	1.2194

(b)

x_n	y_n
1.00	1.0000
1.05	1.0000
1.10	1.0049
1.15	1.0147
1.20	1.0298
1.25	1.0506
1.30	1.0775
1.35	1.1115
1.40	1.1538
1.45	1.2057
1.50	1.2696

13. (a) $h = 0.1$

x_n	y_n
1.00	5.0000
1.10	3.9900
1.20	3.2545
1.30	2.7236
1.40	2.3451
1.50	2.0801

$h = 0.05$

x_n	y_n
1.00	5.0000
1.05	4.4475
1.10	3.9763
1.15	3.5751
1.20	3.2342
1.25	2.9452
1.30	2.7009
1.35	2.4952
1.40	2.3226
1.45	2.1786
1.50	2.0592

(b) $h = 0.1$

x_n	y_n
0.00	0.0000
0.10	0.1005
0.20	0.2030
0.30	0.3098
0.40	0.4234
0.50	0.5470

$h = 0.05$

x_n	y_n
0.00	0.0000
0.05	0.0501
0.10	0.1004
0.15	0.1512
0.20	0.2028
0.25	0.2554
0.30	0.3095
0.35	0.3652
0.40	0.4230
0.45	0.4832
0.50	0.5465

$h = 0.05$

x_n	y_n
0.00	0.5000
0.05	0.5116
0.10	0.5214
0.15	0.5294
0.20	0.5359
0.25	0.5408
0.30	0.5444
0.35	0.5469
0.40	0.5484
0.45	0.5492
0.50	0.5495

(c) $h = 0.1$

x_n	y_n
0.00	0.0000
0.10	0.0952
0.20	0.1822
0.30	0.2622
0.40	0.3363
0.50	0.4053

(e) $h = 0.1$

x_n	y_n
1.00	1.0000
1.10	1.0095
1.20	1.0404
1.30	1.0967
1.40	1.1866
1.50	1.3260

$h = 0.05$

x_n	y_n
0.00	0.0000
0.05	0.0488
0.10	0.0953
0.15	0.1397
0.20	0.1823
0.25	0.2231
0.30	0.2623
0.35	0.3001
0.40	0.3364
0.45	0.3715
0.50	0.4054

$h = 0.05$

x_n	y_n
1.00	1.0000
1.05	1.0024
1.10	1.0100
1.15	1.0228
1.20	1.0414
1.25	1.0663
1.30	1.0984
1.35	1.1389
1.40	1.1895
1.45	1.2526
1.50	1.3315

(d) $h = 0.1$

x_n	y_n
0.00	0.5000
0.10	0.5215
0.20	0.5362
0.30	0.5449
0.40	0.5490
0.50	0.5503

15.

x_n	Euler	Improved Euler
1.0	1.0000	1.0000
1.1	1.2000	1.2469
1.2	1.4938	1.6668
1.3	1.9711	2.6427
1.4	2.9060	8.7989

Exercises 15.5, Page 880

1. (a)

x_n	y_n	x_n	y_n
1.00	5.0000	1.30	2.7236
1.10	3.9900	1.40	2.3451
1.20	3.2545	1.50	2.0801

(b)

x_n	y_n	x_n	y_n
1.00	5.0000	1.30	2.7009
1.05	4.4475	1.35	2.4952
1.10	3.9763	1.40	2.3226
1.15	3.5751	1.45	2.1786
1.20	3.2342	1.50	2.0592
1.25	2.9452		

3. (a)

x_n	y_n
0.00	0.0000
0.10	0.1000
0.20	0.2020
0.30	0.3082
0.40	0.4211
0.50	0.5438

(b)

x_n	y_n
0.00	0.0000
0.05	0.0500
0.10	0.1003
0.15	0.1510
0.20	0.2025
0.25	0.2551
0.30	0.3090
0.35	0.3647
0.40	0.4223
0.45	0.4825
0.50	0.5456

5. (a)

x_n	y_n
0.00	0.0000
0.10	0.0950
0.20	0.1818
0.30	0.2617
0.40	0.3357
0.50	0.4046

(b)

x_n	y_n
0.00	0.0000
0.05	0.0488
0.10	0.0952
0.15	0.1397
0.20	0.1822
0.25	0.2230
0.30	0.2622
0.35	0.2999
0.40	0.3363
0.45	0.3714
0.50	0.4053

7. (a)

x_n	y_n	x_n	y_n
0.00	0.5000	0.30	0.5438
0.10	0.5213	0.40	0.5475
0.20	0.5355	0.50	0.5482

(b)

x_n	y_n	x_n	y_n
0.00	0.5000	0.30	0.5441
0.05	0.5116	0.35	0.5466
0.10	0.5213	0.40	0.5480
0.15	0.5293	0.45	0.5487
0.20	0.5357	0.50	0.5490
0.25	0.5406		

9. (a)

x_n	y_n
1.00	1.0000
1.10	1.0100
1.20	1.0410
1.30	1.0969
1.40	1.1857
1.50	1.3226

(b)

x_n	y_n
1.00	1.0000
1.05	1.0025
1.10	1.0101
1.15	1.0229
1.20	1.0415
1.25	1.0663
1.30	1.0983
1.35	1.1387
1.40	1.1891
1.45	1.2518
1.50	1.3301

11.

x_n	Euler	Improved Euler	Three-Term Taylor
1.0	1.0000	1.0000	1.0000
1.1	1.2000	1.2469	1.2400
1.2	1.4938	1.6668	1.6345
1.3	1.9711	2.6427	2.4600
1.4	2.9060	8.7988	5.6353

13.

x_n	Improved Euler	Three-Term Taylor	True Value
1.00	5.0000	5.0000	5.0000
1.10	5.5300	5.5300	5.5310
1.20	6.1262	6.1262	6.1284
1.30	6.7954	6.7954	6.7992
1.40	7.5454	7.5454	7.5510
1.50	8.3847	8.3847	8.3923

Exercises 15.6, Page 884

1.

x_n	y_n
1.00	5.0000
1.10	3.9724
1.20	3.2284
1.30	2.6945
1.40	2.3163
1.50	2.0533

3.

x_n	y_n
0.00	0.0000
0.10	0.1003
0.20	0.2027
0.30	0.3093
0.40	0.4228
0.50	0.5463

5.

x_n	y_n
0.00	0.0000
0.10	0.0953
0.20	0.1823
0.30	0.2624
0.40	0.3365
0.50	0.4055

7.

x_n	y_n
0.00	0.5000
0.10	0.5213
0.20	0.5358
0.30	0.5443
0.40	0.5482
0.50	0.5493

9.

x_n	y_n
1.00	1.0000
1.10	1.0101
1.20	1.0417
1.30	1.0989
1.40	1.1905
1.50	1.3333

11. $v(5) \approx 35.7678$

13.

1.93	12.50	36.46	47.23	49.00

15.

x_n	y_n
1.00	1.0000
1.10	1.2511
1.20	1.6934
1.30	2.9425
1.40	903.0283

Exercises 15.7, Page 887

1. $y(x) = -x + e^x$; $y(0.2) = 1.0214$, $y(0.4) = 1.0918$, $y(0.6) = 1.2221$, $y(0.8) = 1.4255$

3.

x_n	y_n
0.00	1.0000
0.20	0.7328
0.40	0.6461
0.60	0.6585
0.80	0.7232

5.

x_n	y_n
0.00	0.0000
0.20	0.2027
0.40	0.4228
0.60	0.6841
0.80	1.0297
1.00	1.5569

x_n	y_n
0.00	0.0000
0.10	0.1003
0.20	0.2027
0.30	0.3093
0.40	0.4228
0.50	0.5463
0.60	0.6842
0.70	0.8423
0.80	1.0297
0.90	1.2603
1.00	1.5576

7.

x_n	y_n
0.00	0.0000
0.20	0.0026
0.40	0.0201
0.60	0.0630
0.80	0.1360
1.00	0.2385

x_n	y_n
0.00	0.0000
0.10	0.0003
0.20	0.0026
0.30	0.0087
0.40	0.0200
0.50	0.0379
0.60	0.0629
0.70	0.0956
0.80	0.1360
0.90	1.1837
1.00	0.2384

9.

x_n	y_n
0.00	1.0000
0.10	1.0052
0.20	1.0214
0.30	1.0499
0.40	1.0918

Exercises 15.8, Page 891

1. $y(x) = -2e^{2x} + 5xe^{2x}$; $y(0.2) = -1.4918$, $y_2 = -1.6800$
3. $y_1 = -1.4928$, $y_2 = -1.4919$
5. $y_1 = 1.4640$, $y_2 = 1.4640$
7. $x_1 = 8.3055$, $y_1 = 3.4199$; $x_2 = 8.3055$, $y_2 = 3.4199$
9. $x_1 = -3.9123$, $y_1 = 4.2857$; $x_2 = -3.9123$, $y_2 = 4.2857$

Exercises 15.9, Page 895

1. $y_1 = -5.6774$
   $y_2 = -2.5807$
   $y_3 = 6.3226$

3. $y_1 = -0.2259$
   $y_2 = -0.3356$
   $y_3 = -0.3308$
   $y_4 = -0.2167$

5. $y_1 = 3.3751$
   $y_2 = 3.6306$
   $y_3 = 3.6448$
   $y_4 = 3.2355$
   $y_5 = 2.1411$

7. $y_1 = 3.8842$
   $y_2 = 2.9640$
   $y_3 = 2.2064$
   $y_4 = 1.5826$
   $y_5 = 1.0681$
   $y_6 = 0.6430$
   $y_7 = 0.2913$

9. $y_1 = 0.2660$
   $y_2 = 0.5097$
   $y_3 = 0.7357$
   $y_4 = 0.9471$
   $y_5 = 1.1465$
   $y_6 = 1.3353$
   $y_7 = 1.5149$
   $y_8 = 1.6855$
   $y_9 = 1.8474$

11. $y_1 = 0.3492$
    $y_2 = 0.7202$
    $y_3 = 1.1363$
    $y_4 = 1.6233$
    $y_5 = 2.2118$
    $y_6 = 2.9386$
    $y_7 = 3.8490$

13. $y_0 = -2.2755$
$y_1 = -2.0755$
$y_2 = -1.8589$
$y_3 = -1.6126$
$y_4 = -1.3275$

Exercises 15.10, Page 902

1. $u_{11} = \frac{11}{15}, u_{21} = \frac{14}{15}$
3. $u_{11} = u_{21} = \sqrt{3}/16,$
$u_{22} = u_{12} = 3\sqrt{3}/16$

5. $u_{21} = u_{12} = 12.50,$
$u_{31} = u_{13} = 18.75,$
$u_{32} = u_{23} = 37.50,$
$u_{11} = 6.25, u_{22} = 25.00, u_{33} = 56.25$
7. $u_{14} = u_{41} = 0.5427,$
$u_{24} = u_{42} = 0.6707,$
$u_{34} = u_{43} = 0.6402,$
$u_{33} = 0.4451, u_{44} = 0.9451$
$u_{i,j+1} = \lambda u_{i+1,j} + (1 - 2\lambda)u_{ij} + \lambda u_{i-1,j}$
where $\lambda = k/h^2$

Exercises 15.11, Page 910

1.

Time	$x = 0.25$	$x = 0.50$	$x = 0.75$	$x = 1.00$	$x = 1.25$	$x = 1.50$	$x = 1.75$
0.000	1.0000	1.0000	1.0000	1.0000	0.0000	0.0000	0.0000
0.025	0.6000	1.0000	1.0000	0.6000	0.4000	0.0000	0.0000
0.050	0.5200	0.8400	0.8400	0.6800	0.3200	0.1600	0.0000
0.075	0.4400	0.7120	0.7760	0.6000	0.4000	0.1600	0.0640
0.100	0.3728	0.6288	0.6800	0.5904	0.3840	0.2176	0.0768
0.125	0.3261	0.5469	0.6237	0.5437	0.4000	0.2278	0.1024
0.150	0.2840	0.4893	0.5610	0.5182	0.3886	0.2465	0.1116
0.175	0.2525	0.4358	0.5152	0.4835	0.3836	0.2494	0.1209
0.200	0.2248	0.3942	0.4708	0.4562	0.3699	0.2517	0.1239
0.225	0.2027	0.3571	0.4343	0.4275	0.3571	0.2479	0.1255
0.250	0.1834	0.3262	0.4007	0.4021	0.3416	0.2426	0.1242
0.275	0.1672	0.2989	0.3715	0.3773	0.3262	0.2348	0.1219
0.300	0.1530	0.2752	0.3448	0.3545	0.3101	0.2262	0.1183
0.325	0.1407	0.2541	0.3209	0.3329	0.2943	0.2166	0.1141
0.350	0.1298	0.2354	0.2990	0.3126	0.2787	0.2067	0.1095
0.375	0.1201	0.2186	0.2790	0.2936	0.2635	0.1966	0.1046
0.400	0.1115	0.2034	0.2607	0.2757	0.2488	0.1865	0.0996
0.425	0.1036	0.1895	0.2438	0.2589	0.2347	0.1766	0.0945
0.450	0.0965	0.1769	0.2281	0.2432	0.2211	0.1670	0.0896
0.475	0.0901	0.1652	0.2136	0.2283	0.2083	0.1577	0.0847
0.500	0.0841	0.1545	0.2002	0.2144	0.1961	0.1487	0.0800
0.525	0.0786	0.1446	0.1876	0.2014	0.1845	0.1402	0.0755
0.550	0.0736	0.1354	0.1759	0.1891	0.1735	0.1320	0.0712
0.575	0.0689	0.1269	0.1650	0.1776	0.1632	0.1243	0.0670
0.600	0.0645	0.1189	0.1548	0.1668	0.1534	0.1169	0.0631
0.625	0.0605	0.1115	0.1452	0.1566	0.1442	0.1100	0.0594
0.650	0.0567	0.1046	0.1363	0.1471	0.1355	0.1034	0.0559
0.675	0.0532	0.0981	0.1279	0.1381	0.1273	0.0972	0.0525
0.700	0.0499	0.0921	0.1201	0.1297	0.1196	0.0914	0.0494
0.725	0.0468	0.0864	0.1127	0.1218	0.1124	0.0859	0.0464
0.750	0.0439	0.0811	0.1058	0.1144	0.1056	0.0807	0.0436
0.775	0.0412	0.0761	0.0994	0.1074	0.0992	0.0758	0.0410
0.800	0.0387	0.0715	0.0933	0.1009	0.0931	0.0712	0.0385
0.825	0.0363	0.0671	0.0876	0.0948	0.0875	0.0669	0.0362
0.850	0.0341	0.0630	0.0823	0.0890	0.0822	0.0628	0.0340
0.875	0.0320	0.0591	0.0772	0.0836	0.0772	0.0590	0.0319
0.900	0.0301	0.0555	0.0725	0.0785	0.0725	0.0554	0.0300
0.925	0.0282	0.0521	0.0681	0.0737	0.0681	0.0521	0.0282
0.950	0.0265	0.0490	0.0640	0.0692	0.0639	0.0489	0.0265
0.975	0.0249	0.0460	0.0601	0.0650	0.0600	0.0459	0.0249
1.000	0.0234	0.0432	0.0564	0.0610	0.0564	0.0431	0.0233

3.

Time	x = 0.25	x = 0.50	x = 0.75	x = 1.00	x = 1.25	x = 1.50	x = 1.75
0.000	1.0000	1.0000	1.0000	1.0000	0.0000	0.0000	0.0000
0.025	0.7074	0.9520	0.9566	0.7444	0.2545	0.0371	0.0053
0.050	0.5606	0.8499	0.8685	0.6633	0.3303	0.1034	0.0223
0.075	0.4684	0.7473	0.7836	0.6191	0.3614	0.1529	0.0462
0.100	0.4015	0.6577	0.7084	0.5837	0.3753	0.1871	0.0684
0.125	0.3492	0.5821	0.6428	0.5510	0.3797	0.2101	0.0861
0.150	0.3069	0.5187	0.5857	0.5199	0.3778	0.2247	0.0990
0.175	0.2721	0.4652	0.5359	0.4901	0.3716	0.2329	0.1078
0.200	0.2430	0.4198	0.4921	0.4617	0.3622	0.2362	0.1132
0.225	0.2186	0.3809	0.4533	0.4348	0.3507	0.2358	0.1160
0.250	0.1977	0.3473	0.4189	0.4093	0.3378	0.2327	0.1166
0.275	0.1798	0.3181	0.3881	0.3853	0.3240	0.2275	0.1157
0.300	0.1643	0.2924	0.3604	0.3626	0.3097	0.2208	0.1136
0.325	0.1507	0.2697	0.3353	0.3412	0.2953	0.2131	0.1107
0.350	0.1387	0.2495	0.3125	0.3211	0.2808	0.2047	0.1071
0.375	0.1281	0.2313	0.2916	0.3021	0.2666	0.1960	0.1032
0.400	0.1187	0.2150	0.2725	0.2843	0.2528	0.1871	0.0989
0.425	0.1102	0.2002	0.2549	0.2675	0.2393	0.1781	0.0946
0.450	0.1025	0.1867	0.2387	0.2517	0.2263	0.1692	0.0902
0.475	0.0955	0.1743	0.2236	0.2368	0.2139	0.1606	0.0858
0.500	0.0891	0.1630	0.2097	0.2228	0.2020	0.1521	0.0814
0.525	0.0833	0.1525	0.1967	0.2096	0.1906	0.1439	0.0772
0.550	0.0779	0.1429	0.1846	0.1973	0.1798	0.1361	0.0731
0.575	0.0729	0.1339	0.1734	0.1856	0.1696	0.1285	0.0691
0.600	0.0683	0.1256	0.1628	0.1746	0.1598	0.1214	0.0653
0.625	0.0641	0.1179	0.1530	0.1643	0.1506	0.1145	0.0617
0.650	0.0601	0.1106	0.1438	0.1546	0.1419	0.1080	0.0582
0.675	0.0564	0.1039	0.1351	0.1455	0.1336	0.1018	0.0549
0.700	0.0530	0.0976	0.1270	0.1369	0.1259	0.0959	0.0518
0.725	0.0497	0.0917	0.1194	0.1288	0.1185	0.0904	0.0488
0.750	0.0467	0.0862	0.1123	0.1212	0.1116	0.0852	0.0460
0.775	0.0439	0.0810	0.1056	0.1140	0.1050	0.0802	0.0433
0.800	0.0413	0.0762	0.0993	0.1073	0.0989	0.0755	0.0408
0.825	0.0388	0.0716	0.0934	0.1009	0.0931	0.0711	0.0384
0.850	0.0365	0.0674	0.0879	0.0950	0.0876	0.0669	0.0362
0.875	0.0343	0.0633	0.0827	0.0894	0.0824	0.0630	0.0341
0.900	0.0323	0.0596	0.0778	0.0841	0.0776	0.0593	0.0321
0.925	0.0303	0.0560	0.0732	0.0791	0.0730	0.0558	0.0302
0.950	0.0285	0.0527	0.0688	0.0744	0.0687	0.0526	0.0284
0.975	0.0268	0.0496	0.0647	0.0700	0.0647	0.0495	0.0268
1.000	0.0253	0.0466	0.0609	0.0659	0.0608	0.0465	0.0252

Absolute errors are approximately 2.2×10^{-2}, 3.7×10^{-2}, 1.3×10^{-2}.

5.

Time	$x = 0.25$	$x = 0.50$	$x = 0.75$	$x = 1.00$	$x = 1.25$	$x = 1.50$	$x = 1.75$
0.00	1.0000	1.0000	1.0000	1.0000	0.0000	0.0000	0.0000
0.05	0.5265	0.8693	0.8852	0.6141	0.3783	0.0884	0.0197
0.10	0.3972	0.6551	0.7043	0.5883	0.3723	0.1955	0.0653
0.15	0.3042	0.5150	0.5844	0.5192	0.3812	0.2261	0.1010
0.20	0.2409	0.4171	0.4901	0.4620	0.3636	0.2385	0.1145
0.25	0.1962	0.3452	0.4174	0.4092	0.3391	0.2343	0.1178
0.30	0.1631	0.2908	0.3592	0.3624	0.3105	0.2220	0.1145
0.35	0.1379	0.2482	0.3115	0.3208	0.2813	0.2056	0.1077
0.40	0.1181	0.2141	0.2718	0.2840	0.2530	0.1876	0.0993
0.45	0.1020	0.1860	0.2381	0.2514	0.2265	0.1696	0.0904
0.50	0.0888	0.1625	0.2092	0.2226	0.2020	0.1523	0.0816
0.55	0.0776	0.1425	0.1842	0.1970	0.1798	0.1361	0.0732
0.60	0.0681	0.1253	0.1625	0.1744	0.1597	0.1214	0.0654
0.65	0.0599	0.1104	0.1435	0.1544	0.1418	0.1079	0.0582
0.70	0.0528	0.0974	0.1268	0.1366	0.1257	0.0959	0.0518
0.75	0.0466	0.0860	0.1121	0.1210	0.1114	0.0851	0.0460
0.80	0.0412	0.0760	0.0991	0.1071	0.0987	0.0754	0.0408
0.85	0.0364	0.0672	0.0877	0.0948	0.0874	0.0668	0.0361
0.90	0.0322	0.0594	0.0776	0.0839	0.0774	0.0592	0.0320
0.95	0.0285	0.0526	0.0687	0.0743	0.0686	0.0524	0.0284
1.00	0.0252	0.0465	0.0608	0.0657	0.0607	0.0464	0.0251

Absolute errors are approximately 1.8×10^{-2}, 3.7×10^{-2}, 1.3×10^{-2}.

7.

(a)

Time	$x = 2.00$	$x = 4.00$	$x = 6.00$	$x = 8.00$	$x = 10.00$	$x = 12.00$	$x = 14.00$	$x = 16.00$	$x = 18.00$
0.00	30.0000	30.0000	30.0000	30.0000	30.0000	30.0000	30.0000	30.0000	30.0000
1.00	28.7733	29.9749	29.9995	30.0000	30.0000	30.0000	29.9995	29.9749	28.7733
2.00	27.6450	29.9037	29.9970	29.9999	30.0000	29.9999	29.9970	29.9037	27.6450
3.00	26.6051	29.7938	29.9911	29.9997	30.0000	29.9997	29.9911	29.7938	26.6051
4.00	25.6452	29.6517	29.9805	29.9991	29.9999	29.9991	29.9805	29.6517	25.6452
5.00	24.7573	29.4829	29.9643	29.9981	29.9998	29.9981	29.9643	29.4829	24.7573
6.00	23.9347	29.2922	29.9421	29.9963	29.9996	29.9963	29.9421	29.2922	23.9347
7.00	23.1711	29.0836	29.9134	29.9936	29.9992	29.9936	29.9134	29.0836	23.1711
8.00	22.4612	28.8606	29.8782	29.9898	29.9986	29.9898	29.8782	28.8606	22.4612
9.00	21.7999	28.6263	29.8362	29.9848	29.9977	29.9848	29.8362	28.6263	21.7999
10.00	21.1829	28.3831	29.7878	29.9782	29.9964	29.9782	29.7878	28.3831	21.1829

(b)

Time	$x = 5.00$	$x = 10.00$	$x = 15.00$	$x = 20.00$	$x = 25.00$	$x = 30.00$	$x = 35.00$	$x = 40.00$	$x = 45.00$
0.00	30.0000	30.0000	30.0000	30.0000	30.0000	30.0000	30.0000	30.0000	30.0000
1.00	29.7968	29.9993	30.0000	30.0000	30.0000	30.0000	30.0000	29.9993	29.7968
2.00	29.5964	29.9973	30.0000	30.0000	30.0000	30.0000	30.0000	29.9973	29.5964
3.00	29.3987	29.9939	30.0000	30.0000	30.0000	30.0000	30.0000	29.9939	29.3987
4.00	29.2036	29.9893	29.9999	30.0000	30.0000	30.0000	29.9999	29.9893	29.2036
5.00	29.0112	29.9834	29.9998	30.0000	30.0000	30.0000	29.9998	29.9834	29.0112
6.00	28.8212	29.9762	29.9997	30.0000	30.0000	30.0000	29.9997	29.9762	28.8213
7.00	28.6339	29.9679	29.9995	30.0000	30.0000	30.0000	29.9995	29.9679	28.6339
8.00	28.4490	29.9585	29.9992	30.0000	30.0000	30.0000	29.9993	29.9585	28.4490
9.00	29.2665	29.9479	29.9989	30.0000	30.0000	30.0000	29.9989	29.9479	28.2665
10.00	28.0864	29.9363	29.9986	30.0000	30.0000	30.0000	29.9986	29.9363	28.0864

(c)

Time	$x = 2.00$	$x = 4.00$	$x = 6.00$	$x = 8.00$	$x = 10.00$	$x = 12.00$	$x = 14.00$	$x = 16.00$	$x = 18.00$
0.00	18.0000	32.0000	42.0000	48.0000	50.0000	48.0000	42.0000	32.0000	18.0000
1.00	16.4489	30.1970	40.1561	46.1495	48.1486	46.1495	40.1561	30.1970	16.4489
2.00	15.3312	28.5348	38.3465	44.3067	46.3001	44.3067	38.3465	28.5348	15.3312
3.00	14.4216	27.0416	36.6031	42.4847	44.4619	42.4847	36.6031	27.0416	14.4216
4.00	13.6371	25.6867	34.9416	40.6988	42.6453	40.6988	34.9416	25.6867	13.6371
5.00	12.9378	24.4419	33.3628	38.9611	40.8634	38.9611	33.3628	24.4419	12.9378
6.00	12.3012	23.2863	31.8624	37.2794	39.1273	37.2794	31.8624	23.2863	12.3012
7.00	11.7137	22.2051	30.4350	35.6578	37.4446	35.6578	30.4350	22.2051	11.7137
8.00	11.1659	21.1877	29.0757	34.0984	35.8202	34.0984	29.0757	21.1877	11.1659
9.00	10.6517	20.2261	27.7799	32.6014	34.2567	32.6014	27.7799	20.2261	10.6517
10.00	10.1665	19.3143	26.5439	31.1662	32.7549	31.1662	26.5439	19.3143	10.1665

(d)

Time	$x = 10.00$	$x = 20.00$	$x = 30.00$	$x = 40.00$	$x = 50.00$	$x = 60.00$	$x = 70.00$	$x = 80.00$	$x = 90.00$
0.00	8.0000	16.0000	24.0000	32.0000	40.0000	32.0000	24.0000	16.0000	8.0000
1.00	8.0000	16.0000	24.0000	31.9979	39.7425	31.9979	24.0000	16.0000	8.0000
2.00	8.0000	16.0000	23.9999	31.9918	39.4932	31.9918	23.9999	16.0000	8.0000
3.00	8.0000	16.0000	23.9997	31.9820	39.2517	31.9820	23.9997	16.0000	8.0000
4.00	8.0000	16.0000	23.9993	31.9686	39.0175	31.9686	23.9993	16.0000	8.0000
5.00	8.0000	16.0000	23.9987	31.9520	38.7905	31.9520	23.9987	16.0000	8.0000
6.00	8.0000	15.9999	23.9978	31.9323	38.5701	31.9323	23.9978	15.9999	8.0000
7.00	8.0000	15.9999	23.9966	31.9097	38.3561	31.9097	23.9966	15.9999	8.0000
8.00	8.0000	15.9998	23.9950	31.8844	38.1483	31.8844	23.9950	15.9998	8.0000
9.00	8.0000	15.9997	23.9931	31.8566	37.9463	31.8566	23.9931	15.9997	8.0000
10.00	8.0000	15.9996	23.9908	31.8265	37.7498	31.8265	23.9908	15.9996	8.0000

9.

(a)

Time	x = 2.00	x = 4.00	x = 6.00	x = 8.00	x = 10.00	x = 12.00	x = 14.00	x = 16.00	x = 18.00
0.00	30.0000	30.0000	30.0000	30.0000	30.0000	30.0000	30.0000	30.0000	30.0000
1.00	28.7733	29.9749	29.9995	30.0000	30.0000	30.0000	29.9998	29.9916	29.5911
2.00	27.6450	29.9037	29.9970	29.9999	30.0000	30.0000	29.9990	29.9679	29.2150
3.00	26.6051	29.7938	29.9911	29.9997	30.0000	29.9999	29.9970	29.9313	28.8684
4.00	25.6452	29.6517	29.9805	29.9991	30.0000	29.9997	29.9935	29.8839	28.5484
5.00	24.7573	29.4829	29.9643	29.9981	29.9999	29.9994	29.9881	29.8276	28.2524
6.00	23.9347	29.2922	29.9421	29.9963	29.9997	29.9988	29.9807	29.7641	27.9782
7.00	23.1711	29.0836	29.9134	29.9936	29.9995	29.9979	29.9711	29.6945	27.7237
8.00	22.4612	28.8606	29.8782	29.9899	29.9991	29.9966	29.9594	29.6202	27.4870
9.00	21.7999	28.6263	29.8362	29.9848	29.9985	29.9949	29.9454	29.5421	27.2666
10.00	21.1829	28.3831	29.7878	29.9783	29.9976	29.9927	29.9293	29.4610	27.0610

(b)

Time	x = 5.00	x = 10.00	x = 15.00	x = 20.00	x = 25.00	x = 30.00	x = 35.00	x = 40.00	x = 45.00
0.00	30.0000	30.0000	30.0000	30.0000	30.0000	30.0000	30.0000	30.0000	30.0000
1.00	29.7968	29.9993	30.0000	30.0000	30.0000	30.0000	30.0000	29.9998	29.9323
2.00	29.5964	29.9973	30.0000	30.0000	30.0000	30.0000	30.0000	29.9991	29.8655
3.00	29.3987	29.9939	30.0000	30.0000	30.0000	30.0000	30.0000	29.9980	29.7996
4.00	29.2036	29.9893	29.9999	30.0000	30.0000	30.0000	30.0000	29.9964	29.7345
5.00	29.0112	29.9834	29.9998	30.0000	30.0000	30.0000	29.9999	29.9945	29.6704
6.00	28.8212	29.9762	29.9997	30.0000	30.0000	30.0000	29.9999	29.9921	29.6071
7.00	28.6339	29.9679	29.9995	30.0000	30.0000	30.0000	29.9998	29.9893	29.5446
8.00	28.4490	29.9585	29.9992	30.0000	30.0000	30.0000	29.9997	29.9862	29.4830
9.00	28.2665	29.9479	29.9989	30.0000	30.0000	30.0000	29.9996	29.9827	29.4222
10.00	28.0864	29.9363	29.9986	30.0000	30.0000	30.0000	29.9995	29.9788	29.3621

(c)

Time	x = 2.00	x = 4.00	x = 6.00	x = 8.00	x = 10.00	x = 12.00	x = 14.00	x = 16.00	x = 18.00
0.00	18.0000	32.0000	42.0000	48.0000	50.0000	48.0000	42.0000	32.0000	18.0000
1.00	16.4489	30.1970	40.1562	46.1502	48.1531	46.1773	40.3274	31.2520	22.9449
2.00	15.3312	28.5350	38.3477	44.3130	46.3327	44.4671	39.0872	31.5755	24.6930
3.00	14.4219	27.0429	36.6090	42.5113	44.5759	42.9362	38.1976	31.7478	25.4131
4.00	13.6381	25.6913	34.9606	40.7728	42.9127	41.5716	37.4340	31.7086	25.6986
5.00	12.9409	24.4545	33.4091	39.1182	41.3519	40.3240	36.7033	31.5136	25.7663
6.00	12.3088	23.3146	31.9546	37.5566	39.8880	39.1565	36.9745	31.2134	25.7128
7.00	11.7294	22.2589	30.5939	36.0884	38.5109	38.0470	35.2407	30.8434	25.5871
8.00	11.1946	21.2785	29.3217	34.7092	37.2109	36.9834	34.5032	30.4279	25.4167
9.00	10.6987	20.3660	28.1318	33.4130	35.9801	35.9591	33.7660	29.9836	25.2181
10.00	10.2377	19.5150	27.0178	32.1929	34.8117	34.9710	33.0338	29.5224	25.0019

(d)

Time	$x=10.00$	$x=20.00$	$x=30.00$	$x=40.00$	$x=50.00$	$x=60.00$	$x=70.00$	$x=80.00$	$x=90.00$
0.00	8.0000	16.0000	24.0000	32.0000	40.0000	32.0000	24.0000	16.0000	8.0000
1.00	8.0000	16.0000	24.0000	31.9979	39.7425	31.9979	24.0000	16.0026	8.3218
2.00	8.0000	16.0000	23.9999	31.9918	39.4932	31.9918	24.0000	16.0102	8.6333
3.00	8.0000	16.0000	23.9997	31.9820	39.2517	31.9820	24.0001	16.0225	8.9350
4.00	8.0000	16.0000	23.9993	31.9686	39.0175	31.9687	24.0002	16.0391	9.2272
5.00	8.0000	16.0000	23.9987	31.9520	38.7905	31.9520	24.0003	16.0599	9.5103
6.00	8.0000	15.9999	23.9978	31.9323	38.5701	31.9324	24.0005	16.0845	9.7846
7.00	8.0000	15.9999	23.9966	31.9097	38.3561	31.9098	24.0008	16.1126	10.0506
8.00	8.0000	15.9998	23.9950	31.8844	38.1483	31.8846	24.0012	16.1441	10.3084
9.00	8.0000	15.9997	23.9931	31.8566	37.9463	31.8569	24.0017	16.1786	10.5585
10.00	8.0000	15.9996	23.9908	31.8265	37.7499	31.8269	24.0023	16.2160	10.8012

11. (a) $\psi(x) = \frac{1}{2}x + 20$

(b)

Time	$x=4.00$	$x=8.00$	$x=12.00$	$x=16.00$
0.00	50.0000	50.0000	50.0000	50.0000
10.00	32.7433	44.2679	45.4228	38.2971
20.00	29.9946	36.2354	38.3148	35.8160
30.00	26.9487	32.1409	34.0874	32.9644
40.00	25.2691	29.2562	31.2704	31.2580
50.00	24.1178	27.4348	29.4296	30.1207
60.00	23.3821	26.2339	28.2356	29.3810
70.00	22.8995	25.4560	27.4554	28.8998
80.00	22.5861	24.9481	26.9482	28.5859
90.00	22.3817	24.6176	26.6175	28.3817
100.00	22.2486	24.4022	26.4023	28.2486
110.00	22.1619	24.2620	26.2620	28.1619
120.00	22.1055	24.1707	26.1707	28.1055
130.00	22.0687	24.1112	26.1112	28.0687
140.00	22.0447	24.0724	26.0724	28.0447
150.00	22.0291	24.0472	26.0472	28.0291
160.00	22.0190	24.0307	26.0307	28.0190
170.00	22.0124	24.0200	26.0200	28.0124
180.00	22.0081	24.0130	26.0130	28.0081
190.00	22.0052	24.0085	26.0085	28.0052
200.00	22.0034	24.0055	26.0055	28.0034
210.00	22.0022	24.0036	26.0036	28.0022
220.00	22.0015	24.0023	26.0023	28.0015
230.00	22.0009	24.0015	26.0015	28.0009
240.00	22.0006	24.0010	26.0010	28.0006
250.00	22.0004	24.0007	26.0007	28.0004
260.00	22.0003	24.0004	26.0004	28.0003
270.00	22.0002	24.0003	26.0003	28.0002
280.00	22.0001	24.0002	26.0002	28.0001
290.00	22.0001	24.0001	26.0001	28.0001
300.00	22.0000	24.0001	26.0001	28.0000
310.00	22.0000	24.0001	26.0001	28.0000
320.00	22.0000	24.0000	26.0000	28.0000
330.00	22.0000	24.0000	26.0000	28.0000
340.00	22.0000	24.0000	26.0000	28.0000
350.00	22.0000	24.0000	26.0000	28.0000

Exercises 15.12, Page 916

1.

(a)

Time	$x = 0.25$	$x = 0.50$	$x = 0.75$
0.00	0.1875	0.2500	0.1875
0.10	0.1775	0.2400	0.1775
0.20	0.1491	0.2100	0.1491
0.30	0.1066	0.1605	0.1066
0.40	0.0556	0.0938	0.0556
0.50	0.0019	0.0148	0.0019
0.60	−0.0501	−0.0682	−0.0501
0.70	−0.0970	−0.1455	−0.0970
0.80	−0.1361	−0.2072	−0.1361
0.90	−0.1648	−0.2462	−0.1648
1.00	−0.1802	−0.2591	−0.1802

(b)

Time	$x = 0.4$	$x = 0.8$	$x = 1.2$	$x = 1.6$
0.00	0.0032	0.5273	0.5273	0.0032
0.10	0.0194	0.5109	0.5109	0.0194
0.20	0.0652	0.4638	0.4638	0.0652
0.30	0.1318	0.3918	0.3918	0.1318
0.40	0.2065	0.3035	0.3035	0.2065
0.50	0.2743	0.2092	0.2092	0.2743
0.60	0.3208	0.1190	0.1190	0.3208
0.70	0.3348	0.0413	0.0413	0.3348
0.80	0.3094	−0.0180	−0.0180	0.3094
0.90	0.2443	−0.0568	−0.0568	0.2443
1.00	0.1450	−0.0768	−0.0768	0.1450

(c)

Time	$x = 0.1$	$x = 0.2$	$x = 0.3$	$x = 0.4$	$x = 0.5$	$x = 0.6$	$x = 0.7$	$x = 0.8$	$x = 0.9$
0.00	0.0000	0.0000	0.0000	0.0000	0.0000	0.5000	0.5000	0.5000	0.5000
0.04	0.0000	0.0000	0.0000	0.0000	0.0800	0.4200	0.5000	0.5000	0.4200
0.08	0.0000	0.0000	0.0000	0.0256	0.2432	0.2568	0.4744	0.4744	0.2312
0.12	0.0000	0.0000	0.0082	0.1126	0.3411	0.1589	0.3792	0.3710	0.0462
0.16	0.0000	0.0026	0.0472	0.2394	0.3076	0.1898	0.2108	0.1663	−0.0496
0.20	0.0008	0.0187	0.1334	0.3264	0.2146	0.2651	0.0215	−0.0933	−0.0605
0.24	0.0071	0.0657	0.2447	0.3159	0.1735	0.2463	−0.1266	−0.3056	−0.0625
0.28	0.0299	0.1513	0.3215	0.2371	0.2013	0.0849	−0.2127	−0.3829	−0.1223
0.32	0.0819	0.2525	0.3168	0.1737	0.2033	−0.1345	−0.2580	−0.3223	−0.2264
0.36	0.1623	0.3197	0.2458	0.1657	0.0877	−0.2853	−0.2843	−0.2104	−0.2887
0.40	0.2412	0.3129	0.1727	0.1583	−0.1223	−0.3164	−0.2874	−0.1473	−0.2336
0.44	0.2657	0.2383	0.1399	0.0658	−0.3046	−0.2761	−0.2549	−0.1565	−0.0761
0.48	0.1965	0.1410	0.1149	−0.1216	−0.3593	−0.2381	−0.1977	−0.1715	0.0800
0.52	0.0466	0.0531	0.0225	−0.3093	−0.2992	−0.2260	−0.1451	−0.1144	0.1300
0.56	−0.1161	−0.0466	−0.1662	−0.3876	−0.2188	−0.2114	−0.1085	0.0111	0.0602
0.60	−0.2194	−0.2069	−0.3875	−0.3411	−0.1901	−0.1662	−0.0666	0.1140	−0.0446
0.64	−0.2485	−0.4290	−0.5362	−0.2611	−0.2021	−0.0969	0.0012	0.1084	−0.0843
0.68	−0.2559	−0.6276	−0.5625	−0.2503	−0.1993	−0.0298	0.0720	0.0068	−0.0354
0.72	−0.3003	−0.6865	−0.5097	−0.3230	−0.1585	0.0156	0.0893	−0.0874	0.0384
0.76	−0.3722	−0.5652	−0.4538	−0.4029	−0.1147	0.0289	0.0265	−0.0849	0.0596
0.80	−0.3867	−0.3464	−0.4172	−0.4068	−0.1172	−0.0046	−0.0712	−0.0005	0.0155
0.84	−0.2647	−0.1633	−0.3546	−0.3214	−0.1763	−0.0954	−0.1249	0.0665	−0.0386
0.88	−0.0254	−0.0738	−0.2202	−0.2002	−0.2559	−0.2215	−0.1079	0.0385	−0.0468
0.92	0.2064	−0.0157	−0.0325	−0.1032	−0.3067	−0.3223	−0.0804	−0.0636	−0.0127
0.96	0.3012	0.1081	0.1380	−0.0487	−0.2974	−0.3407	−0.1250	−0.1548	0.0092
1.00	0.2378	0.3032	0.2392	−0.0141	−0.2223	−0.2762	−0.2481	−0.1840	−0.0244

3. (a)

Time	$x = 0.2$	$x = 0.4$	$x = 0.6$	$x = 0.8$
0.00	0.5878	0.9511	0.9511	0.5878
0.05	0.5808	0.9397	0.9397	0.5808
0.10	0.5599	0.9059	0.9059	0.5599
0.15	0.5256	0.8505	0.8505	0.5256
0.20	0.4788	0.7748	0.7748	0.4788
0.25	0.4206	0.6806	0.6806	0.4206
0.30	0.3524	0.5701	0.5701	0.3524
0.35	0.2757	0.4460	0.4460	0.2757
0.40	0.1924	0.3113	0.3113	0.1924
0.45	0.1046	0.1692	0.1692	0.1046
0.50	0.0142	0.0230	0.0230	0.0142

(b)

Time	$x = 0.2$	$x = 0.4$	$x = 0.6$	$x = 0.8$
0.00	0.5878	0.9511	0.9511	0.5878
0.03	0.5860	0.9482	0.9482	0.5860
0.05	0.5808	0.9397	0.9397	0.5808
0.08	0.5721	0.9256	0.9256	0.5721
0.10	0.5599	0.9060	0.9060	0.5599
0.13	0.5445	0.8809	0.8809	0.5445
0.15	0.5257	0.8507	0.8507	0.5257
0.18	0.5039	0.8153	0.8153	0.5039
0.20	0.4790	0.7750	0.7750	0.4790
0.23	0.4513	0.7302	0.7302	0.4513
0.25	0.4209	0.6810	0.6810	0.4209
0.28	0.3879	0.6277	0.6277	0.3879
0.30	0.3527	0.5706	0.5706	0.3527
0.33	0.3153	0.5102	0.5102	0.3153
0.35	0.2761	0.4467	0.4467	0.2761
0.38	0.2352	0.3806	0.3806	0.2352
0.40	0.1929	0.3122	0.3122	0.1929
0.43	0.1495	0.2419	0.2419	0.1495
0.45	0.1052	0.1701	0.1701	0.1052
0.48	0.0602	0.0974	0.0974	0.0602
0.50	0.0149	0.0241	0.0241	0.0149

(c)

Time	$x = 0.1$	$x = 0.2$	$x = 0.3$	$x = 0.4$	$x = 0.5$	$x = 0.6$	$x = 0.7$	$x = 0.8$	$x = 0.9$
0.00	0.3090	0.5878	0.8090	0.9511	1.0000	0.9511	0.8090	0.5878	0.3090
0.01	0.3089	0.5875	0.8086	0.9506	0.9995	0.9506	0.8086	0.5875	0.3089
0.02	0.3084	0.5866	0.8074	0.9492	0.9980	0.9492	0.8074	0.5866	0.3084
0.03	0.3077	0.5852	0.8055	0.9469	0.9956	0.9469	0.8055	0.5852	0.3077
0.04	0.3066	0.5832	0.8027	0.9436	0.9922	0.9436	0.8027	0.5832	0.3066
0.05	0.3052	0.5806	0.7991	0.9394	0.9878	0.9394	0.7991	0.5806	0.3052
0.06	0.3036	0.5775	0.7948	0.9343	0.9824	0.9343	0.7948	0.5775	0.3036
0.07	0.3016	0.5737	0.7897	0.9283	0.9761	0.9283	0.7897	0.5737	0.3016
0.08	0.2994	0.5695	0.7838	0.9214	0.9688	0.9214	0.7838	0.5695	0.2994
0.09	0.2968	0.5646	0.7772	0.9136	0.9606	0.9136	0.7772	0.5646	0.2968
0.10	0.2940	0.5592	0.7697	0.9049	0.9515	0.9049	0.7697	0.5592	0.2940
0.11	0.2909	0.5533	0.7616	0.8953	0.9414	0.8953	0.7616	0.5533	0.2909
0.12	0.2875	0.5468	0.7527	0.8848	0.9303	0.8848	0.7527	0.5468	0.2875
0.13	0.2838	0.5398	0.7430	0.8735	0.9184	0.8735	0.7430	0.5398	0.2838
0.14	0.2798	0.5323	0.7326	0.8613	0.9056	0.8613	0.7326	0.5323	0.2798
0.15	0.2756	0.5242	0.7215	0.8482	0.8919	0.8482	0.7215	0.5242	0.2756
0.16	0.2711	0.5157	0.7097	0.8344	0.8773	0.8344	0.7097	0.5157	0.2711
0.17	0.2663	0.5066	0.6972	0.8197	0.8618	0.8197	0.6972	0.5066	0.2663
0.18	0.2613	0.4970	0.6841	0.8042	0.8456	0.8042	0.6841	0.4970	0.2613
0.19	0.2560	0.4869	0.6702	0.7879	0.8284	0.7879	0.6702	0.4869	0.2560
0.20	0.2505	0.4764	0.6557	0.7708	0.8105	0.7708	0.6557	0.4764	0.2505
0.21	0.2447	0.4654	0.6406	0.7530	0.7918	0.7530	0.6406	0.4654	0.2447
0.22	0.2387	0.4539	0.6248	0.7345	0.7723	0.7345	0.6248	0.4539	0.2387
0.23	0.2324	0.4420	0.6084	0.7152	0.7521	0.7152	0.6084	0.4420	0.2324
0.24	0.2259	0.4297	0.5914	0.6953	0.7311	0.6953	0.5914	0.4297	0.2259
0.25	0.2192	0.4170	0.5739	0.6746	0.7094	0.6746	0.5739	0.4170	0.2192
0.26	0.2123	0.4038	0.5558	0.6533	0.6870	0.6533	0.5558	0.4038	0.2123
0.27	0.2052	0.3902	0.5371	0.6314	0.6639	0.6314	0.5371	0.3902	0.2052
0.28	0.1978	0.3763	0.5179	0.6088	0.6402	0.6088	0.5179	0.3763	0.1978
0.29	0.1903	0.3620	0.4982	0.5857	0.6158	0.5857	0.4982	0.3620	0.1903
0.30	0.1826	0.3473	0.4780	0.5620	0.5909	0.5620	0.4780	0.3473	0.1826
0.31	0.1747	0.3323	0.4574	0.5377	0.5654	0.5377	0.4574	0.3323	0.1747
0.32	0.1666	0.3170	0.4363	0.5129	0.5393	0.5129	0.4363	0.3170	0.1666
0.33	0.1584	0.3013	0.4148	0.4876	0.5127	0.4876	0.4148	0.3013	0.1584
0.34	0.1500	0.2854	0.3928	0.4618	0.4856	0.4618	0.3928	0.2854	0.1500
0.35	0.1415	0.2692	0.3705	0.4356	0.4580	0.4356	0.3705	0.2692	0.1415
0.36	0.1329	0.2527	0.3478	0.4089	0.4299	0.4089	0.3478	0.2527	0.1329
0.37	0.1241	0.2360	0.3248	0.3818	0.4015	0.3818	0.3248	0.2360	0.1241
0.38	0.1152	0.2190	0.3015	0.3544	0.3726	0.3544	0.3015	0.2190	0.1152
0.39	0.1061	0.2019	0.2778	0.3266	0.3434	0.3266	0.2778	0.2019	0.1061
0.40	0.0970	0.1845	0.2539	0.2985	0.3139	0.2985	0.2539	0.1845	0.0970
0.41	0.0878	0.1669	0.2298	0.2701	0.2840	0.2701	0.2298	0.1669	0.0878
0.42	0.0785	0.1492	0.2054	0.2415	0.2539	0.2415	0.2054	0.1492	0.0785
0.43	0.0691	0.1314	0.1808	0.2126	0.2235	0.2126	0.1808	0.1314	0.0691
0.44	0.0596	0.1134	0.1561	0.1835	0.1929	0.1835	0.1561	0.1134	0.0596
0.45	0.0501	0.0953	0.1311	0.1542	0.1621	0.1542	0.1311	0.0953	0.0501
0.46	0.0405	0.0771	0.1061	0.1247	0.1312	0.1247	0.1061	0.0771	0.0405
0.47	0.0309	0.0588	0.0810	0.0952	0.1001	0.0952	0.0810	0.0588	0.0309
0.48	0.0213	0.0405	0.0557	0.0655	0.0689	0.0655	0.0557	0.0405	0.0213
0.49	0.0116	0.0221	0.0305	0.0358	0.0377	0.0358	0.0305	0.0221	0.0116
0.50	0.0020	0.0038	0.0052	0.0061	0.0064	0.0061	0.0052	0.0038	0.0020

5.

Time	$x = 10$	$x = 20$	$x = 30$	$x = 40$	$x = 50$
0.00000	0.1000	0.2000	0.3000	0.2000	0.1000
0.20045	0.1000	0.2000	0.2750	0.2000	0.1000
0.40089	0.1000	0.1938	0.2125	0.1938	0.1000
0.60134	0.0984	0.1688	0.1406	0.1688	0.0984
0.80178	0.0898	0.1191	0.0828	0.1191	0.0898
1.00223	0.0661	0.0531	0.0432	0.0531	0.0661
1.20268	0.0226	−0.0121	0.0085	−0.0121	0.0226
1.40312	−0.0352	−0.0635	−0.0365	−0.0635	−0.0352
1.60357	−0.0913	−0.1011	−0.0950	−0.1011	−0.0913
1.80401	−0.1271	−0.1347	−0.1566	−0.1347	−0.1271
2.00446	−0.1329	−0.1791	−0.2072	−0.1719	−0.1329
2.20491	−0.1153	−0.2081	−0.2402	−0.2081	−0.1153
2.40535	−0.0920	−0.2292	−0.2571	−0.2292	−0.0920
2.60580	−0.0801	−0.2230	−0.2601	−0.2230	−0.0801
2.80624	−0.0838	−0.1903	−0.2445	−0.1903	−0.0838
3.00669	−0.0932	−0.1445	−0.2018	−0.1445	−0.0932
3.20713	−0.0921	−0.1003	−0.1305	−0.1003	−0.0921
3.40758	−0.0701	−0.0615	−0.0440	−0.0615	−0.0701
3.60803	−0.0284	−0.0205	0.0336	−0.0205	−0.0284
3.80847	0.0224	0.0321	0.0842	0.0321	0.0224
4.00892	0.0700	0.0953	0.1087	0.0953	0.0700
4.20936	0.1064	0.1555	0.1265	0.1555	0.1064
4.40981	0.1285	0.1962	0.1588	0.1962	0.1285
4.61026	0.1354	0.2106	0.2098	0.2106	0.1354
4.81070	0.1273	0.2060	0.2612	0.2060	0.1273
5.01115	0.1070	0.1955	0.2851	0.1955	0.1070
5.21159	0.0821	0.1853	0.2641	0.1853	0.0821
5.41204	0.0625	0.1689	0.2038	0.1689	0.0625
5.61249	0.0539	0.1347	0.1260	0.1347	0.0539
5.81293	0.0520	0.0781	0.0526	0.0781	0.0520
6.01338	0.0436	0.0086	−0.0080	0.0086	0.0436
6.21382	0.0156	−0.0564	−0.0604	−0.0564	0.0156
6.41427	−0.0343	−0.1043	−0.1107	−0.1043	−0.0343
6.61472	−0.0931	−0.1364	−0.1578	−0.1364	−0.0931
6.81516	−0.1395	−0.1630	−0.1942	−0.1630	−0.1395
7.01561	−0.1568	−0.1915	−0.2150	−0.1915	−0.1568
7.21605	−0.1436	−0.2173	−0.2240	−0.2173	−0.1436
7.41650	−0.1129	−0.2263	−0.2297	−0.2263	−0.1129
7.61695	−0.0824	−0.2078	−0.2336	−0.2078	−0.0824
7.81739	−0.0625	−0.1644	−0.2247	−0.1644	−0.0625
8.01784	−0.0526	−0.1106	−0.1856	−0.1106	−0.0526
8.21828	−0.0440	−0.0611	−0.1091	−0.0611	−0.0440
8.41873	−0.0287	−0.0192	−0.0085	−0.0192	−0.0287
8.61918	−0.0038	0.0229	0.0867	0.0229	−0.0038
8.81962	0.0287	0.0743	0.1500	0.0743	0.0287
9.02007	0.0654	0.1332	0.1755	0.1332	0.0654
9.22051	0.1027	0.1858	0.1799	0.1858	0.1027
9.42096	0.1352	0.2160	0.1872	0.2160	0.1352
9.62140	0.1540	0.2189	0.2089	0.2189	0.1540
9.82185	0.1506	0.2030	0.2356	0.2030	0.1506
10.02230	0.1226	0.1822	0.2461	0.1822	0.1226

Note: Time is expressed in milliseconds.

Exercises 15.13, Page 924

1. $2, \begin{bmatrix} 1 \\ 1 \end{bmatrix}$ 3. $14, \begin{bmatrix} 1 \\ 3 \end{bmatrix}$ 5. $10, \begin{bmatrix} 1 \\ 1 \\ 0.5 \end{bmatrix}$

7. 7 and 2   9. 4, 3, and 1
11. Approximately 0.2087
13. (c) $A^{-1} = \dfrac{1}{4}\begin{bmatrix} 3 & 2 & 1 \\ 2 & 4 & 2 \\ 1 & 2 & 3 \end{bmatrix}$; (d) 0.59;

   (e) Approximately $9.44 EI/L^2$

Chapter 15 Review Exercises, Page 926

1. 1.6751
3. Midpoint rule: 0.3074, trapezoidal rule: 0.3160, Simpson's rule: 0.3099
5. 130 ft^2
7. All isoclines $y = cx$ are solutions of the differential equation.

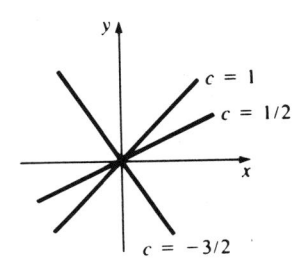

9. **Comparison of Numerical Methods with $h = 0.1$**

x_n	Euler	Improved Euler	3-Term Taylor	Runge–Kutta
1.00	2.0000	2.0000	2.0000	2.0000
1.10	2.1386	2.1549	2.1556	2.1556
1.20	2.3097	2.3439	2.3453	2.3454
1.30	2.5136	2.5672	2.5694	2.5695
1.40	2.7504	2.8246	2.8277	2.8278
1.50	3.0201	3.1157	3.1198	3.1197

Comparison of Numerical Methods with $h = 0.05$

x_n	Euler	Improved Euler	3-Term Taylor	Runge–Kutta
1.00	2.0000	2.0000	2.0000	2.0000
1.05	2.0693	2.0735	2.0735	2.0736
1.10	2.1469	2.1554	2.1556	2.1556
1.15	2.2329	2.2459	2.2462	2.2462
1.20	2.3272	2.3450	2.3454	2.3454
1.25	2.4299	2.4527	2.4532	2.4532
1.30	2.5410	2.5689	2.5695	2.5695
1.35	2.6604	2.6937	2.6944	2.6944
1.40	2.7883	2.8269	2.8278	2.8278
1.45	2.9245	2.9686	2.9696	2.9696
1.50	3.0690	3.1187	3.1198	3.1197

11. **Comparison of Numerical Methods with $h = 0.1$**

x_n	Euler	Improved Euler	3-Term Taylor	Runge–Kutta
0.50	0.5000	0.5000	0.5000	0.5000
0.60	0.6000	0.6048	0.6050	0.6049
0.70	0.7095	0.7191	0.7195	0.7194
0.80	0.8283	0.8427	0.8433	0.8431
0.90	0.9559	0.9752	0.9759	0.9757
1.00	1.0921	1.1163	0.1172	1.1169

Comparison of Numerical Methods with $h = 0.05$

x_n	Euler	Improved Euler	3-Term Taylor	Runge–Kutta
0.50	0.5000	0.5000	0.5000	0.5000
0.55	0.5500	0.5512	0.5512	0.5512
0.60	0.6024	0.6049	0.6049	0.6049
0.65	0.6573	0.6609	0.6610	0.6610
0.70	0.7144	0.7193	0.7194	0.7194
0.75	0.7739	0.7800	0.7802	0.7801
0.80	0.8356	0.8430	0.8431	0.8431
0.85	0.8996	0.9082	0.9083	0.9083
0.90	0.9657	0.9755	0.9757	0.9757
0.95	1.0340	1.0451	1.0453	1.0452
1.00	1.1044	1.1168	1.1170	1.1169

13. $h = 0.2$: $y(0.2) \approx 3.2$
    $h = 0.1$: $y(0.2) \approx 3.23$

15. $u_{11} = 0.8929, u_{21} = 3.5714, u_{31} = 13.3929$
17. (b) $144x^4 - 72x^3 - 131x^2 + 36x + 32 = 0$
(c) $-0.6742, -0.4830, 0.7638, 0.8934$. The point of reflection has the approximate coordinates $(0.7638, 0.6455)$.

Exercises 16.1, Page 936
1. $3 + 3i$ **3.** 1 **5.** $7 - 13i$
7. $-7 + 5i$ **9.** $11 - 10i$ **11.** $-5 + 12i$
13. $-2i$ **15.** $-\frac{7}{17} - \frac{11}{17}i$ **17.** $8 - i$
19. $\frac{23}{37} - \frac{64}{37}i$ **21.** $20i$ **23.** $\frac{102}{5} + \frac{116}{5}i$
25. $\frac{7}{130} + \frac{9}{130}i$ **27.** $x/(x^2 + y^2)$
29. $-2y - 4$ **31.** $\sqrt{(x-1)^2 + (y-3)^2}$
33. $x = -\frac{9}{2}, y = 1$ **35.** $\frac{\sqrt{2}}{2} + \frac{\sqrt{2}}{2}i, -\frac{\sqrt{2}}{2} - \frac{\sqrt{2}}{2}i$
37. $11 - 6i$

Exercises 16.2, Page 941
1. $2(\cos 0 + i \sin 0)$ or $2(\cos 2\pi + i \sin 2\pi)$
3. $3\left(\cos \frac{3\pi}{2} + i \sin \frac{3\pi}{2}\right)$ **5.** $\sqrt{2}\left(\cos \frac{\pi}{4} + i \sin \frac{\pi}{4}\right)$
7. $2\left(\cos \frac{5\pi}{6} + i \sin \frac{5\pi}{6}\right)$
9. $\frac{3\sqrt{2}}{2}\left(\cos \frac{5\pi}{4} + i \sin \frac{5\pi}{4}\right)$
11. $-\frac{53}{2} - \frac{5}{2}i$ **13.** $5.5433 + 2.2961i$
15. $8i; \frac{\sqrt{2}}{4} - \frac{\sqrt{2}}{4}i$
17. $30\sqrt{2}\left(\cos \frac{25\pi}{12} + i \sin \frac{25\pi}{12}\right); 40.9808 + 10.9808i$
19. $\frac{1}{2\sqrt{2}}\left(\cos\left(-\frac{\pi}{4}\right) + i \sin\left(-\frac{\pi}{4}\right)\right); \frac{1}{4} - \frac{1}{4}i$
21. -512 **23.** $\frac{1}{32}i$ **25.** $-i$
27. $w_0 = 2, w_1 = -1 + \sqrt{3}i, w_2 = -1 - \sqrt{3}i$
29. $w_0 = \frac{\sqrt{2}}{2} + \frac{\sqrt{2}}{2}i, w_1 = -\frac{\sqrt{2}}{2} - \frac{\sqrt{2}}{2}i$
31. $w_0 = \frac{\sqrt{2}}{2} + \frac{\sqrt{6}}{2}i, w_1 = -\frac{\sqrt{2}}{2} - \frac{\sqrt{6}}{2}i$
33. $\pm\frac{\sqrt{2}}{2}(1 + i), \pm\frac{\sqrt{2}}{2}(1 - i)$
35. $32\left(\cos \frac{13\pi}{6} + i \sin \frac{13\pi}{6}\right), 16\sqrt{3} + 16i$
37. $\cos 2\theta = \cos^2\theta - \sin^2\theta, \sin 2\theta = 2 \sin \theta \cos \theta$

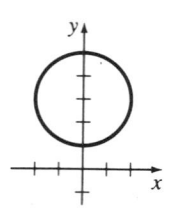

Exercises 16.3, Page 944
1. [graph: vertical line $x = 5$]
3. [graph: horizontal line $y = -3$] 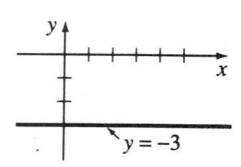
5. [graph: circle centered at origin]
7. [graph: circle centered at $(4, -3)$]
9. Domain 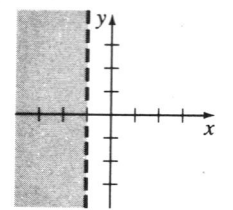
11. Domain 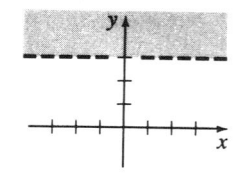
13. Domain 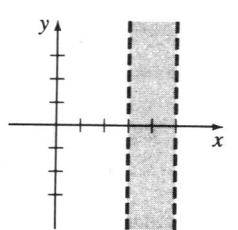
15. Not a domain 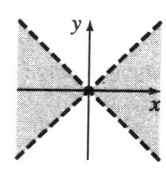
17. Not a domain 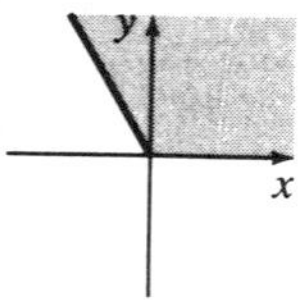
19. Domain

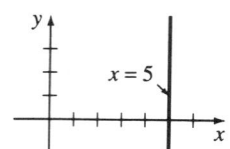

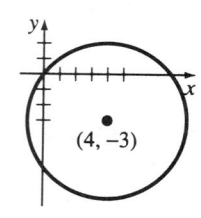

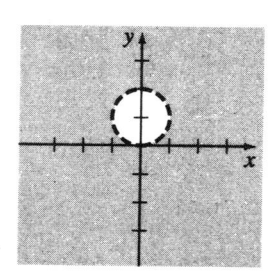

A-86

21. Domain

23. The line $y = -x$
25. The hyperbola $x^2 - y^2 = 1$

Exercises 16.4, Page 951

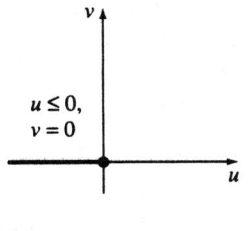

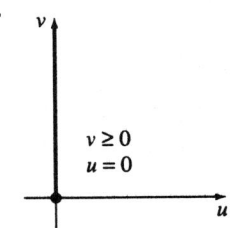

1. $u = \dfrac{v^2}{16} - 4$

3. $u \leq 0$, $v = 0$

5. $v \geq 0$, $u = 0$

7. $f(z) = (6x - 5) + i(6y + 9)$
9. $f(z) = (x^2 - y^2 - 3x) + i(2xy - 3y + 4)$
11. $f(z) = (x^3 - 3xy^2 - 4x) + i(3x^2y - y^3 - 4y)$
13. $f(z) = \left(x + \dfrac{x}{x^2 + y^2}\right) + i\left(y - \dfrac{y}{x^2 + y^2}\right)$
15. $-4 + i$; $3 - 9i$; $1 + 86i$
17. $14 - 20i$; $-13 + 43i$; $3 - 26i$ **19.** $6 - 5i$
21. $-4i$ **27.** $12z^2 - (6 + 2i)z - 5$
29. $6z^2 - 14z - 4 + 16i$ **31.** $6z(z^2 - 4i)^2$
33. $\dfrac{8 - 13i}{(2z + i)^2}$ **35.** $3i$ **37.** $2i, -2i$
41. $x(t) = c_1 e^{2t}$ and $y(t) = c_2 e^{2t}$, the streamlines lie on lines through the origin.
43. $y = cx$, the streamlines are lines through the origin.

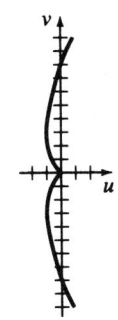

45.

Exercises 16.5, Page 957

15. $a = 1, b = 3$ **21.** $f'(z) = e^x \cos y + ie^x \sin y$
23. $f(z) = x + i(y + C)$
25. $f(z) = x^2 - y^2 + i(2xy + C)$
27. $f(z) = \log_e(x^2 + y^2) + i\left(2 \tan^{-1} \dfrac{y}{x} + C\right)$
29.

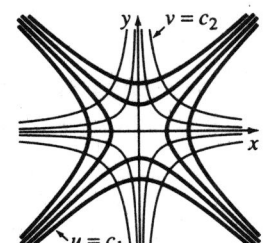

31. The x-axis and the circle $|z| = 1$

Exercises 16.6, Page 966

1. $\dfrac{\sqrt{3}}{2} + \dfrac{1}{2}i$ **3.** $e^{-1}\left(\dfrac{\sqrt{2}}{2} + \dfrac{\sqrt{2}}{2}i\right)$
5. $-e^{\pi}$ **7.** $-1.8650 + 4.0752i$
9. $0.2837 - 0.9589i$ **11.** $-0.9659 + 0.2588i$
13. $e^y(\cos x - i \sin x)$
15. $e^{x^2 - y^2}(\cos 2xy + i \sin 2xy)$
23. $1.6094 + i(\pi + 2n\pi)$
25. $1.0397 + i\left(\dfrac{3\pi}{4} + 2n\pi\right)$
27. $1.0397 + i\left(\dfrac{\pi}{3} + 2n\pi\right)$
29. $2.1383 - \dfrac{\pi}{4}i$ **31.** $2.5649 + 2.7468i$
33. $3.4657 - \dfrac{\pi}{3}i$ **35.** $1.3863 + i\left(\dfrac{\pi}{2} + 2n\pi\right)$
37. $3 + i\left(-\dfrac{\pi}{2} + 2n\pi\right)$ **39.** $e^{(2 - 8n)\pi}$
41. $e^{-2n\pi}(0.2740 + 0.5837i)$ **43.** e^2
47. No; no; yes

Exercises 16.7, Page 971
1. 10.0677 3. $1.0911 + 0.8310i$
5. $0.7616i$ 7. -0.6481 9. -1
11. $0.5876 + 1.3363i$
15. $\frac{\pi}{2} + 2n\pi - i\log_e(2 \pm \sqrt{3})$
17. $\left(-\frac{\pi}{2} + 2n\pi\right)i$ 19. $\frac{\pi}{4} + n\pi$
21. $2n\pi \pm 2i$

Exercises 16.8, Page 974
1. $n\pi + (-1)^{n+1} i\log_e(1 + \sqrt{2})$ 3. $n\pi$
5. $2n\pi \pm i\log_e(2 + \sqrt{3})$ 7. $\pm\frac{\pi}{3} + 2n\pi$
9. $\frac{\pi}{4} + n\pi$ 11. $(-1)^n \log_e 3 + n\pi i$

Chapter 16 Review Exercises, Page 975
1. $0; 32$ 3. $-\frac{7}{25}$ 5. $\frac{4}{5}$ 7. False
9. $0.6931 + i\left(\frac{\pi}{2} + 2n\pi\right)$ 11. $-0.3097 + 0.8577i$
13. False   15. $3 - \frac{\pi}{2}i$ 17. $58 - 4i$
19. $-8 + 8i$
21.

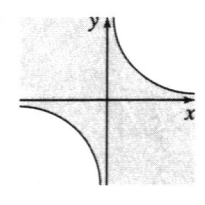

23.

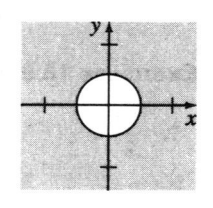

25. An ellipse with foci $(0, -2)$ and $(0, 2)$
27. $1.0696 - 0.2127i$, $0.2127 + 1.0696i$,
    $-1.0696 + 0.2127i$, $-0.2127 - 1.0696i$
29. $5i$ 31. The parabola $v = u^2 - 2u$
33. $1, -1$ 35. Pure imaginary numbers
37. $f'(z) = (-2y - 5) + 2xi$

Exercises 17.1, Page 983
1. $-28 + 84i$ 3. $-48 + \frac{736}{3}i$ 5. $(2 + \pi)i$
7. πi 9. $-\frac{7}{12} + \frac{1}{12}i$ 11. $-e - 1$
13. $\frac{3}{2} - \frac{\pi}{4}$ 15. 0 17. $\frac{1}{2}i$ 19. 0
21. $\frac{4}{3} - \frac{5}{3}i$ 23. $\frac{4}{3} - \frac{5}{3}i$ 25. $\frac{5\pi e^5}{12}$
27. $6\sqrt{2}$ 31. $-11 + 38i; 0$
33. Circulation $= 0$, net flux $= 4\pi$
35. Circulation $= 0$, net flux $= 0$

Exercises 17.2, Page 989
9. $2\pi i$ 11. $2\pi i$ 13. 0
15. $2\pi i; 4\pi i; 0$ 17. $-8\pi i; -6\pi i$
19. $-\pi(1 + i)$ 21. $-4\pi i$ 23. $-6\pi i$

Exercises 17.3, Page 995
1. $-2i$ 3. $48 + 24i$ 5. $6 + \frac{26}{3}i$ 7. 0
9. $-\frac{7}{6} - \frac{22}{3}i$ 11. $-\frac{1}{\pi} - \frac{1}{\pi}i$ 13. $2.3504i$
15. 0 17. πi 19. $\frac{1}{2}i$
21. $11.4928 + 0.9667i$ 23. $-0.9056 + 1.7699i$

Exercises 17.4, Page 1002
1. $8\pi i$ 3. $-2\pi i$ 5. $-\pi(20 + 8i)$
7. $-2\pi; 2\pi$ 9. -8π 11. $-2\pi e^{-1}i$
13. $\frac{4}{3}\pi i$ 15. $-5\pi i; -5\pi i; 9\pi i; 0$
17. $-\pi(3 + i); \pi(3 + i)$ 19. $\pi(\frac{8}{3} + 12i)$
21. 0 23. $-\pi i$

Chapter 17 Review Exercises, Page 1004
1. True   3. True   5. 0 7. $\pi(6\pi - i)$
9. True   11. 0 if $n \neq -1$, $2\pi i$ if $n = -1$
13. $-\frac{7}{2}$ 15. $\frac{136}{15} + \frac{88}{3}i$ 17. 0
19. $-14.2144 + 22.9637i$ 21. $2\pi i$
23. $-\frac{8}{3}\pi i$ 25. $\frac{2}{5}\pi i$ 27. 2π 29. $2n\pi i$

Exercises 18.1, Page 1012
1. $5i, -5, -5i, 5, 5i$ 3. $0, 2, 0, 2, 0$
5. Converges   7. Converges   9. Diverges
11. $\lim_{n\to\infty} \text{Re}(z_n) = 2$ and $\lim_{n\to\infty} \text{Im}(z_n) = \frac{3}{2}$
13. Series converges to $1/(1 + 2i)$.   15. Divergent
17. Convergent, $-\frac{1}{5} + \frac{2}{5}i$
19. Convergent, $\frac{9}{5} - \frac{12}{5}i$
21. $|z - 2i| = \sqrt{5}$, $R = \sqrt{5}$
23. $|z - 1 - i| = 2$, $R = 2$
25. $|z - i| = 1/\sqrt{10}$, $R = 1/\sqrt{10}$
27. $|z - 4 - 3i| = 25$, $R = 25$
29. The series converges at $z = -2 + i$.

Exercises 18.2, Page 1019
1. $\sum_{k=1}^{\infty} (-1)^{k+1} z^k$, $R = 1$
3. $\sum_{k=1}^{\infty} (-1)^{k-1} k(2z)^{k-1}$, $R = \frac{1}{2}$
5. $\sum_{k=0}^{\infty} \frac{(-1)^k}{k!}(2z)^k$, $R = \infty$
7. $\sum_{k=0}^{\infty} \frac{z^{2k+1}}{(2k+1)!}$, $R = \infty$
9. $\sum_{k=0}^{\infty} \frac{(-1)^k}{(2k)!}\left(\frac{z}{2}\right)^{2k}$, $R = \infty$
11. $\sum_{k=0}^{\infty} \frac{(-1)^k}{(2k+1)!} z^{4k+2}$, $R = \infty$

13. $\sum_{k=0}^{\infty} (-1)^k (z-1)^k, R = 1$

15. $\sum_{k=0}^{\infty} \frac{(z-2i)^k}{(3-2i)^{k+1}}, R = \sqrt{13}$

17. $\sum_{k=1}^{\infty} \frac{(z-1)^k}{2^k}, R = 2$

19. $\frac{\sqrt{2}}{2} - \frac{\sqrt{2}}{2 \cdot 1!}\left(z - \frac{\pi}{4}\right) - \frac{\sqrt{2}}{2 \cdot 2!}\left(z - \frac{\pi}{4}\right)^2 + \frac{\sqrt{2}}{2 \cdot 3!}\left(z - \frac{\pi}{4}\right)^3 + \cdots, R = \infty$

21. $e^{3i} \sum_{k=0}^{\infty} \frac{(z - 3i)^k}{k!}, R = \infty$

23. $z + \frac{1}{3} z^3 + \frac{2}{15} z^5 + \cdots$

25. $\frac{1}{2i} + \frac{3}{(2i)^2} z + \frac{7}{(2i)^3} z^2 + \frac{15}{(2i)^4} z^3 + \cdots, R = 1$

27. $2\sqrt{5}$

29. $\sum_{k=0}^{\infty} (-1)^k (z + 1)^k, R = 1;$

$\sum_{k=0}^{\infty} \frac{(-1)^k}{(2+i)^{k+1}} (z - i)^k, R = \sqrt{5}$

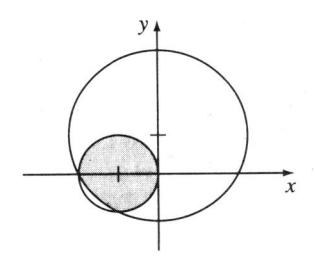

31. (a) The distance from z_0 to the branch cut is one unit.
(c) The series converges within the circle $|z + 1 - i| = \sqrt{2}$. Although the series converges in the shaded region, it does not converge to (or represent) Ln z in this region.

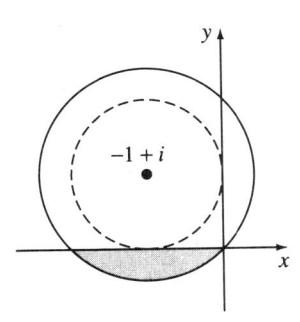

33. $1.1 + 0.12i$ **35.** $\frac{2}{\sqrt{\pi}} \sum_{k=0}^{\infty} \frac{(-1)^k}{(2k+1)k!} z^{2k+1}$

Exercises 18.3, Page 1028

1. $\frac{1}{z} - \frac{z}{2!} + \frac{z^3}{4!} - \frac{z^5}{6!} + \cdots$

3. $1 - \frac{1}{1! \cdot z^2} + \frac{1}{2! \cdot z^4} - \frac{1}{3! \cdot z^6} + \cdots$

5. $\frac{e}{z-1} + e + \frac{e(z-1)}{2!} + \frac{e(z-1)^2}{3!} + \cdots$

7. $-\frac{1}{3z} - \frac{1}{3^2} - \frac{z}{3^3} - \frac{z^2}{3^4} - \cdots$

9. $\frac{1}{3(z-3)} - \frac{1}{3^2} + \frac{z-3}{3^3} - \frac{(z-3)^2}{3^4} + \cdots$

11. $\cdots - \frac{1}{3(z-4)^2} + \frac{1}{3(z-4)} - \frac{1}{12} + \frac{z-4}{3 \cdot 4^2} - \frac{(z-4)^2}{3 \cdot 4^3} + \cdots$

13. $\cdots - \frac{1}{z^2} - \frac{1}{z} - \frac{1}{2} - \frac{z}{2^2} - \frac{z^2}{2^3} - \cdots$

15. $\frac{-1}{z-1} - 1 - (z-1) - (z-1)^2 - \cdots$

17. $\frac{1}{3(z+1)} - \frac{2}{9} - \frac{2(z+1)}{3^3} - \frac{2(z+1)^2}{3^4} - \cdots$

19. $\cdots - \frac{1}{3z^2} + \frac{1}{3z} - \frac{1}{3} - \frac{z}{3 \cdot 2} - \frac{z^2}{3 \cdot 2^2} - \cdots$

21. $\frac{1}{z} + 2 + 3z + 4z^2 + \cdots$

23. $\frac{1}{z-2} - 3 + 6(z-2) - 10(z-2)^2 + \cdots$

25. $\frac{3}{z} - 4 - 4z - 4z^2 - \cdots$

27. $\cdots + \frac{2}{(z-1)^3} + \frac{2}{(z-1)^2} + \frac{2}{z-1} + 1 + (z-1)$

Exercises 18.4, Page 1032

1. Define $f(0) = 2$.
3. $-2 + i$ is a zero of order 2.
5. $-i$ and i are zeros of order 1; 0 is a zero of order 2.
7. $2n\pi i, n = 0, \pm 1, \ldots,$ are zeros of order 1.
9. Order 5   **11.** Order 1
13. $-1 \pm 2i$ are simple poles.
15. -2 is a simple pole; $-i$ is a pole of order 4.
17. $(2n+1)\pi/2, n = 0, \pm 1, \ldots,$ are simple poles.
19. 0 is a pole of order 2.
21. $2n\pi i, n = 0, \pm 1, \ldots,$ are simple poles.
23. Nonisolated

Exercises 18.5, Page 1038

1. $\frac{2}{5}$ **3.** -3 **5.** 0
7. $\text{Res}(f(z), -4i) = \frac{1}{2}, \text{Res}(f(z), 4i) = \frac{1}{2}$
9. $\text{Res}(f(z), 1) = \frac{1}{3}, \text{Res}(f(z), -2) = -\frac{1}{12},$ $\text{Res}(f(z), 0) = -\frac{1}{4}$

11. Res($f(z)$, -1) = 6, Res($f(z)$, -2) = -31,
    Res($f(z)$, -3) = 30
13. Res($f(z)$, 0) = $-3/\pi^4$, Res($f(z)$, π) = $(\pi^2 - 6)/2\pi^4$
15. Res($f(z)$, $(2n+1)\pi/2$) = $(-1)^{n+1}$, $n = 0, \pm 1, \pm 2, \ldots$
17. 0; $2\pi i/9$; 0   19. πi; πi; 0   21. $\dfrac{\pi}{3}$
23. 0   25. $2\pi i \cosh 1$ 27. $-4i$ 29. $6i$
31. $-\dfrac{\pi}{3} + \dfrac{\pi}{3}i$

Exercises 18.6, Page 1046

1. $\dfrac{4\pi}{\sqrt{3}}$ 3. 0   5. $\dfrac{\pi}{\sqrt{3}}$ 7. $\dfrac{\pi}{4}$
9. $\dfrac{\pi}{6}$ 11. π 13. $\dfrac{\pi}{16}$ 15. $\dfrac{3\pi}{8}$
17. $\dfrac{\pi}{2}$ 19. $\dfrac{\pi}{\sqrt{2}}$ 21. πe^{-1} 23. πe^{-1}
25. πe^{-3} 27. $\dfrac{\pi e^{-\sqrt{2}}}{2\sqrt{2}}(\cos\sqrt{2} + \sin\sqrt{2})$
29. $-\dfrac{\pi}{8}\left(\dfrac{e^{-3}}{3} - e^{-1}\right)$

Chapter 18 Review Exercises, Page 1048

1. True   3. False   5. True   7. True
9. $\dfrac{1}{\pi}$ 11. $|z - i| = \sqrt{5}$
13. $1 + \sum_{k=1}^{\infty} \dfrac{(\sqrt{2})^k \cos\dfrac{k\pi}{4}}{k!} z^k$
15. $-\dfrac{i}{z^3} + \dfrac{1}{2!z^2} + \dfrac{i}{3!z} - \dfrac{1}{4!} - \dfrac{i}{5!}z + \cdots$
17. $\cdots + \dfrac{1}{5!(z-i)^3} - \dfrac{1}{3!(z-i)} + (z-i)$
19. $\dfrac{2}{3} + \dfrac{8}{9}z + \dfrac{26}{27}z^2 + \cdots$;
    $\cdots - \dfrac{1}{z^3} - \dfrac{1}{z^2} - \dfrac{1}{z} - \dfrac{z}{3} - \dfrac{z^2}{3^2} - \dfrac{z^2}{3^3} - \cdots$;
    $\dfrac{2}{z^2} + \dfrac{8}{z^3} + \dfrac{26}{z^4} + \cdots$;
    $-\dfrac{1}{z-1} - \dfrac{1}{2} - \dfrac{z-1}{2^2} - \dfrac{(z-1)^2}{2^3} - \cdots$
21. $\dfrac{404\pi}{81}i$ 23. $\dfrac{2\pi}{\sqrt{3}}i$
25. $(\pi + \pi e^{-2}\cos 2)i$ 27. $-\pi i$
29. $\dfrac{9\pi^3 + 2}{\pi^2}i$ 31. $\dfrac{7\pi}{50}$ 33. $\pi\left(\dfrac{90 - 52\sqrt{3}}{12 - 7\sqrt{3}}\right)$

Exercises 19.1, Page 1055

1. The line $v = -u$ 3. The line $v = 2$
5. Open line segment from 0 to πi
7. The ray $\theta = \tfrac{1}{2}\theta_0$ 9. The line $u = 1$
11. The fourth quadrant
13. The wedge $\pi/4 \leq \text{Arg } w \leq \pi/2$
15. The circle with center $w = 4i$ and radius $r = 1$
17. The strip $-1 \leq u \leq 0$
19. The wedge $0 \leq \text{Arg } w \leq 3\pi/4$
21. $w = -i(z - i) = -iz - 1$
23. $w = 2(z - 1)$ 25. $w = -z^4$
27. $w = e^{3z/2}$ 29. $w = -z + i$

Exercises 19.2, Page 1063

1. Conformal at all points except $z = \pm 1$
3. Conformal at all points except $z = \pi i \pm 2n\pi i$
5. Conformal at all points outside the interval $[-1, 1]$ on the x-axis
7. The image is the region shown in Figure 19.11(b). A horizontal segment $z(t) = t + ib$, $0 < t < \pi$, is mapped onto the lower or upper portion of the ellipse
$$\dfrac{u^2}{\cosh^2 b} + \dfrac{v^2}{\sinh^2 b} = 1$$
according to whether $b > 0$ or $b < 0$.
9. The image of the region is the wedge $0 \leq \text{Arg } w \leq \pi/4$. The image of the line segment $[-\pi/2, \pi/2]$ is the union of the line segments joining $e^{i\pi/4}$ to 0 and 0 to 1.
11. $w = \cos(\pi z/2)$ using H-4 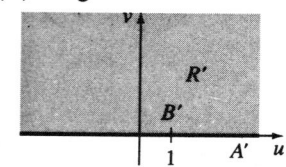
13. $w = \left(\dfrac{1+z}{1-z}\right)^{1/2}$ using H-5 and $w = z^{1/4}$ 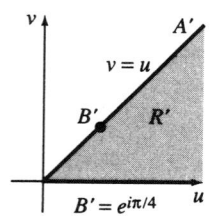
15. $w = \left(\dfrac{e^{\pi/z} + e^{-\pi/z}}{e^{\pi/z} - e^{-\pi/z}}\right)^{1/2}$ using H-6 and $w = z^{1/2}$

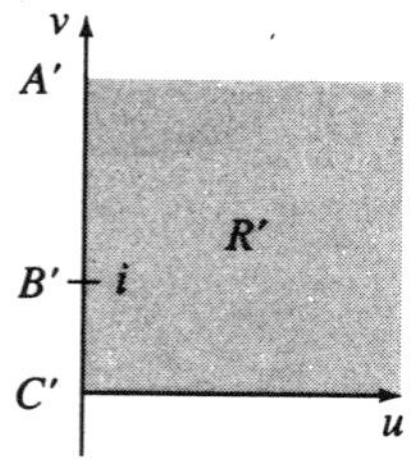

17. $w = \sin(-i \operatorname{Ln} z - \pi/2)$, $A'B'$ is the real interval $(-\infty, -1]$

19. $u = \dfrac{1}{\pi} \operatorname{Arg}(z^4)$ or $u(r, \theta) = \dfrac{4}{\pi}\theta$

21. $u = \dfrac{1}{\pi} \operatorname{Arg}\left(i\dfrac{1-z}{1+z}\right) = \dfrac{1}{\pi}\tan^{-1}\left(\dfrac{1 - x^2 - y^2}{2y}\right)$

23. $u = \dfrac{1}{\pi}[\operatorname{Arg}(z^2 - 1) - \operatorname{Arg}(z^2 + 1)]$

25. $u = \dfrac{10}{\pi}[\operatorname{Arg}(e^{\pi z} - 1) - \operatorname{Arg}(e^{\pi z} + 1)]$

Exercises 19.3, Page 1072

1. $T(0) = \infty$, $T(1) = i$, $T(\infty) = 0$; $|w| = 1$ and the line $v = \tfrac{1}{2}$; $|w| \geq 1$
3. $T(0) = -1$, $T(1) = \infty$, $T(\infty) = 1$; the line $u = 0$ and the circle $|w - 1| = 2$; the half-plane $u \leq 0$
5. $S^{-1}(w) = \dfrac{-w-1}{-w+i} = \dfrac{w+1}{w-i}$, $S^{-1}(T(z)) = \dfrac{(1+i)z - 1}{2z + i}$
7. $S^{-1}(w) = \dfrac{-w+2}{-w+1} = \dfrac{w-2}{w-1}$, $S^{-1}(T(z)) = \dfrac{3}{z}$
9. $w = -2\dfrac{z+1}{z-2}$
11. $w = \dfrac{2z}{z - 1 - 2i}$
13. $w = \dfrac{i}{2}\dfrac{z-1}{z}$
15. $w = 3\dfrac{(1+i)z + 1 - i}{(-3+5i)z - 3 - 5i}$
17. $u = \dfrac{1}{\log_e 2}\log_e\left|\dfrac{z+2}{z-1}\right|$. The level curves are the images of the circles $|w| = r$, $1 < r < 2$, under the linear fractional transformation $T(w) = (w+2)/(w-1)$. Since the circles do not pass through the pole at $w = 1$, the images are circles.
19. Construct the linear fractional transformation that sends $1, i, -i$ to $0, 1, -1$.
21. Simplify $T_2(T_1(z)) = \dfrac{a_2\left(\dfrac{a_1 z + b_1}{c_1 z + d_1}\right) + b_2}{c_2\left(\dfrac{a_1 z + b_1}{c_1 z + d_1}\right) + d_2}$.

Exercises 19.4, Page 1078

1. First quadrant
3.

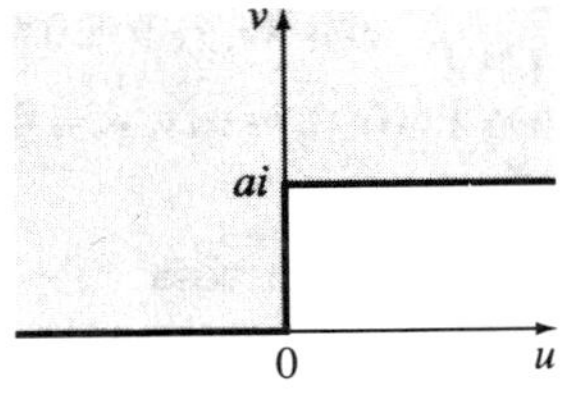

5. $f'(z) = A(z+1)^{-1/2}z^{-1/2}(z-1)^{-1/2}$ for some constant A
7. $f'(z) = A(z+1)^{-1/3}z^{-1/3}$ for some constant A

9. Show that $f'(z) = \dfrac{A}{(z^2 - 1)^{1/2}}$ and conclude that $f(z) = \cosh^{-1} z$.
11. Show that $f'(z) \to A/z$ as $w_1 \to \infty$ and conclude that $f(z) = \operatorname{Ln} z$.
13. Show that $f'(z) \to A(z+1)^{-1/2}z(z-1)^{-1/2} = Az/(z^2-1)^{1/2}$ as $u_1 \to 0$.

Exercises 19.5, Page 1085

1. $u = \dfrac{1}{\pi}\operatorname{Arg}\left(\dfrac{z-1}{z}\right) - \dfrac{1}{\pi}\operatorname{Arg}\left(\dfrac{z}{z+1}\right)$
3. $u = \dfrac{5}{\pi}[\pi - \operatorname{Arg}(z-1)] + \dfrac{1}{\pi}\operatorname{Arg}\left(\dfrac{z}{z+1}\right) - \dfrac{1}{\pi}\operatorname{Arg}\left(\dfrac{z+1}{z+2}\right)$
5. $u = \dfrac{y}{\pi}\left\{1 + \dfrac{y^2 - x^2}{y}\left[\tan^{-1}\left(\dfrac{x-1}{y}\right) - \tan^{-1}\left(\dfrac{x}{y}\right)\right] + x\log_e\left[\dfrac{(x-1)^2 + y^2}{x^2 + y^2}\right]\right\}$
7. $u = \dfrac{1}{\pi}\operatorname{Arg}\left(\dfrac{z^2 - 1}{z^2}\right) + \dfrac{5}{\pi}\operatorname{Arg}(z^2 + 1)$
9. $u = \dfrac{1}{\pi}\operatorname{Arg}\left(\dfrac{(1-i)z - (1+i)}{1 - z}\right) - \dfrac{1}{\pi}\operatorname{Arg}\left(\dfrac{1-z}{-(1+i)z + 1 - i}\right)$
11. $u(0,0) = \tfrac{1}{3}$, $u(-0.5, 0) = 0.5693$, $u(0.5, 0) = 0.1516$
13. Show that $u(0, 0) = \dfrac{1}{2\pi}\displaystyle\int_{-\pi}^{\pi} u(e^{it})\, dt$.
15. $u(r, \theta) = r\sin\theta + r\cos\theta$ or $u(x, y) = x + y$

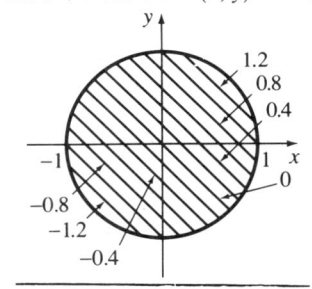

Exercises 19.6, Page 1092

1. $g(z) = e^{-i\theta_0}$ is analytic everywhere and $G(z) = e^{-i\theta_0}z$ is a complex potential. The equipotential lines are the lines $x\cos\theta_0 + y\sin\theta_0 = c$.

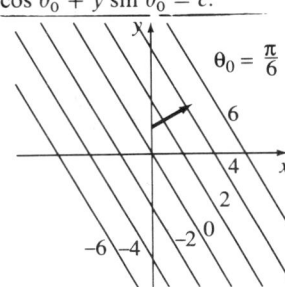

A-91

3. $g(z) = 1/z$ is analytic for $z \neq 0$ and $G(z) = \text{Ln } z$ is analytic except for $z = x \leq 0$. The equipotential lines are the circles $x^2 + y^2 = e^{2c}$.

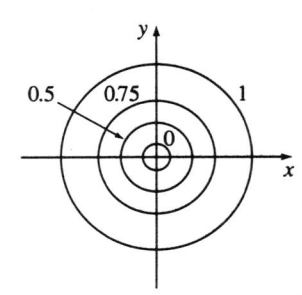

5. $\phi = \dfrac{4}{\pi} \text{Arg } z$ or $\phi(r, \theta) = \dfrac{4}{\pi} \theta$, and $G(z) = \dfrac{4}{\pi} \text{Ln } z$ is a complex potential. The equipotential lines are the rays $\theta = \dfrac{\pi}{4} c$ and $\mathbf{F} = \dfrac{4}{\pi}\left(\dfrac{x}{x^2 + y^2}, \dfrac{y}{x^2 + y^2}\right)$.

7. The equipotential lines are the images of the rays $\theta = \theta_0$ under the successive transformations $\zeta = w^{1/2}$ and $z = (\zeta + 1)/(-\zeta + 1)$. The transformation $\zeta = w^{1/2}$ maps the ray $\theta = \theta_0$ to the ray $\theta = \theta_0/2$ in the ζ-plane, and $z = (\zeta + 1)/(-\zeta + 1)$ maps this ray onto an arc of a circle which passes through $z = -1$ and $z = 1$.

9. (a) $\psi(x, y) = 4xy(x^2 - y^2)$ or, in polar coordinates, $\psi(r, \theta) = r^4 \sin 4\theta$. Note that $\psi = 0$ on the boundary of R.
   (b) $\mathbf{V} = \overline{4z^3} = 4(x^3 - 3xy^2, y^3 - 3x^2 y)$
   (c)

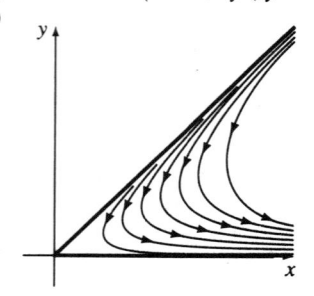

11. (a) $\psi(x, y) = \cos x \sinh y$ and $\psi = 0$ on the boundary of R.
    (b) $\mathbf{V} = \overline{\cos z} = (\cos x \cosh y, \sin x \sinh y)$
    (c)

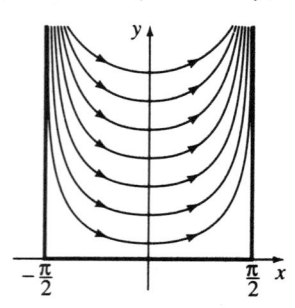

13. (a) $\psi(x, y) = 2xy - \dfrac{2xy}{(x^2 + y^2)^2}$ or, in polar coordinates, $\psi(r, \theta) = (r^2 - 1/r^2) \sin 2\theta$. Note that $\psi = 0$ on the boundary of R.
    (b) $\mathbf{V} = 2\bar{z} - 2/\bar{z}^3$
    (c)

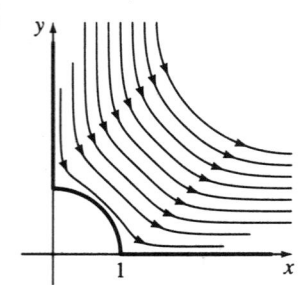

15. (a) $f(t) = \pi i - \dfrac{1}{2}[\log_e|t + 1| + \log_e|t - 1| + i \text{Arg}(t + 1) + i \text{Arg}(t - 1)]$ and so
$$\text{Im}(f(t)) = \begin{cases} 0, & t < -1 \\ \pi/2, & -1 < t < 1 \\ \pi, & t > 1 \end{cases}$$
Hence, $\text{Im}(G(z)) = \psi(x, y) = 0$ on the boundary of R.
   (b) $x = -\dfrac{1}{2}[\log_e|t + 1 + ic| + \log_e|t - 1 + ic|]$
   $y = \pi - \dfrac{1}{2}[\text{Arg}(t + 1 + ic) + \text{Arg}(t - 1 + ic)]$, for $c > 0$
   (c)

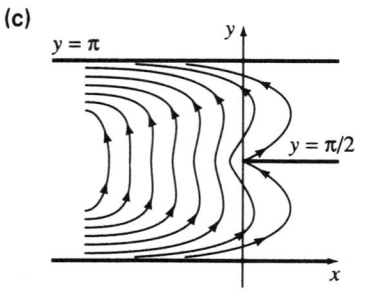

17. (a) $f(t) = \dfrac{1}{\pi}((t^2 - 1)^{1/2} + \cosh^{-1} t) = \dfrac{1}{\pi}((t^2 - 1)^{1/2} + \text{Ln}(t + (t^2 - 1)^{1/2}))$ and so
$$\text{Im}(f(t)) = \begin{cases} 1, & t < -1 \\ 0, & t > 1 \end{cases} \text{ and } \text{Re}(f(t)) = 0 \text{ for}$$
$-1 < t < 1$. Hence, $\text{Im}(G(z)) = \psi(x, y) = 0$ on the boundary of R.
   (b) $x = \text{Re}\left[\dfrac{1}{\pi}\left(((t + ic)^2 - 1)^{1/2} + \cosh^{-1}(t + ic)\right)\right]$
   $y = \text{Im}\left[\dfrac{1}{\pi}\left(((t + ic)^2 - 1)^{1/2} + \cosh^{-1}(t + ic)\right)\right]$
   for $c > 0$

(c)

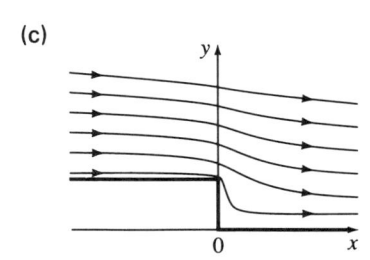

19. $z = 0$ in Example 5; $z = 1$, $z = -1$ in Example 6
21. The streamlines are the branches of the family of hyperbolas $x^2 + Bxy - y^2 - 1 = 0$ that lie in the first quadrant. Each member of the family passes through $(1, 0)$.
23. *Hint:* For z in the upper half-plane,
$$k[\text{Arg}(z - 1) - \text{Arg}(z + 1)] = k\,\text{Arg}\left(\frac{z-1}{z+1}\right).$$

Chapter 19 Review Exercises, Page 1096

1. $v = 4$ **3.** The wedge $0 \le \text{Arg}\, w \le 2\pi/3$
5. True   **7.** $0, 1, \infty$ **9.** False
11. The image of the first quadrant is the strip $0 < v < \pi/2$. Rays $\theta = \theta_0$ are mapped onto horizontal lines $v = \theta_0$ in the w-plane.
13. $w = \dfrac{i - \cos \pi z}{i + \cos \pi z}$

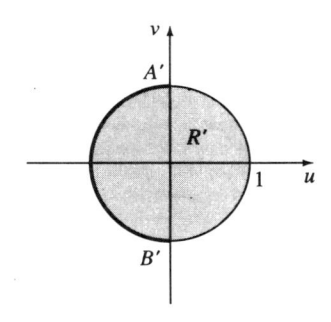

15. $u = 2 - 2y/(x^2 + y^2)$
17. (a) Note that $\alpha_1 \to 0$, $\alpha_2 \to 2\pi$, and $\alpha_3 \to 0$ as $u_1 \to \infty$.
  (b) *Hint:* Write $f(t) = \frac{1}{2}A[\log_e|t + 1| + \log_e|t - 1| + i\,\text{Arg}(t + 1) + i\,\text{Arg}(t - 1)] + B$.
19. $G(z) = f^{-1}(z)$ maps R to the strip $0 \le v \le \pi$, and $U(u, v) = v/\pi$ is the solution to the transferred boundary problem. Hence, $\phi(x, y) = (1/\pi)\text{Im}(G(z)) = (1/\pi)\psi(x, y)$, and so the equipotential lines $\phi(x, y) = c$ are the streamlines $\psi(x, y) = c\pi$.

Appendix I Exercises, Page A-4

1. $24;\ 720;\ 4\sqrt{\pi}/3;\ -8\sqrt{\pi}/15$ **3.** 0.297
5. $\Gamma(x) > \displaystyle\int_0^1 t^{x-1}e^{-t}\,dt > e^{-1}\int_0^1 t^{x-1}\,dt = \dfrac{1}{xe}$
for $x > 0$. As $x \to 0^+$, $1/x \to +\infty$.

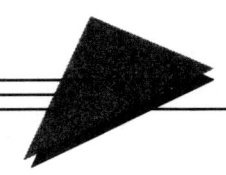

INDEX

Abel, Niels Henrik, 126, 846
Abel's formula, 125
Absolute convergence, 1009
  of a power series, 278
Absolute error, 872
Acceleration, 483
  normal component of, 490
  tangential component of, 490
Adams-Bashforth/Adams-Moulton
    method, 885
Adjoint matrix, 416
Amplitude of vibrations, 177
Analytic function, 950–951
Analyticity at a point, 951
Angle between two vectors, 342
Angle preserving mappings, 1056
Annihilator differential operator, 150
Annulus, 944
Antiderivative of a complex function, 991
Applications of differential equations, 81,
    92
  atomic physics, 82–84
  beams, 249
  biology, 27, 28, 81, 92
  carbon dating, 83
  chemistry, 95
  continuous compound interest, 28, 89
  cooling, 23, 85
  deflection of beams, 25–27, 249–250,
    289, 754
  discharge through an orifice, 25
  electric circuits, 22, 24, 86, 196
  geometric problems, 13, 76
  mixtures, 87
  spring-mass systems, 18, 172
  temperature distributions, 771–772, 774,
    777, 784
  vibrating string, 773, 780
Arc, 677
Area
  as a double integral, 542, 557
  of a parallelogram, 353
  of a surface, 567
Argument of a complex number, 937
  principal, 937
Arithmetic modulo 2, 458
Asymptotic stability, 689

Augmented matrix, 391
Autonomous systems, 674
Auxiliary equation, 131, 272
Axis of symmetry, 25

Back substitution, 390
Backward difference, 893
Banded matrix, 899
BASIC programs, A-20–A-27
Basis of a vector space, 370
  standard, 371
Beams
  static deflection of, 25–27, 249–250
  vibrating, 783
Beats, 198
Bendixson negative criterion, 709
Bernoulli, Jakob, 66
Bernoulli's differential equation, 67
Bessel, Friedrich Wilhelm, 307
Bessel functions
  differential recurrence relations for, 311,
    756
  of first kind, 309
  graphs of, 310
  modified, 317
  properties of, 311
  of second kind, 310
  spherical, 312–313
Bessel's differential equation, 307
  parametric, 311
  solution of, 307
Biharmonic function, 1066
Binormal, 490
Bits, 459
Boundary
  conditions, 774–775
  point, 897, 944
  of a set, 944
Boundary-value problem, 108, 776
  numerical solution of, 892, 896
Bounding theorem for complex integrals,
    981
Box product, 353
Branch
  of the complex logarithm, 964
  cut, 965
Buckling of a thin column, 289, 754

Canonical form, of a system of linear
    differential equations, 620
Carbon dating, 83–84
Cartesian coordinates, 332
Catenary, 25
Cauchy, Augustin-Louis, 271
Cauchy-Euler differential equation, 271
  method of solution, 272, 276
Cauchy-Goursat theorem, 985
  for multiply connected domains,
    986–987
Cauchy-Riemann equations, 952
Cauchy-Schwarz inequality, 348
Cauchy's inequality, 1001
Cauchy's integral formula, 997
  for derivatives, 999
Cauchy's residue theorem, 1035
Cauchy's theorem, 984
Cayley, Arthur, 365, 376
Cayley-Hamilton theorem, 437
Center
  of curvature, 493
  of mass, 548, 585
Central difference, 893
  approximations for derivatives, 892–893,
    896
Centripetal acceleration, 484
Centroid, 548, 586
Chain rule of partial derivatives, 498–499
Change of variables
  in a definite integral, 605
  in a double integral, 608
  in a triple integral, 611
Characteristic equation, 131, 432
Characteristic functions, 748
Characteristic values, 430, 748
Characteristic vectors, 430
Chebyshev's differential equation, 763
Chemical equations, 397–398
Chemical reactions, 95–98
Circle
  in complex plane, 942
  of convergence, 1011
  of curvature, 493
Circulation, 530, 982
Clairaut, Alexis Claude, 66
Clairaut's differential equation, 69

A-95

Classification of a differential equation
 as linear or nonlinear, 5
 by order, 5,
 by type, 4
Closure axioms, 367
Codes, 454, 459
Code words, 459
Cofactor, 401
Commutator of two matrices, 470
Complementary error function, 819
Complementary function, 122, 636
Complete set of functions, 731
Complete solution, 11
Complex form of Fourier series, 737
Complex function, 945
 analytic, 950–951
 continuous, 948
 differentiable, 949
 domain of, 945
 entire, 951
 exponential, 959
 hyperbolic, 969
 inverse hyperbolic, 973
 inverse trigonometric, 971–972
 limit of, 947
 logarithmic, 962
 as a mapping, 945–946, 1051
 polynomial, 948
 rational, 949
 trigonometric, 967
 as a two-dimensional fluid flow, 946
Complex numbers, 932
 addition of, 932
 argument of, 937
 conjugate of, 933
 division of, 933–934, 937
 equality of, 932
 imaginary part of, 932
 modulus of, 935
 multiplication of, 932, 937
 polar form of, 936, 961
 powers of, 938
 pure imaginary, 932
 real part of, 932
 roots of, 939–940
 subtraction of, 932
 triangle inequality for, 935
 vector interpretation, 935
Complex impedance, 961
Complex integral, 978
Complex plane, 935
Complex potential, 1088
Complex powers, 965
 principal value of, 965
Complex sequence, 1007
Complex series, 1008
Complex velocity potential, 1090
Component of a vector on another, 343
Conformal mapping, 1056–1057
 and the Dirichlet problem, 1056–1063
Conformal mappings, table of, A-9–A-19
Conjugate
 harmonic functions, 956
 of a complex number, 933
Connected set, 944

Conservative vector field, 538
Consistent system of linear algebraic
 equations, 388
Continuing method, 885
Continuity
 equation, 603
 of a function of a complex variable, 948
Continuous compound interest, 28
Contour, 978
 indented, 1044
 integral, 978
Convergence
 of a Fourier-Bessel series, 758
 of a Fourier integral, 829
 of a Fourier-Legendre series, 735
 of a Fourier series, 760
 of an integral, 1041
 of a power series, 278
Convolution
 integral, 234
 theorem, 235
Cooling, Newton's Law of, 23, 86
Coordinate planes, 333
Coplanar vectors, 354
Coupled springs, 261
Coupled systems, 656
Cover-up method, 222–223
Cramer, Gabriel, 427
Cramer's rule, 427–428
Crank-Nicholson method, 908
Critical damping, 182
Critical loads, 754
Critical points for linear systems, 676
 center, 684
 degenerate nodes, 683
 locally stable, 681
 saddle point, 682
 stable node, 681
 stable spiral point, 685
 unstable, 681
 unstable node, 682
 unstable spiral point, 685
Critical points, for plane autonomous
 systems
 globally stable, 708
 stable, 689
 unstable, 689
Critical speeds, 754
Critically damped circuit, 199
Cross product, 349
 properties of, 350
Cross ratio, 1070
Cryptography, 454
Curl of a vector field, 517
 physical interpretation of, 519
Curvature, 489
Curves
 closed, 521
 piecewise smooth, 521
 positive direction on, 521
 simple closed, 521
 smooth, 521
Cycle, 677
Cylindrical coordinates, 588–589, 810
 Laplacian in, 810

D'Alembert, Jean le Rond, 784
D'Alembert's solution, 784
Damped amplitude, 186
Damped motion, 181
Damping constant, 181
Damping factor, 182
Decay, radioactive, 81
Decoding a message, 455, 459
Definite integral, 522
Deflation, method of, 922
Deflection curve, 21, 26
Deformation of contours, 986
Degenerate system of differential equations,
 622
Del (∇) operator, 502–503
DeMoivre, Abraham, 939
DeMoivre's formula, 939
Derivative (differentiation)
 of complex exponential functions, 959
 of a complex function, 949, 954
 of complex hyperbolic functions, 970
 of complex inverse hyperbolic functions,
 974
 of complex inverse trigonometric
 function, 973
 of complex logarithm, 965
 of complex trigonometric functions, 968
 rules of, 949–950
Determinants, 400
 cofactor, 401
 evaluating by row reduction, 410
 expansion by cofactors, 157, 404
 minor, 401
 properties of, 406–412
Diagonalizable matrix, 445
Diagonalizability, criterion for, 446–447
Diagonally dominant matrix, 901
Diagonal matrix, 383
Difference equation, 893
 replacement for heat equation, 903–904
 replacement for Laplace's equation, 896
 replacement for wave equation, 912
Difference quotients, 892
Differential
 for a function of several variables, 533–
 534
 operator, 148–149, 502
 recurrence relation, 311–312
 of surface area, 568
Differential equation, 4
 of a family of curves, 13
 linear, 5
 nonlinear, 6
 ordinary, 4
 partial, 4–5
Diffusion equation, 776
Dimension of a vector space, 371
Dirac, Paul Adrian Maurice, 256
Dirac delta function, 256, 258
 Laplace transform of, 257
Dirichlet, Peter G. L., 774
Dirichlet condition, 774
Dirichlet problem, 786, 1060
Direction
 angles, 342

cosines, 342
field, 867
vector, 356
Directional derivative
  for functions of two variables, 504–505
  for function of three variables, 506–507
Discharge through an orifice, 25
Discretization error, 887
Distance formula, 334
Distributions, 258
Divergence of a vector field, 518
  physical interpretation of, 517–518, 601
Divergence theorem, 597
Domain, 944
Dominant eigenvalue, 917
  eigenvector, 917
Dot product, 339
  properties of, 340
  in terms of matrices, 438–439
  as work, 346
Double cosine series, 746
Double eigenvalues, 804
Double integral, 542
  evaluation of, 544–545
  in polar coordinates, 554–555
Double sine series, 746, 797
Driving force, 189
Dulac negative criterion, 711

Effective spring constant, 180, 207
Eigenfunctions, 748–749
Eigenvalues of a boundary-value problem, 748–749
Eigenvalues of a matrix, 430
  complex, 435
  real, 438
Eigenvectors of a matrix, 430
  complex, 435
  orthogonal, 439
Elastic curve, 26
Electrical networks, 264
Electrical vibrations, 199
Elementary matrix, 399
Elementary operations on a linear system, 389
Elementary row operations, 392
  notation for, 393
Elimination methods, 392
Elliptical helix, 475
Elliptic partial differential equation, 769
Embedded beam, 783
Emf, 23
Encoding a message, 455, 459
Entire function, 951
Equation of motion, 174
Equidimensional equation, 271
Equilibrium position, 173
Error(s)
  discretization, 887
  function, 819
  in midpoint rule, 856
  percentage, 872
  propagation, 887
  relative, 872
  round-off, 887

in Simpson's rule, 862
in trapezoidal rule, 858
truncation, 887
Escape velocity, 99
Essential singularity, 1029
Euler, Leonhard, 132
Euler load, 754
Euler-Mascheroni constant, 311
Euler's formula, 132
Euler's methods, 871, 874
  for systems, 891
Evaluation of real integrals by residues, 1039
Exact differential equation, 53
Expansion of a function
  in a complex Fourier series, 737
  in a cosine series, 739
  in a Fourier series, 734
  in a Fourier-Bessel series, 758
  in a Fourier-Legendre series, 760
  in a sine series, 739
  in terms of orthogonal functions, 729–730
Explicit finite difference method, 904
Exponential function, 958–959
  derivative of, 959
  fundamental region for, 960
  period of, 960
  properties of, 960
Exponential order, 211
Exponents of a singularity, 295
Extreme displacement, 174

Family of solutions, 10
Farrari, Lodovico, 846
Fick's law, 91
Finite difference equation, 893
Finite difference method
  explicit, 904
  implicit, 908
First harmonic, 782
First moments, 585
First normal mode, 782
First-order chemical reactions, 95
First-order differential equations
  applications of, 76, 81, 92
  solutions of, 39, 46, 52, 58, 66, 70
First standing wave, 782
First translation theorem, 224–225
Five-point approximation for Laplace's equation, 896
Flow
  of heat, 771–772
  steady-state fluid, 1089
Flux, 573, 982
Forced motion, 189
Forward difference, 893
Fourier, Jean-Baptiste Joseph, 733
Fourier-Bessel series, 758
Fourier coefficients, 733
Fourier cosine transform, 835
  operational properties of, 837
Fourier integral, 828–829, 1043
  complex form, 833
  conditions for convergence, 829

cosine form, 830–831
sine form, 830–831
Fourier-Legendre series, 760
Fourier series, 733–734
  complex, 737
  conditions for convergence, 735
  cosine, 739
  generalized, 731
  sine, 739
  in two variables, 746, 797
Fourier sine transform, 835
  operational properties of, 836
Fourier transform(s), 835–836
Free end conditions, 254
Free motion
  damped, 181
  undamped, 174
Frequency, 174
Frobenius, Ferdinand Georg, 292
Frobenius, method of, 292
Frobenius' theorem, 292
Full-wave rectification of sine, 242
Function
  complex, 945
  error, 819
  even, 737
  even, properties of, 738
  excitation, 123
  forcing, 123
  three or more variables, 494
  two variables, 495
Fundamental frequency, 782
Fundamental mode of vibration, 782
Fundamental matrix, 637
Fundamental region, 960
Fundamental set of solutions
  of a linear differential equation, 117
  of a system of linear differential equations, 634
Fundamental theorem
  of algebra, 1002
  for contour integrals, 992

Galloping Gertie, 195
Gamma function, 308, A-3–A-5
Gauss, Karl Friedrich, 392
Gaussian elimination, 392
Gauss-Jordan elimination, 392
Gauss-Siedel iteration, 398, 899–900
Gauss' theorem, 597
Generalized factorial function, A-3
Generalized Fourier series, 729–731
Generalized functions, 258
General solution
  of a differential equation, 11, 119, 121
  of a system of linear differential equations, 634, 636
Generating function, 317
Geometric series, 1008
Gibbs, Josiah Willard, 741
Gibbs' phenomenon, 740–741
Globally stable critical point, 708
Gompertz curve, 95
Gradient, 502–503
  geometric interpretation, 510–512

Gradient (*continued*)
  field, 538
Great circles, 594
Green's theorem, 559
Growth and decay, 81

Half-life, 82
Half-range expansions, 741
Half-wave rectification of sine, 242
Hamilton, William Rowan, 365
Hamming (7, 4) code, 460
Hamming (8, 4) code, 464
Hanging wire, 24
Harmonic functions, 955
  tranformation theorem for, 1061
Heat equation
  derivation one dimensional equation, 771–772
  one dimensional, 771
  solution of, 777
  two dimensional, 795
Heaviside, Oliver, 223
Heaviside function, 226
Heaviside's expansion theorem, 222–223
Hermite, Charles, 289
Hermite's differential equation, 289, 755
Heun's formula, 874
Homogeneous boundary conditions, 749
Homogeneous differential equation
  ordinary, 46, 114, 131
  partial, 766
Homogeneous functions of degree n, 46
Homogeneous systems of algebraic equations, 388, 396
  nontrivial solutions of, 397
  trivial solution of, 396
Homogeneous systems of linear differential equations, 641
  complex eigenvalues, 645–649
  distinct real eigenvalues, 641–645
  repeated eigenvalues, 649–654
Hooke, Robert, 172
Hooke's law, 172
Hyperbolic partial differential equation, 769

Identity matrix, 384
i, j vectors, 328
i, j, k vectors, 336
Imaginary axis, 935
Impedance, 202
Implicit solution, 7
Impressed force, 123, 189
Improved Euler method, 874
Incompressible fluid, 519
Inconsistent system of linear algebraic equations, 388
Indented contours, 1044
Independence of path, 534, 990
Indicial equation, 295
Indicial roots, 295
Initial conditions, 35
Initial value problem, 35, 106
  for systems of linear differential equations, 630

Inner product
  of two functions, 726
  of two vectors, 339
Input, 123
Insulated boundaries, 775
Integral
  equation, 247
  transform, 835
Integrating factor, 58, 59
Integrodifferential equation, 248
Interest, compounded continuously, 28
Interior mesh points, 893, 897
Invariant region, 712
Inverse Fourier
  cosine transform, 835
  sine transform, 835
  transform, 835
Inverse hyperbolic functions, 973
  derivatives of, 974
Inverse integral transform, 835
Inverse Laplace transform, 217
Inverse of a matrix, 414
  finding, by the adjoint method, 416
  finding, by elementary row operations, 419
Inverse power method, 924
Inverse transform, 835
Inverse trigonometric functions, 971–972
  derivatives of, 973
Invertible matrix, 414
Irregular singular point, 290
  at ∞, 306
Irrotational flow, 519
Isoclines, 867
Isogonal trajectories, 80
Isolated singularity, 1016, 1020
Iterated integral, 544

Jacobi, Carl Gustav Jacob, 607
Jacobian, 607
Jacobian matrix, 692
Jordan, Wilhelm, 392

Kepler's second law of planetary motion, 91
Kernel of a transform, 835
Kirchhoff's point and loop rules, 395–396
Kirchhoff's second law, 199, 264
Kutta, Martin W., 880

Lagrange's identity, 356
Laguerre, Edmond, 306
Laguerre's differential equation, 306, 754
Laplace, Pierre Simon Marquis de, 209
Laplace transform
  behavior as $s \to \infty$, 221
  convolution, 234
  convolution theorem, 235
  definition of, 209
  of derivatives, 233–234
  differentiation of, 232
  of Dirac delta function, 257
  existence of, 210–211
  of integrals, 235
  inverse of, 217, 834–835
  linearity of, 210

operational properties, 224
of a partial derivative, 821
of periodic functions, 237
tables of, A-6–A-8
translation theorems of, 224, 229
Laplace's equation, 771, 784
Laplacian
  in cylindrical coordinates, 810
  in polar coordinates, 802
  in rectangular coordinates, 771
  in spherical coordinates, 813
Latitude, 594
Laurent series, 1020–1021
Laurent's theorem, 1021
Law of Malthus, 103
Law of mass action, 97–98
Least squares
  lines, 466
  method of, 465
Legendre, Adrien Marie, 307
Legendre polynomials, 315
  generating function for, 317, 318
  graphs of, 316
  properties of, 316
  recurrence relation for, 318
Legendre's differential equation, 307
  solution of, 313
Level curves, 494
Level surfaces, 495
Libby, W., 83
Liebman's method, 900
Limit cycle, 714
Lineal elements, 867
Linear combination, 328
Linear dependence
  of functions, 110
  of solution vectors, 632
  of vectors, 370
Linear equations, 387
  systems of, 387–388
Linear fractional transformation, 1066
Linear independence
  of functions, 110
  of solutions, 117
  of solution vectors, 632–633
  of vectors, 370
Linearization, 688, 691
Linear operation, 209
Linear operator, 210
Linear ordinary differential equations, 5, 105
  applications of, 81
  complementary function for, 122
  first-order, 58
  general solution of, 119, 121
  higher order, 105, 135
  homogeneous, 114
  nonhomogeneous, 114, 120
  particular solution of, 120
  superposition principle for, 115, 122
Linear second-order partial differential equations, 766
  homogeneous, 766
  nonhomogeneous, 766
  solution of, 766
  superposition principle for, 768

Linear system(s), 123
  of algebraic equations, 387
  of differential equations, 620, 627
Line integrals
  in the complex plane, 978
  evaluation of, 523–525, 527–528
  fundamental theorem for, 535
  independent of path, 534
  in the plane, 522–523
  in space, 527
Line segment, 357
Lines in 3-space
  parametric equations, 357
  symmetric equations, 358
  vector equation, 356
Liouville, Joseph, 750
Liouville's theorem, 1002
Logarithmic decrement, 189
Logarithm of a complex number, 962
  branch cut for, 965
  derivative of, 965
  principal branch of, 964
  principal value of, 963
  properties of, 964
Logistic equation, 28, 92
  modified, 98
Logistic function, 92–93
  graph of, 93–94
Longitude, 594
Lotka, A. J., 100
Lotka and Volterra, equations of, 100, 702, 704
L-R series circuit, differential equation of, 86
L-R-C series circuit, differential equation of, 199

Magnification in the z-plane, 1053
Malthus, law of, 103
Malthus, Thomas R., 103
Mass, 548, 585
  center of, 548, 585
  action, law of, 98
Mathematical model, 17
Matrices, 376
  addition of, 378
  adjoint, 416
  associative law of, 380
  augmented, 391
  banded, 899
  characteristic equation of, 432
  column, 378
  definition of, 377
  derivative of, 625–626
  determinant of, 400
  diagonal, 383
  diagonalizable, 445
  diagonally dominant, 901
  difference of, 379
  distributive law of, 381–382
  dominant eigenvalue of, 917
  eigenvalues of, 430
  eigenvectors of, 430
  elementary, 399
  equality of, 378
  exponential, 668

  fundamental, 637
  identity, 384
  integral of, 625–626
  lower triangular, 383
  nonsingular, 414
  order of, 377
  orthogonal, 440
  orthogonally diagonalizable, 449
  product of, 380
  reduced row-echelon form, 393
  row-echelon form, 392
  scalar, 383
  scalar multiple of, 379
  singular, 415
  size, 377
  skew-symmetric, 413
  sparse, 899
  square, 377
  symmetric, 384, 437
  transpose of, 382
  triangular, 383
  tridiagonal, 909
  upper triangular, 383
  vector, 377–378
  Wronski, 638
  zero, 383
Matrix form of a linear system, 421
Maxwell's equations, 521
Meander function, 241
Mechanical resonance, 193, 197
Meridian, 594
Mesh, 897
Message, 459
Method of deflation, 922
Method of Frobenius, 293
Midpoint of a line segment, 335
Midpoint rule, 855
  error bound for, 856
Milne's method, 886
  instability of, 887
Minor, 401
Mixed partial derivatives, 497, 500
  equality of, 497–498, 500
Mixtures, 87
ML-inequality, 981
Möbius strip, 571
Modified Bessel function of the first kind, 317
Modified logistic equation, 98
Modulus of a complex number, 935
Moments of inertia, 549, 586
  polar, 552, 558
Multiplicity of eigenvalues, 433, 649–653
Multiply connected
  domain, 984
  region, 536
Multistep methods, 885

n-dimensional vector, 365
Neighborhood, 942
Neumann, C. G., 310, 774
Neumann condition, 774
Neumann's function, 310
Newton-Cotes formulas, 864
Newton's law of cooling, 23, 85
Newton's method, 846–847

Newton's second law of motion, 18, 172
Nodal line, 809
Nonhomogeneous linear differential equation
  ordinary, 114
  partial, 766
Nonhomogeneous systems of algebraic equations, 388
Nonhomogeneous systems of linear differential equations, 658
Nonlinear differential equations, 6
Nonsingular matrix, 414
Norm
  of a function, 727
  of a partition, 542, 569, 583
  of a vector, 328
Normal form, 620
Normalized set of functions, 728
Normal line, 514
Normal modes, 781
Normal plane, 491
Normal vector to a plane, 359
n-parameter family of solutions, 10
nth term test, 1009
Numerical methods
  Adams-Bashforth/Adams-Moulton, 885
  Crank-Nicholson method, 908
  deflation method, 922
  Euler method, 871
  finite-difference methods, 892–893, 896, 904, 908, 912–913
  Gauss-Seidel method, 899–900
  Heun's formula, 874
  Improved Euler's method, 874
  inverse power method, 924
  midpoint rule, 854–855
  Milne's method, 886
  Newton's method, 846–847
  power method, 918
  Runge-Kutta methods, 881
  Simpson's rule, 860–861
  three-term Taylor method, 878
  trapezoidal rule, 857–858

Octants, 333
Odd function, 737
  properties of, 738
One-to-one transformation, 607
Open disk, 942
Open set, 942
Operational properties of the Laplace transform, 224
Operator, differential, 148
Order of a differential equation, 5
Ordinary differential equation, definition of, 4
Ordinary point, 281
Orientation of a surface, 571
Orthogonal curves, 76
Orthogonal eigenvectors, 439
Orthogonal functions, 726
Orthogonally diagonalizable matrix, 449
Orthogonal matrix, 425, 440–441
  constructing an, 442
Orthogonal set, 727
  with respect to a weight function, 730

Index

Orthogonal trajectories, 77
Orthogonal vectors, 341
Orthonormal set of functions, 727
Osculating plane, 491
Output, 123
Overdamped circuit, 199
Overdamped free motion, 182
Overtones, 782

Parabolic partial differential equation, 769
Parallel vectors, 324, 350
Parametric Bessel equation, 311
Parity check bits, 459
Parity check equations, 461
Parity check matrix, 462
Partial derivatives, 496
  higher-order, 497
Partial differential equation
  elliptic, 769
  homogeneous, 766
  hyperbolic, 769
  linear, 766
  nonhomogeneous, 766
  parabolic, 769
  separable, 766
Particular solution, 10
Path independence, 534
  tests for, 536, 539
Pauli spin matrices, 470
Peano, Guiseppe, 38
Pendulum motion, 20, 202, 699
Percentage relative error, 872
Periodic boundary conditions, 755
Periodic driving force, 743
Periodic extension, 735–736
Periodic functions, Laplace transform of, 237
Periodic solution of a plane autonomous system, 677
Phase angle, 177
Phase-plane method, 695
Picard, Charles Emile, 36
Picard's method of iteration, 74
Piecewise continuous functions, 210–211
Piecewise smooth curve, 521
Pitch, 386, 475
Planar transformation, 1051
Plane autonomous system, 675
  types of solutions of, 676
Planes
  Cartesian equation, 359–360
  normal vector to, 359
  vector equation, 359
Plucked string, 774, 783
Poincaré-Bendixson theorems, 714, 716
Poisson, Simeon-Denis, 791
Poisson integral formula, 806
  for unit disk, 1083
  for upper half-plane, 1082
Poisson's partial differential equation, 791
Polar form of a complex number, 936, 961
Polar moment of inertia, 552, 558
Pole
  of order n, 1029
  simple, 1029
Population growth, 27

Positive direction on a curve, 521
Potential
  complex, 1088
  complex velocity, 1090
  function, 538, 1088
Power method, 918
Power series
  circle of convergence, 1011
  differentiation of, 1013
  integration of, 1013
  radius of convergence, 278, 1011
  represents a continuous function, 1013
  review of, 278–279
  solutions, 279, 283
Predator-prey, 702
Predictor-corrector methods, 875, 885
Prime meridian, 594
Principal argument, 937
Principal axes of a conic, 453
Principal branch, 964
Principal normal, 490
Principal nth root, 940
Principal part of a Laurent series, 1028
Principal value
  of a complex power, 965
  of an integral, 1041
  of logarithmic function, 963
Probability integral, 819
Propagation error, 887
Projection of a vector onto another, 344
Pure resonance, 193

Quadratic form, 451
Quasi frequency, 186
Quasi period, 186

Radial vibrations, 807
Radioactive decay, 81–84
Radius of convergence, 278
Radius of gyration, 550, 586
Rational roots, 136
Ratio test, 278
Rayleigh quotient, 919
R-C series circuit, differential equation of, 86
Reactance, 202
Reactions, chemical, 95
Real axis, 935
Reciprocal lattice, 355
Rectangular pulse, 240
Rectified sine wave, 242
Rectifying plane, 491
Recurrence relation, 280
Reduced row-echelon form, 393
Reduction of order, 127
Region
  closed, 944
  in the complex plane, 944
  invariant, 712
  multiply connected, 536
  open, 536
  simply connected, 536
  type I(II), 544
Regular singular point, 290
  at ∞, 306
Regular Sturm-Liouville problem, 749–750
  properties of, 750

Relative error, 872
Residues
  evaluation of integrals by, 1035–1037, 1039–1046
  at a pole of order n, 1033
  at a simple pole, 1033–1034
Residue theorem, 1035
Resonance
  curve, 197
  electrical, 201
  mechanical, 193, 197
Response
  of a system, 86, 123
  transform, 251
Ricatti, Jacobo Francesco, 66
Ricatti's differential equation, 68
Riemann mapping theorem, 1073
Right-hand rule, 332, 349
Robin, Victor G., 774
Robin condition, 774
Rodrigues, Olinde, 318
Rodrigues' formula, 318
Roll, 386
Rotating string, 21
Rotation in the z-plane, 1053
Roundoff error, 887
Row-echelon form, 392
Runge, Carl D. T., 880
Runge-Kutta methods
  first-order, 881
  fourth-order, 880–881
  second-order, 880–881
  for systems, 889

Sawtooth function, 241
Scalar matrix, 383
Scaling, 921
Schrödinger's equation, 306
Schwarz-Christoffel formula, 1075
Second moments, 549, 586
Second-order boundary-value problems, 108, 892
Second-order chemical reactions, 96
Second translation theorem, 229
Self-adjoint form, 752–753
Self-orthogonal family of curves, 80
Separable differential equations
  ordinary, 39
  partial, 766
Separation constant, 767
Series, 278
  absolutely convergent, 278
  circuits, 22–23, 86
  convergent, 1008
  Fourier, 733–734, 739
  Fourier-Bessel, 755, 758
  Fourier-Legendre, 760
  geometric, 1008
  power, 278
  solutions of ordinary differential equations, 279, 283, 292
Shaft through the earth, 32
Sifting property, 258
Simple closed curve, 521
Simple harmonic electrical vibrations, 199
Simple harmonic motion, 174

Index

Simple pendulum, 20, 202
Simple pole, 1029
Simply connected
   domain, 984
   region, 536
Simply supported beam, 783
Simpson's rule, 861
   error bound for, 862
Sine series in two variables, 746, 797
Single-step method, 885
Singular matrix, 415
Singular point of a complex function, 1020
   essential, 1029
   isolated, 1016, 1020
   nonisolated, 1020
   pole, 1029
   removable, 1029
Singular point of a differential equation, 281
   at ∞, 306
   irregular, 290
   regular, 290
Singular Sturm-Liouville problem, 752
Sink, 519, 1094
Skew-symmetric matrix, 413
Sliding bead, 700–701
Slope field, 867
Smooth curve, 477, 521
Smooth surface, 566
Solenoidal vector field, 519
Solution of a differential equation, 6
   complete, 11
   existence and uniqueness of, 36
   explicit, 7
   general, 11, 119, 121
   implicit, 7
   n-parameter family of, 10
   particular, 10
   singular, 10
   trivial, 7
Solution of a system of differential equations, 165, 629
   general, 634, 636
   particular, 635
Solution space, 371
Source, 519, 1094
Space curve, 474
   length of, 480
Sparse matrix, 899
Spherical Bessel functions, 312
Spherical coordinates, 591–593, 813
   Laplacian in, 813
Spring-mass systems, 18, 172–198
Square norm, 727
Square wave, 241
Stability criteria
   for linear systems, 687
   for plane autonomous systems, 692
Staircase function, 240
Standard basis, 371
Standing waves, 781–782, 809
Starting methods, 885
Stationary point, 676
Steady-state
   current, 87
   fluid flow, 1089

   solution, 191, 790
   temperature, 774
Stefan's law of radiation, 99
Stokes, George G., 576
Stokes' theorem, 576
Stream function, 1090
Streamlines, 947, 1090
Streamlining, 1090–1091
String of length n, 459
Sturm, Charles Jacques François, 750
Sturm-Liouville problem
   orthogonality of solutions, 750–751
   properties of, 750
   regular, 749
   singular, 752
Subspace, 369
Substitutions, 70
Successive mappings, 1054
Superposition principle
   for Dirichlet's problem for a rectangular plate, 787
   for linear ordinary differential equations, 115, 122
   for linear partial differential equations, 768
   for systems of linear differential equations, 630
Surface
   area, 567
   integral, 569
Sylvester, James Joseph, 365
Symmetric matrix, 384, 437
   eigenvalues of, 438
Syndrome, 462
Synthetic division, 136
Systems of linear differential equations, solutions of, 620, 629–630, 634–635, 641–653, 656, 658, 661, 664
   Laplace transform method, 260
   matrices, use of, 626, 629–640
   operator method, 165
Systems of linear equations, 387–388
   consistent, 388
   homogeneous, 388, 396
   inconsistent, 388
   nonhomogeneous, 388
   solution of, 388
   solving a, 389

Tangent plane, 512
   equation of, 512
   vector equation of, 512
Taylor's theorem, 1015
Telegraph equation, 777
Thermal diffusivity, 772
Third-order chemical reactions, 99
Three-term Taylor method, 877–878
Torque, 349
Trace, 361
   of a matrix, 681
Tractrix, 31, 100
Trajectories
   isogonal, 80
   orthogonal, 77
Transfer function, 251
Transform pair, 835

Transient
   solution, 191, 201, 790
   term, 87, 191
Translation
   theorems for Laplace transform, 224, 229
   in the z-plane, 1053
Transpose of a matrix, 382
Trapezoidal rule, 858
Traveling waves, 784
Tree diagrams, 499
Triangle inequality, 348
Triangular wave, 242
Triple integral, 583
   in cylindrical coordinates, 590
   evaluation of, 583
   in spherical coordinates, 593
Triple scalar product, 352
Triple vector product, 353
Trivial solution, 7
Trivial vector space, 367
Truncation error, 887
Twisted shaft, 202
Two-point boundary-value problem, 108
Type I (II) region, 544

Underdamped circuit, 199
Underdamped motion, 182
Undetermined coefficients, method of, 139, 152
   for systems of linear differential equations, 658
Uniqueness theorems, 36, 107
Unit impulse, 255
Units
   cgs, 19
   for circuits, 23
   engineering system, 19
   SI, 19
Unit step function, 226
Unit vector, 328
Unstable formula, 887

Van der Pol differential equation, 715
Variables, separable, 39
Variation of parameters, 157
   for systems of linear differential equations, 661–662
Vector(s)
   addition, 324–325, 326
   angle between, 342
   components of, 326, 335, 343, 365
   coplanar, 354
   cross product of, 349, 351
   difference of, 324
   differential operator, 502
   direction, 356
   direction angles of, 342
   direction cosines of, 342
   dot product of, 339, 340, 366
   equal, 324, 326
   equation for a line, 356
   equation for a plane, 359
   free, 324
   function, 474
   horizontal component of, 329

Vectors (*continued*)
   inner product, 339
   length of, 328, 366
   linear combination of, 328
   magnitude of, 328, 366
   multiplication by scalars, 324, 326, 335
   negative of, 324, 326
   normal, 359
   norm of, 328
   orthogonal, 341
   parallel, 324
   in a plane, 324
   position, 325, 335
   projection, 344
   properties of, 327
   right-hand rule, 332–333, 349
   scalar multiple of, 324, 326
   scalar product, 339
   as solutions of systems of linear differential equations, 629
   subtraction of, 326
   sum of, 324
   three-dimensional, 335
   triple scalar product, 352
   triple vector product, 353
   two-dimensional, 325
   unit, 328
   vertical component of, 329
   zero, 324, 328, 335, 366

Vector fields, 515
   and analyticity, 1086–1087
   conservative, 538, 540
   curl of, 517
   divergence of, 518
   flux of, 573, 982
   irrotational, 519
   solenoidal, 519
   velocity, 515, 675
Vector functions, 474
   continuity of, 476
   derivative of, 477, 479
   integrals of, 480
Vector space, 366–367
   finite dimensional, 371
   infinite dimensional, 371
Velocity, 483
   potential, complex, 1090
Verhulst, P. F., 92
Vibrating string, 773, 780
Vibrations, spring-mass systems, 172–198, 261
Volterra, Vito, 100
Volterra integral equation, 247
Volume
   using double integrals, 543, 546
   of a parallelepiped, 353
   using triple integrals, 585
Vortex, 1094

Wave equation
   derivation of one-dimensional equation, 773
   one-dimensional, 771
   solution of, 780
   two-dimensional, 796
Weight, 20
Weight function, 730
Word, 459
Work
   as a dot product, 346
   as a line integral, 529
Wronski, Joseph Maria Hoëne, 112
Wronskian, 112, 633
   Abel's formula for, 125
Wronski matrix, 638

Yaw, 386

Zeros of
   of the complex cosine and sine, 968–969
   of the complex hyperbolic cosine and sine, 970
   of a function, 1030
   of order n, 1030
Zero matrix, 383
Zero-state response, 251
Zero vector space, 367
z-plane, 935

Table of Laplace Transforms

	$f(t)$	$\mathcal{L}\{f(t)\} = F(s)$
1.	1	$\dfrac{1}{s}$
2.	t	$\dfrac{1}{s^2}$
3.	t^n	$\dfrac{n!}{s^{n+1}}$, n a positive integer
4.	$t^{-1/2}$	$\sqrt{\dfrac{\pi}{s}}$
5.	$t^{1/2}$	$\dfrac{\sqrt{\pi}}{2s^{3/2}}$
6.	t^α	$\dfrac{\Gamma(\alpha+1)}{s^{\alpha+1}}$, $\alpha > -1$
7.	$\sin kt$	$\dfrac{k}{s^2+k^2}$
8.	$\cos kt$	$\dfrac{s}{s^2+k^2}$
9.	$\sin^2 kt$	$\dfrac{2k^2}{s(s^2+4k^2)}$
10.	$\cos^2 kt$	$\dfrac{s^2+2k^2}{s(s^2+4k^2)}$
11.	e^{at}	$\dfrac{1}{s-a}$
12.	$\sinh kt$	$\dfrac{k}{s^2-k^2}$
13.	$\cosh kt$	$\dfrac{s}{s^2-k^2}$
14.	$\sinh^2 kt$	$\dfrac{2k^2}{s(s^2-4k^2)}$
15.	$\cosh^2 kt$	$\dfrac{s^2-2k^2}{s(s^2-4k^2)}$
16.	$e^{at}t$	$\dfrac{1}{(s-a)^2}$
17.	$e^{at}t^n$	$\dfrac{n!}{(s-a)^{n+1}}$, n a positive integer

	$f(t)$	$\mathcal{L}\{f(t)\} = F(s)$
18.	$e^{at}\sin kt$	$\dfrac{k}{(s-a)^2+k^2}$
19.	$e^{at}\cos kt$	$\dfrac{s-a}{(s-a)^2+k^2}$
20.	$e^{at}\sinh kt$	$\dfrac{k}{(s-a)^2-k^2}$
21.	$e^{at}\cosh kt$	$\dfrac{s-a}{(s-a)^2-k^2}$
22.	$t\sin kt$	$\dfrac{2ks}{(s^2+k^2)^2}$
23.	$t\cos kt$	$\dfrac{s^2-k^2}{(s^2+k^2)^2}$
24.	$\sin kt + kt\cos kt$	$\dfrac{2ks^2}{(s^2+k^2)^2}$
25.	$\sin kt - kt\cos kt$	$\dfrac{2k^3}{(s^2+k^2)^2}$
26.	$t\sinh kt$	$\dfrac{2ks}{(s^2-k^2)^2}$
27.	$t\cosh kt$	$\dfrac{s^2+k^2}{(s^2-k^2)^2}$
28.	$\dfrac{e^{at}-e^{bt}}{a-b}$	$\dfrac{1}{(s-a)(s-b)}$
29.	$\dfrac{ae^{at}-be^{bt}}{a-b}$	$\dfrac{s}{(s-a)(s-b)}$
30.	$1-\cos kt$	$\dfrac{k^2}{s(s^2+k^2)}$
31.	$kt-\sin kt$	$\dfrac{k^3}{s^2(s^2+k^2)}$
32.	$\cos at - \cos bt$	$\dfrac{s(b^2-a^2)}{(s^2+a^2)(s^2+b^2)}$
33.	$\sin kt\sinh kt$	$\dfrac{2k^2 s}{s^4+4k^4}$
34.	$\sin kt\cosh kt$	$\dfrac{k(s^2+2k^2)}{s^4+4k^4}$

$f(t)$	$\mathscr{L}\{f(t)\} = F(s)$
35. $\cos kt \sinh kt$	$\dfrac{k(s^2 - 2k^2)}{s^4 + 4k^4}$
36. $\cos kt \cosh kt$	$\dfrac{s^3}{s^4 + 4k^4}$
37. $\delta(t)$	1
38. $\delta(t - a)$	e^{-as}
39. $\mathscr{U}(t - a)$	$\dfrac{e^{-as}}{s}$
40. $J_0(kt)$	$\dfrac{1}{\sqrt{s^2 + k^2}}$
41. $\dfrac{e^{bt} - e^{at}}{t}$	$\ln \dfrac{s - a}{s - b}$
42. $\dfrac{2(1 - \cos at)}{t}$	$\ln \dfrac{s^2 + a^2}{s^2}$
43. $\dfrac{2(1 - \cosh at)}{t}$	$\ln \dfrac{s^2 - a^2}{s^2}$
44. $\dfrac{\sin at}{t}$	$\arctan\left(\dfrac{a}{s}\right)$
45. $\dfrac{\sin at \cos bt}{t}$	$\dfrac{1}{2} \arctan \dfrac{a + b}{s} + \dfrac{1}{2} \arctan \dfrac{a - b}{s}$
46. $\dfrac{1}{\sqrt{\pi t}} e^{-a^2/4t}$	$\dfrac{e^{-a\sqrt{s}}}{\sqrt{s}}$
47. $\dfrac{a}{2\sqrt{\pi t^3}} e^{-a^2/4t}$	$e^{-a\sqrt{s}}$
48. $\operatorname{erfc}\left(\dfrac{a}{2\sqrt{t}}\right)$	$\dfrac{e^{-a\sqrt{s}}}{s}$
49. $2\sqrt{\dfrac{t}{\pi}} e^{-a^2/4t} - a \operatorname{erfc}\left(\dfrac{a}{2\sqrt{t}}\right)$	$\dfrac{e^{-a\sqrt{s}}}{s\sqrt{s}}$
50. $e^{ab}e^{b^2 t} \operatorname{erfc}\left(b\sqrt{t} + \dfrac{a}{2\sqrt{t}}\right)$	$\dfrac{e^{-a\sqrt{s}}}{\sqrt{s}(\sqrt{s} + b)}$
51. $-e^{ab}e^{b^2 t} \operatorname{erfc}\left(b\sqrt{t} + \dfrac{a}{2\sqrt{t}}\right) + \operatorname{erfc}\left(\dfrac{a}{2\sqrt{t}}\right)$	$\dfrac{be^{-a\sqrt{s}}}{s(\sqrt{s} + b)}$
52. $e^{at}f(t)$	$F(s - a)$
53. $f(t - a)\mathscr{U}(t - a)$	$e^{-as}F(s)$
54. $f^{(n)}(t)$	$s^n F(s) - s^{(n-1)}f(0) - \cdots - f^{(n-1)}(0)$
55. $t^n f(t)$	$(-1)^n \dfrac{d^n}{ds^n} F(s)$
56. $\displaystyle\int_0^t f(\tau)g(t - \tau)\, d\tau$	$F(s)G(s)$